University Chemistry

University Chemistry

Based on Common Core Syllabus of Universities of Andhra Pradesh

Volume I

Syllabus of B.Sc. I Year (Theory and Practicals)

Series Editor

C. Parameswara Murthy

Authors

C. Parameswara Murthy
M.Sc., Ph.D.
Professor of Chemistry
University College of Engineering
Osmania University, Hyderabad (A.P.)

Syed Fazal Mehdi Ali
M.Sc., Ph.D.
Principal
Rishi Junior & Degree College
Hyderabad (A.P.)

D. Ashok
M.Sc., Ph.D.
P.G. College of Science
O.U. Saifabad, Hyderabad (A.P.)

PUBLISHING FOR ONE WORLD

NEW AGE INTERNATIONAL (P) LIMITED, PUBLISHERS

New Delhi • Bangalore • Chennai • Cochin • Guwahati • Hyderabad
Jalandhar • Kolkata • Lucknow • Mumbai • Ranchi
Visit us at www.newagepublishers.com

Published by New Age International (P) Ltd., Publishers
First Edition: 1995
Reprint: 2008

Branches:

- 36, Malikarjuna Temple Street, Opp. ICWA, Basavanagudi, **Bangalore**. ✆ (080) 26677815
- 26, Damodaran Street, T. Nagar, **Chennai**. ✆ (044) 24353401
- Hemsen Complex, Mohd. Shah Road, Paltan Bazar, Near Starline Hotel, **Guwahati**. ✆ (0361) 2543669
- No. 105, 1st Floor, Madhiray Kaveri Tower, 3-2-19, Azam Jahi Road, Nimboliadda, **Hyderabad.** ✆ (040) 24652456
- RDB Chambers (Formerly Lotus Cinema) 106A,1st Floor, S.N. Banerjee Road, **Kolkata**. ✆ (033) 22275247
- 18, Madan Mohan Malviya Marg, **Lucknow**. ✆ (0522) 2209578
- 142C, Victor House, Ground Floor, N.M. Joshi Marg, Lower Parel, **Mumbai**. ✆ (022) 24927869
- 22, Golden House, Daryaganj, **New Delhi**. ✆ (011) 23262370, 23262368

ISBN : 978-81-224-0742-6

Rs. 395.00

C-08-06-2650

Printed in India at Mohanlal Printers, Delhi.

PUBLISHING FOR ONE WORLD
NEW AGE INTERNATIONAL (P) LIMITED, PUBLISHERS
4835/24, Ansari Road, Daryaganj, New Delhi-110002
Visit us at **www.newagepublishers.com**

Preface

The undergraduate chemistry courses are put to many demands and severe constraints interms of space and time. Only proper selections of topics, extent of their coverage and a mention of their relevance is most essential. Chemistry is conventionally taught through three separate text books devoted to the three main branches, viz., Organic, Physical and Inorganic. But this space age generations are overwhelmingly career oriented and have no time or inclination to go through these exhaustive books and job down relevant material. Further it is also an extensive proposition to buy all the books at the very first year. Therefore rather than keeping critisizing the changing times we wish to capitalize and give what they want through this series. From the decades of experience that we gained in teaching various courses and from the suggestions from friends and colleagues of different universities we present the current series.

In this series we attempt not just to introduce the chemical concepts in a conventional way but try to give an insight into the subject in a self-explanatory, lucid, sequential and developmental fashion. In writing this first volume we have kept the standards of an average intermediate student in mind, and concentrated on block- by-block construction of chemistry, maintaining continuity at every stage. For this reason we have marginally deviated from the common core syllabus and gave details like group properties of s-block element, p-block elements, importance of hydrogen as a fuel, semi-conductors etc. All these changes are incorporated only after a thorough discussion with the subject colleagues and friends.

The salient features in this series are :

(i) The style is conversational but concise.

(ii) Number of applications are listed and discussed, where ever possible.

(iii) Detailed chapter-wise topics and sub-topics are given in the contents.

(iv) Simple analogies are used to clarify the abstract concepts.

(v) Numerous problems are worked out in the text, step by step.

(vi) Numerous problems of varying difficulties are also given at the end of physical chemistry chapters, along with suitable hints wherever necessary, for drilling and appreciation of the subject, answers are provided for these unsolved problems.

(vii) Problems based on organic structural details and conversions are also presented at the end of each organic chapter.

(viii) New terms are introduced in bold type face and definitions and important statements are presented in italics.

(ix) A brief theoretical back-ground of the practicals, and proforma for tabulating the data obtained are also presented in the practical sections.

(x) As we are still in a transition period between CGS and SI units and because the questions paper still contain problems with both the unit systems, we did not develop preference to any system, however in Appendix A we gave conversion factors of these units.

(xi) As our students are overwhelmingly exam-oriented, we have given a number of questions from different question papers at the end of each chapter to help them in judging the pattern of questions which normally appear in the exams.

(xii) Due emphasis is given to EAMCET and other entrance examinations by incorporating Multiple Choice Questions at the end of the each Chapter.

In spite of all the care taken to present a good textbook, I don't deny the entry of some mistakes unnoticingly. I humbly accept any mistakes pointed by esteemed readers. I would also appreciate receiving suggestions for any betterment.

I take this opportunity to record the inspiration given by my parents, Ms. Subba Lakshmi and Dr. Lakshmi Narayana, service rendered by my wife Ms. Sarvani and the omni patience shown by my children Kalyani and Srinivas and other family members in my depressing and disgusting hours.

C. PARAMESWARA MURTHY

Acknowledgement

We have a great pleasure in expressing our sincere thanks to those who have contributed in many ways in one successful completion of this project.

First, we record our appreciation to our family members who constantly encouraged, bearing the inconveniences brought about by rushed deadlines.

We thank the Registrar, O.U., for according permission to take-up this venture.

We are thankful to Mr. Machwe, Ms Supriya Bhale Rao, Mr. Govindan, Mr. Krishna Murthy and other executives of Wiley for their constant support.

We thank Mr. V. Venkatesh & Ms. K. Jayanti for their neat typing and Mr. Satyanarayana for his neat art-work and Mr M. Venugopal of Emvee Computer Services for his support.

I thank Prof D.C. Reddy, Principal, University College of Engineering, O.U., Hyderabad, for providing an inspiring academic atmosphere.

Finally, we thank our friend Prof. C. Ratnam of Andhra University, our colleagues, mainly Dr. G. Veerabhadram and Dr M. Vithal, who helped in shaping this book with their thoughtful reviews and criticisms, proof reading, providing question papers and other material.

C. PARAMESWARA MURTHY

S. FAZAL MEHDI ALI

D. ASHOK

Contents

Part I - INORGANIC CHEMISTRY

Part III – PHYSICAL CHEMISTRY

Part IV – PRACTICAL CHEMISTRY

Part I

Inorganic Chemistry

1

CHEMICAL BONDING

1-1 CONCEPT OF CHEMICAL BONDING

Chemical bonds may be defined as the attraction between atoms that hold them together forming various compounds. The elements lose their original identity, as soon as they are involved in the chemical bonding. Their strength determine the chemical properties of the compounds like addition, substitution behaviours and also determine the heat changes involved with these processes. The radical changes in the chemical behaviour which an atom experiences by knocking out different number of electrons, forming ions of different charges suggest that only extra nuclear electrons are the seat of chemical reactivity. Several studies revealed that only electrons present in the outer most orbit are responsible for the chemical reactivity of the elements.

The behaviour associated with the removal of electrons and addition of electrons is described by their **ionization potential and electron affinity**. Chemical bonds can be divided into three broad categories (a) ionic bonds (b) covalent bonds (c) coordinate covalent bonds. Not all bonds exactly fit in these categories and in several cases we notice intermediate behaviour.

1-1-1 The Ionic Bond

Ionic bonds are formed by the mutual coulombic attractions of oppositely charged ions. These ionic states are assigned by the transfer of electrons from an atom that perferentially loses the electron(s) to another atom which readily accepts the electron(s). Generally metals react with non-metals to form ionic compounds. In these compounds metals always lose electrons, forming positively changed ions called cations and these electrons are gained by the non-metals forming negatively charged ions or anions.

Thus when lithium, Li, and chlorine, Cl, react they form an ionic compound LiCl, which contains Li^+ and Cl^- ions. The electronic configuration of the Li and Cl atoms are

Li $1s^2\ 2s^1$ and

Cl $1s^2\ 2s^2\ 2p^6\ 3s^2\ 3p^5$

As Li being a metal,it loses an electron and this electron is gained by Cl represented by following processes.

$$Li\ (1s^2\ 2s^1) \longrightarrow Li^+\ (1s^2) + e-$$

$$Cl\ (1s^2\ 2s^2\ 2^6\ 3s^2\ 3p^5) + e \longrightarrow Cl^-\ (1s^2\ 2s^2\ 2p^6\ 3s^2\ 3p^6)$$

$$Li^+ + Cl^- \longrightarrow Li\,Cl$$

Similarly NaCl is formed by the combination of Na^+ and Cl^- ions.

$$Na\ (1s^2\ 2s^2\ sp^6\ 3s^1) \longrightarrow Na^+\ (1s^2,\ 2s^2\ 2p^6) + e^-$$

$$Cl\ (1s^2\ 2s^2\ 2p^6\ 3s^2\ 3p^5) + e^- \longrightarrow Cl^-\ (1s^2\ 2s^2\ 2p^6\ 3s^2\ 3p^6)$$

$$Na^+ + Cl^- \longrightarrow NaCl$$

or $$\overset{\times}{Na} + \cdot\ddot{Cl}: \longrightarrow Na\overset{\times}{\cdot}\ddot{Cl}:$$

W. Kossel (1916) made a series of such observations and suggested that the tendency of elements to gain an electron or lose an electron is because all the elements try to achieve an inert gas configuration. Thus in the formation of LiCl, Li^+ takes the helium (He) atom configuration and Cl^- assumes argon (Ar) configuration. In case of NaCl the cation achieves Ne configuration while Cl^- takes the Ar configuration thereby they get stability.

This is also true with the second group and third group elements.

$$Mg\ (1s^2\ 2s^2\ 2p^6\ 3s^2) \qquad Mg^{2+}\ (1s^2\ 2s^2\ sp6) + 2e^-$$
(Ne configuration)

$$O\ (1s^2\ 2s^2\ 2p^4) + 2e^- \qquad O^{2-}\ (1s^2\ 2s^2\ 2p^6)$$
(Ne configuration)

In attempting to explain these reactions, we face the question why metals preferentially lose electrons?, why non-metals are always on the receiving end? and do they have any restrictions in these electron transfers? In all the above cases *the main factor that forces metal or non-metal to lose or gain an electron is the ease with which it can attain a noble gas configuration.*

The factor that controls the formation of ionic compounds like NaCl, LiCl, MgO is primarily the change in potential energy. Always the elements possess a higher potential energy and by a down hill change of potential energy they form the ionic compounds, simultaneously liberating the potential energy loss in the form of heat energy. So of the two ionic processes, formation of cations and formation of anions, even if one reaction is endothermic the other reaction must have a higher exothermicity to make the overall process exothermic. Metals like Na, Li, Mg have few outer most electrons and it is easier to lose an electron than acquiring a number of electrons to attain noble gas configuration. Formation of cations is an endothermic process. Loss of an electron from its outermost orbit has lower ionization potential compared to inner orbit electrons, which are already in a noble gas core.

The electron affinity of non metals, *i.e.,* capability of addition of electrons into valence orbitals controls the formation of anions. Generally addition of first electron is exothermic while formation of anions with -2 or higher negative charge is always an endothermic process. But we know a number of bivalent and multivalent ionic compounds. *The endothermicity involved in the formation of cations and higher valent anions suggest that ionization energy and electron affinity are not the only controlling factors.* The neutral atoms have only Van der Waals forces of attraction and are very weak. On the other hand if the ions are brought together they are bound by electrostatic attraction resulting in a large decrease in potential energy and this compensates any overall endothermicity in the formation of cations and anions. *This lowering of potential energy due to electrostatic attraction is called the* **lattice energy**. This is the driving force behind the formation of ionic compounds of bivalent or higher valent anions and cations, formation of which is usually endothermic.

Both in the formation of cations and anions, the final configuration is restricted to a noble gas configuration because large potential energy changes are involved in breaking into a noble gas core. Except helium all the noble gases have eight electrons in their outer shell. This led to formulation of **octet rule** which states that *"when metals and non metals of the 'B' groups of periodic table react, they often tend to gain or lose electrons until their octet or eight electrons in the outer shell is filled.*

1-1-2 Representation of the Chemical Bond

Gilbert Lewis developed a method for the representation of the ionic bonds using notation known as Lewis symbols. The group number of the periodic table there by the number of electrons in the outer most orbit is taken in cognizance in defining these symbols. Lewis symbols for the pre and post transition elements, which are of importance for the ionic bond formation, are given in the following table.

Group	IA	IIA	IIIB	IVB	VB	VIB	VIIB	0
Symbol	$X\cdot$	$\cdot X\cdot$	$\cdot\dot{X}\cdot$	$\cdot\underset{\cdot}{\dot{X}}\cdot$	$\cdot\underset{\cdot}{\dot{X}}:$	$:\underset{\cdot}{\dot{X}}:$	$\cdot\underset{\cdot\cdot}{\dot{X}}:$	$:\underset{\cdot\cdot}{\ddot{X}}:$

Lewis symbols are highly helpful in writing the Lewis structures or electron dot formulas that speak about the number of valence electrons, charge of the element. Thus combination of sodium of group IA and chlorine of group VIIB to give sodium chloride may be represented as

$$\overset{\times}{Na} + \cdot\underset{\cdot\cdot}{\ddot{Cl}}: \longrightarrow Na^{+}\,[\,\cdot\,\underset{\cdot\cdot}{\ddot{Cl}}:\,] \text{ or } Na^{+} + [\,\overset{\times}{\cdot}\underset{\cdot\cdot}{\ddot{Cl}}:\,]^{-}$$

This formula suggests that by the transfer of an electron to chlorine atom, sodium cation and chloride anion are formed and are bound by an ionic bond. The brackets around chloride indicates that the electron of sodium atom is now the property or part of chloride anion. The eight electrons around Cl also suggests that the octet rule is satisfied. Similarly in the formation of $MgCl_2$, the Lewis structure may be written as

$$Mg^{\times}_{\times} + 2\,\ddot{Cl}: \longrightarrow Mg^{2+},\ 2[\times \ddot{Cl}:]^{-} \text{ or } Mg^{2+} + 2[\times \ddot{Cl}:]^{-}$$

1-1-3 The Covalent Bond

In many instances we may not be able to attain the noble gas configuration, that is completion of an octet, by an outright electron transfer of valence electrons from a metal to a non metal resulting in an ionic bond. ***This is due to highly energetically unfavourable endothermic condition accompanied in the removal of electrons from the outer-most shell. In these cases an octet could be achieved by the sharing of electrons between the atoms, which is an energetically low profile process and a bond obtained by this manner is called a*** **covalent bond** ***or*** **electron pair bond.**

The covalent bonds are formed between atoms with little or no electronegativity difference. We have a number of examples like H_2, Cl_2, N_2, etc., where covalent bond is formed between like atoms having a zero electronegativity difference. For example, when we have two H atoms they readily combine to give hydrogen molecule.

$$H\ (1s^1) + H\ (1s^1) \longrightarrow H_2$$

Expressing in Lewis symbols

$$H\cdot + H\cdot \longrightarrow H:H \quad \text{or} \quad H-H$$

The pair of electrons are equally shared between the atoms, thus attaining the noble gas, He, structure. The pair of electrons between the atoms is also represented by putting a dash between the atoms as represented by the example H - H.

An atom like C having four electrons in its outermost orbit can share them with four H atoms forming methane, CH_4, as represented below

$$\cdot\dot{\underset{\cdot}{C}}\cdot + 4\ {\times}H \longrightarrow H \overset{\displaystyle H}{\underset{\displaystyle H}{\overset{\times}{\underset{\times}{\times C \times}}}} H \quad \text{or} \quad CH_4 \quad \text{or} \quad H-\overset{\displaystyle H}{\underset{\displaystyle H}{\overset{|}{\underset{|}{C}}}}-H$$

Nitrogen belonging to VB group requires only three electrons to attain an inert gas configuration. Hence it combines with three hydrogen atoms having comparable electronegativity

$$\cdot\ddot{N}\cdot + 3H\times \longrightarrow H \times \ddot{N} \times H \ (\text{with } \times H \text{ below N}) \quad \text{or} \quad (H)(\ddot{N})(H),\ (H) \text{ below}$$

The three dashes on nitrogen represent three single bonds and nitrogen is completing its octet and hydrogen is attaining the helium configuration.

It is also possible for two atoms to share a pair of electrons each. For example in case of CO_2, the dot structure may be given as

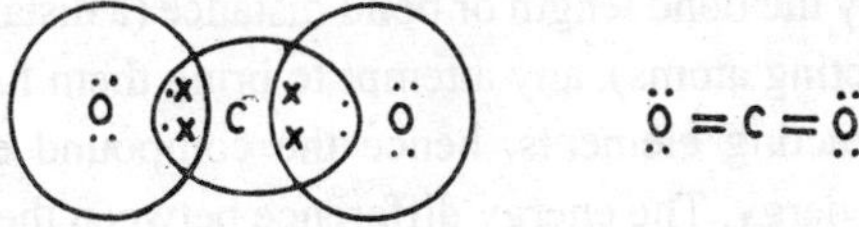

The two pairs of sharing electrons are represented by two dash lines and called a **double bond.**

Similarly a triple bond is formed by sharing of three electrons by any two atoms.

$$.\ddot{N}. + .\ddot{N}. \longrightarrow :N{\equiv}N: \quad \text{or} \quad (:N\ (:::)\ N:)$$

1-1-4 Break down of the Octet Rule in Ionic Compounds

The octet rule as presented by Kossel strictly holds good for the metals in groups IA and II A and aluminium. The octet rule does not work well with cations of transition elements and post transition elements (the few metals like Pb, Sn, etc., of groups III B, IVB, VB, VIB and VIIB) which have generally outer shell electronic configuration not the same as a noble gas. Transition elements exhibit variable valency and the half filled or completely filled *d* orbitals are considered to be most stable states. So a transition element always loses electrons from its outer *d*-subshell giving a half filled *d*-subshell. For example, zinc loses two 4*s* electrons giving Zn^{2+} to give a fully filled *d*-subshell.

$$Zn\ ([Ar]\ 3d^{10}\ 4s^2) \longrightarrow Zn^{2+}\ ([Ar]\ 3d^{10}) + 2e^-$$

Similarly tin loses four electrons to attain a $4d^{10}$ configurations

$$Sn([Kr]\ 4d^{10}\ 5s^2\ 5p^2) \longrightarrow Sn^{4+}\ ([Kr]4d^{10}) + 4e^-$$

Incase of iron, the Fe^{3+} is the most stable state with half filled d-subshell.

$$Fe([Ar]3d^6\ 4s^2 \longrightarrow Fe^{3+}\ ([Ar]3d^5) + 3e^-$$

When we have a totally filled *d*-subshell or half filled *d*-subshell, the extra ordinary stability achieved by an ion is compared with the stability of a noble gas configuration and is called a **pseudo-noble gas configuration.** *The elements which can achieve this pseudo-noble gas configuration form ionic bonds and they do not obey the octet rule.*

1-1-5 Energy Profile of Formation of a Covalent Bond

As expected formation of a covalent bond is also accompanied by a fall in the energy. When the atoms are far apart, the total energy of these atoms is just the arithmetic sum of the energies of individual atoms. As they are brought closer, the energy of the system falls to a minimum because the electron cloud of one atom is coming under the influence of nucleus of another atom and vice-versa. Once the nuclei are separated by the bond length or bond distance (a distance equal to the sum of the radii of the reacting atoms), any attempt to bring them further close amounts to compressing of reacting elements, hence the compound experiences a sharp increase of potential energy. The energy difference between the total energy of free atoms and the minimum energy on the energy profile is called **bond energy** (Fig 1-1).

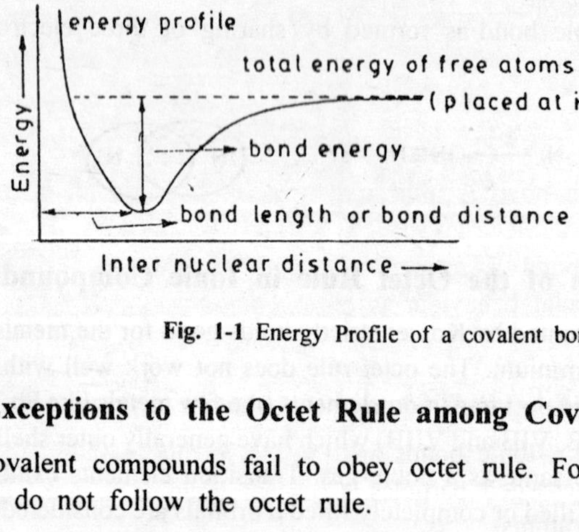

Fig. 1-1 Energy Profile of a covalent bond

1-1-6 Exceptions to the Octet Rule among Covalent Compounds

Many covalent compounds fail to obey octet rule. For example, Be and B chlorides do not follow the octet rule.

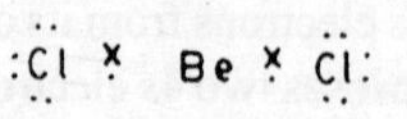

Beryllium chloride

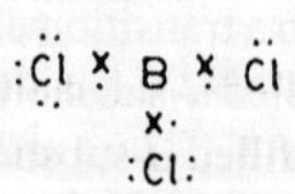

Boron chloride

The metal atoms in Beryllium chloride and Boron chloride have four and six electrons respectively. Exceptions are also known with central atoms having ten and twelve electrons.

```
       :Cl:                      :F:
  :Cl   |   Cl:             :F    |    F:
      \ | /                     \ | /
        P                         S
      /   \                     / | \
  :Cl       Cl:             :F    |    F:
                                 :F:
```

Thus P and S atoms in the above compounds have 10 and 12 bonding electrons in the outermost shell.

1-1-7 The Coordinate Covalent Bond

The co-ordinate covalent bond is formed by the sharing of a lone pair of electrons of an electron rich atom with another atom which does not contribute any electrons towards bonding. Thus the nitrogen in ammonia, NH_3 shares its lone pair of electrons with electron deficient boron trichloride BCl_3 to form an addition compound.

```
    H          Cl                  ·H   Cl              H   Cl
    |          |                    |   |               |   |
H — N:   +     B — Cl  ———————►  H — N : B — Cl   or   H — N ——► B — Cl
    |          |                    |   |               |   |
    H          Cl                   H   Cl              H   Cl
```

The coordinate covalent bond is indicated by a small arrow ($\longrightarrow$) showing the direction of flow of electrons. But it should be noted that it is only a notation as once the bond is formed it doesn't matter where the electrons came from and no special behaviour is observed for the coordinated electron pair or bond.

1-1-8 Comparison of Properties of Ionic and Covalent Bonds

1. Nature of the bond : Ionic bonds are formed by the coulombic attraction of distinct electrically charged particles. Covalent bond is due to sharing of electron pairs between atoms.

2. Nature of crystals : Considering an ionic crystal like sodium chloride we find that the ions Na^+ and Cl^- are arranged in a regular fashion. Each Na^+ is surrounded by six OCl^-, vice versa. Hence no specific Na^+ belongs to a specific Cl^- but in a crystal the number of positively charged ions and negatively charged ions are same. This leads to a single grown crystal where all ions are fixed in an array by the coulombic forces.

In case of a covalent solid like I_2, two iodine atoms are bound together by sharing of an electron pair. These molecules are only kept together in a crystal by the weak Van der Waals forces and are referred to as molecular crystal. The weak forces of attraction result in the formation of relatively small crystals.

3. Melting & boiling points : The melting and boiling points of covalent compounds are generally lower than the ionic compounds. This is the result of weak Van der Waals forces in the covalent molecules against the strong coulombic attractions in the ionic compounds, Table 1-3,

Table 1-1 m.p. and b.p. of some ionic and covalent compounds

Ionic compounds	**m.p°C.**	**b.p°C**
NaCl	801	1413
$SnCl_2$	246	623
Al_2O_3	2050	2250
Covalent compounds		
CCl_4	-23	77
$SnCl_4$	-33	114
AlI_3	191	360

4. Electrical conductivity : Both in the molten state and when dissolved in a polar solvent, an ionic compound behaves as a good conductor of electricity, while covalent compounds are generally bad or nonconductors of electricity. This can be understood by considering a deficit of ions that migrate the electrons.

5. Solubility : Ionic substances are generally soluble in polar solvents and insoluble in non-polar solvents. Thus NaCl is soluble in polar solvents like H_2O, but is insoluble in a non- polar solvents like CCl_4. A covalent compound on the contrary is more soluble in nonpolar solvents, as if to say *like solutes dissolve in like solvents.*

The solubility of ionic compounds in polar solvents is due to solvation (aquation when we use water as solvent) of the ions of solute by solvent molecules.

6. Rates of the chemical reactions : The ionic reactions are almost instantaneous. When NaCl is added to $AgNO_3$ solution, AgCl is immediately precipitated. Reaction between covalent substances proceed relatively slow. Thus chlorination of methane is a very slow process.

$$CH_4 + Cl_2 \longrightarrow CH_3Cl + HCl$$

1-2 CHEMICAL BONDING AND MOLECULAR SHAPES, VSEPR THEORY

Molecules are arranged in different ways in the three dimensional space giving the molecules different shapes. The shapes determine the various physical and chemical properties of molecules. The concept of shapes is also highly relevant in controlling the complicated biological processes. Enzymatic activities are decided by a particular arrangement of atoms and even a small variation in the shape of the molecule destroys the enzymatic activity. Similarly genes, that carry hereditary properties, also have specific arrangement of atoms in the three dimensional space.

The ionic bond is a result of electrostatic attraction between oppositely charged ions. As the charge of an atom is not localized along any particular direction the ionic bonds are non-directional and hence we will not discuss about the shapes of ionic compounds. In the covalently bound substances, the bonds are directional as they are formed by simple sharing of atoms and hence assume different shapes. Hence in the following discussion we are concerned only with the shapes of covalent compounds.

In case of beryllium chloride, the Be atom has two valence electrons which are shared with two chlorine atoms. The dot formula or Lewis structure of $BeCl_2$ is

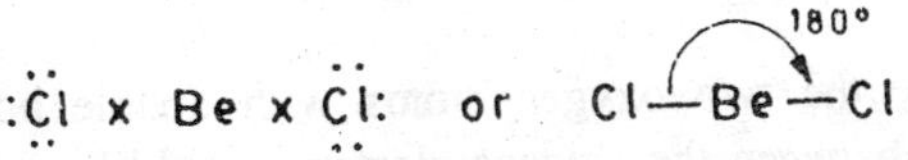

$$:\ddot{\underset{..}{Cl}}\times Be \times \ddot{\underset{..}{Cl}}: \quad \text{or} \quad Cl - Be - Cl \; (180°)$$

At the first sight we see no restrictions imposed on the Cl – Be – Cl angle and get an impression that $BeCl_2$ can take any shape. The first attempt to predict the molecular structure was made by a simple theory known as **Valence Shell Electron Pair Repulsion theory (VSEPR Theory)**. *According to VSEPR theory the geometrical arrangement of atoms around a central atom in a molecule is determined solely by the repulsions among the valence electrons of the central atom and the arrangement is that of minimum repulsive forces between the electrons in the valence (outer most) orbital.*

As stated in the VSEPR theory the shape or geometrical arrangement of a covalent molecule is determined by the

(i) Electrons in the valence orbital (outer most shell) of the central atoms.

(ii) An arrangement where the repulsive forces between the valence electrons is minimum.

On the basis of involvement of type of valence electrons, namely the *bonding electron pairs* and *the lone pair of electrons* we can group the molecules into two classes : (A) Where geometry of the covalent molecule is determined only by the bonding electron-pairs, (B) Where geometry of the covalent molecules is determined by both bonding electron pairs and lone pair of electrons.

A. Geometry of covalent molecules involving participation of only bonding electrons : In the above cited example of $BeCl_2$, the central atom, Be, has two pairs of valence electrons which show a mutual repulsion. A minimum repulsion is achieved when they are placed on both sides of beryllium. Thus $BeCl_2$ assumes a linear structure with a Cl-Be-Cl angle equal to 180°.

$$:\ddot{Cl}-Be-\ddot{Cl}: \quad (180°)$$

Similarly the Lewis structure of CO_2 is written as :

$$\ddot{O} \overset{\times}{\underset{\times}{:}} C \overset{\times}{\underset{\times}{:}} \ddot{O} \quad \text{or} \quad O=C=O \quad (180°)$$

carbon is bound by the two oxygen atoms with double bonds. Here also a minimum repulsion between the valence electrons could be achieved by placing the oxygen atoms 180° apart.

In case of boron trichloride we have three pairs of valence electrons around the central atom 'B'. Applying the VSEPR theory to this compound the repulsive forces could be minimized when the valence electrons are separated by 120°, thus giving BCl3, a planar triangular structure.

Cl
|
B 120°
Cl   Cl

Some more representative examples are given in Table 1-2.

Geometry of covalent molecules involving both bonding electrons and lone pair of electrons : In case of ammonia, NH_3, we expect it to have planar triangular structure akin to BCl_3, as three hydrogen atoms are covalently bound to the central nitrogen. But here the nitrogen, in it's outermost orbital, has in addition to three valence electrons, a lone pair of electrons in the 2*s* orbital. This lone pair, being a part of valence shell, also participates in the mutual repulsions game of the other three valence electron pairs. Hence NH_3 assumes a pyramidal structure.

Table 1-2 Geometry of covalent molecules involving only bonding electron pairs

Molecule	*Bond angle*	*Geometry*
CO_2, $HgCl_2$, $BeCl_2$, BeF_2, $ZnCl_2$, $MgCl_2$	180°	Linear Cl-Be-Cl
BCl_3, BF_3, $AlCl_3$	120°	Plane triangle
CH_4, $SiCl_4$	109.5°	Tetrahedral (a pyramid shape, having four triangular faces)
SF_6	90°	Octahedral (Consists of two square pyramids fused together at the square base. It is a eight sided figure with six corners)
$[Ni(CN)_4]^{2+}$	90°	Square planar

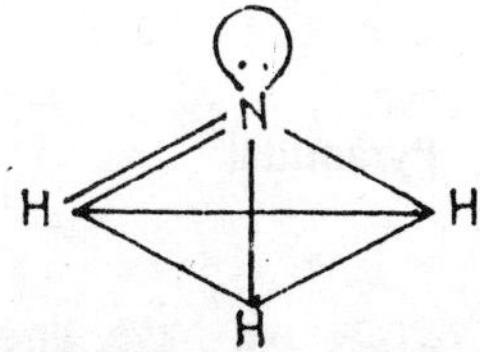

Similar SO_2 instead of taking a linear structure as shown by the dot formula

:O S :O:

takes a V-shaped structure and minimizes the repulsions between the lone pair of electrons and the covalently bonded electron pairs.

Some representative examples are given in Table 1-3

In general the order of repulsive forces between the lone pair of electrons (lp) and bonding electrons (be) is :

lp – lp > lp – be > be-be repulsive forces

Table 1-3 Geometry of covalent molecules involving both bonding electron pairs and Ione electron pairs

Molecule	*Bond angle*	*Geometry*	*Structure*
XeF_2, I_3^-	180o	Linear	F—Xe—F; I—I—I
SO_2	< 120º	V-Shaped	O–S–O, 120°
H_2O	104.5º	V-Shaped	H–O–H, 104.5°
NH_3	107º	Pyramidal	N with H, H, H
XeF_4	90º	Square planar →	Xe with F, F, F, F
ClF_3	90º	T-shaped	F—Cl—F, F
BrF_5	90º	Square pyramidal →	Br with F, F, F, F, F

VSEPR model successfully generates reliable prediction of geometries of molecular structures and is largely based on the Lewis structures of molecules involving pairing of valence electrons of two atoms and repulsions generated by non-bonding electrons. The main drawback of Lewis theory is its failure to explain *why exactly a chemical bond exists inspite of a chance for pairing of electrons or to put in other way, how the molecules share the bonding electron with each other.* Lewis theory does not differentiate a single bond between H atoms in H_2 and I atoms in I_2. Yet experimental observations proved that they have different bond energies, 436 kJ mol^{-1} for H-H and 151 kJ mol^{-1} for I – I, and bond lengths, 0.74A° for H_2 and 2.66 A° for I_2. These and many other descripencies cannot be explained by Lewis theory. Explanation for these discripencies were provided by two theories, viz., Valence bond theory and Molecular orbital theory.

1-3 THE VALENCE BOND THEORY

The valence bond theory was put forward by W. Heitler and F. London in 1927 to explain the stability of covalent bond of hydrogen molecule. Later J.C. Slator and Linus Pauling extended it to explain other covalent bonds.

This theory retains the space distribution of atoms given by VSEPR theory and combines the quantum mechanical approach of various shapes of atomic orbitals and argues that a bond is formed by the overlap of these orbitals. The basic postulates of this theory are :

1. When two atoms form a covalent bond, there is a overlap of atomic orbitals of the valence electrons.
2. The electrons are localized and concentrated between the nuclei.
3. Only two electrons with their spins paired may be shared by one set of overlapping orbitals.
4. The strength of the bond is directly proportional to the extent of overlap of the atomic orbitals. The greater the degree of overlap, the stronger is the bond.
5. The potential energy of the atoms is lowered during the overlap or bond formation and this is achieved by a suitable positioning of atoms to have maximum overlap.

To understand VB theory let us start with the simple molecule H_2. Each H atom has one valence electron.

$$H^{\cdot} \quad + {}^{\cdot}H \longrightarrow H:H \quad \text{or} \quad H-H$$

Lewis structure of H_2

From quantummechanical approach each valence electron is said to have situated in 1*s* orbital and valence bond theory postulates the covalent bond formed between H – H atoms as due to overlap of 1*s* orbitals of H atoms

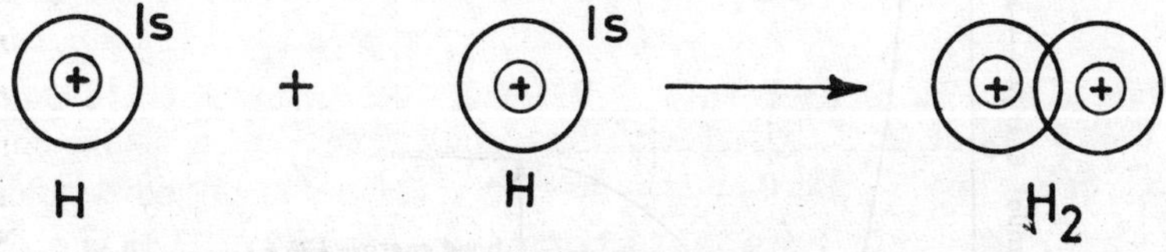

Valence bond model of H atom

HCl is formed by the overlap of 1*s* orbital of H atom and 3*p* orbital of Cl atom.

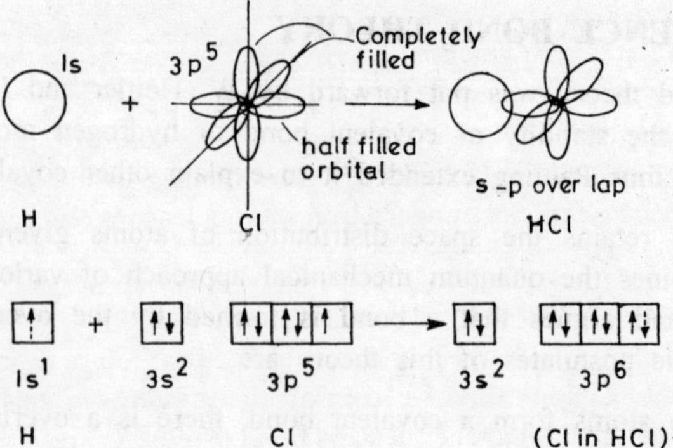

This is an example of overlap between a *s* orbital and a *p* orbital and called *s-p* overlap.

Similarly Cl_2 is formed by the overlap of 3*p* atomic orbitals of Cl atoms involving a *p-p* overlap.

p-p over lap

3 p_x of Cl + 3 p_x of Cl → Cl_2

Which factor(s) favors the overlap of atomic orbitals ? The answer to this question could be found by studying the energitics of a reaction. Formation of a molecule (say H_2) from its atoms (H) is an exothermic process.

$$H + H \longrightarrow H_2 + \text{heat}$$

Therefore a molecule may be said to have lesser amount of energy than its constitutent atoms. Initially when two H atoms are far apart, there is no

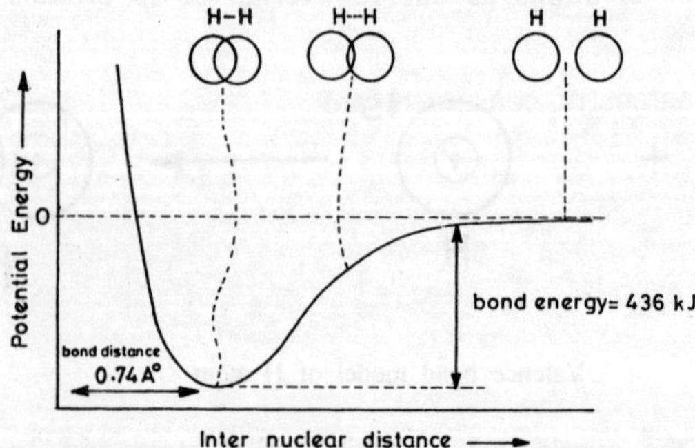

Fig. 1-2 Changes in Potential energy as two H atoms are brought closer

interaction and are at a zero potential energy level. As we bring them closer, the potential energy gradually decreases to a minimum value corresponding to formation of H_2 molecule. This state corresponds to substantial overlap of 1*s* atomic orbitals. Any attempt to further compress the molecule would streeply rise the potential energy because of electron–electron and nuclear-nuclear repulsions. This is shown in Fig. 1-2

The concept of overlap of atomic orbitals explains well the changes in bond energy and bond length for different diatomic molecules. As we discussed already H-H, H-Cl and Cl-Cl atoms involve overlap of different orbitals and thereby expected to have different bond energies and bond lengths as given in Table 1-4.

Table 1-4 Dependence of bond energies and bond lengths on the type of bond overlap

Property \ Molecule	H_2	HCl	Cl_2
Type of overlap	*S-S* overlap	*S-P* overlap	*P-P* overlap
Bond energy (kJ/mol)	435	431	243
Bond length Å	0.74	1.36	1.99

Problem 1 : Describe the structure of the H_2O molecule by the valence bond method.

Step 1 : Electron arrangement of constitutent atoms

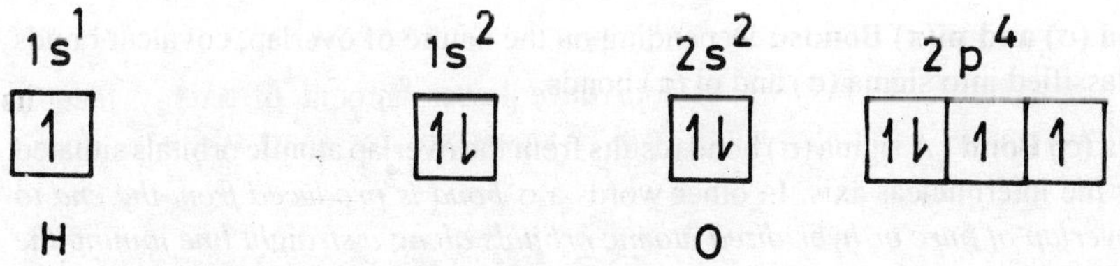

Step 2 : Orbital diagram of constituent atoms

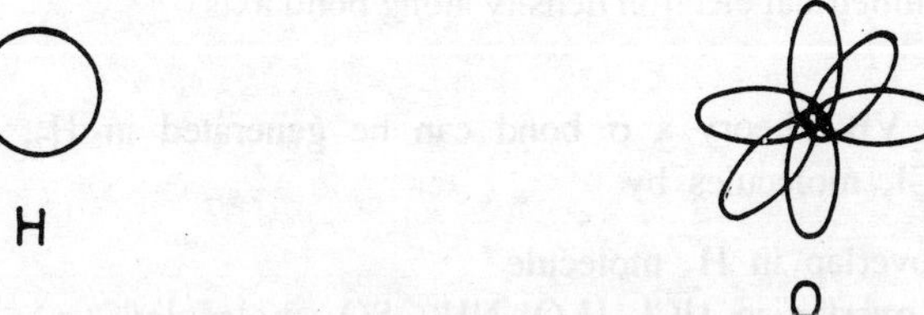

Step 3 : Combination of atomic orbitals as per chemical reaction

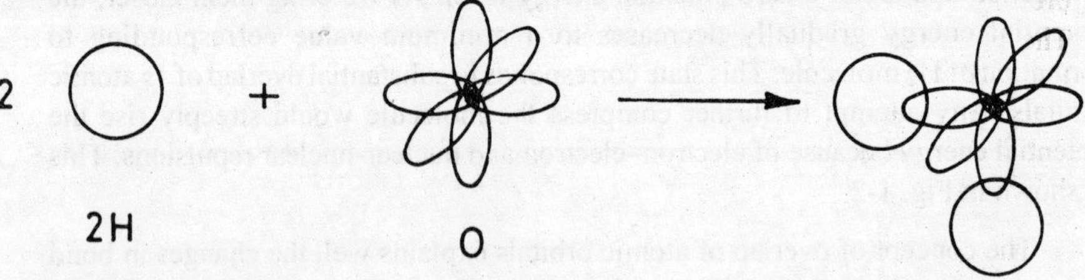

Step 4 : Conclusions

(i) H_2O is a trigonal molecule with a H-O-H bond angle of 90°

(ii) O-H bonds are formed by s-p overlap.

1-3-1 Limitations of Valence Bond Theory : From generalization of Problem 1 we notice that if bonding in a molecule involves two or three *p* orbitals of a central atom, the bond angles are 90°. The experience is contrary to this expectations. The measured bond angles for H_2S, H_2O and NH_3 are 92°, 104.5°, 107° respectively, but not as predicted to be 90°.

In addition to inadequacies discussed in case of molecules like H_2S, H_2O and NH_3, valence bond theory in its original form fails to explain molecules like CH_4 with four equally energitic bonds with equal bond lengths and having a H-C-H bond angle of 109.5°. An explanation to this trivial nature of CH_4 is provided by incorporating concept of hybrid orbitals in original valence bond theory.

1-3-2 Classification of Covalent Bond in the Light of Valence Bond Theory :

Sigma (σ) and pi(π) Bonds : Depending on the nature of overlap, covalent bonds are classified into sigma (σ) and pi (π) bonds.

Sigma (σ) Bond : A sigma (σ) bond results from the overlap atomic orbitals situated along the internuclear axis. In other words a *σ-bond is produced from the end to end overlap of pure or hybridized atomic orbitals along a straight line joining the nuclei of the bonded atoms.* Probability of finding the electron pair involved in the bond formation is maximum in the region between the nuclei. Therefore, it consists of maximum, symmetrical electron density along bond axis.

As discussed in VB Theory a σ bond can be generated in H_2, HCl, H_2O, NH_3, SO_2 and Cl_2 molecules by

(i) *s-s* overlap in H_2 molecule
(ii) *s-p* overlap in HCl, H_2O, NH_3, SO_2 molecules
(iii) *p-p* overlap in Cl_2 molecule

Pi (π) bond : A pi(π) bond results from the sidewise overlap of *pure p* orbitals. The hybrid orbitals do not play any role in the π-bond formations. The electrons density is symmetrically placed above and below the bond axis.

p + p → π bond

Difference between σ and π bonds are given in the first chapter of section 'B'

1-4 MOLECULAR ORBITAL THEORY

Hund and Mulliken originally proposed the molecular orbital theory and was later developed into its present form by Jones and Coulson. In valence bond theory, a covalant bond is said to have formed by overlap of atomic orbital of valence electron, the extent of overlap determining the strength of the bond. Here completely filled and inner orbitals do not participate in any kind of overlap and hence do not contribute to any stability of either the covalent bond or the molecule. This is clearly an understatement of the facts, since the nuclei of approaching atoms to form a covalent bond, mutually affect the orbitals of almost all the electrons. In molecular orbital theory all the atomic orbitals of the participating atoms get mixed up to give equivalent number of new orbitals called **molecular orbitals** which now belong to the molecule as a whole. In other words *a molecular orbital is formed as a result interaction of atomic orbitals of the bonding atoms and are associated with the entire molecule.*

1-4-1 Comparision between Molecular Orbitals and Atomic Orbitals :

Similarities :

1. The molecular orbitals (MOs) follow the same rules as the atomic orbitals in their filling, following established aufbau rule, Hund's rule and Pauli exclusion principle. Therefore like an AO, MO also can accommodate a maximum of two electrons.

2. The number of AOs of participating atoms are equal to the number of MOs.

3. Like AOs, MOs also have different energies and different shapes

4. Similar to AO the incoming electron always occupies MO having lowest energy.
5. Similar to AOs, MOs can also be thought to possess a wave length and designated by certain wave function ψ.

Dissimilarities :

An atomic orbital is influenced only by the respective nuclei while MO is influenced by all the participating nuclei. Therefore AO is unicentric while MO is polycentric.

The wave function describing a molecular orbital is generally obtained by employing one of the following methods.

1. Linear combinations of atomic orbitals method.

2. United atom method

Out of these the first method is rather simple and we will first give a detailed account of it. A brief account of the second method is given at the end of the chapter.

1-4-2 LCAO Method :

According to De Broglie theory of wave particle duality, electrons may be thought to possess both particle as well as wave nature. Therefore electrons in any atomic orbital may be totally described by a wave function (ψ) with a fixed wave length and amplitude. *Waves of similar nature interact with each other producing a resultant wave that has an enhanced amplitude or a diminished amplitude, called* **constructive interference** *and* **destructive interference,** *respectively, Fig. 1-3.*

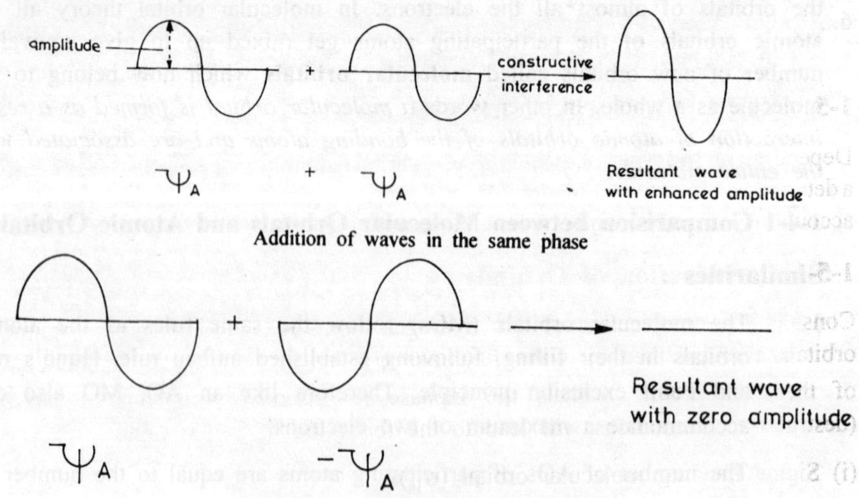

Fig. 1-3 Linear combinations of two waves

When we consider these waves as our electron waves, the resultant wave is comparable to molecular orbital. Thus the molecular orbitals are said to have formed by the linear combination of atomic orbitals (LCAO). *The molecular orbital formed as an result of constructive interference is termed as* **"bonding molecular orbital"** *and the MO formed as a result of destruction interference is termed* **"antibonding molecular orbital".** Therefore linear combination of any two atomic orbitals generates equal number of molecular orbitals, that is orbitals are never destroyed. All the above discussed facts are listed as postulates of MO theory.

1-4-3 Basic Postulates of MO Theory

The rules pertaining to formation of MOs and the manner in which electrons are assigned to them are given by the following postulates.

1. The number of MOs produced is equal to the number of AOs combined.
2. When two AOs combine; of the two MOs produced, one is at a lower energy than the original AOs and called bonding MO and the other is at a higher energy and called antibonding MO
3. Electrons normally seek the lowest energy MOs, i.e., bonding MO, available to them in a molecule.
4. Obeying Pauli exclusion principle, each MO accommodates a maximum of two electrons in it.
5. Obeying Hund's rule, electrons enter MOs of identical energies singly before they pair up
6. Formation of a bond between atoms requires that the number of electrons in bonding MOs exceed the number of antibonding MOs

1-5 TYPES COMBINATIONS OF AOS

Depending on the types of available AOs various combinations result. We present a detailed account of combinations of *s-s, s-p* and *p-p* (axial and lateral) and a passing account of *p-d* and *d-d* and non bonding combinations.

1-5-1 Combination of Orbitals

Consider the combination of two hydrogen atoms A and B whose 1*s* atomic orbitals are described by wave functions ψ_A and ψ_B. On linear combination of these 1*s* atomic orbitals the inphase (constructive) and out of phase (destructive) interactions will lead to the formation of two σ bonds

(i) Sigma bonding molecular orbital (σ_{1s}).

(ii) Sigma antibonding molecular orbital (σ^*_{1s})

A pictorial representation of formation of these two bonds is made in Fig. 1-4.

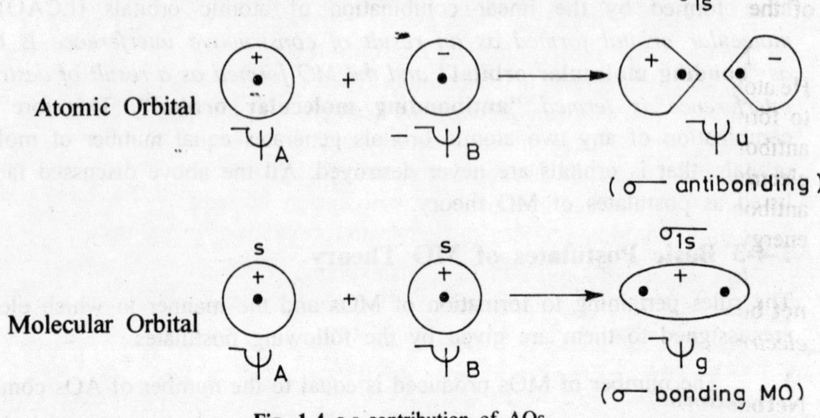

Fig. 1-4 *s-s* contribution of AOs

The bonding molecular orbital function ψ_g leads to increased electron density between the nuclei and therefore is at a lower energy than the original atomic orbitals. Conversely the antibonding molecular orbital function ψ_u leads to two lobes of opposite sign giving a zero probability of finding the electron between the nuclei. ψ_u is at a higher energy level when compared to corresponding atomic orbitals Fig. 1-5

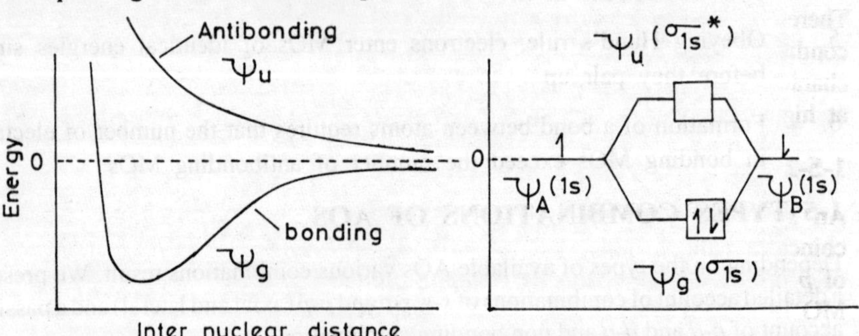

Fig. 1-5 Energy build up of MOs

In the molecular orbital wave functions the subscript g stands for ***gerade*** *(German : even) and the subscript u stands for* **ungerade** *(German : Uneven). When a molecule is rotated about the inter nuclear axis and the line perpendicular to it, if the sign of lobes is unaltered, the orbital is gerade and if the sign changes it is ungerade.*

Energy levels of s-s-orbitals and stabilization : The energy of the bonding molecular orbital is lower than the atomic orbital by ΔE called **stabilizations energy.** Similar to atomic orbitals, a molecular orbital can also accommodate only two electrons. Thus during formation of H_2 molecule each

H atom in its $1s$ state contributes a stabilization energy of ΔE making a total saving of energy of $2\Delta E$. The total stabilization energy of $2\Delta E$ corresponds of the bond energy and accounts for the bond formations.

Now consider the hypothetical case of formation of He_2 molecule. Here each He atom has two electrons in its Is orbital. When $1s^2$ atomic orbitals of He combine to form He_2, two electrons occupy the bonding MO and two more occupy the antibonding MO. The stabilization energy of $2\Delta E$ achieved by the two bonding electrons is offset by the destabilization energy of $2\Delta E$ attained by the electrons of antibonding MO. Since overall there is no energy change, there is no saving of energy in the formation of He_2 and hence no formation of He_2 molecule.

Mathematically, this conclusion is explained by defining a new term called **net bond order.** *It is equal to one half of the difference between the number of electrons in bonding MO and antibonding MO.*

$$\textbf{Net bond order} = \frac{(\text{no. of } e^- \text{ in bonding MOs}) - (\text{no. of } e^- \text{ in antibonding MOs})}{2}$$

According to above definition

Net bond order for $H_2 = \frac{2-0}{2} = 1$

Therefore a stable bond is formed between H atoms.

Net bond order for $He_2 = \frac{2-2}{2} = 0$

Therefore He-He bond is not stable and He_2 does not exist under normal conditions. The molecular orbitals formed from $2s$ orbitals have the same characters as those derived from $1s$ orbital, except for the fact that they are at higher energy.

1-5-2 *s-p* Combination of Atomic Orbitals

An s orbital combines with a p orbital only when the axis of p orbital coincides with the inter nuclear axis. A bonding MO is formed when a lobe of p orbital overlaps with the s-atomic orbital of *similar sign.* An antibonding MO is formed when a lobe of p orbital overlaps an s orbital of *unlike sign,*

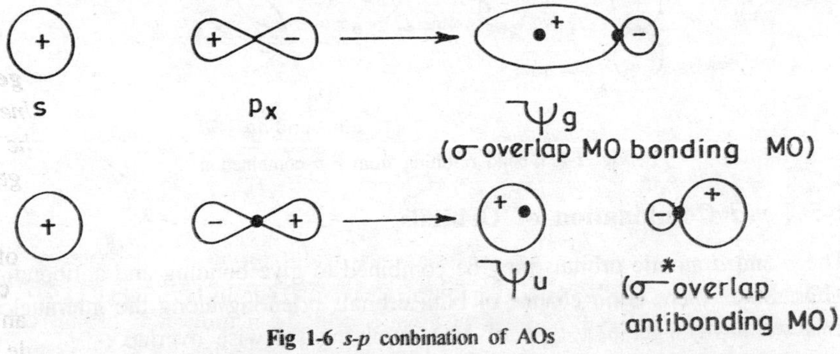

Fig 1-6 *s-p* conbination of AOs

1-5-3 *p-p* Combination of Orbitals

There are two possibilities for combinations of two *p* orbitals. The combinations of *p*- orbitals could occur along inter nuclear axis yielding a σ bond or perpendicular to the inter nuclear axis yielding a π bond.

σ overlap of *p-p* orbitals or Axial combination of *p-p* orbitals : If the *internuclear axis is assumed to fall with x-axis,* then the combination of two p_x atomic orbitals will yield a bonding or antibonding sigma molecular orbital. As mentioned in VB theory this type of overlap is called *head to head overlap* or *end to end or end on overlap,* Fig. 1-7

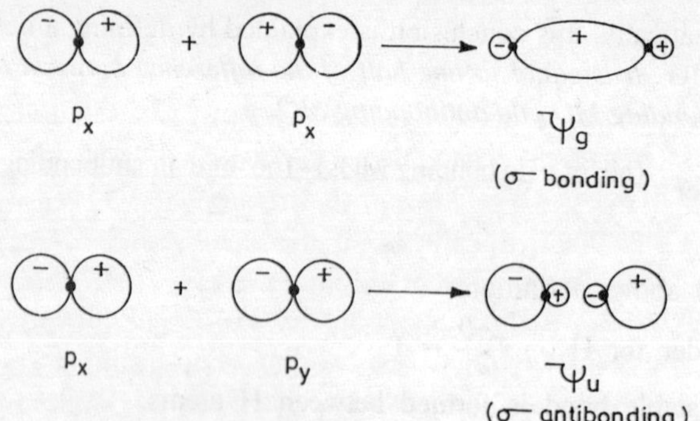

Fig. 1-7 σ bond resulting from *p-p* conbination

π overlap of *p-p*-orbitals or Lateral overlap of *p-p* orbitals : Combination of two *p* orbitals situated perpendicular to internuclear axis yield π bonding and π antibonding MOs, Fig. 1-8

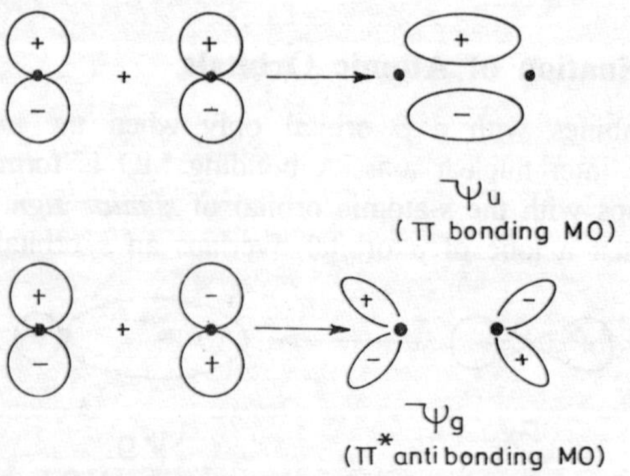

Fig. 1-8 π-bond resulting from *p-p* combination.

1-5-4 *p-d* Combination of Orbitals

The *p* and *d* atomic orbitals may be combined to give bonding and antibonding MOs. Since there is no chance of both orbitals orienting along the internuclear axis we always get a π type of MO involving sidewise overlap.

Bonding of this kind occurs in oxides, oxyacids of phosphorous and sulphur and in transition metal compounds. Orbital diagrams are not given here because of their complexity.

1-5-5 *d-d* Combinations of Orbitals

The bonding and antibonding MOs formed by the overlap of two *d*-orbitals are called δ and δ* overlaps respectively. On rotating these δ bonds about internuclear axis, the sign of the lobes changes four times, compared with two times in case of π bond and no change in case of σ bond.

1-5-6 Nonbonding Combinations of Orbitals

As discussed already a bonding MO results from the combinations of lobes of like signs (constructive interference) and antibonding MOs result from the combinations of lobes of unlike sign (destructive interference). Further we also know that a bonding MO, owing to the stabilization, is at a lower energy while an antibonding MO is at a higher energy than the total energy of corresponding AOs. When the combination of AOs bring no change in overall energy the situation is termed non-bonding. For example let us consider a lateral combinations of *s* and *p* orbitals, Fig 1-9. In this overlap the stabilization occuring from(+) and(+) overlap above internuclear axis is destabilized by (+)

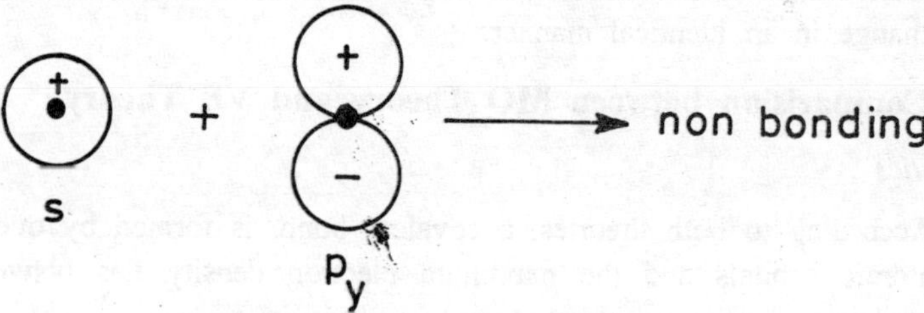

Fig 1-9 Non bonding combination of s and p AOs

and(-)overlap below the internuclear axis, thus leading to nonbonding. Some other nonbonding combinations of AOs are shown in Fig. 1-10.

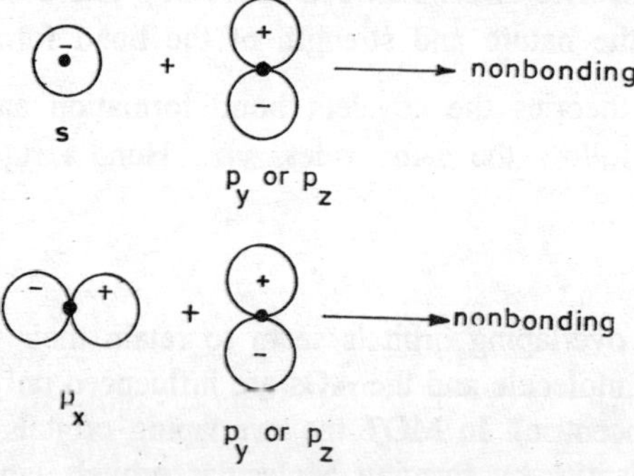

Fig. 1-10 Some nonbonding combinations

1-6 RULES FOR LINEAR COMBINATION OF ATOMIC ORBITALS (LCAO)

A detailed examination of different overlap patterns of AOs and corresponding energy patterns showed that effective combinations of AOs follow three rules.

1. The energy of AOs must be of same order. Thus, in the formations of a homonuclear diatomic molecule like H_2, 1*s* orbital of one atom does not combine with 2*s* orbital of another atom (excited H atom) of the same kind. Similarly combinations of atomic orbitals of the same element is also forbidden. However such combinations between hetero atoms are allowed provided the energies of AOs are of the same magnitude.

2. The overlap of AOs must be to a considerable extent. This implies that the atoms must be close enough for effective overlap. Higher the overlap higher is the strength of bond formed. The bond order is

   s - s overlap > *s - p* overlap > *p - p* axial overlap > p - p lateral overlap

3. The combining AOs must have the same symmetry. In order to produce bonding and antibonding MOs the symmetry of both participating AOs must either remain unchanged when rotated about internuclear axis or change in an identical manner.

1-6-1 Comparision between MO Theory and VB Theory

Similarities :

1. According to both theories, a covalent bond, is formed by overlap of atomic orbitals and the maximum electron density lies between the nuclei.

2. Both the theories require the overlaping, AOs must have comparable energies and same symmetry about the molecular axis.

3. In both the theories directional characters of *p* and *d* orbitalsare important in deciding the nature and strength of the bond formed.

4. In both the theories the covalent bond formation and order of filling the orbitals follow the same rules, viz., Hund's rule, Pauli exclusion principle.

Differences

1. In VBT the overlaping orbitals seem to retain their identity even after formation of molecule and the AOs are influenced only by the respective nuclei (monocentric). In MOT the overlaping orbitals totally loose their individual identity by forming Molecular orbitals which are influenced by both the nuclei (polycentric)

2. Antibonding visualised by MO Theory can have two paired electrons in their antibonding orbitals but VBT treats them as only unpaired electrons.

3. Phenomena of resonance has important role only in VBT and it has no place in MOT.

4. Properties like paramagnetism of O_2, etc., are accountable only by MOT.

5. The total wave function of VBT does not involve ionic terms (components) which are considered in MOT.

1-7 EXAMPLES OF MO TREATMENT

Lewis Theory and VB method basically require a pair of electrons for the formation of a covalent bond. One electron molecular ions like H_2^+, involving one electron bond as detected spectroscopically cannot be accounted by the above approaches but could be successfully explained by MO treatment. To highlight the beauty of MO treatment we now present few examples of homonuclear and heteronuclear diatomic molecules.

1-7-1 Homo Nuclear Diatomic Molecules

H_2^+, molecule ion : Detected spectroscopically, when H_2 gas under reduced pressures is subjected to electric discharge.

(i) It is formed by the combinations of H atom and a proton (H^+)

(ii) The molecule ion has a single electron and occupies σ_{1s} bonding MO, so there is some stabilization.

(iii) Since No of electrons in bonding MO > antibonding MO

$$\text{Bond order} = \frac{1}{2}(1\text{-}0) = \frac{1}{2}$$

Therefore there is a *single electron* or half bond in H_2^+.

H_2 molecule : This is a stable molecule

(i) It is formed by the combination of two H atoms each one having electron in 1*s* orbital.

(ii) The two incoming electrons enter $\sigma 1s$ bonding MO.

(iii) $$\text{Bond order} = \frac{(2-0)}{2} = 1$$

Therefore a regular covalent σ *bond* is formed.

(iv) The energy order $\sigma 1S^2$

He_2^+ molecule ion : Observed only spectroscopically when helium gas is passed through an electric discharge.

(i) It is formed by the combination of a helium atom and a He^+ ion.

(ii) In the MO two electrons enter $\sigma 1s$ bonding MO and one electron in $\sigma^* 1s$ antibonding MO. Therefore stabilization energy slightly overweighs destabilization energy.

(iii) Bond order $= \frac{2-1}{2} = \frac{1}{2}$

Therefore, as in H_2^+, in He_2^+ too has a half bond formation.

(iv) The energy order : $\sigma 1s^2\ \sigma^* 1s^1$

He_2 molecule : This molecule is nonexistent.

(i) It is formed by the combination of two He atoms each containing two electrons in $1s$ orbital

(ii) Two of the incoming electrons enter σ_{1s} bonding MO and other two into $\sigma^* 1s$ antibonding MO.

(iii) Bond order $= \frac{2-2}{2} = 0$

The stabilization and destabilization energies are equal and opposite and cancel each other.

Therefore, no bond is produced.

Molecular orbital diagrams of the above discussed four molecules is given in figure 1-11.

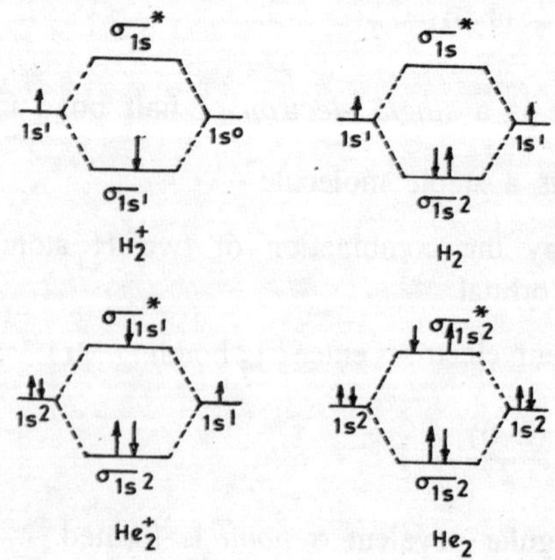

Fig. 1-11 Molecular orbital diagrams of H_2^+, H_2, He_2^+, He_2

B_2 molecule : The electron configuration of boron (Z = 5) is $1s^2\ 2s^2\ 2p^1$.

(i) When B atoms combine to give B_2 we need to accommodate 10 electrons in MO

(ii) The first eight *s* electrons are placed in the σ MOs. Four electrons occupy $\sigma 1s^2$ & $\sigma 2s^2$ bonding MO and the rest four occupy $\sigma^* 1s^2$ & $\sigma^* 2s^2$ antibonding MO.

(iii) $\pi 2p$ orbitals are at lower energy than $\sigma 2p_x$ orbitals. So remaining two $2p$ electrons enter the degenerated (equal energy) $\pi 2p_y$ and $\pi 2p_z$ orbitals. Obeying Hund's rule, each of these degenerated orbitals is occupied by a single electron. Because of these unpaired electrons, B2 is paramagnetic. So the MO build up is represented as

$$\sigma 1s^2,\ \sigma^* 1s^2\ \sigma 2s^2,\ \sigma^* 2s^2,\ \begin{cases} \pi\ 2p_y^1 \\ \pi\ 2p_z^1 \end{cases}$$

The orbital diagram is given in Fig. 1-12

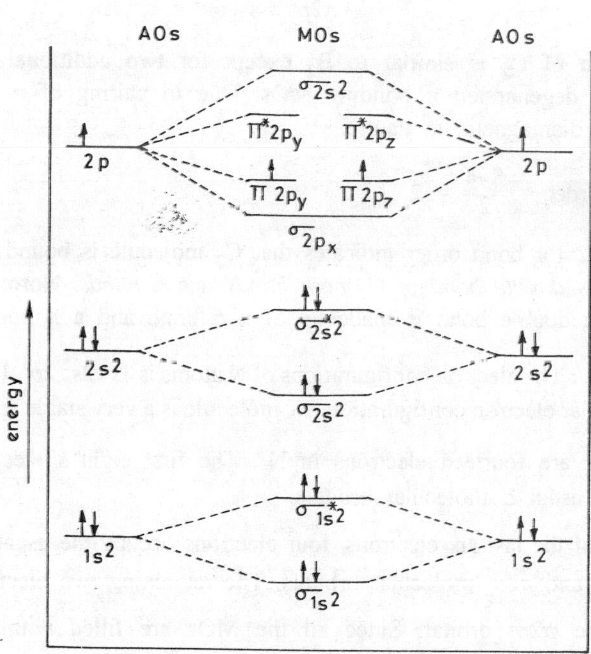

Fig. 1-12 MO diagram of B_2

(iv) Net bond order = $\frac{6-4}{2} = 1$

Therefore, a single bond is formed between B atoms. It is interesting to note that this single bond is a π bond (because of two electrons in π MO) and not a σ bond as found normally to have been.

C2 molecule : The electrcnic configuration of carbon (Z =6) is $1s^2 2s^2 p^2$ c2 molecules have been detected in the carbon vapor.

(i) When C atoms combine to give C2, we need to accommodate 12 electrons in MO.

(ii) The 1s & 2s electrons occupy $\sigma 1s^2$ & $\sigma 2s^2$ bonding MOs and $\sigma^* 1s^2$ & $\sigma^* 2s^2$ antibonding MOs

(iii) Out of the four *2p* electrons, two each go to degenerated energy levels $\pi 2p_y^2$ and $\pi 2p_z^2$ bonding MOs. The order of energy levels of MOs is represented as

$$\sigma 1s^2,\ \sigma^* 1s^2,\ \sigma 2s^2,\ \sigma^* 2s^2,\ \begin{cases} \pi 2p_y^2 \\ \pi 2p_z^2 \end{cases}$$

MO diagram of C_2 is similar to B_2 except for two additional electrons in each of the degenerated π bonding MOs. Due to pairing of π electrons C_2 molecule is diamagnetic in nature.

(iv) Bond order $= \dfrac{8-4}{2} = 2$

A value of 2 for bond order indicates that C_2 molecule is bound by a double bond. *In this double bond of C_2 both bonds are π bonds.* Normally in other molecules, a double bond is made up of a σ bond and a π bond.

N_2 Molecule : The electron configurations of N atoms is $1s^2\ 2s^2\ 2p^3$. Unlike B_2 and C_2 with similar electron configuration, N_2 molecule is a very stable species.

(i) There are fourteen electrons in N_2. The first eight *s* electrons occupy their usual σ molecular orbitals.

(ii) Out of the last six electrons, four electrons occupy the usual degenerated $\pi 2p_y^2$ and $\pi 2p_z^2$ molecular orbitals. The remaining two electrons then go to the $\sigma 2p_x^2$ orbital. Since all the MOs are filled with fully paired electrons, N_2 molecule is diamagnetic. The MO build up is represented as

$$\sigma 1s^2,\ \sigma^* 1s^2,\ \sigma 2s^2,\ \sigma^* 2s^2,\ \sigma 2p_x^2\ ,\ \begin{cases} \pi 2p_y^2 \\ \pi 2p_z^2 \end{cases}$$

(iii) Bond order $= \frac{10-4}{2} = 3$

The high stability of N_2 is due to its triple bond comprised of one sigma bond and two pi bonds. This is the pattern for all stable triple bonds.

O_2 molecule : The oxygen atom has an electronic configuration of $1s^2\ 2s^2\ 2p^4$

(i) There are sixteen electrons in O_2 molecule. The eight s electrons do not involve in any bond formation since equal number (i.e. 4) occupy the bonding and antibonding molecular orbitals.

(ii) The extra two 2p electrons gained in going from N_2 to O_2 occupy the π^*2p_y and π^*2p_z antibonding MOs and have parallel spins as per Hund's rule. Consequently O_2 molecule has two unpaired electrons and is paramagnetic.

This was the first success of MO Theory over VB Theory which failed to account for paramagnetism of O_2.

According to VB approach O_2 was given the resonance structures.

$$\ddot{\underset{\cdot\cdot}{O}}=\ddot{\underset{\cdot\cdot}{O}} \longleftrightarrow \cdot\ddot{\underset{\cdot\cdot}{O}}-\ddot{\underset{\cdot\cdot}{O}}\cdot$$

Out of these two the first structure fails to account for paramagnetism while the second structure fails to accommodate the well recognized double bond character and octet rule. The MO build up of O_2 is represented as

$$\sigma 1s^2,\ \sigma^* 1s^2,\ \sigma 2s^2,\ \sigma^* 2s^2,\ \begin{Bmatrix} \pi 2p_y^2 \\ \pi 2p_z^2 \end{Bmatrix},\ \sigma\ 2p_x^2,\ \begin{Bmatrix} \pi^* 2p_y^1 \\ \pi^* 2p_z^1 \end{Bmatrix}$$

(iii) Bond order $= \frac{10-6}{2} = 2$

Since $\pi*2p_y^1$ and $\pi*2p_z^1$ are half filled they cancel half the effect of $\pi 2p_y^2$ and $\pi 2p_z^2$ respectively, resulting in two half pi bonds. Therefore the double bond expected from a bond order of two may be thought as sum of

$$1\sigma + \frac{1}{2}\pi + \frac{1}{2}\pi = 2$$

O_2^- ion : The superoxide ion, O_2^-, contains seventeen electrons. The electron build up process is similar to O2. The extra electron occupies either $\pi*2p_y$ or $\pi*2p_z$. The order of MOs is thus represented as

$$\sigma 1s^2,\ \sigma^* 1s^2,\ \sigma 2s^2,\ \sigma^* 2s^2,\ \sigma 2p_x^2,\ \begin{Bmatrix} \pi 2p_y^2 \\ \pi 2p_z^2 \end{Bmatrix},\ \begin{Bmatrix} \pi^* 2p_y^2 \\ \pi* 2p_z^1 \end{Bmatrix}$$

Therefore the ion has one unpaired electron and is paramagnetic.

The effects of bonding $\pi 2p_y^2$ and antibonding $\pi^* 2p_y^2$ cancel totally and bonding $\pi 2p_z^2$ and $\pi^* 2p_z^1$ half cancel. Therefore the bond order is the sum of $1\sigma + \frac{1}{2}\pi = 1\frac{1}{2}$. The same value is obtained from general formula for bond order.

$$\text{Bond order} = \frac{10-7}{2} = 1\frac{1}{2}$$

F_2 molecule : The electronic configuration of F atom is $1s^2\ 2s2\ 2p^5$, thus F_2 has eighteen electrons. The extra two electrons gained in going from O_2 to F_2 occupy the $\pi^* 2p_y$ and $\pi^* 2p_z$ antibonding orbitals. The MO build up is thus shown as :

The bonding $1s$, $2s$, $2p_y$, $2p_z$ orbitals are cancelled by the equivalent antibonding orbitals. Thus we have only a σ bond between fluorine atoms. This is also true from bond order.

$$\text{Bond order} = \frac{10-8}{2} = 1$$

The single bond and diamagnetic characters of F_2 are supported by suitable experimental evidences

1-7-2 Heteronuclear Diatomic Molecules

The LCAO rules are same for both homonuclear and heteronuclear molecules. In case of homonuclear molecules the bonding and antibonding MOs are equally, symmetrically spaced above and below the corresponding AOs

In heteronuclear molecules the MO diagram is not symmetric due to the difference in the electronegativities of the participating atoms. The electrons in the bonding MO spend more time near the more electronegative atom and therefore is closer to the energy level of its AO. Similarly the electrons in the antibonding MO are closer to the AO of less electronegative atoms.

To highlight the applicability of MO to heteronuclear molecules, we now discuss formation of NO and CO molecules.

NO molecule : The electron configuration of N and O are $1s^2\ 2s^2\ 2p^3$ and $1s^2\ 2s^2\ 2p^4$ respectively. The N atom has 7 electrons and the O atom has 8 electrons, making 15 electrons in the NO molecule.

(i) The O atom is more electronegative than N atom. Therefore the AOs of oxygen atom are slightly at a lower energy (more stabilized) than corresponding atomic orbitals of nitrogen atom.

(ii) The build up of MOs of NO molecule is

$$\sigma 1s^2,\ \sigma^* 1s^2,\ \sigma 2s^2,\ \sigma^* 2s^2,\ \sigma 2p_x^2,\ \begin{cases}\pi 2p_y^2 \\ \pi 2p_z^2\end{cases},\ \begin{cases}\pi^* 2p_y^1 \\ \pi^* 2p_z^o\end{cases}$$

(iii) The order of energy in MO is shown in Fig 1-13 Due to presence of one unpaired electron the NO molecule is paramagnetic.

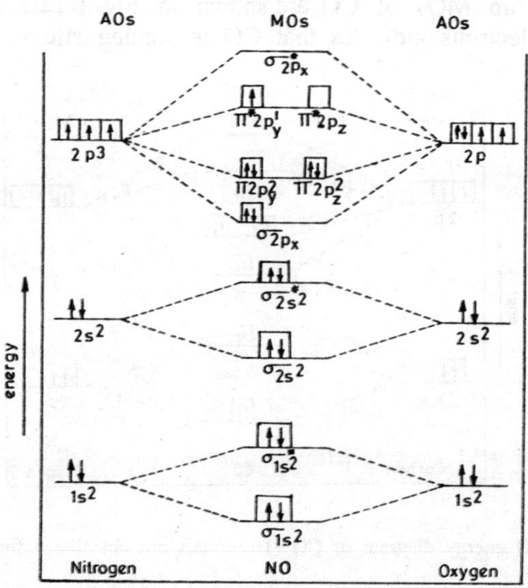

Fig. 1-13 The MO build up in NO

(iv) Net bond order $= \dfrac{10-5}{2} = 2\dfrac{1}{2}$

A bond order of $2\frac{1}{2}$ indicates that NO molecule is less stable than N_2. This is confirmed from their bond dissociations energies of 626.6 kJ mol^{-1} and 941.3 kJ mol^{-1} for NO and N_2 respectively.

(v) The energy gap of AOs of N and O is 250 kJ. This large difference makes the combination of AOs to give heteronuclear molecule NO less effective than corresponding homonuclear molecules (N_2 and O_2) which have a zero energy gap.

(vi) Removal of one electron to give NO^+ increases the bond order to 3 and results in a stronger N-O bond.

CO molecule : The electron configuration of C and O atoms are $1s^2\ 2s^2\ 2p^2$ and $1s^2\ 2s^2\ 2p^4$ respectively. The MO of carbon monoxide should therefore contain 14 electrons.

(i) Oxygen atom being more electronegative than carbon should have more stabilized AOs and are placed at a lower energy level than corresponding AOs of carbon.

(ii) the order of energy levels of MOs are

$$\sigma 1s^2,\ \sigma^* 1s^2,\ \sigma 2s^2,\ \sigma^* 2s^2,\ \sigma 2p_x^2,\ \begin{matrix} \pi 2p_y^2 \\ \pi 2p_z^2 \end{matrix}$$

(iii) The build up MOs of CO are shown in Fig. 1-14 . The absence of unpaired electrons indicates that CO is diamagnetic.

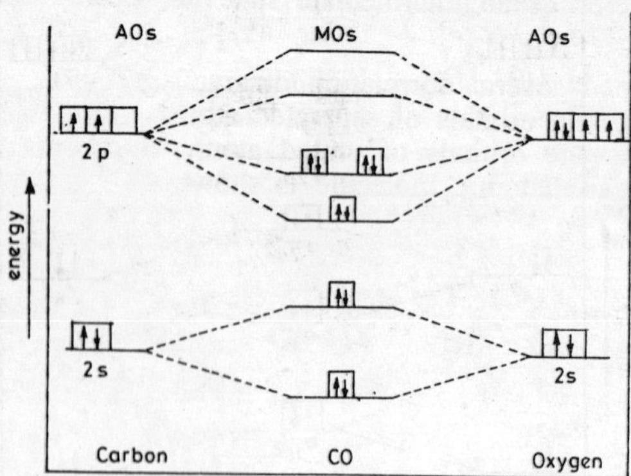

Fig. 1-14 MO energy diagram of CO (1s orbitals are not shown for simplicity)

(iv) bond order $= \dfrac{10-4}{2} = 3$

Therefore, CO molecule has a triple bond. This fact is reflected by its high bond dissociation energy (1069 kJ mol^{-1}). A comparative account of bond orders and bond dissociation energies of few of the above discussed diatomic molecules is given in Table 1-5

Table 1-5 Relation between bond orders and bond dissociations energies of simple diatomic molecules

Molecule	Bond order	Bond dissociation energy (kJ/mol)
CO	3	1069
N_2	3	941
NO	$2\frac{1}{2}$	627
O_2	2	493
H_2	1	458
He_2^+	$\frac{1}{2}$	269
He_2	0	0

1-8 UNITED ATOM METHOD

In LCAO method we mainly discussed the factors that favor in bringing two atoms from infinity to an equilibrium position where atoms are separated by molecular bond length. On the contrary, *in the united atom method we start with a hypothetical united atom where the nuclei are totally superimposed and then go to their equilibrium position.* A united atom is said to have same number of energy levels (orbitals) as a normal atom, but contains electrons from both the atoms. To accommodate the discrepancies arising out of too many electrons in few energy levels, some electrons are promoted to higher energy levels of the united atom. Further, these energy levels are also thought to have some bearing to the nuclear charge arising out of superimposition of both the nuclei. Thus the molecular orbitals formed in the equilibrium position may be considered as an intermediate state between united atom and atomic orbitals.

Mulliken obtained several correlation diagrams by joining the energy levels, starting from atomic orbitals of separated atoms, via the molecular orbitals of the molecule, to the orbitals of united atom. Mulliken's correlation diagram of a homonuclear diatomic molecule is shown in Fig. 1-15

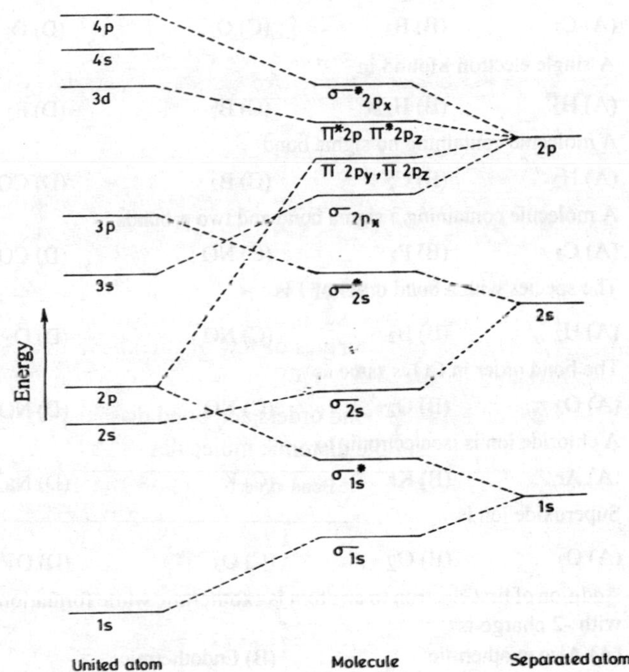

Fig. 1-15 Mulliken correlation diagram of a homonuclear diatomic molecule

1-9 MULTIPLE CHOICE QUESTIONS

1-9-1 Pure π-bond is formed as a result of

(A) Axial overlap of AOs
(B) lateral overlap of AOs
(C) Simultaneous axial and lateral overlap of AOs
(D) Separate axial and lateral overlaps of AOs

1-9-2 Which one of the following overlaps gives strongest covalent bond

(A) *s-s* (B) *s-p* (C) *p-p* (D) *d-d*

1-9-3 If x-axis is the molecular axis, which of the following overlaps yields a pi bond

(A) s-p_y (B) p_x–p_x (C) p_x–p_y (D) p_y–p_y

1-9-4 If x axis is the molecular axis which of the following is a nonbonding combination

(A) s & p_x (B) s & p_y (C) p_x & p_x (D) p_y & p_y

1-9-5 Mole2cule having two pi bonds without a sigma bond

(A) CO_3 (B) NO (C) B_2 (D) C_2

1-9-6 In which set both the molecules are paramagnetic

(A) B_2, O_2 (B) N_2, F_2 (C) C_2, O_2 (D) N_2, O_2

1-9-7 Which of the following molecules is diamagnetic

(A) C_2 (B) B_2 (C) O_2 (D) O_2

1-9-8 A single electron isfound in

(A) H_2^+ (B) H_2 (C) B_2 (D) F_2

1-9-9 A molecule containing no sigma bond

(A) H_2 (B) C_2 (C) B_2 (D) CO

1-9-10 A molecule containing a sigma bond and two π bonds

(A) C_2 (B) F_2 (C) NO (D) CO

1-9-11 The species with a bond order of 1 is

(A) H_2^+ (B) B_2 (C) NO^+ (D) O_2

1-9-12 The bond order in CO is same as

(A) O_2 (B) O_2+ (C) NO (D) NO+

1-9-13 A chloride ion is isoelectronic to

(A) Ar (B) Kr (C) K (D) Na^+

1-9-14 Superoxide ion is

(A) O_3^- (B) O_2^- (C) O_3^{--} (D) O_3^{--}

1-9-15 Addition of first electron to an atom is exothermic while formation of an anion with -2 charge is

(A) Also exothermic (B) Endothermic
(C) Electronegativity (D) Leads to thermal decomposition

1-9-16 The lowering of potential energy due to electrostatic attraction is called

(A) electron affinity (B) Ionisation energy
(C) Electronegativity (D) Lattice energy

1-9-17 The element although forming an ionic bond but not following octet rule

(A) Cl (B) Sn (C) Br (D) Na

1-9-18 Compound containing a covalent bond

(A) NaCl (B) $MgCl_2$ (C) Cl_2 (D) KCl

1-9-19 Covalent compound in crystalline state

(A) Br_2 (B) I_2 (C) CCl_4 (D) NaCl

1-9-20 The pair of compounds not having linear structure

(A) CO_2 & SO_2 (B) H_2O & SO_2 (C) $BeCl_2$ & CO_2 (D) CO_2 & H_2O

1-9-21 The pair of compounds with a square planar shape

(A) CH_4 & SiH_4 (B) $[Ni(CN)_4]^{2+}$ & $[XeF_4]$

(C) XeF_4 & CH_4 (D) SiH_4 & $[Ni(CN)4]^{2+}$

1-9-22 An example of π-p-p overlap

(A) CO_2 (B) P_4 (C) S_4 (D) H_2O

1-9-23 When two atoms are brought close to each other there is a change in

(A) Kinetic Energy (B) Ionisation Energy

(C) Potential Energy (D) Nothing happens

1-9-24 The bond angle HNH in NH3 is 107° but the predicted bond angle is

(A) 104.5^{o} (B) 92° (C) 90° (D) 109°

1-9-25 Waves of similar nature interact with each other producing a resultant wave that has an enhanced amplitude, then it is called as

(A) Constructive Interference (B) Non Interfercence

(C) Covalent Interference (D) Destructive Interference

1-9-26 An antibonding molecular orbital is denoted as

(A) σ p (B) σ (C) π (D) σ *

1-9-27 A wave function is denoted by

(A) σ (B) π (C) ψ (D) δ

1-9-28 The wave function subscript g stands for

(A) gerade (B) generate (C) gene (D) gentic

1-9-29 The one half of the difference between the number of electrons in bonding MO and antibonding MO is termed as

(A) Non bonding (B) Net bond order

(C) Electron deficient (D) Antibonding

1-9-30 The net bond order of He_2 is

(A) 1 (B) 1.5 (C) O (D) 2.5

1-9-31 No. of s-electrons in B_2 are

(A) 2 (B) 4 (C) 6 (D) 8

1-9-32 Oxygen molecule is paramagnetic it has a

(A) Single bond (B) Double bond (C) Triple bond (D) Three electron bond

1-9-33 No. of electrons present in Nitric oxide molecule

(A) 7 (B) 8 (C) 15 (D) 18

1-9-34 On removing an electron from NO the bond order changes to

(A) 2.7 (B) 3 (C) 1.5 (D) 0

1-9-35 The method in which the nuclei are considered to be totally super imposed is

(A) LCAO (B) AO (C) MO (D) ΨAM

1-10 SHORT AND ESSAY-TYPE QUESTIONS

1-10-1 (a) Differentiate MO and AO.

(b) What do you understand by bonding and antibonding molecular orbitals ? Why are they so called ?

1-10-2 How the formation of a covalent bond is explained by valence bond theory?

1-10-3 What is molecular orbital theory ? How does it differ from valence bond theory ?

1-10-4 What do you understand by the terms bonding and antibonding molecular orbitals. Why are they so called ? Illustrate your answer taking H_2 molecule as a typical example.

1-10-5 Explain the significance of bond order. Suggest its usefulness in comparing the strengths of chemical bonds.

1-10-6 Explain the paramagnetic properties of O_2. Discuss the bond order of O_2 in terms of Lewis theory and MO theory and explain why the results are different.

1-10-7 Explain the unique bonding nature of B_2 and C_2 molecules.

1-10-8 Compare the relative strengths of F_2 and F_2^+ with the help of MO method.

1-10-9 Linear combination of homonuclear AO is a facile process than similar heteronuclear AOs. Explain the statement.

1-10-10 Explain why the bond order of N_2 is greater than N_2^+ but the bond order of O_2 is less than that of O_2^+.

1-10-11 Differentiate the MO diagrams of NO and NO^+. Explain their magnetic properties.

1-10-12 Write a brief account of united atom theory.

1-10-13 Compare and contrast VB and MO theories of covalent bonding.

1-10-14 Write an account of *s-p* combination of orbitals. Explain why an *s*-orbital of one atom can not combine with a *p* orbital perpendicular to the inter nuclear axis.

1-10-15 With the help of MO theory account for the following

(i) The bond dissociation energy of NO^+ is higher than of NO.

(ii) The bond length of CO^+ is more than that of CO.

(iii) Ionization energy of N_2 is larger than that of O_2.

(iv) The bond lengths of O_2, O_2^+, O_2^- lie in the order $O_2^+ < O_2 < O_2^-$.

Answers

1-9-1 B	1-9-2 A	9-9-3 D	1-9-4 B	1-9-5 D	1-9-6 A
1-9-7 B	1-9-8 A	1-9-9 C	1-9-10 D	1-9-11 B	1-9-12 D
1-9-13 A	1-9-14 B	9-9-15 B	1-9-16 D	1-9-17 B	1-9-18 C
1-9-19 B	1-9-20 B	1-9-21 B	1-9-22 A	1-9-23 C	1-9-24 C
1-9-24 C	1-9-25 A	1-9-26 D	1-9-27 C	1-9-28 A	1-9-29B
1-9-30 C	1-9-31 D	1-9-32 C	1-9-33 C	1-9-34 B	1-9-35 D

2

THE ELEMENTS OF GROUP IIIB

THE BORON FAMILY

					He
B	C	N	O	F	Ne
Al	Si	P	S	Cl	Ar
Ga	*Ge*	As	*Se*	Br	Kr
In	Sn	Sb	Te	I	Xe
Tl	Pb	Bi	Po	At	Rn

2-1 INTRODUCTION TO THE P-BLOCK ELEMENTS

In the p-block elements the differentiating electron enters the p-subshell. The block comprises of IIIB, IVB, VB, VIB, VIIB and zero groups in which progressively 1 to 6 electrons are filled in the p-orbital. In all these elements the **ns** orbitals are completely filled and hence we can write the general electronic configuration as $ns^2np^{1\text{-}6}$.

Hence we can define p-block elements as those elements in which the differentiating electron enters the p-orbital and gradually fills it to its maximum capacity of six electrons and constitute III B and 0 groups as given in table 2-1

Table 2-1 p-Block elements in the periodic table

IIIB	*IVB*	*VB*	*VIB*	*VIIB*	*0*
					He
B	C	N	O	F	Ne
Al	Si	P	S	Cl	Ar
Ga	*Ge*	*As*	*Se*	Br	Kr
In	Sn	Sb	Te	I	Xe
Tl	Pb	Bi	Po	*At*	Rn

The p-block is the most typical amongst all the blocks of the periodic table. It consists of nonmetals, metals and metalloids, which exist under normal temperature conditions in all three states of matter, viz., solid or liquid or gas. On careful observation of the p-block, we notice that the elements present above the bold line are non-metals, below the bold line including Al are metals and the elements lying between the lines are *metalloids* or semi conductors. This division of the p-block shows a marked variation in the properties of the elements within the same group.

Amongst the sixteen nonmetallic elements, present above the diagonal, ten of them are gases. They are; helium, nitrogen, oxygen, fluorine, neon, chlorine, argon, krypton, xenon and radon. With regards to the reactivity of these gaseous elements of the p-block except He, Ne, Ar, Kr and Rn, rest are all chemically active and are responsible for the formation of numerous compounds. Carbon exists in soft forms like Charcoal, lamp black, as well as in hard crystalline forms like diamond or graphite. Similarly phosphorous, its diagonal neighbour also exists in different forms such as red, white, black, violet, yellow phosphorous. sulphur in its crystalline form exists as monoclinic and rhombic states and in soft states as plastic or milk of sulphur. Selenium also has two forms one red and the other grey. Therefore, it is found that out of the six remaining non metals four of them exists in polymorphic forms. Bromine, one of the three liquid elements of the periodic table at room temperature, and iodine, the real sublimiable crystalline solid, are highly reactive elements.

The elements below and on the diagonal line are all in solid state at room temperature. Gallium has a low melting point and melts even at body temperature.

The metallic character increases going down the groups with a simultaneous increase in the basicity of the elements. Another important feature enjoyed by p-block elements is inert pair effect operative from B to Tl, Si to Pb, As to Bi and Te to Po. This property makes a clear gradation of a metalloid changing to a metal on descending down in each of the groups. Individually all the metalloids distinguish themselves by exhibition of polymorphism as the nonmetals do and this property, with tin as an exception, is not found in the metals. Tin exhibits four allotropes at different temperatures, grey below 18°C, white between 18°C to 161°C and rhombic above 161°C. The allotropic conversions in tin is known as tin disease or tin pest. Germanium has two forms, the nonmetallic light grey form and the metallic dark grey form. Arsenic exists as grey and yellow non metallic allotropes and a black metallic allotrope, antimony also has two similar allotropic modifications a yellow nonmetallic and a highly explosive metallic grey allotrope. In general we can conclude that polymorphism is a very common feature of the p- block elements, with a very few exceptions, and is totally absent in the other blocks of periodic table.

The general trends in the physical properties exhibited by the p- block elements is shown in the figure 4-1. A slight irregularity may be present in a few of the properties, due to the inert pair effect or due to the sheilding by the d-electrons, but this figure gives a comprehensive picture of the properties at a glance. An attempt has been made to explain these generalized trends in the following section, overlooking these irregularities, which are explained in detail in the individual group discussions.

2-2 GENERAL TRENDS IN PHYSICAL PROPERTIES OF p-BLOCK ELEMENTS

2-2-1 Atomic And Ionic Radii

The atomic and ionic radii of the elements increase going down a group, the only exception being gallium, whose atomic radii is the same as that of aluminium. This discrepency arises because of the sheilding by the d-orbital electrons. In each of the period, the atomic radii decreases as we proceed in a period from left to right, with the exception of zero group elements, as shown in table 1-2.

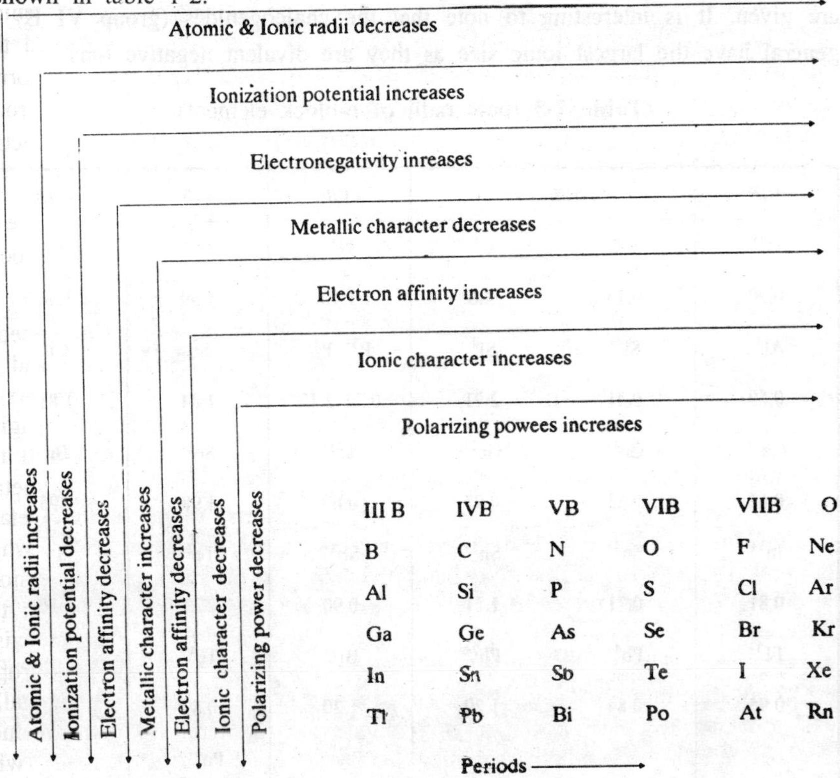

Fig. 1-1 General trends of physical properties exhibited by the p-block elements (The slight irregularities are overlooked).

Table 1-2 The atomic radii values in A units

IIIB	IVB	VB	VIB	VIIB	Zero
B 0.80	C 0.77	N 0.74	O 0.74	F 0.72	Ne 1.60
Al 1.25	Si 1.17	P 1.10	S 1.04	Cl 0.99	Ar 1.91
Ga 1.25	Ge 1.22	As 1.21	Se 1.41	Br 1.14	Kr 2.00
In 1.50	Sn 1.41	Sb 1.40	Te 1.37	I 1.33	Xe 2.20
Tl 1.55	Pb 1.47	Bi 1.46	Po 1.64	At –	Rn –

The largest size of the zero group element in each period is due to the completely filled p-orbital.

With regards to the ionic radii, it is found that every element will form ions of variable oxidation states and hence it will not be possible to have a clear cut generalization. In the table2-3, ionic radii values for the stable ions are given. It is interesting to note that the chalcogenides (group VI B) in general have the largest ionic size as they are divalent negative ions.

Table 2-3 Ionic radii of p-block elements

IIIB	*IVB +*		*VB*	*VIB*	*VIIB*
B^{3+}	C^{4+}	C^{4-}	N^{3-}	O^{2-}	F^-
0.20	0.15	2.6	1.40	1.39	1.36
Al^{3+}	Si^{4+}	Si^{4-}	P^{5+} P^{3-}	S^{2-}	Cl^-
0.50	0.41	2.71	0.34 2.12	1.84	1.81
Ga^{3+}	Ge^{4+}	Ge^{2+}	As^{3+}	Se^{2-}	Br^-
0.62	0.53	0.93	0.69	1.98	1.95
In^{3+}	Sn^{4+}	Sn^{2+}	Sb^{3+}	Te^{2-}	I^-
0.81	0.71	1.21	0.90	2.21	2.16
Tl^{3+}	Pb^{4+}	Pb^{2+}	Bi^{3+}	Te^{4+}	
0.95	0.84	1.20	1.20	0.89	
				Po^{4+}	
				1.02	

It is a known fact that the cations are smaller than their respective atoms and the anions are larger than their respective atoms. While the cations shrink with an increase of charge, the anions assume larger size with increase in their charge. Thus the trivalent negative ions are larger than divalent and movovalent negative ions, but it is other way round in the case of cations. All this can be explained on the basis of the nuclear charge. A comparison of ionic radii of the isolectronic ions of the same period indicates that the ionic radii decreases as we go from IIIB to VIIB.

For a comparison of atomic radii with ionic radii two graphs are drawn (Fig 2-2 a&b), where in we see that the cations are smaller than their atoms and the anions larger than their atoms.

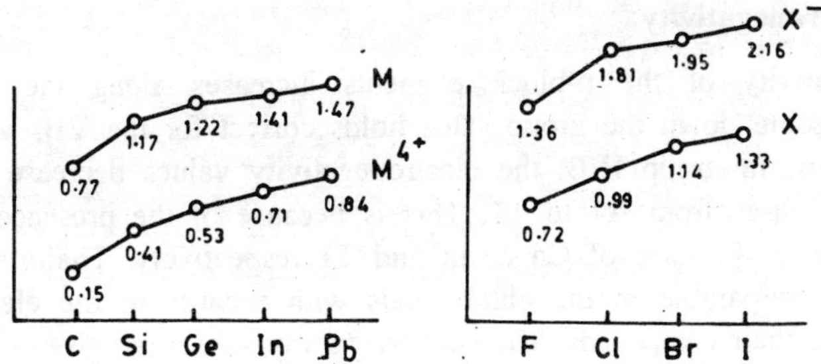

Fig. 2-2 A camparison of cationic and anionic sizes with their respective atoms, (a) Carbon family (b) Halogens.

2-2-2 Ionization Potential

The ionization potentials of the p-block elements are found to decrease going down the groups except in the case of gallium, whose ionization potential is the same as that of the preceeding element Al with a value of 6.0 eV. this is once again because of the presence of the d-block elements. In the periods it is found that the ionization potential of the VB group is more than that of the VIB group, and this is because of the stable configuration due to half filled p-orbital in the nitrogen family. The ionization potential values are given in table 2-4.

Table 2-4 Ionization potential values (ev) of p-block elements

IIIB	*IVB*	*VB*	*VIB*	*VIIB*	*O*
					He 24.6
B 8.3	C 11.3	N 14.6	O 13.6	F 17.4	Ne 21.6
Al 6.0	Si 8.2	P 11.0	S 10.4	Cl 13.0	Ar 15.7
Ga 6.0	Ge 8.1	As 10.0	Se 9.8	Br 11.8	Kr 14.0
In 5.8	Sn 7.3	Sb 8.6	Te 9.0	I 10.4	Xe 12.1
Tl 6.1	Pb 7.4	Bi 8.0	Po –	At –	Rn 10.8

2-2-3 Electronegativity

Electronegativity of the p-block elements increases along the period and decreases going down the group. This holds correct for the VB, VIB and the VIIB groups. In group IIIB, the electronegativity values decrease from B to A1 but increases from A1 to T1. This is because of the presence of d- and f- electrons in the case of Ga & In and T1 respectively. Thallium is highly basic and comparable to the alkalimetals with regards to the electropositive character of their compounds. The electronegativity values are given the table 4-5.

From electronegativity values we can deduce that the metallic character increases going down the group, so will the ionic character. The compounds of Tl^{+}, Pb^{++}, Bi^{+++} are more ionic than their preceeding elements of the respective groups. As the halogens are highly electronegative, the covalency is prominently exhibited in the molecular halogens F_2, Cl_2, Br_2 and I_2, but when combine with highly electropositive elements they form ionic compounds CsF, NaC1 etc. When they combine with elements with high electronegativity HC1, F_2O etc., form polar cvalent compounds.

Table 2-5 The electronegativity values of p-block elements.

IIIB	*IVB*	*VB*	*VIB*	*VIIB*	*O*
					He
B 2.0	C 2.5	N 3.0	O 3.5	F 4.0	Ne
A1 1.5	Si 1.8	P 2.1	S 2.5	C1 3.0	Ar
Ga 1.6	Ge 1.8	As 2.0	Se 2.4	Br 2.8	Kr
In 1.7	Sn 1.8	Sb 1.9	Te 2.1	I 2.5	Xe
T1 1.8	Pb 1.8	Bi 1.9	Po 2.0	At 2.2	Rn

2-2-4 Electron Affinity

The process of electron affinity is an exact reversal of ionization potential. Both these processes are for neutral isolated gaseous atoms. *In electron affinity an electron is added and the energy in the form of heat is released from the atom and it is an exothermic process. In ionization potential, an electron is removed and the energy is required to do so is absorbed in the atom and hence it is an endothermic process.* The energy released when an electron is added makes the atom a negatively charged ion (1st electron affinity) and further addition of electrons to the formation of uninegative ion give dinegative, trinegative etc., ions as in the case of O^{2-}, S^{2-}, N^{3-}, P^{3-}. The first electron enters a neutral atom without much opposition, but the second electron must be forced onto an already negative ion and this process requires more energy as it has to overcome the repulsive forces between two like charged species, the uninegative ion and the electron. But when the electron adds against all odds, the second electron affinity of the atom dips below zero and attains a negative sign, showing that the process has become endothermic instead of exothermic. The values of electron affinities of p- block elements are given in table 2-6.

On examining the table 2-6 we find that the non metals of the second period have electron affinities that are less exothermic than the elements of the same group just below them in the third period, this is because in a given group size increase as we move down the group. Apparently, the crowding of the electrons in the outer shell of a second period element makes the mutual repulsions of the electrons substantially greater than in the outer shell of a third period element.

Therefore, even though the electron that is added to a second period element gets close to the nucleus than one added to a third period element, the greater repulsion in the smaller atoms leads to a lesser exothermicity than a larger atom. Therefore we notice in the table 2-6 that the electron affinity increases upto third period and then gradually decreases down the group from the third period onwards, and it always increases along a period.

Table 2-6 Electronaffinity values of p-block elements.

IIIB	*IVB*	*VB*	*VIB*	*VIIB*	*0*
					He –
B 0.30	C 1.26	N 0.09	O 1.48 (– 7.3)	F 3.62	Ne –
Al 0.46	Si 1.39	P 0.75	S 2.07 (–3.4)	Cl 3.60	Ar –
Ga 0.31	Ge 1.24	As 0.80	Se 2.02 (–4.2)	Br 3.54	Kr –
In 0.31	Sn 1.24	Sb 1.05	Te 1.96	I 3.08	Xe –
Tl 0.31	Pb 1.14	Bi 1.14	Po 1.89	At 2.81	Rn –

2-2-5 Valency, oxidation number and oxidation state

Valency is defined as the combining capacity of an element. Oxidation number is said to be the formal charge left on an atom after the sharing of its valency electrons and hence it is a property of covalency. For example, carbon after sharing its four lone electrons with, say chlorine to give CCl_4, will be said to acquire a formal +δ charge on it and in turn each of the four chlorines will have a formal -δ charge. These are said to be formal charges as the real charges are attained only after loosing or gaining of the electrons. Therefore, carbon exhibits an oxidation number of 4 while chlorine exhibit a valency or oxidation number of 1.

Oxidation state is the change which takes place in an atom after loosing or gaining of electrons. When electrons are lost the atom gains positive charge and on gaining electrons it attains negative charge, hence oxidation state is a property of ionicity. Oxidation state will always carry either a positive or a negative charge before a number eg in NaCl the oxidation state of sodium is +1 and chlorine is – 1, similarly in $MgBr_2$, the oxidation state of magnesium is +2 and bromine is -1. Hence one must represent the sign of the charge, (+) or (–), of the ion present after the formation of a compound.

In the table 1-7, the oxidation numbers or the elements with respect to hydrogen and oxygen are given for the second and the third period of elements.

Table 2-7 Oxidation numbers of 2nd and 3rd period elements

2nd period	LiH 1	BeH_2 2	B_2H_6 3	CH_4 4	NH_3 3	H_2O 2	HF 1
	Li_2O 1	BeO 2	B_2O_3 3	CO_2 4	N_2O_5 5	O_2 0	O_2F_2 1
3rd Period	NaH 1	MgH_2 2	AlH_3 3	SiH_4 4	PH_3 3	H_2S 2	HCl 1
	Na_2O 1	MgO 2	Al_2O_3 3	SiO_2 4	P_2O_5 5	SO_3 6	Cl_2O_7 7

The oxidation number increases from 1 to 4 and decreases back to 1 with respect to hydrogen, but with respect to oxygen the oxidation number increases from 1 to 7 in each of the two periods, the other periods are not considered, as the transition elements lie in between the two sets of these representative blocks, but a similarity do exists.

Each group of the p-block exhibits variable oxidation numbers or states, depending on the type of compounds they form.

Table 2-8 Oxidation numbers of p-block elements

IIIB	*IVB*	*VB*	*VIB*	*VIIB*	*0*
					He
B **3**	C (2) **4**	N 1, 2, **3**, 4, **5**	O **2**	F **1**	Ne –
Al **3** (1)	Si (2) **4**	P **3**, (4), 5	S 2, 4, 6	Cl **1**, 3, 4	Ar –
Ga **3** (1)	Ge 2, **4**	As **3**, **5**	Se 2, 4, 6	Br 1, 3, 4, 5, 6	Kr –
In **3** (1)	Sn **2**, **4**	Sb **3**, **5**	Te 2, 4, 6	I **1**, 3, 5, 7	Xe –
Tl **1** (3)	Pb **2**, 4	Bi **3**, 5	Po 2, **4**, 6	At –	Rn –

In the table 2-8, oxidation numbers of the stable valence states are given in **bold** print, the unstable in light print and the doubtful ones in the parenthesis. The oxidation states of each group are dealt in the individual chapters. It is very interesting to note, that, the maximum number of oxidation states in the periodic table are exhibited by nitrogen. The eleven variable oxidation states of nitrogen are as follows :

NH_3	N_2H_2	NH_2OH	N_2	N_2O	N_2O_2	N_2O_3	N_2O_4	N_2O_5	N_2O_6	HNO_4
–3	–2	–1	0	+1	+2	+3	+4	+5	+6	+7

2-3 CHEMISTRY OF p-BLOCK ELEMENTS

The nonmetals of the p-block elements are responsible for the formation of a wide spectrum of chemicals and it is difficult to assume chemistry without them.

2-3-1 Oxides

Almost all the elements combine with oxygen to give oxides, which fall in three categories viz., basic, amphoteric and acidic oxides. The most metallic elements form strongly basic oxides, metalloids form amphoteric oxides and an increase in the non- metallic character increases their tendency to form acidic oxides. Since the metallic character gradually decreases across a period the nature of oxides formed gradually changes from basic to acidic nature via amphoteric behaviour.

Na_2O	MgO	Al_2O_3	SiO_2	P_2O_5	SO_3	Cl_2O_7
Strongly basic	Less basic	Amphoteric acidic	Weakly acidic	Acidic	Strongly acidic	Strongly acidic

On moving down any group the basic character increases because of increase in the size of atoms and thus metallic character.

CO_2	SiO_2	GeO_2	SnO	PbO
Acidic	Less acidic	Amphoteric	Basic	More basic

2-3-2 Hydrides

We are already familiar with the ionic hydrides formed by the alkali and alkaline earth metals in which hydrogen exists a H^- ion.

Most of the p-block elements form hydrides, however of covalent nature. The hydrides of non-metals are more stable than the metal hydrides. Thus the stability of hydrides across a period, increases and decreases, down the group and the acid strength increases. Among all hydrides HI forms strongest acid solution in water. For members of nitrogen family, the order of stability is NH_3, PH_3, AsH_3, SbH_3, BiH_3.

Table 2-9 Some hydrides of p-block elements

Group IIIB	*IVB*	*VB*	*VIB*	*VIIB*
B_2H_6	CH_4	NH_3	H_2O	HF
$(AlH_3)_2$	SiH_4	PH_3	H_2S	HCl
Ga_2H_6	GeH_4	AsH_3	H_2Se	HBr
InH_3	SnH_4	SbH_3	H_2Te	HI
TlH_3	PbH_4	BiH_3	H_2Po	–

2-3-3 Halides

All the p-block elements form halides. those formed by non metallic elements are covalent. As we proceed from left to right in a period the metallic character decreases, hence the ionic character of halides decreases and covalent character increases. $SbCl_5$ is partially ionic while $TeCl_4$ is covalent. In the case of halides of elements with variable oxidation states, the halides of lower oxidation state are ionic and in higher oxidation states they tend to be covalent.

$PbCl_2$	$PbCl_4$
Ionic	Covalent

The size of the halide ion also plays a role in the determination of the nature of halide, as it effects the property of polarisability, Polarisability increases with increase in the size of the halide resulting in covalent character.

AlF_3	$AlCl_3$	$AlBr_3$	AlI_3
Ionic	Intermediate	Covalent	Covalent

The solubility of halides of a metal decreases in the following order.

$$MI > MBr > MCl > MF.$$

2-4 CHEMISTRY OF IIIB GROUP

The first member of this family being B, the IIIB group is also called as boron family. Boron (B), Aluminium (A1), Gallium (Ga), Indium (In), Thallium (T1) constitute the group IIIB. Since the differentiating electron in these elements enters the p-subshell they come under p-block elements. the general electronic configuration of this family is ns^2np^1.

Boron is a metalloid and other members are typical metals. Boron has abnormally high ionization potential due to its small size which is also the cause of the reduced metallic character. Other members A1, Ga, In, T1 have large atomic radii and are therefore more metallic and form ionic compounds.

2-5 OCCURRENCE

Boron occurs in nature as borates, borax $Na_2B_4O_7$. $10N_2O$. Aluminium is the 3rd most abundant element in the earths crust. Widely distributed ores being aluminium silicate, felspar ($KA1Si_3O_8$), bauxite ($A1_2O_3.xH_2O$) and Cryolite (Na_3A1F_6). Other members of this family Ga, In, T1 are found mainly as minor components in a variety of compounds and hence are available in trace amounts.

2-6 ELECTRONIC CONFIGURATION

All the elements of IIIB group have three electrons in their outermost shell (two s-electrons and one p-electron). Their general electronic configuration can be represented as ns^2np^1.

Element	At.No.	Electronic configuration
Boron (B)	5	$1s^22s^22p^1$
Aluminium (A1)	13	$1s^22s^22p^63s^63p^1$
Gallium (Ga)	31	$1s^22s^22p^63s^23p^64s^23d^{10}4p^1$
Indium (In)	49	$1s^22s^22p^63s^23p^64s^23d^{10}4p^65s^24d^{10}5p^1$
Thallium (T1)	81	$1s^22s^22p^63s^23p^64s^23d^{10}4p^65s^24d^{10}5p^66s^25d^{10}4f^{14}6p^1$

2-7 PHYSICAL CHARACTERISTICS OF GROUP IIIB

All the elements of group IIIB are solids and except B all are metals. A consolidated table of physical properties is given in Table 2-10. With respect to the properties of these elements we observe an increasing trend as we go down the group, in atomic weight, atomic volume, atomic and ionic sizes, and densities. On the other hand properties such as hardness, B.P., M.P., ionization potential, heats of sublimation etc., show a decreasing trend.

Table 2-10 Physical constants of group IIIB elements

Element	B	A1	Ga	In	T1
Physical appearance	Black solid	Silvery white metal	Silvery white metal	Silvery white metal	Silvery white metal
Atomic number	5	13	31	49	81
Atomic weight	10.81	26.98	69.7	114.82	204.37
Oxidation states	+3	+3	+3(+1)	+3(+1)	(+3)+1
Nature of Compounds	Covalent	Covalent	ionic	Ionic	Ionic
Ionization I	8.28	5.98	5.98	5.76	6.11
Potentials II	25.09	18.70	18.81	16.83	20.26
(eV) III	37.82	28.25	30.52	27.82	29.62
Sum of the 3 I.Ps.(eV)	71.19	52.93	55.31	50.41	55.99
Atomic radii Å	0.82	1.18	1.26	1.44	1.48
Ionic radii Å	0.20	0.50	0.62	0.80	0.95
Covalent radii Å	0.8	1.25	1.25	1.50	1.55
Density g/cc	2.34	2.70	5.9	7.31	11.83
Heat of Sublimation	564	324	273	241	179
Melting point (K)	2202	666	31.8	157	303
Boiling point (K)	2823	2723	2570	2275	1730
Electronegativity	2.1	1.5	1.6	1.7	1.8

2-7-1 Density, Melting and Boiling Points, and Heats of Sublimation

Density gradually increases down the group with an exception of A1. There is no regularity in the melting points, but however boiling points follow a more regular decreasing trend down the group. The same type of order follows in the case of heats of sublimation.

2-7-2 Size of Atoms and Their Ions

There is a gradual increase in the size of atoms of the boron family. B is small enough to deviate in some of its properties from the rest of the family. The covalent radii of group IIIB does not increase regularly from B to T1 because of the intervention of ten d-electrons present in the penultimate shell. In the case of Ga and In, because of poor shielding capacity of d-electrons, the outer electrons, are more effectively drawn towards the nucleus making the covalent radius smaller than would otherwise be expected. Similar is the case of

Tl, the intervention of $4f^{14}$ electrons, having much poorer shielding capacity, cause reduction in the covalent radius instead of an increase.

2-7-3 Ionization Potentials

The outermost shell of IIIB group elements contain two s- electrons and one p-electron. As p-electron is held less firmly than s-electron, the first I.P. of these elements is low when compared to the first I.P. of the s-block elements of the same period. The 2nd and 3rd I.P. values are higher.

A close look at the 1st I.P. of IIIB elements show that it is the highest for B, then decreases sharply at Al and remains the same at Ga and again decreases in In and increases for the last element Tl. This *anomalous* behaviour is explained as follows. In the case of B the I.P. is highest because of its small size and least shielding effect. From B to Al there is an appreciable change in the size and also in the effective screening by the intervening s and p-electrons. As we step further down from Al to Ga, there are ten d-electrons in its penultimate shell and these electrons are poorer shielders in comparison to s- and p- electrons. Hence the outermost electrons effectively, and as such the I.P. almost remains practically the same even though atomic number increases from Al to Ga. In the case of In also a similar situation arises and there is a negligible decrease in I.P. In the case of Tl the intervening $4f^{14}$ electrons in the antipenultimate shell have much smaller shielding effect than other electrons. Hence the $6s^2$ electrons in Tl are shielded more firmly by the nucleus and hence an increase in ionization potential.

2-7-4 Oxidation States

All the elements of this group show an oxidation state of +3, however inert pair effect plays a vital role in this family. As we proceed from B to Tl we find that the ns^2 electrons become more stable and inert, hence they do not take part in chemical reactions. Boron and aluminium exhibit stability in +3 state, gallium and indium exhibit a stable +2 state in compounds like $Ga^+[GaCl_4]^-$ and $In^+[InCl_4]^-$ along with stable +3 state. Thallium exhibits only one stable oxidation state of +1, because **inert pair effect is** highly effective in it. The compounds of thallium are analogues to those of the alkali metal compounds. The thallium hydroxide is as alkaline as sodium hydroxide, the crystal lattice of thallium chloride as well as sodium chloride are face centered cubic and the other thallium halides have cesium chloride crystal lattice.

2-7-5 Inert Pair Effect

The occurance of an oxidation state which is two less than the group valency is sometimes referred to as the **inert pair effect**, but it is important to recognize that this is a description and not an explanation. This phenomenon is quite general among the heavier elements of the p-block, specially the post transition elements or the B-sub groups of the periodic table. It is found, that, in the p-block elements the higher oxidation states become unstable and the lower oxidation states become stable on going down the groups.

As seen in the previous section boron and aluminium are stable in +3 oxidation state whereas thallium is stable in +1 oxidation state, similarly in the IV B group it is found that lead and tin are more stable in their +2 state rather than the +4 oxidation state. The +4 oxidation state compounds of these elements undergo reduction to forn +2 stable state compounds.

The term inert pair effect is somewhat misleading since it implies that, the energy required to involve the ns^2 electrons in bonding increases in the sequence Al < Ga < In < Tl. The explaination lies rather in the decrease in bond energy with increase in size from Al to Tl so that the energy required to involve the s electrons in bonding is not compensated by the energy released in forming the two additional bonds. Thus in Tl the $6s^2$ electrons become so inert that only the $6p^1$ electron participates in the bond formation. The inertness of the $6s^2$ electrons in thallium is also explained on the basis of their sheilding by the $5d^{10}4f^{14}$ electrons too.

Therefore, due to the inert pair effect it is found that, in the elements of the III B group and IV B group, a general rule which is observed is, an element is more electropositive in its lower than in the higher oxidation state. The lower oxides and hydroxides are more basic and the higher oxides and hydroxides are more acidic, hence B_2O_3 is an acidic oxide whereas Tl_2O is a basic oxide, similarly, $B(OH)_3$ or H_3BO_3 is an acid whereas TlOH is a strong base.

2-7-6 Metallic Character

Boron is a metalloid or almost a non metal. The other members of the family Al to Tl are typical metals and the metallic character increases down the group. The 1st ionization potentials of these elements fall in a close range. Crystalline form of B is hard, inert and a poor conductor of electricity similar to the covalent crystals, whereas the other elements are good conductors of heat and electricity.

2-7-7 Ionic Behaviour

The high heat of hydration, high lattice energies and high ionization potentials of the family indicate that the compounds formed are covalent. The compounds of boron and aluminium are purely covalent, gallium and indium compounds are partially ionic, but thallium compounds are fully ionic.

2-4-8 Oxidation Potentials

Although the standard oxidation potentials of IIIB elements are high, they are not as high as s-block elements. B does not form an electropositive ion in aqueous solution. aluminium is a strong reducing agent. It is employed for reducing some oxides which are not reduced by carbon. Ionization energy of Al is less than C.

2-8 EXTRACTION OF BORON AND ALUMINIUM

Boron is a non-metal, aluminium is a widely occurring metal and Ga, In and Tl are raw metals. Hence no common general principles of isolation could be designed for group IIIB elements. We shall in brief discuss the principles of isolation of B, Al and the rest being free elements.

2-8-1 Extraction of Boron

The element was isolated in 1807 by Davy. It does not occur in free state in nature. It occurs in the combined form in the following minerals.

(i) Boric acid H_3BO_3: this occurs in the volcanic regions in the form of steam called soffioni.

(ii) Borax $Na_2B_4O_7 10H_2O$ found in India, Tibet, Ceylon and California.

(iii) Boracite, $2Mg_3B_8O_{15}$ $MgCl_2$ occurs in salt deposits of Stassfurt.

(iv) Colemanite $Ca_2B_6O_{11}$. $5H_2O$ found in Asia minor and America.

(v) Boranatrocalcite $Ca_2B_4O_7$ $NaBO_2$. $8H_2O$ is found in Chile.

(vi) Kermate or Razoilite $Na_2B_4O_7$. $4H_2O$, occurs in Majave desert in America in huge deposits.

Methods of preparation

A. Laboratory preparation : On a small scale B is best isolated by reducing Borontrioxide in the presence of Mg.

$$B_2O_3 + 3Mg \longrightarrow 3MgO + 2B$$

It can also be prepared by heating potassium fluoroborates KBF_4 with K

$$KBF_4 + 3K \longrightarrow B + 4KF$$

B. Large scale manufacture : The chief minerals used for the extraction of B are Borax $Na_2B_4O_7$ or Colemanite $Ca_2B_6O_{11}$. Colemanite is first converted to Borax by boiling with sodium carbonate.

$$Ca_2B_6O_{11} + 2Na_2CO_3 \longrightarrow 2CaCO_3 + Na_2B_4O_7 + 2NaBO_2$$

$CaCO_3$ being insoluble settles down and crystals of Borax separate from the mother liquor. The following two steps are important in isolating B from Borax.

Step 1 : Conversion of Borax into Boron trioxide or Boric anhydride : Borax is treated with concentrated hydrochloric acid whereby sparingly soluble boric acid separates out slowly.

$$Na_2B_4O_7 + 2HCl \longrightarrow H_2B_4O_7 + 2NaCl$$

$$H_2B_4O_7 + 5H_2O \longrightarrow \underset{\text{(Orthoboric acid)}}{4H_3BO_3}$$

Orthoboric acid when strongly heated is converted into boron trioxide.

$$2H_3BO_3 \longrightarrow B_2O_3 + 3H_2O.$$

Step 2 : Reduction of Boron Trioxide to Boron : The amorphous form of B is obtained by the reduction of B_2O_3 by Na, K, or Mg metal powders.

$$B_2O_3 + 3Mg \longrightarrow 3MgO + 2B$$

The mixture is heated to bright redness the residue obtained is treated with conc. HCl and boiled to dissolve MgO and unreacted boric acid. Now the residue is dark brown powder of amorphous boron, which is washed with water and dried.

C. Preparation of high grade boron : Pure crystalline form of B is obtained by passing a mixture of BBr_3 vapour and H_2 over electrically heated filament of tungsten at 1200°C. It can also be obtained by subjecting the mixture of BCl_3 vapour and H_2 to the action of high voltage electric arc (=83,000 volts).

$$2BBr_3 + 3H_2 \xrightarrow[\text{1470K}]{\text{Electric arc}} 2B + 6HBr$$

2-8-2 Extraction of Aluminium

It is most widely distributed element in the earths crust (7.3%) the chief ores are (a) Bauxite $Al_2O_3.2H_2O$. (b) Cryolite Na_3AlF_6 (c) Corundum Al_2O_3.

In general aluminium is extracted from Bauxite. The process of extraction involves the following steps.

Step 1 : *Purification of Bauxite :* Bauxite usually contains oxides of Fe, Ti and Si as impurities. Hence the first step in the extraction of Al is formation of its oxide by any one of the following methods. Other oxides are removed by Bayer's process but silica is removed by Serpeck's process.

(a) Bayer's process : The crushed ore is heated at low temperature to convert ferrous to ferric oxide, then it is digested with concentrated NaOH solution. Aluminium dissolves forming $NaAlO_2$ (Sodium meta aluminate) leaving behind Fe_2O_3, SiO_2 unaffected and removed by filtration.

$$Al_2O_3.2H_2O + 2NaOH \longrightarrow 2NaAlO_2 + 3H_2O$$

The filtrate constitute $NaAlO_2$ which is diluted and some crystals of $Al(OH)_3$ are added as seedlings. This solution is agitated for about 36 hours where by slow hydrolysis of sodium aluminate results in the formation of $Al(OH)_3$.

$$NaAlO_2 + 2H_2O \longrightarrow Al(OH)_3 + NaOH$$

Aluminium hydroxide is filtered, washed and ignited to give Al_2O_3. The filtrate is concentrated to be reused.

$$2Al(OH)_3 \xrightarrow{\Delta} Al_2O_3 + 3H_2O$$

(b) Serpeck's process : This method is employed for ores containing large quantities of SiO_2 as impurity, as it also reacts with the base forming silicates. One of the distinct advantages of this process being the formation of ammonia as valuable byproduct.

In this process Al_2O_3 is heated at 1800°C with C in a current of N_2 whereby Al is converted into AlN

$$Al_2O_3 + 3C + N_2 \longrightarrow 2AlN + 3CO$$

Aluminium nitride thus formed is hydrolysed by water to give the hydroxide, which on ignition yields Al_2O_3.

$$AlN + 3H_2O \longrightarrow Al(OH)_3 + NH_3$$

Step 2 : The second main step is electrolytic reduction of alumina, Al_2O_3. Chemical methods failed to reduce alumina as Al reacts with the reducing agent itself. For example, reduction with carbon results in the formation of Al_4C_3. Similarly, it cannot be reduced electrolyticaly in aqueous solution because of high reduction potential of aluminium -1.70V, compared to -0.83V for the reduction (or Electrolysis) of water liberating only H_2 at the cathode.

Hence the best way to obtain Al is by electrolysis of fused Alumina. Pure alumina has very high melting point 2050°C. To bring down the fusion temperature to 1000°C cryolite is added to alumina. Addition of CaF_2 still lowers the fusion temperature to about 900°C. Yet another function of the cryolite is to increase the electrical conductivity of the electrolyte. Iron tank lined with carbon, function as cathode and a series of graphite rods dipped in the molten electrolyte act as anode. The heat loss due to radiation is checked by a layer of powdered coke over the charge. This also reduces the action of oxygen on the carbon rods (anode). Though the electrode reactions are very complicated can be given as follows :

$$4\,AlF_3 \longrightarrow 4Al^{3+} + 12F^-$$

At cathode :

$$4\,Al^{3+} + 12e^- \longrightarrow 4Al$$

At anode :

$$12F^- \longrightarrow 6F_2 + 12e^- \quad \text{(i)}$$

$$2Al_2O_3 + 6F_2 \longrightarrow 4AlF_3 + 3O_2 \quad \text{(ii)}$$

The oxygen evolved at the anode either escapes as such or reacts with carbon rods and forms CO or CO_2 the relative proportion of which depends upon the temperature. The products being gases escape into the atmosphere. The AlF_3 formed ionizes and repeats the cycle, thus more and more amount of Al_2O_3 is reduced.

The Al that gets discharged at the cathode remains at the bottom of the cell in molten condition and can be tapped out at regular intervals. The aluminium is 99% pure.

Electrolytic refining of aluminium : This is done by Hoopes process the electrolytic cell is made of an iron box lined with carbon and acts as anode. This refining cell consists of three layers.

The bottom anode layer consists of the alloy of Al, Cu and Si. The middle layer is of the cryolite and barium fluoride and the top cathode layer is of pure metal.

On electrolysis Al ions from fused mixture rise to the surface and are discharged at the cathode. As the Al ions get discharged at cathode an equivalent amount of Al from alloy present at the bottom layer goes into the middle layer. the impurities of Cu, Si etc., are left behind as anodic mud. Pure form of Al gets collected in the top layer and is tapped out periodically.

2-9 CHEMICAL REACTIVITY OF IIIB ELEMENTS

2-9-1 Action with air, water

Boron does not react with water. However it reacts with air at high temperatures. It burns forming boron trioxide and a little amount of boron nitride, and burns when heated in oxygen.

$$4B + 3O_2 \longrightarrow 2B_2O_3$$

$$2B + N_2 \longrightarrow 2BN$$

Aluminium when comes in contact with oxygen or air it forms a protective film of its oxide on its surface, which makes the metal passive against corrosion. When heated in a current of oxygen it burns with a brilliant light forming its oxide. It is an exothermic reaction.

$$4Al + 3O_2 \longrightarrow 2Al_2O_3 \quad \Delta H = -3229 \text{ kJ}$$

Aluminium does not react with cold water. Saline or saltish waters react with aluminium, hot saline waters react rapidly. Boiling water gets decomposed with aluminium, the reaction with water in both the cases is :

$$2Al + 6H_2O \longrightarrow 2Al(OH)_3 + 3H_2$$

Gallium only gets superficially oxidised with oxygen indium has almost reaction with air or oxygen and thallium forms Tl_2O only

$$4Ga + 3O_2 \longrightarrow 2Ga_2O_3$$

$$4Tl + O_2 \longrightarrow 2Tl_2O$$

Thallium loses its lustre in contact with air or water. In moist air the surfaces gets coated with TlOH.

2-9-2 Action with Acids and Alkalies

Boron reacts with hot concentrated nitric and sulphuric acids to form boric acid.

$$B + 3HNO_3 \longrightarrow H_3BO_3 + 3NO_2$$

$$2B + 3H_2SO_4 \longrightarrow 2H_3BO_3 + 3SO_2$$

It reacts with alkalies to form borates and liberates hydrogen.

$$2B + 6NaOH \longrightarrow 2Na_3BO_3 + 3H_2$$

Aluminium does not react with nitric acid whether dilute or concentrated, neither cold nor hot. Dilute sulphuric acid also has no reaction with aluminium. Hot concentrated sulphuric acid reacts with aluminium to produce sulphurdioxide.

$$2Al + 6H_2SO_4 \longrightarrow Al_2(SO_4)_3 + 3SO_2 + 6H_2O$$

Aluminium dissolves in hydrochloric acid and the reaction increases with the increase of temperature and concentration of the acid.

$$2Al + 6HCl \longrightarrow 2AlCl_3 + 3H_2$$

Strong solutions of the alkalies react with aluminium to produce aluminates or metaaluminates with the liberation of hydrogen.

$$2Al + 6NaOH + 6H_2O \longrightarrow 2Na_3Al(OH)_6 + 3H_2$$

$$2Al + 2KOH + 6H_2O \longrightarrow 2KAl(OH)_4 + 3H_2$$

2-9-3 Oxides and Hydroxides

All the elements of group III B form oxides of the general formula M_2O_3 in which the elements are in +3 oxidation state e.g., B_2O_3, Al_2O_3, Ga_2O_3, In_2O_3. Thallium gives a lower oxide, Tl_2O. Owing to their important role in the industry, an account of the preparation and properties of these compounds is discussed here.

A. Preparation & properties of B_2O_3 Boron trioxide or Boric oxide can be obtained by burning boron in oxygen or by the fusion of orthoboric acid.

$$4B + 3O_2 \longrightarrow 2B_2O_3$$

$$2H_3BO_3 \longrightarrow B_2O_3 + 3H_2O$$

It is usually obtained as a glassy material whose structure consists of randomly oriented three dimensional net work of BO_3 groups, each oxygen atom uniting two boron atoms.

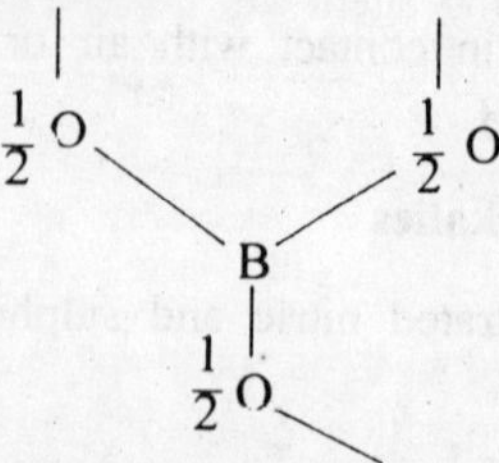

Fig. 2.3 Structure of B_2O_3 whose empirical formula is $BO_{1.5}$.

Borontrioxide reacts with water forming **orthoboric acid**

$$B_2O_3 + 3H_2O \longrightarrow 2H_3BO_3$$

When fused with metallic oxides it forms **borate glasses** which are often colored; this is the basis of the **borax bead test** for the transition metal ion detection in the qualitative analysis.

$$CuO + B_2O_3 \longrightarrow Cu(BO_2)_2$$

$$MnO + B_2O_3 \longrightarrow Mn(BO_2)_2$$

Boric oxide is an amphoteric oxide and it combines with acidic, neutral and basic oxides as well as with alkalies.

$$B_2O_3 + P_2O_5 \longrightarrow 2\ BPO_4$$

$$B_2O_3 + Na_2O \longrightarrow 2\ NaBO_2$$

$$B_2O_3 + 6KOH \longrightarrow 2\ K_3BO_3 + 3H_2O$$

B. Preparation & properties of Al_2O_3 : Alumina or Aluminium trioxide, is obtained by either exposing aluminium for a long period to atmosphere or oxygen; or by heating the metal in a current of air or oxygen.

$$4Al + 3O_2 \longrightarrow 2Al_2O_3$$

Al_2O_3 is also obtained by the dehydration of aluminium hydroxide or by healing any of its oxy salts.

$$2Al(OH)_3 \longrightarrow Al_2O_3 + 3H_2O$$

$$Al_2(SO_4)_3 \longrightarrow Al_2O_3 + 3SO_3$$

It occurs naturally as Bauxite which is usually contaminated with silica or ferric oxide which gives a pale yellow colouration to alumina. When it is freshly prepared it reacts with acids as well as bases thus showing its amphoteric character

$$Al_2O_3 + 6HCl \longrightarrow 2AlCl_3 + 3H_2O$$

$$Al_2O_3 + 2NaOH + 3H_2O \longrightarrow 2NaAl(OH)_4$$

It is used as a dehydrating agent for alcohols to give corresponding alkenes.

$$\underset{\text{ethyl alcohol}}{C_2H_5OH} \xrightarrow[300^{\circ}C]{Al_2O_3} \underset{\text{ethylene}}{C_2H_4} + H_2O$$

As a finely divided powder it is used as an adsorbent in column chromatography, e.g., for the separation of chlorophylls.

The hard form of aluminium trioxide is known as corundum, it is an abrasive and can be obtained on heating alumina to 2000°C.

Alumina in its crystalline form when present in combination with other metal ions in nature becomes precious, with Cr^{3+} it is ruby, with Co^{2+} it is sapphire. When these are prepared artificially they become semi precious stones.

C. Preparation & properties of Ga_2O_3 : Ga_2O_3 is also an amphoteric oxide obtained either by direct combination of the elements Ga and O_2 or by thermal decomposition of its nitrates, sulphates etc.

$$4Ga + 2O_3 \longrightarrow 2Ga_2O_3$$

$$Ga_2(SO_4)_3 \longrightarrow Ga_2O_3 + 3SO_3$$

$$4Ga(NO_3)_3 \longrightarrow 2Ga_2O_3 + 12NO_2 + 3O_2$$

$Ga(OH)_3$ is gelatinous white solid and amphoteric in its behaviour. Although it is amphoteric oxide it has an enhanced acidic character as revealed by its solubility in aqueous ammonia.

D. Preparation & properties of In_2O_3 : is prepared by the thermal decomposition of its hydroxide, nitrate or sulphate.

$$2In(OH)_3 \longrightarrow In_2O_3 + 3H_2O$$

$$In_2(SO_4)_3 \longrightarrow In_2O_3 + 3SO_3$$

It is yellow solid, with predominant basic characters. Hence In_2O_3 is soluble in weak acids. We can explain the basic character on the basis of increased size and reduced electronegativity over the other family members.

E. Preparation and Properties of Tl_2O_3 : Tl_2O_3 is obtained by dehydrating its hydroxide.

$$2Tl(OH)_3 \xrightarrow{\text{heat}} Tl_2O_3 + 3H_2O$$

It is dark brown solid. Predominantly basic in character.

Decomposes around 100°C to Tl_2O which combines with water to give a yellow TIOH, a strong base.

2-9-4 Electron deficiency and acceptor behaviour of boron

Electron deficient compounds are those compounds in which the central atom has deficiency of electrons around it and they are short of electrons to complete their octet, by accepting electron pairs. These compounds are usually the Lewis acids e.g., BF_3, AlH_3, $AlCl_3$ etc., with structures as follows :

```
  F           H           Cl
  ..          ..          ..
 F:B         H:Al        Cl:Al
  ..          ..          ..
  F           H           Cl
```

The boron hydrides are very typical examples of electron deficient compounds in which, even though there is a shortage of electrons around boron, they form a series of compounds called **boranes**. Much work on the study of these compounds was done by Alfred Stock and his collaborators.

Over twenty five neutral boranes B_nH_m and an even number of borane anions $B_nH_m^{x-}$ have been found to the present. All these boranes are classified according to their structure and their stoichiometry into 5 series.

i) Closo-boranes (Greek-Claro meaning cage) have complete closed polyhedral clusters of n boron atoms, with a general formula of $B_nH_n^{2-}$ where n = 6-12.

ii) Nido-boranes (Latin - nidus = nest) have non closed structures in which the B_n clusters occupy 'n' number of corners of (n+1) cornered polynedron. The general formula of nido boranes is $B_nH_n^{+4}$ example : B_2H_6,, B_5H_9, B_6H_{10}, whereas the anionic nido boranes have a general formula of B_nH_{n+3}, these are formed by the removal of one of the bridging hydrogen from compounds like say B_5H_9 to give B_5H_8 similarly $B_{10}H_{13}$, B_4H_7 and B9H12. Another set of nidoboranes with general formula B_nH_{n+2} are also known, where two hydrogens are removed from the neutral molecule to give a divalent negative anion, for example $B_{10}H_{12}^{2-}$ or $B_{11}H_{13}^{2-}$ etc.

iii) Arachno boranes (Greek-arachne = a spider's web) have even more open clusters in which the B atoms occupy n contiguous corners of an (n+2) cornered polyhedron. These boranes have general formula of B_nH_{n+6} example B_4H_{10}, B_5H_{11}, B_6H_{12}, B_8H_{14} n- B_9H_{15}, iso - B_9H_{15} etc. The anionic arachno boranes have the general formula of B_nH_{n+5} and $B_nH_{n+4}^{2-}$ eg. B_2H_7, B_3H_8, B_5H_{10} B_6H_{11} and $B_{10}H_{14}^{2-}$ etc.

iv) Hyphoboranes (Greek - hyphe - net) have the most open clusters in which the B atoms occupy n corners of an (n+3) cornered polyhedron.

They have a general formula of $B_nH_n^{+8}$, example B_8H_{16} and $B_{10}H_{18}$ are Hypho boranes and they also form several adducts.

(v) Conjuncto (Latin - Conjuncto - I join together) have structures formed by linking two (or more) of the preceding types of cluster together the general formula of conjuncto boranes is B_nH_m. The conjuncto boranes are sub divided into five different structure types of interconnected borane clusters, they have the following characteristics.

a) fusion by sharing a single common B atom eg. $B_{15}H_{32}$

b) formation of a direct 2-centre B-B σ bond between 2 clusters eg B_8H_{18} or $(B_4H_9)_2$, $B_{10}H_{16}$ or $(B_5H_8)_2$ etc.

c) fusion of two clusters through two boron atoms at a common edge eg $B_{13}H_{19}$, $B_{14}H_{18}$, $B_{14}H_{20}$ etc.

d) fusion of two clusters through three boron atoms at a common face only soluated compounds of this type are known to be present eg. $(MeCN)_2B_{20}H_{16}$ no neutral or anionic species are known.

e) more extensive fusion involve four boron atoms in various configurations eg $B_{20}H_{16}$ or $B_{20}H_{18}^{2-}$ are classified under this type.

Boranes in general are extremely reactive and several of them are spontaneously flammable in air. The Arachno boranes tend to be more reactive than nido boranes. The reactivity also diminishes with increasing molecular weight. Closo borane anions are exceptionally stable and their chemical behaviour has suggested the term "three dimensional armoticity". Boron hydrides are extremely versatile chemical reagents. Nearly all boranes are highly toxic when inhaled or absorbed through the skin.

Boranes are usually named by indicating the number of Batoms with a latin prefix and the number of hydrogen atoms by an arabic number in parenthesis eg B_5H_9, penta borane (9) or B_5H_{11}, pentaborane (11). Names for anions end in **ates** rather than **ane** and they specify both the number of H and B atoms alongwith the change present on the anion eg. $B_3H_8^{-1}$ octahydrotriborate.

The nomenclature can further be written in detail by the optional inclusion of the term closo, nido etc. in italics eg. decahydro *closo* decaborate (2-) for $B_{10}H_{10}^{2-}$, similarly $B_{15}H_{23}$ can be written as tridodeca *conjuncto* penta decaborane.

Amongst all these, the simplest is the **diborane (B_2H_6)**, whose study is of importance. All the higher boranes are inter convertible and the starting material is the diborane. The general method adopted by Stock for the preparation of diboranes involve the treatment of magnesium boride Mg_3B_2 by an acid giving a mixture of boranes. This method is now seldom used as it is difficult to separate the boranes from the mixture. We will restrict our discussion to diborane.

2-9-5 Diborane :

Diborane is prepared by treating Mg_3B_2 with H_3PO_4 (Stock's original method) to get a mixture of diboranes which on heating gives B_2H_6.

$$Mg_3B_2 + H_3PO_4 \longrightarrow \text{Mixture of Boranes} \xrightarrow{\text{Heat}} B_2H_6$$

It is also obtained by the reduction of BCl_3 with either $LiAlH_4$ or with hydrogen in the presence of a silent electric discharge and a catalyst.

$$3LiAlH_4 + 4BCl_3 \longrightarrow 2B_2H_6 + 3LiCl + 3AlCl_3$$

$$6H_2 + 2BCl_3 \xrightarrow[\text{Cu - Cr } 450^\circ C]{\text{Silent electric discharge}} B_2H_6 + 6HCl$$

The best method for the manufacture of diborane is by the decomposition of sodium borohydride with BF_3 or by treating it with H_2SO_4 or H_3PO_4

$$3NaBH_4 + 4BF_3 \longrightarrow 3NaBF_4 + 2B_2H_6$$

$$2NaBH_4 + H_2SO_4 \longrightarrow Na_2SO_4 + 2H_2 + B_2H_6$$

$$2NaBH_4 + 2H_3PO_4 \longrightarrow B_2H_6 + 2NaH_2PO_4 + 2H_2$$

Properties of B_2H_6

(i) Diborane is a gas and can be liquified. It has a disagreeable foul smell with a toxic effect.

(ii) All the boranes get decompose at red heat to boron and hydrogen.

$$B_2H_6 \xrightarrow{\text{red heat}} 2B + 3H_2$$

(iii) The diborane when ignited in air or oxygen, burn with a green flame.

(iv) The diborane gets absorbed by alkalies

$$B_2H_6 + 2KOH \longrightarrow 2KOBH_3 + H_2$$

(v) Diborane is decomposed by water to form orthoboric acid.

$$B_2H_6 + 6H_2O \longrightarrow 2H_3BO_3 + 6H_2$$

(vi) All the hydrides react readily with halogens under controlled conditions, when a mixture of partially halogenated products are obtained.

$$B_2H_6 + Cl_2 \longrightarrow B_2H_5Cl + HCl$$

$$B_2H_6 + 3Cl_2 \longrightarrow 2BCl_3 + 3H_2$$

(vii) By the reaction of alkyl compounds, alkyl derivatives of diborane are obtained. In this reaction alkyl group replaces the hydrogen of diborane. This replacement is possible only to an extent of four hydrogen atoms.

$$B_2H_6 + CH_3Cl \rightarrow B_2H_5CH_3 + HCl \xrightarrow{CH_3Cl} B_2H_4(CH_3)_2 + HCl$$

$$\downarrow CH_3Cl$$

$$B_2H_2(CH_3)_4 + HCl \xleftarrow{CH_3Cl} B_2H_3(CH_3)_3 + HCl$$

(vii) Different organic reagents react with diborane under different conditions.

Tertiary amines react with diboranes to form boron amine complexes which are quite stable.

$$B_2H_6 + 2(CH_3)_3\,N \longrightarrow 2BH_3\;N(CH_3)_3$$

It combines with ethers and thio ethers to form adducts.

$$B_2H_6 + (CH_3)_2O \longrightarrow (CH_3)_2OB_2H_6$$

$$B_2H_6 + (C_2H_5)_2S \longrightarrow (C_2H_5)_2SB_2H_6$$

It combines with pyridine to form a coordinated pyridine salt of diborane.

$$C_5H_5N: + B_2H_6 \longrightarrow C_5H_5\;N{:}\!\rightarrow B_2H_6$$

(viii) Diborane combines with carbon monoxide under pressure to form an adduct.

$$B_2H_6 + 2CO \longrightarrow 2BH_3CO.$$

(ix) Diborane is a powerful reducing agent. Some functional groups that are reduced by diborane are :

RCHO	RCH_2OH
RCN	RCH_2NH_2

(x) Diborane undergoes hydroboration reactions with olefinic or acetylinic compounds to form trialkyl boranes.

The alkyl boranes are not isolated as such, instead are converted into

(a) hydrocarbons by reacting them with aqueous acids or carboxylic acids

(b) alcohols by treating with alcoholic H_2O_2

(c) Ketones or carboxylic acids by the oxidation with chromic acid.

$$BR_3 \xrightarrow{H_3O^+} 3RH$$

$$BR_3 + 3CH_3COOH \longrightarrow 3RH + (CH_3CO)_3\;B$$

$$B(RCH_2CH_2)_3 + 3H_2O_2 \longrightarrow 3RCH_2CH_2OH + H_3BO_3$$

$$B(CH_3CH_2)_3 + H_2CrO_4 \longrightarrow 3CH_3COOH$$

The bridging bonds in B_2H_6 are readily cleaved, even by weak ligands to give either symmetrical (homolytic) or unsymmetrical (heterolytic) cleavage products

$$B_2H_6 + 2L \longrightarrow 2L \rightarrow BH_3$$

a) Symmetrical (homolytic) cleavage.

$$B_2H_6 + 2L \longrightarrow [L{:}\rightarrow BH_2]^+ + [BH_4]^-$$

b) Unsymmetrical (heterolytic) cleavage.

The symmetrical cleavage is the more common mode. The ligand, H^+ is a special case since it gives the symmetrical tetrahedral borohydride $[BH_4]^+$ ion, isoelectronic with CH_4, the $[BH_4]$ ion itself provides a rare example of a ligand that can be unidentate, bidentate or tridentate as illustrated in Fig.

MePh2P, MePh2P, MePh2P, Cu, H, B, H, H, H; 1.18 A°, 1.33 A°, 1.17 A°, 1.7 A°, 2.28 A°, 121.7°

Cu.......B 2.52 A°

$Cu+(MePh_2P)_3$ BH_4 complex

Secondly the BH_4 shows four or five coordination number which is quite unusual of an element with such a low atomic number. The chemistry of this ion is studied in the later sections.

xii) All the boranes react with ammonia under different conditions to form some interesting compounds.

$$B_2H_6 + NH_3 \xrightarrow[\text{Low temperature}]{\text{Excess } NH_3} B_2H_62NH_3 \text{ (Diborane amine)}$$

$$B_2H_6 + NH_3 \xrightarrow[\text{high temperature}]{\text{Excess } NH_3} (BN)_x \text{ (Boron nitride)}$$

$$B_2H_6 + NH_3 \xrightarrow[\text{very high temperature}]{\text{1:2 mixture}} B_3N_3H_6 \text{ (Borozole)}$$

Boron nitride has a structure similar to graphite and borozole is commonly known as inorganic benzene because of its analogous behaviour to benzene with respect to the isoelectronic structures which are planar. In both the compounds, C in benzene N and B in borozole, are all in sp_2 hybrid forms. In benzene there is a π-bond present but in borozole there is a dative bond present between nitrogen and boron.

Borozole can be best prepared by the reduction of the borontrichloro derivative with ammonium chloride

$$3BCl_3 + 3NH_4Cl \xrightarrow[-9HCl]{\text{heat}} (BClNH)_3 \xrightarrow{3NaBH_4} B_3N_3H_6 + 3NaCl + \frac{3}{2} B_2H_6$$

The physical properties of borozole can be compared with those of benzene in the table below :

Name	Borazole/Borazine	Benzene
Formula	$B_3N_3H_6$	C_6H_6
Molecular weight	80.53	78.12
Melting Point	-58°C	6°C
Boiling Point	55°C	80°C
Heat of vapourization kcal/mole	7.0	7.4
Density g/ml	0.81	0.81
Bond lengths C–H Å	-	1.08
Bond lengths C–C Å	-	1.42
Bond lengths B–H Å	1.20	–
Bond lengths B–N Å	1.44	–
Bond lengths N–H Å	1.02	-
Surface Tension dynes/cm	31.1	31.0

The chemical properties suggests that borozole has little aromatic character. It reacts with H_2O, CH_3OH and HCl to yeild 1:3 adducts with the elimination of hydrogen.

$$B_3N_3H_6 + 3H_2O \xrightarrow{[O]} [BH(OH)NH_2]_3 \xrightarrow{10^\circ C} [B(OH)NH]_3 + 3H_2$$

$$B_3N_3H_6 + 3HCl \longrightarrow B_3N_3H_9Cl_3$$

At 50°C the hydrochloric acid present on the adduct compound looses its hydrogen

$$B_3N_3H_9Cl_3 \xrightarrow[-3H]{50^\circ C} B_3N_3H_3Cl_3$$

On methylation of borozole hydrogen gets displaced to give a thermally stable $B_3N_3(CH_3)_6$. This compound can react with $[Cr(CO)_3\ (CH_3CN)_3]$ to give the complex $[CrB_3N_3(CH_3)_6(CO)_3]$ which is closely related to hexamelthyl complex of benzene, $[CrC_6(CH_3)_6\ (CO)_3]$.

Thermolysis of borozole produces polymeric and polymeric analogues of nephthalene, biphenyl etc.

Naphthalene analogue of Borazole　　Biphenyl analogue of Borazole

The structures of these compounds are as seen in Fig.2-2 a, b, c, d.

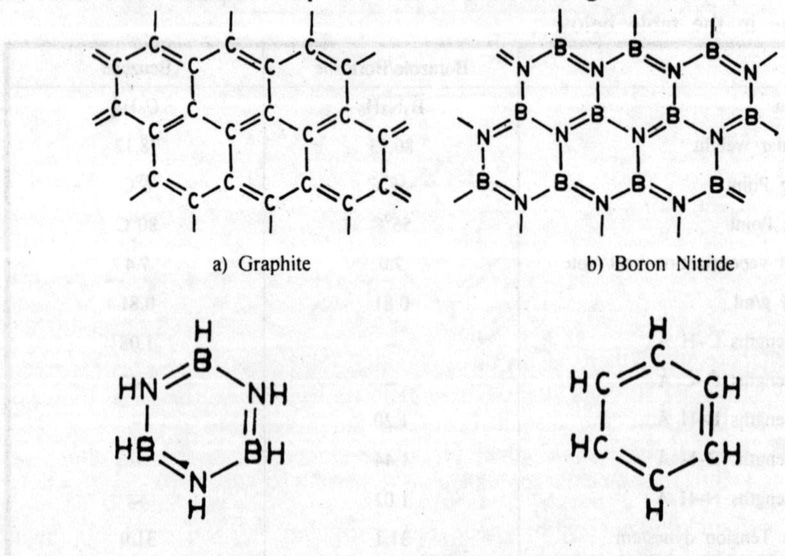

Fig.2-2 Structures of Boron compounds and their carbon analogues

Structure of Diborane : The structure of diborane is of great interest to a chemist. It has a molecular formula B_2H_6 analogous to C_2H_6 (ethane), hence it is expected that diborane also, should have a structure similar to ethane i.e.,

```
    H   H
    |   |
H — C — C — H
    |   |
    H   H
```

But to have such a structure it requires 14 electrons, 12 for B – H bondings and 2 for B – B bonding, which the compound does not possess. According to the electronic configuration of $B(1s^2 2s^2 2p^1)$, it can provide only 3 electrons from its valence shell for bond formation of the type :

```
    H   H
    |   |
H — B — B — H
    |   |
    H   H
```

and there is no possibility for the B-B bonding to take place. As mentioned before the compound being electron deficient, it is not capable of providing enough number of electrons for forming covalent bonds. The Raman spectra and the infrared spectral data assign the following structure to diborane;

1·77 Å
1·33 Å
1·19 Å
97°
121·5°

It is seen here, that, there is no direct bonding between B – B, instead they are linked together through hydrogens which act as a bridge between the two borons. On the plane the distance between the two boron atoms is 1.77Å. The B-H terminal distance is equal for all the four such hydrogens which is 1.19Å. The B–H distance is equivalent to 1.33Å in the B–$\hat{H}$–B arrangement.

It is found by nuclear magnetic resonance and electron diffraction that the four terminal hydrogens are in a different environment from the other two hydrogen atoms. The two borons and the four terminal hydrogen atoms lie in the same plane while the other two bridging hydrogen atoms lie, one above and the other below, in a plane perpendicular to the plane of the molecule. Hence four hydrogens are of one type and the other two are of different type. This is confirmed by the fact that diborane cannot be methylated beyond four hydrogens, to give finally $B_2H_2(CH_3)_4$, without breaking the molecule into $B(CH_3)_3$. It is found that the B– H distance of the terminal hydrogens is the same as the bond length measured in a non-electron deficient compound, and therefore the electron deficiency lies with bridging groups. Moreover the B–H distance in the bridge is abnormally long.

Let us now consider how such a bonding is possible and what makes the bridge. The B–H–B bond is a three centre electron pair bond, i.e., one electron pair binds three atoms: B, H and B. The three centre BHB orbital may be considered to result from the combination of one s-orbital and all the three p-orbitals of boron including the empty p_z orbital, thus giving a pseudo-sp^3 hybridization. They give a delocalized orbital containing one pair of electrons but covering all the three nuclei, during which the hydrogen gets enveloped by the empty p_z orbital & that of its own, forming a banana bond as shown in Fig.2-5.

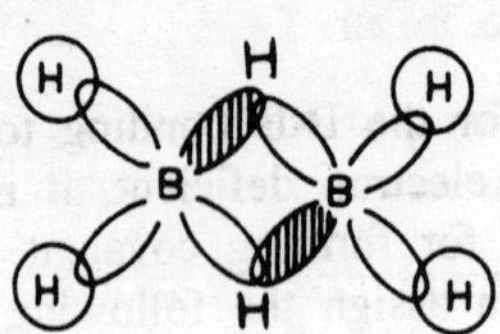

Fig.2-5 The shaded portion is the empty $2p_z$ orbital which helps in the 3 centre electron pair bond.

According to the bond angles it is found that the H$\hat{B}$H (terminal) is equal to 121.5°. So a sp^2 hybridization is expected, where we can think of the 2s, $2p_x$ and $2p_y$ orbitals getting involved. The empty $2p_z$ orbital is also expected to be present on the boron atom, while the enveloping of hydrogen takes place, the two electrons (one from boron and the second from hydrogen) make their way into the empty $2p_z$ orbital and thus create a three centre electron pair bond. The support for this discussion can be taken from the H$\hat{B}$H (bridging internal angle) of 97°, which we can suggest to be, because of a distorted sp^3 hybridization, involving the empty $2p_z$ orbital, thus resulting in the formation of a τ (tau) bond or a banana bond. It is a typical case of dual hybridization in the boron atom.

A tau bond or a banana bond is a typical bond formed in very unusual circumstances such as in diborane, where two bonds are formed using only two electrons. It is also referred to as a three centered electron pair bond. As the bond is formed due to strain, it is shown in the form of a banana, hence the name banana bond, Fig. 2-6.

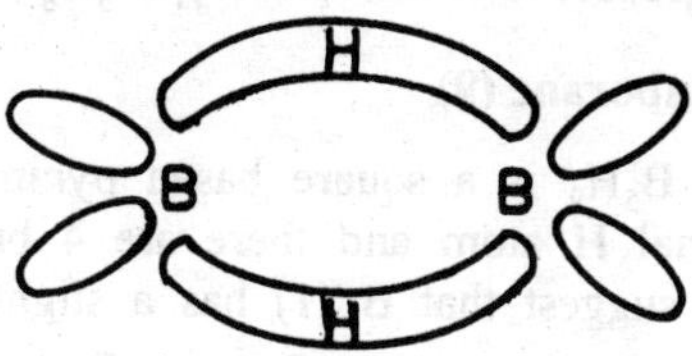

Fig. 2-6 Banana bond

2-9-6 Chemistry of Higher Boranes

Chemistry of Pentaborane (9), B_5H_9

It is a colourless, volatile liquid, boils at 60°C, it is chemically very reactive and spontaneously flammable in air.

It reacts with Lewis bases to form adducts, some of which have been recognized as belonging to a new series of hypho-borane derivatives B_nH_{n+8}.

PMe_3 (trimethyl phosphine) give an adduct, on reaction with B_5H_9; to form $[B_5H_9(PMe_3)_2]$, which is formally analogous to $[B_5H_{11}]^{2-}$ alongwith the (unknown) B_5H_{13}.

B_5H_9 also acts as a weak Bronsted acid and from proton competition reactions with other boranes and borane anions, it has been established that acidity increases with increasing size of the borane cluster and that arachno boranes are more acidic than the nido-boranes.

nido : $B_5H_9 < B_6H_{10} < B_{10}H_{14} < B_{16}H_{20} < B_{18}H_{22}$

arachno : $B_4H_{10} < B_5H_{11} < B_6H_{12}$ and $B_4H_{10} > B_6H_{10}$

Accordingly B_5H_9 can be deprotonated at low temperatures by loss of H to give B_5H_8, providing a sufficiently strong base such as lithium alkyl or an alkalimetal hydride is used to produce metallo borane clusters.

$$B_5H_9 + LiCH_3 \longrightarrow LiB_5H_8 + CH_4$$

$$B_5H_9 + NaH \longrightarrow NaB_5H_8 + H_2$$

Cluster expansion and cluster degradation reactions are a feature of many polyhedral species. An example of cluster expansion reaction is

$$LiB_5H_8 + \frac{1}{2} B_2H_6 \xrightarrow[-78o\ (fast)]{(C_2H_5)O_2} LiB_6H_{11} \xrightarrow{HCl} B_6H_{12}$$

The cluster degradation reaction can be

$$[NMe_4]\ B_5H_8 + 6C_3H_7OH \longrightarrow [NMe_4]\ B_3H_8 + 2\ B\ (C_3H_7O)_3 + 3H_2$$

Structure of nido-pentaborane (9)

The structure of B_5H_9 is a square based pyramid of B atoms each of which carries a terminal H atom and there are 4 bridging H atoms around the base. Calculations suggest that B (1) has a slightly higher electron

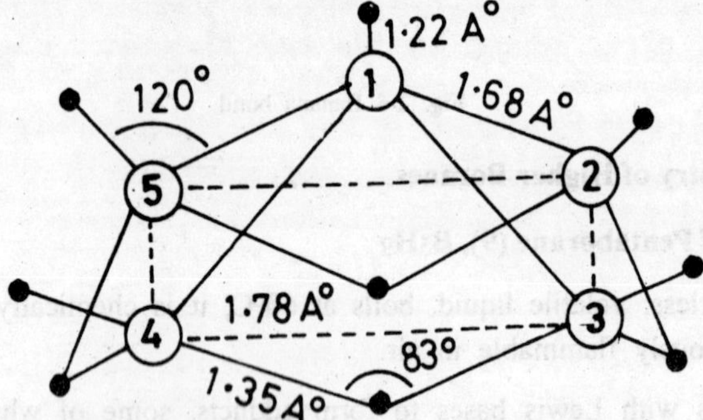

density than the basal borons and that the H_μ (bridging) is slightly more positive than the H_t (terminal). The square base has a 1.78Å length for the B–B distance. The pyramidal apex B is 1.68Å away from the basal B atoms, the B–H_t distance is 1-22Å whereas B–H_μ distance is 1.35A. the H_t B B angle is 120° and the BH_μ B angle is of 83°.

Structure of arachno-pentaborane (11) :

The structure of B_5H_{11} is a five cornered polyhedron as seen below, not much of details are available about this compound.

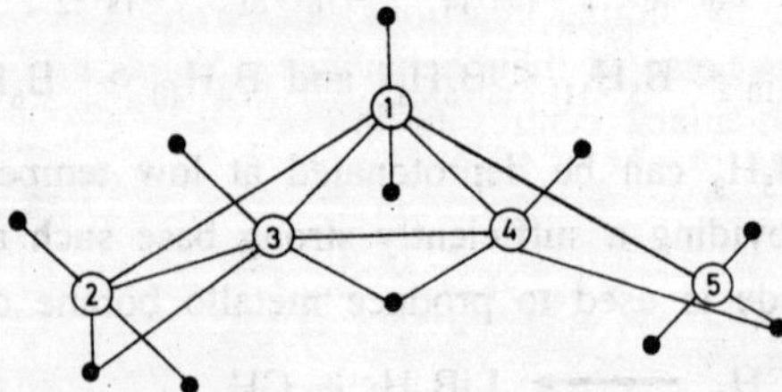

Structure of tetraborane : B_4H_{10} has almost a planar structure. Each of the four boron and the four H_μ contribute 1 electron to the B–H–B bonds, ie., 4 pairs of electrons for the 4 (3 centre) bonds. The two hinge BH_t groups each have 1 remaining electron and an orbital which intereact to give a 2 centre B–B bond.

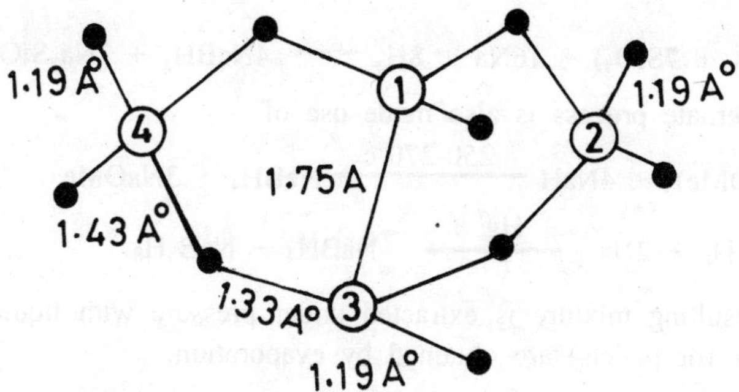

2-9-7 Compounds closely related to Boranes :

Boron, as seen in the earlier section, can combine with hydrogen to form boranes which has defenite stiochiometric formulae and structures. Boron combines with hydrogen to form a certain class of borox hydrides which are either closely related to boranes or they are derivatives of different boranes.

They are characteristic of their anionic forms or they are also coordination derivatives of different boronhydrides with special reference to the borohydride and polyhedral borohydrides.

Closely related to the boranes are a large number of anionic species which are commonly referred to as hydroborates. The simplest of the hydroborates is the tetrahydroborate ion BH_4

This anionic species is commonly referred to as borohydride but its salts are designated as tetrahydroborates. Formally these can be regarded as a product of Lewis acid – base interaction between BH_3 and H to give a coordinated $[H_3B \rightarrow H]^-$ion.

The first tetrahydroborate of Aluminium was reported in 1940 by Schlesinger and Brown. It was prepared by the reaction of trimethylaluminium and diborane at 80°c.

$$Al(CH_3)_3 + 2B_2H_6 \longrightarrow Al(BH_4)_3 + B\ (CH_3)_3$$

Later on the tetrahydroborates of lithium, beryllium, magnesium, zinc etc. were prepared by similar reactions, these products were made use of in organic synthesis because of their behaviour as versatile nucleophillic reducing agents which could attack centres of low electron density electrophiles,

The preparation of tetrahydroborates can also be done by the direct reaction of a MH with either B_2H_6 or BX_3 at room temperatures.

$$2LiH + B_2H_6 \xrightarrow{Et_2O} 2LiBH_4$$

$$4\ NaH + BX_3 \xrightarrow{Al_2Et_6} NaBH_4 + 3NaX$$

Industrially these are manufactured by Bayer's process to give quantitative yeilds Borax (or ulexite) quartz, sodium metal and gaseous hydrogen are reacted under pressure at 400° to 500°C to get a 98% yeild.

$$(Na_2B_4O_7 + 7SiO_2) + 16Na + 8H_2 \longrightarrow 4NaBH_4 + 7Na_2SiO_2$$

An alternate process is also made use of

$$B(OMe)_3 + 4NaH \xrightarrow{250\text{-}270^oc} NaBH_4 + 3NaOMe$$

$$2B_2H_6 + 2Na \xrightarrow{Hg} NaBH_4 + NaB_3H_8$$

The resulting mixture is extracted under pressure with liquid ammonia, pure powder (or pellets) are obtained by evaporation.

Number of reactions are used to prepare different tetrahydroborates under different conditions as follows :

$$NaBH_4 + LiCl \xrightarrow[\text{amine}]{\text{isopropyl}} LiBH_4 + NaCl$$

$$2NaBH_4 + BeCl_2 \xrightarrow{125^o} Be(BH_4)_2 + 2NaCl$$

$$8LiBH_4 + 2TiCl_4 \xrightarrow{20^o} 2Ti(BH4)3 + 8LiCl + B_2H_6 + H_2$$

$$In(CH_3)_3 + 2B_2H_6 \xrightarrow{-40oc} In\ (BH_4)_3 + B\ (CH_3)_3$$

The metallic tetrahydroborates may have M–H–B bridges as seen in the structure below

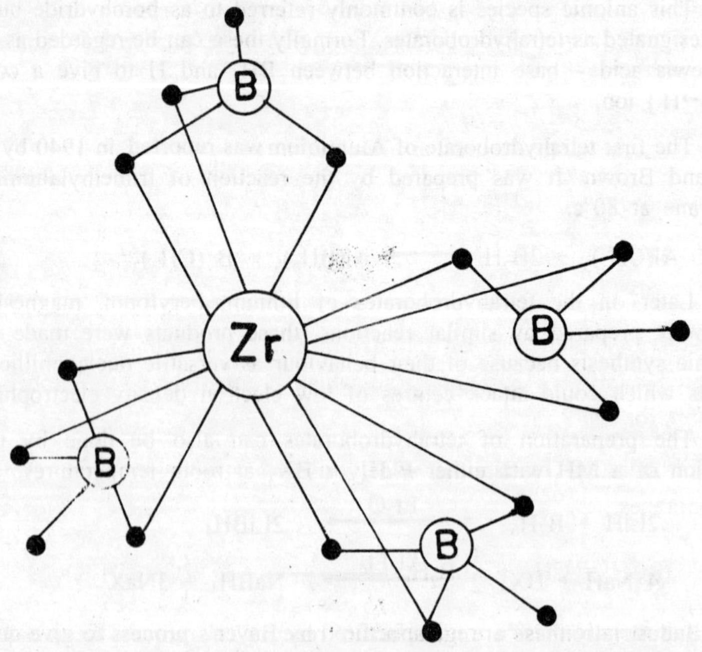

The structure of $Zr^{IV}(BH_4)_4$

The alkali metal tetrahydroborates are of much importance specially $LiBH_4$ and $NaBH_4$.

In the chemical industry $LiBH_4$ plays a very vital role in the reduction of esters to alcohols, whereas $NaBH_4$ reduces ketones, acid chlorides and aldehydes, if functional groups like –CN, – NO_2 etc are present, the BH_4 ion does not touch these groups and thus selective reaction takes place. Apart from its reducing property, the BH_4 ion is also used in the bleaching of wood pulp and electroless (chemical) plating of metals. $NaBH_4$ is used for the deposition of nickel on metal or nonmetallic substrates to prevent coerosion and to make nickel coating hard. this type of coating is applied to electronics. During this process about 5% by weight of the deposit is boron, the chemical reaction taking place is

$$10NiCl_2 + 8NaBH_4 + 17NaOH + 3H_2O \longrightarrow (3Ni_3B + Ni) + 20NaCl + 5NaB(OH)_4 + 17.5\ H_2$$

It is found that 1kg $NaBH_4$ precipitates about 2.2 kg of $(3Ni_3B+Ni)$ or $(Ni_{10}B_3)$ which is sufficient to cover an area of $10.7m^2$ with 25 μ_m, thickness.

$NaBH_4$ is a convenient source of diborance by its reaction with BF_3 or by protonation.

$$3NaBH_4 + 4BF_3 \xrightarrow{\text{diglyme}} 3NaBF_4 + 2B_2H_6$$

$$2NaBH_4 + 2HCl \longrightarrow 2NaCl + 2H_2 + B_2H_6$$

Because of their strong reducing nature nonmetallic halides can converted to their hydrides. (eg. $SiCl_4$ SiH_4)

Sodium tetrahydroborate is the best source for the **hydroboration reactions.**

The addition of B_2H_6 to alkenes and alkynes in either solvents at room temperatures is referred to as a **hydroboration reaction.** Such reactions were discovered by Brown & Subba Rao which brought them, the Nobel prize in chemistry for 1979.

The hydroboration reaction is widely used in the quantitative synthesis of organoboranes.

$$3RCH = CH_2 + \frac{1}{2} B_2H_6 \longrightarrow B(CH_2CH_2R)_3$$

Hydroboration is a regiospecific reaction in which the boron shows preferential attachment to the least substituted C atom (An anti Markownikoff's).

The protonolysis of the organoborane by refluxing it with an anhydrous carboxylic acid yeilds the alkane corresponding to the initial alkene, the oxidative hydrolysis with alkaline hydrogen peroxide yeilds the corresponding primary alcohol.

$$B\ (CH_2CH_2R)_3 \xrightarrow{EtCOOH} 3RCH_2CH_3$$

$$B\ (CH_2CH_2R)_3 \xrightarrow{H_2O_2/NaOH} 3RCH_2\ CH_2OH$$

Tertiary alcohols can be prepared by the hydroboration reaction by the transfer of alkyls to carbonmonoxide

$$BR_3 + CO \xrightarrow[110^oc]{diglyme} R_3CBO \xrightarrow[H_2O_2]{NaOH} R_3C.OH$$

If the reaction is conducted in water instead of diglyme a ketone is formed of chain length (2n+1) C atoms and the other variants yeild primary alcohols or aldehydes with (n+l) C atoms

$$BR_3 + CO \xrightarrow[H_2O]{100^o} RB\ (OH)\ CR_2OH \xrightarrow{NaOH} R_2CO$$

$$BR_3 + CO \xrightarrow[NaBH_4]{45o} \text{Intermediate} \xrightarrow[\text{compound}]{KOH} RCH_2OH$$

$$BR_3 + CO \xrightarrow[THF]{25^o} \text{Intermediate} \xrightarrow[\text{compound}]{H_2O_2} ROH + RCHO$$

Diborane being an electrophillic reducing agent prefers attaching a molecule at a position of high electron density. Reductions can occur either with or without bond rapture or by the removal of oxygen, hence nitrites are reduced to amines, schiff's bases to secondary amines, oximes to alkylhydroxyl amines, by hydroboration reactions.

$$RC \equiv N \xrightarrow{BH_4} RCH_2NH_2$$

$$RC_6H_4CH = NC_6H_4R' \xrightarrow{BH_4} RC_6H_4CH_2NHC_6H_4R'$$

$$RR'C = NOH \xrightarrow{BH_4} RR'CHNHOH$$

By hydroboration, oxygen atoms are removed either with or without the addition of hydrogen atoms to a molecule. Thus phosphire oxide reduces to phosphine whereas aromatic nitroso compounds are reduced to amines.

There are certain reactions of diborane, which are not classified under the hydroboration reactions for example, the thermolysis of it with $NaBH_4$ under controlled conditions or its reaction with $LiAlH_4$. These reactions produces a group of ions with higher symmetry in which there is a closed **polyhedral** framework of linked boron atoms.

$$5B_2H_6 + 2NaBH_4 \xrightarrow[NEt_3]{180^o} Na_2B_{12}H_{12} + 13H_2$$

$$5B_2H_6 + 2LiAlH_4 \xrightarrow{-160^o} Li_2B_{10}H_{10} + LiB_{12}H_{12} + ... \text{ (not balanced)}$$

These **polyhedral borohydrides** as they are referred to are classified as closo boranes, they have a general formula of $(BH)_n^{2-}$ where n can have values of 6 to 12. The polyhedral borohydrides are chemically very stable possessing extensive electron delocalization and show aromatic behaviour in their reactions. Although all the species from $B_6H_6^{2-}$ to $B_{12}H_{12}^{2-}$ are easily prepared and most of them are stable even above 600°C the most accessible and best known are the $B_{10}H_{10}^{2-}$ and $B_{12}H_{12}^{2-}$ ions and their salts.

They are extremely stable towards nucleophiles but are moderately susceptible to electrophilic attack. The hydrogen at the 1,10 positions is easily substituted forming numerous isomers giving equal number of substituted compounds eg. $B_{10}H_9X^{2-}$ shows 2 geometrical isomers, $B_{10}H_8X_2^{2-}$ has 7 geometrical isomers whereas $B_{10}H_7X_3^{2-}$ shows as many as 16 isomers. Many of these isomers do exist. In the case of $B_{12}H_{12}^{2-}$ only one isomer for $B_{12}H_{11}X^{2-}$, three isomers exist $B_{12}H_{10}X_2^{2-}$ as ortho, para & meta isomers and five isomers for $B_{12}H_9X_3^{2-}$. Hence the number of compounds of $B_{10}H_{10}^{2-}$ out number that of $B_{12}H_{12}^{2-}$.

Oxidation of closo $B_{10}H_{10}^{2-}$ with aqueous solutions of Fe^{3+} or Ce^{4+} results in the formation of conjuncto $B_{20}H_{18}^{2-}$ which can be considered to be a dimerization reaction alongwith deprotonation. If the oxidation is carried out in a two phase system with Fe^{3+} ion or at 0°C with Ce^{4+}, the intermediate H bridged species $B_{20}H_{19}^{3-}$ is formed and it can be isolated. Reduction of $B_{20}H_{18}^{2-}$ with Na/NH_3 yeilds another conjuncto species $B_{20}H_{18}^{4-}$.

Alongwith these heavier polyhedral borohydride anions, a number of smaller ions are reported and studied.

Pyrolysis of alkali metal salts of $B_3H_8^-$ ion in vaccum at 200 – 230°C results in the formation of a mixture of salts of BH_4, B_9H_9, $B_{10}H_{10}$, $B_{12}H_{12}$, The salts of $B_9H_9^{2-}$ can be isolated from this mixture. By the air oxidation of $Na_2B_9H_9$ in its sodium salt dimethoxyethane solution $Na_2B_8H_8$ if formed which is another species of the family. In turn, the air oxidation of $B_8H_8^{2-}$ ion in 1, 2 dimethoxyethane gives salts of $B_7H_7^{2-}$ and $B_6H_6^{2-}$ as major products alongwith a mixture of other species but in small quantities. The structure of $Cs_2B_7H_7$, has been found to be a pentagonal bipyramidal a well known salt of the borohydride group of compounds and that of $B_6H_6^{2-}$ ion is found to be an octahedral.

The polyhedral borohydrides forms a number of complex species with metals forming M-B links and/or B-B links as well as M-H-B links. The best example is the combination of Mn(CO)4 with B3H8 ion, as seen in the figure below :

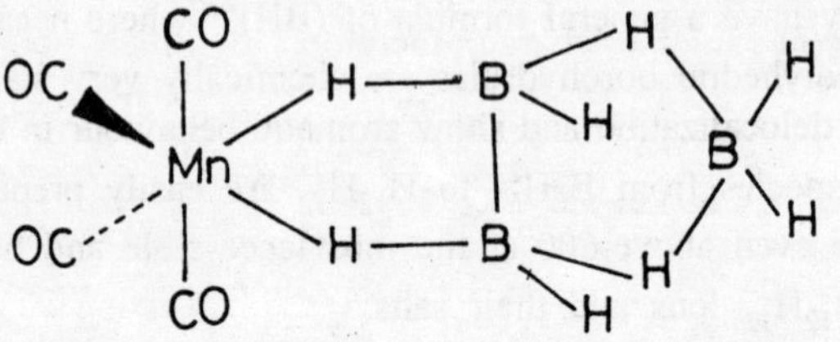

The B_3H_8 has itself a triangular structure, in which each boron holds two hydrogen atoms and one side of the triangle consists of a B–B bond but the other two sides are single hydridic bridges.

The B_3H_8 can be obtained from tetraborane by the following reaction.

$$B_4H_{10} + 2NH_3 \longrightarrow B_3H_8 + H_3NBH_2NH_3^+$$

$$2B_4H_{10} + 2NaH \longrightarrow B_2H_6 + 2B_3H_8^- + 2Na^+$$

$$B_4H_{10} + NaBH_4 \longrightarrow B_3H_8^- + Na^+$$

$$B_4H_{10} + 2Na \xrightarrow[Hg]{ether} 2Na^+ + 2B_3H_8^- + \frac{1}{2} B_2H_6 + (BH)_x$$

Diborane reacts with sodium amalgam to form sodium salt of B_3H_8 ion

$$2B_2H_6 + 2Na(Hg) \longrightarrow NaBH_4 + NaB_3H_8$$

2-9-8 Carboranes : Carboranes or the carbaboranes are mixed hydrides of carbon and boron having both carbon and boron atoms in an electron deficient skeltal framework. These are considered to be obtained by the substitution of carbon for boron in polyhedral clusters of boranes. It is possible to interact certain transitional metal ions which get attached to these polyhedral clusters to form metallocarboranes.

The carboranes are compounds with basic structural units containing a number of C and B atoms arranged on the vertices of a triangulated polyhedron. Their structures are closely related to those of the isoelectronic boranes, eg nido B_6H_{10} provides the basic cluster structure for the four carboranes viz CB_5H_9, $C_2B_4H_8$, $C_3B_3H_7$ and $C_4B_2H_6$, this is possible, because, there is a successive replacement of a basal B atom by a C atom being compensated by the removal of one bridging H. Carboranes have the general formula $[(CH)_a(BH)_m\ H_b]^{c-}$ with **a** number of **CH** units and **m** number of **BH** units present at the polyhedral vertices **b** number of extra **H** atoms which are either the bridging (Hμ) or endo H_{endo} ie. tangential to the surface of the polyhedron as distinct from the axial H_t atoms specified in the CH and BH groups. The H_{endo} occur in BH_2 groups which can be specified as BH_tH_{endo}, It is found that the number of electrons available for the skeletal bonding is 3e from each CH unit, 2e from each BH unit, one from each of Hμ and H_{endo} and c electron from the anionic charge, therefore the total number of skeletal bonding electron pairs $= \frac{1}{2}(3a + 2m + b + c)$

$$= n + \frac{1}{2}(3a + b + c)$$

where n = (3a+m) is the number of occupied vertices of the polyhedron.

Closo - Carborane structures have (n+1) pairs of skeltal bonding electrons (ie a + b + c = 2)

nido - carborane structures have (n+2) pairs of skeltal bonding electrons (ie. a + b + c = 4)

arachno - Carborane structures have (n+3) pairs of skeltal bonding electrons (ie. a + b + c = 6)

If **a** has a value of zero the compound will be a borane or borane anion rather than a carborane. If b = O there will be no Hμ or H_{endo} hydrogens, this is observed in all the closo-carboranes with the exception of l-C B_5H_7 which contains a triply bridging Hμ over one B_3 face of the octahedron.

Closo carboranes are the most stable of the carboranes hence they out number the other types. These are colorless volatile liquids or solids. They can be prepared from an alkeyne and a borane by pyrolysis or by passing a silent electric discharge through the reactants. A mixture of light closo carboranes are obtained.

$$B_5H_9 \xrightarrow[600^o]{C_2H_2} 1,\ 5\text{-}C_2B_3H_5 + 1,\ 6\text{-}C_2B_4H_6 + 2,\ 4\text{-}C_2B_5H_7$$

At lower temperature the same reactants produces nido carbonanes

$$\text{nido- } B_5H_9 \xrightarrow[200^o]{C_2H_2} \text{nido- } 2,\ 3\ \text{-}C_2B_4H_8$$

Pyrolysis of nido or arachno carboranes or their reaction in the presence of silent electric discharge leads to the formation of closo carboranes either by the loss of hydrogen or disproportionation.

$$C_2B_2H_6 \longrightarrow C_2B_2H_4 + H_2$$

$$2C_2\ B_2H_6 \longrightarrow C_2BH_3 + C_2B_3H_5 + 2H_2$$

Cluster expansion reactions with diborane provide an alternative route to the preparation of closo carboranes which are intermediate produces and these reactions leads to more open structures of nido and arachno carboranes.

$$\text{closo } 1,7-C_2B_6H_8 + \frac{1}{2}B_2H_6 \longrightarrow \text{closo } 1,6-C_2B_7H_9 + H_2$$

$$\text{closo } 1,6-C_2B_7H_9 + \frac{1}{2}B_2H_6 \longrightarrow \text{closo } 1,6-C_2B_8H_{10} + H_2$$

$$\text{closo } 1,6\ C_2B_8H_{10} \xrightarrow[OH^- + H_2O]{\text{base hydrol}} \text{arachno} - 1,3\ C_2B_7H_{12} + B(OH)_3$$

$$\text{nido} - 1,7\ C_2B_9H_{12} \xrightarrow[6H_2O+6e^-]{\text{chromic acid}} \text{arachno} - 1,3, - C_2B_7H_{13} + B(OH)_3 + 5H^+$$

In carboranes the terminal H atoms attached to B undergo facile electrophilic substitution.

$$1,6\text{–}C_2B_7H_9 \xrightarrow{Br_2/AlCl_3} 8Br\text{–}1,6\text{–}C_2B_7H_8 + HBr$$

$$1,10\text{–}C_2B_8H_{10} \xrightarrow{8Cl_2/CCl_4} 1,10\ (CH)_2\ B_8Cl_8 + 8HCl$$

Both 1, 2 – $C_2B_{10}H_{12}$ are degraded by strong bases such as methoxide ion in alkaline solutions :

$$1,2\text{–}C_2B_{10}H_{12} + CH_3O^- + 2CH_3OH \xrightarrow[40^\circ C]{KOH,\ CH_3OH} B(OCH_3)_3 + 1,2 - C_2B_9H_{12} + H_2$$

Although closo-carboranes are stable to high temperatures and common reagents they can be specifically converted to nido- carborane anions by the reaction of strong bases similar to the above reaction.

$$1,7 - C_2B_{10}H_2 + EtO^- + 2EtOH \xrightarrow{85^\circ C} 7,9 - C_2B_9H^-_{12} + H_2 + B(OEt)_3$$

Similarly deprotonation of the monoanions by NaH removes the bridge proton to give nido-dianions 7, 8 – $B_2B_9H_{11}^{2-}$ and 7, 9 – $C_2B_9H_{11}^{2-}$

The most important reaction of the $C_2B_9H_{12}$ ion is, that with NaH

$$C_2B_9H_{12}^- + NaH \longrightarrow Na^+C_2H_9H_{11}^{2-} + H2$$

The $C_2B_9H_{11}^{2-}$ contains an open pentagonal face, with five approximately Sp^3 orbitals directed towards the vacant vertex of the icosahedroal ion, these five orbitals combine to give three bonding and two antibonding molecular orbitals. When the bonding orbitals are filled with six available electrons the situation is similar to that in ferrocene and it becomes possible for the dicarbollide ion to coordinate with the partially filled d-orbitals of the transition metals forming sandwich compounds. These compounds of the carboranes with transition metals are referred to as metallocarboranes.

The metallocarboranes can be prepared by six methods.

i) Coordination of the metal using nido carborane anions as ligands : Reaction of $C_2B_9H_{11}$ when reacted with $FeCl_2$ in tetrahydro furan in the absence of air. and moisture then a sandwich compound is formed

$$2C_2B_9H_{11}^{2-} + Fe^{2+} \longrightarrow [Fe^{2+}(n\text{-}C_2B_9H_{11})_2]^{2-}$$

On the oxidation of this compound Fe^{2+} oxidizes to Fe^{3+} forming the corresponding Fe^{3+} complex.

When the reaction is carried out in the presence of NaC_2H_5 then a mixed ligand purple sandwich compound is formed.

$$C_2B_9H_{11}^{2-} + C_5H_5 + Fe^{2+} \xrightarrow{THF} Fe^{3+}(n\text{-}C_5H_5)(n\text{-}C_2B_9H_{11})$$

ii) Polyhedral expansion : In the reactions, there is a 2 - electron reaction of a closo carborane with a strong reducing agent such as sodium naphthalide in tetrahydrofuran followed by a trasition metal salt solution.

$$\text{closo } 1,7\text{ - }C_2B_6H_8 \xrightarrow[CoCl_2]{2e^-/THF} [Co(C_5H_5)(C_2B_6H_8)] + NaC_5H_5$$

$$\xrightarrow[CoCl_2 + NaC_5H_5]{2e^-/THF} [\{Co(C_5H_5)\}_2(C_2B_6H_8)]$$

iii) Polyhedral subrogation : Replacement of a BH vertex by a metal vertex without changing the number of verticies in the cluster is termed polyhedral subrogation.

$$[Co(n\text{-}C_5H_5)(C_2B_{10}H_{12})] \xrightarrow[2)\ Co^{2+}C_5H_5^-]{1)\ OH^-} [\{Co(n\text{-}C_5H_5)\}_2(C_2B_9H_{11})]$$

iv) Thermal metal transfer : This method is the pyrolysis of preexisting metallocarboranes to that of a closo carboranes at high temperatures

$$[(C_5H_5)_2Co]^+ + [2,3\text{ - }C_2B_8H_{10})_2Co]^- \xrightarrow[\text{Hot tube}]{525^o} (C_5H_5)_2\ CoC_2B_8H_{10}.$$

v) Direct oxidative insertion of a metal centre : Nucleophillic zero valent derivatives of Ni, Pd and Pt insert directly into the closo carborane clusters which involves a net transfer of electrons from the metal to the cage.

$$Pt(PEt_3)_3 + (2,3\text{ - }Me_2\text{ - }2,3\text{ - }C_2B_9H_9) \xrightarrow[\text{Petrol}]{-30^oC} PtEt_3 +$$

$$[1 - \{Pt(PEt_3)_2\}\ \text{-}2,4\ (MeC)_2B_9H_9]$$

The chemistry of metallocarboranes is developing everyday and we can expect reactions of unusual type also during these research works.

2-9-9 Boric Acid/Ortho boric acid

Orthoboric acid is formed when the boron halides are hydrolysed or when dilute hydrochloric acid is added to a solution of borax.

$$Na_2B_4O_7 + 2HCl + 5H_2O \longrightarrow 4H_3BO_3 + 2NaCl$$

It can also be obtained by the action of super heated water on boron nitride

$$BN + 3H_2O \longrightarrow H_3BO_3 + NH_3$$

It is obtained as a white solid on crystallization.

Properties: Ortho boric acid is a weak monobasic acid and in aqueous solution the boron atom completes its octet by removing OH from the water molecule.

$$B(OH)_3 + 2H_2O \longrightarrow B(OH)_4^- + H_3O^+$$

It therefore functions as a Lewis acid and not as a proton donor. When heated orthoboric acid first gets converted to metaboric acid and then boron trioxide.

$$H_3BO_3 \xrightarrow[-H_2O]{100^\circ C} HBO_2 \xrightarrow[-H_2O]{160^\circ C} H_2B_4O_7 \xrightarrow[-H_2O]{\text{red heat}} B_2O_3$$

The structure of orthoboric acid is based on the planar $B(OH)_3$ units, individual units being joined together by hydrogen bonds to give a two dimensional sheet. The hydrogen bonds linking the oxygens of different $B(OH)_3$ groups are unsymmetrical, i.e., they are not midway between the oxygen atoms. The layers of two dimensional sheets are held together by weak Van der Waals forces, and this accounts for the cleavage of the solid into flakes.

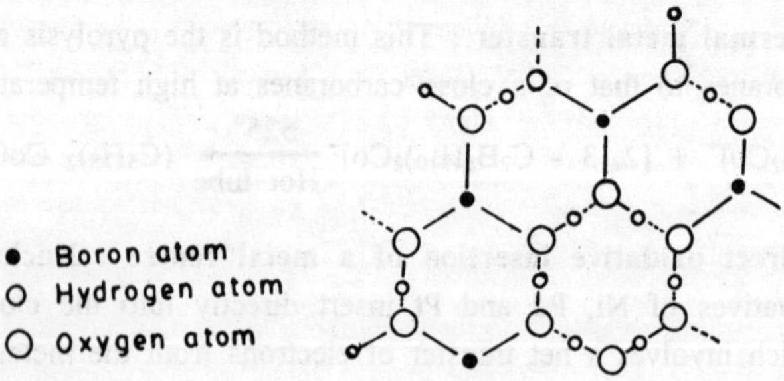

● Boron atom
○ Hydrogen atom
◯ Oxygen atom

Structure of Orthoboric acid

2-9-10 Borax/Sodium tetraborate ($Na_2B_4O_7.10H_2O$)

Borax or sodium tetraborate occurs naturally as tincal with 55% of borax content in it. Borax is extracted from tincal by dissolving the mass in water and evaporating the solution until crystals separate.

Borax can be prepared by boiling calcium borate with sodium carbonate.

$$Ca_2B_6O_{11} + 2Na_2CO_3 \longrightarrow 2CaCO_3 + Na_2B_4O_7 + 2NaBO_2$$

The calcium carbonate is removed by filtration and the solution is evaporated until crystals of borax separate out. The $NaBO_2$ is also converted to borax by passing carbon dioxide in the solution.

$$4NaBO_2 + CO_2 \longrightarrow Na_2B_4O_7 + Na_2CO_3$$

Properties : Borax is obtained as large colourless crystals with ten molecules of water of crystallization. It is highly soluble in hot water and sparingly soluble in cold water. Borax on heating looses its water of crystallization and it swells, due to removal of water, on complete removal it becomes a glassy solid of anhydrous borax. Fused borax can combine with many metal oxides to form coloured glasses, this property is made use of in the borax bead test in qualitative analysis for the detection of transition metal ions. The colour of the beads is given in Table 2-10

$$B_2O_3 + CuO \longrightarrow Cu(BO_2)_2$$

Table 2-10 Color of the transition metal meta borates obtained by borax bead test

Atomic Number	*Metal*	*Oxidising Flame*		*Reducing Flame*	
		Hot	*Cold*	*Hot*	*Cold*
22	Titanium	Yellow	Colourless	Grey	Pale Violet
23	Vanadium	Yellow	Greenish Yellow	Brown	Emerald Green
24	Chromium	Yellow	Green	Green	Green
25	Manganese	Violet	Amethyst	Colourless	Colourless
26	Iron	Yellowish brown	Yellow	Green	Green
27	Cobalt	Blue	Blue	Blue	Blue
28	Nickel	Violet	Reddish Brown	Grey	Grey
29	Copper	Green	Blue	Colourless	Opaque Red
42	Molybdenum	Yellow	Colourless	Brown	Brown
58	Cerium	Orange Red	Colourless	Colourless	Colourless
74	Tangstun	Yellow	Colourless	Yellow	Yellow Brown
79	Gold	Rose Violet	Rose Violet	Red	Violet
92	Uranium	Yellow	Pale Yellow	Green	Bottle Green

Borax is used in the glass industry and in manufacture of enamels, glazes, optical glass, soap and for drying of oils. It is also used for the stiffening of candle wicks and for the glazing of paper. It is an antiseptic too and as a flux it is used in soldering.

2-9-11 Aluminium hydride

$(AlH_3)_n$ is called **alane,** it can be prepared by the action of lithium hydride with $(AlH_3)_n$ excess of aluminium chloride in ether.

$$3n\ LiH + n\ AlCl_3 \longrightarrow 3n\ LiCl + (AlH_3)_n$$

or By the action of $AlCl_3$ with AlH_4 in an ethereal solution.

$$4n\ AlCl_3 + 3n\ AlH_4 \longrightarrow 4(AlH_3)_n + 3n\ AlCl_4$$

Properties : It is a polymetric compound with bridging H atoms and constitute electron deficient molecules.

It is a white solid, which decomposes to give out H_2 at strong heat. It reacts with B_2H_6 to form aluminium borohydride.

$$2(AlH_3)_n + 3n\ B_2H_6 \longrightarrow 2n\ Al(BH_4)_3$$

It reacts under pressure at 130°C with ethylene to give a hydrocarbon chain. It hydrolyses to produce polyethylene.

$$Al\begin{cases} H \\ H \\ H \end{cases} + n\ \underset{\text{ethylene}}{C_2H_4} \longrightarrow Al\begin{cases} CH_2 - CH_2 - CH_2 \text{ -------- } CH_3 \\ CH_2 - CH_2 - CH_2 \text{ -------- } CH_3 \\ CH_2 - CH_2 - CH_2 \text{ -------- } CH_3 \end{cases}$$

Structure of $(AlH_3)_n$: It is a polymeric hydride with Al surrounded by 6 H atoms octahedrally. In brief the structure contains a net of Al–H–Al bridges. The neighbouring aluminium atoms are held by two 3 centered electron pair bonds above and below the plane of net work of Al atoms.

2-9-12. Alkyls of Aluminium : The alkyls of aluminium have gained great industrial importance because of their use as catalysts in the polymerization of ethylene as well as propylene.

The alkyls of aluminium are also widely used both as reducing agents as well as alkylating agents for the transition metal complexes.

Preparation of alkyls of aluminium : It is obtained by the reduction of alkyl-Mg compounds by Al.

$$2Al + 3R_2\ Mg \longrightarrow 2R_3Al \text{ or } (R_3Al)_2$$

By the action of alkylmagnesium halide with aluminium chloride, aluminium alkyl halides are obtained.

$$RMgCl + AlCl_3 \longrightarrow RAlCl_2,\ R_2AlCl,\ R_3Al$$

More direct methods available for large scale production are by the action of AlH_3 or $LiAlH_4$ on unsaturated hydrocarbons, C_nH_{2n}.

$$AlH_3 + 3C_nH_{2n} \longrightarrow Al\ (C_nH_{2n+1})_3$$

$$Li\ AlH_4 + 4C_nH_{2n} \longrightarrow Li\ [Al(C_nH_{2n+1})_4]$$

There is no direct combination of Al and H_2 to give AlH_3 but in the presence of alkyl groups it gives a **dialkyl**

$$Al + 1/2H_2 + 2\ R \longrightarrow AlR_2H$$

The dialkyl can further react to give trialkylaluminium

$$AlR_2H + R' \longrightarrow AlR_2R'$$

Therefore direct interaction of Al, H_2 and olefines can be used to give either the dialkyl hydrides or the trialkyls.

The lower alkyls of Al are reactive liquids and catch fire in air, and explode in water. Other derivatives are spontaneously inflammable and are sensitive to air and moisture. The structure of trimethyl aluminium is

Each aluminium atoms supplies an sp^3 hybrid orbital and so does the C atom.

The main acheivement in the organoaluminium chemistry was the discovery of alkene insertion reactions by K. Ziegler. In which two processes were discovered i)**growth reactions** to synthesize unbranched long chain primary alcohols and alkenes and ii) low pressure polymerization of ethene and propene in the presence of organometallic mixed catalysts, for which Ziegler & Natta were jointly awarded the Noble prize for chemistry in 1963.

A typical Ziegler Natta catalyst can be made by mixing $TiCl_4$ and $Al_2(C_2H_5)_6$ in heptane : partial reduction to Ti^{3+} and alkyl transfer occur and a brown suspension forms which rapidly absorbs and polymerizes ethene even at room temperature and atmospheric pressure. Typical industrial conditions required are 50-150°c and 10 mm pressure $Al_2(CH_3)_2$ can be prepared on a laboratory scale by the reaction of $Hg(CH_3)_2$ on Al at ~ 90°c. Similarly $Al_2(C_6H_5)_6$ can be prepared by using $Hg(C_6H_5)_2$ on the industrial scale Al is alkylated by means of any alkylhalide or by alkenes in the presence of H_2

$$2Al + 3RCl \xrightarrow[AlCl_3/AlR_3]{I_2} R_3Al_2Cl_3 \rightleftharpoons \frac{1}{2} R_4Al_2Cl_2 + \frac{1}{2} R_2 Al_2Cl_4$$

On adding NaCl, $R_2Al_2Cl_4$ gets removed as a complex, $2NaAlCl_3R$ and enables $R_4Al_2Cl_2$ to be distilled from the mixture, Reaction with Na yeilds a trialkyl eg :

$$3(CH_3)_4\ Al_2Cl_2 + 6\ Na \longrightarrow 2Al_2\ (CH_3)_6 + 2\ Al + 6NaCl$$

Higher trialkyls are more readily prepared on an industrial scale by the alkene route in which H_2 adds to aluminium in the presence of AlR_3 to give a dialkylaluminium hydride which gets readily adds to the alkene

$$2Al + 3H_2 + 2Al_2(C_2H_5)_6 \xrightarrow{150^o} \{6C_2H_5AlH\} \xrightarrow{6C_2H_4} 3Al_2(C_2H_5)_6$$

Polyethene produced at surface of such a catalyst is 85-95% crystalline. The probable mechanism of the catalyst is

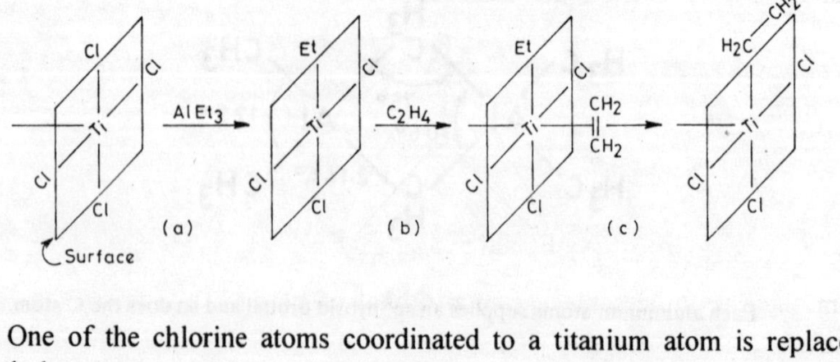

Al Et3 → (a) C2H4 → (b) → (c)

Surface

a) One of the chlorine atoms coordinated to a titanium atom is replaced by an ethyl group

b) then, because the titanium atom on the surface of the solid has a vacant coordination site a molecule of ethylene (ethene) can attach itself.

c) migration of the ethyl group to the ethylene by the cis insertion process occurs.

The result of this cis-insertion is that a vacant site is left behind and this can be occupied by another ethylene molecule, steps (a) & (b) gets repeated in a chain producing polyethene.

The discovery of this catalyst and its commercial importance has opened new vistas in the industrial chemistry.

2-9-13 Halides of Boron :

Boron trifluoride : Boron trifluoride (BF_3) can be prepared by the action of B_2O_3 with CaF_2 in the presence of sulphuric acid. The $CaSO_4$ obtained is another important product. The gaseous BF_3 is collected in gas jars

$$B_2O_3 + CaF_2 + H_2SO_4 \longrightarrow BF_3 + CaSO_4 + H_2O$$

another method of obtaining BF_3 is by the action of B_2O_3 with KHF_4.

$$B_2O_3 + KHF_4 \longrightarrow K(B_4O_6F) + BF_3$$

BF_3 has an unoccupied orbital and hence can readily unite with electron donors of which flouride ion and water are the most important. It first gets hydrolysed with water to form boric acid with the liberation of F^- ion which further combines with BF_3 to form fluoborate ion, which is not easily hydrolyzed.

$$4BF_3 + 3H_2O \longrightarrow H_3BO_3 + 3H^+$$

The proton combines with the fluoborate ion forming fluoboric acid HBF_4 which cannot be isolated but will be present only in solution. The fluoborate ion forms stable anhydrons salts, for example : KBF_4

BF_3 is a colourless gas with an equilateral triangular structure with fluorine present at the three corners. As the molecule has very short bondlengths with high bond energy, the unshared electron pairs

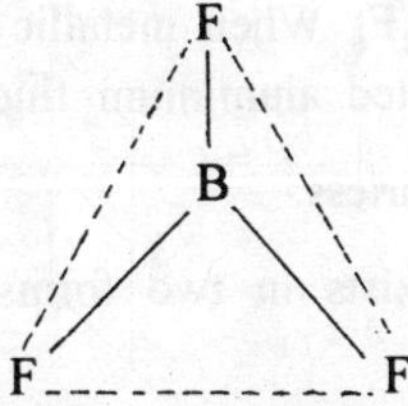

from fluorine utilizes the vacant p orbital of boron. This makes the molecule a strong electron acceptor and in turn a strong Lewis acid. The symmetry of the molecule with no intermolecular attractions, makes BF_3 volatile.

Boron trichloride : Boron trichloride (BCl_3) can be prepared by heating either a metal boride or boron in a current of chlorine. It can also be prepared by the action of chlorine at high temperature on a mixture of carbon and boric oxide.

$$M_3B + 3Cl_2 \longrightarrow BCl_3 + 3MCl$$

$$2B + 3Cl_2 \longrightarrow 2BCl_3$$

$$2B_2O_3 + 6Cl_2 + 3C \longrightarrow 4BCl_3 + 3CO_2$$

Boron trichloride can be hydrolysed with water and it can also get ammonolysed.

$$BCl_3 + 3H_2O \longrightarrow H_3BO_3 + 3HCl$$

$$BCl_3 + 6NH_3 \longrightarrow B(NH_2)_3 + 3NH_4Cl$$

BCl_3 can coordinate with a Cl^- ion forming BCl_4^- ion, BCl_3 is a planar triangular molecule similar to that of BF_3.

Diborontetrachloride : A liquid subchloride of boron, B_2Cl_4 is obtained by passing BCl_3 vapour through 1-2 mm pressure through a glow discharge between mercury electrodes. Its structure is of interest, in the vapour and liquid phase the two BCl_2 groups are in planes perpendicular to each other.

Cl Cl
120° B 1·7 Å B
Cl 1·75 Å Cl

2-9-14 Halides of Aluminium

Aluminium reacts with all the four important halogens forming AlF_3, $AlCl_3$, $AlBr_3$ and AlI_3

Preparation & properties of Al_2F_6 When metallic Al, or its oxide Al_2O_3 is dissolved in excess of HF, hydrated aluminium fluoride gets deposited.

In the pure form it is colourless.

The hydrate $AlF_3 . 3H_2O$ exists in two forms, a soluble form and an insoluble form.

It dissolves in HF and forms fluoroaluminic acid

$$3HF + AlF_3 \longrightarrow H_3AlF_6$$

The mineral cryolite (Na_3AlF_6) is a sodium salt of $H_3\ AlF_6$.

Preparation & properties of AlCl3

The crystal hydrate form of $AlCl_3$, gets deposited on dissolving the metal in excess of HCl.

$$6HCl + 2Al + 12H_2O \longrightarrow Al_2Cl_6\ 12H_2O + 3H_2$$

On heating the hydrate decomposes to give alumina

$$Al_2Cl_6{+}12H_2O \longrightarrow Al_2O_3 + 6HCl + 9H_2O$$

Anhydrous $AlCl_3$ is prepared by heating Al in a dry stream of Cl_2 or HCl gas

$$2Al + 3Cl_2 \longrightarrow 2AlCl_3$$

$$2Al + 6HCl \longrightarrow 2AlCl_3 + 3H_2$$

It can also be obtained by heating Al_2O_3 either with carbon and chlorine stream or with carbonyl chloride

$$Al_2O_3 + 3C + 3Cl_2 \longrightarrow 2AlCl_3 + 3CO$$

$$Al_2O_3 + 3COCl_2 \longrightarrow 2AlCl_3 + 3CO_2$$

$AlCl_3$ is white and solid subliming at 183°C and it is highly hygroscopic. It fumes strongly in the presence of moisture due to formation of HCl and attains a yellow color.

$$2AlCl_3 + 6H_2O \longrightarrow 2Al(OH)_3 + 6HCl$$

It is polar covalent compound and hence is soluble in alcohol, acetone, ether, carbon tetrachloride etc.

$AlCl_3$ combines with donor molecules such as NH_3 to form an adduct.

$$AlCl_3 + :NH_3 \longrightarrow Cl_3AlNH_3$$

It forms complexes with ethers, aldehydes etc.

Vapour density measurements at 670 K correspond to a dimer forming Al_2Cl_6.

It is used as a catalyst in the organic synthesis (Friedal Crafts reaction).

$AlCl_3$ is widely used in the manufacture of drugs and pharmaceuticals etc.

Preparation of $AlBr_3$ and AlI_3 : Both the compounds are prepared by heating the metal in the presence of Br_2 or I_2 vapours.

$$2Al + 3Br_2 \longrightarrow 2AlBr_3$$

$$2Al + I_2 \longrightarrow 2AI_3$$

$AlBr_3$ resembles $AlCl_3$ in its properties. AlI_3 reacts with CCl_4 giving $AlCl_3$ and CI_4

$$4AlI_3 + 3CCl_4 \longrightarrow 4AlCl_3 + 3CI_4$$

2-9-15 Characteristic differences between B and Al halides :

The characteristic differences between B and Al halides is that former exists as a monomer and the latter exists as a dimer (Al_2Cl_6) in non polar solvents and in vapour phase.

Such bridged structures are common for electron deficient compounds. Boron trihalides do not dimerize. This is due to the small size of boron and its inability to expand its coordination sphere.

Halides of Ga, In, Tl : The trihalides of Ga, In and Tl are covalent but the covalent character decreases from Ga to Tl with a simultaneous increase in ionic character.

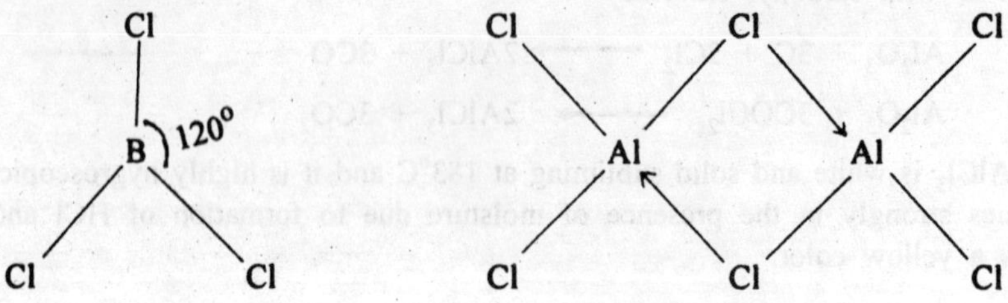

Structure of BCl_3 and Al_2Cl_6 (Dimeric $AlCl_3$)

$GaCl_3$ is a white solid with MP = 780°C and BP = 6800°C and dimerizes to Ga_2Cl_6. $InCl_3$ does not dimerize and is relatively more ionic than $GaCl_3$. TlCl does not dimerize and is most ionic.

2-10 DIAGONAL RELATIONSHIP OF BORON WITH SILICON

Boron has unique chemistry with only few resemblances in common with Al and the rest of the group III elements. It shows many resemblances in common with Si and provides one of the clearest examples of diagonal relationship. Some of the common features are listed below :

(i) Occurrence : Both the elements *i.e.* B and Si do not occur in free state in nature. They are available as oxo compounds *e.g.* Borates and Silicates.

(ii) Non metallic character : Both of them have non metallic character, they have similar physical properties and exist in both amorphous and crystalline forms.

(iii) Covalence : The charge to radius ratio for both B and Si is high and comparable hence their chemistry is mainly covalent.

(iv) Both the elements have comparable

(a) Electronegativities (B = 2.0 and Si = 1.8)

(b) Densities g/cc B = 2.34 Si = 2.35

(c) Boiling temperatures B = 2550°C Si = 2680°C.

(v) Both the elements form hydrides from their magnesium salts.

$$Mg_3B_2 + 6HCl \longrightarrow B_2H_6 + 3MgCl_2$$

$$Mg_2Si + 4HCl \longrightarrow SiH_4 + 2MgCl_2$$

These hydrides are gases and burn spontaneously in air or oxygen.

(vi) Both the oxides are polymeric macro molecules and are soluble in alkaline solutions.

(vii) Chlorides of B and Si hydrolyse to give

$$BCl_3 + 3H_2O \longrightarrow B(OH)_3 + 3HCl$$

$$SiCl_4 + 4H_2O \longrightarrow Si(OH)_4 + 4HCl$$

(viii) The oxides and hydroxides of both these elements are feebly acidic and react with alcohol in the presence of conc. H_2SO_4 to form volatile products.

$$B(OH)_3 + 3ROH \longrightarrow B(OR)_3 + 3H_2O$$

$$Si(OH)_4 + 4ROH \longrightarrow Si(OR)_4 + 4H_2O$$

B-O bond and Si-O bonds are so strong that hydroxides prefer to release H^+ and behave like weak acids.

(ix) Both these elements are rather inert and are almost stable at room temperature. They do not react with H_2O, but are reactive only with conc H_2SO_4.

2-11 SIMILARITIES BETWEEN B AND Al

B and Al resemble each other in some respects and also behave differently in other respects.

(a) Outer electronic configuration is same for both elements $ns^2\ np^1$, without any 'd' electrons in them.

(b) Both of them have a tendency to share 3 outermost electrons and hence show non variable +3 oxidation state.

(c) B has a greater tendency to form covalent compounds than aluminium which exhibits ionic nature in AlF_3. BX_3 and $AlCl_3$ behave as electron deficient molecules and act as Lewis acids by combining with electron rich compounds.

$H_3N: + BX_3 \longrightarrow H_3N:\!\rightarrow BX_3$

$H_3N: + AlCl_3 \longrightarrow H_3N:\!\rightarrow AlCl_3$

(d) Both B and Al react with alkalies to liberate H_2

$2B + 6NaOH \longrightarrow 2Na_3BO_3 + 3H_2$

Sodium borate

$2Al + 2NaOH + 2H_2O \longrightarrow 2NaAlO_2 + 3H_2$

Sodium meta aluminate

(e) On heating to high temperature (above 700°C) in air or oxygen atmosphere, both the elements form trioxides.

$4B + 3O_2 \longrightarrow 2B_2O_3$

$4Al + 3O_2 \longrightarrow 2Al_2O_3$

(f) On heating in a current of Cl_2 gas both these elements form chlorides which are covalent, volatile, hygroscopic and are easily hydrolyzed.

$2B + 3Cl_2 \longrightarrow 2BCl_3$

$BCl_3 + 3H_2O \longrightarrow H_3BO_3 + 3HCl$

Boric acid

$2Al + 3Cl_2 \longrightarrow 2AlCl_3$

$AlCl_3 + 3H_2O \longrightarrow Al(OH)_3 + 3HCl$

(g) Both the elements form nitrides when the elements are heated in a current of N_2

$2B + N_2 \longrightarrow 2BN$

$2Al + N_2 \longrightarrow 2AlN$

Both these nitrides hydrolyse with the liberation of NH_3

$BN + 3H_2O \longrightarrow H_3BO_3 + NH_3$

$AlN + 3H_2O \longrightarrow Al(OH)_3 + NH_3$

$BN + 3NaOH \longrightarrow Na_3BO_3 + NH_3$

$AlN + NaOH + H_2O \longrightarrow NaAlO_2 + NH_3$

(h) Both these elements form similar organic compounds.

$2BCl_3 + 3\ Zn(CH_3)_2 \longrightarrow 2B(CH_3)_3 + 3ZnCl_2$

$AlCl_3 + 3NaH + 3C_2H_4 \longrightarrow Al(C_2H_5)_3 + 3NaCl.$

2-12 MULTIPLE-CHOICE QUESTIONS

2-12-1 Ionization potential of Al and Ga is

a) 8.28 & 5.98 eV c) 5.98 & 5.76 eV

b) 5.98 & 5.98 eV d) 8.28 & 5.76 eV

2-12-2 Thalium is stable in the oxidation state of

a) + 2 b) + 3 c) + 1 d) + 4

2-12-3 Colemanite is an ore of

a) Aluminium b) Gallium c) Thallium d) Boron

2-12-4 Borax is

a) $Na_2B_4O_710H_2O$ c) $Na_3BO_37H_2O$

b) $Ca_2B_6O_{11}$ d) H_3BO_3

2-12-5 In Bayer's Process for the extraction of Aluminium the ore used is

a) Carnalite b) Bauxite c) Corundum d) Cryolite

2-12-6 Aluminium reacts with H_2SO_4 to give $Al_2(SO_4)_3$ with SO_2 whereas boron reacts with H_2SO_4 to give alongwith SO_2 :

a) H_3BO_3 b) $B_2(SO_4)_3$ c) Na_3BO_3 d) $Na_3BO_3+H_2$

2-12-7 In the compound $Ga^+[GaCl_4]$; the individual Ga shows an oxidation state of

a) +1 b) +3 c) +2 d) +4

2-12-8 The banana bond is a

a) Single electron bond b) Two electron bond
c) A Tau bond d) Three electron bond

2-12-9 Diborane is

a) $B_2H_4X_2$ b) B_2H_6X c) B_2H_5M d) B_2H_6

2-12-10 The general formula for closoboranes is

a) $B_nH_n^{2-}$ b) $B_nH_n^{4-}$ c) $B_nH_n^{1-}$ d) $B_nH_n^{3-}$

2-12-11 The nido boranes amongst the following pairs are

a) B_4H_{10} & B_5H_9 b) B_2H_6 & B_5H_9

c) B_2H_6 & B_4H_{10} d) B_4H_{10} & B_5H_9

2-12-12 Which one is not an ionic nido borane

a) $B_5H_8^-$ b) $B_9H_{12}^-$

c) $B_5H_9^-$ d) $B_{10}H_{13}^-$

2-12-13 An arachnoborane pair amongst the following is

a) $B_2H_7^-$ and $B_3H_8^-$ b) B_2H_8 and B_6 H_{12}

c) B_8H_{16} and B_8 H_{14} d) $B_{10}H_{14}$ and B_3 H_9

2-12-14 The Hypho borane amongst the following is

a) $B_{10}H_{18}$ b) B_8H_{14} c) $B_{10}H_{14}$ d) B_9H_{15}

2-12-15 Fusion of two clusters through two boran atoms at a common edge produces a borone of the type

a) Hypho b) closo c) nido d) conjuncto

2-12-16 $B_3N_3H_6$ is

a) Borazine b) Borazene c) Borazone d) Boraxane

2-12-17 $B_3N_3H_6$ resembles in many properties to

a) Benzoin b) Benzene c) Benzidine d) Benzyne

2-12-18 The surface tension of benzene and Benzine are

a) less than each other b) more than each other
c) equal to each other d) quite different

2-12-19 The boron-boron distance in diborane is

a) 1.70 Å b) 1.77 Å c) 1.66 Å d) 1.19 Å

2-12-20 Which is the correct nomenclative for the compound B_5H_9

a) penta borane b) 9-penta borane

c) nona penta borane d) penta borane-9

2-12-21 The tetrahydroborate ion is

a) B_4H_9 b) $B_4H_8^{2-}$ c) BH_4^{2-} d) BH_4

2-12-22 Hydroboration reaction, is a reaction between

a) B_2H_6 & alkynes b) B_2H_6 & benzene
c) benzene & alkynes d) B_2H_6 & silicones

2-12-23 Hydroboration reaction is

a) condensation reaction b) Polymerisation reaction

c) Regiospecific reaction d) substitution reaction

2-12-24 Usually the polyhedral borohydrides are formed by

a) Hyphoboranes b) Closoboranes

c) Conjunto boranes d) None of them

2-12-25 Carboranes are compounds of

a) Carborundum b) Carbon & Alkenes

c) Carbon & Silicon d) Carbon & Boron

2-12-26 Metallocarboranes can be obtained by

a) Polyhedral decomposition b) Polyhedral fusion

c) Polyhedral condensation d) Polyhedral Subrogation

2-12-27 During the borax bead test, the compound formed is

a) Metallo metaborates b) Metallo borides
c) Metallo boxoxides d) Meta carborates

2-12-28 In a Ziegler-Natta catalyst the compounds used are

a) $TiCl_4 + Al_2(C_2H_5)_6$ b) $TiCl_4 + C_2H_5OH$
c) $Al_2(C_2H_5)_6 + C_2H_5OH$ d) $C_2H_{\bar{5}} + TiCl_4$

2-13 SHORT AND ESSAY TYPE QUESTIONS

2-13-1 Discuss the general behaviour of the IIIB group of the periodic table.

2-13-2 Discuss the oxidation states of the III B group elements with special reference to the Inert pair effect.

2-13-3 a) What are the important minerals of boron?

b) How is Boron extracted from borax

2-13-4 a) What are main ores of aluminium ?

b) How is Aluminium extracted by ? Serpeck's process

2-13-5 Discuss the action of air, water, alkalies & acids on the III B group elements

2-13-6 What are the different forms of boranes

2-13-7 Discuss the preparation, structure and bonding in diborane

2-13-8 How are tetrahydroborates prepared. Discuss their industrial importance & their structures

2-13-9 What is meant by hydroboration reaction ?

2-13-10 Discuss the Chemistry of polyhedralborohydrides

2-13-11 What are carboranes? How are they synthesized ?

2-13-12 What are the different methods used for preparing metallocarboranes?

2-13-13 Write notes on the following

a) Alkyls of Aluminium b) Aluminium hydrides
c) Ziegler-Natta Catalysts d) Aluminium Halides

2-13-14 Discuss the halides of boron

2-13-15 Write notes on the following
a) Diagonal relationship of boron with silicon
b) Similarities & differences between boron and aluminium

2-13-16 What is meant by an electron defficient compound? Describe the structure of such a compound formed by Aluminium. Explain clearly the bonding involved in it. (O.U.. Apr. 90. 88)

2-13-17 What are electron deficient molecules

(i) Give the structure of diborane and borazole

(ii) Explain the nature of the following hydroxides.
$B(OH)_3$, $Al(OH)_3$, $Tl(OH)_3$ (K.U., Oct. 1990)

2-13-18 What is electron deficient bond? Describe the preparation, properties and structure of diborane. (N.U., Apr. 1989)

2-13-19 What are electron deficient molecules? Explain the bonding and structure taking two examples of electron deficient molecules. (K.U., May. 1988)

2-13-20 Account for the following :

(i) Boron does not form trivalent cationic species

(ii) Thallium exists in a stable + 1 oxidation state (O.U., Oct. 1989)

2-13-21 What is borax ? Mention any two of its important properties and any two uses. (O.U., Apr. 1988)

2-13-22 Write an explanatory note on a 3 centred bonds. (O.U., Apr. 1988)

2-13-23 Give the preparation and two important chemical properties of diborane. Write a note on the structure and nature of bonding in diborane.(O.U., Oct. 1988)

2-13-24 How is diborane prepared? Discuss its structure. (A.U., Apr. 1988)

2-13-25 Write short notes on (a) structure of diborane (b) Aluminothermic process. (K.U., Oct. 1988)

2-13-26 Describe the proportion of borax. (K.U., Apr. 1989)

2-13-27 Describe the extraction, characteristic properties and purification of aluminium. (K.U., Apr. 1989)

2-13-28 'Boron trichloride and aluminium trichloride' among these two which is a monomer and which is a dimer? Explain. (A.U., Apr. 1988)

2-13-29 Describe the digonal relationship between beryllium and alumnium. (K.U., May. 1988)

2-13-30 Explain the diagonal relationship between B and Si.

2-13-31 Describe the physical and chemical properties of boron and silicon with reference to their diagonal relationship. (N.U., Oct. 1989)

2-13-32 Give any two properties to show the similarity between boron and aluminium and two properties in which these two elements differ. (O.U., Apr. 1990)

ANSWERS 2-12

2-12-1 B	**2-12-2** C	**2-12-3** D	**2-12-4** A	**2-12-5** B	**2-12-6** A
2-12-7 C	**2-12-8** C	**2-12-9** D	**2-12-10** A	**2-12-11** B	**2-12-12** C
2-12-13 B	**2-12-14** A	**2-12-15** D	**2-12-16** A	**2-12-17** B	**2-12-18** C
2-12-19 B	**2-12-20** D	**2-12-21** D	**2-12-22** A	**2-12-23** C	**2-12-24** C
2-12-25 D	**2-12-26** D	**2-12-27** A	**2-12-28** A		

3

ELEMENTS OF GROUP IV B
THE CARBON FAMILY

					He
B	C	N	O	F	Ne
Al	Si	P	S	Cl	Ar
Ga	Ge	As	Se	Br	Kr
In	Sn	Sb	Te	I	Xe
Tl	Pb	Bi	Po	At	Rn

3-1 INTRODUCTION

The elements that follow carbon constitute the IV B group or the carbon family. They are carbon (C), silicon (Si), germanium (Ge), tin (Sn) and lead (Pb). Carbon forms tthe most important constituent of organic matter and opens the doors of a new branch of chemistry known as organic chemistry where as silicon is the main constituent of inorganic matter commonly found as sand etc. In this group a change from nonmetallic to metallic gradations with increase in atomic number is well illustrated. Carbon and silicon are distinct non metals, germanium is a metalloid and tin and lead are distinct metals.

3-2 OCCURRENCE

Carbon and silicon occur abundantly in nature. Carbon is the only element in this group that occurs free in nature as coal, diamond, graphite. In the combined state in nature carbon is present in the form of natural gas, petroleum crude and various types of organic matter. They form the main source of energy. Silicon is the 2nd most abundant element in the earth crust (28%), it is also the 2nd most abundant element after oxygen in human body (18%) and it is the main constituent of rocks, in the form of silicon and silicates. Germanium occurs in traces in silver and zinc ores. Tin and lead occur in concentrated ores such as tin stone (SnO_2) and galena (PbS).

3-3 ELECTRONIC CONFIGURATION AND OXIDATION STATES

The general electronic configuration of the valence shell of this group of elements is represented as ns^2np^2. With this general electronic configuration these elements are expected to show similar properties. But because of the differences in the number of electrons present and saturation of the penultimate shells of these elements viz., carbon with 2 electrons, Si with 8 electrons, germanium with 18 electrons, tin and lead each with 18 electrons in the penultimate shells with a scope of filling further 14 electrons in their 4f and 5f orbitals (maximum capacity of 4th and 5th shells being 32 and 50 electrons respectively) it is clear that none of these elements have a similarity in the electronic configuration, and hence are not expected to show close resemblances in the group properties as in the other groups, especially IA and IIA.

With $ns^2 np^2$ electronic configuration of their valence shells they are capable of attaining the noble gas octet configuration either by gaining 4 electrons or by losing 4 electron. C^{4+} ion is isoelectronic with helium and has a low charge density, therefore non existent. This would require an abnormally high ionization potential of 147.9 eV. When carbon gains 4 electrons it would be isoelectronic with argon configuration and such a process is also energetically not viable except in metal hydrides, Be_2C and $A1_4C_3$. Instead carbon shows a covalence of four by promoting 2s electrons to 2p level and undergoes sp^3 hybridization, resulting in 4 orbitals of equivalent energy, directed towards the 4 corners of a tetrahedron. Though the ionization potential is much less in the case of Si to give Si^{4+} than that is required to give C^{4+} ion, silicon also prefers to show a covalence of 4 in it compounds similar to that of carbon.

While carbon and silicon uniformly show a valency of four, Ge, Sn and Pb exhibit both +2 and +4 oxidation states. The stability of +4 state decreases and that of the +2 increases from Ge to Pb. This is because of the inert pair effect, which increases on descending the group. The unstable +2 state of Germanium gets oxidized to +4 state easily, hence the Ge^{2+} compounds are strong reducing agents. The inert pair effect is much greater in tin, hence in tin the +2 is highly stable and +4 state gets easily reduced to +2. Sn^{4+} compounds are strong oxidizing agents. In lead the inert pair effect is so marked that only +2 state is highly stable, the +4 compounds are very unstable and are proven oxidizing agents.

$$Ge^{2+} \longrightarrow Ge^{4+} + 2e^-$$

$$Sn^{4+} \longrightarrow Sn^{2+} + 2e^-$$

$$Pb^{4+} \longrightarrow Pb^{2+} + 2e^-$$

Unstable state stable state

3-4 PHYSICAL PROPERTIES OF THE CARBON FAMILY

The physical properties are listed in table 3-1.

3-4-1 The Ionization Potential

The ionization potential of the IVB group decreases on descending the group. The ionization potential of silicon indicates that it is capable of forming positive ions more easily than carbon but it forms mainly covalent compounds. The only elements which are likely to give a large electronegativity difference leading to ionic character are fluorine and oxygen in ionic compounds like SnO_2, SnF_4 and PbO_2.

Table 3-1 Physical constants of group IVB elements

Element / Property	Carbon	Silicon	Germanium	Tin	Lead
Symbols	C	Si	Ge	Sn	Pb
Automic number	6	14	32	50	82
Oxidation states	(+2), +4	(+2), +4	+2, +4	+2, +4	+2, (+4)
Ionization Potentials I	11.27	8.15	7.89	7.32	7.41
Ionization Potentials II	24.37	16.34	15.91	14.61	15.04
Ionization Potentials III	47.83	33.47	34.21	30.65	32.09
Ionization Potentials IV	64.44	45.09	45.66	40.71	41.93
Total I.E. (eV)	147.9	103.05	103.67	93.29	96.47
Atomic weight	12.011	28.09	72.60	118.7	207.27
Atomic radii (A)	0.77	1.17	1.22	1.41	1.54
Density (g ml^{-1})	3.51 diamond 2.25 graphite	2.35	5.36	6.00	11.34
Melting point (K)	3853	1698	1233	504.9	600
Boiling point (K)	4473	2953	3103	2960	2024
Electronegativity	2.5	1.8	1.8	1.8	1.8

3-4-2 Electronegativity

Carbon is the most electronegative element of the family. The electronegativity values for Si, Ge, Sn and Pb are same (1.8) for all the elements, this may be due to the presence of *d* and *f* orbitals.

3-4-3 Catenation

Catenation is the property of formation of chains of similar atoms. Carbon, silicon and to some extent germanium, catenate to form many chain compounds. Carbon can catenate to form even multimillion atom catenation compounds. Because of the high bond energy (85 kcals $mole^{-1}$), the C – C bonds are very strong in these homo polar compounds in comparison to Si – Si bond. Hence the catenation in silicon is possible to an extent of Si_6H_{14} unlike limitless carbon-carbon chains.

This bonding is covalent in nature and this property is manifested in hydrides or hydrocarbons. The high C – H bond energy (99 kcals $mole^{-1}$) and high C – C bond energy accounts for a high stability of the saturated hydrocarbons. In contrast the Si – H bond energy (76 kcals $mole^{-1}$) is relatively low and this makes the SiH_4 highly reactive group, restricting the catenation of Si to just 6 atoms.

Germanium can catenate to an extent of Ge_3H_6, tin to an extent of Sn_2H_6 where as lead cannot catenate at all.

In addition to catenation between same elements or inter atomic catenation, the intra atomic catenation is also well recognized. The best example for intra atomic catenation is silicon carbide, (SiC)n, involving catenation between C and Si. Silicon carbide is a very strong and inert compound, and hence used as an abrasive material. It is commonly known as carborundum.

3-4-4 Complex Formation

For the formation of a complex, the complexing metal ion must be capable of accepting electron pairs in its vacant d-orbitals. The size of atom is another important factor in the complexation. Carbon has no empty d-orbitals and hence is incapable of complex formation. Si, Ge, Sn and Pb form stable complexes.

$$SiF_4 + 2F^- \longrightarrow [SiF_6]^{2-}$$

$$SnCl_4 + 2Cl^- \longrightarrow [SnCl_6]^{2-}$$

3-4-5 Allotropy

All the elements of the family exhibit, the characteristic property of allotropy or polymorphism.

Carbon forms two type of allotropes (i) Crystalline and (ii) Amorphous. The crystalline allotropes of carbon are diamond and graphite. In diamond, the sp^3 hybrid carbon has a tetrahedral arrangement with great hardness and with the C – C bond distance of 1.54 A. In graphite the layers of carbon atoms are in a regular hexagonal network with a C – C bond distance of 1.415 A and these hexagonal layers are separated by a distance of 3.35 A, figure 6-1. In graphite the arrangement of the carbon is in Sp^2 hybrid type, holding two electrons in each orbital. The unhybridized electrons in the p-orbital are mobile electrons as in a metal lattice and hence graphite conducts electricity.

Diamonds with a refractive index around 2.42 are used as precious stones in jewellery. Large diamonds are used for even cutting the metals. Diamond powder is a good abrasive material. Artificial diamonds are manufactured by heating graphite to temperatures above 3000°C under pressure of 1×10^5 atmospheres in the presence of a metal catalyst like Fe, Co or Mn.

The type of Graphite manufactured depends on the need for which it has to be used. The lubricative form of it is obtained by heating powdered coke in the presence of silicon for about 24 to 48 hours around 2500°C in an electric furnace. The graphite used for electrodes is made by mixing powdered coke with pitch then it is molded to the required shape. It is heated to 1200°C to remove the volatile substances then for final graphitizing it is further heated to 2500°C.

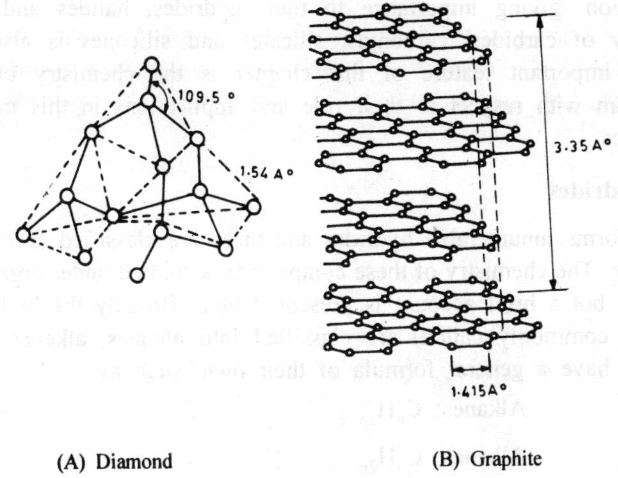

(A) Diamond (B) Graphite

Fig.3-1 Structure of diamond (A) and graphite (B)

The amorphous allotropes of carbon are charcoal, coke, gas carbon, animal charcoal, lamp black etc.

Charcoal is obtained by burning sugars, wood or cellulose and coke is obtained by the hardening of charcoal or from the residue of the distillation of oils, gas carbon is present in the furnaces, animal charcoal is the product obtained by the charring of bones or animal products, lamp black is the product obtained by the incomplete combustion of oils.

3-5 CHEMICAL REACTIVITY OF THE CARBON FAMILY

All the elements of the family of group IVB are quite reactive, carbon forms numerous compounds and combines with almost all the nonmetals and metals to form inorganic group of compounds, it combines with hydrogen alone or with hydrogen and nonmetals like oxygen, halogens, sulphur, phosphorus, nitrogen etc., to form innumerable groups of compounds possessing typical characteristics of their own. Their typical molecular and structural formulae and bonding necessitated the development of a branch of chemistry known as organic chemistry. It will not be improper if we state that whatever is around us and in us are only manifestation of carbon and our very existence is under its umbrella.

Silicon's most important compound is SiO_2 (silica or sand). Its derivatives such as quartz, tridymite, cristobolite, silicones and silicates are the oxides of silicon. Another important series of compounds of silicon are the silanes (the hydrides) and silicon halides. Similarly germanium, tin and lead chemistries are also of equal importance.

The chemistry of important compounds of these elements are studied in this section, giving importance to their hydrides, halides and oxides. The chemistry of carbides, carbonyls, silicates and silicones is also dealt with. Another important feature of this chapter is the chemistry of silicon and germanium with respect to their role and applications in this modern age of electronics.

3-5-1 Hydrides

Carbon forms innumerable hydrides and these are classified according to their behaviour. The chemistry of these compounds is studied under organic chemistry in detail, but a brief account is presented here. Broadly the hydrocarbons (as they are commonly called) are classified into alkanes, alkenes and alkynes. They all have a general formula of their own, such as

Alkanes: C_nH_{2n+2}

Alkenes: C_nH_{2n}

Alkynes: C_nH_{2n-2}

The hydrides are present in either long chains or in cyclic forms. The open chain compounds are called aliphatic compounds and the closed chain cyclic compounds are called alicyclic compounds. Another set of cyclic compounds are aromatic compounds, these are benzene, fused benzene rings like naphthalene and their derivatives such as phenol, benzoic acid, benzaldehyde, naphthol etc. In addition there are substituted hydrocarbons and heterocyclic hydrocarbons also. The structures of a few of the above cited compounds are given here for appreciating the beauty of this class of compounds.

```
     H           H   H
     |           |   |
 H - C - H   H - C = C - H     H - C ≡ C - H
     |
     H
```

(i) Methane (CH_4) (ii) Ethylene (C_2H_4) (iii) Acetylene(C_2H_2)

```
      H2                H   H   H   H   H   H
      C                 |   |   |   |   |   |
  H2C   CH2         H - C - C - C - C - C - C - H
   |     |              |   |   |   |   |   |
  H2C   CH2             H   H   H   H   H   H
      C
      H2
```

(iv) Cyclohexane(C_6H_{12}) (v) n-Hexane (C_6H_{14}) Open chain compound

(vi) Isohexane (C_6H_{14}) (vii) Benzene (C_6H_6) (viii) Naphthalene

Branched chain compound Aromatic compounds

(ix) Phenol (C_6H_5OH) (x) Pyridine (C_5H_5N) (xi) Furan (C_4H_4O)

Derivative of benzene Heterocyclic compounds

Silicon hydrides are called silanes, they have a general formula Si_nH_{2n+2}, these hydrides are analogous to alkanes. The most basic silane is monosilane (SiH_4), its homologues are disilane Si_2H_6 and trisilane (Si_3H_8). The monosilane is prepared by the reduction of silicon tetrachloride with lithium aluminium hydride.

$$SiCl_4 + LiAlH_4 \longrightarrow SiH_4 + AlCl_3 + LiCl$$

It is also obtained by the reaction of magnesium silicide with hydrochloride acid.

$$Mg_2Si + 4HCl \longrightarrow 2MgCl_2 + SiH_4$$

Disilane is analogous to ethane as monosilane is to methane. Similar to ethane the bonding in disilane is between two Si atoms.

Methane Monosilane Ethane Disilane

The silanes are highly reactive when compared to alkanes and hence are less stable. This difference in behaviour between alkanes and silanes is probably due to the difference in electronegativity between Si (1.8) and C (2.5). The electronegativity of H is 2.1. This results in an uneven sharing of the bonding electrons leaving a negative charge on carbon and positive charge on silicon on combining with hydrogen.

$$C^{\delta-}H_4^{\delta+} \qquad Si^{\delta+}H_4^{\delta-}$$

Germanes are similar to silanes in their properties. Monogermane is obtained by the addition of alkali borohydride to GeO_2 or by the reduction of germanium tetrachloride with $LiAlH_4$.

$$GeCl_4 + LiAlH_4 \longrightarrow GeH_4 + LiCl + AlCl_3$$

Digermane (Ge_2H_6), trigermane (Ge_3H_8) are also reported. Germanes upto Ge_9H_{20} are also obtained but are not well characterized.

Stannanes (SnH_4 & Sn_2H_6) and plumbane (PbH_4) are known. SnH_4 can be prepared by reduction of $SnCl_4$ with $LiAlH_4$ in an ethereal solution.

$$SnCl_4 + LiAlH_4 \longrightarrow SnH_4 + LiCl + AlCl_3$$

The stability of the hydrides decreases from C to Pb. This is accounted by three factors :

1) The relative bond energies :

C – C : 326 kJ mole^{-1} Si – Si : 222 kJ mole^{-1}
Ge – Ge : 167 kJ mole^{-1} Sn – Sn : 155 kJ mole^{-1}

2) The relatively high C – H bond energy :

C – H : 414 kJ mole^{1} Si – H : 314 kJ mole^{-1}
Ge – H : 310 kJ mole^{-1} Sn – H : 197 kJ mole^{-1}

3) The carbon atom does not have easily accessible d-orbitals and it cannot expand its octet. On the contrary other members of this group have easy access to d-orbitals and can expand their octet thus explaining their increased reactivity.

As the stability decreases with simultaneous increase in their chemical reactivity, silicones, germanes, stannanes are strong reducing agents and they readily get hydrolysed producing hydrogen when reacted with either an acid or a base or water.

$$SiH_4 + HCl \longrightarrow SiH_3Cl + H_2$$

$$GeH_4 + 2NaOH + H_2O \longrightarrow Na_2GeO_3 + 4H_2$$

$$SiH_4 + 3H_2O \longrightarrow H_2SiO_3 + 4H_2$$

Stannane decomposes at room temperature producing H_2

$$SnH_4 \longrightarrow Sn + 2H_2$$

Hence the reducing property increases from alkanes to silanes.

The thermal stability of the hydrides also decreases from alkanes to plumbanes. Alkane decomposes at 800°C while silane at 440°C, germane at 285°C, stannane at 150°C and plumbane merely at 0°C.

3-5-2 Oxides

The elements of the family form stable oxides, carbon forms at least three oxides CO, CO_2, C_3O_2.

Carbonmonoxide (CO) is formed when carbon is burnt in the presence of a limited supply of oxygen or air.

$$2C + O_2 \longrightarrow 2CO$$

It is obtained by the reduction of carbondioxide with carbon.

$$CO_2 + C \longrightarrow 2CO$$

By the dehydration of formic acid or oxalic acid carbon monoxide is formed.

$$HCOOH \xrightarrow{H_2SO_4} CO + H_2O$$

$$H_2C_2O_4 \xrightarrow{P_2O_5} CO + CO_2 + H_2O$$

CO is a colourless odourless gas, burns in air with a blue flame forming carbondioxide. It is highly poisonous, as it combines with haemoglobin present in the blood more readily than oxygen. In an atmosphere of carbon monoxide the oxygen normally carried by haemoglobin is replaced by CO and the iron present is attacked by CO to form iron carbonyl. The depletion of oxygen content and formation of iron carbonyls clot the blood and immediately arrest the biological reactions leading to death. Ordinary gas masks used for adsorbing gases like chlorine, sulphurdioxide etc. contain active charcoal and are not of much help to work in a CO atmosphere due to poor adsorbtion of CO by charcoal. Hence masks containing a mixture of manganese dioxide and cupric oxide are used which effectively react with CO liberating carbondioxide. These masks are a boon to the mine workers.

The reducing property of carbon monoxide is employed in the extraction of metals like iron and nickel.

$$Fe_2O_3 + 3CO \longrightarrow 2Fe + 3CO_2$$

$$NiO + CO \longrightarrow Ni + CO_2$$

It reacts with many transition elements forming volatile carbonyls. An exclusive study of carbonyls is made later in this chapter. Carbonmonoxide addition to reacting with oxygen combines with sulphur, chlorine etc.

$$CO + S \longrightarrow COS \text{ (Carbonyl sulphide)}$$

$$CO + Cl_2 \xrightarrow{h\nu} COCl_2 \text{ (Carbonyl chloride) (Phosgene)}$$

Carbonmonoxide reacts with sodium hydroxide and hydrogen forming sodium formate and methyl alcohol respectively.

$$\underset{\text{(Fused)}}{NaOH + CO} \longrightarrow HCOONa \text{ (sodium formate)}$$

$$2H_2 + CO \xrightarrow{ZnO/Cr_2O_3} CH_3OH \text{ (methyl alcohol)}$$

Carbonmonoxide could be represented by three resonance structures, Figure 3-2.

$$\underset{(a)}{:\bar{C}:::\overset{+}{O}:} \longleftrightarrow \underset{(b)}{:\bar{C} \equiv \overset{+}{O}:} \longleftrightarrow \underset{(c)}{C = O}$$

Fig 3-2 Resonating structures of carbon monoxide

In carbonmonoxide, carbon is in its sp hybrid state, one of the sp hybrid orbital overlaps with an orbital of oxygen forming a single σ bond. One pair of electrons present in the other sp orbital remains unaffected. The vacant unhybridized p-orbital gets involve in the formation of two π bonds with two p-orbitals of the oxygen atom resulting in a formal (δ–) charge on carbon and a formal (δ+) charge on oxygen.

From the resonance structures of CO, the structure given by fig. 3-2 (b) is more acceptable as the lone pair of electrons on carbon are available for donation to an acceptor like a metal ion. Hence CO can be said to have a triple bond : C ≡ O : with a pair of electrons each on carbon and oxygen atoms, which makes it act as a very good coordinating ligand.

Carbondioxide (CO_2) is formed by the burning of carbon in the presence of unlimited supply of air or oxygen.

$$C + O_2 \longrightarrow CO_2$$

In the atmosphere it is present to about 0.03% by volume, and this concentration is maintained in the atmosphere by the carbon dioxide cycle.

CO_2 is evolved in numerous laboratory reactions.

(i) When sodium carbonate is treated with dil HCl, CO_2 is evolved.

$$Na_2CO_3 + 2HCl \longrightarrow 2NaCl + H_2O + CO_2$$

(ii) When marble chips are treated with a dilute acid, CO_2 is readily evolved.

$$CaCO_3 + H_2SO_4 \longrightarrow CaSO_4 + H_2O + CO_2$$

(iii) When lime stone is heated at 1000°C, CO_2 is produced.

$$CaCO_3 \longrightarrow CaO + CO_2$$

(iv) During fermentation, carbondioxide is obtained.

$$C_6H_{12}O_6 \xrightarrow{\text{enzymes}} 2C_2H_5OH + 2CO_2$$

(v) By the action of heat on bicarbonates, carbondioxide is obtained.

$$2NaHCO_3 \xrightarrow{\text{heat}} Na_2CO_3 + H_2O + CO_2$$

It is a colourless, odourless gas, dissolves in water forming acidic solutions as CO_2 is an anhydride of carbonic acid, it cannot be isolated.

$$CO_2 + H_2O \longrightarrow H_2CO_3 \longrightarrow H^+ + HCO_3^- \text{ or } 2H^+ + CO_3^{--}$$

CO_2 reacts with alkalies forming carbonate salts and when it is in excess bicarbonates are formed. This process is applied for the removal of the permanent hardness of water.

$$Ca(OH)_2 + CO_2 \longrightarrow CaCO_3 + H_2O \xrightarrow{CO_2} Ca(HCO_3)_2$$

Formation of insoluble $CaCO_3$ is used as a test in the qualitative analysis of CO_2.

CO_2 is decomposed to CO at 1500°C.

$$2CO_2 \longrightarrow 2CO + O_2$$

but when CO_2 is passed over red hot coke, no oxygen is formed

$$CO_2 + C \longrightarrow 2CO$$

Carbondioxide gets absorbed by plants in the presence of sunlight during photosynthesis giving starch which on burning produces CO_2 back with the liberation of energy.

$$6CO_2 + 6H_2O + \xrightarrow{\text{Sunlight}} C_6H_{12}O_6 + 6O_2$$

$$C_6H_{12}O_6 + 6O_2 \longrightarrow 6CO_2 + 6H_2O + \text{energy}$$

CO_2 can be easily liquified and is therefore kept under pressure in fire extinguisher. Further compression of CO_2 converts the gas into solid carbondioxide popularly called dry ice. Dry ice is widely used for conducting reactions below 0°C.

Carbondioxide is a linear molecule with three resonating structures, figure 3-3, all with a zero dipole moment, of which structure (c) is more acceptable.

$$\bar{O} - C \equiv \overset{+}{O} \longleftrightarrow \overset{+}{O} \equiv C - \bar{O} \longleftrightarrow :O = C = O:$$

(a) (b) (c)

Fig.3-3 Resonating structures of carbondioxide

Carbon suboxide (C_3O_2) is formed by the dehydration of malonic acid with a strong dehydrating agent like P_2O_5. Another oxide of carbon is $C_{12}O_9$ which is an anhydride of mellitic acid, $C_6(COOH)_6$.

Silicon forms two oxides, monoxide (SiO) and dioxide (SiO_2). The monoxide does not exist at room temperature but is formed by high temperature reduction of SiO_2.

$$SiO_2 + Si \longrightarrow SiO$$

Silicon dioxide is commonly called as silica. Carbondioxide is a discrete molecule, where as silicon dioxide has an infinite three dimensional structure, because silicon cannot form double bonds as carbon does. Thus CO_2 is a gas and SiO_2 is a high melting solid, occurring in a number of different crystalline forms. Each form contains silicon atoms bonded tetrahedrally to four oxygen atoms by means of single Si – O bonds. The three main crystalline forms of SiO_2 are quartz, tridymite and cristobalite, each of which has a high and low temperature modification. The crystal symmetries and equilibrium inter conversion temperatures are represented by the following flow diagram.

α-Quartz (Helical)		α-Tridymite		α-Cristobalite (Bi-axial)		
⇅ 573°C		⇅ 120° - 160°C		⇅ 200° - 275°C		
β-Quartz (Hexagonal)	⇌ 870°C	β-Tridymite (Hexagonal)	⇌ 1470°C	β- Cristobalite (Cubic)	⇌ 1710°C	Liquid SiO_2

These names (quartz, tridymite and cristobalite) are derived either from their properties or place of extraction. Quartz is derived from a polish word, Kwardy, meaning very hard, tridymite in Greek tridymos means three fold, and cristobalite is derived from the name of a place in germany St. Cristoball, where it is available as an earthy deposit.

The difference between these structures is the arrangement of the tetrahedral SiO_2 units. All the three forms occur in nature and the transition between these modifications are quite rapid. Silica melts at 1710°C and has a very low coefficient of expansion hence it is immune to thermal shocks. The restriction of earthquake tremors to a small zone is due to resistance of thermal shocks by silica. Pure silica is transparent and is used for making sophisticated optical lenses, in the electronic industries and for making special laboratory equipments.

A tetrahydrated silica, ($SiO_2 4H_2O$), is called silica gel. It is a hard, granular translucent substance, with a large surface area and has the property of adsorbing liquids, hence it is used as a dehydrating agent and as a catalyst in the hydrolysis of aryl halides to phenols. It is also a supporter catalyst. It is used with vanadium pentoxide, in the oxidation of naphthalene to phthalic acid and sulphur dioxide to sulphur trioxide.

Germanium and tin form dioxides by the oxidation of respective metals at high temperature. Lead dioxide is much less thermally stable and is prepared by the oxidation of Pb^{2+} in alkaline solutions. All the three oxides GeO_2, SnO_2 and PbO_2 have rutile structures.

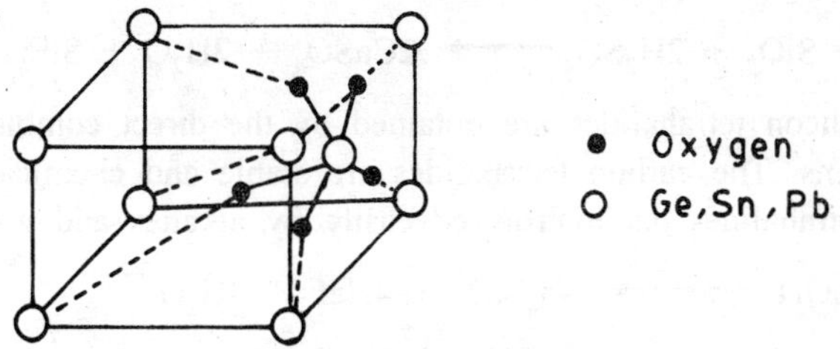

Fig.3-4 Rutile structure of GeO_2, SnO_2, PbO_2

Lead also forms PbO, Pb_2O_3, Pb_3O_4, Pb_7O_{11} and PbO_6. By the thermal decomposition of PbO_2, two non stoichiometric intermediate compounds are formed viz., Pb_7O_{11} and Pb_2O_3. Red lead Pb_3O_4 is formed by heating litharge (PbO) or lead dioxide (PbO_2) in air and this compound is found to possess chains of PbO_6. The structure of Pb_3O_4 is shown in Fig.3-5.

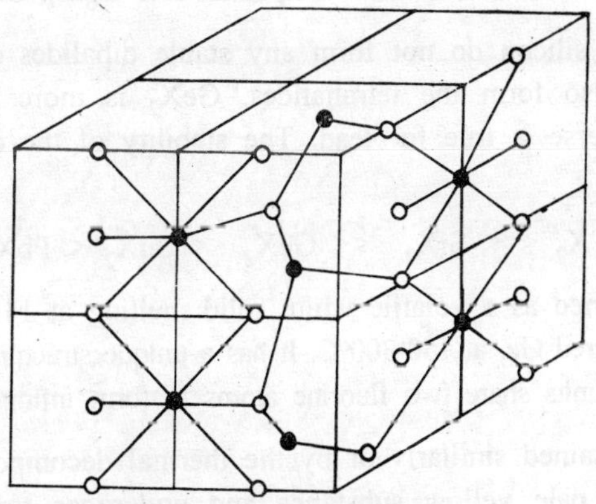

Fig.3-5 Structure of Pb_3O_4

3-5-3 Halides

Carbon forms many halides because of its capability of forming chains, or catenation. Some simple halides are CF_4, CCl_4, CBr_4 and CI_4. Silicon also forms similar halides SiF_4, $SiCl_4$, SiI_4. In addition, some very interesting fluoro carbons like C_2F_4, C_2F_6, C_3F_8, substituted carbon tetrachlorides like CCl_3F, CCl_2F_2, $CClF_3$, and compounds like $(CF_3)_3$ N are also synthesized.

Carbontetrachloride can be prepared by passing Cl_2 into CS_2 in the presence of a catalyst like trace amount of iodine. The mixed halides are separated by azeotropic distillation.

$$CS_2 + 3Cl_2 \longrightarrow \underset{\text{(b.p. 138°C)}}{S_2Cl_2} + \underset{\text{(b.p. 76.7°C)}}{CCl_4}$$

The silicon tetrafluoride is obtained when calcium fluoride is mixed with SiO_2 in the presence of H_2SO_4.

$$2CaF_2 + SiO_2 + 2H_2SO_4 \longrightarrow 2CaSO_4 + 2H_2O + SiF_4$$

The other silicon tetrahalides are obtained by the direct combination of silicon and halogens. The carbon tetrahalides are stable and chemically inert, whereas silicon tetrahalides get hydrolyzed easily by alkalies and water.

$$SiF_4 + 8NaOH \longrightarrow Na_4SiO_4 + 4NaF + 4H_2O$$

$$SiCl_4 + 4H_2O \longrightarrow Si(OH)_4 + 4HCl$$

$$2SiF_4 + 4H_2O \longrightarrow SiO_2 + 2H_3O^+ + SiF^{2-}_6 + 2HF$$

The SiF_6^{2-} is an octahedral complex with sp^3d^2 hybridization. The fluorosilicic acid, H_2SiF_6, which is known only in solution, is a strong acid and forms salts on reacting with metals and their compounds. The salts of the heavy metals are water soluble but those of Na, K, Ba and the lanthanides are sparingly soluble. Silicon forms chain compounds like Si_6Cl_4, Si_3Br_8, Si_5Cl_{12} etc.

Carbon and silicon do not form any stable dihalides as germanium, tin or lead which also form the tetrahalides. GeX_2 is more stable than GeX_4 where as the reverse is true for lead. The stability of the dihalides is in the sequence of

$$CX_2 <<< SiX_2 <<< GeX_2 << SnX_2 < PbX_2$$

GeF_2 is formed as a volatile white solid melting at 110°C by the action of GeF_4 on powdered Ge, at 150-300°C. It has a unique structure in which trigonal pyramidal (GeF_3) units share two fluorine atoms to form infinite spiral chains.

$GeCl_2$ is obtained similarly or by the thermal decomposition of $GeHCl_3$ at 70°C. It is a pale yellow substance and undergoes typical reactions as shown in the following flow chart.

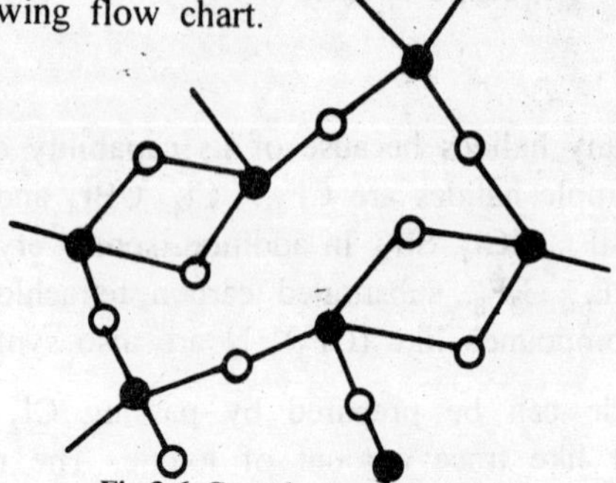

Fig.3-6 Crystal structure of GeF_2

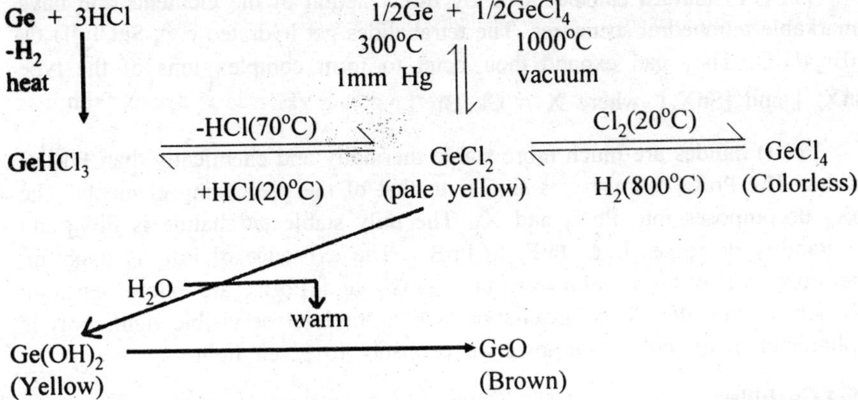

$GeBr_2$ and GeI_2 are obtained by the reduction of their respective tetrahalides. Germanium tetrahalides are readily prepared by direct action of the elements or by the action of aqueous hydrogen halides on germanium oxide.

$$Ge + 2X_2 \longrightarrow GeX_4$$

$$Ge + 4HX \longrightarrow GeX_4 + 2H_2$$

SnF_2, which is obtained as colourless monoclinic crystals by evaporation of a solution of SnO in 40% aqueous HF, is composed of Sn_4F_8 tetramers interlinked by weaker Sn – F interactions. $SnCl_2$ is obtained by dissolving the metal in HCl.

$$Sn + 2HCl \longrightarrow SnCl_2 + H_2$$

$SnCl_2$ in gaseous phase forms bent molecules, but the crystalline material (m.p. 246°C) has a layer structure with chains of corner shared trigonal bipyramidal $SnCl_3$ groups. It forms a dihydrate crystalline complex with hydrogen bonding in it. The three structures of $SnCl_2$ arc shown in fig.3-7

(a) $SnCl_2$ (gaseous)

(b) $SnCl_2$ (Crystalline)

(c) $[SnCl_2(H_2O)_2]$ complex.

Fig.3-7 Structures of $SnCl_2$ in (a) gaseous, (b) crystalline and (c) complex.

Tin (IV) halides can be made by direct action of the elements and have remarkable tetrahedral structures. The tetrahalides get hydrated e.g., $SnCl_4 4H_2O$, $SnBr_4 4H_2O$. They can expand their octet to form complex ions of the type $[SnX_6^{2-}]$ and $[SnX_5]^-$ where X = Cl, Br, I.

Pb(II) halides are much more stable thermally and chemically than Pb(IV) halides. The PbX_2 chemistry is similar to that of the preceeding elements. The PbX_4 decomposes into PbX_2 and X_2. The only stable tetrahalide is PbF_4 and the stability decreases from PbF_4 to $PbBr_4$. The existence of PbI_4 is doubtful. The PbX_2 are ortho rhombic crystals. $PbCl_2$ and $PbBr_2$ are photo sensitive and deposit metallic Pb on irradiation with ultraviolet or visible light, PbI_2 is a photoconductor and decomposes on exposure to green light.

3-5-4 Carbides

The carbides can be divided into four groups (i) Salt like (ii) Interstitial (iii) Covalent (iv) Iron type.

(i) Salt like carbides: The salt like carbides are formed by the most reactive metals of the IA, IIA, IB and lanthanides. These carbides are prepared by heating the metal, its oxide or its hydride with carbon. They are obtained as colourless solids and non conductros of electricity. The salt like carbides hydrolyse to produce acetylene, hence are referred to as acetylides. Beryllium and aluminium carbides, which produce methane on hydrolysis are called **methanides.**

$$MgC_2 + 2H_2O \longrightarrow Mg(OH)_2 + C_2H_2$$

$$Al_4C_3 + 12H_2O \longrightarrow 4Al(OH)_3 + 3CH_4$$

Hence in acetylides, the acetylide (– C = C –) ion is present and in methanides C^{4-} ion is present.

(ii) Interstitial carbides : The interstitial carbides are prepared by the direct reaction of the transition metals with carbon, or the reduction of the oxide with carbon at about 2000°C. They are non-stoichiometric compounds, carbon occupying the spaces between metal atoms in the lattice without too much distortion. They are high melting compounds and can conduct electricity and heat. It is observed, that, metals with 1.3A° cannot take up carbon atoms without causing distortion of the metal lattice. The carbides of this type are TiC, WC, TaC etc. They are used as hard facings for tools and dies. These carbides are obtained by hard pressing of the metal with carbon powder at 2000°C.

(iii) Covalent carbides: Covalent carbides are giant molecules with covalent bonding running completely throughout the structure. These are the binary compounds of carbon with elements of higher electronegativity like boron and silicon, but not hydrogen, chlorine or sulphur. The typical examples of covalent carbides are silicon carbide SiC and boron carbide B_4C_3. Both the carbides are extremely hard. SiC has a structure resembling to that of diamond and is prepared by reducing silica with carbon.

$$SiO_2 + C \longrightarrow SiC + O_2$$

Boron carbide is obtained by reducing

$$B_2O_3 + 3C \longrightarrow B_4C_3 + 3O_2$$

The structure of B_4C_3 is complicated. Both the carbides are thermally highly stable, very hard, chemically inert solids.

(iv) Iron type carbides: They are the carbides of those metals having a metallic radii of less than 1.3Å. The metals like chromium (1.27Å) and nickel (1.24Å) can form such carbides. As the iron group and its neighbouring, elements can form these carbides, they are called iron type carbides, Mn_3C, Fe_3C, Ni_3C have C atoms inside the trigonal prisms formed by the metal atoms.

In chromium carbide, Cr_3C_2, the carbon atoms form chains in the solid. These carbides can easily get decomposed by acids and water.

3-5-5 Cyanides

The chemistry of C – N bond plays an important role in inorganic carbon chemistry. These bondings are present in cyanide, cyanate and thiocyanate ions and their derivatives. The cyanide ion (CN^-) is important chemically as well as it plays an important role in the biological system. It is a derivative of hydrogen cyanide which is present in two tautomeric forms as HCN and HNC. The molecules of HCN are found to be in quite excess when compared to HNC. The CN^- ion and its molecular form $(CN)_2$ behave as a pseudo halogen. The chemistry of cyanide ion and cyanogen are discussed in chapter 6, of halogens. In this section the metal cyanides are studied. The metal cyanides can be classified as ionic, covalent and complex cyanides and their derivatives. The ionic cyanides are formed by the alkali, the alkaline earth metals, lanthanides and thallium. The alkali metal cyanides of sodium, potassium, rubidium have the face centered cubic crystal lattice, i.e., of NaCl, where the cyanide ion replaces the chloride ion. Thallium cyanide has the body centered cubic crystal lattice similar to CsCl.

(i) Ionic cyanides: The ionic cyanides are obtained by treating the metals or their compounds with hydrogen cyanide.

$$2Na + 2HCN \longrightarrow 2NaCN + H_2$$

The cyanide ion is 57% ionic and forms resonance structures.

$$\cdot C :: N: \quad \text{or} \quad :\overset{+}{C}\!:\; :\overset{-}{N}:$$

(ii) Covalent Cyanides : The square planar compounds formed by Cu^{2+}, Ni^{2+}, Pd^{2+} and Pt^{2+} forms covalent cyanides, silver also forms covalent cyanides, it contains infinite chains of silver and cyanides arranged alternately as – Ag – C = N – Ag – C = N – Ag –.

(iii) Complex cyanides : The complex cyanides have discrete $[M_x(CN)_n]^{(2-n)}$ groups formed by the transition metals. Usually they are highly stable species and are often unaffected by even acids. $[Fe(CN_6)]^{4-}$ and $[Fe(CN)_6]^{3-}$ are the most common species of this type. The ions have an octahedral arrangement with sp^3d^2 hybridization.

In all the cyanides it is found that the metal is coordinated to the carbon atom of – CN groups as observed by X-ray diffraction. Pauling suggested a triple bond in ionic cyanides as M – C ≡ N; and resonating structures, M = C = N, M – C ≡ N, for the covalent cyanides.

3-5-6 Carbonyls

These purely diamagnetic compounds of VIII group, VIIA group and of the coinage metals with CO group are formed by the reaction of carbon monoxide with either free metal or metal salts and are characterized by having CO groups bonded directed to the metal atoms. The properties of the carbonyls are those of covalent compounds, carbonmonoxide being purely neutral, the metals, will be in their zero valent states, and the bond character of – C ≡ O is retained.

Preparation of carbonyls: Here we restrict our discussion to a study of the preparation of few important carbonyls only.

Nickel tetracarbonyl is prepared by treating fine pellets of nickel with carbon monoxide at room temperature and ordinary pressures.

$$Ni + 4CO \longrightarrow Ni(CO)_4$$

Finely divided iron when treated with CO at 200°C and 100 atmosphere pressure, iron pentacarbonyl, $Fe(CO)_5$, is formed.

$$Fe + 5CO \longrightarrow Fe(CO)_5$$

This on irradiation with ultraviolet rays $Fe_2(CO)_9$ is formed, such carbonyls are called polynuclear carbonyls.

$$2Fe(CO)_5 \xrightarrow{\text{uv rays}} Fe_2(CO)_9 + CO$$

Some carbonyls are formed by the reduction of metallic salts **under** pressure by carbon monoxide or a metal carbonyl.

$$RuI_3 \xrightarrow{CO} Ru(CO)_2I_2 \xrightarrow[Ag/Cu]{\text{High pressure}} Ru(CO)_5 \text{ Colorless liquid}$$

$$OsO_4 \xrightarrow[1000^\circ C \text{ \& } 50 \text{ atm}]{CO} Os(CO)_5 \text{ Colourless liquid}$$

$$Re_2S_7 \xrightarrow[250^\circ C\ \&\ 200\ atm]{CO} Re_5(CO)_{10} \text{ Colourless crystals}$$

$$MoCl_5 \xrightarrow{Fe(CO)_5} Mo(CO)_6 \text{ Colourless crystals}$$

$$WCl_6 \xrightarrow{Fe(CO)_5} W(CO)_6 \text{ Colourless crystals.}$$

(ii) Structure and bonding in carbonyls : The structure and bonding in carbonyls have been studied by various workers and it is concluded that these compounds, specially the simple ones of the type $M(CO)_n$ have M – C – O bonds to be collinear. The bonding of metal carbonyls is represented schematically in Fig 3-8 to indicate the orbital overlaps leading to M – C – O bonding. The Fig. 3-8a, represents overlap and donation from the lone pair on carbon into a vacant (hybrid) metal orbital to form a σ M – C bond and fig.3-8b represents a π overlap and a donation from a filled 'd_{xz}' or 'd_{yz}' orbital of the metal into a vacant antibonding 'p' orbital on CO to form a π, M – C bond.

Fig.3-8 Orbital overlapping in the metal carbonyl bonding

The metal – carbon bonds are shorter than the expected values, for example the calculated M – C distance is 2.00Å but in almost all the metal carbonyls the M – C distance ranges from 1.82Å depending upon the metal. The angle in the bridging carbonyls for the M – CO – M is around 77°-80°, whereas it is interesting to note that in organic compounds C – CO – C bond angle is 120-124°, which suggests a different bonding trend. The structures of a few carbonyl compounds are given in Figs. 3-9 to 3-12.

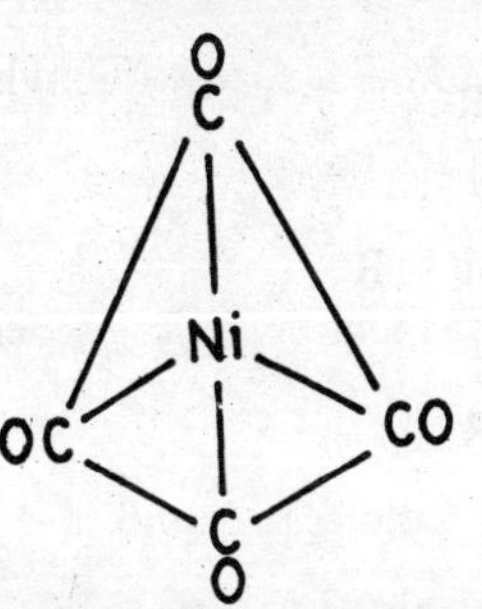

Fig 3-9 Structure of $Ni(CO)_4$. (It is a tetrahedral molecule with Ni-C distance being 1.82Å. It is a mononuclear moelcule.)

Fig 3-10 Structure of $Fe_2(CO)_9$ (The Fe-Fe distance = 2.493Å. It is a dinuclear molecule.)

Apart from usual carbonyls, substituted carbon monoxide ligands form stable complexes. The ligands such as $= CR_2$ (Carbene), $= CR$ (Carbyne) and $- C$ (Carbide) forms carbon monoxide substituted complexes. This development is very recent one (1976-1984) and the complexes are quite stable.

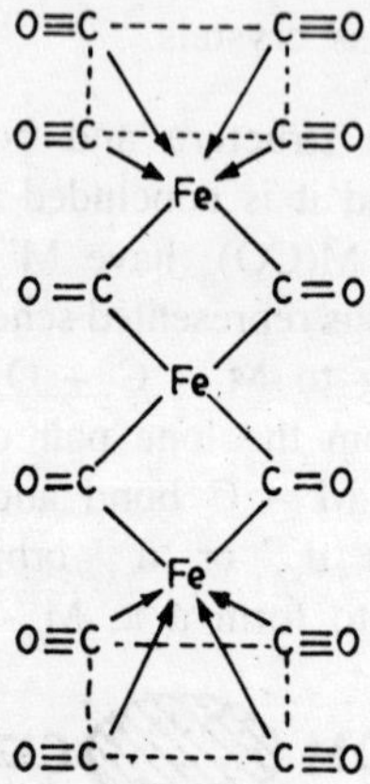

Fig.3-11 Structure of $Fe_3(CO)_{12}$ (It is a trinuclear molecule.)

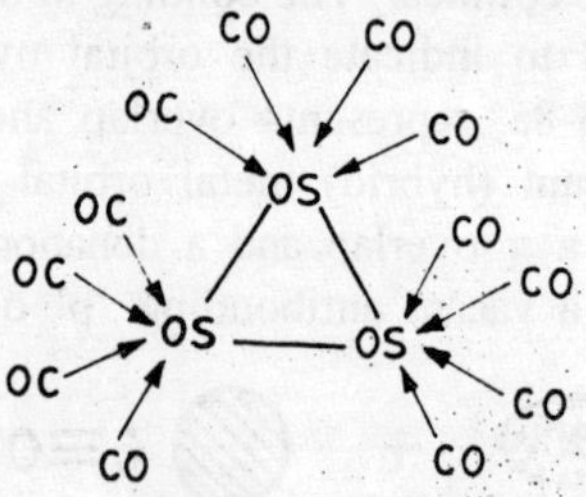

Fig.3-12 Structure of $Os_3(CO)_{12}$ (It is a trinuclear molecule.)

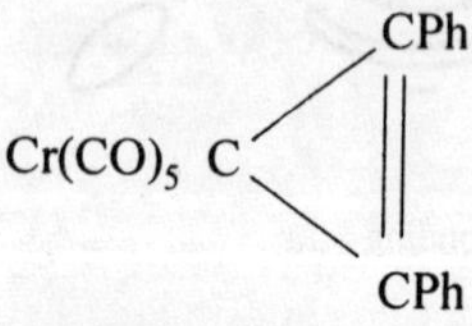

$Cr(CO)_5\ C$ (bonded to CPh≡CPh) or $WBr(CO)_4(\equiv CPh)$

3-5-7 Silicones

Silicones are semi-inorganic polymers used in almost every industry and home in the country today. These polymeric compounds contain alternately linked silicon and oxygen atoms and either alkyl or aryl groups (R) attached to the silicon through covalent bonds, example :

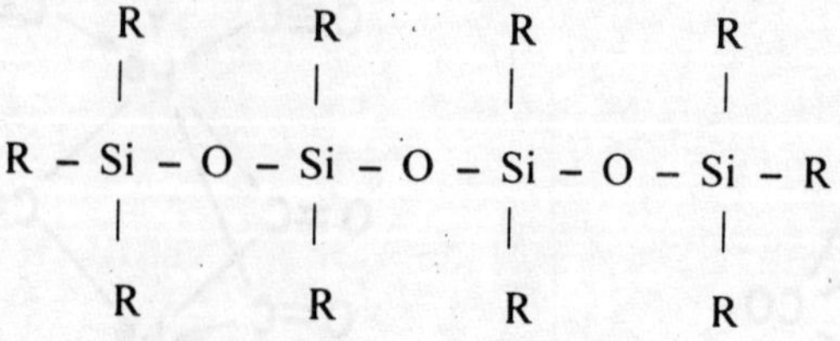

$$R_3Si - O - SiR_2 - O - SiR_2 - O - SiR_3$$

Silicones are prepared by the reaction of chloro alkane or chloro benzene in the silicon in the presence of a catalyst like copper at 300°C. The product obtained is a mixture of both arylchloro silane and alkyl chloro acid

$$3\,Si + 6\,R\,Cl \xrightarrow[300^{o}C]{Cu} R_3SiCl + R_2SiCl_2 + RSiCl_3$$

They are separated by fractional distillation followed by washing

$$2R_3\,SiCl + H_2O \longrightarrow R_3SiOSiR_3 + 2HCl$$

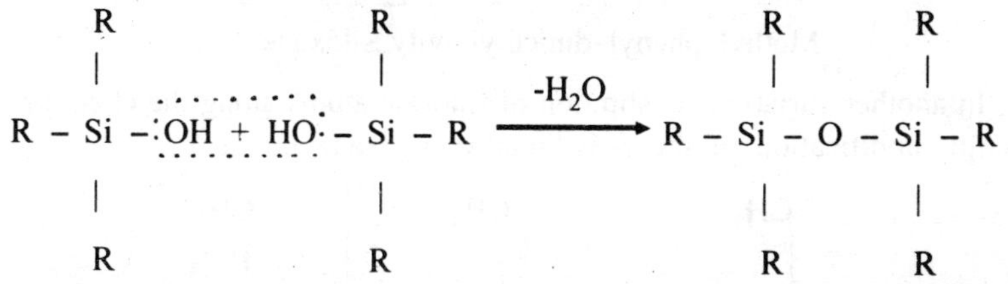

$$R_3Si{-}OH + HO{-}SiR_3 \xrightarrow{-H_2O} R_3Si{-}O{-}SiR_3$$

Silanes

Silanes are highly water repellent compounds. As the silanes are fluids, the silicones assume many different forms such as rubbers, greases, adhesives as reactive monomors, intermediates and elastomers. All these retain some of the characteristics of inorganics while offering a range of properties that are generally associated with organics. Silicones have found applications in practically every industry from the air-craft and missile manufacture to furniture making, as electrical insulators and textile finishes, in pharmaceutical and cosmetic preparation and even in the baking of bread and frying/cooking of food in the nonstick ware. As engineering materials the silicones offer ranges of thermal stability with surfaces and electrical properties that are not obtainable from other polymers and plastics.

The silicones are modified as per the requirement and its use. the basic polydimethyl siloxane structure may be modified and varied in many ways. For example substituting a phenyl group for one of the methyl groups produces a new polymer which extends the thermal stability, flexibility and processability of the silicones.

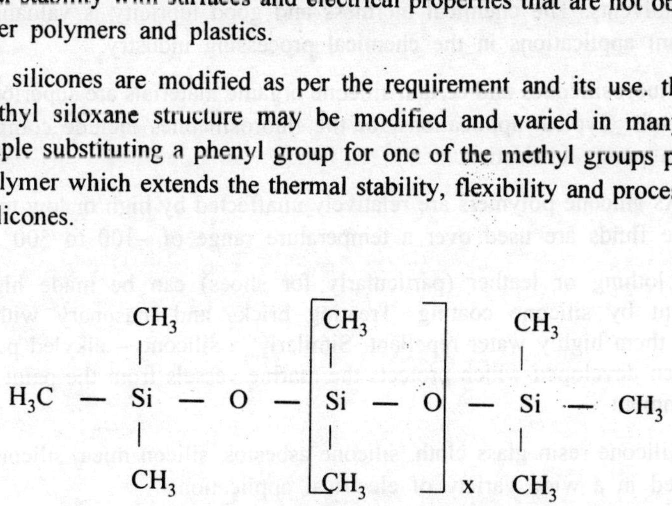

$$H_3C-Si(CH_3)_2-O-\left[Si(CH_3)_2-O\right]_x-Si(CH_3)_2-CH_3$$

Polydimethyl siloxanc

$$\begin{array}{ccccccccccccc} & & CH_3 & & & & CH_3 & & & & CH_3 & & & & CH_3 & & \\ & & | & & & & | & & & & | & & & & | & & \\ H_3C & - & Si & - & O & - & Si & - & O & - & Si & - & O & - & Si & - & CH_3 \\ & & | & & & & | & & & & | & & & & | & & \\ & & CH_3 & & & & CH_3 & & & & C_6H_5 & & & & CH_3 & & \end{array}$$

Methyl phenyl dimethyl poly siloxane

In another variation substitution of fluorine atoms along the chain produces a useful modification of the polydimethys siloxane polymer

$$\begin{array}{ccccccccccc} & & CH_3 & & & & CH_3 & & & & CH_3 & & \\ & & | & & & & | & & & & | & & \\ H_3C & - & Si & - & O & - & Si & - & O & - & Si & - & CH_3 \\ & & | & & & & | & & & & | & & \\ & & CH_3 & & & & CH_2 & & & & CH_3 & & \\ & & & & & & | & & & & & & \\ & & & & & & CH_2 & & & & & & \\ & & & & & & | & & & & & & \\ & & & & & & CF_3 & _x & & & & & \end{array}$$

Introducing the fluorine along the chain in the form of a trifluoropropyl group has a profound effect on solubility and in the case of fluidity, lubricity. Fluorosilicones are insoluble in common aromatic, aliphalic and chlorinated solvents in contrast to the siloxane polymer which is completely soluble in these solvents. The chemical inertness and good lubricity is valuable in many lubricant applications in the chemical processing industry.

Fluorosilicones and certain silicone organic materials are superior lubricants for metals. Typical applications for the fluorosilicones include compressor and vaccum pump lubrication.

As silicone polymers are relatively unaffected by high or low temperatures silicone fluids are used over a temperature range of –100 to 500° F.

Clothing or leather (particularly for shoes) can be made highly water repellent by silicone coating. Treating bricks and masonary with silicones makes them highly water repellant. Similarly, a silicone – alkyled paint system has been developed which protects the marine vessels from the paint destroying environment.

Silicone resin glass cloth, silicone asbestos, silicon mica, silicon laminates are used in a wide variety of electrical applications.

3-5-8 Silicates : Silica when fused with sodium carbonate, alongwith carbon-dioxide, a water soluble glass commonly known as (water glass) is obtained.

The water glass is the starting material for a number of sodium silicates. The most primary silicate anion SiO_4^{4-} is a very strong base which is capable to attract protons from water molecules, which in turn get condensed to form other type of silicate anions, this can be experimented in the laboratory.

$$SiO_2 + 2Na_2CO_3 \longrightarrow Na_4SiO_4 + 2CO_2$$

$$Na_4SiO_4 + 4H_2O \longrightarrow H_4SiO_4 + 4NaOH$$

$$H_4SiO_4 \longrightarrow HSiO_4^{3-} + 3H^+$$

$$2HSiO_4^{3-} \longrightarrow Si_2O_7^{6-} + H_2O$$

(Pyrosilicate ion)

The structures of the silicates were investigated & studied by X- ray crystallographic methods. The first and the formost important results observed were that, in the process involved in the formation of these silicates in nature isomorphous replacement frequently occured. Such as, magnesium (0.66Å), calcium (0.99Å), iron (0.74Å), manganese (0.80Å), chromium (0.84Å) and other bivalent metal ions of almost similar sizes may replace one another mutually to a lesser or greater extent. This may happen only when the ions have the same size and same charge with respect to each other. Similarly trivalent ions will replace each other chromium (0.63Å), manganese (0.66Å), iron (0.64Å), etc. It is also observed that ions with different charges also can replace each other Mg^{2+} (0.66Å) with Fe^{3+} (0.64Å). The presence of different metal ions in the silicates results in different structures, with the SiO_4^{4-} ion remaining as a tetrahedral unit in all of them. They differ from each other only in the way, in which these units are linked together.

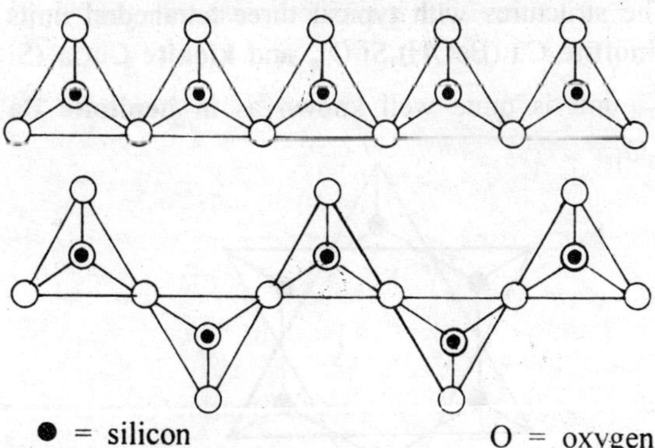

● = silicon O = oxygen

The silicates are broadly divided into five classes. i) The silicates containing the discrete SiO_4^{4-} ion, these are known as **orthosilicates** such as in **zircon**, $ZrSiO_4$, in **berylliumortho silicate** (Be_2SiO_4), in **olivine** $(MgFe)_2SiO_4$, in phenacite (Be_2SiO_4) or **willemite** Zn_2SiO_4.

ii) The silicates containing the **pyrosilicate** ($Si_2O_7^{6-}$) anion, share one oxygen atom between two tetrahedral units, forming a discrete, disilicate ion, which is quite rare, the best example of this type is **thortveitite** $SC_2Si_2O_7$, in which the Si–O–Si bond is linear between staggered tetrahedra. The Si–O–Si bond varies from 180° to 130° in the **lanthanide pyrosilicates** with the increase in size from Lu^{3+} to Nd^{3+}. Another typical pyrosilicate is that of zinc, the mineral **Hemimorphite** has an angle Si–$\hat{O}$–Si of 150°, its X-ray structure shows that the compound has a composition of $Zn_4(OH)_2Si_2O_7$. H_2O but it is commonly written as $2H_2Zn_2SiO_5$. The structure of this compound is seen below :

The cyclic meta silicates have an emprical formula of $[SiO_3]n_2^{n-}$ having 3, 4, 6 or 8 linked tetrahedra. The common amongst them are the 3 and 6 linked ones. The structures with typical three tetrahedral units are quite rare, such as **in aminoffite** $Ca_3(BeOH)_2Si_3O_{10}$ **and kionite** CuCa $(Si_3O_{10})_2H_2O$. The common $Si_3O_9^{6-}$ unit is quite well known as in **benitoite** Ba $TiSi_3O_9$ or in **beryl** $Be_3AlSi_6O_{18}$.

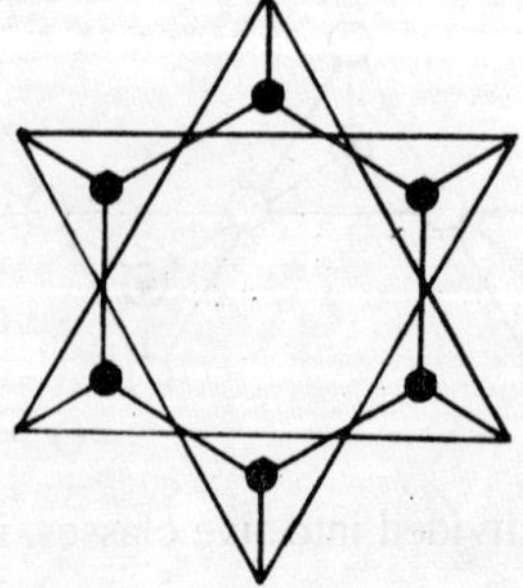

Similarly a tetrasilicate $Ag_{10}Si_4O_{13}$ though reported to be prepared, by the action of, AgO on SiO_2 for 72 hours at 550° C under a very high current of oxygen, its structure is not well defined, whereas the structure of $Si_4O_{12}^{8-}$ is common as in **murite** $[Ba_{10}(CaMnTi)_4\ Si_8O_{24}(Cl,OH,O)_{12}]4H_2O$.

In nature chain metasilicates with emprical formula $[SiO_3^{2-}]_\infty$ are formed by sharing of the corner of SiO_4^- tetrahedra. The examples of this type are, **pyroxene** minerals such as **enstatite** $(Mg_2Si_2O_6)$, **diopride** $(CaMgSi_2O_6)$, **jadeite** $(NaAlSi_2O_6)$ etc. The chain meta silicates are three dimensional substances and when they form **double chains or ribbons,** they form the **amphiboles** or **asbestos** minerals which adopt the $[Si_4O_{11}^{6-}]_\infty$ double chain, as seen below

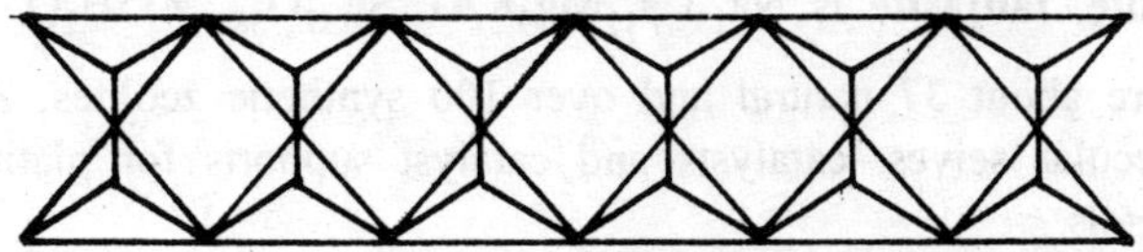

The asbestos is actually a combination of **serpentines** $Mg_6Si_4O_{10}(OH)_8$ and the blue **amphibole material, crocidolite** $[Na_2\ Fe_3^{II}\ Fe_2^{III}\ Si_8O_{22}(OH)_2]$ and the grey brown mineral $[(MgFe)_7Si_8O_{22}(OH)_2]$

Other different type of silicates are, china clay mineral **kaolinite** $Al_2(OH)_4Si_2O_5$, by the replacement of $2Al^{3+}$ by $3Mg^{2+}$ in kaolinite gives the serpentive asbestos mineral **chrysotile** $[Mg_3(OH)_4Si_2O_5]$ on a similar replacement in **pyrophylrite, talc** $[Mg_3(OH)_2Si_4O_{10}]$ is obtained. The structures of a) kaolinite and (b) pyrophyllite are given below.

Si
Al
O
OH

(a) Kaolinite (b) Pyrophyllite

Clays are also classified as silicates they are obtained by weathering or decomposition of rocks or by the weathering of alkaline feldspar

$$2[KAlSi_3O_8] + CO_2 + H_2O \longrightarrow Al_2(OH)_4Si_2O_5 + 4SiO_2 + K_2CO_3$$

Micas are formed when one fourth of the Si^{4+} in pyrophyllite is replaced by Al^{3+} and the negative charge is balance with K^+.

$$[Al_2(OH)_2Si_4O_{10}] \longrightarrow [KAl_2(OH)_2\ (Si_3AlO_{10})]$$

pyrophyllite muscovite (white mica)

Zeolites are another very important form of silicates, they have open aluminosilicate frameworks and they can take up loosely bound water or other small molecules in their structures. Zeolites have structures which are characterized by the presence of systems of interconnected cavities, these can be linked either in one direction giving fibrous crystals or more usually in two or more than two directions giving three dimensional structures. A naturally occuring zeolite, **faujasite** is $Na_{13}Ca_{11}Mg_9KAl_{55}Si_{137}O_{384}\ 235H_2O$.

There are about 37 natural and over 100 synthetic zeolites. Zeolites are used as molecular seives, catalysts and catalyst supports for platinum group and other metals.

Synthetically zeolites are prepared by hydrothermal process. the high silica zeolites are ZSM-5 (Mobil Oil) and silicalite (Union carbide)

3-6 SEMI CONDUCTORS : APPLICATION OF SILICON AND GERMANIUM IN ELECTRONICS

Silicon, Germanium, Selenium and alloys isoelectronic of these elements formed from gallium, arsenic and chalcogenides are of great importance in semiconductor devices. These are applicable in the understanding of photoelectric, thermoelectric, magnetic and diffusional properties of solids. They have been used in diodes, thermistors, photo-electric cells, rectifiers, transistors, phosphours, luminescent materials and computer components.

These elements are rather unusual type of elements with properties some what intermediate of metals and non metals. They are neither good nor bad conductors of electricity, thus taking name : **semiconductors**. The semiconductors are processed to permit an electric current to pass through much readily in one direction than in the opposite direction, i.e., when an alternating current (A.C.) is passed through a semiconductor, the forward surging current is allowed to pass through, whereas the backward surging current is blocked, this accounts to the conversion of A.C. to D.C. Hence these are used in electronic industry, particularly those centered on the inter conversion of electrical and light energy. For example, they are produced commercially as **light emitting diodes (LEDS),** familiar in pocket calculators, wrist watches and alpha numeric output displays of many instruments. They are also used in infrared emitting diodes, injection lasers, infrared detectors, photocathodes, photomultiplier tubes and laser windows.

A diode for example can be had from a crystal of a semiconductor such as silicon or germanium which is grown from a melt which contains a small amount of an impurity, in the form of an element of a higher or lower group in the periodic table. If the element is from the higher group, a *n*-type crystal is obtained and if the element is from the lower group a *p*-type crystal is formed. The *n*-type crystals contain an excess of electrons over those in the pure crystal, whereas the *p*-type crystals contain a deficiency of electrons. A semiconducting diode is then prepared by arranging to have a n-type and p-type junction in a crystal. There are various ways of getting such junctions, for instance a bit of one type of impurity and then a bit of the other type may be added to a crystal as it is being drawn from a melt of the pure material.

Similarly a transistor is a three type unit in which an emitter and collector of one type of semiconducting material are separated from each other by a thin slice of the other type of semiconducting material. Hence the transistors are either of *n-p-n* or *p-n-p* types. The thin slice separating the semiconductor of the same type is of a thinness less than 0.02mm.

An extremely elegant chemical solid state technology has evolved in which crystals of the required properties are deposited, etched and modified to form the appropriate electrical circuits. The LEDs were first prepared in 1961 from an alloy of Ga-As (which is isoelectronic to elemental germanium) by chemical vapour deposition and crystal wafers as large as 20 cm^{-2} were produced commercially. In a LED, the colour of the emitted radiation is determined by the energy band gap (Eg). The wafer thickness can extend from 0.4 mm to 0.5 mm depending on the requirement.

They not only rectify signals but they also amplify the sound system and hence are used in telephone sets as amplifiers. The weak high frequency currents set up in radio and television can be amplified many times over, by using silicon chips or germanium, which is still a better semiconductor than silicon in this matter. This property led to the application of these elements in telecommunications, diodes, triodes etc., which amplify and rectify the signals. These chips operate much better than radio valves consuming very little electric current and without being heated up or worked out. The greatest attraction of the transistor, is that, they can be made as small as pin heads operating on tiny dry batteries, computers, satellites, telecommunication devices, without the conversion of current thus passing electrical energy. The semiconductors normally operate at considerably lower voltages.

The commercially available semiconductors are those of Si, Ge, Se, $GaAs_{1-x}P_x$, GaAs, GaP, GaAsNPTe, $AgGa_2O_4$ etc.

$MgGa_2O_4$ when activated with Mn^{2+}, is used in ultraviolet activated powders as a brilliant green phosphorescence material, in the xerox copying machines.

3-7 MULTIPLE-CHOICE QUESTIONS

3-7-1 The electronegativity of Si, Ge, Sn & Pb is same, the value is

a) 2.5 b) 1.8 c) 2.3 d) 1.3

3-7-2 Silicon can catenate upto the formation of its compound

a) Si_xH_{x+5} b) Si_3H_7 c) Si_6H_{14} d) Si_2H_{16}

3-7-3 Carborundum is

a) $(SiC)_n$ b) Si_2C c) SiC d) B_4C_3

3-7-4 Disilane is analogous to

a) ethane b) methane c) butane d) propane

3-7-5 Silica gel is

a) SiO_2 b) SiO_24H_2O c) SiO_4 d) Si_2O_4

3-7-6 On hydrolysis $SiCl_4$ forms $Si(OH)_4$ whereas CCl_4 forms

a) CO_4 b) CO_24H_2O c) CO_2 d) None of them

3-7-7 Solid CO_2 is popularly known as

a) Mellitic acid b) DryIce c) lime stone d) quartz

3-7-8 MgC_2 on hydrolysis form

a) Methane b) $Mg(CH_3)_2$ c) C_2H_2 d) CH_3OH

3-7-9 Al_4C_3 on hydrolysis form

a) Methane b) $Mg\ (CH_3)_2$ c) C_2H_2 d) CH_3OH

3-7-10 Lanthanide carbides are

a) Ionic carbides b) Covalent carbides
c) Interstitial carbides d) Iron type carbides

3-7-11 The Interstitral carbide is

a) SiC b) TaC c) MgC_2 d) Cr_3C_2

3-7-12 Covalent carbides are formed between

a) non metals and carbon b) nonmetals and a metallic carbides
c) metals & carbon d) metals and nonmetalic carbides

3-7-13 Iron type carbides are formed with metals having an ionic radii of

a) 1.8Å b) 1.74 Å c) 0.94 Å d) 1.3Å

3-7-14 Cyanide ion has a structure of

a) $-C = N^{-}-$ b) $-C = N^{-}$ c) $-C = N^{-}$: d) $= C = N$:

3-7-15 Thallium cyanide has a crystal lattice similar to

a) CaCl b) NaCl c) ZnS d) NaCN

3-7-16 An isocyanide has a bonding of the type

a) M = C = N b) M – C = N c) M – N = C d) M = N = C

3-7-17 Carbonyls of group VIII are

a) Paramagnetic b) Diamagnetic c) Ferromagnetic d) Antimagnetic

3-7-18 Carbonyls are

a) divalent b) trivalent c) zerovalent d) tetravalent

3-7-19 Nickel carbonyl has a structure of

a) trigonal bipyramid b) Octahedral
c) tetragonal d) tetrahedral

3-7-20 A Carbene ligand is

a) = CR b) $= CR_2$ c) – C d) C_6HS

3-7-21 Polymethyl siloxane is an example of

a) silicon b) silicates c) silanes d) silicons

3-7-22 On addition of fluorine to a silicone they become -------- to the aromatic solvents

a) repellent b) stable c) insoluble d) inert

3-7-23 Silicone are ----- to water

a) hydrophillic b) repellent c) soluble d) resistant

3-7-24 The Pyrosilicate ion is

a) $Si_2O_2^{-7}$ b) $Si_2O_7^{4-}$ c) $Si_2O_6^{-7}$ d) $Si_2O_8^{-7}$

3-7-25 In nature isomorphous replacement of metal ions takes place when the metal radii is

a) statement is false b) smaller than each other
c) almost equal to each other d) Bigger than the other

3-7-26 Olivine is

a) Be_2SiO_4 b) $(MgFe)_2SiO_4$ c) $2Na_2SiO_4$ d) Zr SiO_4

3-7-27 Zircon is

a) Be_2Si_4 b) $MgFe_2$ SiO_4 c) Zn_2SiO_4 d) $ZrSiO_4$

3-7-28 Cyclic meta silicate have a general formula of

a) $[(SiO_3)_n]^{2n-}$ b) $[(SiO_3)_n]^{4n-}$ c) $[(SiO_6)_n]^{2n-}$ d) $[(SiO_3)_n]^{6n-}$

3-7-29 Beryl is

a) Be_2SiO_4 b) $Be_3Al_2Si_6O_{18}$ c) Be_4 $(SiO_4)_2$ d) Ca_3 $(Be\ OH)_2$ Si_3O_{10}

3-7-30 Asbestos belongs to the group of

a) Amphiboles b) Serpentines
c) Amphiboles & Serpentines d) Crocidolite

3-7-31 Talc is

a) $Mg_3(OH)_2Si_4O_{10}$ b) $Mg_3(OH)_4Si_2O_5$
c) K $AlSi_3O_8$ d) Al_2 $(OH)_4Si_2O_5$

3-7-32 Zeolites are prepared by the process of

a) Condensation b) Polymerisation
c) Hydrothermal d) Dissociation

3-7-33 $MgGa_2O_4$ when activated with Mn^{2+} produces

a) energy activation b) Fluorosence
c) Phorphorescence d) diodes

3-7-34 Transistors are of

a) *p-n-p* type b) *p-p* type c) *n-p* type d) *p-n-n* type

3-8 SHORT AND ESSAY-TYPE QUESTIONS

3-8-1 Justify the position of C, Si, Ge, Sn and Pb in the same group. (O.U., Oct.1989)

3-8-2 Describe the physical and chemical properties of the elements C, Si, Ge, Sn, Pb and justify their position in the periodic table.

3-8-3 Name the elements of IV-B. Describe the gradation in properties with reference to
(a) Oxidation states(b) Electronegativity
(c) Catenation(d) Metallic character

3-8-4 What is catenation? How does this property make carbon a unique element? Discuss the gradation of this property in group IV-B.

3-8-5 Give a brief account of the allotropic forms of carbon. Illustrate the structure of graphite and diamond.

3-8-6 Account for the following
i) Although covalently bonded, diamond has a high melting point.
ii) SiO_2 is a solid where as CO_2 is a gas. (O.U., Apr.1989, 1990)

3-8-7 Illustrate and explain the "inert pair effect" with suitable examples. (O.U., Oct.1988)

3-8-8 Write the reaction of hydrolysis of silicon tetrahalide. (O.U., Nov. 1990)

3-8-9 Write down the electronic configuration of C and Si. Compare these two elements with respect to
a) Tendency for catenation.
b) Behavior of their tetrahalides on hydrolysis. (O.U., Oct. 1988)

3-8-10 Bring out the clear difference between diamond and grphite. (O.U., Apr. 1988)

3-8-11 Reason the following
a) Alkanes are large in number while silanes are few.
b) Silicon carbide is as hard as diamond.
c) Ge and Sn are reducing agents where as Pb is an oxidizing agent.
d) $PbBr_4$ and PbI_4 do not exist.
e) CO_2 is acidic while PbO_2 is basic.

3-8-12 $SnCl_4$ and $SiCl_4$ behave as Lewis acids but not CCl_4. Explain.

3-8-13 Explain the following :
a) Silicon dioxide is a solid while carbon dioxide is a gas
b) Alkanes are more reactive than silanes. (N.U., Apr. 1989)

3-8-14 Discuss the structures of diamond and graphite.
(A.U., Apr. 1989) (A.U., Jun.1990)

3-8-15 Discuss the structure of diamond. (K.U., Apr. 1989)

3-8-16 Explain why
a) CO_2 is a gas at room temperature while SiO_2 is a solid
b) SiC is as hard as diamond
c) Si has much less tendency to catenation than carbon (K.U., Oct. 1990)

3-8-17 How are carbides classified? Give a comparative account of the physial and chemical properties of different carbides. (N.U., Oct. 1989)

3-8-18 What are carbides? How are they classified? Give one example each. Write the hydrolysis reaction of Magnesium carbide. (O.U., Nov. 1990)

3-8-19 Write a note on metal cyanide complexes. (O.U., Apr.1989)

3-8-20 Write a note on
a) Silicates
b) Carbonyls
c) Cyanides
d) Silicones

3-8-21 Write notes on carbonyls & silicones. (A.U., Dec.1989)

3-8-22 What are carbonyls? How are they prepared? Describe their structures.

3-8-23 What are metal carbonyls? Give their methods of preparation and properties. Describe the structure of nickel carbonyl. (N.U., Oct.1989)

3-8-24 What are carbonyls? Describe the chemistry of Nickel carbonyl.

3-8-25 What are metal carbonyls? Describe the structure and bonding of any one simple carbonyl. (O.U., Oct. 1989)

3-8-26 Write the structure of the following carbonyls a
a) $Fe_3(CO)_{12}$
b) $Os_3(CO)_{12}$
c) $Fe_2(CO)_9$

3-8-27 What are silicones? Give a brief account of their importance in industries.

3-8-28 What are silicones and how they prepared? Give their composition.
(O.U. April, 1990 & Oct 88)

3-8-29 How are silicates prepared in the laboratory?

3-8-30 Name few ortho-silicates with their composition.

3-8-31 What are cyclosilicates? Discuss their structures with reference to their mineral ores & composition.

3-8-32 Write notes on
a) Asbestos
b) Clays & Mica
c) Zeolites

3-8-33 Discuss the importance of Semi conductors.

3-8-34 Write notes on
a) Transistors
b) Diodes
c) LEDs

3-8-35 Write an Essay on the chemistry of Germanium & silicon as electronic materials.

3-7 ANSWERS

3-7-1 B	**3-7-2** C	**3-7-3** A	**3-7-4** A	**3-7-5** B	**3-7-6** D
3-7-7 B	**3-7-8** C	**3-7-9** A	**3-7-10** A	**3-7-11** B	**3-7-12** A
3-7-13 D	**3-7-14** B	**3-7-15** A	**3-7-16** C	**3-7-17** B	**3-7-18** C
3-7-19 D	**3-7-20** B	**3-7-21** D	**3-7-22** B	**3-7-23** B	**3-7-24** C
3-7-25 C	**3-7-26** B	**3-7-27** D	**3-7-28** A	**3-7-29** B	**3-7-30** C
3-7-31 A	**3-7-32** C	**3-7-33** C	**3-7-34** A		

4

THE ELEMENTS OF GROUP VB
THE TYPICAL GROUP

					He
B	C	N	O	F	Ne
Al	Si	P	S	Cl	Ar
Ga	*Ge*	As	*Se*	Br	Kr
In	Sn	Sb	Te	I	Xe
Tl	Pb	Bi	Po	At	Rn

4.1 INTRODUCTION

The group VB of the periodic table is referred to as a typical group, because nitrogen (N) is a gas, phosphorous (P) is a waxy solid; both N and P are non metals; the third in the group arsenic (As) is a metalliod with more non-metallic properties than metallic, antimony (Sb) is also a metalliod with predominant metallic character with less non-metallic character, the last element bismuth (Bi) is a true metal. Hence all the possible characteristics of physical state is found to be present in this group. It is found that, there occurs a progressive increase in the metallic character as we go from nitrogen to bismuth. The metallic character first appears when the underlying shell of 18 electrons is first encountered with arsenic. The pronounced change from non metallic to metallic behaviour in the family is due to the increase in size and decrease in ionization protential and electronegativity of the elements with increasing atomic number.

The transition from the non metallic to the metallic character down the group is shown not only by the physical properties but also by the chemical properties of the elements and their compounds.

Nitrogen occurs in the free state in the atmosphere to an extent of 75% and also in combined state as ammonia, ammonium salts, amino acids, nitric oxide, nitrogen dioxide, nitrates etc. Phosphorous occurs in combined state in the form of phosphorite; $8Ca_3(PO_4)_2$ $Ca(OH)_2$, apatite; $3Ca(PO_4)_2$ $Ca(FCl)_2$ and is also present in bones as $Ca_3(PO_4)_2$. Arsenic is present as orpiment As_2S_3 and arsenolite As_4O_6. Antimony is found in stibinite Sb_2S_3 and in antimony bloom Sb_2O_3. Bismuth occurs as bismuth glance Bi_2S_3 and bismite Bi_2O_3.

Nitrogen was probably first recognised as a distinct gas by Rutherford in 1772 and its elemental nature was established by Lavoisier in 1774. The word nitrogen is derived from the Greek words nitrogenon meaning = nitre former, this refers to the formation of a nitre salt by the element. Phosphorous was discovered in 1669 by Brand when a micro cosmic salt was obtained by the complete evaporation of urine to dryness. Arsenic, antimony and bismuth were known since the 14th century.

4.2 ELECTRONIC CONFIGURATION

All the elements of the family have a very stable electronic configuration of an exactly half filled p-orbital, which makes the elements very stable. The physical data is given in table 6-1.

Nitrogen (7) N : $1s^2\ 2s^2\ 2p^3$

Phosphorous (15) P : $1s^2\ 2s^2\ 2p^6\ 3s^2\ 3p^3$

Arsenic (33) As : $1s^2\ 2s^2\ 2p^6\ 3s^2\ 3p^6\ 4s^2\ 3d^{10}\ 4p^3$

Antimony (51) Sb : $1s^2\ 2s^2\ 2p^6\ 3s^2\ 3p^6\ 4s^2\ 3d^{10}\ 4p^6\ 5s^2\ 4d^{10}\ 5p^3$

Bismuth (83) Bi : $1s^2\ 2s^2\ 2p^6\ 3s^2\ 3p^6\ 4s^2\ 3d^{10}\ 4p^6\ 5s^2\ 4d^{10}$
$5p^6\ 6s^2\ 4f^{14}\ 5d^{10}\ 6p^3$.

Table 4-1 Physical Data of Elements of Group VB

	Nitrogen	Phosphorous	Arsenic	Antimony	Bismuth
Symbol	N	P	As	Sb	Bi
Molecular state	N_2	P_4	As_4	Sb_4	Bi
Atomic number	7	15	33	51	83
Atomic weight	14.007	30.97	74.92	121.75	208.99
Atomic volume	17.3	17.0	13.1	18.2	21.3
Atomic density (g cm^{-3})	0.81	1.82	5.73	6.02	9.78
Atomic radii (Å)	0.75	1.06	1.19	1.38	1.46
Ionic radii of trivalent ion, Å	1.71	2.12	2.22	2.45	1.20
Ionization Potential [$M \rightarrow M^{3+} + 3e^-$ (eV)]	47.4	30.2	28.0	24.7	25.4
Electronegativity	3.0	2.1	2.0	1.8	1.7
Melting point (°C)	–210.1	44.2	817	630.5	271.5
Boiling point (°C)	–196	280	613 (sublimes)	1440	1420
Oxidation State	–3, + 1, + 2, + 3, + 4, + 5	–3, + 3, + 4, + 5	–3, + 3, + 5	–3, + 3, + 5	+3, +5
Heat of fusion kcal / g atom^{-1}	0.086	0.15	6.62	4.74	2.6

From the electronic configuration it is noticed that either these elements can gain three electrons showing an oxidation state of –3 or loses three or all the five electrons exhibiting an oxidation states of +3 or +5 respectively. It is very interesting to find that nitrogen shows the maximum number of oxidation states when it combines with oxygen. Thus in N_2O (+1); N_2O_2 (+2); N_2O_3 (+3); N_2O_4 (+4) and N_2O_5 (+5) oxidation states are observed, along with the common oxidation state of –3 of the family. +3 and +5 oxidation states are shown by the VB group elements with the halogens and sulphur. The inert pair effect plays an important role in the family. The tendency of the pair of s electrons to remain inert increases on going down the group, allowing only the p-electrons to be used in bonding resulting in stable trivalent state. Thus Bi^{3+} and N^{5+} are very stable, where as Bi^{5+} is highly unstable.

4-3 ATOMICITY OF NITROGEN AND PHOSPHOROUS

Nitrogen has an electronic configuration of $1s^2\ 2s^2\ 2p^3$ which can also be written as $1s^2\ 2s^2\ 2p_x^1\ 2p_y^1\ 2p_z^1$. Two nitrogen atoms combine to form a molecule of nitrogen by the overlapping of the three half filled orbitals, thus forming three bonds.

	$2s^2$	$2p^3$			$3s^2$	
N	↑↓	↑	↑	↑		N
N_2	↑↓	↑↓	↑↓	↑↓	↑↓	N_2

The atomic orbitals overlap to form one σ and two π bonds it can be represented as

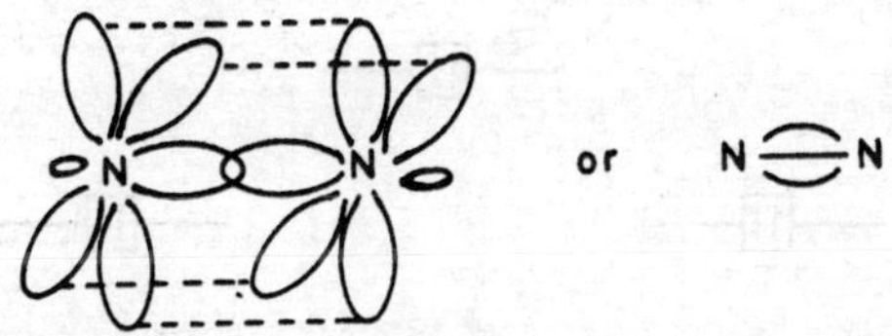

The σ bond is formed by the end on overlapping along the same plane by one of the orbitals, the other two orbitals overlap laterally with the corresponding two orbitals of the other nitrogen atom forming π bonds.

In the nitrogen molecule the bond length is 1.10 Å, the bond energy is 224 kcal $mole^{-1}$. This value shows why the N_2 molecule is so stable and inert despite a very low single bond energy (as calculated) of only 93.8 kcal $mole^{-1}$.

The molecular orbital diagram can explain the atomicity and bonding in nitrogen molecule very easily. The way in which two N atoms combine to form a N_2 molecule is represented in figure 4-1.

The resulting structure for the nitrogen molecule is then given by

$$N_2[1s^2 1s^2 (\sigma\ 2s)^2\ (\sigma{*}2s)^2\ (\sigma\ 2p)^2\ (\pi_y 2p)^2\ (\pi_z 2p)2]$$

The bonding of electrons in the σ2s – bonding orbital is neutralized by the antibonding of the electrons in the σ*2s – orbital, so that six electrons remain in the bonding orbitals forming a triple bond between the two nitrogen atoms.

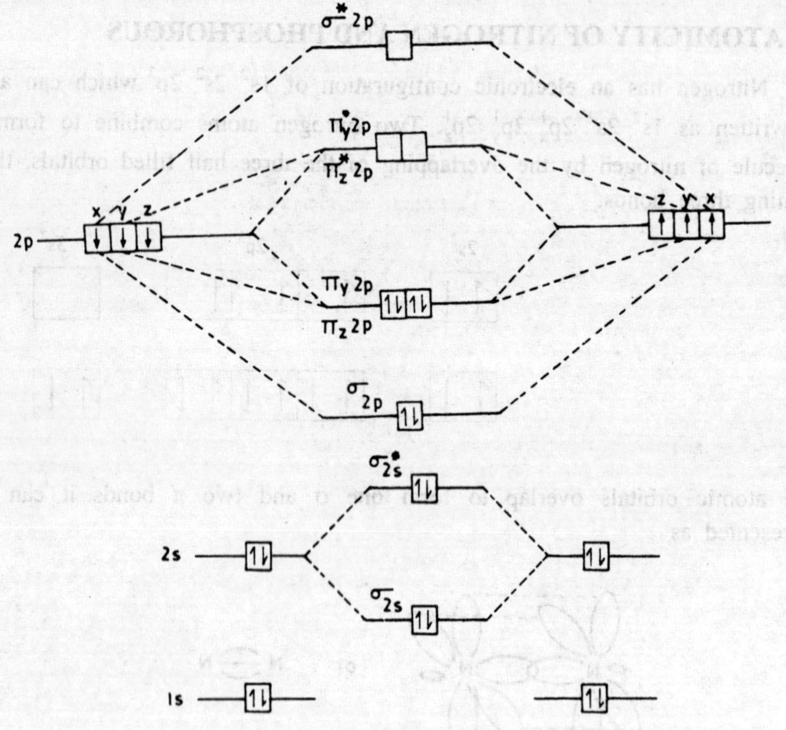

Fig 4-1 Molecular orbital diagram of N_2 molecule

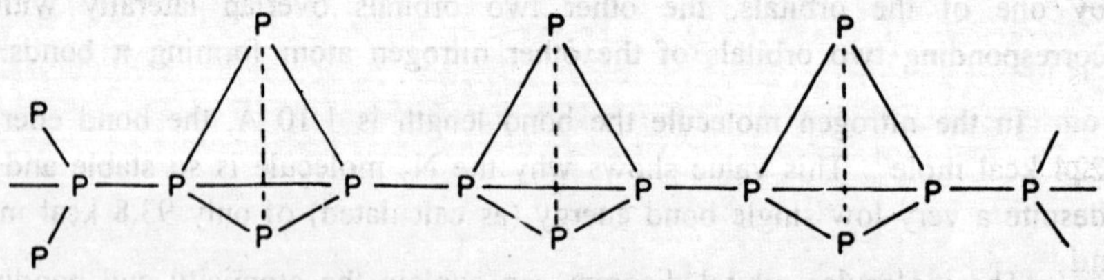

Fig. 4-2 Structure of phosphorous (P_4)

The phosphorous atom is in general thought to be of a tetrahedral molecule, four phosphorous atoms linked tetrahedrally together in a catenated form as seen in Fig.4-2, hence it has a polymeric structure.

As phosphorous is present in many allotropic forms, mainly being white, red and black. There are in all about eleven known modifications of phophorous, some are amorphous, some have definite identity but most of them have unknown structures, the above mentioned tetrahedral molecule (P_4) is that of solid red phosphorous and liquid white phosphorous. The bond distance P – P is 2.12Å and the P – P bond angle is 60°. The very low bond angle is due to the considerable strain and the strain energy of the phosphorous molecule is calculated to be 23 kcal mole^{-1}. Dividing the strain energy between the six bonds present in the molecule it becomes very small, thus making phosphorous a reactive element. The black phosphorous which is the crystalline form of phosphorous consists of corrugated sheets, each P atom is bonded to three other atoms of phosphorous. The structure of black phosphorous is therefore flaky and graphite like. It is the most chemically inactive form of phosphorous. Its structure is given in Fig. 4-3, showing the arrangement of atoms in the corrugated planes forming a crystalline black phosphorous.

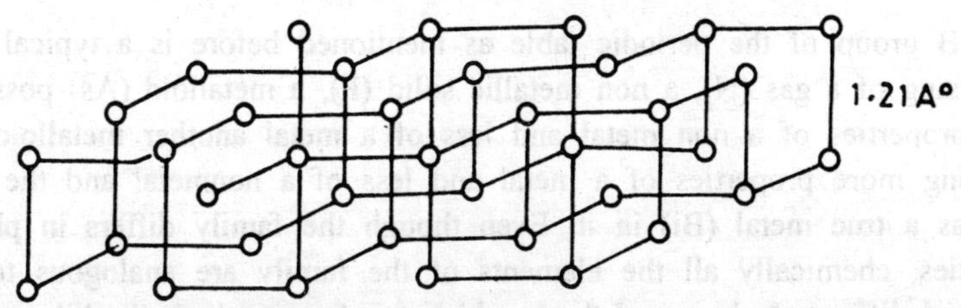

Fig. 4-3 The crystalline structure of P_2 molecule

The molecular orbital diagram of phosphorous (P_2) molecule is similar to the nitrogen (N_2) molecule which is written as

$$P_2\ [1s^2 1s^2 2s^2 2s^2 2p^6 2p^6\ (\sigma 3s)^2 (\sigma^* 3s)^2 (\sigma 3p)^2 (\pi_y 3p)^2 (\pi_z 3p)^2]$$

4-4 ANOMALOUS NATURE OF NITROGEN

It is found in all the groups of the periodic table that the first element differs from rest of the elements of its family. Similarly nitrogen also differs in many respects with regards to phosphorous, arsenic, antimony and bismuth. The anamolous behaviour of nitrogen is due to its small size, high electronegativity and its inability to expand its valency shell beyond the octet. Apart from these characteristics, nitrogen is a diatomic gas with a triple bond between the two atoms. The other elements are all solids showing allotropy. Phosphorous, arsenic and antimony are tetraatomic molecules (P_4, As_4, and Sb_4) whereas Bi is monoatomic. Nitrogen forms multiple bonds whereas the other elements do not, hence nitrogen forms a series of compounds which are no parallel to the other elements of the family. Some of the typical compounds are N ≡ N, N=O, NO_2, NO_3, $(C_6H_5)_2N_2$, – C ≡ N etc. Except for NF_3 the other halides of nitrogen are unstable and explode whereas the halides of other elements are stable. The nitrogen halides can neither donate nor accept an electron pair but the halides of other elements of the family function both as donors and acceptors. The nitrogen halides on hydrolysis produce ammonia but the halides of other V B elements produce hydrogen halides and oxyacids or oxyhalides.

$$NBr_3 + 3H_2O \longrightarrow NH_3 + 3HOBr$$

$$PCl_3 + 3H_2O \longrightarrow 3HCl + H_3PO_3$$

$$PCl_3 + H_2O \longrightarrow 2HCl + POCl$$

The nitrogen trioxide is unstable whereas the trioxides of phosphorous, arsenic, antimony are stable and are dimeric. Bismuth trioxide is monomeric and stable, nitrogen forms five oxides whereas the other elements form either tri or penta oxides, the oxides of nitrogen are N_2O, N_2O_2, N_2O_3, N_2O_4, N_2O_5.

4-5 REACTIVITY OF ELEMENTS

The VB group of the periodic table as mentioned before is a typical group comprising of a gas (N), a non metallic solid (P), a metalloid (As) possessing more properties of a non metal and less of a metal another metalloid (Sb); possesing more properties of a metal and less of a nonmetal and the group also has a true metal (Bi) in it. Even though the family differs in physical properties, chemically all the elements of the family are analogous to each other and differ only here and there, which can be over looked while studying the group relationship of the family.

Hence, to make a group study of the VB group of the periodic table we have chosen to study the halides, oxides, and oxyacids of nitrogen and phosphorous, along with oxyhalides, hydrides and azides in the following sections.

4-5-1 Halides

Trihalides of nitrogen with fluorine and chlorine have been isolated, but for bromine and iodine only the ammonia complexes $NBr_3(NH_3)_6$ and $NI_3(NH_3)_6$ are known. Nitrogen trifluoride is a colourless gas, boiling at –119°C. It is prepared by electrolysing fused NH_4HF_2 in a copper vessel. The crude gas which comprises of the following mixture;

$$HF, N_2, H_2, N_2O$$

$$O_2, O_3, NH_2F \text{ \& } NHF_2$$

is passed over KF to remove HF and over MnO_2 to remove O_3, it is frozen, distilled and then washed with a solution of sodium hydroxide to remove NH_2F and NHF_2. A highly exothermic compound NF_3 with a high thermal stability and chemical inertness more than the other nitrogen halides is obtained. It reacts violently with hydrogen but it cannot get hydrolysed with water.

$$2NF_3 + 3H_2 \longrightarrow N_2 + 6HF$$

The yellow, oily liquid, nitrogen trichloride is formed as a very explosive compound by the action of chlorine or sodium hypochlorite on ammonium chloride in a concentrated aqueous solution.

$$NH_4Cl + 3Cl_2 \longrightarrow NCl_3 + 4HCl$$

It explodes above its boiling point and can easily get hydrolysed.

$$NCl_3 + 3NaOH \longrightarrow NH_3 + 3NaOCl$$

$$NCl_3 + 3H_2O \longrightarrow NH_3 + 3HOCl$$

Difluorodiazine (N_2F_2), tetrafluorohydrazine (N_2F_4), monochloramine (NH_2Cl), hexaamine nitrogen tribromide [$NBr_3(NH_3)_6$], monoaminenitrogentriiodide [NI_3NH_3] are a few derivatives of nitrogen halides.

Phosphorous trifluoride is a stable gas formed by the reaction between phorphorous trichloride and arsenic trifluoride or any other suitable fluoride. PF_3 is colourless, it does not fume in air and gets hydrolysed slowly by water. The other, trihalides are formed by the direct combination of the elements. In an air free retort, phosphorous burns in a current of chlorine to give phosphorous trichloride. It is distilled as a colourless fuming liquid which boils at 76°C.

$$2P + 3Cl_2 \longrightarrow 2PCl_3$$

The tribromide and triiodide of phosphorous are prepared by the same method as that of trichloride under controlled conditions.

The phosphorous trihalides hydrolyse, the ease of hydrolysis increases with the atomic weight of the halogen.

$$PX_3 + 3H_2O \longrightarrow 3HX + H_3PO_3$$

They all react with oxygen and sulphur

$$2PX_3 + O_2 \longrightarrow 2POX_3$$

$$PX_3 + S \longrightarrow PSX_3$$

PX_3 compounds act as very good ligands and form complexes.

$$Ni(CO)_4 + 4PCl_3 \longrightarrow Ni(PCl_3)_4 + 4CO$$

The phosphorous trichloride and tribromide combine with Cl_2 and Br_2 respectively to form corresponding phosphorous pentahalide.

$$PX_3 + X_2 \longrightarrow PX_5$$

The solid pentachloride is made by dropping liquid PCl_3 into dry chlorine. PCl_5 is quite unstable thermally and gets dissociated into trichloride releasing chlorine.

$$PCl_5 \xrightarrow{\text{heating}} PCl_3 + Cl_2$$

It gets reduced by some metals on warming.

$$PCl_5 + Cd \longrightarrow PCl_3 + CdCl_2$$

Phosphorous pentahalides get hydrolysed to phosphoric acid, the reaction takes place in two steps.

$$PX_5 + H_2O \longrightarrow POX_3 + 2HX$$

$$POX_3 + 3H_2O \longrightarrow H_3PO_4 + 3HX$$

Gaseous PCl_5 has a trigonal bipyramidal structure, but in its solid state it contains a tetrahedral PCl_4^+ and an octahedral PCl_6^- ion arrangement. The PBr_5 splits into PBr_4^+ and Br^- ions. The structure of trigonal bypyramidal, PCl_5 can be seen in figure 4-4.

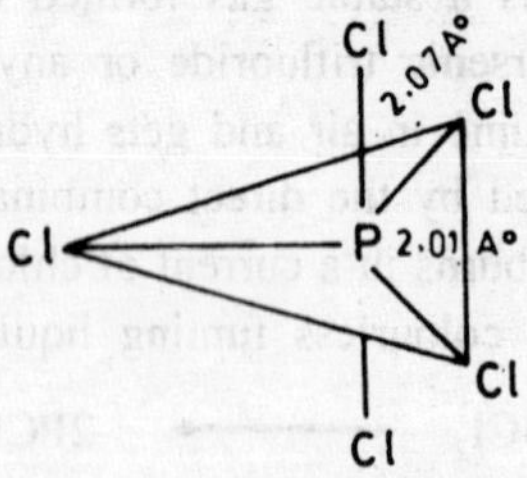

Fig. 4-4 Structure of PCl_5

The P – Cl distance at the apex of the pyramid is 2.07 Å and the distance on the plane of the triangle is 2.01 Å. The bond angle between the apex chlorine, phosphorous and the chlorine on the plane is 90°, whereas the bond angle on phosphorous with respect to the chlorines in the plane of the triangle is 120°.

Nitrogen cannot form a pentahalide because of the absence of suitable empty d-orbitals.

4-5-2 Oxides

Nitrogen forms five oxides, N_2O, N_2O_2, N_2O_3, N_2O and N_2O_5. Nitrogen is the only element in the group to form so many oxides with variable oxidation states. All these are colourless in their solid state except N_2O_3, which is light blue. But a mixture of these gases forms a golden brown colour.

Oxide :	N_2O	N_2O_2/NO	N_2O_3	NO_2/N_2O_4	N_2O_5
Oxidation state	+1	+2	+3	+4	+5

The nitrous oxide, N_2O, is obtained by heating NH_4NO_3 strongly. It decomposes above 600°C and acts as a good oxidising agent. It is not an acid anhydride.

$$2N_2O \longrightarrow 2N_2 + O_2$$

As it is isoelectronic to CO_2, the physical properties of the two compounds show similarities. It has a linear structure which is rendered non polar due to resonance.

$$N^- = N^+ = O \longleftrightarrow N \equiv N^+ - O^-$$

Nitric oxide, N_2O_2 or NO is a colourless gas which gets condensed to a dark blue liquid. It is a paramagnetic molecule with one odd electron in an antibonding π-orbital. It is obtained on a large scale by oxidising ammonia in the presence of a catalyst like platinium above 450°C.

$$4NH_3 + 5O_2 \xrightarrow{\text{Pt, 450°C}} 4NO + 6H_2O$$

In the laboratory it is prepared by the reduction of nittric acid or a solution of a nitrate in an acid. Copper is commonly used for reduction. On heating them together nitric oxide is obtained which is collected in water to get a pale yellow solution.

$$3Cu + 8HNO_3 \longrightarrow 3Cu(NO_3)_2, + 2NO + 4H_2O$$

In the qualitative analysis presence of nitrate or nitric acid is detected by the formation of a brown ring. The brown ring is of $[Fe(NO)]^{++}$ ion complexion formed by the trapping of NO evolved by treating a nitrate or nitric acid in the presence of hot concentrated sulphuric acid by ferrous ion. Thus unstable complex ion decomposes after a while to produce pure NO. The nitric oxide is thermally stable and decomposes only on heating over red hot copper to form nitrogen and copper (II) oxide.

$$2NO + 2Cu \longrightarrow 2CuO + N_2$$

From the X-ray diffraction of nitric oxide it is found to be present in a dimeric orientation. It has a resonating dimeric structure, figure 4-5.

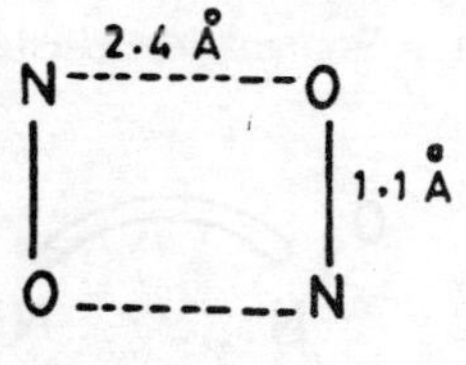

Fig. 4-5 Structure of $(NO)_2$

The resonating structures of nitric oxide are : $N = O : \longleftrightarrow N^+ = O :$ and there is no evidence of diamagnetism in the molecule. On getting dimerized the paramagnetic character increases upto 250°C, above that temperature the paramagnetism becomes virtually constant.

Nitric oxide being an unsaturated compound, reacts with other metal ions to form a. addition complexes, called nitrosyls. In these complexes, nitrogen is the donor atom and in most of them NO gets bonded as NO^+, an ion isoelectronic with CN^- and CO. Coordination compounds containing NO^+ species, do not usually contain more than one NO group. Hence tricarbonylnitrosylcobalt, $[Co(NO)(CO)_3]$ is as stable as nickel tetracarbonyl $Ni(CO)_4$.

But unstable nitrosyls are obtained by the replacement of CO from the metallic carbonyls with nitric oxide. Some nitrosyls reported are those of Fe, Ru and Ni. A few are mentioned here. Nitroniumtrinitrosyl ferrate(II), $NO^+[Fe(NO)_3]$, is a black solid, obtained by heating iron carbonyl in a current of nitric oxide. The ruthenium tetra nitrosyl, $Ru(NO)_4$, is a red crystalline compound obtained on passing nitric oxide in a solution of ruthenium dodecacarbonyl $[Ru_3(CO)_{12}]$. When nitric oxide is passed into a solution of nickel tetracarbonyl in chloroform, $[Ni(NO)_2]$ is obtained. Nitrosyl halides are also reported. When chlorine is reacted with nitric oxide under appropriate conditions, in a copper tube then NOCl is formed. Its structure is found to be an angular one with $Cl\hat{N}O$ = 116°, figure 4-6.

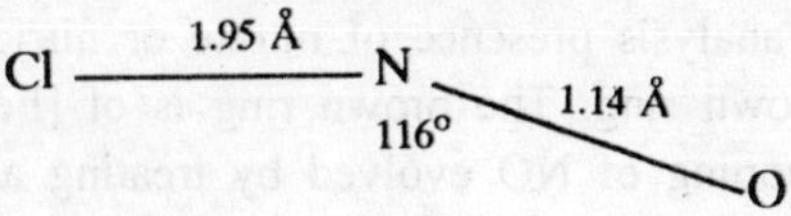

Fig. 4-6 Structure of NOCl

The nitrogen sesquioxide is unstable and can be obtained at 20°C, by the condensation of NO and NO_2 together. It is an anhydride of nitrous acid and exists only in the solid state, its melting point being – 102°C. The sesquioxide (N_2O_3) being a combination of NO and NO_2 as mentioned above, the structure is expected to have ony π bonding between nitrogen atoms as shown in fig. 4-7.

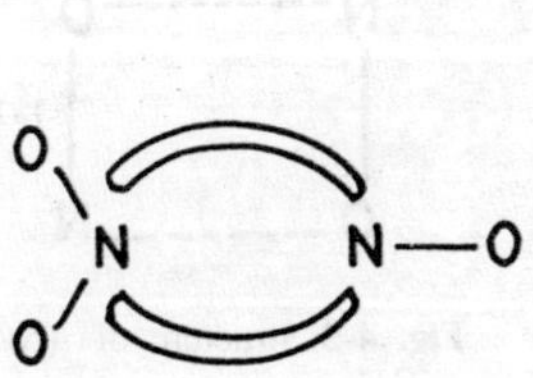

Fig.4-7 Structure of N_2O_3

The spectral study of N_2O_3, shows a bond of π–$\pi*$ transition in the near ultraviolet region which is assigned to a nitroso-nitro molecule with a single π bond between the two (N – N) nitrogen atoms. The nitrite io.., (NO_2^-), present in N_2O_3 is a good coordinating agent because of the lone pairs of electrons present on both oxygen atom as well as the nitrogen atoms. Hence coordination is possible through both the atoms exhibiting linkage isomerism of the type M ← ONO and M ← NO_2.

The nitrogen tetroxide, N_2O_4, is prepared by heating lead nitrate and condensing the gases evolved.

$$2Pb(NO_3)_2 \longrightarrow 2PbO + O_2 + 2N_2O_4$$

It is a reddish brown poisonous gas, and on condensation forms a brown liquid which turns paler on cooling and finally becomes a colourless solid. It melts at 9.3°C and boils at 21.3°C. The monomer NO_2 is angular as in figure 4-8a.

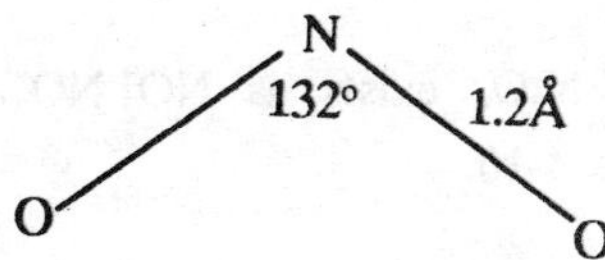

Fig. 4-8a Structure of NO_2

The compound is diamagnetic in its dimeric form. NO_2 being an odd electron molecule its resonance energy is sufficient enough to form a dimer. The lone electrons on the monomers pair up to form a dimer and the X-ray diffraction also shows the molecule to be a dimer, Fig. 4-8b.

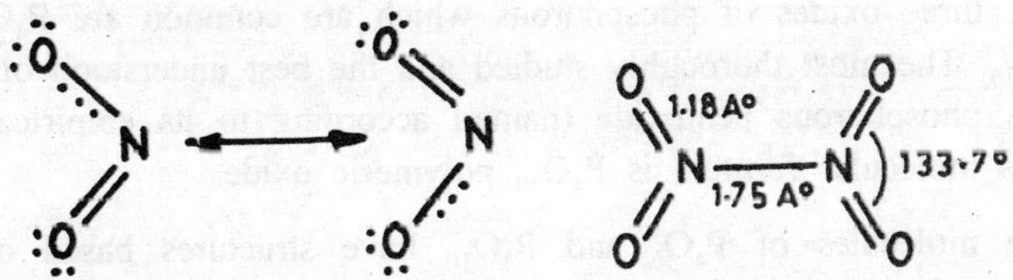

Fig. 4-8b (i) Resonance structure of NO_2 (ii) Structure of N_2O_4

Metals and their salts react with liquid N_2O_4 releasing nitric oxide and forming metallic nitrates

$$Zn + 2N_2O_4 \longrightarrow Zn(NO_3)_2 + 2NO$$

$$2KCl + 2N_2O_4 \longrightarrow 2KNO_3 + 2NO + Cl_2$$

$$NaCl + N_2O_4 \longrightarrow NaNO_3 + NOCl$$

Nitrogen pentoxide is an acid anhydride. On hydrolysis it forms HNO_3 and it is prepared by the dehydration of nitric acid with P_2O_5.

$$N_2O_5 + H_2O \longrightarrow 2HNO_3$$

$$2HNO_3 \longrightarrow N_2O_5 + H_2O$$

It ionises in concentrated H_2SO_4 forming nitronium ion.

$$N_2O_5 + 3H_2SO_4 \longrightarrow 2NO_2^+ + 3HSO_4^- + H_3O^+$$

Its probable structure is given in fig.4-9

O　　　　　　O
　N – O – N
O　　　　　　O

Fig. 4-9 Structure of N_2O_5

In the solid state N_2O_5 exists as $NO_2^+ NO_3^-$. The nitrate ion has six resonance structures, fig. 4-10.

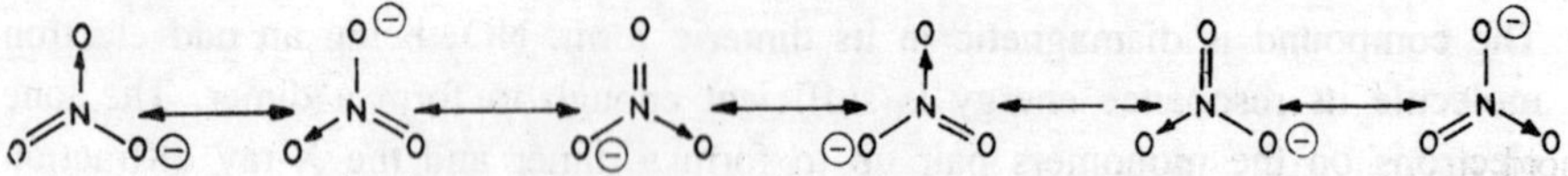

Fig. 4-10 Resonance structures of the nitrate ion

The three oxides of phosphorous which are common are P_4O_6, $(PO_2)_n$ and P_4O_{10}. The most thoroughly studied and the best understood of the three oxides is phosphorous pentoxide (named according to its empirical formula P_2O_5), its molecular formula is P_4O_{10} polymeric oxide.

The molecules of P_4O_6 and P_4O_{10} have structures based on the P_4 tetrahedron. $(PO_2)_n$ is a polymeric oxide.

Phosphorous trioxide is prepared by burning white phosphorous in a slow current of air or in oxygen under reduced pressure. As the vapour is contaminated with the side product, phosphorous pentoxide and it being less volatile is trapped in a glass wool plug. Phosphorous trioxide is then condensed in a U-tube cooled in a freezing mixture. It is a white waxy solid with a melting point 23.8°C and it boils at 173°C. The product obtained is melted in the U-tube using warm water for further purification. The molecular formula of the compound is found to be P_4O_6.

In the P_4O_6 molecule oxygen atoms bridge the edges of the tetrahedron, the P – O distance is 1.65°A and the angles $P\hat{O}P$ and $P\hat{P}P$ are 127.5° and 99° respectively, figure 4-11.

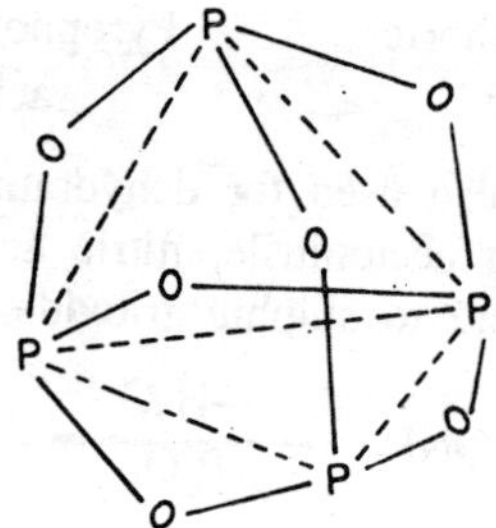

Fig. 4-11 Structure of P_4O_6 molecules.

On heating in a current of oxygen or air the trioxide burns to form pentoxide

$$P_4O_6 + 2O_2 \longrightarrow P_4O_{10}$$

Phosphorous trioxide on dissolving in cold water gives phosphorous acid hence it is an acid anhydride.

$$P_4O_6 + 6H_2O \longrightarrow 4H_3PO_3$$

When dissolved in hot water it produces violently a mixture of phosphoric acid and phosphine.

$$P_4O_6 + 6H_2O \longrightarrow 3H_3PO_4^- + PH_3$$

When P_4O_6 is heated above 210°C, the polymeric oxide $(PO_2)_n$ and red phosphorous are formed. Solid $(PO_2)_n$ sublimes at about 180°C and the density indicates the formula of P_4O_8 in its vapour state. It reacts with water and yields phosphoric and phosphorous acids, and hence it could be a mixed oxide, $P_4O_{10}.P_4O_6$, P_8O_{16} or $(PO_2)_8$.

$$2PO_2 + 3H_2O \longrightarrow H_3PO_4 + H_3PO_3$$

Phosphoric oxide or phosphorous pentoxide is obtained when phosphorous is burnt in excess of oxygen or air.

$$P_4 + 5O_2 \longrightarrow P_4O_{10}$$

It is polymorphic and exist in three crystaline forms, hexagonal, orthorhombic and tetragonal. It is the most stable oxide of phosphorous. Electron diffraction experiments and vapour density determinations indicate that the formula is P_4O_{10} in the vapour. It is the anhydride of phosphoric acid.

$$P_4O_{10} + 6H_2O \longrightarrow 4H_3PO_4$$

It has a strong affinity towards water, and is the most efficient drying agent known. It absorbs moisture; while acting as a drying agent to form metaphosphoric acid first, followed by the pyro and ortho acids.

$$P_4O_{10} \xrightarrow{2H_2O} 4HPO_3 \xrightarrow{2H_2O} 2H_4P_2O_7 \xrightarrow{2H_2O} 4H_3PO_4$$

Phosphoric Oxide — Metaphosphoric acid — Pyrophosphoric acid — Orthophosphoric acid

Phosphoric oxide is also used for dehydrating organic compounds. Thus acetamide is dehydrated to acetonitrile, nitric acid is dehydrated to nitrogen pentoxide and sulphuric acid to sulphur trioxide.

$$CH_3CONH_2 \xrightarrow[P_2O_5]{-H_2O} CH_3CN$$

$$HNO_3 \xrightarrow[P_2O_5]{-H_2O} N_2O_5$$

$$H_2SO_4 \xrightarrow[P_2O_5]{-H_2O} SO_3$$

These processes are reversed on the addition of water. The structure of P_4O_{10} is similar to that of P_4O_6 except that the four atoms of oxygen are coordinated to the four phosphorous atoms as seen in figure 4-12.

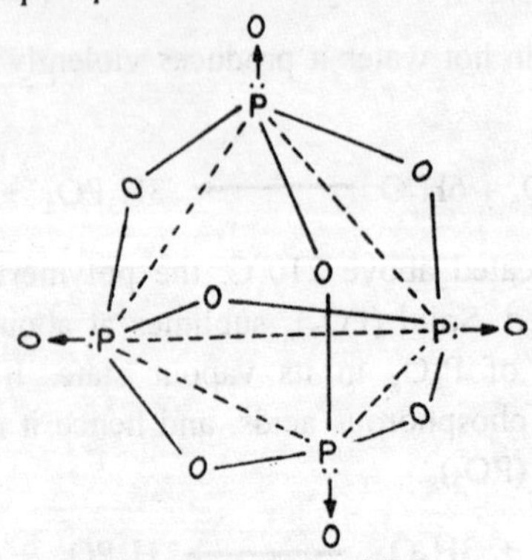

Fig. 4-12 Molecule of P_4O_{10}

4-5-3 Oxyhalides of Phosphorous

The oxyhalides, of phosphorus are POF_3, $POCl_3$ and $POBr_3$. Mixed halides such as POF_2Cl, $POFCl_2$, POF_2Br and $POFBr_2$ are also isolated. Phosphoryl chloride is best made by heating PCl_5 with anhydrous oxalic acid.

$$PCl_5 + H_2C_2O_4 \longrightarrow POCl_3 + 2HCl + CO_2 + CO$$

The mixed halides are made by treating the chloride or bromide with SbF_3. Phosphorous oxychloride hydrolyses slowly with cold water.

$$POCl_3 + 3H_2O \longrightarrow H_3PO_4 + 3HCl$$

It is a tetrahedral molecule. The properties and structures of $POBr_3$ and $POCl_3$ are similar (figure 4-13).

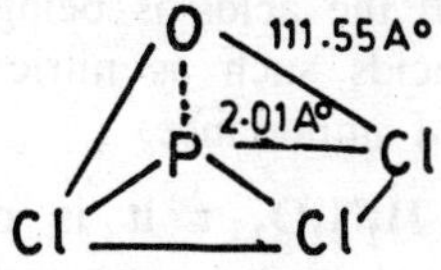

Fig. 4-13 Structure of $POCl_3$

The P = O distance is 1.55 Å and P – Cl is 2.01 Å. $POCl_3$ gets self ionized.

$$2POCl_3 \longrightarrow POCl_{2+} + POCl_{4-}$$

A complex oxohalide is also known which is made from P_2O_5 and PCl_5

```
     O       O
     ||      ||
Cl − P − O − P − Cl
     |       |
     Cl      Cl
```

4-5-4 Oxoacids of Nitrogen and Phosphorous

A large number of oxo acids of nitrogen and phosphorous are known.

(1) Oxoacids of Nitrogen

(i)	Hyponitrous acid	$H_2N_2O_2$	+1
(ii)	Hydronitrous acid	H_2NO_2	+2
(iii)	Nitrous acid	HNO_2	ı3
(iv)	Hyponitric acid	$H_2N_2O_3$	+5
(v)	Nitric acid	HNO_3	+5
(vi)	Pernitric acid	HNO_4	+7

(2) Oxoacids of Phosphorous

(i)	Hypophosphorous acid	H_3PO_2	+1
(ii)	Phosphorous acid	H_3PO_3	+3
(iii)	Hypophosphoric acid	$H_4P_2O_6$	+4
(iv)	Orthophosphoric acid	H_3PO_4	+5
(v)	Pyrophosphoric acid	$H_4P_2O_7$	+5
(vi)	Metaphosphoric acid	HPO_3	+5
(vii)	Peroxophosphoric acid	H_3PO_5 or $H_4P_2O_8$	+7 or +6
(viii)	Tripolyphosphoric acid	$H_5P_3O_{10}$	+5
(ix)	Tetra polyphosphoric acid	$H_6P_4O_{13}$	+5

A brief chemistry of all the acids is being dealt in this section giving a detailed account of few acids such as nitric acid, phosphorous acid and phosphoric acid, which are of importance.

(i) Hyponitrous acid $H_2N_2O_2$: It is obtained by treating sodium hyponitrite with silver nitrate in the presence of acetic acid. The silver hyponitrite obtained, is treated with dil HCl in an ethereal solution to get $H_2N_2O_2$.

$$Na_2N_2O_2 + 2AgNO_3 \longrightarrow Ag_2N_2O_2 + 2NaNO_3$$

$$Ag_2N_2O_2 + 2HCl \longrightarrow H_2N_2O_2 + 2AgCl$$

It is a weak acid and it behaves as a strong reducing agent. It is a dibasic acid and forms two series of salts. $NaHN_2O_2$ and $Na_2N_2O_2$

(ii) Hydronitrous Acid, H_2NO_2 : Only salts of this acid are known and the acid cannot be isolated freely. When sodium nitrite is treated with sodium metal in an ammonical solution hydronitrite is formed.

$$NaNO_2 + Na \xrightarrow{NH_3} Na_2NO_2$$

(iii) Nitrous Acid, HNO_2 : The acid cannot be isolated freely. It is present only in solution that too in a very unstable state. Barium nitrite when treated with sulphuric acid forms, nitrous acid in solution.

$$Ba(NO_2)_2 + H_2SO_4 \longrightarrow 2HNO_2 + BaSO_4$$

$BaSO_4$ can be removed by filteration to get a salt free aqueous solution of the acid.

Nitrous acid can also be prepared by treating either sodium or potassium nitrite with dilute mineral acids.

$$2NaNO_2 + H_2SO_4 \longrightarrow 2HNO_2 + Na_2SO_4$$

$$KNO_2 + HCl \longrightarrow HNO_2 + KCl$$

It is a reducing agent as well as a strong oxidising agent. It can reduce a strong oxidising agent such as potassium permangnate and oxidise a reducing agent like hydrogen iodide.

$$2KMnO_4 + 3H_2SO_4 + 5HNO_2 \longrightarrow K_2SO_4 + 2MnSO_4 + 5HNO_3 + 3H_2O$$

$$2HNO_2 + 2HI \longrightarrow 2H_2O + I_2 + 2NO$$

The salts of nitrous acid are known as nitrites. The most important nitrites are those of sodium and potassium. The nitrite ion is present in two tautomeric forms, as ONO^-, nitrito group and – NO_2^-, nitro group. Both the groups have free lone electron pairs present on nitrogen as well as on oxygen atoms, hence two types of compounds for this ion are reported, e.g.

$$Na \leftarrow O - N = O \longleftrightarrow Na \leftarrow N \genfrac{}{}{0pt}{}{\nearrow\!\!\!= O}{\searrow O}$$

(a) Sodium nitrito compound (b) Sodium nitrite

From these tautomeric forms we may expect the structure of the acid to be;

$$H - N \genfrac{}{}{0pt}{}{= O}{\rightarrow O} \quad \text{or} \quad HO - N = O$$

and the nitrite ion to be

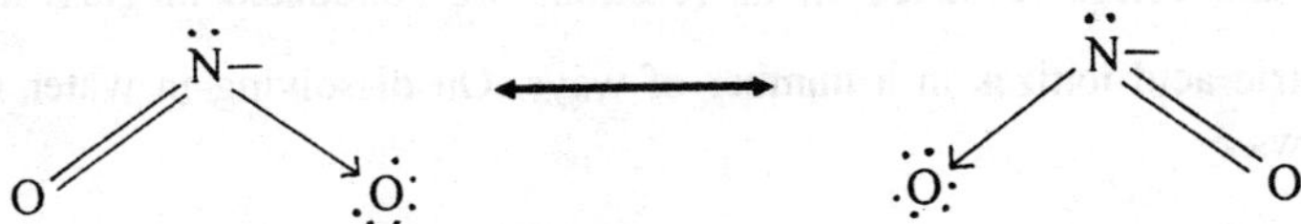

The bond length of N = O is 1.13 Å and N – O is 1.23 Å and the bond angle O$\hat{N}$O is found to be 132°.

(iv) Hyponitric acid $H_2N_2O_3$: It is very unstable but its sodium salts can be made as monohydrates by treating free hydroxylamine in methanol with ethyl nitrate in the presence of sodium methoxide.

$$NH_2OH + C_2H_5NO_3 \longrightarrow HO.NHNO_2 + C_2H_5OH \xrightarrow{2CH_3ONa} Na_2N_2O_3.H_2O$$

It decomposes in the presence of an acid to form nitric oxide.

(v) Nitric Acid HNO_3 : It is the most important of all the oxy acids of nitrogen.

It can be prepared by treating a nitrate salt with concentrated sulphuric acid.

$$NaNO_3 + H_2SO_4 \longrightarrow NaHSO_4 + HNO_3$$

Industrially it is made by the catalytic oxidation of ammonia by Ostwald's process. Required ammonia is obtained by the Haber's process. When a mixture of ammonia and air is passed over a platinum-rhodium catalyst at about 850°C, NO and steam are formed.

$$4NH_3 + 5O_2 \xrightarrow[\text{Catalyst}]{\text{Pt–Rh}} 4NO + 6H_2O$$

The mixture is cooled and the nitric oxide allowed to react with air to form nitrogen dioxide, which when dissolved in water in the presence of air to produce nitric acid.

$$2NO + O_2 \longrightarrow 2NO_2$$

$$4NO_2 + 2H_2O + O_2 \longrightarrow 4HNO_3$$

Any excess NO_2 present in air is removed by passing into sodium hydroxide solution. The acid obtained is of 75-80% purity. This acid is now boiled and distilled to get an acid of 98% purity. Any further dehydration, if required, is achieved by treating the acid with P_2O_5. The solution now obtained has a constant boiling point of 120.5°C which is an azeotropic mixture.

Pure nitric acid is a colourless liquid and boils at 86°C, and freezes at –41.55°C to a crystalline solid. Pure nitric acid can be had by distilling the dilute acid with concentrated sulphuric acid. Fuming nitric acid is obtained by dissolving excess of NO_2 in concentrated nitric acid. Nitric acid is extremely reactive and corrosive hence all its reactions are conducted in glass apparatus.

Nitric acid ionizes in a number of ways. On dissolving in water it ionizes as follows :

$$HNO_3 + H_2O \rightleftharpoons H_3O^+ + NO_3^-$$

Anhydrous nitric acid undergoes ionization in two steps,

$$2HNO_3 \rightleftharpoons H_2NO_3^+ + NO_3^-$$

$$H_2NO_3^+ \rightleftharpoons H_2O + NO_2^+$$

which on summing gives

$$2HNO_3 \rightleftharpoons NO_3^- + NO_2^+ + H_2O$$

Sulphuric acid also ionizes nitric acid.

$$H_2SO_4 + HNO_3 \longrightarrow NO_2^+ + HSO_4^- + H_2O$$

Hence a mixture of concentrated sulphuric acid and nitric acid act as a strong nitrating agent. Nitrobenzene is prepared by using such a mixture.

$$C_6H_6 + HNO_3 \xrightarrow{H_2SO_4} C_6H_5NO_2 + H_2O$$

Nitric acid reacts with metallic oxides, hydroxides and carbonates.

$$Na_2O + 2HNO_3 \longrightarrow 2NaNO_3 + H_2O$$

$$KOH + HNO_3 \longrightarrow KNO_3 + H_2O$$

$$CaCO_3 + 2HNO_3 \longrightarrow Ca(NO_3)_2 + CO_2 + H_2O$$

Nitric acid reacts differently with different metals. The reaction of the acid depends upon its concentration also.

In eq. (i) concentrated acid reacts with copper giving NO_2 whereas in eq. (ii) dilute acid gives NO

$$Cu + 4HNO_3 \longrightarrow Cu(NO_3)_2 + 2H_2O + 2NO_2 \quad (i)$$

$$3Cu + 8HNO_3 \longrightarrow 3Cu(NO_3)_2 + 4H_2O + 2NO \quad (ii)$$

Zinc in cold condition reacts with (i) dilute acid giving either N_2O or N_2 whereas (ii) very dilute acid gives NH_4NO_3

$$4Zn + 10HNO_3 \longrightarrow 4Zn(NO_3)_2 + N_2O + 5H_2O \quad (i)$$

or

$$5Zn + 12HNO_3 \longrightarrow 5Zn(NO_3)_2 + N_2 + 6H_2O$$

$$4Zn + 10HNO_3 \longrightarrow 4Zn(NO_3)_2 + 3H_2O + NH_4NO_3 \quad (ii)$$

Magnesium and manganese give hydrogen.

$$Mg + 2HNO_3 \longrightarrow Mg(NO_3)_2 + H_2$$

$$Mn + 2HNO_3 \longrightarrow Mn(NO_3)_2 + H_2$$

Noble metals like Pt, Au, Rh and Ir cannot react with nitric acid, but aquaregia can attack these metals (Aqua regia contains 1 part of conc. HNO_3 and 3 parts of conc HCl) The aquaregia contains free chlorine in it which reacts with the metals.

$$HNO_3 + 3HCl \longrightarrow 2H_2O + Cl_2 + NOCl$$

$$2Au + 3Cl_2 \longrightarrow 2AuCl_3$$

$$AuCl_3 + HCl \longrightarrow HAuCl_4 \text{ (Aurochloric Acid)}$$

As mentioned before nitric acid in the presence of sulphuric acid acts as a nitrating agent. The nitrating species being the nitronium ion NO_2^+, it is thought that the NO_2^+ ion attacks the benzene ring first with the formation of a positively charged complex, which then loses a proton to give the nitron compound

$$C_6H_6 + NO_2^+ \xrightarrow{\text{slow}} [C_6H_6NO_2]^+ \xrightarrow{\text{fast}} C_6H_5NO_2 + H^+$$

Nitric acid oxidizes cane sugar to oxalic acid.

$$C_{12}H_{22}O_{11} + 36HNO_3 \longrightarrow 6C_2O_4H_2 + 36NO_2 + 23H_2O$$

Nitric acid is a planar molecule having a resonance structure given in figure 4-14.

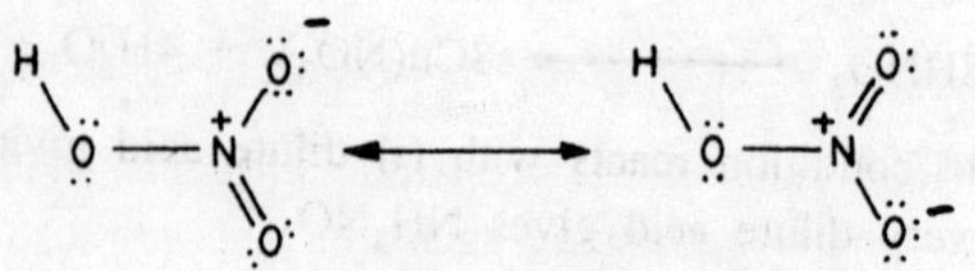

Fig. 4-14 Resonance structure of HNO_3

The bond lengths and bond angles are represented in Fig. 4-15

angle : a = 102°
b = 116°
c = 130°
d = 114°

Length : α = 0.96 Å
β = 1.41 Å
γ = 1.22 Å
μ = 1.22 Å

Fig. 4-15 Structure of Nitric Acid

(vi) Pernitric acid HNO_2 : When anhydrous hydrogen perioxide is mixed with N_2O_5 pernitric acid is obtained.

$$N_2O_5 + H_2O_2 \longrightarrow HNO_2 + HNO_3$$

It is highly explosive acid, hence is very unstable. It is an oxidising agent.

It has a structure similar to that of nitric acid Fig. 4-16, with a peroxo group.

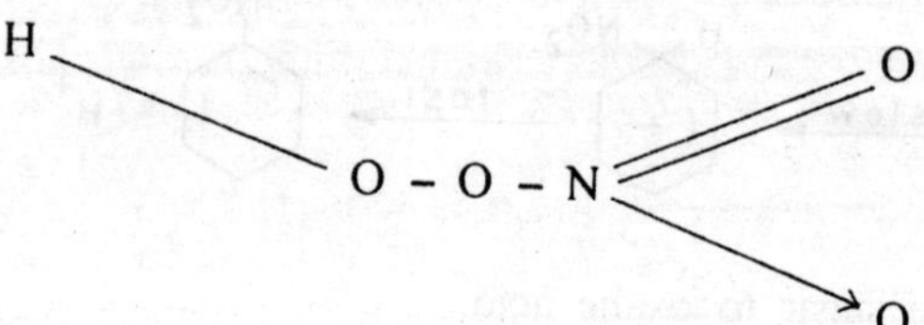

Fig. 4-16 Structure of pernitric acid

(vii) Hypo phosphorous acid H_3PO_2 : The hypo phosphorous acid is made when barium hypophosphite is acidified with dilute sulphuric acid. The barium hypophosphite is obtained when white phosphorous is dissolved in hot baryta water

$$Ba(H_2PO_2)_2 + H_2SO_4 \longrightarrow 2H_3PO_2 + BaSO_4$$

It is a colourless, deliquescent crystalline substance melting at 26.5°C. It is a monobasic acid and is a reducing agent. Its salts are also reducing agents. Its structure is given in Fig. 4-17.

```
          O
          ||
H – O – P – H
          |
          H
```

Fig. 4-17 Structure of Hypophosphorous acid

(vii) Phosphorous acid H_3PO_3 : On hydrolysis of phosphorous trichloride with ice cold water and on crystallization, a solid crystallizes out, which melts at 70°C. Pure acid is deliquescent.

$$PCl_3 + 3H_2O \longrightarrow H_3PO_3 + 3HCl$$

Since it is a dibasic acid its structure is thought to contain only two hydroxyl groups, although its formation by hydrolysis of phosphorous trichloride suggests three. It is thought that an equilibrium exists of the type shown below, which is very largely displaced to the right, Fig 4-18.

```
  OH                O
  |                 ||
: P – OH  ⟷  H – P – OH
  |                 |
  OH                OH
```

Fig. 4-18 Structure of H_3PO_3

Hence phosphorous atom is essentially five covalent. The acid and its salts are reducing agents. The alkali metal phosphites are water soluble but other metal salts are insoluble it undergoes decomposition forming ortho phosphoric acid and phosphine when heated.

$$4H_3PO_3 \longrightarrow 3H_3PO_4 + PH_3$$

It ionises, showing its dibasic nature as follows, giving primary phosphite ion and secondary phosphite ion respectively.

$$H_3PO_3 \longrightarrow H_2PO_3^- + H^+$$
$$H_3PO_3 \longrightarrow HPO_3^{2-} + 2H^+$$

Their salts are also named accordingly e.g. NaH_2PO_3, sodium primary phosphite, $CaHPO_3$, calcium secondary phosphite.

(ix) Hypophosphoric acid $H_4P_2O_6$: The acid is prepared as a trihydroxo compound from the solution obtained when $Na_2H_2P_2O_6$ is poured through a cation exchange column resin.

$$Na_2H_2P_2O_6 + 2H^+ \text{ Resin} \longrightarrow 2\ Na^+ \text{ Resin} + H_4P_2O_6$$

It is a tetrabasic acid, but the common salts of the acid are those in which only two of the hydrogens are replaced. It is a very stable acid and so are the salts.

The structure of the acid, as elucidated from P^{31} n.m.r. spectrum is seen in Fig. 4-19.

```
      HO        OH
       |         |
  H — P — O — P — O
       |         |
       O        OH
```

Fig. 4-19 Structure of Hypophosphoric acid

(x) Orthophosphoric acid, H_3PO_4 : It is the best known oxy acid of phosphorous and it is commonly called as phosphoric acid. It is prepared by adding P_2O_5 to water and boiling it.

$$P_4O_{10} + 6H_2O \longrightarrow 4H_3PO_4$$

The laboratory method of preparing phosphoric acid is to heat white phosphorous with an equimolar solution of nitric acid.

$$P_4 + 20HNO_3 \longrightarrow 4H_3PO_4 + 20NO_2 + 4H_2O$$

On large scale, the acid can be obtained by action of concentrated sulphuric acid with either rock phosphate or bone ash.

$$Ca_3(PO_4)_2 + 3H_2SO_4 \longrightarrow 3CaSO_4 + 2H_3PO_4$$

Calcium sulphate is filtered and the syrupy liquid is concentrated to get 85% pure acid.

It is a tribasic acid, thus forming three series of salts. The ionization takes place as follows :

Ionization constant

$$H_3PO_4 \rightleftharpoons H_2PO_4^- + H^+ \qquad K_1 = 7.5 \times 10^{-3}$$

$$H_2PO_4^- \rightleftharpoons HPO_4^- + H^+ \qquad K2 = 6.2 \times 10^{-8}$$

$$HPO_4^{=} \rightleftharpoons PO_4 + H^+ \qquad K_3 = 1.1 \times 10^{-12}$$

The first dissociation is fairly strong and the third is extremely weak. It gives salts of the type NaH_2PO_4, Na_2HPO_4 and Na_3PO_4.

It has a tetrahedral structure, figure 4-20, angle a = c = 112° and b = 106° the P – OH distance = 1.57 Å and P = O distance = 1.52 Å.

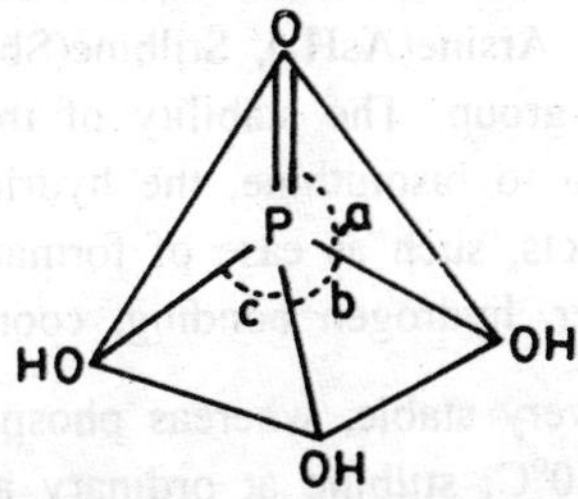

Fig. 4-20 Structure of ortho-phosphoric acid

(xi) Pyrophosphoric acid $H_4P_2O_7$: It is obtained when orthophosphoric acid is heated to 250° – 260°C. It is a colourless solid.

$$2H_3PO_4 \longrightarrow H_4P_2O_7 + H_2O$$

All the hydrogens are replaceable, hence it is a tetrabasic acid, but the common pyrophsophates are $M_2H_2P_2O_7$ and $M_4P_2O_7$. Its structure is seen in Fig. 4-21.

Fig. 4-21 Structure of $H_4P_2O_7$

(xii) Meta phosphoric acid, HPO_3 : It is obtained by the dehydration of ortho phosphoric acid at 316°C. It is a glassy polymeric solid with an emprical formula of HPO_3.

$$H_3PO_4 \longrightarrow HPO_3 + H_2O$$

The structure of polymeric HPO_3 is represented in figure 4-22.

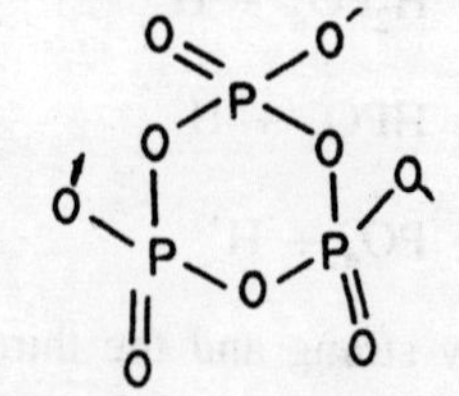

Fig. 4-22 Structure of HPO_3

4-5-5 Hydrides

All the elements form volatile gaseous hydrides of the type MH_3 : Ammonia (NH_3), Phosphine (PH_3), Arsine(AsH_3), Stilbine(SbH_3) and Bismuthine (BiH_3) are the hydrides of the group. The stability of the hydrides decreases down the group from ammonia to bismuthine, the hydrides show a marked change in properties in all respects, such as ease of formation, thermal stability, basic nature, reducing character, hydrogen bonding, coordinate bond formation etc.

Ammonia is thermally very stable, whereas phosphine decomposes at 400°C, arsine decomposes at 230°C, stilbine at ordinary and bismuthine decomposes at room temperature.

Ammonia is a base and forms salts with both weak and strong acids, phosphine is a weaker base than ammonia whereas the other hydrides are not at all basic in character. The reducing character is just the reverse, ammonia is not a reducing agent but BiH_3 is a strong reducing agent. All the hydrides have a lone pair of electrons in them, by which they are expected to form a coordinate bond with metal ions. Except ammonia and phosphine to a lesser extent, none of them is capable of forming coordination compounds. All the hydrides of the family have a pyramidal structure, with the bond angle decreasing from ammonia onwards. The angle in ammonia actually suggests a structure of distorted tetrahedron. The decrease in the bond angles from NH_3 to BiH_3 is because of the repulsion between the bonded electron pairs and the lone pair resting on the elements of VB group.

Ammonia has hydrogen bonding in the liquid state. The other hydrides do not have hydrogen bonding in them. this is because of the high electronegativity of nitrogen (3.0). The electronegativity of other elements is quite less and are therefore not capable of forming hydrogen bonding. Ammonia undergoes auto ionization similar to water.

$$NH_3 + NH_3 \rightleftharpoons NH_4^+ + NH_2^-$$

similar to $$H_2O + H_2O \rightleftharpoons H_3O^+ + OH^-$$

In effect, one ammonia molecule removes a proton from another and this is equivalent to taking a proton from the amide ion, NH_2^-. Amide ion is of course a much better electron donor than an ammonia molecule. In a fair competition with an ammonia molecule, amide ion will always win the proton, consequently, if the NH_4^+ ion comes in contact with the NH_2^- ion, the NH_2^- will at once take back the proton forming two ammonia molecules, therefore, within the bulk of the liquid ammonia there will be ammonium and amide ions present although in very small concentrations.

An amide ion is a better electron donor than OH^- ion as seen by the immediate and complete hydrolysis of NH_2^- ion in water.

$$NH_2^- + H_2O \rightleftharpoons NH_3 + OH^-$$

This means that ammonia molecule does not give up a proton as easily as does water. Hence the autoionization of ammonia is not expected to proceed to as great extent as observed for water. Experimentally it was found that the ionic product of liquid ammonia is 1×10^{-33} which is much less than the ionic product of water, 1×10^{-14}. Liquid ammonia as a solvent deserves much more detailed treatment, hence it's role as a solvent is dealt in section 4-6 of this chapter.

Ammonia can be prepared in the laboratory by heating any ammonium salt along with sodium hydroxide or slaked lime.

$$NaOH + NH_4Cl \longrightarrow NaCl + H_2O + NH_3$$

$$Ca(OH)_2 + 2NH_4Cl \longrightarrow CaCl_2 + 2H_2O + 2NH_3$$

Ammonia is also obtained by heating ammonium sulphate or phosphate and also by heating microcosmic salt (NH_4NaHPO_4).

$$(NH_4)_2SO_4 + 2NaOH \longrightarrow Na_2SO_4 + 2H_2O + 2NH_3$$

$$(NH_4)_3\,PO_4 \longrightarrow HPO_3 + H_2O + 3NH_3$$

$$NH_4Na\,HPO_4 \longrightarrow NaPO_3 + H_2O + NH_3$$

Ammonia is best formed by the reduction of nitric oxide with nascent hydrogen in the presence of a catalyst like platinum.

$$2NO + 5H_2 \xrightarrow{Pt} 2NH_3 + 2H_2O$$

It is manufactured on a large scale by Haber's process, based on LeChatelier's principle. Haber's process is an adaptation of the natural process by which ammonia is formed in the nature.

Nitrogen is obtained from a reaction between producer gas and hydrogen from water gas. This mixture is made to pass over coke which reduces CO present in the producer and water gases, setting free N_2 & H_2. They are mixed in a proportion of 1:3 and the mixture is passed through steam and then over ferric oxide, which acts as a catalyst, converting the carbon monoxide to carbondioxide. CO_2 is removed passing through water under pressure, (25 atm), and any residual carbon monoxide is then removed by ammoniacal solution of cuprous chloride.

$$N_2 + 3H_2 \rightleftharpoons 2NH_3 + 22.4 \text{ kcals}$$

Since the reaction is endothermic and proceeds with a decrease in the volume, the reaction favoured by low temperatures (400-500°C) and high pressures of almost 200 atmospheres. A flow sheet is seen in Fig. 4-23.

Ammonia is a colorless gas with a characteristic odor, highly soluble in water and its aqueous solution is commonly called as ammonium hydroxide, showing its alkaline character.

$$MnSO_4 + 2NH_4OH \longrightarrow Mn(OH)_2 + (NH_4)_2SO_4$$

$$ZnCl_2 + 2NH_4OH \longrightarrow Zn(OH)_2 + 2NH_4Cl$$

$$AgNO_3 + 2NH_4OH \longrightarrow Ag(NH_3)_2NO_3 + 2H_2O$$

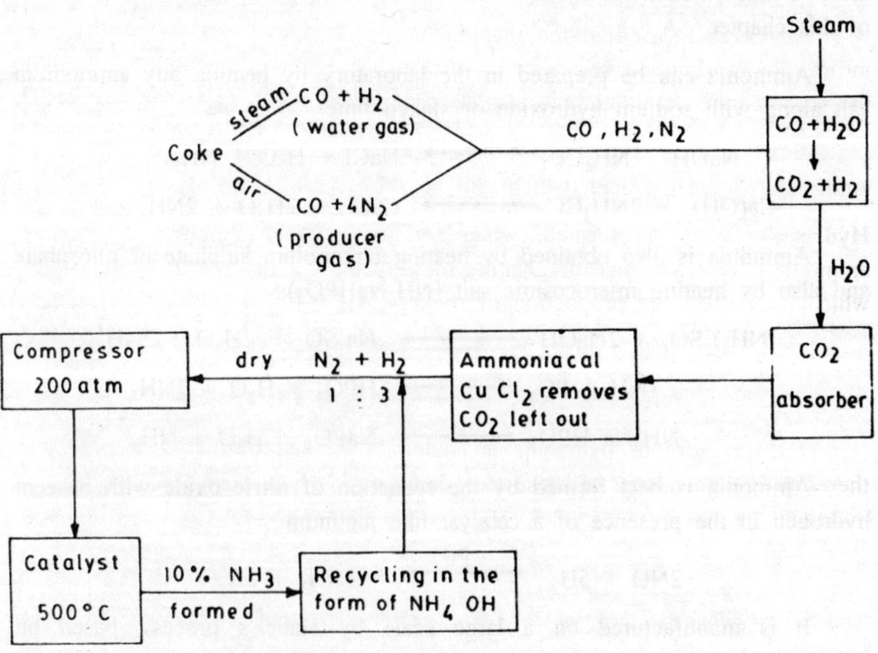

Fig. 4-23 Flow sheet for Haber's process

Gaseous ammonia does not burn in air but it burns in pure oxygen with a yellow flame forming nitrogen.

$$4NH_3 + 3O_2 \longrightarrow 2N_2 + 6H_2O$$

It reacts with Cl_2 producing nitrogen but with excess of Cl_2 produces NCl_3 with HCl it produces NH_4Cl, hence NH_3 is a base.

$$8NH_3 + 3Cl_2 \longrightarrow N_2 + 6NH_4Cl$$

$$NH_3 + 3Cl_2 \longrightarrow NCl_3 + 3HCl$$

$$NH_3 + HCl \longrightarrow NH_4Cl$$

Alkali metals form amides with ammonia and certain other metals form nitrides and hydrogen

$$2Na + 2NH_3 \longrightarrow 2NaNH_2 + H_2$$

$$3Mg + 2NH_3 \longrightarrow Mg_3N_2 + 3H_2$$

Ammonia is used in the preparation of many ammonium salts, which are used for numerous purposes.

Ammonia is used as an excellent non-aqueous solvent.

Ammonia is used in the manufacture of fertilizers and as a refrigerent in ice plants. Ammonia is used in the manufacture of artificial silk and in the manufacture of urea. It is also used to remove greases in dry cleaning.

Nitrogen also forms two other hydrides, viz., hydrazine and hydroxyl amine and hydrazoic acid, which will be studied under the heading of azides.

Hydrazine, N_2H_4 is formed, when sodium hypochlorite solution oxidizes ammonia in the presence of gelatine or glue. It is a colourless fuming liquid which is used in the production of rocket fuel.

$$NH_3 + NaOCl \longrightarrow NH_2Cl + NaOH$$

$$NH_2Cl + 2NH_3 \longrightarrow NH_2NH_2 + NH_4Cl$$

The structure of hydrazine is analogous to that of hydrogen perioxide, therefore the chemistry of these two compounds is related to each other.

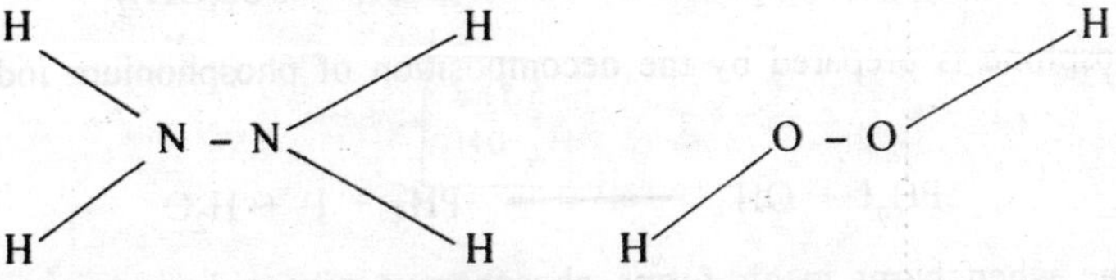

Both the compounds are strong reducing agents in alkaline media as well as in acidic media.

$$N_2H_4 + 2Br_2 \longrightarrow 4HBr + N_2$$

$$N_2H_4 + 2O_2 \longrightarrow 2H_2O_2 + N_2$$

It also acts as an oxidizing agent

$$N_2H_4 + Zn + 2HCl \longrightarrow ZnCl_2 + 2NH_3$$

Hydrazine forms derivatives like phenylhydrazine and 2:4 dinitrophenylhydrazine. These derivatives are used in the preparation of solid derivatives for the identification of aldehydes and ketones.

Hydroxylamine NH_2OH is prepared by the electrolytic reduction of nitric acid using a lead cathode.

$$HNO_3 + 6H^+ + 6e^- \longrightarrow NH_2OH + 2H_2O$$

Hydroxylamine is an unstable monoacid i.e. base. It is a colourless crystalline solid, highly soluble in water, liable to decompose, highly explosive. Its decomposition starts at a temperature of 15°C, hence should be preserved below the decomposition temperature.

$$3NH_2OH \longrightarrow NH_3 + N_2 + 3H_2O$$

$$4NH_2OH \longrightarrow 2NH_3 + N_2O + 3H_2O$$

It form salts called hydroxylamines which are strong reducing agents. Hydroxylamine and its salts reduce Fehling's solutions to Cu(I) oxide, and ammoniacal silver nitrate to silver.

$$4Cu^{++} + 2NH_2OH + 8OH^- \longrightarrow 2Cu_2O + N_2O + 7H_2O$$

$$4Ag^+ + 2NH_2OH + 4OH^- \longrightarrow 4Ag + N_2O + 5H_2O$$

Hydroxylamine is used to prepare crystalline derivatives of aldehydes and ketones called oximes, which have sharp melting points.

Phosphine, PH_3, is usually made by the action of hot, strong alkali on white phosphorous. It is a colourless noxiously smelling gas and is highly poisonous. Impure PH_3 may be prepared by the action of water or dilute acids on metal phosphides.

$$Ca_3P_2 + 6H_2O \longrightarrow 2PH_3 + 3Ca(OH)_2$$

Pure phosphine is prepared by the decomposition of phosphonium iodide with alkali

$$PH_4I + OH^- \longrightarrow PH_3 + I^- + H_2O$$

Phosphine when burnt in air forms phosphorous pentoxide

$$4PH_3 + 8O_2 \longrightarrow P_4O_{10} + 6H_2O$$

It burns in chlorine forming PCl_3

$$PH_3 + 3Cl_2 \longrightarrow PCl_3 + 3HCl$$

Substitution by alkyl or phenyl radicals of phosphine with hydrogen produces very effective ligands, that activate the lone pair of electrons on phosphorous. Compounds like $P(CH_3)_3$, trimethylphosphine, $P(C_2H_5)_3$ triethyphosphine, $P(C_6H_5)_3$, triphenyl phosphine find a very important role in the coordination chemistry as ligands.

4-5-6 Azides

Azides are salts prepared from hydrazoic acid and are potentially very explosive, **Hydrazoic acid or azoimide** was first prepared in 1890 by Curtuis. It was obtained by the careful oxidation of hydrazine with nitric acid.

$$3N_2H_4 + 5[O] \longrightarrow 2HN_3 + 5H_2O$$

Hydrazoic acid is also formed by the action of hydrazine on a solution of nitrogen trichloride in benzene

$$N_2H_4 + NCl_3 \longrightarrow HN_3 + 3HCl$$

It was in 1892, that, hydrazoic acid was first prepared by Wislicenus, using sodamide $NaNH_2$ and ammonia, by heating them together over sodium in a hard glass tube at 200°C. The salt obtained NaN_3 when distilled with sulphuric acid gave HN_3. The salt NaN_3, sodium azide was found to be ionic with $Na^+N_3^-$ ions in it. The hydrazoic acid was found to react with many metals as well as their salts to give their azides.

$$Zn + 3HN_3 \longrightarrow Zn(N_3)_2 + N_2 + NH_3$$

The hydrazoic acid neutralizes ammonia forming ammonium azide.

$$NH_3 + HN_3 \longrightarrow NH_4N_3$$

The azide ion, the cyanate ion and the neutral nitrous oxide molecules are similar, suggesting a linear structure by X-ray diffraction.

[N::N::N], [N::C::O], [N::N::O]

The hydrazoic acid molecule has a hydrogen atom linked by a bond at an angle to the linear azide group

$$\overset{1.14\ \text{Å}}{N \longleftarrow N} \overset{1.25\ \text{Å}}{\longleftarrow N} - H$$

The azide ion has a resonance hybrid of three forms:

$[N::N::N]^-$, or $[:N:N::N:]^-$ or $[:N::N:N:]^-$

or with the alternative formulation of the coordinate link

[:N=N=N:], or [:N–N≡N:] or [:N≡N–N:]

4-6 LIQUID AMMONIA AS A SOLVENT

A brief discussion about the self ionization of ammonia has been done in section 4-5-5 where we have said that ammonia molecules gets self ionized to form NH_4^+ ions and NH_2^- ions and the ionic product constant was found to be 1×10^{-33}. In this section, the non-aqueous solvent chemistry of ammonia is dealt, within the scope of this book. Actually the chemistry of this non aqueous solvent needs volumes to study it in detail.

Liquid ammonia gets ionised similar to water, but its ionisation is very slow and to a very slight extent when compared to water.

$$NH_3 + NH_3 \rightleftharpoons NH_4^+ + NH_2^-$$

$$H_2O + H_2O \rightleftharpoons H_3O^+ + OH^-$$

In water, acids provide hydronium ions and bases provide hydroxyl ions, but in liquid ammonia, acids generate ammonium ions and bases the amide ions. Thus in **liquid ammonia acid-base system** an ammono-acid neutralises an ammono-base to give a salt and ammonia.

$$NH_4^+Cl^- + Na^+NH_2^- \longrightarrow Na^+Cl^- + 2NH_3$$

or $$NH^+{}_4 + NH^-{}_2 \longrightarrow 2NH_3$$

$$H^+Cl^- + Na^+OH^- \longrightarrow Na^+Cl^- + H_2O$$

or $$H^+ + OH^- \longrightarrow H_2O$$

Salts get hydrolysed in water while they get ammonolysed in ammonia.

In water : $SiCl_4 \longrightarrow Si(OH)_4 \longrightarrow SiO_2 + 2H_2O$

In ammonia : $3SiCl_4 \longrightarrow 3Si(NH_2)_4 \longrightarrow Si_3N_4 + 8NH_3$

A similar type of precipitation occurs in both solvents

$$(NH_4)_2S + CuSO_4 \xrightarrow[\text{liq.}NH_4]{\text{water or}} CuS + (NH_4)_2SO_4$$

However, due to difference in solubulities, on mixing $BaCl_2$ and $AgNO_3$ solution AgCl is precipitated in water while $BaCl_2$ is precipated in liquid ammonia.

In water : $BaCl_2 + 2AgNO_3 \longrightarrow Ba(NO_3)_2 + 2AgCl$

In ammonia : $Ba(NO_3)_2 + 2AgCl \longrightarrow BaCl_2 + 2AgNO_3$

The amphoteric character of zinc salts is exhibited similarly by both the solvents.

In water : $Zn^{++} + 2NaOH \longrightarrow Zn(OH)_2 \xrightarrow{\text{excess NaOH}} Na_2\,[Zn(OH)_4]$

In ammonia : $Zn^{++} + 2NaNH_2 \longrightarrow Zn(NH_2)_2 \xrightarrow{\text{excess NaNH}_2} Na_2\,[Zn(NH_2)_4]$

The solubilities of salts in liquid ammonia are markedly different from their solubilities in water. Ammonia is in general not a good solvent for highly ionic compounds. But it is a better solvent for non-polar compounds, when compared to water.

Alkali metals dissolve in ammonia without the evolution of hydrogen, giving blue, strongly conducting solutions. In dilute solutions the cation and anion appears to be solvated also contributing to the colour of the solution. This solution is a useful reducing agent and the reaction sequence is represented by the following equations.

$$Na \rightleftharpoons Na^+ + e^-$$

$$Na^+ + xNH_3 \rightleftharpoons [Na(NH_3)_x]^+$$

$$e^- + yNH_3 \rightleftharpoons [e(NH_3)_y]^-$$

4-7 NITROGEN FIXATION

The assimilation of nitrogen compounds by plants and animals is essential for life. Nitrogen occurs in all proteins, which are the structural units of living matter. With few exceptions, atmospheric nitrogen cannot be used directly by plants but must be fixed as suitable compounds in the soil for absorption by the root system. The exceptions are pod bearing plants or leguminous plants such as peas, beans, lupins and agricultural crops which bear nodules in their roots with which nitrogen fixing bacteria are associated. Mosses, algae and fungi have this power too. Nitrogen is taken up by the leguminous plants from air spaces in soil. Animals acquire combined nitrogen by eating plants or other animals, restoring it back to soil through excreation. After death, saprophytic bacteria break down the remains of plants and animals to ammonium salts, which are oxidized to nitrates and then they are built up into nitrogenous living matter. Some nitrogen is lost by the action of denitrifying bacteria on nitrates, some by the leaching out of salts by rain water, but much important is the loss through farming, the production of crops and the disposal of sewage. The circulation of nitrogen between soil and living matter is thrown out of balance by the requirements of civilizations. The nitrogen cycle in nature is represented in Figure 4-24.

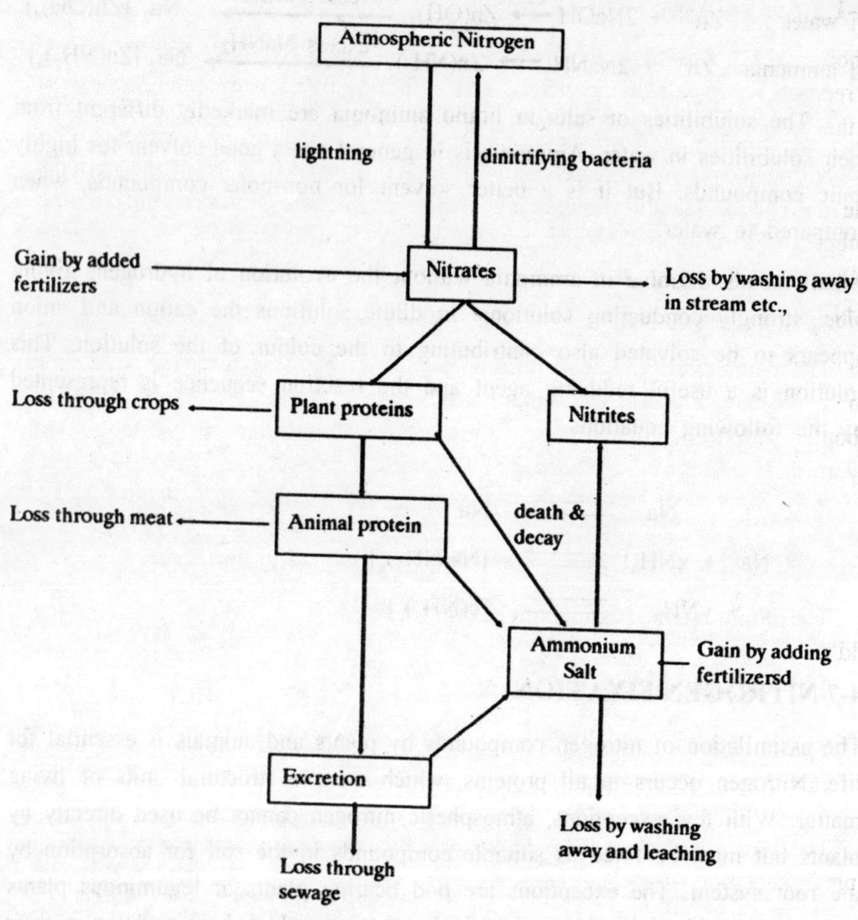

Fig. 4-24 The nitrogen cycle in nature

4-8 Inorganic Polymers containing nitrogen :

In this section we will be studying the chemistry of two of the important inorganic polymers containing nitrogen in them. Before we study about these compounds, S_4N_4 and $P_3N_3Cl_6$. We should know what a polymer means. In simple terms it is considered to be a group of atoms present in their simple multiples such as $(AB)_x$, where the emprical formula of the polymer is AB and x number of integers of AB are chemically bonded together, hence it can be defined as a substance in which many structural units are connected by valuence bonds. Some of the polymers which you are acquinted with are, P_4, S_8, Se_8 or compounds like diamond, graphite or silicones. All the polymers will either have a chain structure or may have ring structures.

4-8-1 Tetra sulphur tetra nitride :

Although S_4N_4 was first prepared by Gregory(1835) its study was not made till recently, until a polymer $(SN)_x$ was discovered and was found to possess all the properties that of a metal, and it even showed superconductivity.

Nitrogen and sulphur are related diagonally to each other in the periodic table and hence should show many similar properties. Nitrogen and sulphur combines to form numerous compounds like S_4N_4, S_5N_6, cyclo S_2N_2, polymeric $(SN)_x$, cyclo S_4N_2, bicyclo $S_{11}N_2$ and a number of higher homologues like $S_{15}N_2$, $S_{16}N_2$, $S_{17}N_2$, $S_{19}N_2$ etc.

The compound S_4N_4 can be prepared by passing a current of ammonia into a warm solution of either SCl_2 or S_2Cl_2 in either benzene or carbontetrachloride to get orange yellow crystals which are stable to air but they melt at 178.2°C.

$$6SCl_2 + 16NH_3 \longrightarrow S_4N_4 + 2S + 12NH_4Cl$$

$$6S_2Cl_2 + 16NH_3 \longrightarrow S_4N_4 + 8S + 12\ NH_4Cl$$

The ammonium chloride on heating with S_2Cl_2 at 160°C, adds on the yield of S_4N_4.

$$6S_2Cl_2 + 4NH_4Cl \longrightarrow S_4N_4 + 8S + 16HCl$$

The sulphur on treating with ammonia also gives S_4N_4.

$$10S + 4NH_3 \longrightarrow S_4N_4 + 6H_2S$$

The H_2S obtained is removed by passing it through silver iodide solution to precipitate out silver sulphide and ammonium iodide.

The S_4N_4 is insoluble and nonreactive towards water, but can get hydrolyse in the presence of both dilute or concentrated alkalies

Dil. : $2S_4N_4 + 6NaOH + 9H_2O \longrightarrow 8NH_3 + Na_2S_2O_3 + 2Na_2S_3O_6$

Conc : $S_4N_4 + 6NaOH + 3H_2O \longrightarrow Na_2S_2O_3 + 2Na_2SO_3 + 4NH_3$

In dilute solutions a trithionate is formed but in a concentrated solution a sulphite is formed.

Halogenacids, substitutes nitrogen from S_4N_4 forming $[S_4N_3]^+ X^-$, in the presence of CCl_4, where $X^- = F^-, Cl^-, Br^-$.

$$3HX + 3S_4N_4 \xrightarrow{CCl4} 3\ [S_4N_3]^+ X^- + NH_3$$

whereas HI produces H_2S, NH_3 and I_2

$$8HI + S_4N_4 \longrightarrow 4H_2S + 4NH_3 + 4I_2$$

On the reaction with ammonia S2N2NH a complex ammoniate compound, is formed which behaves like an acid amide in ammonia solution.

$$S_4N_4 + 2NH_3 \longrightarrow 2S_2N_2NH_3$$

S_4N_4 is found to be an eight membered heterocyclic compound with a cradle configuration. The S-N bond distance in the compound (1.65 Å) is shorter than the supposed covalent radii between sulphur and nitrogen (1.78Å), this may be due to the delocalization of the electrons in the heterocyclic ring. The structure of S_4N_4 cannot be drawn as a fixed one, as it may undergo numerous resonance hybrids, the typical ones can be considered as seen in Fig. 4-25

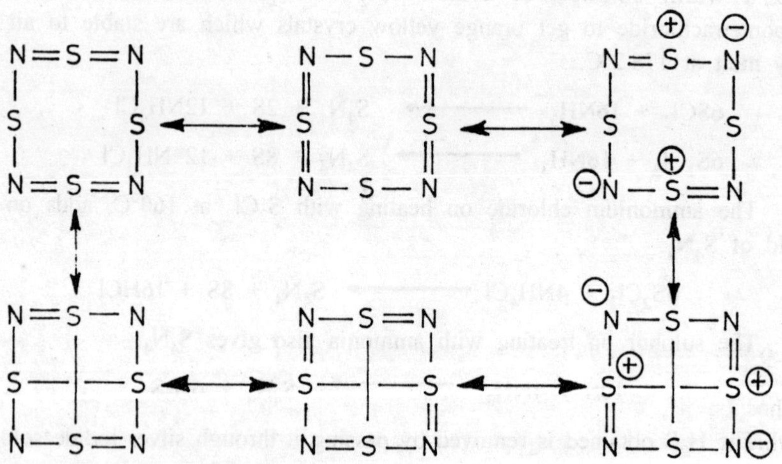

Fig. 4-25 Resoning Structures of S_4N_4

Another structure can also be seen in the figure 4-26 below, it is found that the bridging

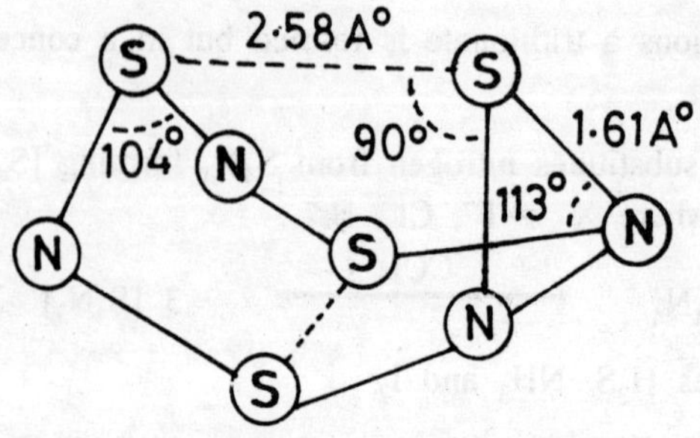

Fig. 4-26 : Structure of S_4N_4

nitrogen is placed at an angle of 113° with respect to the two sulphurs bonding it. The bond angle of N$\hat{S}$S is 90° whereas the N$\hat{S}$N bond angle is 105° as seen the figure, the S-S bond length is 2.58Å, which is an intermediate between the usual S-S bond distance of 2.08Å and the nonbonding vander Waals distance of 3.3Å, this suggests that the bonding in S_4N_4 is weak but is structurally strong. S2N2 is a better coordinating agent than S4N4. The structure and bonding in [Pb(S2N2)NH3] is notably very interesting and it features unequal S-N bonding with variable bond angles, as seen in the figure 4-27.

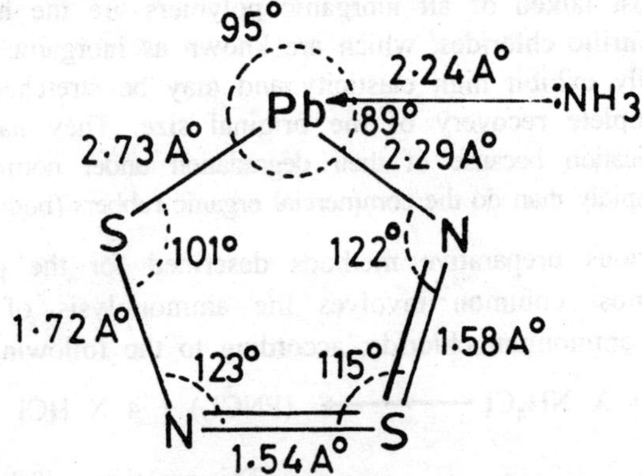

Fig. 4.27 Structure of [Pb(S_2N_2)NH_3]

4-9 POLYMERS OF PHOSPHOROUS

Phosphorous shows variable polymorphs such as white, black, red, etc. The white phosphorous has a discrete P_4 tetrahedral structure, Fig.4-25(a) in solid and liquid states but the vapors show the presence of P_2 molecules also. The black phosphorous exists in four forms, amorphous, rhombohedral, cubic and orthorhombic. The ortho rhombic form is a soft flaky material with a longer structure with continuous net work as shown in Fig. 4.25(b)

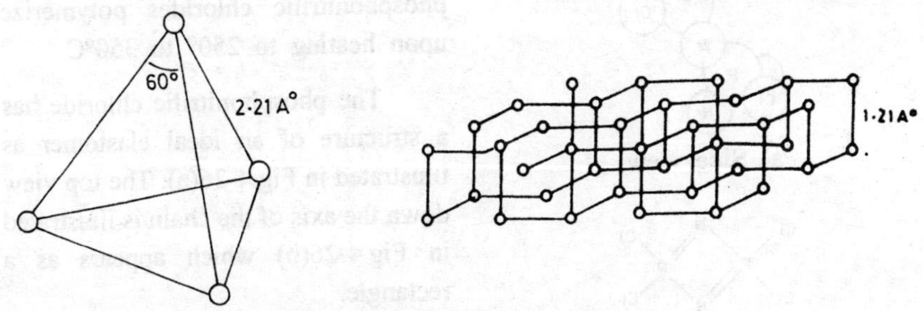

Fig. 4.28 (a) Tetrahedral structure of white phosphorous (b) Ortho rhombic phosphorous

The red phosphorous is highly polymerized allotrope of phosphorus. It is highly insoluble and it is in an amorphous form. On heating its crystalline form is produced.

Cyclic phenylphosphorous compounds have also been discovered and are called **phosphobenzenes**, having a formula $(PhP)_4$. A pentamer $(PhP)_5$ and an hexamer $(PhP)_6$ have also been isolated and their structures determined.

4-9-1 Polymeric phosphonitrilic chlorides :

Probably the most talked of all inorganic polymers are the high molecular weight phosphonitrilic chlorides, which are known as inorganic rubber. These materials generally exhibit high elasticity and may be stretched several fold with almost complete recovery of the original size. They have not found commercial application because of their degradation under normal atmospheric conditions more rapidly than do the commercial organic rubbers (both natural rubber.

Of the various preparative methods described for the phosphonitrilic chlorides, the most common involves the ammonolysis of phosphorous pentachloride by ammonium chloride, according to the following equation.

$$X\ PCl_5 + X\ NH_4Cl \longrightarrow (PNCl_2) + 4\ X\ HCl$$

This reaction may be carried out in a refluxing solvent, such as sym-tetrachloroethane, or by heating the solid mixture in vented tubes. Depending on how the reaction is carried out, the product becomes a mixture of oil and crystals or a buttery mass containing some crystalline material. By differential solubility, fractions of various molecular weights can be obtained. All the lower phosphonitrilic chlorides polymerize upon heating to 250° to 350°C.

The phosphonitrilic chloride has a structure of an ideal elastomer as illustrated in Fig.4-26(a). The top view down the axis of the chain is illustrated in Fig.4-26(b) which appears as a rectangle.

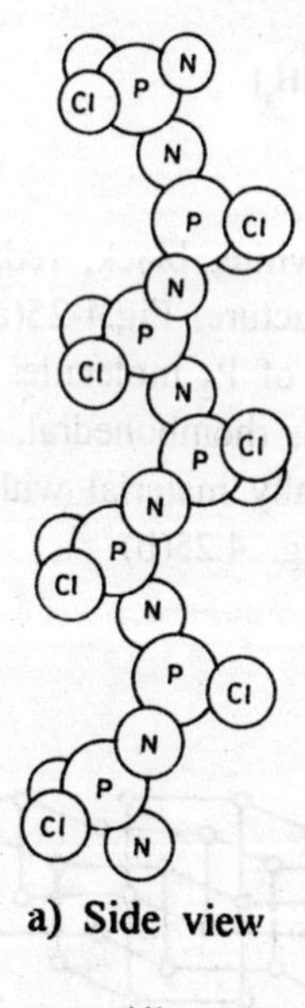

a) Side view

b) Top view

Fig.429 Structure of phosphonitrilic chloride

When highly elastic samples of inorganic rubber are stored in the absence of air, there is no change in its elastic properties. However storage in air results in a considerable increase in the elastic modules and in the brittleness of the samples. This is explained by the action of water in forming oxygen bridges form chain to chain causing a decrease in the lattice length of the polymer and an increase in elasticity, Fig.4-27.

```
   Cl        Cl                        Cl        Cl
   |         |                         |         |
 - P = N  -  P = N -                 - P = N  -  P = N -
   |         |                         |
   Cl        Cl                        Cl
   |                    + H2O                    O        + 2 HCl
   Cl        Cl                        Cl
   |         |                         |
 - P = N  -  P = N -                 - P = N  -  P = N -
   |         |                         |         |
   Cl        Cl                        Cl        Cl

     Linear chains                      Bridged chains
```

Fig 4·30 Linear and bridged chain of polyphosphonitritic chloride.

4-9-2 Polyphosphoryl chloride

The simplest way to prepare the polyphosphoryl chlorides is by combining proper proportions of ortho phosphoryl chloride with phosphorous pentoxide.

$$(n+2)\ POCl_3 + (n-1)P_2O_5\ 3P_nO_{2n-1}\ Cl_{n+2}$$

The polyphosphoryl chlorides are strongly acidic, being rather similar in properties to the chlorides of organic acids such as acetyl chloride. The polyphosphorylchlorides in the presence of moist air, fumes to give hydrogen chloride. They are good chlorinating agents for organic reaction and found commercial use in this area.

4-10 FERTILIZERS

In order to keep up the soil fertility and to get good yeilds of various crops it is necessary to add the essential nutrients to the plants, plants require along with carbon, hydrogen and oxygen, elements like phosphorous, nitrogen, sulphur, potassium, calcium, magnesium, and iron, besides tracer metals like boron, copper, cobalt, zinc, manganese and molybdenum which are essential as the food of the plants. Carbon, hydrogen and oxygen can be had from nature in the form of air and water. However, the other essential elements had to be supplied to the plants through soil. As long as the soil is virgin, it will not require any aritificial supply of these elements, but after continuous cropping for years together, the soil looses all its natural nutrients and the need for the essential elements become a necessity. The most essential plant nutrients are nitrogen, phosphorous and potassium. These nutrients are supplied to the soil through manures and fertilizers. Manures are natural nutrients which are usually biological products like animal dung, blood, bones, fishes crap, animal excreta, guano, horns, decayed leaves, bark, seeds and ashes, whereas fertilizers are chemical compounds usually containing nitrogen, phosphorous and potassium. Fertilizers are classified under three heads.

(1) Nitrogenous fertilizers

(2) Phosphatic fertilizers

(3) Potash fertilizers

4-10-1 Nitrogen Fertilizers

The nitrogenous compounds used as fertilizers are urea, ammonium sulphate, calcium ammonium nitrates, ammonium nitrate, calcium cyanamide, sodium nitrate etc. The highest nitrogen content amongst these is in urea with 42-46% of it. The need of the nitrogenous fertilizers is due to the loss of nitrogen content in the soil, it has to be brought in equilibrium as seen in the nitrogen cycle in the previous section. In this section a few important fertilizers are discussed.

(i) Urea : Urea is the best fertilizer as it liberates ammonia immediately on hydrolysis

$$(NH_2)_2CO + H_2O \longrightarrow 2NH_3 + CO_2$$

carbondioxide is not harmful as it leaves the soil after getting liberated and is absorped by the plant itself through its leaves. The advantage in using urea is it can be stored with no risk of any hazards, but cannot be kept long as it decomposes even at low temperatures. Urea can be prepared on a large scale by combining ammonia with carbondioxide under high pressure, ammonium carbonate formed loses two molecules of water and gets converted into urea.

$$2NH_3 + CO_2 + H_2O \longrightarrow (NH_4)_2CO_3$$

$$(NH_4)_2CO_3 \longrightarrow (NH_2)_2CO + 2H_2O$$

By Haber's process urea is obtained on a large scale by reacting liquid ammonia with gaseous carbondioxide at 125°C and under 220 atomspheres of pressure, ammonium carbonate is formed, which decomposes to give urea.

$$2NH_3 + CO_2 + H_2O \longrightarrow H_4NCOOONH_4$$

$$NH_4COOONH_4 \longrightarrow NH_2CONH_2 + 2H_2O$$

(ii) Ammonium sulphate: is by and large another very important nitrogen fertilizer. Calcium sulphate is treated with aqueous ammonia and carbondioxide. Calcium carbonate is filtered off and ammonium sulphate is crystallized out from the filterate by evaporation.

$$2NH_3 + CO_2 + H_2O + CaSO_4 \longrightarrow (NH_4)_2SO_4 + CaCO_3$$

Ammonium sulphate is also obtained as a by product in the fractional distillation of petroleum products. As a fertilizer ammonium sulphate is used along with super phsophate. The excess use of ammonium sulphate causes adverse effects on the soil as it becomes acidic. Usually this is used only on the germination of seeds and neither before nor after germination.

(iii) Ammonium nitrate and calcium ammonium nitrate : which is also called otherwise, nitro limestone or CAN. Ammonium nitrate being explosive is mixed with lime and pellets of the fertilizer are made which are then protected from moisture by a coating of soap stone. The fertilizer is manufactured by treating ammonia with nitric acid. Both these reactants are byproducts of Ostwald's process.

$$4NH_3 + 5O_2 \xrightarrow{Pt} 4NO + 6H_2O$$

$$4NO + 2O_2 \xrightarrow{800^\circ C} 4NO_2$$

$$4NO_2 + 2H_2O + O_2 \longrightarrow 4HNO_3$$

$$NH_3 + HNO_3 \longrightarrow NH_4NO_3$$

Calcium ammonium nitrate is better fertilizer than ammonium sulphate as calcium prevents the soil acidity and is more soluble.

(iv) Calcium cyanamide : Calcium cyanamide, $Ca(CN)_2$, is produced by heating calcium carbide in a current of purified nitrogen and the mass obtained is pulverised, then is treated with water to decompose any unreacted carbide and to hydrate the lime. The commercial names of this fertilizer are lime nitrogen, nitrolium and nitrolime. Calcium cyanamide is used as a straight fertilizer or in a fixed form. Very less quantities of this fertilizer are enough to fertilize large land masses. It has to be mixed with the soil much before plantation, then sufficient time is given to the fertilizer to decompose before planting. The cyanamide converts itself to urea, ammonia and finally to nitrates, which are essential for the plants.

4-10-2 Phosphatic Fertilizers

The fertilizers calssified under this type are (i) calcium superphosphate and (ii) Phosphatic slag.

(i) Calcium super phosphate or super phosphate of lime is obtained by treating phosphate rock with sulphuric acid. The product formed contains about 14-20% of P_2O_5.

$$Ca_3(PO_4)_2 + 2H_2SO_4 + 3H_2O \longrightarrow CaH_4(PO_4)_2 H_2O. 2CaSO_4 2H_2O$$

it is a dicalcium salt. It may at times contain impurities like calcium fluoride which reacts with sulphuric acid to form HF, in turn HF reacts with SiO_2present as another impurity which forms SiF_4, this gets hydrolysed to hydrofluosilicic acid, because of all these unrequired impurities and their reactions, the fertilizer becomes undesirable in many instances as sulphuric acid, hydrogen fluoride, hydrofluosilicic acid causes damage to the soil. Hence purified samples should be used for better results. Super phosphate can be used for any type of crop and for all sorts of soils. It is mixed with the soil before ploughing. The super phosphate decomposes slowly when compared to the instanteneous decomposing nitrogenous fertilizers.

(ii) The phosphatic slag is a by product of the steel industry. As an impurity, phosphorous present in the ore forms a slag on the addition of a flux such as lime. They combine together to form $Ca_3(PO_4)_2$ which floats on the surface of the molten metal in the Blast furnance. This slag rich in phosphorous content is used as a fertilizer. It is used in the alkaline soils, and not in acidic soils because of its basic character.

4-10-3 Potash Fertilizers

The potash fertilizers are potassium nitrate, potassium chloride and potassium sulphate. The potassium fertilizers are used for the tobacco crop which makes the leaves more resistant to heat, and gives high melting ash. Potassium is always present along with little of magnesium, hence when the potash fertilizers are used indirectly magnesium is also supplied to the soil. Magnesium being a very important constituent of chlorophyll in the green plants, hence potash fertilizers are used for the growth of green vegetables and fruits, specially citrus which contains potassium to a large extent in them.

Another set of fertilizers which are in use are the mixed fertilizers. The mixed fertilizers are nitrogenous, phosphatic and the potash fertilizers mixed together in different proportions depending upon the type of soil, and the plant. These fertilizers give nitrogen, phosphorous as well as potassium to the soil or the plants. Commercially it is called as N.P.K. fertilizers.

The fertilizer industry in India, is one of the major industries in the country. Fertilizers are in great demand to bring about the green revolution in the country. The total requirement of the fertilizers is such, that, the demand cannot be fulfilled even though there are several units all over the country producing tons of fertilizers.

4-11 MULTIPLE-CHOICE QUESTIONS

4-11-1 The only element with maximum number of oxidation states

a) Phosphorous b) Nitrogen c) Arsenic d) Antimony

4-11-2 The element with stable oxidation states of +1 & +2 is

a) Phosphorous b) Nitrogen c) Arsenic d) Antimony

4-11-3 The smallest trivalent ion (1.2Å) is of

a) P^{3+} b) N^{3+} c) Bi^{3+} d) As^{3+}

4-11-4 The only element found to be monoatomic is

a) Phosphorous b) Nitrogen c) Bismuth d) Arsenic

4-11-5 The products obtained on hydrolysis of PX_3 is

a) H_3PO_3 + HX b) H_3PO_3 + P_2O_5

c) HX + HPO_3 d) H_3PO_4 + HX

4-11-6 The products obtained on hydrolysis of PX_5 is

a) H_3PO_3 + HX b) H_3PO_4 + HX

c) HPO_3 + HX d) H_3PO_3 + P_2O_3

4-11-7 The oxidation product of NH_3 is

a) NO b) N_2O c) NO_2 d) N_2O_3

4-11-8 NO on reacting with red hot copper produces

a) CuO + NO_2 b) Cu_2O + N_2O c) Cu_2O + N_2 d) CuO + N_2

4-11-9 The compounds of NO are known as

a) Nitrosyls b) Nitros c) Nitritos d) nitratos

4-11-10 P_4O_6 is an acid anhydride of

a) HPO_3 b) H_3PO_3 c) H_3PO_4 d) $H_4P_2O_7$

4-11-11 P_4O_{10} is an acid anhydride of

a) HPO_3 b) H_3PO_3 c) H_3PO_4 d) HuP_2O_7

4-11-12 $H_5P_3O_{10}$ is

a) Peroxo phosphoric acid b) Meta phosphoric acid

c) Tripoly phosphoric acid d) Hypo phosphoric acid

4-11-13 Pernitric acid is

a) $H_2N_2O_3$ b) HNO_4 c) H_2NO_2 d) $H_2N_2O_2$

4-11-14 Hyponitrous acid is

a) $H_2N_2O_3$ b) HNO_4 c) H_2NO_2 d) $H_2N_2O_2$

4-11-15 Hydronitrous acid is

a) $H_2N_2O_3$ b) HNO_4 c) H_2NO_2 d) $H_2N_2O_2$

4-11-16 Hyponitric acid is

a) $H_2N_2O_3$ b) HNO_4 c) N_2NO_2 d) $H_2N_2O_2$

4-11-17 Pyrophosphoric acid is

a) HPO_3 b) $H_4P_2O_7$ c) $H_4P_2O_6$ d) H_3PO_2

4-11-18 Hypophosphoric acid is

a) HPO_3 b) $H_4P_2O_7$ c) $H_4P_2O_6$ d) H_3PO_2

4-11-19 Hypophosphorous acid is

a) HPO_3 b) $H_4P_2O_7$ c) $H_4P_2O_6$ d) H_3PO_2

4-11-20 Meta phosphoric acid is

a) HPO_3 b) $H_4P_2O_7$ c) $H_4P_2O_6$ d) H_3PO_5

4-11-21 Conc. HNO_3 reacts with Cu to give

a) NO b) NO_2 c) N_2O_3 d) N_2O_5

4-11-22 Dilute HNO_3 reacts with Cu to give

a) NO b) NO_2 c) N_2O_3 d) N_2O_5

4-11-23 Zinc reacts with dil HNO_3 in cold to give

a) N_2O and N_2 b) NH_4NO_3

c) NH_4NO_3 and N_2 d) N_2O and N_2

4-11-24 Zinc reacts with conc. HNO_3 in cold to give

a) N_2O and N_2 b) NH_4NO_3 & N_2 c) NH_4NO_3 (d) NO_2

4-11-25 H_3PO_2 is a

a) dibasic acid b) tribasic acid c) Inert d) monobasic acid

4-11-26 Chemical with which NaOH reacts to produce NH_3

a) NO_2 b) NO c) NH_4Cl d) $Ca(OH)_2$

4-11-27 The source of pure N_2 in the manufacture of ammonia is

a) Fuel gas b) Producer gas

c) Water gas d) Ordinary diesel

4-11-28 Oxidation of ammonia on burning in oxygen produces

a) N_2 b) N_2O c) NO_2 d) N_2O_3

4-11-29 Excess of chlorine when reacted with NH_3 produces

a) N_2 b) NH_4Cl c) NCl_3 d) NCl_3 + HCl

4-11-30 Hydroxylamine is

a) NH_4OH b) N_2H_4 c) NH_2OH d) NH_2Cl

4-11-31 N_2H_4 reacts with NCl_3 to produce HCl alongwith

a) NH_3 b) HN_3 c) NH_2OH d) N_2

4-11-32 The ionic product of ammonia is

a) 10^{-14} b) 10^{-33} c) 10^{-44} d) 10^{-58}

4-11-33 Urea is a

a) Nitrogenous b) Phosphatic c) Potasium d) Uric fertilizer

4-11-34 The phosphatic slag is

a) $CaNa_2PO_4$ b) $Ca_3(PO_4)_2$ c) $CaH_4\ (PO_4)_2$ (d) Na_3PO_4

4-11-35 S_4N_4 reacts with HI to give

a) H_2 b) N_2 c) S_8 d) I_2

4-11-36 Phosphonitrilic chloride is

a) $N_3P_3Cl_6$ b) $N_3S_3Cl_6$ c) N_3PCl_3 d) $N_3P_3Cl_3$

4-12 SHORT AND ESSAY-TYPE QUESTIONS

4-12-1 Why is the VB group of the periodic table referred to as a typical family. Elucidate your answer suitably with logical discussions.

4-12-2 Give the electronic configuration of N, O, As, Sb and Bi. (O.U., Oct. 1985)

4-12-3 Compare the physical and chemical properties of nitrogen family on the basis of their electronic configuration. (S.K.U., Apr. 1988)

4-12-4 Write short notes on allotropic forms of phosphorous. (K.U., Oct. 1990)

4-12-5 Explain inert pair effect with two suitable examples. (O.U., Apr. 1988)

4-12-6 Account for the following: the first ionization potential of nitrogen is more than oxygen. (O.U., Mar. 1989)

4-12-7 Illustrate the anamalous nature of nitrogen in group VB. (O.U., Mar. 1990)

4-12-8 Explain the variations in the properties of the elements N, P, As, Sb and Bi with special reference to hydrides and halides. (O.U., Apr. 1988 & Oct. 1989)

4-12-9 Discuss the trends in the following properties of the elements of the nitrogen family
(i) Ionization potential
(ii) Metallic character
(iii) Oxidation states. (O.U., Oct. 1985)

4-12-10 Compare the hydrides formed by the elements of group VB. (O.U., Oct. 1985)

4-12-11 Explain why

(a) Ionization potential of nitrogen is higher than O_2.

(b) N_2 forms only a trichloride but not a pentachloride. (O.U., Mar. 1986)

4-12-12 Give a comparative account of the chemical behaviour of nitrogen and phosphorous with reference to the properties of their

(a) Electronic configurations (b) Hydrides
(c) Oxyacids (d) Oxides
(O.U., Oct. 1984 & Apr. 1985)

4-12-13 Compare and contrast the chemical behaviour of nitrogen & phosphorous with reference to the properties of their
(a) Hydrides
(b) Oxides
(c) Halies.

4-12-14 How are the following changes brought about
(i) $Ca_3(PO_4)_2$ to $Ca(H_2PO_4)_2$
(ii) NaCl to $Na_2\ CO_3$
(iii) HF to F_2 (O.U., Oct. 1988)

4-12-15 Why nitrogen is least reactive in V group elements? (A.U., Apr. 1988)

4-12-16 Explain the anamolous nature of nitrogen. (A.U., Apr., Dec. 1989)

4-12-17 In what respect nitrogen differs from the rest of the VB group elements?

4-12-18 Give any one method of preparing PCl_5. Give equations to show its reaction with water. Write its structure. Why is it not possible to obtain PI_5? (O.U., Oct. 1988)

4-12-19 Give an account of the preparation and properties of the oxides of nitrogen. Discuss their structure.

4-12-20 Give a brief account of the chemistry of the oxides of nitrogen.

4-12-21 Make a comparative study of the nature of oxides formed by the elements of group VB.
(O.U., Mar. 1989)

4-12-22 Write notes on the oxides & oxyhalides of phosphorous.

4-12-23 Describe the manufacture of nitric acid by Ostwald's process with the physico chemical principles involved in it. (O.U., Apr. 1988)

4-12-24 Write equations for the action of Conc. HNO_3 on I_2 and zinc. (O.U., Apr. 1988)

4-12-25 Describe in detail the reactivity of the following acids
(a) Nitrous acid
(b) Nitric acid (K.U., Apr., 1989)

4-12-26 Briefly compare & contrast the oxyacids of nitrogen and phosphorous.

4-12-27 Describe the preparation, properties and structures of oxyacids of phosphorous.

4-12-28 Explain why concentrated nitric acid turns yellow in sunlight? (A.U., Jun. 1990)

4-12-29 Write a brief note on oxyacids of P. (O.U., Oct. 1987)

4-12-30 Write equations to show the reactions of hot Conc. HNO_3 with
(a) Iodine (b) Copper
(c) H_2S (O.U., Oct. 1985)

4-12-31 Give the outlines of the Ostwald's process for converting ammonia into nitric acid. (O.U., Apr. 1985)

4-12-32 Write equations to show what happens when
(i) Cu turnings are heated with Conc. HNO_3
(ii) NH_3 is passed into a solution of $CuSO_4$. (O.U., Jun. 1985)

4-12-33 Give an account of the preparation and structure of various oxyacids of phosphorous. (O.U., Jun. 1984)

4-12-34 Write equations for the action of conc. HNO_3 on iron & zinc. (O.U., Nov. 1986)

4-12-35 Explain why H_3PO_3 is dibasic and NO_2 is brown at higher temperatures. (O.U., Apr. 1985)

4-12-36 (a) Discuss the physicochemical principle involved and state the necessary conditions for
(i) Synthesis
(ii) Oxidation of ammonia into nitric acid
(b) Write balanced equations for the reaction of
(i) Conc. HNO_3 with sulphur
(ii) dilute HNO_3 with zinc. (O.U., Mar. 1986)

4-12-37 Write a note on hydrides of V group. (O.U., Nov. 1985, Nov. 86)

4-12-38 Write a note on hydrides of V group elements. (A.U., Apr. 1988)

4-12-39 Compare the properties of the hydrides formed by the group VB (nitrogen family) elements. (O.U., Mar. 1990)

4-12-40 Describe the Haber's process for the manufacture of ammonia emphasising the physicochemical principles involved in the reaction. (O.U., Apr. 1985, Nov. 85 & Oct. 87)

4-12-41 What happens when chlorine reacts with excess of ammonia and ammonia reacts with excess of chlorine? Give equations. (O.U., Nov. 1986)

4-12-42 Compare the properties of ammonia with those of phosphine. (O.U., mar. 1989)

4-12-43 Starting the appropriatte conditions, describe what happens when ammonia reacts with
(a) Oxygen
(b) Silver nitriate
(c) Sodium
(d) Chlorine

4-12-44 Discuss the behaviour of ammonia as a solvent. (A.U., Apr. 1988 & A.U., Apr. 1989)

4-12-45 Discuss the various types of reactions taking place in liquid ammonia as a solvent. (N.U., Apr. 1989)

4-12-46 What are the two factors which favour liquid ammonia as a solvent?

4-12-47 Write a note on ammonia, as a non aqueous solvent.

4-12-48 Discuss the fixation of nitrogen in the atmosphere.

4-12-49 Give equations to show that hydroxyl amine and hydrazine behave as oxidising and reducing agents.

4-12-50 Which metals from azides? Describe the preparation, structure and uses of a typical azide. (O.U., Mar. 1989)

4-12-51 Describe the preparation and any two uses of superphosphate of lime. (O.U., Mar. 1985 & Nov. 86 & Mar. 90)

4-12-52 How is the fertilizer urea manufactured? (O.U., Mar. 1989)

4-12-53 Describe the large scale preparation of the following fertilizers
(a) Ammonium sulphate
(b) Calcium cyanamide
(c) Super phosphate of lime

4-12-54 Mention the uses of compounds of nitrogen and phsophorous as fertilizers. (A.U., Apr. 1989)

4-12-55 Discuss the bonding & structure of tetrasulphur tetranitride.

4-12-56 How is S_4N_4 prepared. Write down the reactions of S_4N_4 with dilute and concentrated alkali and halogen acids.

4-12-57 Discuss the bonding and structure in phosphonitrilic chloride.

4-12-58 Write notes on the preparation of polyphosphoryl chloride & phosphonitrilic chloride.

ANSWERS

4-11-1 B	**4-11-2** B	**4-11-3** C	**4-11-4** C	**4-11-5** A	**4-11-6** B
4-11-7 A	**4-11-8** D	**4-11-9** A	**4-11-10** B	**4-11-11** C	**4-11-12** C
4-11-13 B	**4-11-14** D	**4-11-15** C	**4-11-16** A	**4-11-17** B	**4-11-18** C
4-11-19 D	**4-11-20** A	**4-11-21** B	**4-11-22** A	**4-11-23** A	**4-11-24** C
4-11-25 D	**4-11-26** C	**4-11-27** B	**4-11-28** A	**4-11-29** D	**4-11-30** C
4-11-31 B	**4-11-32** B	**4-11-33** A	**4-11-34** B	**4-11-35** D	**4-11-36** A

5

THE ELEMENTS OF GROUP VI B *THE CHALCOGENS*

					He
B	C	N	O	F	Ne
Al	Si	P	S	Cl	Ar
Ga	*Ge*	As	*Se*	Br	Kr
In	Sn	Sb	Te	I	Xe
Tl	Pb	Bi	Po	At	Rn

5-1 INTRODUCTION

The group VI B elements, oxygen (O), sulphur (S), selenium(Se) tellurium(Te) and polonium(Po) are commonly called **Chalcogens** (ore forming) as most of these elements are present in a combined state in different mineral ores. Oxygen is a gas and is the most abundant element of the earth's crust present up to 47% in the atmosphere. Sulphur in the combined state is present in earth's crust up to 0.052%. It is also present in free state. Se, Te and Po are trace elements and are present to an extent of 10^{-7} to 10^{-14}%, of the earth's crust.

Oxygen was first isolated by Scheele in 1772 and was discovered independently by Priestley in 1774. Sulphur was perhaps known to the ancient Egyptians and Babylonians. It was well known to the Greeks and Romans. The Phlogistonists considered it to be a compound of phlogiston and sulphuric acid. It was Lavoisier who recognized it as an element in 1777. Selenium was discovered in 1817, and tellurium in 1832 by Berzellius. Polonium was discovered in 1998 by Madame Curie by precipitation of pitch blend with hydrogen sulphide.

Non metallic character is strongest in oxygen and sulphur which are insulators, selenium and tellurium are semiconductors and polonium is metallic. As all these elements can enter into chemical combination and complete their octets by gaining two electrons to form a negatively charged divalent ion, they are all grouped together in the periodic table, although Po is an exception. Their compounds are practically covalent and as with other non metallic elements the concept of oxidation state has only formal of significance. Oxygen exhibits only – 2 state, but sulphur, selenium, tellurium and polonium have vacant d-orbitals that can be utilized without too great an energy change, hence are able to exhibit +2, +4 and +6 valency states (It is discussed further in the section 5-2).

5-2 ELECTRONIC CONFIGURATION

All the elements of the group have two electrons less than an octet in their outermost shell.

Oxygen (8) (O) : $1s^2\ 2s^2\ 2p^4$

Sulphur (16) (S) : $1s^2\ 2s^2\ 2p^6\ 3s^2\ 3p^4$

Selenium (34) (Se) : $1s^2\ 2s^2\ 2p^6\ 3s^2\ 3p^6 4s^2\ 3d^{10}\ 4p^4$

Tellurium (52) (Te) : $1s^2\ 2s^2 2p^6 3s^2\ 3p^6\ 4s^2\ 3d^{10}\ 4p^6\ 5s^2\ 4d^{10}\ 5p^4$

Polonium (84) (Po) : $1s^2\ 2s^2\ 2p^6\ 3s^2 3p^6\ 4s^2\ 3d^{10}\ 4p^6\ 5s^2\ 4d^{10}\ 5p^6\ 6s^2\ 5d^{10}\ 6p^4$

It is clear from the electronic configuration, that the elements S, Se, Te and Po have d orbitals available for bonding, and they can form four to six bonds by unpairing of electrons. In the ground state, the two unpaired electrons present are involved in the bond formation and expecting sp^3 hybridization they assume a tetrahedral structure with two positions being occupied by the lone pairs of electrons, as in H_2S.

In the ground state

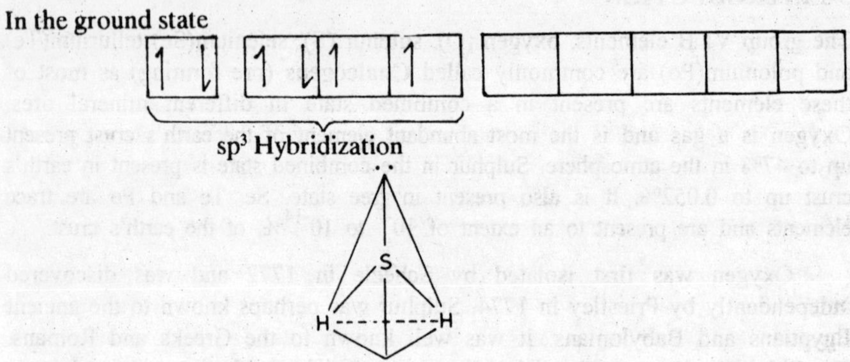

In the first excited state, four unpaired electrons are available by the excitation of a 'p' electron into the d-orbital, thus forming four bonds expecting sp^3d hybridization with a trigonal bipyramidal structure with one position occupied by one pair of electrons, as in $TeCl_4$

In the first excited state

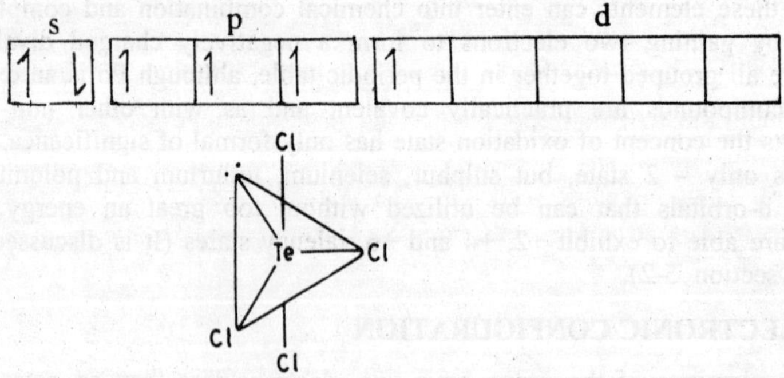

On further excitation of these atoms six unpaired electrons are available forming six bonds expecting sp^3d^2 hybridization with an octahedral structure. such as in SeF_6

In the second excited state

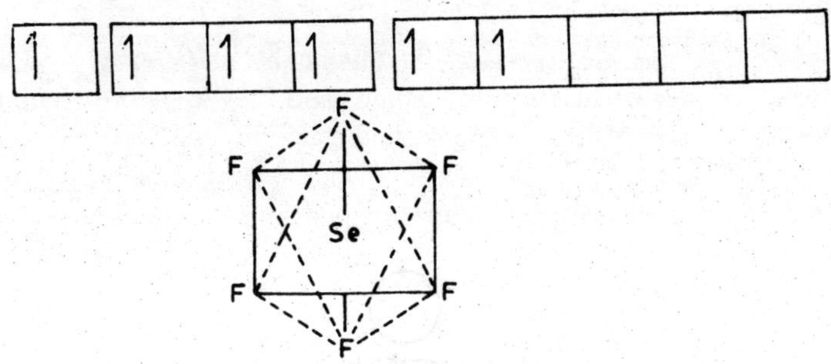

The higher oxidation state compounds are all very volatile as they are all covalent.

5-3 ATOMICITY

Oxygen exists in the form of discrete molecules with a double bond uniting two oxygen automs together, O = O (O_2). It is a paramagnetic molecule in the gaseous, liquid and as well as in solid states. It has a high dissociation energy of 117 kcals mole^{-1}. The valence bond theory in its simplest form would predict the electronic structure to be :Ö = Ö:, which, while accounting for the strong bond, fails to account for the paramagnetism. Alternatively we might write the electronic structure as :O ⩦ O: to account for the presence of two unpaired electrons. The molecular orbital treatment shows the configuration of the O_2 to be $1s^2$, $\sigma{*}1s^2$, $2s^2$, $\sigma{*}2s^2$, $\sigma\ 2p^2{}_x$, $\sigma 2p_y^2$, $2p_z^2$, $\pi^*2p_y^1$, $\pi^*2p_z^1$, which confirms the presence of two unpaired electrons. The molecular orbital diagram of oxygen molecule showing paramagnetism is seen in fig. 5-1.

Sulphur and selenium both exists as octamolecular species S_8 and Se_8 at room temperature. The molecules have puckered ring structures. Tellurium has a spiral chain network with Te - Te distance 2.84 Å and Polonium has a cubic crystalline structure with Po - Po distance being 3.35 Å.

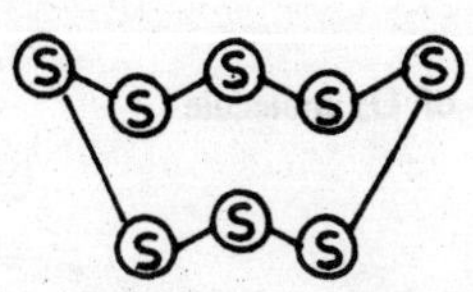

S - S = 2.06 Å
S - Ŝ - S = 107° 54′

Dihedral angle = 98.9°
(between two layers)

Se - Se = 2.33 Å
Se - $\widehat{Se}$ - Se = 105.7°
Dihedral angle = 101°.3'
(between two layers)

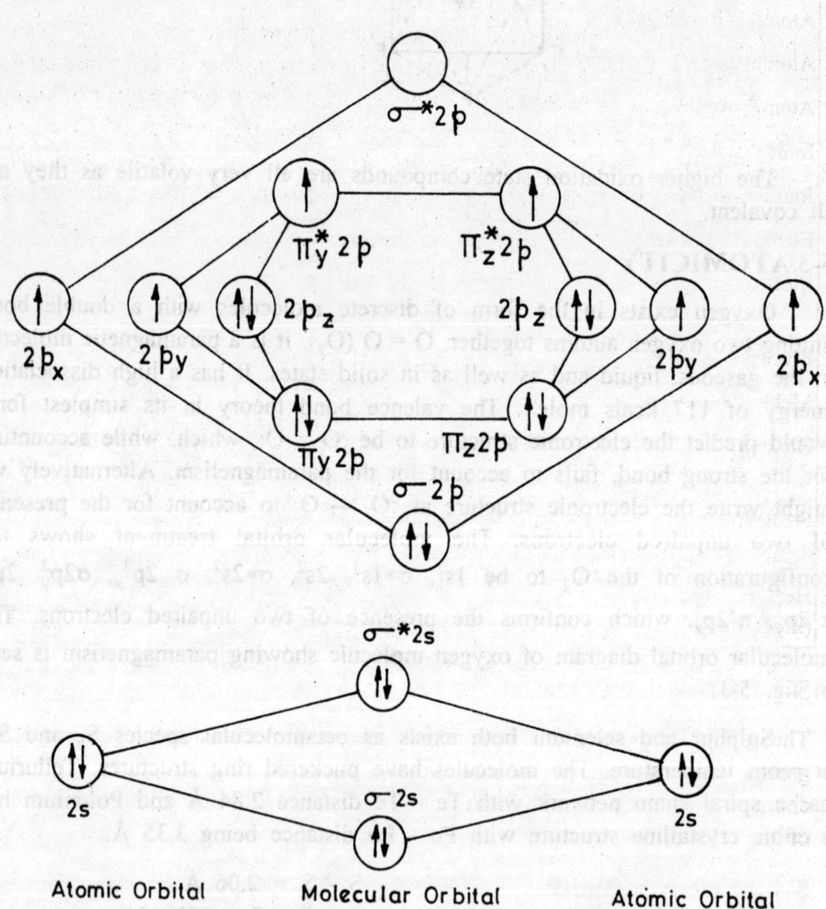

Fig. 5-1 Molecular Orbital diagram of O_2 molecule

5-4 PHYSICAL PROPERTIES

From the table 5-1 it is observed that there is a regular variation in their physical properties, on descending down the group from oxygen to polonium. As expected the atomic volume, radii, density, ionic radii, melting and boiling points, increase gradually with the increase in atomic number, however the melting point of polonium is rather low and there is no explanation to it.

Table 5-1 Physical Data

Property	*O*	*S*	*Se*	*Te*	*Po*
Atomic number	8	16	34	52	84
Atomic weight	15.99	32.064	78.96	127.60	210
Atomic volume (cc)	14.1	15.5	16.5	20.5	22.7
Atomic density (g cm^3)	1.14	2.07	4.79	6.24	9.2
Atomic Radii A	0.73	1.09	1.16	1.35	1.52
Ionic Radii A	1.30	1.84	1.98	2.21	2.3
Ionization potentials (eV)	13.61	10.36	9.75	9.01	8.41
Electron affinity (eV)	1.48	2.07	-	-	-
Electron affinity (M^{2-})	(-7.3)	(-3.4)	(-4.2)	-	-
Electronegativity	3.5	2.5	2.4	2.1	1.76
Melting point (°C)	218.9	119	217.4	449.84	254
Boiling Point (°C)	182.96	444.60	684.8	1390	962
Oxidation states	-2	-2, +2 +4, +6	-2, +2, +4 +6	-2, +2, +4, +6	+2, +4
Heats of atomization (kcals $mole^{-1}$)	59.2	56.9	49.6	46.7	-

5-4-1 Ionization Potential

The ionization potentials of the family decrease systematically going down the group. The chalcogens can lose either 2, 4 or 6 electrons to form positively charged ions even though high energy requirements are involved in the loss of the electrons.

5-4-2 Electronaffinity

The electron affinity values given in the physical data table 5- 1, show that the energy is released on the addition of first electron to oxygen and sulphur. The electron affinity values given in the parenthesis for O, S and Se are the energies released on the addition of the second electron to give dinegative ions. The high repulsive forces operating in the addition of an electron to the metastable uninegative chalcogenide ion to give dinegative ions, arc indicated by the negative sign of electron affinity. The electron affinities are dependent on size and effective nuclear charge, and is calculated with the help of the Born Haber cycle.

5-4-3 Electronegativity

The electronegativity values of the family show that the elements are highly electronegative except Po, which being a metal shows an electropositive character. the electronegativity decreases down the group with oxygen having a value of 3.5, intermediate of fluorine and chlorine values. Hence all the metallic oxides of chalcogenides are ionic and their ionicity is in between the metalic flourides and chlorides. However due to decrease in electronegativity on descending the group, only few sulphides, selenides and tellurides of most electropositive metals have more than 50% ionic character. Thus, except polonium all the elements of this family can exhibit an oxidation state of –2 in their binary compounds with metals.

5-4-4 Polymorphism

The Chalcogens exhibit polymorphism or they exist in more than one allotropic form. Oxygen occurs in two non metallic polymorphs as diatomic oxygen (O_2) and triatomic ozone (O_3). Sulphur exists in two crystalline forms rhombic or α-sulphur and monoclinic or β-sulphur. The former is stable at room temperature and the later is stable above 96ºC. Plastic sulphur or amorphous sulphur and milk of sulphur are the other two polymorphs of sulphur. All these forms of sulphur are non metallic. Selenium exists in two forms grey metallic selenium and red nonmetallic selenium. The former is less stable than the later nonmetallic form of selenium. In tellurium, which also forms two grey allotropes viz., metallic and nonmetallic, but the stability gets reversed. Polonium exists in two allotropic forms α and β, both of them having exclusively metallic character.

5-4-5 Oxidation States

All the elements of the family exhibit the oxidation state of –2 except polonium. Oxygen exhibits only –2 oxidation state. It shows no positive oxidation state except when it combines with fluorine forming OF_2, O_2F_2, OF, and O_2F_3 of which the later two compounds are highly unstable. But as we move down the group positive oxidation states become more stable. The electronic configuration of ns^2np^4 suggests two positive states of +4 and +6. The +4 state is the result of loss of all the four p electrons while the +6 state is caused when both s and p electrons get involved. Excepting oxygen all the other elements of the family have vacant d-orbitals, in the ground state there are only two unpaired electrons which permit the formation of two bonds. In the excited state however the paired up p-electrons get unpaired and get excited to go into the vacant d-orbitals of the same shell, accordingly for four unpaired electrons and six unpaired electrons available for chemical bonding, exhibiting +4 and +6 oxidation states. The +4 state is exhibited in the compounds of S, Se, Te with oxygen and fluorine.

5-4-6 Catenation

Catenation is the phenomenon of formation of chains with repetition of the same atom. Oxygen and sulphur exhibit the property of catenation. The catenation is more prominent in sulphur than oxygen. The other members of this family does not have any tendency to catenate. The catenated compounds, called polyoxides and polysulphides, generally contain two to six O or S atoms and are represented as

$$H-O-O-H,\ H-S-S-H,\ H-S-S-S-H,\ H-S-S-S-S-H.$$

Dichlorosulphanes, represented by the general formula S_xCl_2, where x can take a value as high as 100 are also known.

5-4-7 Anomalous Nature of Oxygen

Oxygen differs from the rest of the chalcogens due to its small size, very high electronegativity, and the absence of d orbitals in the valency shell.

Because of its high electronegativity oxygen is more ionic in its compounds than the rest of the elements of the family. Oxygen is a diatomic molecule, S and Se exist as octa atomic molecules, while Te and Po are only monoatomic.

Oxygen forms hydrogen bonding because of its high electronegativity hence H_2O is a liquid while H_2S, H_2Se, H_2Te are all gases with very low boiling points.

Oxygen is paramagnetic in all the three phases, whereas the other chalcogens are not.

Oxygen molecule is highly stable with a bond energy of 117 kcals $mole^{-1}$. As a result of this oxygen requires high energy of activation to form oxides, hence, oxygen combines with other elements only at high temperatures. That is why, there are less number of oxide ores in nature than sulphide ores.

5-5 CHEMICAL REACTIVITY OF CHALCOGENS

Although it is found that oxygen differs chemically from the rest of the elements of the family, there are many primary reactions which show a similarity of O_2 with S, Se, Te & Po. The chemistry of polonium is not studied extensively because of its radioactive nature. Under this heading the chemistry of these elements with respect to their hydrides, halides, oxides and the oxyacids of chalcogens and peroxyacids of sulphur is being described.

5-5-1 Hydrides

All the chalcogens form binary hydrides of the type H_2Ch (H_2O, H_2O, H_2S, H_2Se,H_2Te & H_2Po). All the chalcogen hydrides are gaseous, except H_2O which is also present as a solid as well as a liquid at ordinary temperatures.

Being gaseous their volatility decreases going down the group. H_2O is the most stable of all the hydrides and the stability also decreases from H_2O to H_2Po.

Water is prepared by passing a mixture of hydrogen and oxygen through, an electric arc or subjecting them to high temperatures. The hydrogen sulphide, hydrogen selenide and hydrogen telluride are obtained by the action of mineral acids (H_2SO_4 or HCl) on corresponding metal chalcogenides.

$$FeS + 2HCl \longrightarrow FeCl_2 + H_2S$$

$$Na_2Se + H_2SO_4 \longrightarrow Na_2SO_4 + H_2Se$$

Hydrogen polonide is obtained by dissolving mangesium foil plated with Po in 0.2 M HCl.

All the hydrides are awfully smelling except H_2O which is colourless, odourless and tasteless. H_2O is a high boiling liquid because of the hydrogen bonding present between its molecules, whereas the other hydrides boil at a very low temperature.

H_2O	100°C
H_2S	–60.75°C
H_2Se	–41.5°C
H_2Te	–1.8°C

All the hydrides dissociate into H^+/HCh^- and $2H^+/Ch^{2-}$. The degree of dissociation increases down the group, showing that all the hydrides are acidic in character and the acidic character increases going down the group. Thus the values of dissociation constant at 25°C are

H_2O	1.0×10^{-14}	H_2S	1×10^{-7}
H_2Se	1.7×10^{-4}	H_2Te	2.3×10^{-3}

H_2O dissociates only at 2000°C while H_2Te decomposes at ordinary temperature, the thermal stability decreases from H_2O to H_2Te. All the hydrides are strong reducing agents except water and the reducing character increases on moving from H_2S to H_2Te. The $H\hat{Ch}H$ bond angle decreases from H_2O to H_2Te present in the tetrahedron structure. Lesser the angle more will be the distortion and hence instability in the molecule.

	Bond angle ($H\hat{Ch}H$)	Bond length (H-Ch)
H_2O	104.5°	0.95 Å
H_2S	92.5°	1.30 Å
H_2Se	90.0°	1.45 Å
H_2Te	89.0°	1.72 Å

Oxygen and sulphur also forms polyoxides and polysulphides with hydrogen, which are less stable than the usual hydrides. Oxygen forms H_2O_2 (H-O-O-H) and sulphur forms H_2S_2 (H-S-S-H), H_2S_3 (H-S- S-S-H) and H_2S_4 (H-S-S-S-S-H). Although H_2O_2 and H_2S_2 have similar structures they differ in their chemical behaviour.

H_2O_2, hydrogen peroxide is obtained by tthe reaction of an alkali metal or alkaline earth metal peroxide with a dilute acid.

$$Na_2O_2 + 2HCl \longrightarrow 2NaCl + H_2O_2$$

$$BaO_2 + H_2SO_4 \longrightarrow BaSO_4 + H_2O_2$$

It is also obtained by the electrolysis of sulphate to give peroxydisulphuric acid ($H_2S_2O_8$), which oxidizes water into peroxide.

$$2SO_4^{-} + H_2O \xrightarrow{\text{electrolysis}} H_2S_2O_8 + 2e^{-}$$

$$H_2S_2O_8 + H_2O \longrightarrow H_2SO_5 + H_2SO_4$$

$$H_2SO_5 + H_2O \longrightarrow H_2SO_4 + H_2O_2$$

H_2O_2 is a strong oxidizing agent and is fairly stable.

H_2S_2 can be prepared in a reaction similar to H_2O_2.

$$Na_2S_2 + 2HCl \longrightarrow 2NaCl + H_2S_2$$

But neither it resembles in its properties with H_2O_2 nor in its stability. There are no reports on the formation of H_2Se_2 and H_2Te_2.

5-5-2 Halides

Sulphur and selenium does not combine with iodine and polonium has no reaction with fluorine, but all the other possible halides of the family are known to exist. Since fluorine is more electronegative than oxygen, their compounds are called fluorides, and as oxygen is more electronegative than chlorine, bromine and iodine, the compounds of oxygen with these halides are called oxides.

While the compounds with chlorine, bromine and iodine exhibit an oxidation state of maximum +4 for the chalcogens, fluorine helps them in attaining a maximum oxidation state of +6.

The halides of the chalcogens are shown in the table 5-2

Table 5-2 Halides of chalcogens

	O	*S*	*Se*	*Te*	*Po*
F	O_2F_2, OF_2	S_2F_2, SF_4	Se_2F_2	TeF_4	-
	O_3F_2, O_4F_2	SF_6, S_2F_{10}	SeF_4, SeF_6	TeF_6	
Cl	Cl_2O, ClO_2	S_2Cl_2	Se_2Cl_2	$TeCl_2$	$PoCl_2$
	Cl_2O_6, Cl_2O_7	SCl_2, SCl_4	$SeCl_4$	$TeCl_4$	$PoCl_4$
Br	Br_2O	S_2Br_2	Se_2Br_2	$TeBr_2$	$PoBr_2$
	BrO_2, BrO_3		$SeBr_4$	$TeBr_4$	$PoBr_4$
I	I_2O_4, I_4O_9, I_2O_5	-	-	TeI_4	PoI_4

The halogen compounds of oxygen are discussed in the subhead (6- 6-2).

The monohalides are obtained by the direct reaction of the elements. Sulphur monochloride (S_2Cl_2) is prepared by passing chlorine over fused sulphur. Selenium monobromide (Se_2Br_2) is obtained by the reduction of tetrabromide with selenium.

$$2S + Cl_2 \longrightarrow S_2Cl_2$$

$$SeBr_4 + 3Se \longrightarrow 2Se_2Br_2$$

The structure of the monohalides is similar to that of hydrogen peroxide.

Cl
104°
1.99 A°
S—S
2.05 A°
1.99 A°
Cl

The dihalides have angular shapes with distorted sp^3 hybrid orbitals.

Sulphur dichloride is prepared by reacting chlorine with sulphur monochloride at O°C. It is a red liquid which dissociates above O°C to give back it's reactants by the backward reaction, hence are highly unstable.

$$S_2Cl_2 + Cl_2 \underset{\text{above } 0^oC}{\overset{\text{at } 0^oC}{\rightleftharpoons}} 2SCl_2$$

Tellurium dihalides are obtained either by the reduction of the tetrahalides with tellurium or by heating the tetrahalide strongly.

$$TeCl_4 + Te \longrightarrow 2TeCl_2$$

$$TeBr_4 \xrightarrow{\text{heating}} TeBr_2 + Br_2$$

All the dihalides hydrolyse readily.

$$2SeX_2 + 3H_2O \longrightarrow Se + H_2SeO_3 + 4HX$$

The tetrahalides of all the types except tetrafluorides are prepared by the direct combination of the elements. SBr_4, SI_4 and SeI_4 are not known.

$$2TeF_6 + Te \xrightarrow{180^\circ C} 3TeF$$

$$Se + 2Cl_2 \longrightarrow SeCl_4$$

Except $TeCl_4$ all the other halides are thermally unstable and they all decompose by water yielding the appropriate dioxides or the corresponding acid and the halogen acid. The $TeCl_4$ has a tetrameric crystalline structure, as seen in the figure below.

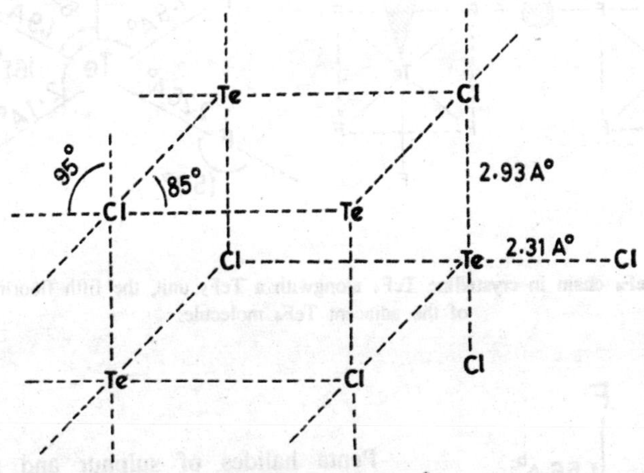

Structure of $(TeCl_4)_4$ in a crystalline form.

The tetra fluorides with one position occupied by a lone pair of electrons in a trigonal bipyramidal structure are seen in thc figure below. The TeF_4 forms a chain with its molecules. They are all capable of forming adducts with compounds like BF_3, AsF_5 etc. through their lone pair of electrons.

F
1.64 A°
1.54 A°
F
S
101.6°
F
173.1°
F

F
1.77 A°
1.68 A°
F
Se
100°
F
169.2°
F

a) SF_4 b) SeF_4

$$SCl_4 + 2H_2O \longrightarrow 4HCl + SO_2$$

$$SeBr_4 + 3H_2O \longrightarrow 4HBr + H_2SeO_3$$

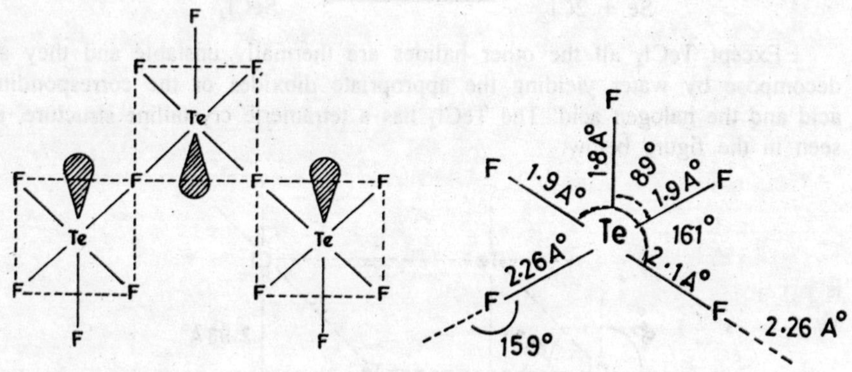

c) TeF_4 chain in crystalline TeF_4 alongwith a TeF_5 unit, the fifth fluorine is of the adjacent TeF_4 molecule.

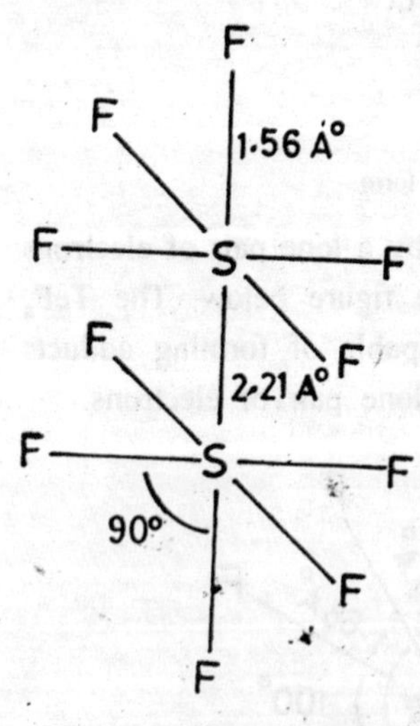

Penta halides of sulphur and tellurium are isolated but no corresponding pentahalide of Se is known. The pentahalides are liquids. S_2F_{10} is inert towards action of water and aqueous alkalies. It appears to have a typical structure of two SF_5 molecules linked through the sulphur atoms.

The hexafluorides are prepared by the direct union of the concerned elements.

$$Se + 3F_2 \longrightarrow SeF_6$$

The hexafluorides of sulphur, selenium and tellurium are all colourless gases, with very low boiling points and are highly covalent. SF_6 is highly inert even at red heat. SeF_6 is active and reacts with ammonia above 200°C and TeF_6 gets hydrolysed easily with water.

$$TeF_6 + 4H_2O \longrightarrow H_2TeO_4 + 6HF$$

They have an octahedral structure with sp^3d^2 hybridization.

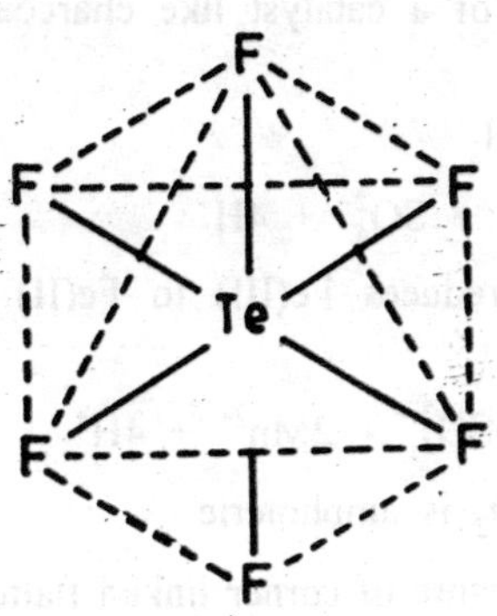

Structure of Octahedral TeF_4

Mixed hexa halides of tellurium are obtained by oxidative fluorination of $TeCl_4$ or $TeBr_4$ to get $TeClF_5$ or $TeBrF_5$ respectively. Mixed halides of the type $TeBr_2Cl_2$ (a yellow solid/ruby red liquid) or $TeBr_2I_2$ (garnet red crystals are reported.

5-5-3 Oxides

The most important oxides of the family are the dioxides and the trioxides. All the elements of the family form such oxides except SeO_3 which has not been isolated yet in its pure form.

The dioxides are obtained by burning the elements in air, this process is usually applicable in a large scale manufacture.

$$S + O_2 \longrightarrow SO_2$$

The dioxides are obtained in the laboratory by treating the elements with hot concentrated nitric acid and heating the initially formed products. Treating a sulphite salt with dilute HCl gives SO_2.

$$SO_3^{--} + 2H^+ \longrightarrow SO_2 + H_2O$$

The dioxides are angular molecules with a lone pair of electrons occupying one corner of the triangle.

Hydrogen chalcogenides reduce the dioxides to their respective chalcogens

$$2H_2S + SO_2 \longrightarrow 2H_2O + 3S$$

$$2H_2Se + SeO_2 \longrightarrow 2H_2O + 3Se$$

$$2H_2Se + SO_2 \longrightarrow 2H_2O + S + 2Se$$

Sulphur dioxide dissolves in water forming sulphurous acid, hence it is an acid anhydride,

$$SO_2 + H_2O \longrightarrow H_2SO_3$$

Similar reactions take place with SeO_2 and TeO_2.

SO_2 dissolves in NaOH to form sodium sulphite and when present in excess quantity results in the formation of bisulphite.

$$SO_2 + 2NaOH \longrightarrow Na_2SO_3 + H_2O$$

$$Na_2SO_3 + H_2O + SO_2 \longrightarrow 2NaHSO_3$$

SO_2 reacts with chlorine in the presence of a catalyst like charcoal to give sulphuryl chloride, SO_2Cl_2.

$$SO_2 + Cl_2 \longrightarrow SO_2Cl_2$$

$$2Fe^{3+} + SO_2 + 2H_2O \longrightarrow 2Fe^{2+} + SO_4^{2-} + 4H^+$$

SO_2 is a strong reducing agent and it reduces Fe(III) to Fe(II) and decolourizes potassium permanganate solution.

$$2MnO_4^- + 5SO_2 + 2H_2O \longrightarrow 5SO_4^{2-} + 2Mn^{2+} + 4H^+$$

SO_2 and SeO_2 are acidic oxides, and TeO_2 is amphoteric.

In the solid state SeO_2 has a polymeric structure of corner linked flattened (SeO_3) pyramides each selenium linked to a pendent like terminal oxygen alone as seen in the figure below :

whereas the TeO_2 has a pseudo trigonal bipyramidal three dimension structure.

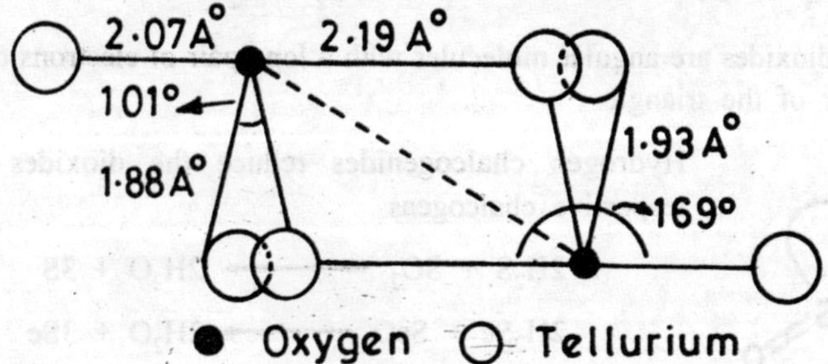

Sulphur trioxide is evolved as a white smoke when some sulphates are strongly heated.

$$Fe_2(SO_4)_3 \longrightarrow Fe_2O_3 + 3SO_3$$

SO_2 is oxidized to SO_3 by oxygen when passed over vanadium pentoxide

$$2SO_2 + O_2 \xrightarrow{V_2O_5} 2SO_3$$

SO_3 is best prepared when a mixture of SO_2 and O_2 is bubbled through concentrated H_2SO_4 over heated platinised asbestos. SO_3 formed gets condensed as a white solid in a receiver cooled in ice and fitted with a sulphuric acid trap to prevent the absorption of moisture.

$$2SO_2 + O_2 \xrightarrow[400^\circ C]{\text{Pt. Catalysed}} 2SO_3$$

Sulphur trioxide exists in polymorphic forms. One form has a trimeric cyclic structure, in which four oxygen atoms are arranged approximately tetrahedrally around each sulphur atom. This form gradually changes over into a linear polymerized structure in the presence of moisture. This strand like form of the solid accounts very neatly for its fibrous needle like appearance.

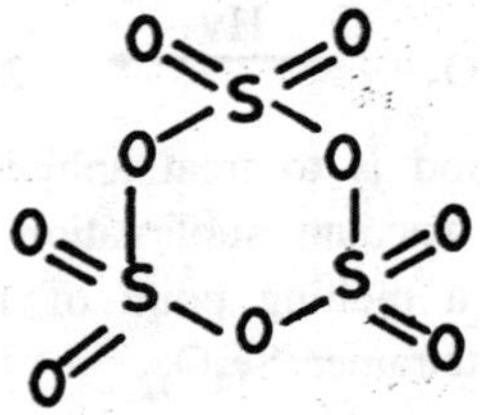

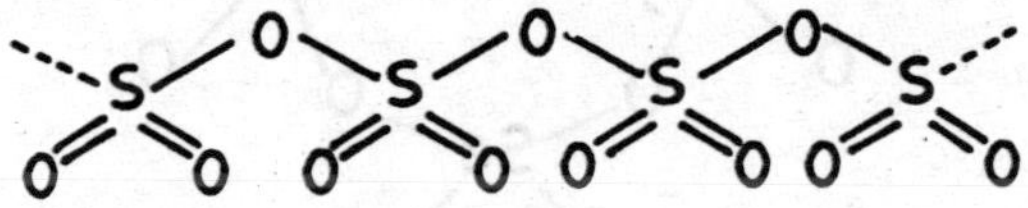

On heating, discrete SO_3 molecules are formed in vapour state, and they have a symmetrical planer structure.

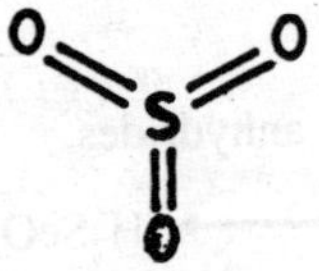

It is a powerful acidic oxide and fumes in moist air and reacts vigorously with water to form sulphuric acid.

$$SO_3 + H_2O \longrightarrow H_2SO_4$$

This sulphuric acid when saturated with sulphur trioxide gives pyrosulphuric acid (also called fuming sulphuric acid or Oleum).

$$H_2SO_4 + SO_3 \longrightarrow H_2S_2O_7$$

With basic oxides it reacts with the liberation of heat to form sulphates.

$$CaO + SO_3 \longrightarrow CaSO_4$$

It oxidizes hydrogen bromide to free bromine.

$$2HBr + SO_3 \longrightarrow H_2O + Br_2 + SO_2$$

SeO_3 is formed by a silent electric discharge on Se and O_2 gases,

$$2Se + 3O_2 \xrightarrow{H\nu} 2SeO_3$$

An improvised method is to treat anhydrons K_2SeO_3 with SO_3, which is refluxed, followed by vacuum sublimation at 120°, a white hygroscopic solid is obtained. It has a melting point of 118°C. The crystal structure is built up from the cyclic tetramer Se_4O_{12}

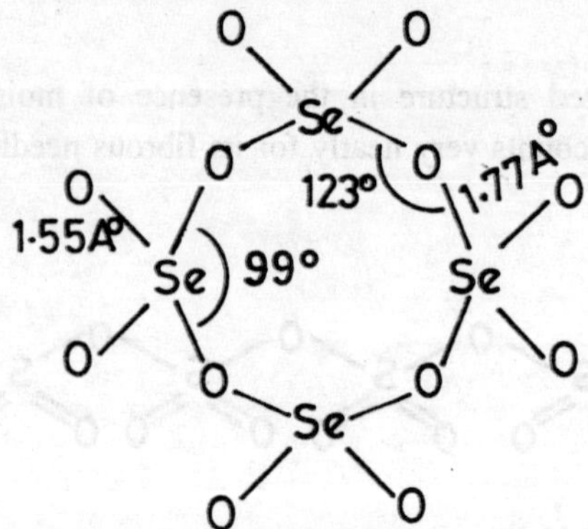

Structure of $(SeO_3)_4$ or Se_4O_{12}

The TeO_3 is prepared by heating tellurium acid (H_6TeO_6) strongly. It is found to be present in two polymorphic forms, a yellow orange, α-TeO_3 and a grey β-TeO_3, the former is an octahedra but the later has a 3D lattice structure.

Both the trioxides are acidic anhydrides

$$SeO_3 + H_2O \longrightarrow H_2SeO_4$$

$$TeO_3 + H_2O \longrightarrow H_2TeO_4$$

5-5-4 Oxyacids and peroxyacids of sulphur

A large number of oxyacids of sulphur are formed, but most of them do not exist in free state, but are manifested as their anions and salts. The oxyacids are classified into five groups.

(a) Sulphuric acids

(b) Sulphurous acids

(c) Sulphoxylic acids

(d) Thinoic acids

(e) Peroxo acids.

The mothods of preparation of these acids dealt with in this section are in a summarised form only, so as to get a brief understanding of them according to the scope of this book.

Sulphuric acid : Sulphuric acid is prepared usually by the contact process. This involves the oxidation of sulphur dioxide to give sulphur trioxide in the presence of iron or iron oxide as a catalyst at a temperature of 450°C under atmospheric pressure. This is based on Le Chatelier's principle, to get the maximum yeild of sulphutrioxide, low temperatures are involved as the reaction is endothermic.

$$2SO_2 + O_2 \rightleftharpoons 2SO_3 - 46.98 \text{ kcals.}$$

The sulphurtrioxide thus formed is dissolved in 98% pure sulphuric acid forming oleum, oleum is diluted to the required consistency to get sulphuric acid.

$$SO_3 + H_2SO_4 \longrightarrow H_2S_2O_7 \text{ (Oleum)}$$

$$H_2S_2O_7 + H_2O \longrightarrow 2H_2SO_4 \text{ (Sulphuric acid)}$$

Sulphurous acid : Sulphurous acid cannot be isolated, hence there is no evidence of its existence, the solution of sulphurdioxide in water which is highly acidic is commonly called sulphurous acid. The salts of this acid are reported as sulphites and bisulphites, hence the formation of this acid in water seems to be possible.

$$SO_2 + H_2O \longrightarrow H_2SO_3$$

Pyrosulphuric acid is obtained by passing sulphur trioxide in concentrated sulphuric acid

$$H_2SO_4 + SO_3 \longrightarrow H_2S_2O_7$$

Permonosulphuric acid also known as **Caro's acid** H_2SO_5 is obtained by grinding crystalline potassium persulphate with concentrated sulphuric acid in the presence of a freezing mixture.

$$K_2S_2O_8 + H_2SO_4 \longrightarrow H_2SO_5 + K_2S_2O_7$$

It is also prepared by the action of hydrogen peroxide with sulphuric acid

$$H_2SO_4 + H_2O_2 \longrightarrow H_2SO_5 + H_2O$$

Another oxyacid of sulphur is **persulphuric acid or disulphuric acid** $H_2S_2O_8$ or **Marshall's acid** is obtained by the electrolysis of a 50% solution of sulphuric acid using platinum electrodes. The electrolysis is conducted by keeping the electrolytic bath in a freezing mixture. Hydrogen escapes from the cathode and a solution of $H_2S_2O_8$ remain in the bath.

At Cathode $\quad H^+ + e^- \longrightarrow H$

$\quad\quad\quad\quad\quad H + H \longrightarrow H_2$

At Anode $\quad 2HSO_4^- \longrightarrow H_2S_2O_8 + 2e^-$

The structural features of a few oxyacids are shown below with a short text about them.

(i) Sulphuric acids : This category comprises of three acids, sulphuric acid, thiosulphuric acid and pyrosulphuric acid. The structures of these acids are given below.

```
      O                 S                  O         O
      ||                ||                 ||        ||
HO – S – OH       HO – S – OH       HO – S – O – S – OH
      ||                ||                 ||        ||
      O                 O                  O         O

Sulphuric acid    Thiosulphuric acid     Pyrosulphuric acid
```

The salts of these acids end with the suffix -ate, such as sulphates, thiosulphates and pyrosulphates.

(ii) Sulphurous acids : This series contain four acids, the sulphurous acid, thiosulphurous acid, pyrosulphurous acid and dithionous acid. The structure of these acids are given below.

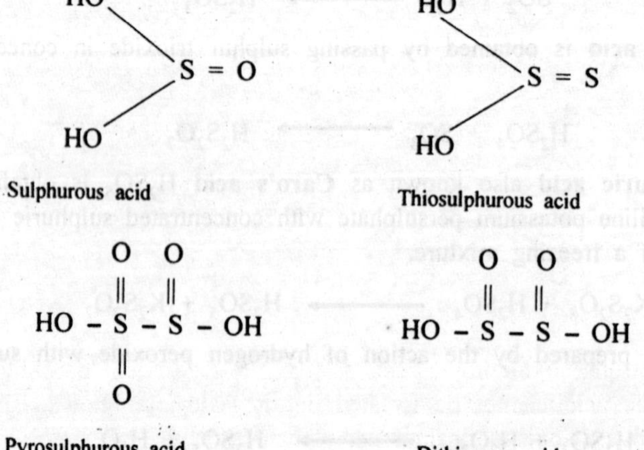

Sulphurous acid

Thiosulphurous acid

Pyrosulphurous acid

Dithionous acid

The salts of these acids end with the suffix -ite such as pyrosulphites and dithionites.

(iii) Sulphoxylic acids : The only sulphoxylic acid known is H_2SO_2.

(iv) Thionic acids : This series contain two acids, dithionic acid and polythionic acid. Their structure are

```
      O   O                     O              O
      ||  ||                    ||             ||
HO  - S - S - OH          HO  - S - (S)n  -  S - OH
      ||  ||                    ||             ||
      O   O                     O              O
```

Dithionic acid   Polythionic acid

(v) The peroxoacids : are the peroxosulphuric acid and the peroxodisulphuric acid

```
            O                     O               O
            ||                    ||              ||
HO - O  -  S - OH         HO  -  S - O - O  -  S - OH
            ||                    ||              ||
            O                     O               O
```

Peroxosulphuric acid   peroxodisulphuric acid

From the structures of the oxy and peroxy acids of sulphur, it is clear that the former set of acids have either an oxygen-sulphur linkage in them, and the later set of acids have an oxygen-oxygen linkage similar to that of hydrogen peroxide. The structure of persulphuric acid is seen below :

It is a strong oxidizing agent similar to hydrogen peroxide. The peroxodisulphuric acid cannot be isolated but its salts are known. The best known salt of this acid is potassium persulphate ($K_2S_2O_8$).

5-6 Application of Selenium in Xerography :

This particular topic has been added to this chapter so as to give the student the idea of how they get a photocopy or a **xerox** copy. What is the chemistry behind this dry process of getting a permanent print of anything whose copy is required.

The term xerography is derived from the greek words xero = dry, graphy = writing. This process was invented by Carlson (1937-42), today it is the most sought after and easiest method of getting a photocopy. The process actually involves seven steps basically, as given below.

1. Sensitization of the photoreceptor : The photoreceptor consist of a vacuum deposited film of amorphous Se of about 50 μm thickness on an aluminium substrate which is sensitized by an electrostatic charge from a coronal discharge of a field with ~ 10^5 V cm^{-1}

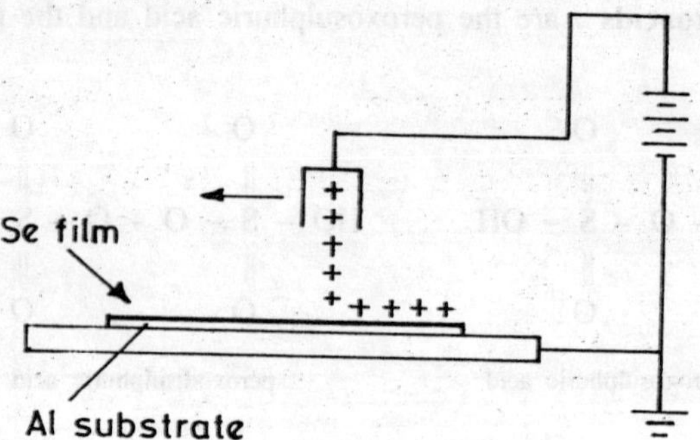

2. The second step involved is, the **exposur and latent image formation,** here the sensitized photoreceptor is exposed to a light and dark image pattern, in the light areas the surface potential of a photoconductor is reduced due to a photoconductive discharge. As the current can only flow perpendicular to the plane of the surface, in this step, it produces an electrostatic potential distribution which replicates the light dark pattern of the image.

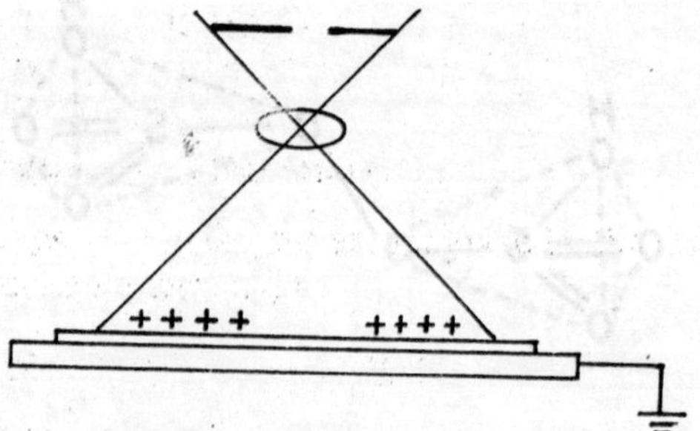

3. Development of the image takes place in this step, which is done using a mixture of black toner spherical particles of 10mμ size in diameter mixed with carrier heads of ~100mμ diameter. The toner particles get charged by friction (triboelectrical property) and are preferentially attracted either by the surface fringe field at light dark boundaries or by the absolute potential in the darker areas, they adhere to the photoreceptor thus forming a visible image corresponding to the latent electrostatic image.

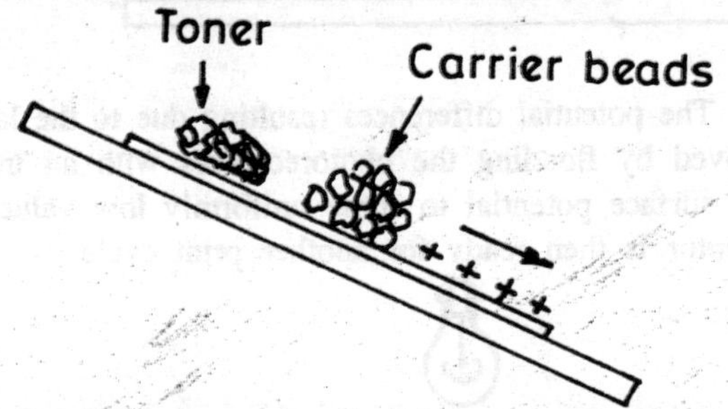

4. Image transfer : The image is transfered electrostatically by the attraction of toner particles

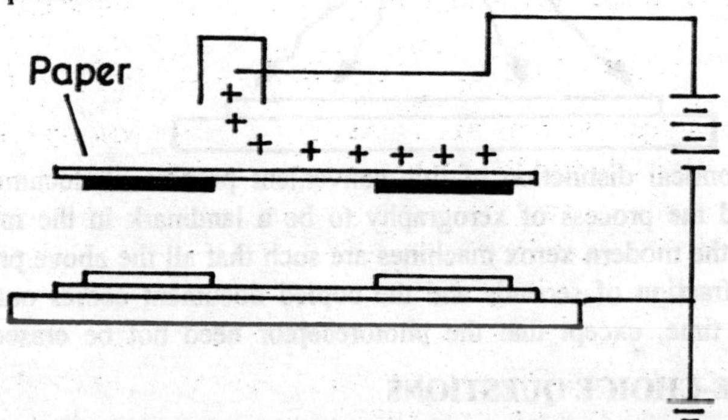

5. Fixing of the print : The powder image formed in the above step is made permanent by either fusing or melting the toner particles on the surface of the paper, this is done either by the heat, heat & pressure or by the solvent vapours.

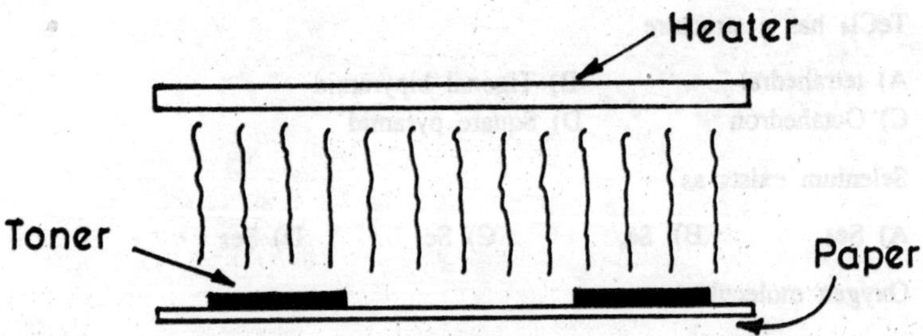

6. Cleaning : Any toner particles which gets adhere on the photoreceptor are cleaned by either a cloth, brush etc or by mechanical electrostatically, in an automatic copier.

7. Image erasure : The potential differences resulting due to the latent image formation are removed by flooding the photoreceptors with an intense light source to drive the surface potential to some uniformly low value of 10^4 V cm^{-1}, the photoreceptor is then ready for another print cycle.

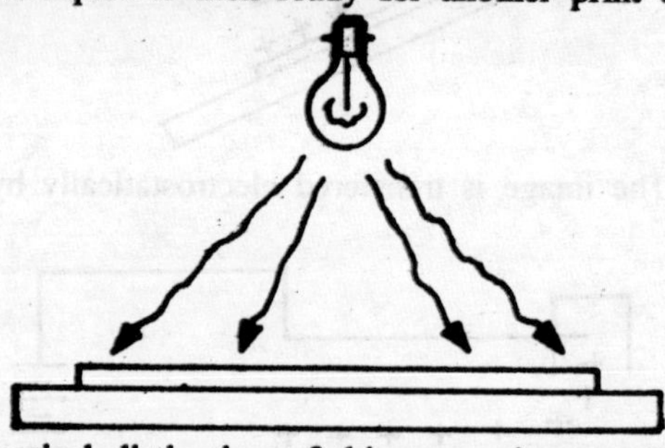

The economical distinction of this convenient process of documentation copying has led the process of xerography to be a landmark in the inventory science. Today the modern xerox machines are such that all the above processes takes place in fraction of seconds and the copied document comes out of the machine in no time, except that the photoreceptor need not be erased.

5-7 MULTIPLE-CHOICE QUESTIONS

5-7-1 The semiconductors of group VI B are

A) S and Se B) Se and Te C) Ge and Se D) Ge and Te

5-7-2 H_2S has a structure

A) linear B) trigonal C) tetrahedral D) trigonal bipyramid

5-7-3 $TeCl_4$ has a structure

A) tetrahedral B) Tigonal bipyramid
C) Octahedron D) Square pyramid

5-7-4 Selenium exists as

A) Se_4 B) Se, C) Se D) Se_8

5-7-5 Oxygen molecule is

A) Paramagnetic B) diamagnetic
C) Ferromagnetic D) None of them

5-7-6 An Oxygen atom is

A) Paramagnetic B) Diamagnetic
C) Ferromagnetic D) None of them

5-7-7 Selenium molecule is

A) Paramagnetic B) diamagnetic
C) Ferromagnetic D) None of them

5-7-8 The most stable chalcogenhydride is

A) H_2Te B) H_2Se C) H_2S D) H_2O

5-7-9 The bond angle in HSH is

A) 89° B) 90° C) 92.5° D) 104.5°

5-7-10 Hydrogen chalcogenides reduce their dioxides to form their respective

A) monoxides B) hydrogen C) hydrides D) chalcogens

5-7-11 SeO_2 has a structure

A) chain B) square planar C) linear D) Tetrahedron

5-7-12 SO_3 exists as

A) dimer B) monomer C) cyclic D) trimeric cyclic

5-7-13 Pyrosulphuric acid is

A) $H_2S_2O_4$ B) H_2SO_4 C) $H_2S_2O_7$ D) H_2SO_5

5-7-14 Marshall's acid is

A) Pyrosulphuric acid B) Permonosulphuric
C) Persulphuric acid D) Ionic acid

5-7-15 Dithonous acid is

A) $H_2S_2O_4$ B) H_2SO_4 C) $H_2S_2O_5$ D) $H_2S_2O_6$

5-7-16 Dithonic acid is

A) $H_2S_2O_4$ B) $H_2S_nO_6$ C) $H_2S_2O_5$ D) $H_2S_2O_6$

5-7-17 Peroxodisulphuric acid

A) $H_2S_2O_5$ B) $H_2S_nO_6$ C) $H_2S_2O_8$ D) $H_2S_2O_6$

5-7-18 $H_2S_2O_5$ is

A) Thiosulphuric acid B) Peroxosulphuric acid
C) Pyrosulphurous acid D) Thiosulphurous acid

5-7-19 $H_2S_3O_3$ is

A) Thiosulphuric acid B) Peroxosulphuric acid
C) Pyrosulphurous acid D) Thiosulphurous acid

5-7-20 $H_2S_2O_2$ is

A) Thiosulphuric acid B) Dithionos acid
C) Dithionic acid D) Thiosulphurous acid

5-7-21 $H_2S_2O_8$ is

A) Pyrosulphurous acid B) Pyrosulphuric acid
C) Persulphuric acid D) Persulphurous acid

5-7-22 TeO_3 is an acidic anhydride of

A) H_2TeO_3 B) H_2TeO_4 C) H_2TeO_5 D) $HTeO_3$

5-7-23 Oleum is

A) $H_2S_2O_8$ B) $H_2S_2O_6$ C) $H_2S_2O_7$ D) $H_2S_2O_4$

5-7-24 Fuming sulphuric acid is

A) Oleum B) Marshall's acid C) Caro's acid D) None of them

5-7-25 S_2F_{10} has an

A) S-F B) F-S C) S-S D) S_2-F bonding

5-8 SHORT AND ESSAY-TYPE QUESTIONS

5-8-1 Discuss the properties of VIB group elements on the basis of their
(a) Electronic configuration
(b) Hydrides
(c) Oxides
(d) Oxidation states
(e) Ionization potential
(f) Electronegativity and catenation. (O.U., Jun. 1984 & Oct. 1985)

5-8-2 (a) Name various elements represented by the electronic configuration ns^2np^4. Discuss the periodicity of the following properties among the elements of oxygen family.
(i) Ionization potential
(ii) Electron affinity
(iii) Oxidation state
(b) Water is a liquid whereas hydrogen sulphide is a gas at room temperature. Explain why?
(c) How do you account for the paramagnetic character of oxygen?
(K.U., Oct. 1988)

5-8-3 Based on the electronic configuration of the chalcogens justify their inclusion in the same group.

5-8-4 Discuss the similarities between oxygen & sulphur and their dissimilarities with selenium, tellurium and polonium. Justify the inclusion of all these elements in the same groups.

5-8-5 Give the electronic configuration of the elements of oxygen family.
(N.U., Apr. 1989)

5-8-6 Explain the trends in physical & chemical properties found to be present in VI B group.

5-8-7 Discuss the group properties of 6th group elements with respect to
(a) Hydrides

(b) Metallic and non metallic character
(c) Electronic configuration and
(d) Oxidation states (O.U., Oct. 1987)

5-8-8 Discuss the structure of oxygen molecule. (A.U., Dec. 1989 & A.U., Jun. 1990)

5-8-9 Explain the magnetic nature of O_2 molecule. (O.U., Oct. 1988)

5-8-10 Explain the paramagnetism in oxygen.

5-8-11 Oxygen can exhibit an oxidation state of –2 only. Why?

5-8-12 How does electronegativity change from oxygen to tellurium? (N.U., Oct. 1989)

5-8-13 What is allotropy? Mention allotropic modifications of sulphur. (N.U., Oct. 1989)

5-8-14 Discuss the allotropy of sulphur. Draw the structure of a sulphur molecule.

5-8-15 Compare and contrast the properties of oxygen and sulphur. (O.U., Oct. 1989)

5-8-16 Give a comparative account of the following properties of VI group elements (i) Hydrides (ii) Oxides (iii) Halides (iv) Oxidation states. (K.U., May 1988)

5-8-17 Discuss in detail about the hydrides of VI B group.

5-8-18 Discuss in brief polymorphism & catenation in sulphur.

5-8-19 Explain the gradation in the nature of the hydrides of oxygen family. (O.U., Oct. 1988)

5-8-20 Write a note on H_2S as a reagent in qualitative analysis. (O.U., Oct. 1987)

5-8-21 Discuss the structure of SO_3. (O.U., Apr. 1985)

5-8-22 Discuss the physicochemical principles involved in the manufacture of H_2SO_4 by contact process. (O.U., Nov. 1986)

5-8-23 Write notes on the oxyacids and peroxyacids of sulphur. (A.U., Dec. 1989)

5-8-24 What is the action of Conc. H_2SO_4 on $MnSO_4$ and Copper? Give balanced equations. (O.U., Nov. 1986)

5-8-25 How is peroxosulphuric acid prepared? Give its structure. (O.U., Apr. 1985 & Oct. 1989)

5-8-26 Give the structure of H_2SO_3, $H_2S_2O_8$ and H_2SO_4. (O.U., Mar. 1986)

5-8-27 Draw the structures of different oxyacids of sulphur.

5-8-28 Classify the oxyacids of sulphur. Name them. Find out the oxidation state of sulphur in them and draw their structures.

5-8-29 Write short notes on oxyacids of sulphur. (O.U., Mar. 1986 & O.U., Mar. 1988)

5-8-30 What are amphoteric oxides and peroxides give one example each? (K.U., May 1988)

5-8-31 Discuss the chemistry of sulphur trioxide.

5-7-1 B **5-7-2 C** **5-7-3 B** **5-7-4 D** **5-7-5 A** **5-7-6 A**

5-7-7 B **5-7-8 D** **5-7-9 C** **5-7-10 D** **5-7-11 A** **5-7-12 D**

5-7-13 C **5-7-14 C** **5-7-15 A** **5-7-16 D** **5-7-17 C** **5-7-18 B**

5-7-19 C **5-7-20 D** **5-7-21 C** **5-7-22 B** **5-7-23 C** **5-7-24 A**

5-7-25 C

					He
B	C	N	O	F	Ne
Al	Si	P	S	Cl	Ar
Ga	*Ge*	As	*Se*	Br	Kr
In	Sn	Sb	Te	I	Xe
Tl	Pb	Bi	Po	At	Rn

6 THE ELEMENTS OF GROUP VIIB *THE HALGOGENS*

6-1 INTRODUCTION

Halogens are the elements of VII B group of the periodic table. The term halogens is derived from the words halos-sea and gemnos produced from. The halogens are fluorine (F), chlorine (Cl), bromine (Br), iodine (I) and astatine (At).

Sodium chloride is the main constituent of the sea water, hence chlorine is present abundantly in nature, fluorine and bromine are present in certain sea weeds and they are extracted from their ashes. Iodine is present in the decayed sea animals as thyroxine. There are many other sources of these elements, but being highly reactive all the elements, except the radioactive astatine; are all present in combined state in different mineral ores. The radioactive astatine (At^{210}) has the highest half life of 8.3 hours. The mineral ores of the elements are discussed later in the same chapter.

Halogens constitute the most homogenous group of the periodic table, displaying a uniform trend in their properties. At room temperature fluorine is a light green gas, chlorine is a green gas, bromine is a red fuming liquid, iodine is shining violet crystalline solid and astatine is a grey metal.

6-2 ELECTRONIC CONFIGURATION AND OXIDATION STATES

The electronic configuration of the elements show the presence of seven electrons in the outermost orbit exhibiting a general electronic configuration of $ns^2\ np^5$, hence all these elements are short of one electron to complete their octet or attain noble gas configuration. Therefore all halogens form univalent, negatively charged ions easily, showing strong electronegative character.

Halogen	At. number	Electronic configuration
F	9	$1s^2 2s^2 2p^5$
Cl	17	$1s^2 2s^2 2p^6 3s^2 3p^5$
Br	35	$1s^2\ 2s^2\ 2p^6\ 3s^2\ 3p^6\ 4s^2\ 3d^{10} 4p^5$
I	53	$1s^2 2s^2 2p^6 3s^2 3p^6 4s^2 3d^{10} 4p^6 5s^2 4d^{10} 5p^5$
At	85	$1s^2 2s^2 2p^6 3s^2 3p^6 4s^2 3d^{10} 4p^6 5s^2 4d^{10} 5p^6 6s^2 4f^{14}$

Univalency is the main characteristic of the family both in ionic as well as in covalent compounds except in combination with oxygen. The stronger metal halides are ionic while those with weak metals and non metals covalent.

Fluorine is the only element of the family exhibiting only one oxidation state of –1, whereas all the other elements show, although rarely, variable oxidation states. Fluorine cannot exhibit oxidation states other than –1 because of its small size, strong electron density around the nucleus creating a high charge resulting in high electronegative character. More over it does not have any empty orbitals to enable itself to get hybridized as the other elements of the family do. Chlorine, bromine and iodine show oxidation states of –1 along with +1, +2, +3, +4, +5, +6 and +7. The positive oxidation states are exhibited when the halogens combine with oxygen.

$$Cl_2O\ (+1),\ ClO\ (+2),\ ClO_2\ (+4)$$

$$I_2O_5\ (+5),\ BrO_3\ (+6),\ Br_2O_7\ (+7)$$

or in the interhalogen compounds,

$$ClF_3,\ BrF_3,\ ICl_3$$

show +3 state in Cl, Br and I respectively.

This is due to vacant d orbitals which allow the excitation of the p-electrons and in +7 state even the s-electrons get excited. The excitation of the electrons is shown below for +3, +5, +7 states.

	s	p	*d*
One unpaired electron accounts for a covalency of 1	[↑↓]	[↑↓][↑↓][↑]	[ ][ ][ ][ ][ ]
Three unpaired electrons account for a covalency of 3 due to first excitation	[↑↓]	[↑↓][↑][↑]	[↑][ ][ ][ ][ ]
Five unpaired electrons account for a covalency of 5 due to second excitation.	[↑↓]	[↑][↑][↑]	[↑][↑][ ][ ][ ]
Seven unpaired electrons accounts for a covalency of 7 due to third excitation.	[↑]	[↑][↑][↑]	[↑][↑][↑][ ][ ]

6-3 GENERAL GROUP TRENDS OF VARIOUS PHYSICAL PROPERTIES

The physical constants are shown in the table 6-1. It is observed that the family of halogens show a regular trend in all the properties except electron affinity. The physical constants are discussed individually under respective headings.

Table 6-1 Physical Properties of Halogens

	Fluorine	*Chlorine*	*Bromine*	*Iodine*	*Astatine*
Symbol	F	Cl	Br	I	At
At Number	9	17	35	53	85
At. Weight	18.9984	35.45	79.90	126.90	–
At. radius X Å	0.64	0.99	1.14	1.33	–
Ionic radius X'Å	1.31	1.81	1.96	2.22	–
Ionization Potential (eV)	17.42	13.01	11.84	10.44	9.5
Electron affinity (eV)	3.62	3.79	3.56	3.28	–
Electronegativity*	4.10	2.83	2.74	2.21	1.96
Boiling point °C	-188	-34.6	58.76	184.25	–
Melting point °C	-223	-103	-7.2	113.5	–
Heat of dissociation kcal. $mole^{-1}$	38	58	46	36	–
Heat of hydration kcals $mole^{-1}$	-122	-89	-81	-72	–
Heat of sublimation kcal. $mole^{-1}$	–	–	7.4	10.4	–
Oxidation potential kcal	-186.5	-147.4	-136.4	-122.6	–
Abundance in earth's crust %	0.075	0.04	0.0002	0.00003	–

* electronegativity values are calculated as per the Alfred- Rochow method Journal of Inorganic and Nuclear Chemistry vol. 5, page 264, 1958.

6-3-1 Atomic Volume, Atomic & Ionic Radii and Density

As expected in this family there is a regular increase in the atomic volume (from 17.1 cc to 25.5 cc), atomic radii (from 0.64 Å to 1.33 Å), ionic radii of the halide ion from fluorine (1.31 Å) increases to iodine (2.22 Å) and the density of the halogens in their liquid state increases (from 1.1 to 3.76 as we move from fluorine to iodine).

6-3-2 Colour

It is interesting to note, that, the color of the liquid halogens is darker when compared to the same element in its vapour state. Fluorine is a pale yellow gas but yellow in liquid form. Chlorine is a greenish-yellow gas where as liquid chlorine is amber yellow. Bromine being a reddish brown liquid has pale orange to red vapours when it fumes, Iodine vapours are violet but it is a shining dark violet solid. The colour of the halogens can be attributed to the absorption of visible light by the molecules, resulting in the excitation of outer electrons to higher energy levels. Smaller the atoms more the energy required for excitation, thus fluorine will need the maximum and iodine minimum in the family. Fluorine absorbs violet light associated with higher energy and appears yellow, where as iodine on the other extreme absorbs yellow and green, low energy radiation and appears violet.

6-3-3 Ionization Potential

The ionization potential (eV) or ionization energy (kcal) of the halogens seems to be high and decreases down the group. The higher ionization potentials are indicated as these elements have very low tendency to loose electrons. The ionization potential value of iodine shows that iodine can even form a unipositive ion which is seen in the formation of I^{3+}.

6-3-4 Electronegativity

Halogens are the most electronegative elements of the periodic table possessing very high electronegativity values. The most electronegative element of the family is fluorine with a value of 4.1 and iodine is the least electronegative with a value of 2.21 almost equal to that of gold (2.4) thus showing somewhat electropositive character in it, this enables it to form I^+ ion and combines with other electronegative radicals.

6-3-5 Electron Affinity and Oxidizing Power

Electron affinity or the tendency of an atom to gain electrons is maximum in chlorine amongst the halogens. the unexpectedly low value for the electron affinity of fluorine may be associated with electron-electron repulsion in the relatively compact 2p sub level. Therefore it appears that chlorine should be a better oxidizing agent than fluorine sine more energy is released when the chlorine atoms acquire an electron than when the fluorine does. However the dissociation energy of fluorine (38 kcal mole^{-1}) is less than that of chlorine (58 kcal mole^{-1}). The low value for the dissociation energy of fluorine may be partially attributed to the repulsions of the non-bonding electrons, but it has generally been assumed that for chlorine, bromine and iodine, hybridization of p and d valence shell orbitals strengthen these bonds.

Oxidation may be regarded as the removal of electrons by the oxidizing agent. Hence the halogens act as oxidizing agents. The strength of an oxidizing agent (that is its oxidation potential) depends on several energy terms and is best represented by a Born-Haber type of cycle.

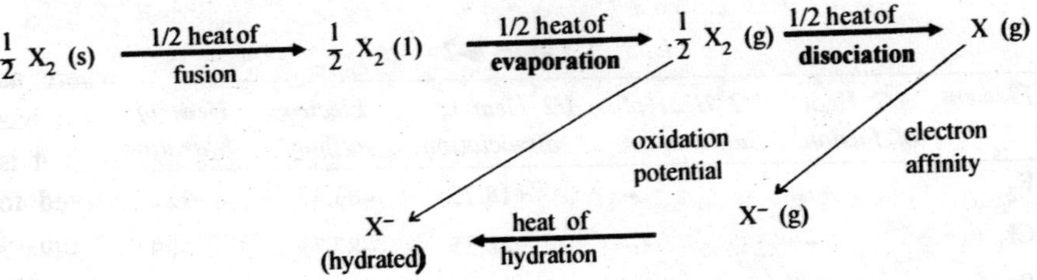

Fluorine and chlorine being gases at room temperatures the heat of fusion and evaporation can be omitted. Hence the oxidation potential is based only on the sum of the heat of dissociation, heat of hydration and electron affinity. The values of heat of fusion, evaporation and dissociation are positive because energy must be supplied. Thus, despite the electron affinity of chlorine being highest, fluorine is the strongest oxidizing agent because of its low heat of dissociation and high heat of hydration. This gives fluorine the capability to replace the higher halogens both in solution as well as in the gaseous state. Similarly chlorine replaces bromine and iodine and bromine replaces iodine from their compounds. In general the lower halogens replace the higher halogens. Hence the oxidizing power decrease from fluorine to iodine; so much so, iodine even gets the capability of acting as a reducing agent. Fluorine oxidizes water to oxygen, whereas oxygen oxidizes iodide ions to iodine.

$$F_2 + H_2O \longrightarrow 2H^+ + 2F^- + 1/2\ O_2 \ -90 \text{ kcal mol}^{-1}.$$

$$2I^- + O_2 \rightleftharpoons I_2 + 2O^- + 25 \text{ kcal mole}^{-1}$$

The electrode potential (E) values supports the decreasing oxidizing power of the halogens, and more over it is clearly evident that the electrode potential is more important for an element to show its oxidizing behaviour rather than its electron affinity.

$F_2 + 2e^- \longrightarrow 2F^-$	E = +2.85 V
$Cl_2 + 2e^- \longrightarrow 2Cl^-$	E = +1.36 V
$Br_2 + 2e^- \longrightarrow 2Br^-$	E = +1.07 V
$I_2 + 2e^- \longrightarrow 2I^-$	E = +0.54 V

The oxidation potential i.e. the strength of the oxidizing agent can be calculated by Hess Law which is given by the expression.

$$E = 1/2H_f + 1/2H_s + 1/2H_d - E.A - H_h$$

where H_f = heat of formation, H_s = heat of sublimation (vaporization), H_d = heat of dissociation, E.A. = electron affinity and H_h = heat of hydration. Knowing the values of E.A., H_f, H_s, H_d and H_h; E can be calculated.

Table 6-2

Element	*1/2 Heat of Fusion*	*1/2 Heat of Sublimation*	*1/2 Heat of dissociation*	*Electron affinity*	*Heat of hydration*	*E kcals*
F_2	–	–	+18.75	–83.47	–122	–186.72
Cl_2	–	–	+28.55	–87.39	–89	–148.39
Br_2	–	+3.7	+23.30	–82.09	–81	–136.09
I_2	+1.8	+5.2	+17.85	–75.63	–72	–122.78

6-3-6 Non Metallic Character

From the high ionization potential and high electronegativity values it is found that the halogens are non-metallic in character, Iodine shows a character of a metalloid in a few cases.

6-4 EXTRACTION OF HALOGENS

Halogens, as mentioned before are extensively present in nature in combined form. The halogens can be extracted from their respective ores or their simple salts. The process is not tedious and simple methods are employed for their extraction. Individual processes are discussed below.

6-4-1 Extraction of Fluorine

Fluorspar (Latin flour = flowing) was described by Agricola as early as 1530, as a solid executing fluorescence and melting into a red liquid. Fluorspar is calcium fluoride, CaF_2. The composition of fluorspar was unknown until it was found that it is a salt of lime and a peculiar acid which was eating away every type of vessel used for identifying it. The work of Keyer (1781) and Wenzel (1783) showed that iron or lead vessels should be used for its isolation. Ampere (1810) suggested that this acid to be a compound of hydrogen with an unknown radical analogous to chlorine. The element was first isolated by Moissan in 1886.

Fluorine does not occur free in nature because of its extremely reactive nature, but is widely distributed in nature in combination with other elements. It is present in traces in biological systems such as bone, teeth, milk, blood, plant and in small quantities in mica, mineral springs, sea water etc. The main ores of fluorine are fluorspar (CaF_2), cryolite ($AlF_3 3NaF$) and fluorapatite [$CaF_2 . 3Ca_3(PO_4)_2$].

Fluorine can be prepared by electrolysis and the chemicals used must be free from traces of water which react rapidly with the free halogen. The electrolyte used is a (1:1) fused mixture of potassium fluoride and hydrogen fluoride. The electrolytic bath of steel serves also as the cathode. The anode is of carbon. Fluorine gets discharged at the anode and hydrogen at the cathode, periodically HF is added to the cell to maintain correct composition of the mixture. Figure 6-1 shows the arrangement of Moissan's cell for the preparation of fluorine.

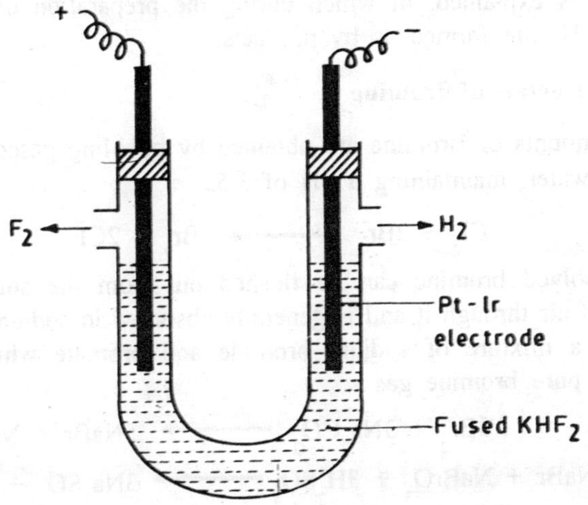

Fig. 6-1 Preparation of fluorine by Moissan's method

6-4-2 Extraction of Chlorine

In 1658 Glauber obtained the "spirit of salt" by distilling common salt with concentated sulphuric acid and dissolving the vapours in water. The other product found was Glauber's salt and sodium sulphate. Priestley (1772), Scheele (1774), Bertholett (1786), Lavoisier (1789) all of them made the primary study of this, the then unknown element. Much later Davy (1810) discovered the greenish gas to be chlorine (Greek Chlorous = pale green). Muriatic acid as named by Priestley was renamed as hydrochloric acid by Davy.

Chlorine as in sodium chloride is found abundantly in sea water, dry salt lakes, brine wells. It is found to be present in nature as sodium or other alkali or alkaline earth salts such as carnallite, $KClMgCl_26H_2O$, horn silver, AgCl, and sylvine, KCl. The natural chlorine consists of 75.4% of Cl^{35} isotope and 24.6% of Cl^{37} isotope. Samples enriched in one or the other of these stable isotopes are obtainable, and the enriched isotope can be used as a tracer. Cl^{36} is an artificial radioactive isotope and is α β emitter with 2×10^8 years half life.

Chlorine can be prepared in the laboratory by the oxidation of concentrated hydrochloric acid using manganese dioxide or potassium permanganate as the oxidizing agents.

$$MnO_2 + 4H^+ + 2Cl^- \longrightarrow Cl_2 + Mn^{++} + 2H_2O$$

$$2MnO^{4-} + 16\ H^+ + 10Cl^- \longrightarrow 5Cl_2 + 2Mn^{2+} + 8H_2O$$

Chlorine can be manufactured on a large scale by Nelson's cell. Nelson cell working is explained. In which during the preparation of sodium hydroxide, Cl_2 and H_2 are formed as by products.

6-4-3 Extraction of Bromine

Large amounts of Bromine are obtained by bubbling gaseous chlorine through the sea water, maintaining a pH of 3.5.

$$Cl_2 + 2Br^- \longrightarrow Br_2 + 2Cl$$

The dissolved bromine can be flushed out from the solution by passing a stream of air through it and is generally absorbed in sodium carbonate solution forming a mixture of sodium bromide and bromate which on acidification liberates pure bromine gas.

$$3Br_2 + 3Na_2CO_3 \longrightarrow 5NaBr + NaBrO_3 + 3CO_2$$

$$5NaBr + NaBrO_3 + 3H_2SO_4 \longrightarrow 3Na_2SO_4 + 3H_2O + 3Br_2$$

To obtain one kilogram of bromine about 9000 litres of sea water is required. The bromine obtained is carefully distilled over potassium bromide to get a sample of cent percent purity in which chlorine is completely removed by this process.

$$2KBr + Cl_2 \longrightarrow 2KCl + Br_2$$

6-4-4 Extraction of Iodine

Iodine may be obtained by heating potassium or sodium iodide with dilute sulphuric acid and manganese dioxide.

$$2KI + MnO_2 + 3H_2SO_4 \longrightarrow I_2 + 2KHSO_4 + MnSO_4 + 2H_2O$$

Iodine can be easily extracted and purified by sublimation.

6-5 USES OF HALOGENS

Halogens and their halides are extensively used on a large scale. In this section a few important uses are mentioned.

(1) Fluorine, chlorine and bromine are used as oxidizing agents.

(2) Chlorine is used as a bleaching agent. Its salt NaClO and hypochlorous acid HClO are also used in the bleaching industry.

(3) The compounds of fluorine and chlorine are used as germicides and insecticides.

(4) Chlorine is used in the sterlization of drinking water.

(5) Compounds like difluorodichloromethane **(Freon)**, polytetrafluoroethylene **(Teflon)** find importance in industry. Freon a non toxic, non inflammable compound is used in air conditioning and refrigeration in place of NH_3 and SO_2. Teflon is a new synthetic substitute for plastic and is resistant to both acids and bases. Its thermal resistance and non-sticky nature are taken advantage in making nonstick cooking utensils.

(6) Phosgene ($COCl_2$), Teargas (CCl_3NO_2), and mustard gas ($ClC_2H_4SC_2H_4Cl$) are used in warfare.

(7) Sodium bromide and potassium bromide are used as sedatives whereas silver bromide is used in photography.

(8) Iodine is extensively used in medicine, as tincture of iodine, iodex (a mixture of methyl salicylate and iodine), as iodoform etc.

(9) Iodine and Bromine along with their salts KI and KBr are important laboratory reagents.

(10) All the halogens are used widely for the preparation of halogen derivatives of organic compounds.

(11) Alkali halides specially potassium iodide are generally used in the infra-red spectroscopy as cells and lenses.

(12) The high dielectric constant and poor thermal conductivity (insulation) allow the sulphur hexafluoride to be used in X-ray and high voltage machines.

6-6 CHEMICAL BEHAVIOUR

All the halogens are highly reactive, combining with almost all the metals and non-metals forming halides and they also react amongst themselves to form interhalogens. The chemical reactivity of the halogens decreases with increasing atomic number.

The main characteristic feature of the halogens is their marked univalency in both ionic and covalent bonding. As the halogens are short of one electron than inert gas configuration, they either accept an electron or share an electron both completing the octet. With strong metals they form ionic halides and with weak metals and non metals covalent halides are formed, except fluorine which forms ionic compounds with mercury(II), tin(IV), bismuth(III) although others form covalent compounds.

6-6-1 Hydrides

Fluorine reacts with hydrogen even in dark, chlorine reacts in the presence of sunlight explodes when brought in contact with hydrogen. Bromine and hydrogen combine when ignited or passed over a heated platinized asbestos catalyst at 200°–300° C, forming in all the cases the hydrogen halides.

$$H_2 + F_2 \longrightarrow 2\,HF$$

$$H_2 + Cl_2 \longrightarrow 2\,HCl$$

$$H_2 + Br_2 \longrightarrow 2\,HBr$$

Iodine combines with hydrogen to give an equilibrium mixture in a reversible reaction forming hydrogen iodide, when heated with a catalyst on platinized asbestos at about 400°C.

$$I_2 + H_2 \rightleftharpoons 2HI$$

The stability of the hydrides decreases from HF to HI. Hydrogen fluoride is a very stable liquid with a hydrogen bonding because of high electronegativity of fluorine. It is usually represented as H – F .... H – F .... H – F. Hydrogen chloride, hydrogen bromide and hydrogen iodide are gases. They are reducing agents and the property decreases. HCl is a mild reducing agent, HBr is stronger than HCl, but HI is the strongest reducing agent amongst all the hydrogen halides.

All the hydrogen halides are acids in aqueous solutions; but in their dry gaseous state they are covalent, in the presence of moisture or water they polarize and form ions to act as acids. The ionization tendency increases from HF to HI, accordingly by the strength of the acids increase from HF to HI.HF is the weakest acid and HI is the strongest.

6-2-2 Oxides

All the halogens form oxides except fluorine which forms oxygen fluoride, as fluorine is more electronegative than oxygen.

a) The oxygenfluorides : Of all the known oxygen fluorides the most stable is OF_2, it is a colourless and a very poisonous gas, it can be obtained by the action of fluorine on an alkali, like NaOH.

$$2F_2 + 2NaOH \longrightarrow 2NaF + OF_2 + H_2O$$

when pure it is highly stable till 200°C, at higher temperatures it dissociates into its elements

$$2OF_2 \longrightarrow 2F_2 + O_2$$

The OF_2 is not linear but instead it is an angular molecule which looks like:

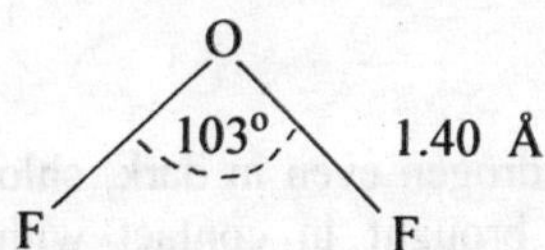

One of the fluorine can be substituted with hydrogen to form HOF, by a photolysed reaction between F_2 and H_2O in frozen form in the presence of solid N_2. Because of its unstability the reaction forms back the product, to avoid.

$$F_2 + H_2O \xrightleftharpoons{N_2} HOF + HF$$

such a thing to happen, the reaction can be carried out at 40°C. The product on cooling further gets rid of HF and H_2O (unreacted) between -50°C to -79°C, the product HOF can be collected at a temperature of -183° as a white solid. It melts at -117° to a pale yellow liquid which boils below room temperature and dissociates into HF and O_2

$$2HOF \longrightarrow 2HF + O_2$$

It reacts vigorously with water

$$4HOF + 2H_2O \longrightarrow 4HF + 2H_2O_2 + O_2$$

It is an angular molecule with an $H\hat{O}F$ angle of 97° and O-F distance of 1.44 Å with O-H distance of 0.96 Å,

O
0.96 Å 97° 1.44Å
H F

The **dioxygen difluoride** can be obtained by passing a silent electric discharge in a 1:1 mixture of O_2 and F_2 kept at a very low pressure of 7-17 mm Hg

$$O_2 + F_2 \longrightarrow O_2F_2$$

When O_2 and F_2 are subjected to a 3 Mev radiation at -195°C, a yellow solid is obtained of O_2F_2

O_2F_2 is less stable than OF_2, O_2F_2 is a powerful oxidizing and fluorinating agent.

$$ClF + O_2F_2 \longrightarrow ClF_3 + O_2$$

$$BrF_3 + O_2F_2 \longrightarrow BrF_5 + O_2$$

$$H_2S + 4O_2F_2 \longrightarrow SF_6 + 2HF + 4O_2$$

The structure of O_2F_2 is similar to that of H_2O_2 as seen in the figure below.

F 1.57Å 1.21Å O—O 109.5° F   H 0.95Å 1.47Å O—O 94.8° H

Trioxygen difluoride (O_3F_2) : is a blood red liquid and tetra oxygen difluoride (O_4F_2), is a red brown solid, both these compounds are stable only at extremely low temperatures. they can be prepared by silent electric discharge. O_3F_2 is considered to be a mixture of O_4F_2 and O_2F_2, as it can exist in a dimeric state, as (O_6F_4).

(b) Oxides of other halogens : Chlorine forms a number of oxides, Cl_2O, ClO, ClO_2, Cl_2O_6 & ClO_4, bromine forms Br_2O, BrO_2, BrO_3, Br_2O_7, whereas iodine forms only I_2O_5. the oxides of chlorine are of importance, hence a detail study of the stable ones among them is done here with a reference to the other halogen oxides, wherever necessary. Chlorine oxides are explosive compounds and in general are prepared by endothermic processes.

Chlorine monoxide (Cl_2O) is formed when dry chlorine is passed over freshly precipitated dry mercury (II) oxide, which is kept ice cooled.

$$2\,Cl_2 + 2HgO \longrightarrow Cl_2O + HgCl_2.HgO$$

It is a brownish yellow gas, getting condensed in a freezing mixture. The orange solid melts at -11.6° C and boils at 2°C, the gas explodes violently dissociating into chlorine and oxygen.

$$2\,Cl_2O \longrightarrow 2Cl_2 + O_2$$

It is hence a strong oxidizing agent. It is an acidic anhydride and forms hypochlorous acid on dissolving in water.

$$Cl_2O + H_2O \longrightarrow 2\,HClO$$

With NaOH, Cl_2O form hypochlorites. The structures of F_2O, Cl_2O and Br_2O are similar and are related to tetrahedron.

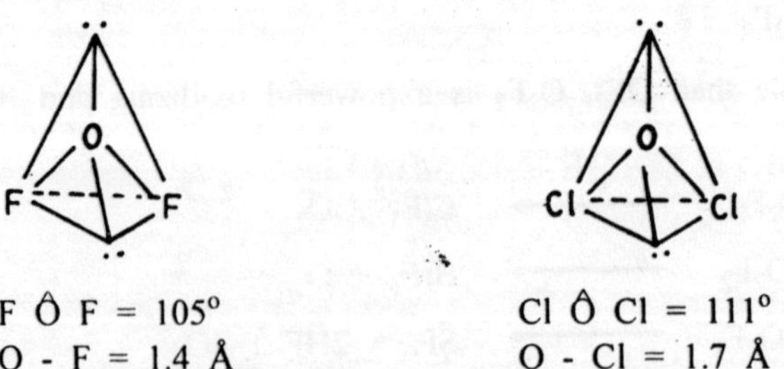

$F\hat{O}F = 105°$ $Cl\hat{O}Cl = 111°$

O - F = 1.4 Å O - Cl = 1.7 Å

The two unpaired electrons from oxygen bind with the halogen atoms and two lone pairs of electrons are present on two apexes of the tetrahedron. They all exhibit sp^3 hybridization.

Chlorine dioxide (ClO_2) is prepared by the action of nitrogen dioxide on sodium chlorate solution.

$$NaClO_3 + NO_2 \longrightarrow NaNO_3 + ClO_2$$

Chlorine dioxide, ClO_2 an orange yellow gas, is highly explosive and has a distinctive odour. It is a powerful oxidizing agent and dissolves in water to form a fairly stable solution of chloric acid and hydrochloric acid.

$$6\ ClO_2 + 3H_2O \longrightarrow 5HClO_3 + HCl.$$

Aqueous solutions of alkalis when treated with ClO_2 forms chlorite and chlorate salts, hence chlorine dioxide is a mixed acid anhydride.

Chlorine dioxide is a paramagnetic gas containing an odd number of electrons. as the odd electron molecules are highly reactive, the reactivity of chlorine dioxide is a typical example of it. The bond lengths are quite shorter than in a single bond and has a resonance structure, involving two coordinate bonds and a three electron bond, resonating with either of the two bonds.

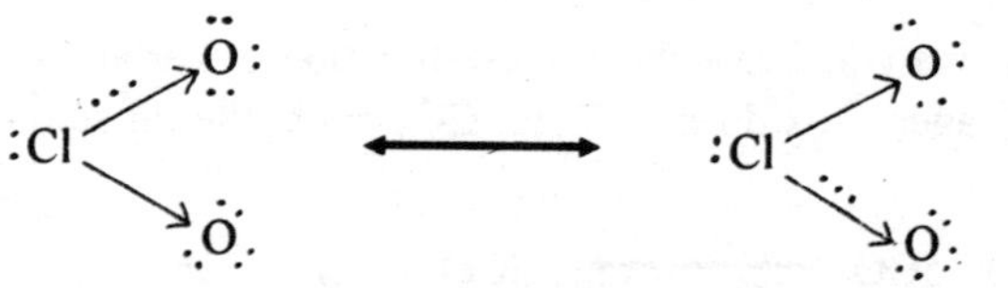

$O\hat{Cl}O = 117°$

$Cl - O = 1.49 Å$

Chlorine hexaoxide (Cl_2O_6) is a red liquid which can be prepared by the action of chlorine dioxide on ozonized oxygen. The liquid dissociates on heating, it is a strong oxidizing agent and explodes on contact with grease. With alkalis it forms chlorates and perchlorates showing its mixed anhydride character.

$$Cl_2O_6 + 2NaOH \longrightarrow NaClO_3 + NaClO_4 + H_2O$$

As it has no unpaired electrons it is diamagnetic, and it is probably in equilibrium with ClO_3, which is formed on heating Cl_2O_6 mildly.

$$Cl_2O_6 \xrightarrow{\text{heat}} 2ClO_3$$

Chlorine heptoxide, (Cl_2O_7), is a dehydrated product of perchloric acid with P_2O_5 and it is distilled off under reduced pressure.

$$4HClO_4 + 2P_2O_5 \longrightarrow 2Cl_2O_7 + 4HPO_3$$

It is a colourless oily liquid boiling at 80°C, but slowly decomposes on standing. It hydrolyses with even a slight amount of moisture forming perchloric acid, hence it is a perchloric acid anhydride.

$$Cl_2O_7 + H_2O \longrightarrow 2HClO_4$$

The structures of Cl_2O_6 and Cl_2O_7 are not known definitely but are suggested to be as

Cl_2O_6 Cl_2O_7

The oxides of bromine are analogous to those of chlorine, the only oxide of iodine found to be reported is I_2O_5, iodine pentoxide which is a white solid, is obtained by the dehydration of iodic acid by heating to 170ºC.

$$2HIO_3 \longrightarrow I_2O_5 + H_2O$$

Hence it is an anhydride of iodic acid as the reaction reverses on slight contact with moisture.

It is an important analytical reagent in the estimation of carbon monoxide. It being an oxidizing agent, oxidizes CO to CO_2 with the liberation of I_2 which is titrated.

$$I_2O_5 + 5CO \longrightarrow 5CO_2 + I_2$$

Its structure is presumably planar

O ← I — O — I → O (each I also bonded → O)

With iodine undergoing sp^2 hybridization with respect to the bridging oxygen and forming two coordinate bonds with the two terminal oxygens.

6-6-3 Reactivity with Metals and Non-metals

Almost all the metals, non metals and most of their compounds react with the halogens, directly or indirectly forming stable halides with an exception of helium, neon and argon. Halides generally are amongst the most important and common compounds. Halides are classified into binary halides, covalent halides, oxohalides, hydroxo halides and complex halides depending mainly on the type of metal, nonmetal or their compounds.

The binary halides are metal halides of predominantly ionic character with some covalency in them in the case of few metals. A few metal halides are intermediate of ionic and covalent character too. Basically the binary ionic halides are those in which the crystal lattice consists of discrete ions rather than the definite molecular units as present in covalent halides. The ionic halides are of the alkalimetals, alkaline earth metals (with the exception of Li & Be) and of most of the lanthanides and a few actinides. By the increase of the charge/radius ratio the ionicity decreases to give place to covalency. Another factor to determine the ionic or covalent character of the halide is a metal with a lower valency forms ionic halides and with a higher valency forms covalent halide. For example $PbCl_2$ is ionic, $PbCl_4$ is covalent, $SnCl_2$ is ionic, $SnCl_4$ is covalent. In some cases the halogen itself plays a vital role e.g., AlF_3 is ionic, $AlCl_3$ show some extent of covalency, $AlBr_3$ and AlI_3 are covalent.

Most of the halides are water soluble and forms hydrated metal ions and the halide ions, but the solubility depends upon the halogen also which is in the order of iodide > bromide > chloride > fluoride.

The binary halides are formed by the direct combination of the metal or non metal with halogen or with halogen halides.

$$2M + X_2 \longrightarrow 2MX$$

$$2M + 2HX \longrightarrow 2MX + H_2$$

$$2K + F_2 \longrightarrow 2KF$$

$$Mg + 2HCl \longrightarrow MgCl_2 + H_2$$

$$2P + 3Cl_2 \longrightarrow 2PCl_3$$

$$2S + Br_2 \longrightarrow S_2Br_2$$

$$Hg + I_2 \longrightarrow HgI_2$$

The covalent halides with discrete molecules in crystalline state or some consisting of separate molecules held together by Van der Waals force have low lattice energy and are hence very volatile with low melting & boiling points when compared to ionic halides of high melting and boiling points. Most of the electronegative elements and the metals in very high oxidation states form covalent halides, the most important covalent halides are the hydrogen halides, and those of B, C, P, As, Sb, Si, Ge, Sn, S, Se, Te.

All the halides have the ability of forming ligands with both ionic and covalent halides such as $[SnCl_6]^{2-}$, $[KI_3]^{2-}$, $[FeCl_4]^-$, $[IF_6]^-$ etc.

6-6-4 Reactivity with Water

Fluorine displaces oxygen from water (Some hydrogen peroxide and ozone are formed as well).

$$2H_2O + 2F_2 \longrightarrow 4HF + O_2$$

Chlorine is moderately soluble in water, 3.04 volumes dissolving in 1 volume of water at 0°C. The yellowish green solution formed contains acids, the latter being responsible for the strong bleaching and germicidal power of wet chlorine.

$$H_2O + Cl_2 \longrightarrow HCl + HClO$$

Bromine is slighly soluble in water, 3.6 volumes dissolving in 100 volumes of water at 20ºC. No chemical reaction is expected. the bromine water is a good oxidizing agent but not as powerful as chlorine water. The bromine water is useful in the laboratory for bromination reactions in the presence of acetic acid which helps in retaining bromine in solution. In sunlight bromine hydrolysis.

$$2\ H_2O + 2Br_2 \longrightarrow 4\ HBr + O_2$$

Iodine is almost insoluble in water.

6-6-5 Reactivity with Ammonia

Chlorine and bromine reacts with concentrated ammonia vigorously to produce nitrogen and the respective ammonium halide.

$$8NH_3 + 3Cl_2 \longrightarrow N_2 + 6NH_4Cl$$

$$8\ NH_3 + 3Br_2 \longrightarrow N_2 + 6NH_4Br$$

Excess chlorine when passed into aqueous ammonia a dangerously explosive nitrogen trichloride separates out as an oil.

$$2\ NH_3 + 6\ Cl_2 \longrightarrow 2NCl_3 + 6HCl$$

Excess bromine forms ammonium perbromide, NH_4Br_3, and in excess iodine gives a brown precipitate of nitrogen iodide, which is explosive when dry.

$$5\ NH_3 + 3I_2 \longrightarrow 3NH_4I + NH_3NI_3$$

6-6-6 Reactivity with hydrocarbons

Fluorine decomposes hydrocarbons.

$$CH_4 + 2F_2 \longrightarrow 4\ HF + C$$

Under controlled conditions, hydrogen is replaced by fluorine or fluorides

$$C_6H_6 + 6F_2 \xrightarrow[260^\circ C]{\text{Copper catalyst}} C_6F_6 + 6HF$$

Some interesting reactions of electrolytic fluorination are also performed

$$2\ (C_2H_5)O + 5F_2 \longrightarrow 2\ (C_2F_5)O + 5H_2$$

$$C_8H_{18} + 9F_2 \longrightarrow C_8F_{18} + 9H_2$$

$$(C_4H_9)_3N + 9F_2 \longrightarrow (C_4F_9)_3N + 9H_2$$

In direct sunlight a mixture of chlorine and methane explodes.

$$CH_4 + 2Cl_2 \longrightarrow 4\ HCl + C$$

Chlorine and bromine but not iodine will bring about substitution of hydrogen atoms in saturated paraffin hydrocarbons in diffused day light, HCl is formed at each stage.

$$CH_4 \xrightarrow[-HCl]{+Cl_2} CH_3Cl \xrightarrow[-HCl]{+Cl_2} CH_2Cl_2 \xrightarrow[-HCl]{+Cl_2} CHCl_3 \xrightarrow[HCl]{+Cl_2} CCl_4$$

Methane, Monochloro Methane, Dichloro Methane, Tricholro Methane, Tetrachloro Methane

Chlorine, bromine and in some cases iodine will form addition products with unsaturated hydrocarbons.

$$\begin{array}{c} CH_2 \\ \| \\ CH_2 \end{array} + Br_2 \longrightarrow \begin{array}{c} CH_2\ Br \\ | \\ CH_2\ Br \end{array}$$

Ethylene — 1, 2 Dibromoethane

Wax burns with a red flame in chlorine atmosphere.

$$C_xH_y + y/2\ Cl_2 \longrightarrow xC + yHCl$$

6-6-7 Reactivity with Alkalies

With excess of cold dilute sodium hydroxide, chlorine forms a solution containing sodium chloride and sodium hypochlorite, similarly bromine forms bromide and hypobromite similar products with iodine are formed.

$$Cl_2 + 2NaOH \longrightarrow NaCl + NaClO + H_2O$$

With hot concentrated alkali, chloride and chlorate, bromide and bromate, iodide and iodate are formed respectively

$$3Cl_2 + 6NaOH \longrightarrow 5NaCl + NaClO_3 + 3H_2O$$

6-6-8 Oxidation of Compounds of Metals

Chlorine and bromine water will oxidize solutions of iron(II) salts to the iron(III) state

$$2\ Fe^{2+} + Cl_2 \longrightarrow 2Fe^{3+} + 2Cl^-$$

Potassium ferrocyanide is oxidized to potassium ferricyanide by chlorine or bromine.

$$2K_4Fe(CN)_6 + Br_2 \longrightarrow 2K_3Fe(CN)_6 + 2KBr$$

Similarly Sn(II) compounds are oxidized to Sn (IV) compounds with Cl_2, Br_2 & I_2.

$$Sn^{2+} + I_2 \longrightarrow Sn^{4+} + 2I^-$$

6-6-9 Hydrogen Fluoride

Hydrogen fluoride is made by the reaction between calcium fluoride and sulphuric acid.

$$CaF_2 + H_2SO_4 \longrightarrow 2HF + CaSO_4.$$

The insoluble $CaSO_4$ is removed by filtration.

Hydrogen fluoride is also formed by the direct combination of hydrogen and fluorine, which combine violently even in dark.

$$H_2 + F_2 \longrightarrow 2\ HF$$

In vapour state hydrogen fluoride is in monomeric state at high temperatures, but at lower temperatures; because of the hydrogen bonding present in the compound; it is present as dimeric H_2F_2 or even in associated $(HF)_n$ state. In solid form zig-zag chains are present, Fig. 6-2

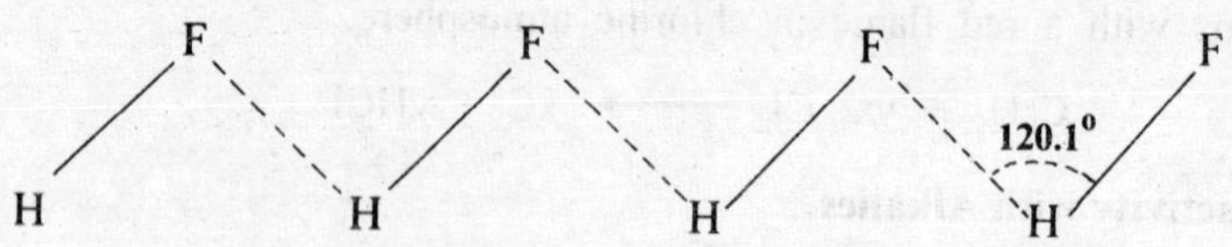

Fig. 6-2 Associated structure of hydrogen fluoride.

It is highly soluble in water, the dilute aqueous solutions of hydrogen fluoride behave as weak acids whereas 5-15 M solutions of its behave as a strong acid because of its ionization into H_3O^+, HF_2^- and more complex species $H_2F_3^-$, $H_3F_4^-$, etc. Liquid HF is somewhat less acidic than Conc. H_2SO_4. The acid strength is such that HNO_3 behaves as a base when reacted with liquid HF,

$$HF + HNO_3 \longrightarrow H_2NO_3^+ + F^-$$

Hydrogen fluoride reacts with metals and their compounds forming metal fluorides.

$$2Na + 2HF \longrightarrow 2NaF + H_2$$

$$KOH + HF \longrightarrow KF + H_2O$$

$$CaO + 2HF \longrightarrow CaF_2 + H_2O$$

$$MgCO_3 + 2HF \longrightarrow MgF_2 + CO_2 + H_2O$$

Many metal and non metal fluoride salts formed are electron deficient, hence act as Lewis acids. They are strong F^- acceptors and also combine with Lewis bases such as NH_3, H_2O etc.

$$2B + 3F_2 \longrightarrow 2BF_3$$

$$2BF_3 \xrightarrow{NH_3} [BF_3NH_3] \qquad 2BF_3 \xrightarrow{F^-} [BF_4]^-$$

Hydrogen fluoride reacts with silica hence cannot be stored in glass.

$$Na_2SiO_3 + 6HF \longrightarrow Na_2SiF_6 + 3H_2O$$

$$SiO_2 + 4HF \longrightarrow SiF_4 + 2H_2O$$

$$SiF_4 + 2HF \longrightarrow H_2SiF_6 \text{ (Fluoro silicic acid)}$$

This property is used for etching glass.

Hydrogen fluoride is used as a non-aqueous solvent. It is one of the most water like solvents amongst the non aqueous solvents, as it possess fairly similar solvent properties except for a greater dissociating power of hydrogen fluoride than water.

The non-aqueous solvent chemistry of hydrogen fluoride was initially studied by K. Fredenhagen (1928) and was extended by other workers. In this section we will restrict ourselves only to a few classic reactions of HF.

(i) It is found that many organic compounds such as alcohols, aldehydes, ketones, ethers, acids, acid anhydrides, carbohydrates and a few nitrogen compounds dissolve in liquid hydrogen fluoride to give conducting solutions, for example :

$$CH_3OH + HF \longrightarrow (CH_3OH.H)^+ + F^-$$

$$(C_2H_5)_2O + HF \longrightarrow [(C_2H_5)_2O.H]^+ + F^-$$

$$CH_3COOH + HF \longrightarrow (CH_3COOH.H)^+ + F^-$$

(ii) Hydrocarbons are insoluble in hydrogen fluoride.

(iii) Inorganic compounds dissolve in hydrogen fluoride indicating that it is a good solvent for ionic compounds also.

$$SbF_5 + HF \longrightarrow SbF_6 + H^+$$

$$H_2O + HF \longrightarrow H_3O^+ + F^-$$

$$H_2SO_4 + 2HF \longrightarrow \underset{\text{Fluorosulphuric acid}}{HOSO_2F} + H_3O^+ + F^-$$

Precipitation of such compounds which are not possible in aqueous solutions are conducted in liquid HF.

$$KClO_4 + HF \longrightarrow HClO_4 + KF$$

$$Ba(ClO_4)_2\ 2AgF \longrightarrow 2AgClO_4 + BaF_2$$

6-6-10 Oxy-acids of Chlorine

Fluorine cannot form any oxy acid. Chlorine, bromine and iodine form oxyacids of the general type HOX, HXO_2 HXO_3, HXO_4. The chemistry of these acids and their salts is very complicated as many of them are known only in solution or as their salts, solutions of all of these acids can be obtained by reaction of the free halogens with water or aqueous bases.

In aqueous solution saturated with chlorine contains an appreciable concentration of **hypochlorous acid (HOCl)**. When freshly precipitated HgO in water is shaken with Cl_2, the hypochlorous acid is formed.

$$2HgO + H_2O + 2Cl_2 \longrightarrow HgOHgCl_2 + 2HOCl$$

Bromine and iodine also forms their respective hypobromous & hypoiodous acids.

Hypochlorous acid is the most stable of the other hypohalous acids. Sodium hypochlorite which is used for bleaching cotton is prepared by the electrolysis of brine. The chlorine liberated at the anode reacts with the hydroxyl ions before it escapes the cell.

at anode $$2\,Cl^- \longrightarrow Cl_2 + 2\text{ electrons}$$

$$Cl_2 + 2NaOH \longrightarrow NaOCl + NaCl + H_2O$$

At cathode the sodium metal formed due to electrolysis reacts with water to give sodium hydroxide

at cathode $$2Na^+ + 2\text{electrons} \longrightarrow 2Na$$

$$2Na + 2H_2O \longrightarrow 2NaOH + H_2$$

Chlorous acid ($HClO_2$) is the only definite **halous acid** formed. It is stronger than HOCl and its salts are called chlorites. It cannot be isolated and is known only to be present in solution. $NaClO_2$ is the best known salt of it and is used in textile industry.

The known **halic acids** are $HClO_3$, $HBrO_3$ and HIO_3. The acids are all strong oxidizing agents

$$NaClO_3 + 3Na_2SO_3 \longrightarrow NaCl + 3Na_2SO_4$$

In aqueous solution the halic acids behave as strong acids.

$$HXO_3 + H_2O \longrightarrow H_3O^+ + XO_3$$

The **chloric acid** is formed by the action of chlorine on hydroxides or by the electrolysis of hot chloride solutions.

$$6OH^- + 3Cl_2 \longrightarrow ClO_3^- + 5Cl^- + 3H_2O$$

Potassium chlorate decomposes in two different forms.

$$2\ KClO_3 \longrightarrow 2\ KCl + 3O_2 \text{ (at very high temperature)}$$

$$4\ KClO_3 \longrightarrow 3\ KClO_4 + KCl \text{ (at 400°C)}$$

The chlorate ion (ClO_3^-) has a tetrahedral structure with one position occupied by a tone pair of electron.

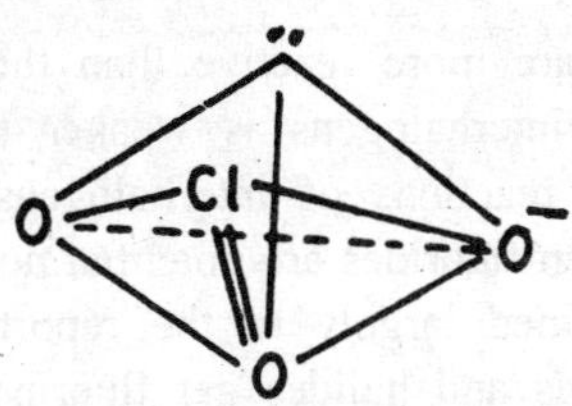

Perchloric acid, $HClO_4$, can be prepared by vacuum distillation of a mixture of potassium perchlorate and concentrated sulphuric acid.

$$2KClO_4 + H_2SO_4 \longrightarrow K_2SO_4 + 2HClO_4$$

It is a liquid and starts decomposing at it is boiling point of 90ºC. It explodes when comes in contact with any organic matter.

It is a strong acid in aqueous solutions, its monohydrate $HClO_4.\ H_2O$ is represented as $H_3O^+.\ ClO_4^-$. The acid is powerful oxidizing agent. The alkali metal perchlorates can be prepared by the controlled heating of the corresponding chlorates. The perchlorate ion is tetrahedral.

6-6-11 Interhalogens

The binary and ternary combinations of the halogens amongst themselves results in the formation of compounds called interhalogens. They are all covalent substances and polarized in such a way that a fractional negative charge resides on the lighter halogen atom(s). With the exception of BrCl, ICl, ICl_3, and IBr, the compounds are all halogen fluorides such as ClF, ClF_3, ClF_5. BrF, BrF_3, BrF_5, IF_5 and IF_7. Hence the inter halogens may be classified into four types, XX′, XX'_3, XX'_5 and XX'_7 where X is the heavier halogen and X′ is the lighter halogen. The only two ternary interhalogen compounds are $IFCl_2$ and IF_2Cl.

All the compounds are formed due to the thermal combination of the halogens.

$$Cl_2 + F_2 \longrightarrow 2ClF - 13.5 \text{ kcals (gas)}$$

$$I_2 + 5F_2 \longrightarrow 2IF_5 - 57.7 \text{ kcals (liquid)}$$

$$Br_2 + 3F_2 \longrightarrow 2BrF_3 - 61.1 \text{ kcals (liquid)}$$

$$Br_2 + 5F_2 \longrightarrow 2BrF_5 - 102.5 \text{ kcals (liquid)}$$

$$I_2 + Cl_2 \longrightarrow 2ICl + 4.2 \text{ kcals (liquid)}$$

$$I_2 + 7Cl_2 \longrightarrow 2ICl_7 - 229.' \text{ kcals (gas)}$$

The formation of these compounds is not surprising in view of large differences in electronegativity within the group and the ability of the heavier halogens to exhibit multi valencies.

The interhalogens are more reactive than the halogens; except fluorine, because X-X′ bond of interhalogens is weaker than the X-X bond of the halogens. The chemical reactions of interhalogens are almost similar to the halogens. The interhalogen fluorides are potential non aqueous ionizing solvents. Their chemistry is confined largely to the reports of fluorination reactions. Many metals, their oxides and halides get fluorinated or form adducts.

$$3M + 2XF_3 \longrightarrow 3MF_2 + X_2$$

$$10MO + 4XF_5 \longrightarrow 10MF_2 + 2X_2 + 5O_2$$

$$NaCl + XF \longrightarrow NaF + XCl$$

$$KI + IF_5 \longrightarrow KI_2\ F_6$$

$$BrF_3 + SbF_3 \longrightarrow SbBrF_6$$

The iodine monochloride is known as Wij's reagent which is used in the estimation of iodine number in unsaturated fats and oils. It also forms polyhalides with alkali metal halides.

$$NaBr + ICl \longrightarrow NaIBrCl$$

Another property of ICl is, it can chlorinate as well as iodinate depending on the conditions of the reactions

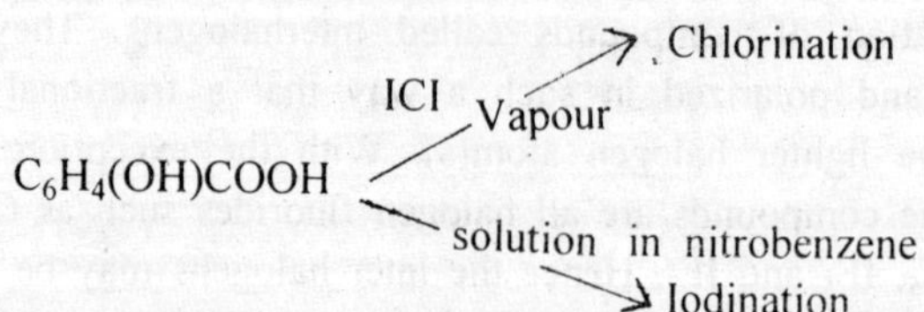

Structures of Interhalogens : The structures of BrF, ClF, BrCl, ICl and IBr are known from spectroscopic and electron diffraction studies of their vapours. They are linear molecules. The bond lengths in gaseous state for these molecules is

Br-F 1.756A	Br-Cl 2.138 A	I-Br 2.470 A
Cl-F 1.628A	ICl 2.321 A	

It is found that greater the electronegativity difference, shorter the bond will be. The thermal stability is in the order if IF > BrF > ClF > ICl >

IBr > Br Cl, this is too dependent of electronegativity difference, more polar the covalent bond stabler the compound will be.

The structure of IF_3 is unknown, but ClF_3 and BrF_3 have been studied by X-ray diffraction as solids, as well as by microwave spectroscopy for the vapours, the results by both the methods are in good agreement with each other. These structures are described as trigonal bipyramids, in which two equatorial positions are occupied by non bonding pairs of electrons.

F —— Cl —— F
1.698Å
89.4°
1.598Å
F

Three possible structures are predicted but it is not certain, which of the three will be formed.

Another suggested mechanism, shows the formation of a dimer, this may be because of the presence of two lone pair of electrons

The F^{19} n.m.r. spectra of BrF_3 shows a similar structure.

From the temperature dependence of the F^{19} n.m.r. spectrum it has been concluded, that there is a rapid fluorine exchange between the apical and equatorial positions. The compounds show an sp^3d hybridization with three unpaired electrons forming bonds with three fluorine atoms. The thermal stability of XX_3A type is of the order for BrF_3 > ClF_3 > ICl_3

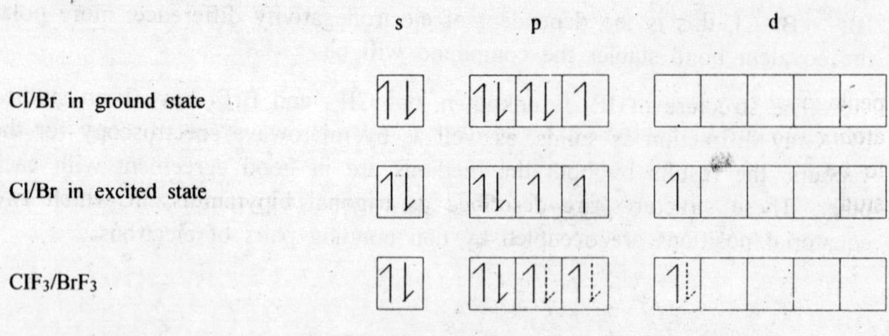

sp^3d hybridization

Iodine trichloride which is present as ICl & Cl_2 in its vapourstate, has a structure in the solid state, similar to dimeric ClF_3, in the form of I_2Cl_6

Cl Cl Cl
I I
Cl Cl Cl

Both (BrF_5, IF_5) the pentahalides have a square pyramidal structure with a pair of electrons occupying the vacant octahedral site. The F(apical) - $\widehat{Br}$ - F (equatorial) angles are all in the range of 80-87°

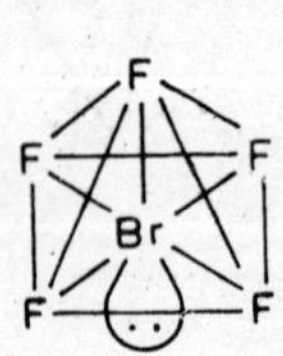

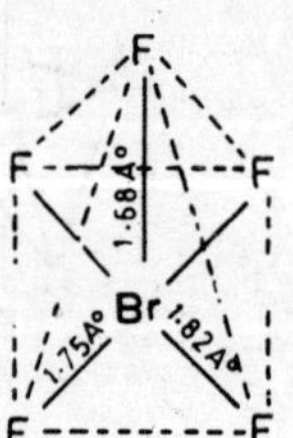

But when a position is provided for the lone pair of electrons as per the expected sp^3d^2 hybridization the structure is expected to be an octahedron with one position occupied by a lone pair.

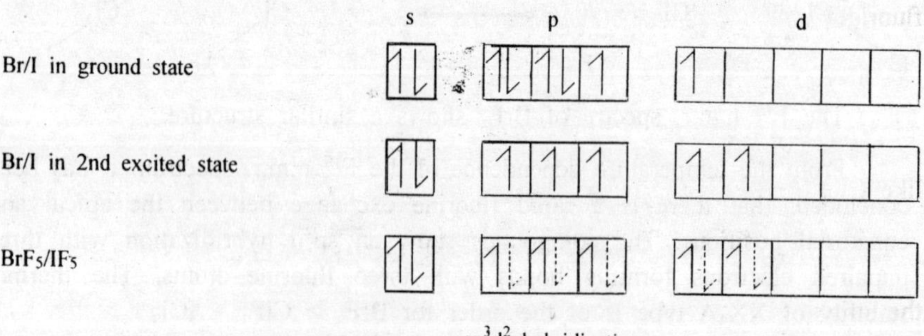

sp^3d^2 hybridization

The BrF_5 is most reactive. IF_5 is a liquid and is a good conductor

The structure of IF_7 molecule as shown in the figure is found to be a pentagonal bipyramid. The fluorine atoms are grouped round the central iodine atom. The electron diffraction study has shown the I-F (apical) bond distance to be 1.77 Å and the equatorial distance being 1.85 Å. The infrared spectral study has suggested the same structure to rhenium and osmium heptafluorides.

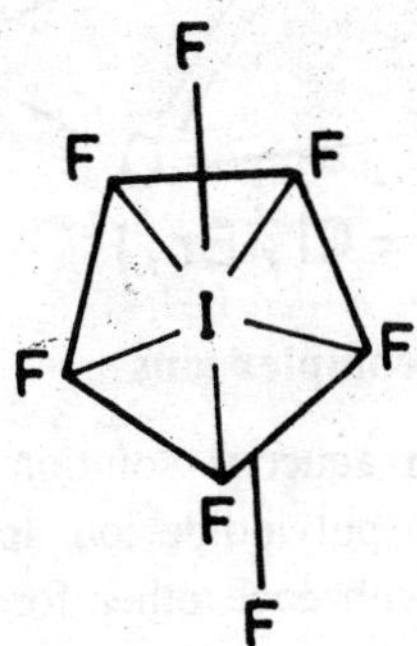

The hybridized orbital approach shows an sp^3d^3 hybridization of the iodine atom.

	s	p			d				
I in ground state	↑↓	↑↓	↑↓	↑					
I in 3rd excited state	↑	↑	↑	↑	↑	↑	↑		
IF_7	↑↓	↑↓	↑↓	↑↓	↑↓	↑↓	↑↓		

sp^3d^3 hybridization

Apart of these, several oxofluorides are reported such as ClO_2F (Chloryl fluoride) which can be obtained by the following reaction

$$6KClO_3 + 10\ BrF_3 \longrightarrow 6KBrF_4 + 2Br_2 + 3O_2 + 6ClO_2F$$

Similarly BrO_2F and IO_2 F are also be found. the perhalogen compounds like ClO_3F, BrO_3F, IO_3F can also be prepared

$$2HIO_4 + 2F_2 \xrightarrow{HF} 2IO_3F + 2HF + O_2$$

the structure of XO_3F is found to be a tetrahedron with X as its centre.

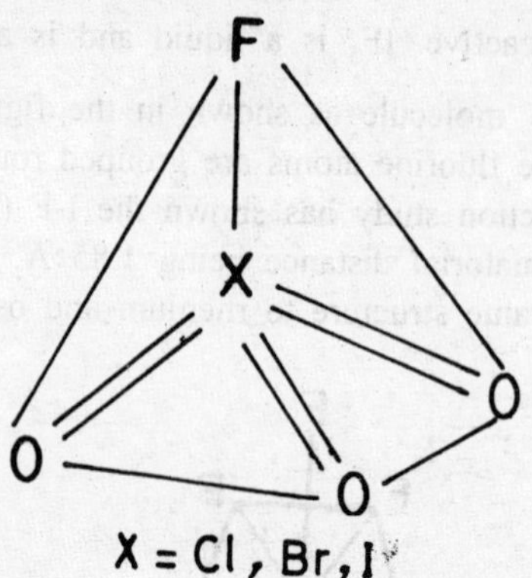

6-6-12 Polyhalides and their complex ions

Iodine when dissolved in an aqueous solution of potassium iodide it forms KI_3, the I_3 ion is said to be a polyiodide ion. Iodine and iodide ions, specially have a tendency to interact with each other forming polyiodide species of the type I_n^-.

The polyhalide ions can be formed not only in aqueous media but can also be formed in non-aqueous media too, such as in solvents like CH_3OH, CH_3CN, C_2H_5OH etc.

The polyhalides can be prepared by the addition of an halide ion to an interhalogen compound.

$$ICl_3 + Cl^- \longrightarrow ICl_4^-$$

$$IBr + Cl^- \longrightarrow IBrCl^-$$

$$IF_5 + F^- \longrightarrow IF_6^-$$

$$BrF_3 + F^- \longrightarrow BrF_4^-$$

The stability of a polyhalide depends upon the size of the central atom, large the central atom stabler the polyhalide will be, sometimes to make a polyhalide more stable a larger cation species is introduced. The thermal stability of a few triatomic polyhalides is in the order of

$$I_3^- > IBr_2^- > ICl_2^- > I_2Br^- > BrCl_2^- > Br_2Cl^-$$

A few polybromides and polychlorides are also reported such as Br_3 and $(NEt_4)Cl_3$. The stability of the polyhalides decreases in the order of $I_3^- > Br_3^- > Cl_3 > I^-$ no polyfluorides are reported.

The polyhalides can be classified as follows, depending on the number of atoms.

Anions			Cations		
Triatomic	*Pentatomic*	*Heptatomic*	*Triatonic*	*Pentatomic*	*Heptatomic*
ICl_2^-,	ICl_4^-	IF_6^-	ClF_2^+	ClF_4^+	ClF_6^+
IBr_2^-	$ICl_3\ F^-$	BrF_6^-	Cl_2^+	BrF_4^+	BrF_6^+
$IBrCl^-$	IF_4^-, BrF_4^-		BrF_2^+	IF_4^+	IF_6^+
$BrCl_2^-$	ClF_4^-, I_5^-		ICl_2^+		
ClF_2^-,			I_2Cl^+		
Cl_3^-, Br_3^-, I_3^-			$IBrCl^+$		
			I_2Br^+		

The general method of preparation of polyhalide compounds is by the addition of a halogen or interhalogen to a solution containing a halide ion or in rare cases a solid halide. The solvent must be resistant to chemical attack by any of the halogens or interhalogens and it must dissolve the halide but should not solvolyze the desired product, hence water is avoided and the usual solvents used are methanol, ethanol, acetic acid etc.

$NR_4I + I_2 \longrightarrow [NR_4][I]_n$ where R = (Me, Et Pr, Bu)

n = 3, 5, 7, 9

$NMe_3H^+ + Br_3^- \longrightarrow [NMe3H]\ [Br3]$

$NH_4^+ + IBiCl \longrightarrow [NH4I]\ [BiCl]$

$I_2Cl^+ + AlCl_4^- \longrightarrow [I_2Cl]\ [AlCl_4]$

$ClF_6^+ + PtF_6 \longrightarrow [ClF_6]\ [Pt\ F_6]$

$BrF_6^+ + Sb_2F_{11}^- \longrightarrow [BrF_6]\ [Sb_2F_{11}]$

$IF_6^+ + BF_4^- \longrightarrow [IF_6]\ [BF_4]$

The ions ClF_6^+ and BrF_6^+ cannot be thought of as the derivatives of ClF_7 and BrF_7 formed by removing one F^- ion, but IF_6^+ is formed IF_7 by a similar process.

As ClF_7 and BrF_7 cannot be formed, hence ClF_6 and BrF_6^+ cannot be formed by the same process of IF_6^+. But instead there can be obtained by the following reactions :

$$ClF_5\ (\text{excess}) + PtF_6 \longrightarrow ClF_6^+\,PtF_6^- + ClF_4^+\,PtF_6^-$$

$$BrF_5\ (\text{excess}) + KrF^+AsF_6^- \longrightarrow Kr + BrF_6^+ + AsF_6^-$$

The polyhalides being quite bulky molecules, their structures are also quite elaborated, as seen for the compound $[BrF_4]\ [Sb_2F_{11}]$

av1·8A° F
2·49 A°
96°
Br
173°
2·24 A°
171°
1·86 A°
1·94 A°
173°
2·15 A°
Sb
mean of five Sb–F 1·75 A°
mean of six Sb–F 1·9 A°

even a simple I_5^- ion, has such a structure, as in Me_4NI_5,

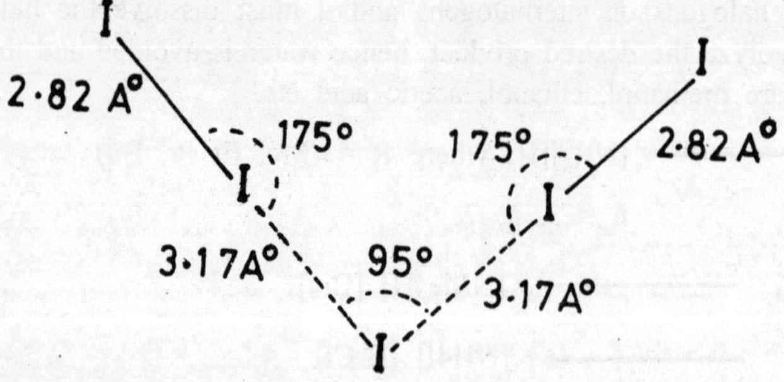

6-6-13 Pseudohalogens

The term pseudohalogen was first suggested by Birckenbach and Kellerman to a molecule consisting of more than two electronegative atoms which in the free state resembles halogens in their chemical behaviour. The principle radicals in this typical category are cyanogen, $(CN)_2$, oxycyanogen, $(OCN)_2$, sulphocyanogen, $(SCN)_2$, selenocyanogen, $(SeCN)_2$; azidocarbondisulphide., $(SCSN_3)_2$ and azides N_3^-. They are also called Halogenoids.

The pseudohalogens should possess the following characters:

(i) They should be volatile with symmetrical structures containing two radicals e.g. $-C \equiv N=$.

(ii) They should be capable of combining with metals to form salts.

$$M + (OCN)_2 \longrightarrow M(OCN)_2$$

(iii) Their Ag, Hg and Pb salts should be insoluble in water.

(iv) Their acids should exist similar to hydrogen halides. e.g. HN_3.

(v) The pseudohalogens should form inter pseudohalogens or mixed pseudohalogens, such as CNN_3, CNSCN, OCNSCN etc.

(vi) The pseudohalogens should combine with halogens, e.g., ClN_3, CNF, OCNCl, SeCNBr, etc.

(vii) The pseudohalogen should form covalent compounds analogous to covalent halides such as $C(CN)_4$, $C_2H_4(SCN)_2$ etc.

(viii) Moreover, the pseudohalogens should get oxidized by a suitable oxidizing agent.

6-6-14 Basic Character of Iodine

The basic character of the family increases as we go down a group in the periodic table. Hence iodine assumes basic character. The evidence of its basicity is the formation of polar covalent compounds by it. With the other halogens iodine forms the interhalogens which on ionization form positively charged iodine ions and negatively charged F, Cl and Br^- ions. Positive iodine compounds are well established. On electrolysis of ICl in the molten state it is found that at anode both I_2 and Cl_2 are liberated, but at cathode only I_2 is liberated, showing its positive ion formation, probably the accompanying reactions is

$$2ICl \longrightarrow I^+ + ICl_2^-$$

Similarly on the electrolysis of $I(CH_3COO)_3$, it is found that a reaction of the following type occurs.

$$I(CH_3COO)_3 \longrightarrow I^{3+} + 3CH_3COO^-$$

Another interesting reaction of iodine is the formation of a basic radical when I_2 is oxidized with conc. HNO_3 which in the presence of acetic anhydride and phosphoric acid gives iodine phosphate, thereby confirming that iodine is basic in its behaviour.

$$I + Ac_2O + H_3PO_4 \longrightarrow IPO_4$$

Cationic iodine exists in sulphuric acid solutions. In dilute HCl^+ iodine forms tripositive I^{3+} ion, when iodine is dissolved in it, whereas when dissolved in concentrated oleum, monopositive I^+ ion is formed.

6-7 ANOMALOUS BEHAVIOUR OF FLUORINE

As it is observed, that the first element of a group always behaves differently from other members of the family, following the same tradition, fluorine, the first member of the halogen family, also shows characteristic properties of its own, which are not similar to chlorine, bromine, iodine and astatine. Some of these difference are as follows :

i) Fluorine is an element with the highest electronegativity in its family, but also amongst the other elements of the periodic table. It has a small size. Hence it exhibits only oxidation state of –1, whereas the other members of the family exhibit oxidation states of +1, +3, +5 and even +7 also.

ii) Fluorine is the most reactive element of the family having a very low F-F bond energy, with a value of 38.5 kcal per mole. The X-X bond energy in Cl_2, Br_2 and I_2 is quite high.

iii) Because of the absence of empty d-orbitals in fluorine, it cannot form hybridized orbitals like chlorine, bromine and iodine do. Hence fluorine cannot form polyhalides ions like Cl_3^-, Br_3^- and I_3^-.

iv) Fluorine combines with hydrogen forming hydrogen fluoride, the molecules of hydrogen fluoride are associated together intermolecularly with hydrogenbonding. HCl, HBr, HI cannot form hydrogen bonding. This ability of formation of hydrogen bonding by HF makes it exsit as a liquid with a boiling point of 19°C. The other hydrogen halides are all gases.

v) The high electronegativity and smaller size of fluorine also result in stronger attractive forces between the constitutents of HF. Hence HF is a weak acid while the HCl, HBr and HI are strong acids.

vi) the metal fluorides behave differently when compared with the metal chlorides, bromides and iodides. For example AgF is soluble in water while the other AgX (X = Cl, Br, I) are water insoluble. On the contrary CaF_2 is insoluble in water whereas CaX_2 (X = Cl, Br, I) are quite water soluble and find the application as descicants.

vii) The only natural isotope of fluorine, F^{19}, has a nuclear magnetic moment, and hence plays a significant role in NMR studies, whereas the other halogens are not useful for this study.

viii) The small size and high electronegativity of flourine allows other metals to attain their maximum coordination number. Thus we observe compounds like IF_7, SF_6 SiF^{-2} etc. The large size of other halides do not permit the achievement of maximum coordination number.

ix) A few fluorides are more stable than their respective other halides. NF_3 is a stable compound while NCl_3 is an explosive.

x) The small size of fluoride ion permits the formation of ions like HF_2^- having a hydrogen bonded structure (H-F-H). This anion combines with K^+ forming KHF_2. $KHCl_2$, $KHBr_2$ and KHI_2 are non existent.

xi) fluorides are highly ionic whereas chlorides, bromides and iodides have more covalent characters. For example AIF_3 and SnF_4 are ionic but their corresponding chlorides, bromides and iodides are covalent.

xii) Fluorine resembles oxygen in its chemical reactivity and physical properties, whereas none of the other halogens show any similarity with it.

For example :

a) Property of hydrogen bonding is exhibited by HF and H_2O.

b) The sulphur shows an oxdation state of +6 in compounds like SO_3 and SF_6.

c) Both oxygen and fluorine combine with carbon directly.

None of these properties are exhibited by other halogens.

6-8 MULTIPLE-CHOICE QUESTIONS

6-8-1 Thyroxine contains

A) astatire B) fluorine C) iodine D) chlorine

6-8-2 The radioactive element is

A) astatine B) fluorine C) iodine D) chlorine

6-8-3 Voilet crystals are of

A) astatine B) bromine C) iodine D) chlorine

6-8-4 $1s^2 2s^2 2p^6 3s^2 3p^6 4s^2 3d^{10} 4p^5$ is the electronic configuration of

A) astatine B) bromine C) iodine D) chlorine

6-8-5 The element showing a lone oxidation state

A) bromine B) chlorine C) astatine D) fluorine

6-8-6 The oxidation state of +7 in a halogen is due to the excitation of the electrons of the outermost orbital. It is due to the

A) second B) third C) first D) fourth excitation

6-8-7 The ionic radii of 1.96 A is of

A) fluorine B) chlorine C) bromine D) iodine

6-8-8 Chlorine has an atomic radii of

A) 1.31A B) 1.44A C) 2.22A D) 0.99A

6-8-9 Ionisation potential of iodine is

A) 9.5 ev B) 10.4 ev C) 11.84 D) 17.42

6-8-10 The value of 3.79 is the electron affinity

A) fluorine B) chlorine C) bromine D) iodine

6-8-11 Oxidation potential is based only on the sum of :

A) lonisation potential, electron affinity, heat of hydration.

B) Electronegativity, Ionisation potential, electron affinity.

C) heat of dissociation, heat of hydration, electronegativity.

D) heat of dissociation, heat of hydration, electron affinity.

6-8-12 The oxidation potential can be calculated by

A) Hess law B) Pauli principle C) Aufbau principle D) Jahn-Teller effect

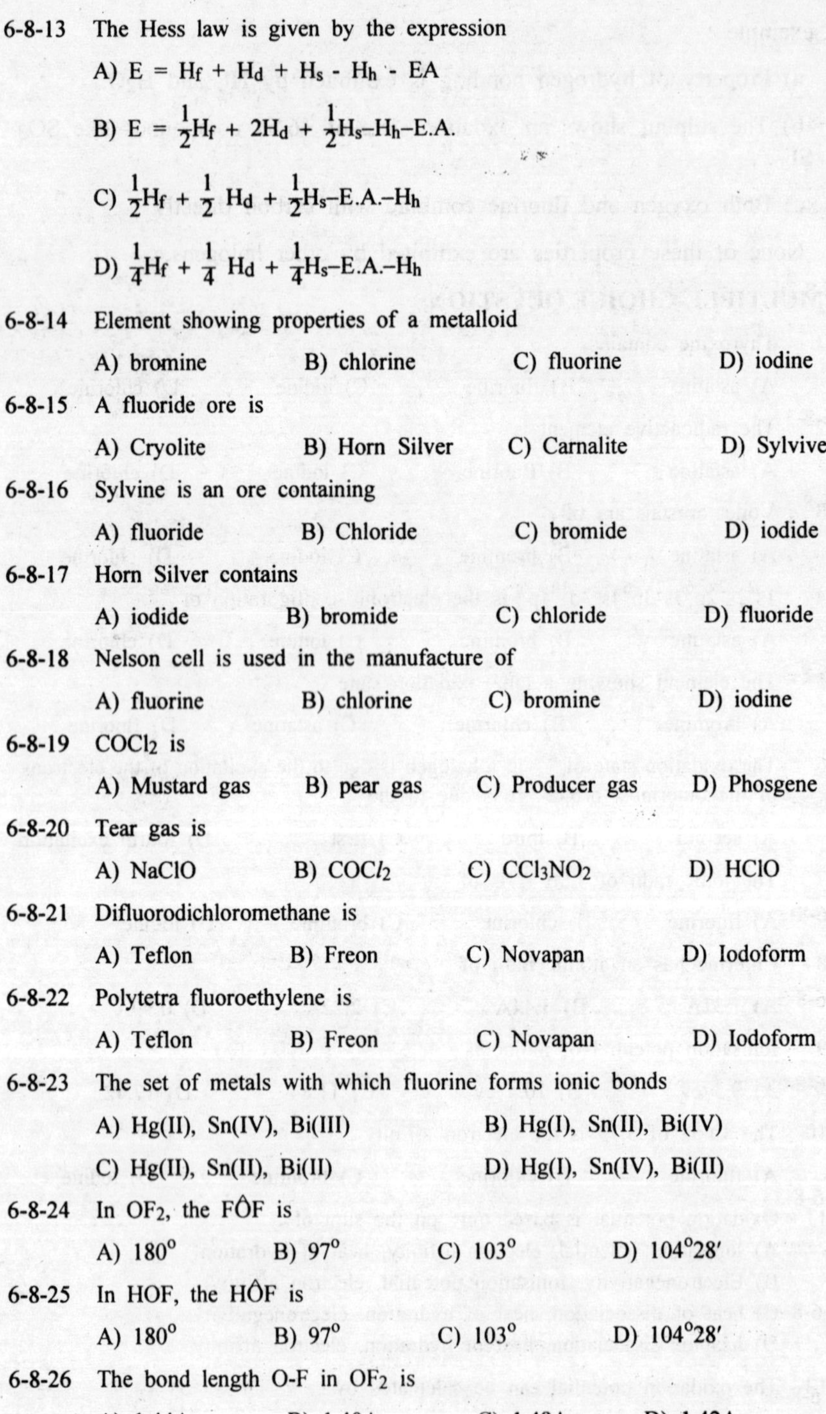

6-8-13 The Hess law is given by the expression

A) $E = H_f + H_d + H_s - H_h - EA$

B) $E = \frac{1}{2}H_f + 2H_d + \frac{1}{2}H_s - H_h - E.A.$

C) $\frac{1}{2}H_f + \frac{1}{2}H_d + \frac{1}{2}H_s - E.A. - H_h$

D) $\frac{1}{4}H_f + \frac{1}{4}H_d + \frac{1}{4}H_s - E.A. - H_h$

6-8-14 Element showing properties of a metalloid

A) bromine B) chlorine C) fluorine D) iodine

6-8-15 A fluoride ore is

A) Cryolite B) Horn Silver C) Carnalite D) Sylvive

6-8-16 Sylvine is an ore containing

A) fluoride B) Chloride C) bromide D) iodide

6-8-17 Horn Silver contains

A) iodide B) bromide C) chloride D) fluoride

6-8-18 Nelson cell is used in the manufacture of

A) fluorine B) chlorine C) bromine D) iodine

6-8-19 $COCl_2$ is

A) Mustard gas B) pear gas C) Producer gas D) Phosgene

6-8-20 Tear gas is

A) $NaClO$ B) $COCl_2$ C) CCl_3NO_2 D) $HClO$

6-8-21 Difluorodichloromethane is

A) Teflon B) Freon C) Novapan D) Iodoform

6-8-22 Polytetra fluoroethylene is

A) Teflon B) Freon C) Novapan D) Iodoform

6-8-23 The set of metals with which fluorine forms ionic bonds

A) Hg(II), Sn(IV), Bi(III) B) Hg(I), Sn(II), Bi(IV)

C) Hg(II), Sn(II), Bi(II) D) Hg(I), Sn(IV), Bi(II)

6-8-24 In OF_2, the $F\hat{O}F$ is

A) 180° B) 97° C) 103° D) $104^o28'$

6-8-25 In HOF, the $H\hat{O}F$ is

A) 180° B) 97° C) 103° D) $104^o28'$

6-8-26 The bond length O-F in OF_2 is

A) 1.44A B) 1.40A C) 1.48A D) 1.42A

6-8-27 Cl_2O is an acid anhydride of

A) $HClO_3$ B) $HClO$ C) $HClO_4$ D) HCl

6-8-28 The acidic anhydride of $HClO_4$ is

A) ClO_2 B) Cl_2O C) Cl_2O_7 D) Cl_2O_6

6-8-29 Chlorite salts are derivatives of

A) Cl_2O B) Cl_2O_6 C) Cl_2O_7 D) ClO_2

6-8-30 The pair of salts formed from the reaction of Cl_2O_6 with NaOH are

A) Perchlorates & Chlorates B) Chlorates & Chlorites

C) Chlorites & Hypochlorites D) Hypochlorites & perchlorates

6-8-31 The structure of I_2O_5 is

A) Tetragonal B) trigonal bipyramid C) Planar D) Tetrahedron

6-8-32 The solubility of a metal halide in water is in the order of :

A) iodide bromide chloride fluoride
B) fluoride chloride bromide iodide
C) chloride bromide iodide fluoride
D) bromide iodide fluoride chloride

6-8-33 $PbCl_2$ is ionic, $PbCl_4$ is

A) also ionic B) covalent

C) coordinate compound D) has vander Waals forces

6-8-34 AlF_3 is

A) Ionic B) covalent
C) coordinate compound D) has vander Walls forces

6-8-35 Chlorine when dissolved in water forms

A) $HClO$ B) $HClO_2$ C) HCl_2O D) $HClO_4$

6-8-36 A solution of bromine in water is

A) $HBrO$ B) $HBrO_2$ C) $HBrO_4$ D) HBr

6-8-37 $2NH_3 + 6Cl_2$ gives

A) $N_2 + 6NH_4Cl$ B) $2NCl_3 + 6\ HCl$
C) $N_2 + 2NCl_3$ D) $2NCl_3 + 6NH_4Cl$

6-8-38 $8NH_3 + 3Cl_2$ gives

A) $N_2 + 6\ NH_4Cl$ B) $2NCl_3 + 6HCl$
C) $N_2 + 2NCl_3$ D) $2NCl_3 + 6NH_4Cl$

6-8-39 $5NH_3 + 3I_2$ gives

A) $N_2 + 6\ NH_4I$ B) $3NH_4I + N_2$ C) $N_2 + 2NI_3$ D) $3NH_4I + NH_3NI_3$

6-8-40 In the presence of sunlight chlorine reacts with metane to give

A) $HCl + C$ B) $C_2H_6 + Cl_2$ C) $C_3H_8 + Cl$ D) $CHCl_3 + H_2$

6-8-41 Chlorine reacts with cold dilute NaOH solution to give

A) NaCl + NaClO + Cl_2O B) NaCl + $NaClO_2$ + Cl_2O

C) $NaClO_2$ + Cl_2O + H_2O D) NaCl + NaClO + H_2O

6-8-42 Chlorine reacts with hot concentrated NaOH solution to give

A) NaCl + $NaClO_3$ + H_2O B) NaCl + $NaClO_2$ + Cl_2O

C) $NaClO_2$ + Cl_2O + H_2O D) NaCl + $NaClO_2$ + Cl_2

6-8-43 Nitronium ion is formed when HF is reacted with

A) HNO_2 B) H_2NO_2 C) HNO_3 D) HNO_4

6-8-44 Methanol when treated with hydrogen fluoride gives one of the products as

A) CH_3F B) CH_3OF C) $(CH_3OH\text{-}H)^+$ D) C_6H_5F

6-8-45 Hypobromous acid is

A) HBrO B) $HBrO_2$ C) $HBrO_3$ D) $HBrO_4$

6-8-46 Perchloric acid is

A) $HClO_3$ B) $HClO_4$ C) $HClO_2$ D) HCl_2O

6-8-47 In aqueous solutions the halic acids behave as strong acids because of the formation of

A) HXO_3^+ B) H_3O^+ C) OH^- D) H^+

6-8-48 An interhalogen amongst the following is

A) BrF_3^+ B) BrF_5^- C) BrF_5 D) $KBrF_5$

6-8-49 The possible structure of ClF_3 is

A) trigonal bipyramid B) triangular planar
C) tetrahedron D) planar

6-8-50 ClF_3 can form a

A) trimer B) tetramer C) polymer D) dimer

6-8-51 Chloryl fluoride is

A) ClO_2F B) ClO_3F C) ClOF D) $ClFO_4$

6-8-52 The structure of BrO_3F is

A) trigonal bipyramid B) tetrahedron
C) Octahedron D) tetragon

6-8-53 $1F_7$ has a structure of

A) Octahedron B) heptagon
C) pentagonal bipyramid D) trigonal bipyramid

6-8-54 Sulphocyanogen is

A) SeCN B) SCN^- C) $(SeSN)_2$ D) $(SCN)_2$

6-8-55 An interpseudohalogen amongst these is

A) OCNCl B) $(C\ N)_4$ C) CNN_3 D) $(OCN)_2$

6-8-56 The analogue of methane amongst these is

A) $C_4(OCN)$ B) $C_3H(CN)$ C) $C(CN)_4$ D) CH_3Br

6-8-57 Halogenoid amongst these is

A) CNSCN B) $(CN)_2$ C) Br D) $BrCl_3$

6-8-58 Iodine can form one amongst the following pairs of ions

A) I_3^- and I^{3+} B) I_3^{3+} and I^{3-} C) I^- and I^{3-} D) I^{3+} and I^{2+}

6-8-59 A polyhalide amongst these is

A) $BrCl_3$ B) IBr_3 C) $3Br^-$ D) Br_3^-

6-8-60 Compound formed on reacting I_2Cl^+ with $AlCl_4^-$ will have its composition as :

A) $AlCl_5I_2$ B) $[I_2Cl]$ $[AlCl_4]$
C) $[I_2Cl_3]$ $[AlCl_3]$ D) Al $[Cl_4I_2]Cl$

6-9 SHORT AND ESSAY-TYPE QUESTIONS

6-9-1 Discuss the gradation of properties of halogens with reference to
a) Colour
b) Electronaffinity
c) State
d) Action with alkalies. (O.U., Oct. 1988)

6-9-2 Give the structure of ClF_3 and IF_5. (O.U., Apr. 1988)

6-9-3 Discuss the method of preparation of Fluorine. (O.U., Apr. 1988)

6-9-4 Write short notes on the reactivity of pseudohalogens. (O.U., Oct. 1987, Apr. 1985 & May 1989)

6-9-5 Chlorine has the highest electron affinity but fluorine is most reactive. Explain. (O.U., Oct. 1987 & Apr. 1987)

6-9-6 HF is a liquid and HCl is a gas at room temperature. Explain. (O.U., Apr. 1987)

6-9-7 Give the method of preparation of fluorine. In what respects does it differ from other halogens? (O.U., Apr. 1987)

6-9-8 What are inter halogens compounds ? Give two examples. Write the structures of the inter halogen compounds that iodine & fluorine form. (O.U., Apr. 1987, Mar. '89, '90, K.U., Nov. '90 and Oct. '89, May '88)

6-9-9 HF is a weaker acid in aqueous solution than other hydrogen halides. Explain. (O.U., Apr. 1987)

6-9-10 Describe the preparation, geometry and hybridization of ICl_5 and IF_7. (O.U., Nov. 1986)

6-9-11 Discuss the gradation in the properties of the elements of group VII B (i) Hydrides (ii) Oxides. (O.U., Mar. 1986. & Mar. 1990)

6-9-12 Comment on the basic character of iodine and suitably illustrate it. (O.U., Oct. 1989)

6-9-13 Make a comparative study of the halides formed by the elements of the halogen family. (O.U., Oct. 1989)

6-9-14 Why does iodine show basic properties? Name two salts in which cationic iodine is present. (O.U., Mar. 1989)

6-9-15 Account for the

A) Dissociation energy of F_2 is lower than Cl_2

B) Interhalogen compounds are diamagnetic (O.U., Oct. 1989)

6-9-16 Make a comparative study of the elements of group VIIB with reference to

(i) Colour

(ii) Electrode potential

(iii) Electronegativity

(iv) Oxidation states

(v) Reaction with hydrogen. (O.U., Oct. 1988)

6-9-17 Outline the principle of the electrolytic process used for the preparation of fluorine. (O.U., Oct. 1988)

6-9-18 What is electron affinity? Why do the halogens have high electron affinity values? (O.U., Apr. 1985 & Apr. 1988)

6-9-19 Why does fluorine show an oxidation state of -1 only? (O.U., Oct. 1984)

6-9-20 Why is the electron affinity of fluorine less than that of chlorine? (O.U., Oct 1984 & Apr. 1984)

6-9-21 How do you explain BrCl is covalent whereas NaCl is ionic? (O.U., Apr. 1984)

6-9-22 Write about two hydrides of halogens with a stress on physical state, reactivity and acidic properties. (A.U., May 1988)

6-9-23 Write a note of the oxides of halogens. (A.U., Oct. 1989)

6-9-24 How do you explain different physical states and colour of halogens? (A.U., May 1988)

6-9-25 Taking fluorine as an example, show that the first element in a group behaves quite differently from its group elements. (O.U., Nov. 1986)

6-9-26 Why does iodine from tri,penta and hepta fluorides only? (O.U., Nov. 1985)

6-9-27 How is chlorine heptaoxide prepared? Give two of its important properties and write its structure. (O.U., Oct. 1985)

6-9-28 What are the difficulties encountered in the preparation of fluorine and how are they remedied? (O.U., Apr. 1985 & Oct. 1985)

6-9-29 Write equations to show the action of fluorine with dilute and concentrated potassium hydroxide solution. (O.U., Oct. 1985)

6-8-30 Give the structure of ClF_3. (O.U., Oct. 1985)

6-9-31 How is fluorine prepared?

a) In what respect does fluorine differ from the rest of the elements in the VII group?

b) Write the reaction of F with H_2O.

c) What are the uses of F. (O.U., Apr. 1985)

6-9-32 What is the action of fluorine on

a) water?

b) $KClO_3$? Give equations. (O.U., Jun. 1984)

6-9-33 Describe the preparation and the structure of ClF_3. (O.U., Jun. 1984)

6-9-34 Write the electronic structure of all the oxyacids of chlorine. Give the properties of perchloric acid as an acid and as an oxidant. (O.U., Nov. 1984)

6-9-35 Give a concise comparative account of chemistry of the halogens with reference to general variations in properties with increasing atomic number.

6-9-36 Discuss the chemistry of fluorine and any two of its compounds.

6-9-37 Oxygen combines with Cl_2, Br_2 & I_2 to form oxides but combines with F_2 to form fluorides. Explain.

6-9-38 Give an account of the chemistry of the fluorides.

6-9-39 "The halogens form a well defined family of elements as indicated by their position in the periodic table". Discuss.

6-9-40 Write in the form of reactions what happens when
(1) Cl_2 reacts with hot NaOH, & cold NaOH
(2) Cl_2 reacts with H_2S.
(3) Cl_2 reacts with acetylene
(4) Cl_2 reacts with aqueous solution of KBr

6-9-41 Compare and contrast the reactions of fluorine and iodine.

6-9-42 Write an account of the oxyacids of chlorine.

6-9-43 Account for the high reactivity of fluorine when compared to other halogens.

6-9-44 Draw the structures of the four types of oxyacids formed by halogens. Give a write up about the chemistry of any two oxyacids of chlorine.

6-9-45 Justify the position of halogens in the same group of the periodic table.

6-9-46 Write a comparative account of the oxides and oxyacids of VII B group elements. (A.U., 1978)

6-9-47 What are interhalogen compounds? How are they classified and prepared? Mention the important uses of ICl. (A.U., 1980)

6-9-48 Enumerate the properties and reactions of periodic acid. (S.V.U., 1980)

6-9-49 In F_2O, the bond angle is less than tetrahedral angle while in Cl_2O and Br_2O it is greater than tetrahedral angle. Explain. (A.U., 1985)

6-9-50 Discuss the structures of BrF_3 and IF_5. (A.U., 1985)

6-9-51 Discuss the structures of ClO_4 and IF_3 on the basis of hybridization. (O.U., 1983)

6-9-52 Give the hybridization and shape of the following IF_5 and IF_7. (N.U., 1987)

6-9-53 Give the names and formulae of any two oxyacids of chlorine and calculate the oxidation number of chlorine in each. (N.U., 1988)

6-9-54 Write notes on

a) Interhalogens compounds

b) Pseudohalogens (N.U., Oct. 1989)

6-9-55 Explain why halogens are oxidizing agents? (N.U., Oct. 1989)

6-9-56 Which halogen acid is the strongest? Why? (N.U., Apr. 1989)

6-9-57 Write about the hydrides of halogens with a stress on their physical state reactivity and acidic property.

6-9-58 What are interhalogen compounds? Write a note on structural formulae of interhalogen compounds. (S.K.U., Apr. 1988)

6-9-59 How do you explain different physical states and colours of halogens. (A.U.. Apr. 1988)

6-9-60 Give an account of interhalogen compounds. (A.U., Dec. 1989 & K.U., Oct. 1990)

6-9-61 Explain the basic character of iodine. (A.U., Dec. 1989)

6-9-62 Give an account of pseudohalogens. (A.U., Dec. 1989)

6-9-63 Write short notes on oxyacids of chlorine (A.U., Jun. 1990)

6-9-64 Name and give structures of chief oxides and oxyacids of chlorine. Describe the methods of preparation and properties of hypochlorous acid. (S.K.U., Apr. 1988)

9-9-65 Justify the position of fluorine, chlorine and bromine in the same sub-group of the periodic table outline the chief points of similarity of fluorine and oxygen. (K.U., Oct. 1988)

ANSWERS

6-8-1 C	**6-8-2** A	**6-8-3** C	**6-8-4** B	**6-8-5** D	**6-8-6** B
6-8-7 C	**6-8-8** D	**6-8-9** B	**6-8-10** B	**6-8-11** D	**6-8-12** A
6-8-13 C	**6-8-14** D	**6-8-15** A	**6-8-16** B	**6-8-17** C	**6-8-18** B
6-8-19 D	**6-8-20** C	**6-8-21** B	**6-8-22** A	**6-8-23** A	**6-8-24** C
6-8-25 B	**6-8-26** B	**6-8-27** B	**6-8-28** C	**6-8-29** D	**6-8-30** A
6-8-31 C	**6-8-32** A	**6-8-33** B	**6-8-34** A	**6-8-35** A	**6-8-36** D
6-8-37 B	**6-8-38** A	**6-8-39** D	**6-8-40** A	**6-8-41** D	**6-8-42** A
6-8-43 C	**6-8-44** C	**6-8-45** A	**6-8-46** B	**6-8-47** B	**6-8-48** C
6-8-49 A	**6-8-50** D	**6-8-51** A	**6-8-52** B	**6-8-53** C	**6-8-54** D
6-8-55 C	**6-8-56** C	**6-8-57** B	**6-8-58** A	**6-8-59** D	**6-8-60** B

7

THE ELEMENTS OF THE ZERO GROUP *THE NOBLE GASES*

					He
B	C	N	O	F	Ne
Al	Si	P	S	Cl	Ar
Ga	*Ge*	As	*Se*	Br	Kr
In	Sn	Sb	Te	I	Xe
Tl	Pb	Bi	Po	At	Rn

7-1 INTRODUCTION

The noble gases of the zero group of the periodic table are helium (He), neon (Ne), argon (Ar), krypton (Kr), xenon (Xe) and radon (Rn). As these elements form chemical compounds less readily than the other elements of the periodic table they are named **noble elements.** Except helium, which has an electronic structure containing a duplet, $1s^2$, the others have an electronic structure containing an octet. Hence none of these elements show any oxidation state, and therefore the group to which these elements belong to is also known as **the zero group.**

The noble gases were not discovered at the time when Demitri Mendeleev formulated his periodic table and therefore these gases do not find a place in Mendeleev's periodic table. Lord Rayleigh discovered the presence of these gases in the atmosphere in 1894, the discovery being an off shoot of his experiments related to the nitrogen's presence in the air. Rayleigh found that the density of nitrogen prepared from ammonia was 1.2505 g dm^{-3}, while that prepared from air was 1.2572 g dm^{-3}. By passing nitrogen obtained from the air repeatedly over heated magnesium (which forms magnesium nitride with nitrogen), Ramsay obtained traces of a gas, which on spectroscopic examination, showed no signs of nitrogen but contained new lines corresponding to new element argon, (Greek : inert). The detection of helium (Greek : helios = sun) in the atmosphere in trace amounts by Kaiser in 1895, was followed by isolation of another inert gas neon, (Greek : new) by Ramsay and Travers in 1898 during the fractional distillation of impure liquid argon. The less volatile fractions from liquid air were soon shown by the same workers to contain two other new elements krypton (Greek : hidden) and xenon (Greek : stranger) in 1898. The family was rendered complete by the discovery of radon as a product of radioactive disintegration.

Two other significant developments in helium chemistry also merit attention. The identity of ionized helium, called as the α-particle, was established in 1903 by Ramsay and Soddy. In 1907, Cady and McFarland reported the presence of 1.84% of helium by volume in certain natural gases.

Helium is unique because it only forms a solid under pressure and it exists in two liquid phases. Liquid He^3 appears to be normal liquid, but He^4 has very peculiar properties which are usually not expected of any known substance. At a temperature of about 4K it forms a liquid, known as He I that possesses properties of a normal liquid. If this is further cooled under a pressure of 1 atmosphere at a temperature of about 2K results in the formation of another super fluid form of helium, known as He II, which is in equilibrium with He I. The physical properties change abruptly. He II has a viscosity at least as low as most sensitive methods find it difficult to measure, which is 1/100th of that of gaseous hydrogen, and the liquid flows apparently without any friction. If the levels in two concentric vessels are unequal it will flow from one to the other, even going uphill if necessary. The specific heat, surface tension and compressibility are also quite abnormal. Its thermal conductivity increases to a very large extent and is almost 800 times greater than that of copper. Some researchers have referred to He II as a fourth state of matter. This abnormal behaviour of helium has got no explanation for it yet. From the temperature-pressure phase diagram, Fig. 7- 1, of helium near absolute zero it is noticed, that, there is an absence of a triple point and also the solid, liquid and gaseous phases do not coexist in equilibrium.

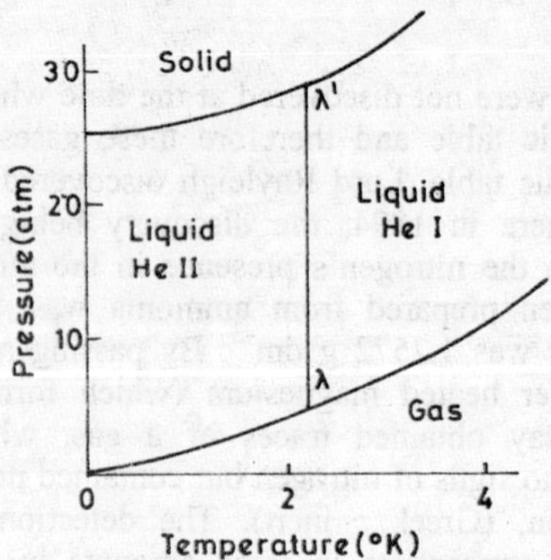

Fig. 7-1 Phase diagram of He near absolute zero

The other noble gases do not possess any such peculiar properties, although on the whole, these gases behave differently from the other gases of the periodic table.

7-2 POSITION IN THE PERIODIC TABLE

These mono atomic gases which are often called by different names such as **rare gases, inert gases, noble gases** etc. were allotted a place between the highly electronegative halogens and the highly electropositive alkali metals, thus forming a bridge between the two most chemically reactive groups of the periodic table. Hence these gases are often referred to as the **bridging elements** also.

Group VII B	*Group O*	*Group IA*
H	He	Li
F	Ne	Na
Cl	Ar	K
Br	Kr	Rb
I	Xe	Cs
At	Rn	Fr

These bridging elements are placed at the extreme end of the periodic table at the termination of the p-block and hence also take the name **terminal elements.** This position is in conformity with their atomic structure, gaseous nature and some non-metallic properties, although the periodic properties which follow a regular uniform trend in other members of the periodic table alter so abruptly after the halogens that there is little similarity worth mentioning. Even though the inert gases are nonmetals like the preceding VIIB group elements, (the halogens), they do not form diatomic molecules as fluorine (F_2), chlorine (Cl_2) etc., does. On the other hand, the succeeding alkali metals have nothing common and are quite different both physically and chemically from the noble gases. Thus, none of the properties of these elements have any resemblance with their immediate neighbours of group VIIB or group IA. This also negates any of our contention of categorize them as elements of transition between the highly electronegative halogens and the highly electropositive alkali elements but we can call them to be the bridging elements of the periodic table.

Owing to their *inertness and the complete octet showing zero valency, Ramsay,* who made an extensive work in their discovery and chemistry, introduced a new group called **zero group** and placed them just after the halogens due to their nonmetallic gaseous behaviour. Later Moseley, the father of modern periodic table, when arranged the elements as per their atomic number also approved this position of these gases.

7-3 GENERAL GROUP TRENDS OF PHYSICAL PROPERTIES

The noble gases form a very homogenous group of the periodic table as observed from the regular trends in their physical properties presented in table 7-1.

The uniatomic state of the noble gases is proved to be correct from the ratio of the specific heats which correspond with the ideal value of 1.667 for mono atomic gases.

From regularly increasing melting point and boiling point values it is clear that the interatomic van der Waals forces increase on going from He to Rn. Due to low melting point and hence negligible cohesive forces helium exists as a normal liquid (He I) and a super fluid (He II) state.

Table 7-1 Physical properties of zero group elements

Property	*He*	*Ne*	*Ar*	*Kr*	*Xe*	*Rn*
At. No.	2	10	18	36	54	86
At. Wt.	4.0	20.18	39.94	83.7	131.3	222
Van der Waals Radius (A°)	1.2	1.6	1.9	2.0	2.2	–
At. volume (gm^3)	31.8	16.8	28.5	32.2	42.9	50.5
Ionization Potential (eV)	24.5	21.5	15.7	13.9	12.1	10.7
Density (g gm^{-3})	0.126	1.204	1.65	2.6	3.06	4.4
Heat of fusion (kcal $mole^{-1}$)	0.033	0.08	0.28	0.34	0.55	–
Heat of vaporization (kcal $mole^{-1}$)	0.025	0.405	1.60	2.24	3.10	3.6
Ratio of specific heats (r=Cp/Cv)	1.65	1.64	1.65	1.69	1.67	–
Solubility in water (mol lit^{-1} 20°C)	13.8	14.7	37.9	73	110	–
Melting Point (K)	0.9	24	84	116	161	202
Boiling point (K)	4.0	27	87	121	164	211

The ionization potentials of the noble gases as well as the van der Waals radii are found to be highest in their own period. It is because of a very stable electronic structure with completely filled orbitals.

7-4 ELECTRONIC CONFIGURATION

The electronic configuration of the noble gases shows that they have a very stable structure with completely filled valency shells, which make these elements chemically inert. They have very little tendency to form any sort of chemical bonding by the normal methods. The compounds of these elements are prepared under very specific conditions with very limited number of elements. That is why, very few compounds of these gases are known.

Noble gas	*Atomic number*	*Electronic configuration*
Helium (He)	2	$1s^2$
Neon (Ne)	10	$1s^2\ 2s^2\ 2p^6$
Argon (Ar)	18	$1s^2\ 2s^2\ 2p^6\ 3s^2\ 3p^6$
Krypton (Kr)	36	$1s^2\ 2s^2\ 2p^6\ 3s^2\ 3p^6\ 4s^2\ 3d^{10}\ 4p^6$
Xenon (Xe)	54	$1s^2\ 2s^2\ 2p^6\ 3s^2\ 3p^6\ 4s^2\ 3d^{10}\ 4p^6\ 5s^2\ 4d^{10}\ 5p^6$
Radon (Rn)	86	$1s^2\ 2s^2\ 2p^6\ 3s^2\ 3p^6\ 4s^2\ 3d^{10}\ 4p^6\ 5s^2\ 4d^{10}\ 5p^6\ 6s^2\ 5d^{10}\ 4f^{14}\ 6p^6$

7-5 PRINCIPLES OF ISOLATION

All the noble gases are present in the atmosphere except radon which is a product of radioactive distintegration of radium and it has physical and chemical properties appropriate to its position as the last member of the zero group.

Helium is the second most abundant element in the universe (65% H_2 and 23% He), formed as a result of its synthesis from hydrogen, but being too light to be retained by the earths gravitational field, all the primordial helium has been lost into outer atmosphere.

$$_1H^2 + {}_1H^2 \longrightarrow {}_2He^4 + 26.72 \text{ MeV}$$

Terrestrial helium is the result of radioactive α- decay argon, $_{18}Ar^{40}$, is obtained from electron capture by $_{19}K^{40}$. The noble gases make up about 1% of the earth's atmosphere in which their major component is Ar. Smaller concentrations are occluded in igneous rocks, but the atmosphere is the principal commercial source of Ne, Ar, Kr and Xe, which are obtained as bye-products of the liquifaction and separation of air. Some Ar is also obtained from synthetic ammonia plants in which it accumulates after entering as an impurity in the nitrogen and hydrogen feeds.

The volume percent of the noble gases in the atmosphere is as follows :

He : 0.0005%

Ne : 0.0016%

Ar : 0.93%

Kr : 0.1 to 1.0 × 10^{-5} %

Xe : 0.1 to 1.0 × 10^{-6} %

Hence radioactive sources and air are the common sources of the inert gases. They are isolated from these sources.

7-5-1 Isolation of crude inert gas mixture

An inert gas mixture is isolated from liquid air by adopting one of the following methods.

Rayleigh's Method : In this process a 50 litre glass globe is taken with eleven volumes of oxygen and nine volume of air in it. A discharge of 6000-8000 volts is passed into it using two heavy platinum electrodes. A continuous steam of NaOH solution is introduced into this bulb, which is drawn out after it absorbs the oxides of nitrogen.

$$N_2 + O_2 \rightleftharpoons 2NO$$

$$2NO + O_2 \rightleftharpoons 2NO_2$$

$$2NO_2 + 2NaOH \longrightarrow NaNO_2 + NaNO_3 + H_2O$$

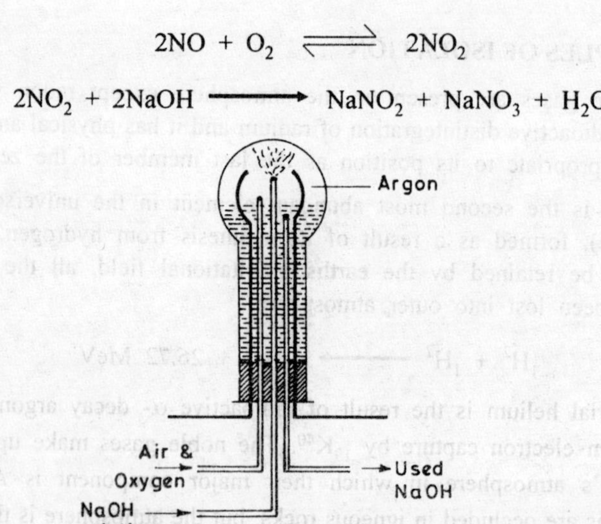

Fig. 7-2 Rayleigh's method for the preparation of Argon

The excess oxygen is removed by passing the residual gas through potassium sulphide, figure 7-2. the gas now left behind contains a mixture of all the inert gases, with argon present in plenty and hence it is often referred to as crude argon.

Rayleigh-Ramsay Chemical Method : This method was first applied by Rayleigh and Ramsay. The crude fraction of the mixture of rare gases is obtained by removing the water vapour by passing over phosphorous pentoxide; carbon dioxide by passing over solid caustic soda; oxygen when passed over heated copper and nitrogen by passing it over heated magnesium. The residual gases are a mixture of the inert gases.

7-5-2 Fractional Adsorption

The crude argon obtained from the above two processes is now subjected to fractional adsorption by passing it over activated coconut charcoal. This method is also referred to as **Dewar's Charcoal Method.** The activated coconut charcoal is kept in a bulb of liquid air maintained at a temperature of -120° C, at this temperature all gases except helium and neon get adsorbed over the charcoal. The unadsorbed gases are passed over activated charcoal in a separate bulb maintained at a temperature of - 180°C. At this temperature neon is adsorbed while helium remains unadsorbed and is collected. The adsorbed neon is collected by warming the charcoal slowly to a temperature of - 120°C.

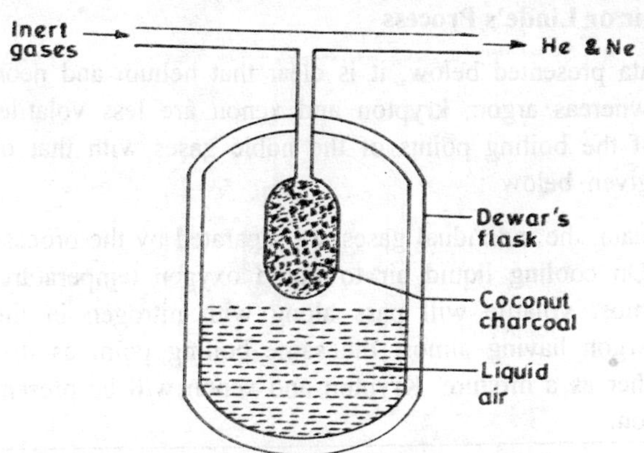

Fig. 7-3 Dewar's fractional adsorption apparatus for rare gases

The charcoal containing the mixture of argon, krypton and xenon is now cooled in a separate Dewar's flask to -120°C. Argon liquifies and gets released while krypton and xenon are still adhered to the charcoal surface. The charcoal is now heated to - 80°C, where krypton gets released and is collected. The temperature is now raised to 0°C in order to free xenon from the charcoal. Thus all the noble gases are separated by this method, the Dewar's activated charcoal apparatus is shown in Fig. 7-3.

The above adsorption process is summarized in Flow sheet 7-1.

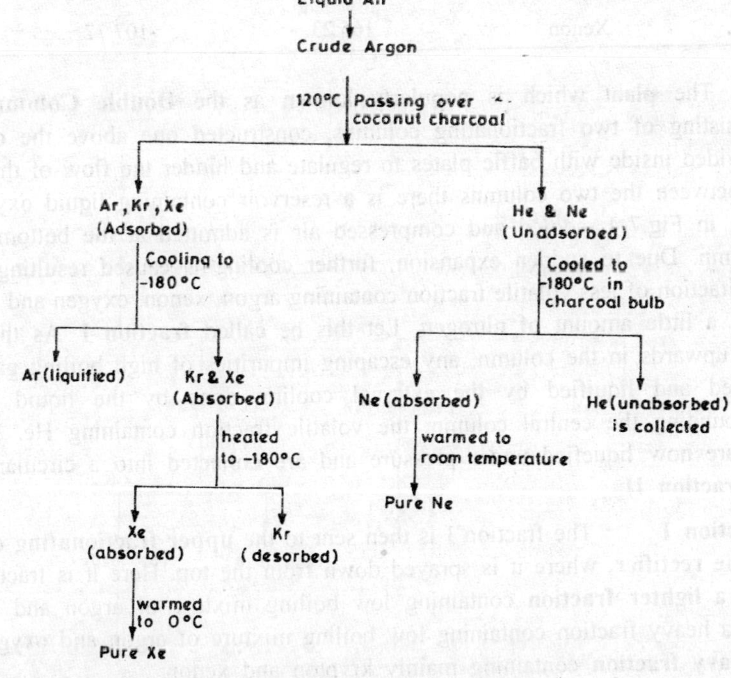

Flow sheet 7-1 Adsorption process of rare gases

7-5-3 By Liquifaction of air or Linde's Process

From the boiling point data presented below, it is clear that helium and neon are most volatile gases, whereas argon, krypton and xenon are less volatile. A comparative acctount of the boiling points of the noble gases with that of nitrogen and oxygen, is given below :

Based on the above data, the individual gases are separated by the process of selective liquifaction. On cooling liquid air to liquid oxygen temperature, helium and neon being most volatile will pass along with nitrogen in the highly volatile fraction. Argon having almost the same boiling point as that of oxygen, will pass together as a mixture. Krypton and xenon will be present in the least volatile fraction.

Element	*Boiling point*	
	^oK	^oC
Helium	4.13	-268.87
Neon	27.18	-245.82
Nitrogen	77.20	-195.80
Argon	87.15	-185.85
Oxygen	90.02	-182.98
Krypton	120.11	-152.89
Xenon	165.23	-107.77

The plant which is popularly known as the **Double Column plant** consisting of two fractionating columns, constructed one above the other, is provided inside with baffle plates to regulate and hinder tne flow of the gases. In between the two columns there is a reservoir containing liquid oxygen, as seen in Fig.7-4. cooled and compressed air is admitted at the bottom of the column. Due to sudden expansion, further cooling is caused resulting in the liquifaction of less volatile fraction containing argon, xenon, oxygen and krypton with a little amount of nitrogen. Let this be called **fraction I**. As the gases rise upwards in the column, any escaping impurities of high boiling gases get cooled and liquified by the external cooling done by the liquid oxygen surrounding the central column. the volatile fraction containing He, Ne and N_2 are now liquefied under pressure and are collected into a circular jacket as **fraction II.**

Fraction I : The fraction I is then sent to the **upper fractionating column of the rectifier,** where it is sprayed down from the top. Here it is fractionated into a **lighter fraction** containing low boiling mixture of argon and oxygen and a heavy fraction containing !ow boiling mixture of argon and oxygen and a **heavy fraction** containing mainly krypton and xenon.

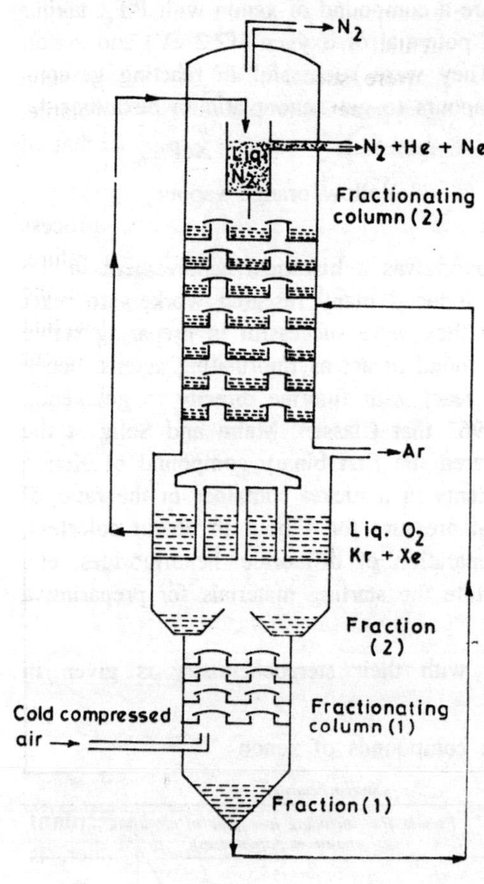

Fig. 7-4 Linde's double column plant

Lighter fraction : Oxygen present in the lighter fraction is either burnt in the presence of hydrogen or removed by passing over hot copper.

$$O_2 + 2H_2 \longrightarrow 2H_2O$$

$$O_2 + 2Cu \longrightarrow 2CuO$$

Thus pure argon is isolated from the lighter fraction.

Heavier fraction : The heavier, high boiling fraction contains mainly Kr and Xe. These two gases have a boiling point difference of 42° and hence are separated by fractionation.

Fraction II :

Fraction II has N_2, Ne and He. The nitrogen is removed by cooling the fraction to liquid nitrogen temperature followed by removal of trace amounts by heating with hot magnesium to give magnesium nitride. The left out mixture of neon and helium is then cooled with liquid hydrogen which solidifies neon leaving helium free as a gas.

Thus all the required noble gas fractions are obtained.

7-6 COMPOUNDS OF NOBLE GASES

The discovery of the noble gases was a direct result of their unreactive nature and early unsuccessful attempts to induce chemical reaction establishes the belief of their inertness. In 1933 Linus Pauling predicted that KrF_6 and XeF_6 should be syntesized. D.M. Yost and A.L. Kaye unsuccessfully attempted to prepare these compounds. Peacock was successful in preparing $ArBF_3$ as orange crystals, but they decomposed immediately after its isolation. In 1962, PtF_6 was prepared and also its dioxygen complex $[O_2]^+$ $[PtF_6]^-$ or O_2PtF_66. This made Bartlett and Lohmann to prepare a compound of xenon with PtF_6, taking into consideration that the ionization potential of oxygen (12.2 eV) and xenon (12.13 eV) are almost the same. They were successful in reacting gaseous xenon with platinum hexafluoride vapours to get xenonplatinum hexafluoride.

$$Xe(g) + PtF_6 \rightleftharpoons XePtF_6$$

Red vapour   Yellow orange vapour

The preparation of this compound was a historical achievement in the chemistry of the noble gases. This induced many research workers to react other hexafluorides, with xenon and they were successful in preparing stable compounds. These compounds were found to act as fluorinating agents, hence it was thought that xenon can also react with fluorine directly to get xenon fluorides. It was, on 2nd August 1962 that Classen, Malm and Selig at the Argonne National Laboratories prepared the first binary compound of Xenon with fluorine. They took these elements in a nickel container in the ratio of 1:5 and heated at 400ºC and 6 atm. pressure for 1 hour to obtain colorless, XeF_4 crystals. This lead to the preparation of difluoride, hexafluorides, etc. These binary xenon fluorides constitute the starting materials for preparing a number of xenon compounds.

A list of Xenon compounds with their stereochemistry is given in Table 7-2.

Table 7-2 Some compounds of xenon

Compound	*Oxidation State of Xenon*	*M.P°C*	*Stereochemistry*	
			Actual	*Pseudo (i.e. including lone pair of electrons, shown in parentheses)*
1	2	3	4	5
XeF_2	+2	129	Linear	Trigonal bipyramidal (3) F—Xe—F
XeF_4	+4	117	Square planar	Octahedral (2)
XeF_6	+6	49.6	Distorted octahedron	Pentagonal bipyramidal (1)
$[XeF_5]^+ AsF_6^-$	+6	130.5	Square pyramidal	Octahedral
$CsXeF_7$, Cs_2XeF_8	+6	50d	—	—

$[NO_2]^+$ $[XeF_8]^{2-}$	+6		Antiprismatic	Lone pair inactive	
$XeOF_4$	+6	-46	Square pyramidal	Octahedral (1)	
XeO_2F_2	+6	30.8	See-Saw	Trigonal bipyramidal (1)	
$CsXeOF_5$	+6	-	Distorted tetrahedral	oc-Capped octahedral (1)	
$KXeO_3F$	+6	-	Square pyramidal	Octahedral (1)	
XeO_3	+6	Explodes	Pyramidal	Tetrahedral	
XeO_3F_2	+8	-54.1	Trigonal bipyramidal	Trigonal bipyramidal	
XeO_4	+8	-35.9	Tetrahedral	No lone electrons	
XeO_6	+8	-	Octahedral	No lone electrons	
Ba_2XeO_6	+8	300d	Octahedral	No lone electrons	

It is not possible to study all these compounds, so we will restrict our study to few representative samples like; XeF_2, XeF_4, XeF_6, $XeOF_2$, $XeOF_4$, XeO_2F_2, XeO_3, XeO_4, based on their preparation, structure and bonding.

7-6-1 Xenondifluoride (XeF_2)

It is prepared by photochemical combination of xenon with fluorine in the presence of mercury as a sensitizer.

$$Xe + F_2 \xrightarrow[h\nu]{\text{Mercury}} XeF_2$$

On heating a mixture of xenon and fluorine in the molecular ratio of 1 : 3 in a nickel tube at 400°C followed by sudden cooling also gives colourless XeF_2.

$$Xe + F_2 \xrightarrow[Ni]{400^oC} XeF_2$$

If an electric discharge is passed in a 1 : 3 mixture of Xe and F_2 then also XeF_2 is obtained.

$$Xe + F_2 \xrightarrow{\text{Electric discharge}} XeF_2$$

XeF_2 is a colourless crystalline solid, dissolving in HF and melting at -140°C. It is a very mild fluorinating agent. It reacts with benzene to form fluorobenzene liberating xenon.

$$C_6H_6 + XeF_2 \longrightarrow C_6H_5F + HF + Xe$$

XeF_2 is a strong oxidizing agent. It oxidizes hydrochloric acid to chlorine, and iodine to iodine fluoride in the presence of boron trifluoride.

$$2HCl + XeF_2 \longrightarrow Cl_2 + Xe + 2HF$$

$$I_2 + XeF_2 \xrightarrow{BF_3} 2IF + Xe$$

Structure of XeF_2 : As per the valence bond theory SeF_2 has a trigonal bipyramidal structure, Fig. 7-5. Fluorine being highly electronegative,

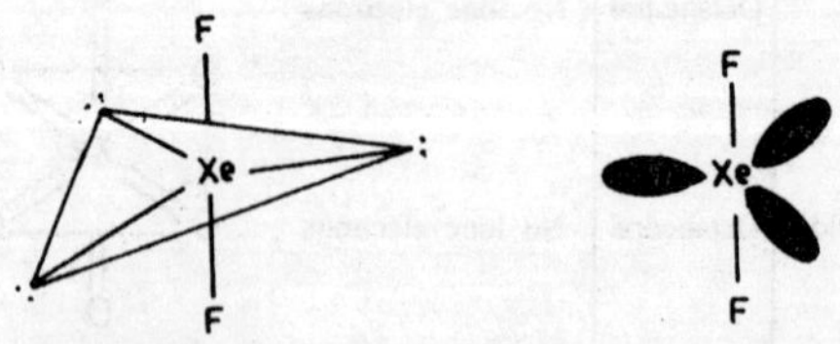

Fig. 7-5 Shape of XeF_2

considerable amount of polarity might be expected. Thus polarity of XeF_2 is found as :

$$\overset{-0.5\delta}{F} — \overset{+1\delta}{Xe} — \overset{-0.5\delta}{F}$$

the structure can be studied by the Valence bond concept given in Fig.7-5. Xe has electronic configuration.

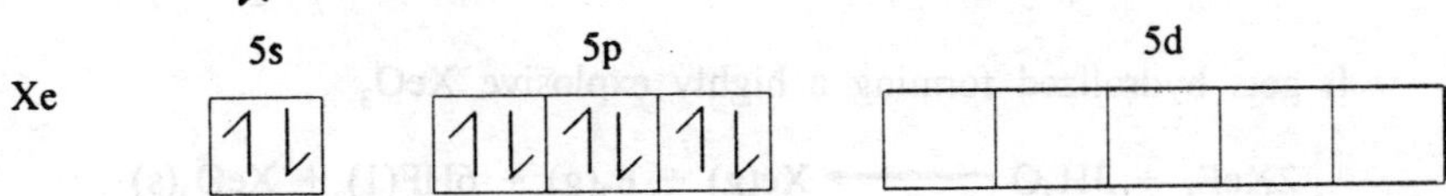

In the complex formation, one electron pair gets unpaired and the fluorine with two lone electrons get paired with them forming a covalent bond. The hybridization exhibited is sp^3d,

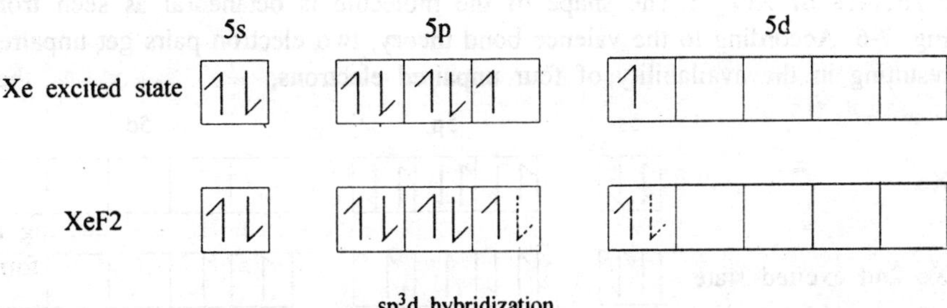

confirming the structure to be trigonal bipyramid with three electron pairs present on the apex of the triangle with xenon and fluorine lying diagonal to each other, showing the compound to be a three center, four electron σ, i.e., a delocalized σ MO (molecular orbital) array over the linear FXeF molecule.

7-6-2 Xenon tetrafluoride (XeF_4)

It is prepared by heating a mixture of xenon and fluorine, in a ratio of 1 : 5, at a temperature of 400°C in a nickel vessel under a pressure of 5 to 6 atmospheres.

$$Xe + 2F_2 \xrightarrow[400^oC]{Ni} XeF_4$$

When a silent electric discharge is passed in a 1 : 2 mixture of Xe and F_2 at a very low temperature of -80°C, XeF_4 is liberated.

$$Xe + 2F_2 \xrightarrow[-80^oC]{\text{electric discharge}} XeF_4$$

It is a colourless, crystalline solid decomposing at 77°C and sublimes very readily. It is a mild oxidizing agent, and gets reduced with hydrogen

$$XeF_4 + 2H_2 \longrightarrow Xe + 4HF$$

It reacts with boron trichloride to form boron trifluoride.

$$3XeF_4 + 4BCl_3 \longrightarrow 4BF_3 + 3Xe + 6Cl_2$$

It gets hydrolized forming a highly explosive XeO_3

$$2XeF_4 + 3H_2O \longrightarrow Xe(g) + F_2(g) + 6HF(1) + XeO_3(s)$$

It is capable of forming adducts with compounds like SbF_5

$$XeF_4 + SbF_5 \longrightarrow XeF_4.SbF_5\ [XeF_3]^+\ [SbF_6]^-$$

Structure of XeF_4 : The shape of the molecule is octahedral as seen from Fig. 7-6. According to the valence bond theory, two electron pairs get unpaired resulting in the availability of four unpaired electrons,

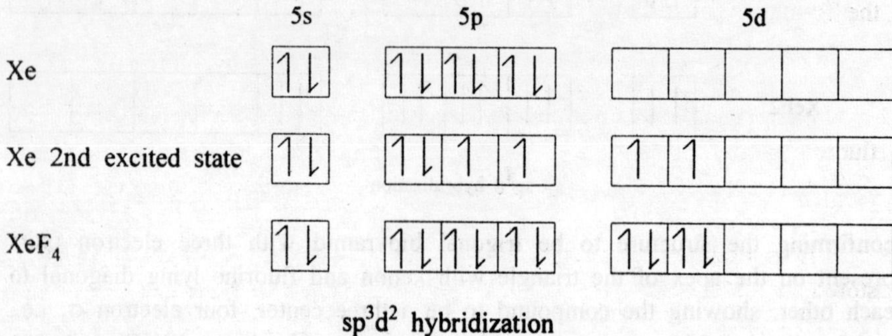

sp^3d^2 hybridization

which combine covalently with four fluorine atoms. The fluorine atoms are present on the square planar structure of the octahedral with two lone electron pairs occupying the apex of the octahedral. It shows a sp^3d^2 hybridization.

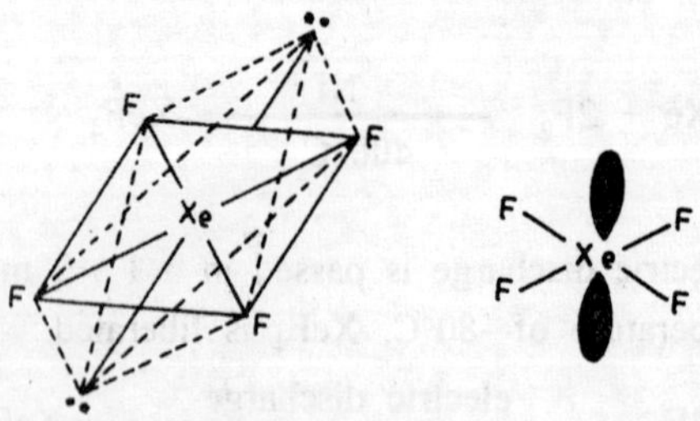

Fig. 7-6 Shape of XeF_4

7-6-3 Xenon hexafluoride XeF_6

When excess of fluorine is heated with xenon, (ratio of 1 : 20), at a pressure of 50 atmospheres in a nickel vessel, XeF_6 is formed.

$$Xe + 3F_2 \xrightarrow[50\ atm]{Ni} XeF_6$$

It is prepared by the interaction of XeF_4 with F_2 under pressure.

$$XeF_4 + F_2 \longrightarrow XeF_6$$

When a silent electric discharge is passed through a mixture of xenon and fluorine at a low temperature, XeF_6 is formed.

$$Xe + 3F_2 \longrightarrow XeF_6$$

It is a crystalline solid and is highly volatile. It melts at 46°C and the vapours of this compound has greenish yellow colour.

XeF_6 dissolves in hydrofluoric acid giving a conducting solution due to the formation of a complex $XeF_5 . HF_2$.

$$XeF_6 + HF \longleftrightarrow XeF_6HF \longleftrightarrow [XeF5]^+ \quad [HF_2]^-$$

Same complex is formed by the reaction between xenondifluoride and fluorine.

$$XeF_2 + 2F_2 \longrightarrow [XeF_5]^+ . [HF_2]^-$$

It is highly reactive and it even reacts with SiO_2, hence it cannot be stored in glass vessel.

$$2XeF_6 + SiO_2 \longrightarrow 2XeOF_4 + SiF_4$$

It undergoes reduction with hydrogen to give hydrogen fluoride.

$$XeF_6 + 3H_2 \longrightarrow Xe + 6HF$$

Similar to XeF_4, on hydrolysis, it produces the highly explosive XeO_3

$$XeF_6 + 3H_2O \longrightarrow XcO_3 + 6\ HF$$

Structure of XeF6 : According to the general belief and the usual trend observed in many compounds, XeF6 should have been an octahedron, but it is not so. The shape of the molecule is a pentagonal bipyramid, with a slight distortion due to the presence of a lone electron pair, as seen in Fig. 7-7. The reasoning given for the latter shape is its tendency to form an adduct, which is possible only when it coordinates

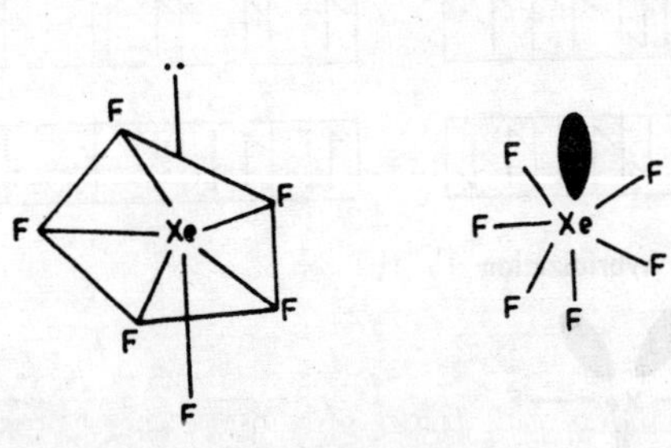

Fig. 7-7 Structure of XeF6

through its electron pair with the reacting molecule like CsF, SbF_5 etc.

$$CsF + XeF_6 \longleftrightarrow CsF.XeF_6 \longleftrightarrow [XeF_5]^+pt\ .\ [CsF_2]^-$$

The compound formation can take place covalently. The xenon gets excited, resulting in the unpairing of three electron pairs contributing six lone electrons, which gets paired up with six electrons of the fluorine atoms. The hybridization is of sp^3d^3 type.

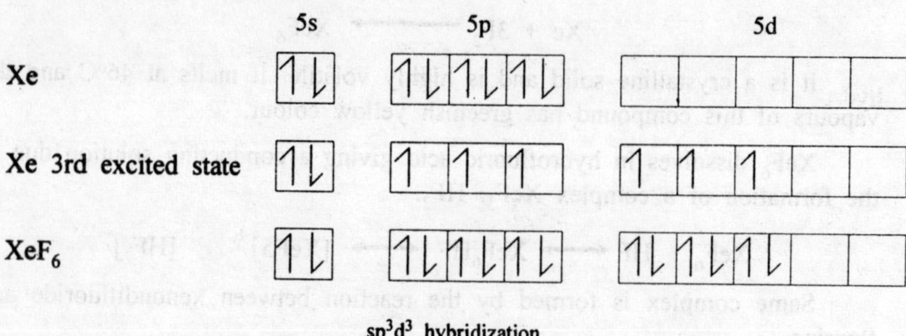

sp^3d^3 hybridization

7-6-4 Xenon oxydifluoride, $XeOF_2$

This is prepared by the partial hydrolysis of XeF_4 at a low temperature, but is an unstable compound.

$$XeF_4 + H_2O \longrightarrow XeOF_2 + 2HF$$

Structure of $XeOF_2$: $XeOF_2$ should have a trigonal bipyramid structure as it shows an sp^3d hybridization. Two of the electron pairs of p-orbital gets unpaired contributing four lone electrons. The two fluorine atoms form two sigma bonds, and the oxygen forms a double bond, with one sigma and one pi bond,

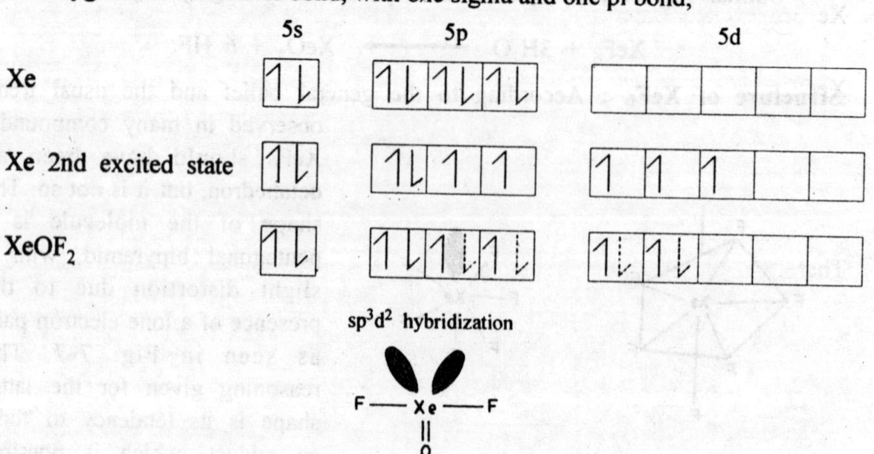

Fig. 7-8 Structure of $XeOF_2$

7-6-5 Xenon oxytetrafluoride, $XeOF_4$

In very small quantities $XeOF_4$ is prepared by the partial hydrolysis of xenon hexafluoride.

$$XeF_6 + H_2O \longrightarrow XeOF_4 + 2HF$$

In quantitative amount it is prepared by heating Xe and F_2 in a ratio of 1 : 4 in the presence of oxygen at 230°C.

$$2Xe + 4F_2 + O_2 \longrightarrow 2XeOF_4$$

It is a colourless compound that melts at -40°C, and gets reduced by hydrogen very easily.

$$XeOF_4 + 3H_2 \longrightarrow Xe + H_2O + 4HF$$

It gets hydrolysed with water giving xenon dioxydifluoride.

$$XeOF_4 + H_2O \longrightarrow XeO_2F_2 + 2HF$$

It reacts with silica also to form xenon trioxide.

$$XeOF_4 + SiO_2 \longrightarrow XeO_3 + SiF_4$$

Structure of $XeOF_4$: With regard to the structure of $XeOF_4$, it is found that on excitation of Xe, three of the p-electrons are promoted to the d-orbital, thus generating six lone electrons available for bond formation. Four of them form sigma bonds with the four fluorine atoms and two of them form a double bond with oxgygen (containing one sigma and one pi bond). The five sigma bonds formed indicate a sp^3d^2 hybridization.

	5s	5p			5d				
Xe	↑↓	↑↓	↑↓	↑↓					
Xe 3rd excited state	↑↓	↑	↑	↑	↑	↑	↑		
$XeOF_4$	↑↓	↑↓	↑↓	↑↓	↑↓	↑↓	↑↓		
		(σ bonds)			(π bonds)				

sp^3d^2 hybridization

Therefore, the structure of the molecule is octahedron as seen in Fig. 7-9.

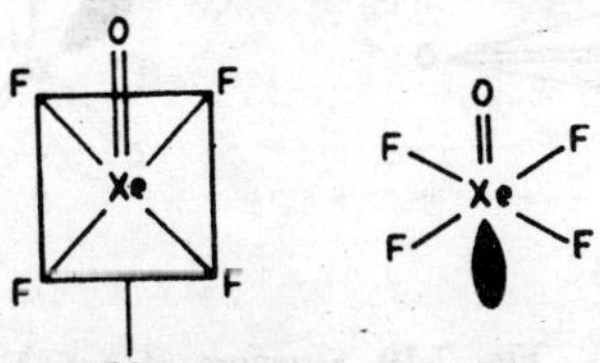

Fig. 7-9 Structure of $XeOF_4$

7-6-6 Xenon dioxydifluoride XeO_2F_2

It is prepared by the hydrolysis of $XeOF_4$ or by the reaction of $XeOF_4$ with xenon trioxide.

$$XeOF_4 + H_2O \longrightarrow XeO_2F_2 + 2HF$$

$$XeOF_4 + XeO_3 \longrightarrow 2XeO_2F_2$$

It is colourless crystalline solid which melts at 31°c. It gets hydrolysed easily with water to form XeO_3.

$$XeO_2F_2 + H_2O \longrightarrow XeO_3 + 2HF$$

Structure of XeO_2F_2 : As in the previous case and in this compound also Xe excites three of its p-electrons to d-orbital, making six unpaired electrons available for bond formation. Two of them combine with fluorine, and four of them form two double bonds with two oxygen atoms, each double bond having one sigma and one pi bond. Hence it is observed to have a sp^3d hybridization. Therefore the possible structure of the compound should be trigonal bipyramid, but it seems to be distorted because of the presence of the double bonds and a lone pair of electrons in one of the equatorial positions.

	5s	5p			5d				
Xe	↑↓	↑↓	↑↓	↑↓					
Xe 3rd excited state	↑↓	↑	↑	↑	↑	↑	↑		
XeO_2F_2	↑↓	↑↓	↑↓	↑↓	↑↓	↑↓	↑↓		
		(4 σ bonds)			(2 π bonds)				

sp^3d hybridization

The structure of XeO_2F_2 is given in the Fig. 7-10.

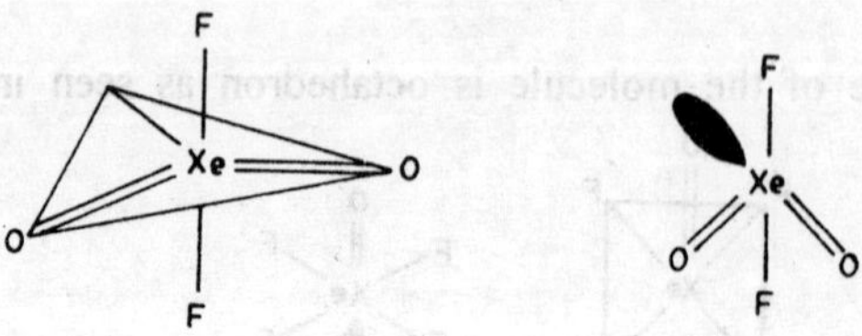

Fig. 7-10. Structure of XeO_2F_2

7-6-7 Xenon trioxide XeO3

Xenon trioxide is obtained as a white non-volatile compound when XeF_6 or XeF_4 is hydrolysed.

$$XeF_6 + 3H_2O \longrightarrow XeO_3 + 6HF$$

$$2XeF_4 + 3H_2O \longrightarrow XeO_3 + Xe + 6HF + F_2$$

It is highly stable in aqueous solutions and is highly explosive. It explodes violently when dry even on a slight rubbing or pressure.

It is a strong oxidizing agent and oxidizes Pu^{3+} to Pu^{4+} in the presence of an acid.

$$6PuCl_3 + XeO_3 + 6HCl \longrightarrow 6PuCl_4 + Xe + 3H_2O$$

When reacted with a strong base, disproportionation occurs forming perxenates.

$$2XeO_3 + 4NaOH + 6H_2O \longrightarrow Na_4XeO_6.8H_2O + Xe + O_2$$

The perxenate solutions are powerful oxidizing agents.

The stable perxenate salts can be precipitated as $Na_4XeO_36\text{-}8H_2O$, $Na_4XeO_6.6H_2O$, $Ba_2XeO_6.5H_2O$. It is found by X-ray crystallography that XeO_6 is present in an octahedral form.

Structure of XeO_3 : The xenon gets excited to an extent of three p-electrons going into d-orbital, making available six electrons for bond formation. The three oxygen atoms form double bonds with Xe, therefore there are 3 sigma bonds and 3 pi bonds, showing a hybridization of sp^3 type.

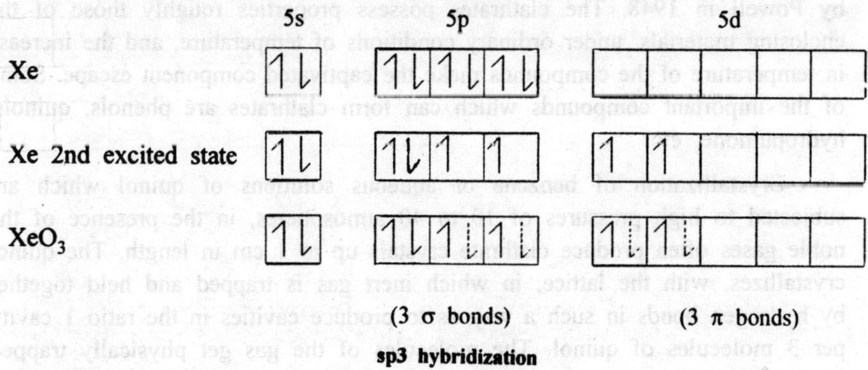

The molecule has a tetrahedral shape as seen in Fig. 7-11.

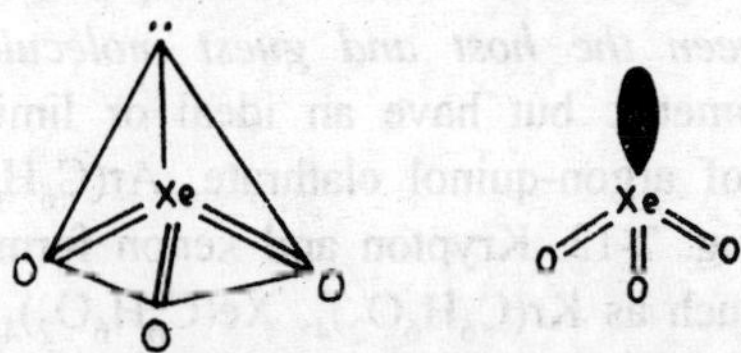

Fig. 7-11 Structure of XeO_3

7-6-8 Xenon tetraoxide, XeO_4

It is prepared by treating barium perxenate with concentrated sulphuric acid.

$$Ba_2XeO_6 + 2H_2SO_4 \longrightarrow XeO_4 + 2BaSO_4 + 2H_2O$$

It is highly explosive compound and explodes at temperatures as low as –40°C. The molecule has a tetrahedral shape as shown in Fig. 7-12.

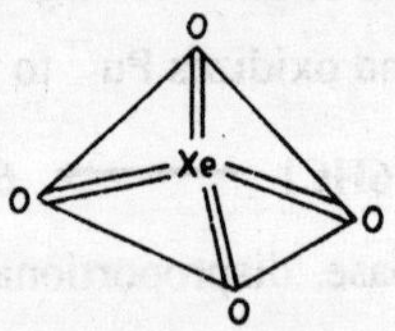

Fig. 7-12 Structure of XeO_4

7-7 CLATHRATES

Certain solid molecular compounds have been prepared in which a molecule of one component is completely enclosed by one or more molecules of a second component in such a way that its escape is prevented, unless the forces which bind its surrounding together are not overcome. Such **cage compounds** *have been called* **clathrate compounds.** They were first prepared and reported by Powell in 1948. The clathrates possess properties roughly those of the enclosing materials, under ordinary conditions of temperature, and the increase in temperature of the compounds make the captivated component escape. Some of the important compounds which can form clathrates are phenols, quinols, hydropuinone, etc.

Crystallization of benzene or aqueous solutions of quinol which are subjected to high pressures of 10 to 40 atmospheres, in the presence of the noble gases often produce clathrate crystals up to 1 cm in length. The quinol crystallizes, with the lattice, in which inert gas is trapped and held together by hydrogen bonds in such a way as to produce cavities in the ratio 1 cavity per 3 molecules of quinol. The molecules of the gas get physically trapped in these cavities, and there being only weak van der Waals forces acting between the *guest* inert gas molecules and host quinol molecules, no *chemical bond is formed between the host and guest molecules.* The clathrates are therefore non stoichiometric but have an ideal or limiting composition. The limiting composition of argon-quinol clathrate, $Ar(C_6H_6O_2)_3$ is 1 : 3 and its structure is given in Fig. 7-13. Krypton and xenon form clathrates with quinol having compositions such as $Kr(C_6H_6O_2)_4$, $Xe(C_6H_6O_2)_4$ etc. once formed they have considerable stability and have a storage life of even few years, but the gas is released on dissolution or melting.

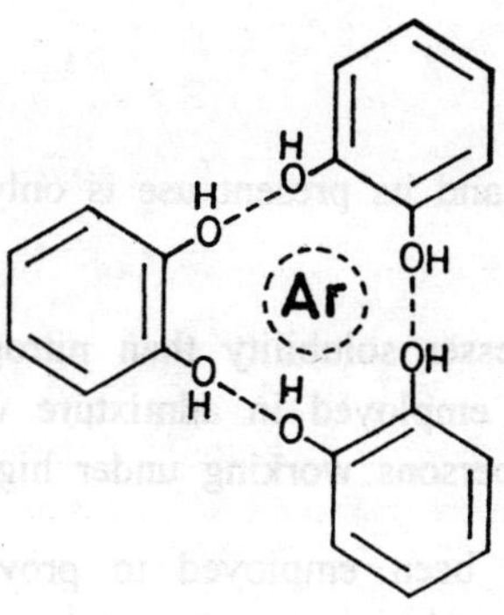

Fig. 7-13 : Ar $(C_6H_6O_2)_3$

Noble gases also form clathrates with water, when frozen under high pressure of the gas. They have the ideal composition of $[(NG)_8(H_2O)_{48}]$. *Clathrates are formed only by Ar, Kr and Xe but not by He or Ne.* The small size of He and Ne molecules do not permit the inert gas to fit into the cage. A comparable phenomenon occur when synthetic zeolites are cooled under a high pressure of the gases like Ar and Kr, where they get capitulated in the aluminosilicate frameworks of the zeolites. Samples containing upto 20% by weight of Ar have been obtained

Uses of clathrate compounds

(i) Clathrates provide a means of storing noble gases.

(ii) Clathrates also provide a means of holding the various radioactive isotopes of Kr and Xe which are produced in nuclear reactors. Thus radioactive Kr^{85} isotope produced by the fission of uranium in the nuclear reactors is selectively trapped and used as a compact energy source in small atomic batteries.

(iii) The fact that He and Ne do not form clathrate compounds is advantageously used in separating these gases from the other three inert gases, viz., Ar, Kr, Xe, present in an inert gas mixture.

7-8 USES OF NOBLE GASES

The general uses of the noble gases are, they are used as coolants in high thermal reactors. They are used as a flow gas in gas-liquid chromatography and for the deaeration of solutions and as a general inert diluent to create the inert atmosphere.

Uses of Helium :

(i) Helium is used as a non-flammable gas.

(ii) It has a lifting capacity of approximately 1 kg per m^3, hence for quite a long time, helium was used in airships, but, now this has become outdated and its present use is only in lifting meteorological baloons.

(iii) Because of its lesser solubility than nitrogen in the blood stream, helium has been employed in admixture with oxygen by deep sea divers and with persons working under high pressures.

(iv) Helium has also been employed to provide inert atmosphere for handling active metals.

(v) In combination with argon it is used in metallurgical processes such as welding.

(vi) Helium mixed with oxygen is used as artificial breathing mixture, for asthma patients.

(vii) Because of high thermal conductivity, low viscosity, low density and extra ordinary inertness, helium is used as a coolant in the nuclear reactors.

(viii) The inertness and low boiling points allow liquid helium to be used in studying low temperature reactions and scientific observations like super conductivity.

Uses of Neon :

(i) Neon is used in discharge tubes used for illumination and decorative lightings.

(ii) Different colors can be obtained by mixing with other noble gases and these colored discharge lamps find application in neon signs, used for advertising purposes. In these tubes neon is subjected to a discharge of about 1000 volts at a pressure of 2 mm. Neon normally produces a brilliant orange red color. When mixed with mercury vapour it emits green or blue elumination depending upon the proportion of the mixture.

(iii) The neon lights are visible through long distances, and can penetrate through thick fogs.

(iv) Neon lights have a very important use in the strodoscope (an instrument for observing, measuring, or adjusting, vibration, rotation, or the like; by using a stroke light that make the moving object appear stationary).

Uses of Argon :

(i) Argon is used extensively in gas filled incandescent lamps, it is preferred over nitrogen because of its inertness.

(ii) It is also used in producing a complete inert atmosphere for welding and metallurgical purposes.

(iii) Many Geiger-Mueller counter tubes contain argon in them.

(iv) It has been used in thermoionic valves and fluorescent tubes.

Uses of Krypton and Xenon

Krypton and Xenon because of their rarity have been used only to a limited extent.

(i) They are better substitutes for argon in the incandescent lamps.

(ii) They are used in discharge tubes for producing high speed flash of bluish light which is used in quick photography, the flash some times last as short as one fifty thousandth (1/50000) part of a second.

(iii) Liquid xenon is used in the detectors employed for the detection of gamma photons and neutral mesons.

Uses of Radon :

(i) Radon has been used in the treatment of malignant tumors.

(ii) It is also used as a radioactive source in testing metal castings.

(iii) It is used as a substitute for X-rays in industrial radiography.

Uses of compounds of Xenon

(i) The xenon compounds find applications of their own. Xenon fluorides are said to prolong the life time of tungsten filaments in incandescent lamps.

(ii) They may also prove useful for fluorinating unsaturated hydrocarbons.

(iii) Xenon trioxide is used as a powerful oxidant. It has been used in the laboratories to estimate alcohols, carboxylic acids and amino acids.

7-9 MULTIPLE-CHOICE QUESTIONS

Place a tick mark before the correct answer.

7-9-1 The zero group gases have an electronic configuration of

(A) ns^2np^6 (B) $ns^2np^6(n-1)d^{10}$
(C) ns^1np^6 (D) $ns^2nd^{10}(n-1)p^8$

7-9-2 The only element of the zero group which will not form any definite compound is

(A) Xenon (B) Neon (C) Radon (D) Krypton

7-9-3 The shape of Xenon tetrafluoride is

(A) Square planar (B) Tetrahedral
(C) Octahedron (D) Distorted octahedron

7-9-4 The number of lone pairs of electrons in xenon hexafluoride is

(A) One (B) Two (C) Three (D) Four

7-9-5 One of the noble gases which can be obtained from radioactive minerals by the action of conc. H_2SO_4 is

(A) He (B) Ar (C) Kr (D) Rn

7-9-6 Xe is obtained as the main product by the hydrolysis of

(A) XeF_2 (B) $XeFO_4$
(C) XeF_6 (D) All these compounds

7-9-7 Structure of Xenon hexafluoride is

(A) Octahedron (B) Pentagonal bipyramid
(C) Trigonal bipyramid (D) Hexagon

7-9-8 The Cp/Cv ratio of a rare gas is

(A) 4.33 (B) 3.44 (C) 1.66 (D) 2.55

7-9-9 The rare gas that is not found in atmosphere

(A) Helium (B) Xenon (C) Krypton (D) Radon

7-9-10 The forces which are between the atoms of the rare gases are of

(A) Covalent (B) Ionic (C) Van der Waals (D) Hydrogen

7-9-11 In the clathrate compounds of rare gases, the gaseous atoms are held by

(A) Covalent bonding (B) Hydrogen bonding
(C) Covalent coordinate bonding (D) π-bonding

7-9-12 The rare gas which cannot form an octet

(A) He (B) Ne (C) Ar (D) Kr

7-9-13 The stable electronic configuration of He is

(A) $1s^1 2s^1$ (B) $1s^2$ (C) $1s^0 2s^2$ (D) $1s^2 2s^1$

7-9-14 The radioactive element in the family of rare gases is

(A) Kr (B) Xe (C) Rn (D) He

7-9-15 The hybridization observed in XeF_4 is

(A) sp^2d^2 (B) sp^3 (C) sp^3d^2 (D) $d\ sp^2$

7-10 SHORT AND ESSAY-TYPE QUESTIONS :

7-7-1 (a) How are zero group elements separated by Dewar's method ?

(b) Give the structures of XeF_4 and XeO_3 and mention the hybridization in these compounds. (O.U., Apr. 1991)

7-7-2 What are clathrate compounds ? How are they prepared ? (O.U., Apr. 1991)

7-10-3 Write an account of the preparation, structure and bonding of fluoride and oxyfluorides of xenon. (A.U., Dec. 1989)

7-10-4 Write briefly about the clathrate compounds of noble gases. (A.U., Dec. 1989)

7-10-5 Explain the structure of xenon hexafluoride. (A.U., Jun. 1990)

7-10-6 How are XeF_2, XeF_4, XeF_6 prepared ? Write their structures ? (O.U., Oct. 1990)

7-10-7 Describe the conditions under which noble gases can form compounds ? (O.U., Oct. 1990)

7-10-8 (a) Prove that noble gases are not so inert ?

(b) Justify the position of noble gases in the periodic table ?

(c) Describe the uses of noble gases. (K.U., Apr. 1990)

7-10-9 Give the structure of xenon tetrafluoride.

7-10-10 Write short notes on special behaviour of helium.

7-10-11 (a) Discuss the position of noble gases in the periodic table.

(b) Describe the separation of noble gases by Dewar's absorption method. (O.U., Apr. 1990)

7-10-12 Give the preparation of xenon tetrafluoride. (O.U., Apr. 1989)

7-10-13 How are zero group elements isolated from liquid air using Linde's process?

7-10-14 Why are the zero group elements known as the noble gases ?

7-10-15 Discuss the position of the noble gases in the periodic table. Why are they called the bridging elements ?

7-10-16 Discuss the preparation properties and bonding of the following :

A) XeF_2 B) XeF_6 C) XeO_3 D) $XeOF_4$

7-10-17 Discuss the abnormal behaviour of helium.

7-10-18 Write the electronic configuration of the noble gases.

7-10-19 What type of bonding is to be expected of noble gases? Discuss taking into consideration, at least four xenon fluorides, or oxyfluorides.

7-10-20 What prompted Bartlett to prepare xenon compounds? Discuss the preparation & properties of fluorides of xenon.

7-10-21 Give the structures of the following :
A) XeF_6 and B) XeO_4 (A.U., Apr. 1989 & S.K.U. Apr. 1990)

7-10-22 How are zero group elements isolated from air. Write a note on clathrate compounds of noble gases (A.U., Jun. 1989)

ANSWERS

7-9-1 A **7-9-2** C **7-9-3** C **7-9-4** A **7-9-5** D **7-9-6** B
7-9-7 B **7-9-8** C **7-9-9** D **7-9-10** C **7-9-11** B **7-9-12** A
7-9-13 B **7-9-14** C **7-9-15** C

8
ANALYTICAL CHEMISTRY

8-1 INTRODUCTION

Analytical chemistry is the practical branch of chemistry that gives a scientific treatment of methods employed for the determination of the chemical composition of matter. The methods are numerous for analyzing a particular compound or mixture and each method has a principle of its own. When we talk about composition, we encounter the questions what is present and how much is present. Basing on these questions analytical chemistry is divided into two branches viz., **qualitative analysis** (*what is present*) and **quantitative** analysis (*How much is present*). A broad subclassification is presented in flow chart - I. We will restrict our study to the principles of **titrimetric, garvimetric** and **semimicro** analysis.

Flow Chart - I

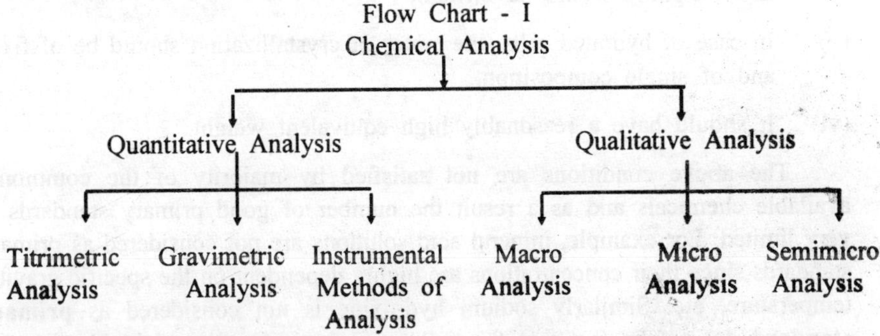

8-1-1 Quantitative Analysis

It deals with the determination of the amount of the components present in a unit quantity of sample and can be estimated either by titrimetric (also called volumetric) or gravimetric or instrumental methods of analysis.

8-1-2 Qualitative Analysis

It deals with finding the composition, of the species present or its structural details. This is achieved composition of the either by consuming large amounts of a substance or by taking small amounts of the substance or by using extremely small amounts of substance. Accordingly these methods are respectively called **macro** analytical method, **semimicro** analytical method and **micro** analytical method.

8-2 VOLUMETRIC METHODS OF ANALYSIS

It deals with the volumes of two reagent solutions of which one is of known concentration and the other of unknown concentration. The solution whose concentration is known accurately is referred to as a **standard solution**. A standard solution can be prepared by taking a fixed amount of a **primary** standard substance and dissolving it in a known amount of a solvent. Though weight and volume percent solutions may be considered as standard substance in quantitative analytical methods, chemists usually express the concentrations in normality, molarity and molality.

A primary standard substance must comply the following conditions

(i) It should be readily obtained in a pure form and should not be too expensive. Established methods should be available for testing its purity.

(ii) It should be easy to handle and should not pose any stringent restrictions in its handling.

(iii) It should be sufficiently stable when exposed to constituents of atmosphere.

(iv) It should be non-hygroscopic and non-efflorescent; otherwise, drying and weighing would be difficult.

(v) In case of hydrated salts, the water of crystallization should be of fixed and of single composition.

(vi) It should have a reasonably high equivalent weight.

The above conditions are not satisfied by majority of the commonly available chemicals and as a result the number of good primary standards is very limited. For example, mineral acid solutions are not considered as primary standards since their concentrations are highly dependent on the specific gravity, temperature, etc. Similarly sodium hydroxide is not considered as **primary standard** due to the fact that the pellets are hygroscopic and highly reactive with CO_2 present in the atmosphere. Substances like KCl, $AgNO_3$, Na_2CO_3, oxalic acid, Sodium oxalate satisfy the above mentioned conditions and are called primary standard substances. Here oxalic acid ($C_2H_6O_4$), inspite of having two molecules of water (water of crystalization) with every oxalic acid molecule is considered a primary standard since the ratio of H_2O is always fixed. Substances such as alkali hydroxides, mineral acids, hypo, potassium permanganate, ammonium thiocyanate, many hydrated salts which do not comply the above mentioned rules are called **secondary standard substances.** A solution of a secondary standard substance may be standardised by titrating against a suitable solution of primary standard substance and then may be used as a standard solution for further analysis.

The standard solution is usually taken in the burette. The process of adding the standard solution from the burette until the chemical reaction is just complete is known as **titration.** *The term **titrimetric analysis** is used for the estimation of solutions only while the term **volumetric analysis** also covers gas analysis that involves measurement of volumes of gaseous species produced* *or consumed.* A standard solution or a solution which has already been standardized is known as **titrant** and the substance of unknown concentration whose estimation is being carried out is known as **titrate**. When the titrant and titrate react in a conical flask, the point of exact completion of the reaction is called the **end point** or **point of equivalence.** The completion of the reaction should be clearly visualised by our eye which is achieved by either the usage of an **external indicator** or **a self indicator.** An external indicator is a foreign substance which when added to the titrant shows a distinct color change at or very close to the equivalence point.

For example, in the titration of a strong acid with a strong base phenolphthalein is used as an external indicator which changes its color at the end point to pink from colorless. **An internal (self) indicator** is a chemical substance which is also one of the reactant and shows a distinct color change on the completion of the reaction. For example, in the titration of oxalate ions by permanganate, the equivalence point is indicated by a color change from colorless to pink due tó presence of slight excess of $KMnO_4$.

Wherever it is difficult to find a suitable indicator, the volumetric analysis is performed taking help of instrumental techniques. Some of these methods are

(i) Potentiometric titrations
(ii) Conductometric titrations
(iii) Amperometric titrations
(iv) pH metric titrations
(v) Spectrophotometric titrations

The titrimetric method of analysis is the simplest of all the analytical methods, as it involves simple procedures often requiring least expensive apparatus like burettes, pipettes, standard flasks, conical flasks and beakers.

8-2-1 Classification of volumetric analysis

The volumetric analysis may be classified into following four classes depending upon the type of reaction taking place during titration.

1. Acid-base or Neutralization reactions
2. Oxidation - Reduction or Redox reactions
3. Precipitation reactions
4. Complexometric reactions.

8-3 ACID-BASE OR NEUTRALIZATIOꟿ REACTIONS

In this class of reactions, an acid and a base react to give a salt and water.

$$\underset{\text{acid}}{HCl} + \underset{\text{base}}{NaOH} \longrightarrow \underset{\text{salt}}{NaCl} + \underset{\text{water}}{H_2O}$$

Since the acidic and basic properties are due to free H^+ and OH^- ions the basic neutralization reactions may be written as

$$\underset{\text{(from acid)}}{H^+} + \underset{\text{(from base)}}{OH^-} \longrightarrow H_2O$$

Water is a very weak electrolyte and ionizes feebly. Therefore, in the absence of any interference of the salt formed, the pH of the reactions mixture remains at 7 (neutral pH) at equivalance point of the titration. The acid-base titrations could be between

(i) Strong acid and strong base

(ii) Strong acid and weak base (NH_3, Na_2CO_3, $Na_2B_4O_7$, etc.)

(iii) Weak acid (CH_3COOH, $H_2C_2O_4$, etc.) and strong base

The end points employed in neutralization titrations are ordinarily based upon the abrupt change in pH that occurs in the vicinity of the equivalence point.

In case of weak acid vs weak base titrations due to lack of any sharp pH change at the equivalance point, they are not included in the volumetric titrations listed above.

The dependence of neutralization reactions on ionic product pH and proper selections of indicators necessitates a detailed study of these parameters. We therefore present first a concise account of these parameters and then give an account of various titration curves obtained in acid-base titrations.

8-3-1 Ionic Product of Water

Water has a negligible but definite conductance of 3.8×10^{-8} S cm^{-1}. This is due to **Auto ionization or Self ionization** of water molecules, which is often simplified and written as

$$H_2O \rightleftharpoons H^+ + OH^-$$

Since the autoionization of water is a reversible process, the equilibrium constant expression takes the form

$$K = \frac{[H^+]\,[OH^-]}{[H_2O]} \tag{8.1}$$

The pure water concentration $[H_2O]$ is, of course, constant and is equal to 55.56 mol L^{-1} at 25°C.

Multiplying both sides of above expression with this value we get,

$$K_w = K\,[H_2O] = [H^+]\,[OH^-] \tag{8.2}$$

where K_w is a constant equal to 1×10^{-14} mol^2L^{-2} at 25°C and is called **dissociation constant or ionic product of water.** In pure water, the concentrations of hydrogen and hydroxyl ions are equal, that is $[H^+] = [OH^-]$. Hence

$$[H^+] = [OH^-] = \sqrt{1\times10^{-14}} = 1 \times 10^{-7} \text{ mol L}^{-1}$$

Ionic product expression (8.1) is also obeyed closely by all dilute aqueous solutions. In such solution $[H_2O]$ is essentially unaffected by the chemical reactions that occur. Acidic solutions contain an excess of H^+ ions over OH^- ions and basic solutions contain an excess of OH^- ions over H^+ ions. For example, an acid solution containing 10^{-2} M hydrogen ions has 10^{-12} M hydroxyl ions.

$$[OH^-] = \frac{10^{-14}}{10^{-2}} = 10^{-12} \text{ mol L}^{-1}$$

Similarly, when the molar concentration of OH^- is 0.001, the concentration of H^+ ions is

$$[H^+] = \frac{10^{-14}}{10^{-3}} = 10^{-11} \text{ mole L}^{-1}$$

So in aqueous solutions of acids and bases, both H^+ and OH^- are present. In acid solution $[H^+]$ is greater than 1×10^{-7} M, while in basic solution $[H^+]$ is less than 1×10^{-7} M.

8-3-2 The pH and pOH Scale

We often encounter, the acid and base solutions of different concentrations. The acidity and basicity or the strength of a solution is measured and compared in terms of the free hydrogen ion concentrations. For example, a 0.001 M HCl has a hydrogen ion concentration of 0.001 M or 1.0×10^{-3} M and a 0.001 M NaOH has a hydrogen ion concentration of 1.0×10^{-11} M. If we talk within the limits of 1 normal acid to 1 normal base, all the concentrations are always expressed in numbers involving negative exponents. When we have to compare two such numbers, like 1.23×10^{-8} and 9.64×10^{-7}, we first look for the exponents of 10 and then the value before the exponent. This slows down the comparision and puts a non-mathematical man in confusion. To simplify the situation the Danish biochemist S.P.L. Sörensen in 1909 introduced the pH scale, [where p stands for potenz, (German : power) of H^+].

The pH of a solution is defined as negative logarithm of hydrogen ion concentration (mol / lit). Mathematically

$$\mathbf{pH = - \log [H^+]}$$

Therefore, knowing $[H^+]$ of a solution we can compute its pH or knowing pH we can calculate $[H^+]$ of the solution.

Thus pure water having a $[H^+] = 10^{-7}$ mol L^{-1} has a pH = -log 10^{-7} = 7. Therefore pH of an acid solution is always less than 7. For example, the pH of a 10^{-3} M. HCl is equal to

$$pH = -\log 10^{-3} = 3$$

Similarly an aqueous solution of a base has always a pH greater than 7.

By common practice, pH scale is restricted to a range 0 to 14, that is to say from $[H^+] = 1$ M to 10^{-14} M or $[OH^-] = 10^{-14}$ M to 1 M or $[H^+] = 1$ M to $[OH^-] = 1$ M. Out side this range, concentrations will not have negative exponents and are simply given in molar concentration.

To calculate $[H^+]$ from pH, we can use the modified pH expression as

$$\mathbf{[H^+] = 1 \times 10^{-pH}}$$

Extending the same approach to hydroxide ions, it is possible to define pOH as a *negative logarithm of hydroxide ion concentration* (mol L^{-1}).

$$\boxed{\mathbf{pOH = -\log [OH^-]}}$$

Similar to interdependence of $[H^+]$ and $[OH^-]$ expressed by equation (8.2), it is possible to arrive a relation between pH and pOH. Taking logarithm on both sides of equation (8.2), we get,

$$\log K_w = \log [H^+] + \log [OH^-]$$

Multiplying throughout with -1 gives

$$-\log K_w = -\log [H^+] - \log [OH^-]$$

Extending the concept of pH and pOH, -log K_w may be replaced by pK_w.

Therefore,

$$\boxed{\mathbf{pK_w = pH + pOH}}$$

Because $K_w = 1 \times 10^{-14}$

$$pK_w = -\log 10^{-14} = 14$$

Hence

$$\boxed{\mathbf{pH + pOH = 14}}$$

Therefore, a knowledge of pH is sufficient to know pOH.

$$pOH = 14 - pH$$

We can easily recognize the importance of pH scale by considering the pH values of some common substance, Fig. 8-1

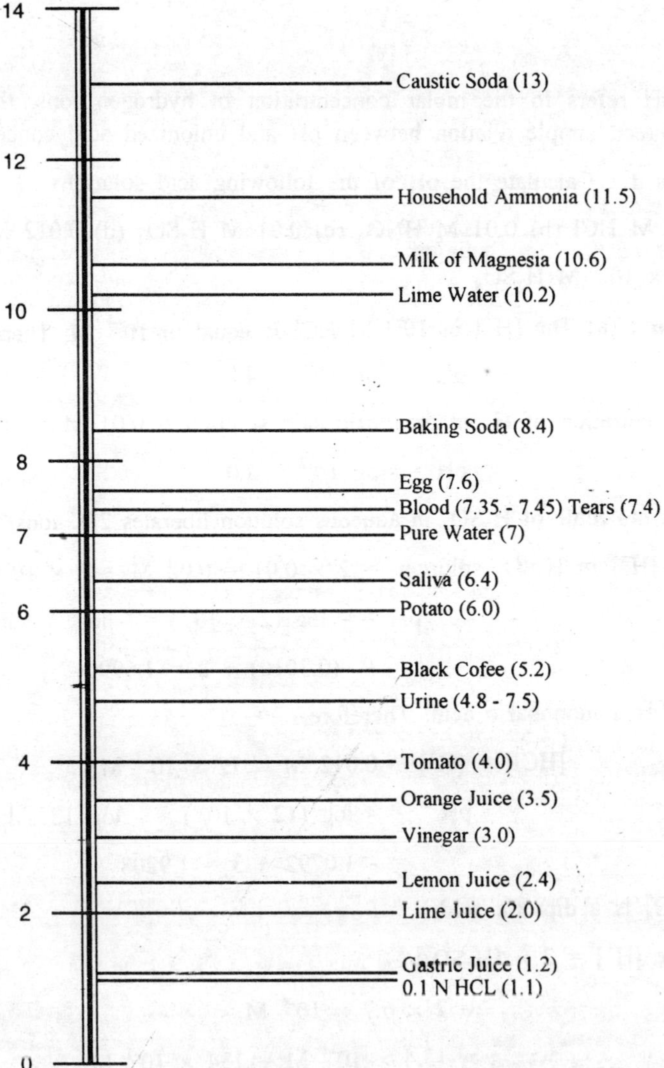

Fig. 8-1 The pH scale and pH values of some common substances

While reading the pH scale two factors have to be kept in mind,

(i) *A small change in* pH *means large change in* $[H^+]$. Though lime juice and tomato juice seem to differ only by 2 pH units, actually hydrogen ion concentration of former is greater by 100 times over the latter. This means, seemingly small pH change refers to a very large change in the actual hydrogen ion concentration.

(ii) pH *refers to* $[H^+]$ *concentration but not unionized acid concentration.* In case of strong acids, due to 100% dissociation, the $[H^+]$ and acid concentrations are one and the same. But, if we consider a weak acid only a small percentage of molecules exist in dissociated form. Because pH refers to the molar concentration of hydrogen ions, there is no direct, simple relation between pH and unionized acid concentration.

Problem 1 : Calculate the pH of the following acid solutions

(a) 10^{-4} M HCl (b) 0.01 M HNO_3 (c) 0.01 M H_2SO_4 (d) 0.012 M HCl (e) 6.7×10^{-4} M H_2SO_4.

Solution : (a) The $[H^+]$ in 10^{-4} M HCl is equal to 10^{-4} M. Therefore.

$$pH = -\log 10^{-4} = 4.0$$

(b) Concentration of H^+ in the nitric acid solution = 0.01 M = 10^{-2} M

$$pH = -\log 10^{-2} = 2.0$$

(c) Each molecule of H_2SO_4 in aqueous solution liberates $2H^+$ ions. Therefore,

$$[H^+] \text{ in } H_2SO_4 \text{ solution} = 2 \times 0.01 = 0.02 \text{ M} = 2 \times 10^{-2} \text{ M}$$

$$pH = -\log (2 \times 10^{-2}) = -\log 2 - \log 10^{-2}$$

$$= -(0.3010) + 2 = 1.6990$$

(d) HCl is a monobasic acid. Therefore,

$$[HCl] = [H^+] = 0.012 \text{ M} = 12 \times 10^{-3} \text{ M}$$

$$pH = -\log (12 \times 10^{-3}) = -\log 12 - \log 10^{-3}$$

$$= -1.0792 + 3 = 1.9208$$

(e) H_2SO_4 is a diprotic acid.

Therefore $[H^+] = 2 \times [H_2SO_4]$

$$= 2 \times 6.7 \times 10^{-4} \text{ M}$$

$$= 13.4 \times 10^{-4} \text{ M} = 134 \times 10^{-5} \text{ M}$$

$$pH = -\log (134 \times 10^{-5})$$

$$= -\log 134 - \log 10^{-5}$$

$$= -2.1271 + 5$$

$$= 2.8729$$

Problem 2 : Calculate pH and pOH of the following completely ionized bases (a) 0.015 M NaOH, (b) 0.01 M $Ba(OH)_2$.

Solution : (a) One molecule of NaOH ionizes to give one molecule of OH^-

Therefore $[NaOH] = [OH^-] = 0.015 \text{ M} = 15 \times 10^{-3}$ OH^-

$$\textbf{pOH} = -\log [OH^-] = -\log (1.5 \times 10^{-2})$$

$$= -\log 15 - \log 10^{-3} = -1.1761 + 3.0$$

$$pOH = 1.8239$$

$$pH = 14 - pOH = 14 - 1.8239 = 12.1761$$

(b) Each molecule of $Ba(OH)_2$ on dissociation yields two OH^- ions.

$$[OH^-] = 2 \times [Ba(OH)_2] = 2 \times 0.01 \text{ M} = 0.02 \text{ M} = 2 \times 10^{-2} \text{ M}$$

$$pOH = -\log [OH^-] = -\log (2 \times 10^{-2})$$

$$= -\log 2 - \log 10^{-2} = -0.3010 + 2 = 1.6990$$

$$pH = 14 - 1.6990 = 12.3010$$

Problem 3 : Calculate the [H+] in the solution with pH (a) 4.032 (b) 8.123

Solution

(a) $[H^+] = 1 \times 10^{-pH}$ or $\log [H^+] = -pH$

$\log H^+ = -4.032 = \bar{5}.968$

$[H^+] = \text{anti log } (\bar{5}.968) = 9.29 \times 10^{-5}$ M

(b) $\log [H^+] = -8.123 = \bar{9}.877$

$[H^+] = \text{antilog } (\bar{9}.877) = 7.534 \times 10^{-9}$ M

Problem 4 : What would be the pH of the mixture formed by mixing 40 mL of 0.015 M HCl and 50 mL of 0.03 M NaOH

Solution : Number of moles of HCl taken $= \dfrac{40 \times 0.015}{1000} = 0.6 \times 10^{-3}$ mole

Number of moles of NaOH taken $= \dfrac{50 \times 0.03}{1000} = 1.5 \times 10^{-3}$ moles.

Since 1 mole of HCl is neutralized by 1 mole of NaOH, 0.6×10^{-3} moles of HCl consumes 0.6×10^{-3} moles of NaOH

Amount of NaOH left unneutralized $= 1.5 \times 10^{-3} - 0.6 \times 10^{-3}$

$= 0.9 \times 10^{-3}$ mole.

Volume of total mixture = 40 + 50 = 90 ml

Therefore,

90 ml of solution contains 0.9×10^{-3} moles of NaOH
1000 ml of solution contains ?

$$= \frac{1000}{90} \times 0.9 \times 10^{-3} = 1 \times 10^{-2} \text{ M}$$

Concentration of NaOH left out = 1×10^{-2} M

pOH of the mixture = - log 10^{-2} = 2.0

pH of the mixture = 14 - 2 = 12.0

Problem 5 : To a 50 ml aqueous sodium hydroxide solution of concentration 0.1 M, 0.063 g of oxalic acid ($C_2O_4H_2$ $2H_2O$) is added. Find the pH of the mixture obtained after neutralization.

Solution : Molecular weight of oxalic acid = 126

126 g of oxalic acid is equal to 1 g mole

0.063 of oxalic acid is equal to ?

$$\frac{0.063}{126} = 0.5 \times 10^{-3} \text{ moles}$$

Amount of NaOH present in 50 ml of alkali solution $= \dfrac{50 \times 0.1}{1000}$

$= 5.0 \times 10^{-3}$ mole

0.5 mole of oxalic acid is neutralized by 1.0 mole of NaOH

0.5×10^{-3} mole of oxalic acid is neutralized by ?

$$\frac{0.5 \times 10^{-3}}{0.5} \times 1.0 = 1.0 \times 10^{-3} \text{ moles of NaOH}$$

Amount of NaOH left unneutralized = Total moles - Moles of NaOH of NaOH reacted with oxalic acid

$= 5 \times 10^{-3} - 1 \times 10^{-3}$

$= 4 \times 10^{-3}$ moles of NaOH

Volume of the solution is unchanged since 0.063g of oxalic acid crystals occupy negligible volume. Therefore,

50 ml of mixture contains 4 x 10^{-3} moles of unneutralized NaOH
1000 ml of mixture contains ?

$$= \frac{1000}{50} \times 4 \times 10^{-3} = 8 \times 10^{-2} \text{ M}$$

Concentration of unneutralized NaOH = 8 x 10^{-2} M

pOH = -log [OH-] = - log (8 x 10^{-2})

= -0.9031 + 2 = 1.1

pH = 14 - 1.1 = 12.9

8-3-3 Indicators

Acid-base indicators are weak organic acids or bases that undergo color changes when converted from essentially undissociated molecular state to completely dissociated salts.

or

Acid-base indicators are weak organic acids or bases that on ionization give ions with different color from that of the undissociated molecules.

An indicator not only has its characteristic color but also changes the color depending on pH of the medium and pK_{In} of the indicator.

$$K_{In} = \frac{[H+]\,[HIn]}{[HIn]} \;;\; pK_{In} = -\log(K_{In})$$

pK_{In} is called **indicator constant.**

Taking logarithm on both sides

$$\log K_{In} = \log H^{+} + \log \frac{[In]}{[HIn]}$$

Multiplying with -1 gives

$$-\log K_{In} = -\log H^{+} - \log \frac{[In]}{[HIn]}$$

$$\boxed{\mathbf{pK_{In} = pH - \log \frac{[In]}{[HIn]}}}$$

or $$pH = pK_{In} + \log \frac{[In]}{[HIn]}$$

Therefore the color shown by an indicator depends on the ratio $[In^-]/[HIn]$, which in turn depends on the pH of the solution. Intermediary colors are shown in the pH range of $pK_{In} \pm 1$.

When $[In^-] / [HIn] > 10$, the indicator exists predominantly in the base ion-form and takes the color of In^-.

When $[In^-] / [HIn] < 0.1$, the indicator exists predominantly undissociated form HIn and assumes its color.

A few commonly used acid-base indicators are listed in Table 8-1 along with the pK_{In} values and pH range.

Table 8-1 Some important Acid-Base indicators

Indicator	*Color change*		pK_{In}	*pH at color change*
	Acid	*Base*		
Thymol blue	Red	Yellow	1.7	1.2 - 2.8
Methyl yellow	Red	Yellow	-	2.9 - 4.0
Methyl orange	Red	Orange	3.5	3.2 - 4.4
Methyl Red	Red	Yellow	5.1	4.4 - 6.2
Bromthymol blue	Yellow	Blue	7.0	6.0 - 7.6
Litmus (Azolitmin)	Red	Blue	6.5	5.5 - 7.5
Phenol red	Yellow	Red	7.9	6.8 - 8.4
Phenolphthalein	Colorless	Pink	9.6	8.3 - 10.0
Alizarin yellow	Yellow	Violet	-	10 - 12
Trinitrobenzene	Colorless	Brown	13.5	12 - 14

8-3-4 Determination of pH Using Indicator Solution

An approximate value of the pH of a solution can be found using different indicators covering the total pH range. For this the experimental solution is taken in different small test tubes and a drop of different indicators covering the whole pH range are added to each test tube. The pH of the solution will be one in which one of the indicators gives a basic color and the next indicator with higher pK_{In} gives acid color. For example, a solution of pH = 8 gives a basic color with litmus and bromothymol blue having $pK_{In} \approx 6.8$ while it is acidic and is colorless to phenolphthalein having $pK_{In} = 9.3$ and pH range between 8.3 - 10.0. In recent times a suitable combination of indicators called **universal indicator** is used which exhibit different colors at different pH values. A comparison with standard pH color strip provided with the universal indicator solution establishes the approximate pH of the experimental solution. To find exact pH one has no alternative but to employ suitable instrumental methods.

8-3-5 Choice of Indicators

The choice of indicator lies at the pH equivalence point of a tritration as discussed below.

(a) Titration of a strong acid by a strong base

In the strong acid-strong base titrations, the salt produced does **not** hydrolyze and the solution is exactly neutralized at pH = 7.0. This is called **equivalence** point of the titration. When we plot the pH of the solution versus **volume of** base added we get a curve illustrated by Fig. **8-2**, called a **titration curve.**

Since pH scale is a logarithmic scale, despite of linear variation of $[H^+]$ with volume of base added, the pH variation gives a sigmoidal (S-shaped) curve. At equivalence point the pH rises very sharply, approximately 6 pH units, for an addition of just 0.1 mL of the base. Before and after equivalence point, the pH change is negligible. The indicators whose color change falls along the steep portion of the titration curve are all suitable to act as indicators for this titration. Therefore, all indicators having their pK_{In} in the range of 5-9, in other words whose color changes in the pH range 4-10 are suitable for these titrations. Indicators like thymol blue and alizarin yellow are not suitable to act as indicators as the former, changes color too soon (at pH = 2.0) and the latter too late (at pH = 12).

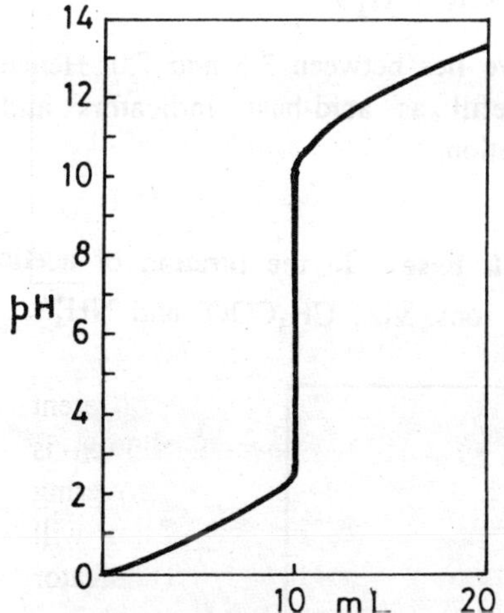

Fig. 8-2 Titration curve of strong acid (10 mL of 0.1 M HCl) by a strong base (0.1) M NaOH)

Fig. 8-3 Titration curve of acetic acid (10 mL, 0.1M by 0.1 M NaOH)

(B) Titration of a weak acid by a strong base : In the titration of weak acid, say acetic acid, by a strong base, acetate ions are produced. The acetate ions being strong conjugate base are hydrolyzed to give acetic acid and hydroxyl ions.

$$CH_3COOH + NaOH \longrightarrow CH_3COO^- + Na^+ + H_2O$$
$$CH_3COO^- + H_2O \rightleftharpoons CH_3COOH + OH^-$$

Therefore, the resultant solution obtained is basic in nature and the pH at the equivalence point is greater than 7, and lies around (8.7), Fig.8.3.

Since the pH range of phenolphthalein exactly matches the equivalence point pH of the titration curve, it is a most suitable indicator in weak acid vs strong base titrations. Indicators like methyl orange or methyl red having lower pH range are not suitable as Indicators, in this type of titrations.

(C) Titration of a weak base by a strong acid : In the titration of NH_4OH, a weak base, by a strong acid, HCl, a strong conjugate acid ammonium ion is formed.

$$\underset{\text{Weak base}}{NH_4OH} + HCl \longrightarrow \underset{\substack{\text{Strong}\\\text{conjugate}\\\text{acid}}}{NH_4^+} + Cl^- + H_2O$$

NH_4^+, strong conjugate acid, readily undergoes hydrolysis to give an acidic solution.

$$NH_4^+ + H_2O \rightleftharpoons NH_3 + H_3O^+$$

The equivalence point of this titration curve lies between 3.5 and 7.0. Hence methyl orange and methyl red are useful as acid-base indicators and phenolphthalein does not work in this titration.

(D) Titration of a weak acid by a weak base : In the titration of acetic acid by NH_4OH we get two hydrolyzable ions, viz., CH_3COO^- and NH_4^+.

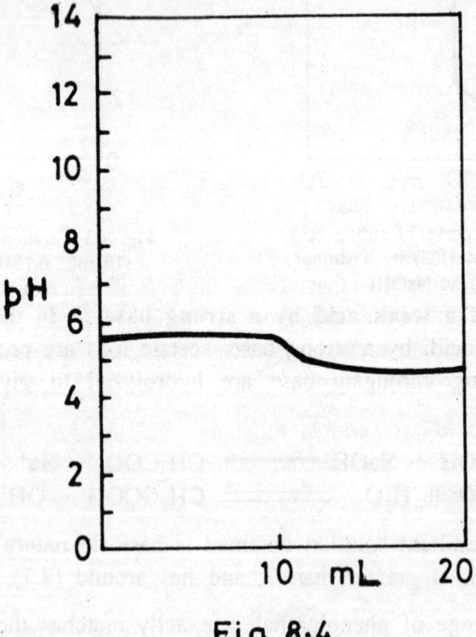

Fig. 8-4 Titration curve of weak acid (10 mL of 0.1 CH_3COOH) by a weak base (0.1 M NH_4OH)

$$CH_3COOH + NH_4OH \rightleftharpoons CH_3COO^- + NH_4^+ + H_2O$$

Weak acid — Weak base — Strong conjugate base — Strong conjugate acid

The acid & base both being weak the upper and lower limits of pH lie between 4.5 - 8.0. Even near the equivalence the pH variation is gradual between 6.5 to 7.5. Due to absence of steep rise in pH at the equivalence point, it is not possible to have an indicator that can give meaningful titre values. Phenol red having a wörking pH range 6.5 - 8 gives approximate titre values, (Fig.8-4).

8-3-6 Theories of Indicators : There are two theories that explain the color changes of indicators during acid-base titrations.

(A) Ostwald's theory or Ionic theory : Most of the discussion given earlier for characterizing an indicator belongs to this theory. The main features of this theory are as follows :

(i) All the acid-base indicators are either weak acids or weak bases. They ionize partially to give H^+ or OH^- as in the case of simple weak acids and bases.

$$\underset{\substack{\text{weak acid}\\\text{(Bronsted acid)}}}{HIn} \rightleftharpoons \underset{\text{conjugate base}}{H^+ + In^-}$$

$$\underset{\substack{\text{weak base}\\\text{(Bronsted base)}}}{InOH} \rightleftharpoons \underset{\text{conjugate acid}}{OH^- + In^+}$$

(ii) All the indicators with $pK_{In} < 7$ are called acid indicators and with $pK_{In} > 7$ are called base indicators. In other words, the acid indicators undergo ionization in acidic medium and base indicators in basic medium. According to this definition methyl orange, methyl violet, methyl red, thymol blue, etc., are acid indicators and phenolphthalein, alizarin yellow, nitrobenzene, etc., are base indicators.

(iii) A conjugate acid-base couple must possess different colors to act as an indicator. In other words the ionized and unionized forms of indicator should possess different colors.

$$\underset{\text{Color A}}{HIn} \rightleftharpoons \underset{\text{Color B}}{H^+ + In^-}$$

or

$$\underset{\text{Color C}}{InOH} \rightleftharpoons \underset{\text{Color D}}{OH^- + In^+}$$

Colours corresponding to acidic and basic solutions of different indicators are given in Table 8-1.

(iv) While titrating a weak acid (or base), the indicator should be such that its pH range falls in the pH equivalence range. Following this rule we cannot use methyl orange and methyl red as indicators for the titration of acetic acid by a strong base.

(B) Quinonoid Theory : Though ionic theory was successful in correlating the indicator constant (pK_{In}) and indicator pH range to the pH of the titration mixture and existence of various ionic forms, it has no answer as to what makes any indicator acquire its color. An answer to this part of the question was provided by **quinonoid theory.**

The following features explain the quinonoid theory

i) Organic compounds which have benzene ring (**benzenoid** structures) are colorless or very lightly colored, while those having quinonoid structure with extended conjugation are highly colored.

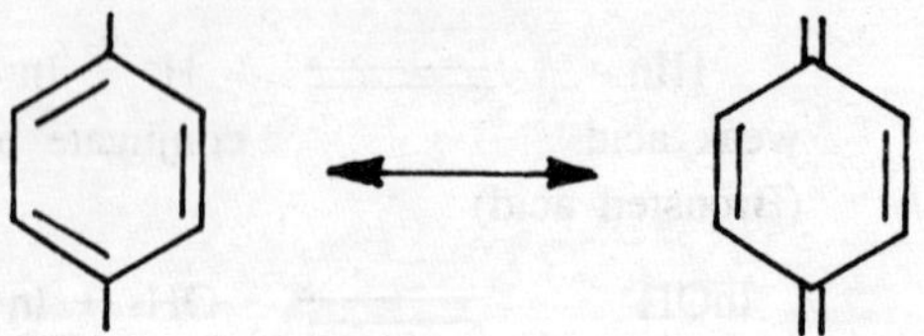

Benzenoid structure Quinonoid structure

ii) The benzenoid and quinonoid structures are two tautomeric forms and are at dynamic equilibrium.

iii) The concentration of a particular form depends on the pH condition of the solution. Therefore the color of the indicator solution varies with pH. For example, in case of phenolphthalein; the structures are

(Benzenoid)
Colourless

(Quinonoid
Red

Similarly for methyl orange;

$$\underset{\text{(Quinonoid) Red}}{(H_3C)_2\overset{+}{N}{=}C_6H_4{=}N{-}NH{-}C_6H_4{-}SO_3^-} \rightleftharpoons \underset{\text{(Benzenoid) Yellow}}{(H_3C)_2N{-}C_6H_4{-}N{=}N{-}C_6H_4{-}SO_3^-}$$

Problem 6 : The indicator constant of phenolphthalein is 9.3. If this indicator is added to 1×10^{-4} M NaOH solution, find the ratio of concentration of phenolphthalate anion to phenolphthalein.

Solution : $\quad pH = pK_{In} + \log \frac{[In^-]}{[HIn]}$

NaOH dissociates completely

$$pOH = -\log(1 \times 10^{-4}) = 4.0$$

$$pH = 14 - 4 = 10.0$$

$$10.0 = 9.3 + \log \frac{[In^-]}{[HIn]}$$

$$\log \frac{[In^-]}{[HIn]} = 0.7$$

$$\frac{[In^-]}{[HIn]} = \text{anti} \log 0.7 = 5.01$$

Problem 7 : The indicator range of phenolphthalein lies in the pH range 8.3 - 10.0. Assuming that the neutral point of the indicator is in the centre of the range, calculate the ionization constant of the indicator.

Solution : Since pH scale is a negative logarithmic scale of hydrogen ion concentration the centre of range cannot be determined by averaging pH values of the range. So it is essential to calculate $[H^+]$ of corresponding ranges to get the middle value which lies very close of pK_A.

$[H^+]$ corresponding to pH 8.3 = 5.01×10^{-9} mol l^{-1}

$[H^+]$ corresponding to pH 10.0 = 1×10^{-10} mol l^{-1}

$$\text{Average } [H^+] = \frac{5.01 \times 10^{-9} + 1 \times 10^{-10}}{2} = 2.56 \times 10^{9}$$

$$H\,In \rightleftharpoons H^+ + In^-$$

$$K_{In} = \frac{[H^+][In^-]}{[HIn]}$$

pH = pK_{In} when the salt (In^-) and acid (HIn) concentrations are equal; i.e., $[In^-]$ = [HIn]. Therefore,

$$K_{In} = [H^+] = 2.56 \times 10^{-9}$$

$$pK_{In} = 8.6$$

The indicator constant for phenolphthalein = 8.6

8-4 REDOX TITRATIONS

Oxidation-Reduction or Redox reactions are characterized by transfer of electrons between the reacting species. Loss of electrons or increase in oxidation number of a substance leads to its oxidation whereas reduction is accompanied by a decrease in oxidation number or gain of electrons. In any oxidation-reduction reaction the number of electrons lost by one species is equal to the number gained by the other. The the substance containing an atom which shows a decrease in oxidation number as a result of reaction is called *oxidizing agent.* Similarly a *reducing agent* is a substance having an atom which shows an increase in oxidation number.

The above terms could be well understood considering a reaction between ferric chloride and stannous chloride.

$$2FeCl_3 + SnCl_2 \longrightarrow SnCl_4 + 2FeCl_2$$

$$2Fe^{3+} + Sn^{2+} \longrightarrow Sn^{4+} + 2Fe^{2+} \text{ (ionic representation)}$$

In this reaction stannous ion loses two electrons to form stannic ion and these electrons are readily accepted by two ferric ions to form ferrous ions. Ferric ions which show a decrease in its oxidation number is called oxidizing agent and stannous ion which increases its oxidation number, is called a reducing agent. The oxidation and reduction components of the redox reaction are written as two half reactions indicating species that gains electrons and species that loses electrons.

Oxidation half reaction :

$$Sn^{2+} \rightleftharpoons Sn^{4+} + 2e^-$$

Reduction half reaction :

$$2Fe^{3+} + 2e^- \rightleftharpoons 2Fe^{2+}$$

Overall reaction is the sum of these two half reactions. The feasibility and spontaneity of a half reaction is decided by corresponding standard electrode potentials. The electrode potentials of few important half reactions are given in Table 8-2.

Table 8-2 Standard electrode potentials of few half reactions

Reduction half reaction			E^o *in V*
$MnO_4^- + 8H^+ + 5e^-$	⟶	$Mn^{2+} + 4H_2O$	1.52
$Cl_2 + 2e^-$	⟶	$2Cl^-$	1.36
$Cr_2O_7^{2-} + 14H^+ + 6e^-$	⟶	$2Cr^{3+} + 7H_2O$	1.33
$Hg_2^{2+} + 2e^-$	⟶	2Hg	0.79
$Fe^{3+} + e^-$	⟶	Fe^{2+}	0.77
$I_2 + 2e^-$	⟶	$2I^-$	0.54
$Cu^{2+} + e^-$	⟶	Cu^+	0.153
$Sn^{4+} + 2e^-$	⟶	Sn^{2+}	0.15

From the selected list of redox potential presented in Table 8-2, we can suggest that MnO_4^- ion having highest reduction potential, is the most powerful oxidizing agent and is capable of oxidizing all other ions. Similarly $Cr_2O_7^{2-}$ is capable of oxidizing ions that have lower reduction potential viz., Hg, Cu^+, Fe^{2+}, I^-, Sn^{2+} but cannot oxidize Mn^{2+} or Cl^- that have higher potential. In the titrimetric estimation of Fe^{2+} by permanganate absence of any chloride salt or HCl should be ensured, to avoid any interference due to the oxidation of Cl^-. In this situation dichromate oxidation, having slightly lower potential, is most suitable as $Cr_2O_7^{2-}$ is unreactive to Cl^- but oxidizes Fe^{2+}.

Standard potentials speak only about the feasibility of a reaction, but do not give any information about the rate of the reaction. In some redox titrations which proceed at a slow rate, a catalyst is added to the reaction mixture to force the reaction to proceed at a reasonable velocity.

8-4-1 Oxidation - Reduction Indicators : Two types of indicators are encountered in Redox titrations.

1. Specific Indicators : *A specific indicator is a substance that reacts with one of the participants of the titration to produce a color change.* In the iodometric titrations, **starch** is often used as a specific indicator. **Amylose** in starch binds to I_3^- to produce a dark-blue colored complex. This complex often serves to signal the end point in redox titrations in which iodine is either produced or consumed. **Potassium thiocyanate (KSCN)** is another specific indicator used in the titration of Fe(III). The end point is indicated by the disappearence of red color due to reduction of iron (III) thiocyanate complex to give colorless iron (II) thiocyanate.

2. True redox indicators : The color properties of some substances depend on the potential of the system. For example, diphenyl amine in dilute acid is violet in the oxidized state (above 0.76 V) and colorless in the reduced state (below 0.76 V). The potential at which a color transition will occur depends upon, E° - standard potential of indicator and is called **transition potential.**

The true redox indicators are not specific to a particular system but the behavior depends only upon change in potential of the redox titrations. So they have a much wider scope of applications than do the specific indicators. The color changes and transitions potentials of some true redox indicators are listed in Table 8-3.

Table 8-3 Indicator properties of some redox indicators

Indicator	*Color*		*Transition potential (V)*	*Medium*
	Oxidized	*Reduced*		
1. 5-Nitro 1, 10 phenanthroline iron (II) complex	pale blue	red violet	1.25	1M H_2SO_4
2. 1, 10 phenanthroline iron (II) complex	pale blue	red	1.11	1M H_2SO_4
3. diphenyl amine	violet	colorless	0.76	dil. acid
4. methylene blue	blue	colorless	0.53	1 M acid
5. indigo tetrasulfonate	blue	colorless	0.36	1 M acid
6. phenosafranine	red	colorless	0.28	1 M acid

In the titration of iron (II) with cerium (IV) solutions at the equivalence point the potential of the system changes from 0.95 V to 1.25 V. Any indicator having a transition potential in this range should be satisfactory. 1, 10 phenanthroline iron (II) complex having a transition potential of 1.11 V should therefore be an ideal candidate for this purpose.

8-4-2 Some typical redox titrations

To demonstrate the potential of the redox titrations we will now discuss oxidation reactions involving $KMnO_4$, $K_2Cr_2O_7$ and ceric sulphate.

A. Titrations involving potassium permanganate (Permangnometry) : $KMnO_4$ is one of the most powerful and widely used oxidising agents employed in various volumetric titrations. The intense color of permanganate allows it to be used as a self indicator and eliminates the use of an external indicator. However, for very dilute permanganate solutions diphenylsulfonic acid or orthophenanthroline iron (II) complex are used as indicators. The tendency of permanganate to oxidize chloride ions makes it unusable in the presence of chloride salts and HCl. The reactions need to be conducted in acid medium and acidification with H_2SO_4 is often preferred. In solutions with $[H^+] \geq 0.1$ N the common reduction product is manganese (II) ion.

$$MnO_4^- + 8H^+ + 5e^- \rightleftharpoons Mn^{2+} + 4H_2O \qquad E^o = 1.55 \text{ V}$$

In weakly acidic (pH > 4), neutral and weakly alkaline solutions MnO_2 is the most common reduction product.

Weakly acid or neutral, or weakly alkaline $\quad MnO_4^- + 4H^+ + 3e^- \rightleftharpoons MnO_2 + 2H_2O, \quad E^o = 1.66$ V

Strongly basic (> IN) $\quad MnO_4^- + e^- \rightleftharpoons MnO_4^{2-}, \quad Eo = 0.56V$

The decomposition of permanganate is catalysed by light, heat, acids, bases, Mn^{2+} and MnO_2. MnO_2 shows an autocatalytic effect and therefore its presence should be avoided while preparing a standard $KMnO_4$ solution. The photochemical decomposition could be avoided by storing the permanganate solutions in colored flasks.

Owing to autocatalytic decomposition due to heat and light effects potassium permanganate is not considered as primary standard. It is usually standardised either with a standard solution of oxalic acid or its sodium salt.

B. Titrations involving Potassium Dichromate (Dichrometry) : $K_2Cr_2O_7$ is a less powerful oxidizing agent than KMnO4 having a redox potential of 1.33 V. However, its main advantages lie in its

(i) availability in a pure form

(ii) less expensive than Ce^{4+} salts

(iii) thermal stability.

(iv) indefinite shelf life

(v) nonreactivity with Cl^-.

The main limitations with $K_2Cr_2O_7$ is its lesser oxidising characteristics and slowness of some of its reactions. Since it is a primary standard, its solutions could be readily prepared and stored for long periods. The oxidation reaction proceeds in acid medium following the reaction.

$$\underset{\text{(Orange)}}{Cr_2O_7^{2-}} + 14H^+ + 6e^- \longrightarrow \underset{\text{(Green)}}{Cr^{3+}} + 7H_2O, \qquad E^o = 1.33V$$

Diphenylamine sulphonate is used as external indicator in dichrometric titrations. Potassium dichromate is mainly used in the determinations of iron. The ferrous iron can be titrated directly in the presence of phosphoric acid and sulphuric acid. The main purpose of using phosphoric acid lies in lowering of Fe^{3+}/Fe^{2+}

electrode potential, achieved by the formation of $[Fe\ (HPO_4)]^+$ complex ion. In addition dichromate is used in the titration of calcium, persulphate, manganese dioxide, nitrite, etc.

C. Titrations involving Ceric Sulphate : Quadrivalent cerium, Ce^{4+}, is a very strong oxidising agent with a reduction potential of 1.7 V in perchlorate media, 1.66 V in nitrate media and 1.44 V in sulphate media.

$$Ce^{4+} + e \rightleftharpoons Ce^{3+} \qquad E^o : \text{Depends on anions of acid used}$$

1.7 V in $HClO_4$

1.66 V in HNO_3

1.44 V in H_2SO_4.

Similar to dichromate, Ce^{4+} can be used only in acidic medium. It has an edge over potassium permanganate in terms of (a) its stability (b) slightly higher oxidation potential (c) definite stiochiometry of redox reactions because of single electron reduction (d) does not oxidize chloride at a detectable rate and therefore titrations could be carried on in hydrochloric acid medium.

The disadvantages of Ce^{4+} incomparision to permanganate ions are (i) Ce^{4+} cannot be used in neutral and basic media. (ii) It is not sufficiently intense in color to serve as self indicator (iii) It is about 50 times more expensive than permanganate. Its advantages are, Ceric ammonium nitrate of primary standard quality is commercially available and standard solutions can be prepared directly by weight. Other lower grade reagents could be standardised with arsenious oxide, sodium oxalate, ferrous ammonium sulphate, etc.

D. Titrations involving Iodine (Iodometry) : The redox titrations involving iodine are of two types

(i) in which a standard solutions of iodine is used to titrate easily oxidized substances and has limited application since iodine is a weak oxidant.

(ii) in which the iodine liberated in the reaction solution through oxidation of iodide is titrated with a strong reductant like hypo and computed.

The former is called direct or **iodimetry** titration and the later is called indirect or **iodometry** titration. In both the cases the basic reaction is

$$I_2 + 2e^- \rightleftharpoons 2I^-$$

The Iodide ion (I^-) formed will readily react with free iodine to form a triiodide ion (I^-_3).

$$I^- + I_2 \rightleftharpoons I_3^-$$

$$(I_3^- + e^- \longrightarrow 3I^-, \quad E^o = 0.536\ V)$$

Therefore overall reaction is written as .

Hence it is a weak oxidising agent. This low oxidizing power brings it a distinct advantage over many other oxidizing agents by bringing selectivity in its behaviour.

With stannous ions $Sn^{2+} + I_2 \rightleftharpoons Sn^{4+} + 2I^-$

with sulphide $H_2S + I_2 \rightleftharpoons S + 2H^+ + 2I^-$

with sulphite $SO_3^{2-} + I_2 + H_2O \rightleftharpoons SO_4^{2-} + 2H^+ + 2I^-$

with thiosulphate $2S_2O_3^{2-} + I_2 \rightleftharpoons S_4O_6^{2-} + 2I^-$

With weaker reducing agents iodine reacts in neutral or faintly acidic media as given in the following reactions.

$$H_3AsO_3 + I_2 + H_2O \rightleftharpoons H_3AsO_4 + 2H^+ + 2I^-$$

$$Sb_2O_3 + 2I_2 + 2H_2O \rightleftharpoons Sb_2O_5 + 4H^+ + 4I^-$$

$$2[Fe(CN)_6]^{4-} + I_2 \rightleftharpoons 2[Fe(CN)_6]^{3-} + 2I^-$$

Iodine hydrolyses in alkaline solutions

$$I_2 + OH^- \rightleftharpoons HOI + I^-$$

The hypoiodite (HOI^-) so formed disproportionates in part to give iodate (IO_3^-) and iodide as shown by the following equation

$$3HOI + 3OH^- \rightleftharpoons IO_3^- + 2I^- + 3H_2O$$

The occurence of this reaction complicates the stiochiometry of the reactions and lead to serious errors in iodine titration. Therefore iodometry/iodimetry titrations can not be performed in alkaline solutions.

The reduction potential of iodine is markedly increased in the presence of ions that complex with iodide formed. For example, in the presence of Hg(II), which forms stable iodide complex, iodine oxidize arsenious acid (H_3AsO_3) quantitatively to arsenic acid (H_3AsO_4).

Aqueous suspension of starch is the widely used indicator in iodimetric titrations. Starch is comprised of two components viz; α-amylose and β-amylose. Iodine is absorbed on to β-amylose and imparts a deep blue color to the reaction mixture. Potato, rice and arrow root starches predominently contains β-amylose and can be used as indicators. Corn starch is rich in α- amylose and hence is unsuitable as indicator. Starch when added to a solution containing a high concentration of iodine decomposes to a product whose indicator properties are not entirely reversible. Hence starch indicator should be added only at the approach of equivalence point indicated by pale brown color of the titrate.

Iodine - hypo (sodium thiosulphate) reaction is the most popularly used iodometric reaction. The reaction is written as

$$2S_2O_3^{2-} + I_2 \rightleftharpoons S_4O_6^{2-} + 2I^-$$

or $$2S_2O_3^{2-} + I_3^- \rightleftharpoons S_4O_6^{2-} + 3I^-$$

The reaction takes place tracing the following steps.

step (i) $S_2O_3^{2-} + I_3^- \rightleftharpoons S_2O_3I^- + 2I^-$

step (ii) $2S_2O_3I^- + I^- \rightleftharpoons S_4O_6^{2-} + I_3^-$

step (iii) $S_2O_3I^- + S_2O_3^{2-} \rightleftharpoons S_4O_6^{2-} + I^-$

During the titration Thiosulphate anion reacts with I_3^- to give colorless intermediate $S_2OI_3^-$. This intermediate either reacts through step (ii) to produce I_3^- thus leading to reappearence of color on starch or reacts with $S_2O_3^{2-}$ to give unreactive tetrathionate ion $S_4O_6^{2-}$. At the equivalence point the reaction terminates through step (iii).

8-5 PRECIPITATION TITRATIONS

Volumetric methods based on the formation of a sparingly soluble compound are called precipitation titrations. Although many precipitation reactions are known, only very few of them could be used in volumetric analysis that possess following characteristics.

1. The precipitate formed is practically insoluble.
2. The precipitation is rapid
3. The titration values are not disturbed or changed due to adsorption effects even when coprecipitation occurs
4. The end point is easily detectable.

The most important of the precipitation reactions is based on the precipitation of silver following the equation

$$Ag^+ + X^- \longrightarrow AgX$$

where X is chloride, bromide, iodide, cyanide or thiocyanate

The involvement of silver in volumetric titrations and its study is known as **argentometry.**

Different methods are adopted while working with precipitation reactions.

8-5-1 No indicator method or Turbidometric technique : In this technique the titrimetric reactions are followed employing light scattering phenomena.

For example, when sodium chloride is added to a solution containing silver ions, the turbidity of the titrant increases till equivalence point and thereafter turbity remains constant due to lack of coagulation.

8-5-2 The Volhard's method : In this method silver ions are titrated by ammonium thiocyanate in the presence of ferric chloride serving as indicator. During the titrations initially silver ions react with ammonium thiocyanate to form colorless precipitate of silver thiocyanate. After precipitation of silver ions, that is at the equivalence point, thiocyanate reacts with ferric ions to give a deep blood red coloration due to formation of ferric thiocyanate. Volhard's method can be used in the estimation of chloride ions in acidic media through an indirect precipitation titration. An excess of known volume of $AgNO_3$ is added to chloride solution and the unreacted silver ions are back titrated by conventional Volhard's method. This method suffers from complication arising by the reaction.

$$AgCl + SCN^- \rightleftharpoons Ag\,(SCN) + Cl^-$$

Therefore this titration consumes more titrant and chloride concentration appears high by about 2%. This error can be eliminated by either filtering the AgCl precipitate before back titration or by adding little nitrobenzene to adhere to AgCl precipitate and protect it from thiocyanate added in the back titration.

8-5-3 Mohr's Method : This is the best known method of argentometry titrations. Halides are directly titrated with silver nitrate in the presence of potassium chromate serving as indicator.

The principle of precipitation is same as in Volhard method. Silver ions react with halide ions to form AgX precipitate and at the end point, i.e. when the precipitation is complete an excess drop of the titrant will react with the indicator to form Ag_2CrO_4, a brick red precipitate. The titrate should be neutral or slightly alkaline, but not very alkaline, to avoid precipitation of silver as silver hydroxide. Further, with too much of acid, the end point is delayed due to lowering of CrO_4^{2-} concentration due to reversible reaction.

$$H^+ + CrO_4^{2-} \rightleftharpoons HCrO_4^-$$

The precipitation titration could also be employed for the estimation of Cl^- Br^- and I^- using mercuric nitrate as titrant instead of silver nitrate. These titrations are then called **mercurimetric** titrations. Mohr's method is accurate, time saving and applicable even to low Cl^- concentrations.

Organic indicators like sodium salts of rhodizonate and hydroxy quinone are also used in the Mohr titrations. Sulphate ions are estimated by titrating them with $BaCl_2$ using these organic indicators. The end point is indicated by the formation of red barium salt of the indicator ions. Similarly fluoride ions are estimated with $Th(NO_3)_4$ using sodium alizarin sulphate (Alizarin-S)as indicator.

8-5-4 Adsorption indicators : Precipitation titrations also employ a new class of indicators called **adsorption indicators.** The colour change in these indicators is caused by their adsorption on the precipitate particles. For example, when $AgNO_3$ is added to NaCl in the presence of **fluorescein indicator**, (Fl^-) the end point is indicated by a colour change from light yellow to rose pink. On standing we notice that the colour is actually that of the precipitate and the supernatant liquid is colourless.

Before the end point (when Cl^- is present in excess) : The precipitate particles are adsorbed by chloride ions and do not allow the adsorption of indicator

$$(Ag\ Cl)\ Cl^- + Fl^- \longrightarrow \text{No adsorptions}$$

At the end point (when there is definte Ag^+ concentration) : The precipate particles are adsorbed by Ag+ which allow the adsorption of indicator

$$(AgCl)\ Ag^+ + Fl^- \longrightarrow \underset{\text{pink}}{(Ag\ Cl)\ (Ag^+\ Fl^-)}$$

Fluorescein is a typical example of **anionic indicator**. In case of **cationic indicators** like methyl violet (MV) the scheme of adsorption will be

$$(Ag\ Cl)\ Cl^- + MV^+ \longrightarrow \underset{\text{Adsorption}}{(Ag\ Cl)\ (Cl^-\ MV^+)}$$

$$(Ag\ Cl)\ Ag^+ + MV^+ \longrightarrow \text{No adsorption}$$

So with cationic indicators, we have to titrate $AgNO_3$ with NaCl.

The adsorption indicators are not suitable with precipitates that coagulate in the presence of highly charged ions, eg. Al (III). Such titrations can be performed in the presence of protective colloids, which minimize or eliminate the problem of coagulation.

8-6 COMPLEXOMETRIC TITRATIONS

Metal ions can act as electron pair acceptors and form compounds in the presence of suitable electron donors. These compounds are called **coordination compounds or complex compounds** and the donor species is called **ligand.** Ammonia, aniline, oxalate ion, chloride ion, cyanide ion, water molecule are some of the most common ligands. Some molecules have two or more donor groups and are called **chelates.** A chelating agent that has two donor groups is called **bidentate** ligand, while one with three donor groups is called **tridentate** ligand. Quadridentate, quinque dentate and hexadentate chelating agents are also encountered.

Reactions involving complex ions or undissociated neutral complex compounds fall under complexometric titrations. In complexometric titrations it is important to know the condition of equilibrium and corresponding equilibrium contant. For this consider an equilibrium between a metal ion M, a ligand D and their 1 : 1 complex MD.

$$MD \rightleftharpoons M + D \qquad K_d = \frac{[M]\,[D]}{[MD]}$$

where k_d is called **dissociation or instability constant**. The reciprocal of K_d gives the **formation** or **stability constant** K_f. More the stability constant higher is the stability of the complex formed.

8-6-1 EDTA Titrations : Ethylenediaminetetraacetic acid (EDTA in short) is one of the versatile ligands used in many complexometric titrations. It is also called with several other trade names such as versene, complexane (III), Trilon B, Nullapon, etc.

$$\begin{matrix} H^{1}OOC\text{-}CH_2 & & & & CH_2 - COOH^{3} \\ & \searrow & N - CH_2 - CH_2\ N & \nearrow & \\ H^{2}OOC\text{-}CH_2 & \nearrow & & \searrow & CH_2 - COOH^{4} \end{matrix}$$

Fig. 8.5 Structure of EDTA

It is a weak acid with pk_a values of 2.0, 2.67, 6.16 and 10.26. These values indicate that first two protons are lost easily than the remaining two. Thus it has four carboxylic protons and two nitrogen atoms a total of six sites as donor groups. Because of poor solubility, the disodium salt of EDTA is often employed as ligand, in which two of the sites are occupied by sodium atoms making it tetradentate ligand. The tetra sodium salt is not satisfactory owing to extensive hydrolysis in aqueous media.

Nitrilotriacetic acid (abbreviated NTA) is another common ligand employed in complexometric titrations.

$$\begin{matrix} H\ OOC^{-}CH_2 & & CH_2 - COOH \\ & H - \overset{+}{N} - & CH_2COO^{-} \\ & & CH_2 - COOH \end{matrix}$$

Selections of metal ion indicators

The EDTA titrations are dependent of metal ion indicators as the accuracy of the titration is dependent on the sharp color change at the equivalence point. The indicators used are all chelating agents themselves as EDTA and can and form either 1 : 1, 2 : 1 or 1 : 2 complexes. Eriochrome black-T, Murexide, Solochrom dark blue, Sulphane black F, Xylenol orange, pyrocatecol violet, Eriochrome Red B, etc. are used as metal ion indicators and except for the first two the rest are all alkaline in their character. Eriochrome black T is used in the titrations of zinc, calcium, magnesium, manganese, nickel and aluminium, where as copper is estimated using fast sulphon black F. Nickel is estimated using either murexide or bromopyrogallol red indicator. Xylenol orange is used in the estimation of bismuth, thorium, lead, cobalt, zinc and cadmium.

Titration methods using EDTA : Three different procedures are popularly employed in complexometric titrations using EDTA.

1. Direct - titration method : Direct titration of metal ions with EDTA solution is carried out in the presence of suitable indicator. For example, in the titration of magnesium and other divalent ions by EDTA the indicator commonly employed is Eriochrome black-T. The titration is carried out in moderately alkaline medium.

$$Mg^{2+} + H_2Y^{2-} \rightleftharpoons Mg\ Y^{2-} + 2H^+$$

Erichrome black - T is doubly dissociated in alkaline and neutral solutions to give blue colored HIn^{2-}. This ion forms a red colored complex in the presence of Mg^{2+}

$$\underset{\text{blue}}{Mg^{2+} + HIn^{2-}} \rightleftharpoons \underset{\text{red}}{Mg\ In^- + H^+}$$

In the titration of Mg^{2+}, initially the solution is red colored due to formation of MgIn. At the equivalence point the magnesium ion concentration is sufficiently low (high pMg) to destabilize $MgIn^-$ to produce free blue colored HIn^{2-} and therefore the color changes to blue from red. Further, EDTA being a stronger chelate than Eriochrome black-T, at the equivalence point EDTA also plays a significant role in destabilizing the red $MgIn^-$. The formation of metal hydroxide due to presence of OH^- is avoided by adding a little amount of tartarate or citrate salt.

2. Back titration method : When the metal ion is incapable of forming a colored ion with the indicator anions or the rate of complexation with EDTA is slow or the buffer interferes with the direct titration, a back titration method is often employed. In this technique excess amount of EDTA solution is added to the titrate along with a buffer and the uncomplexed EDTA is then titrated against a standard solution of magnesium ion using Eriochrome black-T as the indicator. To carry out this titration it is essential that the metal-EDTA complex must be more stable than magnesium-EDTA complex, otherwise the back titration could displace the metal ion. The end point is given by the color change from blue to red.

3. Displacement or substitution titration : In the displacement titrations the metal ion is first treated with an excess of magnesium or zinc complex of EDTA; if the metal forms more stable complex with EDTA than the magnesium or zinc complex, the following displacement reaction occurs.

$$MgY^{2-} + M^{2+} \rightleftharpoons MY^{2-} + Mg^{2+}$$

The amount of magnesium which gets freed corresponds to the concentration of metal ion and is estimated by titrating with a standard solution of EDTA in the presence of Eriochrome black T. This type of titration is often utilized in the estimation of Ca^{2+} present in hard water.

$$Ca^{2+} + MgY^{2-} \rightleftharpoons CaY^{2-} + Mg^{2+}$$

Applications of EDTA in analysis : Some of the most popular applications of EDTA are

(i) Determination of magnesium by direct titration

(ii) Determination of calcium by substitution titration

(iii) Determination of calcium by back titration

(iv) Determination of calcium and magnesium present in hard water.

(v) Selective titration of various ions by judicious choice of pH

Mg^{2+}, Cr^{2+}, Ca^{2+}, Ba^{2+} can be titrated at pH = 11

Mn^{2+}, Fe^{2+}, Co^{2+}, Ni^{2+} can be titrated at pH = 4.0 - 7.0

Cr^{3+}, Fe^{3+}, Zn^{2+}, Cd^{2+}, Al^{3+}, Pb^{2+}, Cu^{2+}, Ti^{2+} and V^{4+}

can be titrated at pH = 1.0 - 4.0

8-7 GRAVIMETRIC ANALYSIS

A gravimetric method is a quantitative analytical tool in which the analysis is completed by a weighing operation. This is carried out in two ways.

(i) The substance to be analysed is isolated (separated) from the other constituents in the form of an insoluble precipitate and the analysis is completed by determining the weight of this precipitate. This method of analysis is called **precipitation method**.

(ii) The substance to be determined is isolated by distillation and either the sample obtained on condensation of vapors is directly weighed or the weight loss in the left out sample is weighed and computed for the weight of distilled substance. This method of analysis is called **volatilization method**. Of the two methods, precipitation method is most widely used in gravimetric analysis.

8-7-1 Precipitation method

The precipitation of a metal ion is effected by suitable choice and addition of anion. The formed precipitate should be *almost insoluble* in the media in which it got precipitated and in common washing reagents like water, alcohol, acetone, etc. Another important aspect of selection of precipitate is its physical nature that allows easy separation through filteration technique. The size of the particles of the precipitate should be large enough and crystalline to allow filteration through a sintered crucible or a Whatmann filter paper. Finally the last aspect that must be borne in mind is that the precipitate must be pure enough or convertible into a pure substance of definite chemical composition.

8-7-2 Optimum conditions for precipitation

Following parameters often effect the quality and quantity of the precipitate.

(a) To get good yield of the precipitate, the solubility product should be less than 10^{-8}

(b) The precipitation should be carried out in dilute solution as it minimises errors due to coprecipitation.

(c) A slow mixing of reagents is preferred as it helps proper growth of precipitate particles which allow effective filteration.

(d) The precipitate should be initially washed with dilute solution instead of pure water (solvent) in order to avoid complications arising out of peptization.

(e) The solubility of the precipitate directly depends on the temperature of the reaction media. Therefore if reaction mixture is heated to allow completion of reaction, it should be cooled to room temperature before performing filteration.

(f) Heating any precipitate to a certain temperature alongwith the solvent is called **digestion**. Crystalline precipitates should be digested in order to avoid absorbed impurities.

(g) The presence of common ion in solutions containing the precipitate will help in suppressing the solubility of the precipitate. For example, AgCl is quantitatively precipitated in the presence of slight excess common ion Cl^-. Far excess of common ion may sometimes increase the solubility of precipitate by making it more polar.

$$\underset{\text{less soluble}}{AgCl} + xCl^- \longrightarrow \underset{\text{more soluble}}{AgCl_2^- \text{ or } AgCl_3^{-}}$$

Hence this has to be kept in mind while adding an electrolyte with a common ion.

8-7-3 Coprecipitation

A precipitate formed in solution by the combination of suitable reagents is invariably contaminated to a greater or smaller extent by the ions present in the solution. This is known as **coprecipitation**.

The coprecipitation may occur through different mechanisms such as (i) surface adsorption (ii) isomorphic inclusion (iii) nonisomorphic inclusion (iv) Occlusion.

(i) Surface adsorption : Contaminant in the form of ions is adsorbed on the surface of the solid and the counter ions of opposite charge present in the solution are adsorbed immediately adjacent to the particle. Greater the surface area available more is the amount of adsorption. Therefore, *gelatinous precipitates, having maximum surface area, show maximum coprecipitation while microcystalline precipitates show least coprecipitation.*

(ii) Isomorphic inclusion : When the coprecipitated compound has *same dimensions* and chemical composition as precipitate, then this compound can replace precipitate molecules/ions in its crystal structure without causing appreciable strain or distortion. This *mixed crystal formation is known as isomorphic inclusion.*

This type of coprecipitation has no answer except for preventing the occurence by removing the troubling ion before precipitation.

(iii) Non-isomorphic inclusion : When minute amounts of non-isomorphic substance (having different dimension and composition) is carried out of solution by the precipitate, the crystals formed will have the impurities uniformly distributed. This type of coprecipitation is called non-isomorphic inclusion. Similar to isomorphic inclusion, there is no way that we can separate the impurity except for taking steps to prevent such inclusion.

(iv) Occlusion : *Occlusion is a type of coprecipitation in which the soluble impurity is entrapped within the crystal structure of the precipitate.* When precipitation is carried slowly the impurities will have sufficient time to escape from the precipitate before they become entrapped. Prolonged digestion of the precipitate is effective in eliminating contamination by occlusion. The property of occlusion is governed by Paneth - Fajans-Hahn rule of adsorption according to which *"the less soluble the salt the more it will occlude"*. Obeying this law it was found that barium nitrate, being less soluble will occlude more in comparison with highly soluble barium chloride during precipitation of barium sulphate.

8-7-4 Post precipitation

Post precipitation is a special type of introducing contamination, where the impurity precipitates after all the major part of the analytical precipitate has formed. In otherwords *it is the precipitation of a sparingly soluble impurity on the surface of main precipitate.* This can be well understood considering following example. When a solution of calcium chloride is treated with oxalic acid solution a precipitate of calcium oxalate is formed. But in the presence of magnesium ions as impurity, the precipitation of calcium oxalate induces the precipitation of magnesium oxalate causing error in the estimation. The longer the time of contact greater is the post precipitation.

Post precipitation is common in the sulphides. For example, when copper sulphide is precipitated in the presence of zinc, the CuS precipitate should be filtered away immediately or else zinc postprecipitates and interferes with the analysis.

The main differences between post precipitation and coprecipitation are listed below :

Postprecipitation	*Coprecipitation*
1. Contamination increases with time of contact of supernatant liquid and main precipitate.	1. Contamination decreases with time of contact of supernatant liquid and main precipitate.
2. Agitation of liquid enhances postprecipitation.	2. Agitation of liquid has no effect.
3. Extent of contamination is large.	3. Extent of contamination is small.

8-7-5 Estimation of sulphate as soluble sulphate

This example serves as a typical example to understand the care and pitfalls encountered in gravimetric analysis. The analysis is based on the precipitation of soluble sulphate with barium ion.

$$Ba^{2+} + SO_4^{2-} \longrightarrow BaSO_4$$

The barium sulphate is then filtered, washed with water, dried by igniting and weighed.

Procedure : To 20mL of the given sulphate solution in a 400mL beaker, add 1 mL of conc. HCl and 200 mL water and heat the solution till it boils. Now add 20mL of 5% solution of $BaCl_2$ slowly, drop wise, with constant stirring. Digest the precipitate for 30 minutes. The precipitate formed is allowed to settle for 10 to 15 minutes. A few drops of the precipitant ($BaCl_2$ solution) is added to the clear supernatant solution to test the completion of the precipitation. Any precipitation prompts, further addition of 5 mL of $BaCl_2$. After digestion test the solution once again for the completion of the precipitation with a few drops of barium chloride.

The precipitate is now filtered through a Whatmann filter paper No.40. The precipitate on the filter paper is washed with hot water till chloride is completely removed. The absence of chloride in the filtrate is ensured by adding few drops of $AgNO_3$ solution to it. Transfer all the precipitate into the funnel, with the help of a glass rod, a wash bottle and a policeman.

Now allow the precipitate to drain out all water, dry it in a hot air oven at 120°C. The filter paper with the precipitate is then folded and placed into a previously ignited, weighed and cooled silica glass crucible. The crucible is placed in an inclined position on a clay pipe triangle, with half covered lid and incinerated over a low bunsen flame placed beneath the crucible. The flame is gradually increased and finally the crucible is heated strongly. In the incineration process at 600°C, barium sulphate is reduced partially to sulphide. This sulphide is converted to sulphate by heating the incinerated sample with a few drops of sulphuric acid. The crucible is now cooled and kept in a desiccator. The cooled crucible with the incinerated product, $BaSO_4$, is then weighed.

The weight of sulphate is calculated from the weight of the barium sulphate thus obtained.

Calculations

Weight of empty crucible = x g

Weight of crucible + $BaSO_4$ = y g

Weight of $BaSO_4$ present in 20 mL of solution = $(y - x)$ g = z g

Weight of $BaSO_4$ present in 1000 mL of solution = $\frac{1000 \times z}{20}$

233.36 g of $BaSO_4$ contains 96 g of SO_4^{2-}

$\frac{(1000 \times x)}{20}$ grams of $BaSO_4$ contains $\frac{z \times 1000 \times 96}{20 \times 233.36}$ g of SO_4^{2-} per litre

Superficially, the proposed method sounds simple and straight forward. Infact, however, many inteferences are encountered in actual practice due to coprecipitation of anions and cations. Table 8-4 summerizes some of the common interference problems and their reasons.

Table 8-4 Some Common interferences and reasons in the estimations of soluble sulphate as $BaSO_4$

Nature of Interference	*Reasons*	*Solution*
1. Interference yielding low yields of precipitate		
a. Coprecipitation of ammonium ion	Amonium ion leaves the precipitate as volatile $(NH_4)_2SO_4$.	NH_4^+ should be avoided
b. Coprecipitation of sulphuric acid	H_2SO_4 occluded is removed during ignition	Excess mineral acid is present
c. Coprecipitation of iron	–	Iron should be removed prior to estimation of SO_4^{2-}
d. Partial reduction of $BaSO_4$ to BaS	Filter paper is charred too rapidly	

2. **Interference yielding high $BaSO_4$**		
a. Precipitation of $BaCO_3$ and $Ba_3(PO_4)_2$	Absence of mineral acid	
b. Coprecipitation of NO_3^-, CrO_4^{2-}		Avoid the ions

8-8 SEMIMICRO ANALYSIS

Qualitative analysis of a salt or mixture of salts can be performed on either in a macro scale or semimicro scale or micro scale. *Macroscale analysis is done using quite appreciable amounts in normal sized apparatus.* The weights of the substance can vary from 0.5 to 1.9 g and volume of solutions can be of the order of 10mL. In semimicro analysis 0.05g of the salt is utilized for the reactions in a volume of 1 mL or slightly less. Micro analysis technique is also termed as milligram technique and employ 0.005g or 5mg of substance in $\simeq$ 0.1 mL of the solutions. From a careful observation of these techniques we notice that *although the weight of the salt is reduced, the salt to volume ratio is retained.* Microanalysis can be performed on a one tenth scale in comparision to the semimicro technique. However it is very difficult to draw a clear distinct line between all the three techniques. The microscale analysis is done for non-interfering radicals by performing very specific spot tests and flame tests (These topics alongwith other related techniques are dealt with in Vol II of this series). The main difficulty lies in the unambiguos inference of the observation and the result needs be accepted with caution till the student acquires perfection.

In semimicro technique the apparatus used is of relatively small size in comparison to the macroscale of analysis with which the student is already acquainted. Semi-micro analysis needs less amount of chemicals, work space and time. Hence it is always preferred over the macroscale analysis.

All the radicals are grouped into **anions** (*acidic radicals*) and **cations** (*basic radicals*). These radicals are further categorized into different classes depending on the type of reactions they undergo with some specific group reagents.

8-8-1 Classification of anions or acid radicals

The acid radicals cannot be classified into systematic groups as done in case of basic radicals. However, for the sake of convenience, the anions are classified into three groups depending upon their reaction with mineral acid.

Group	*Radical*	*Group reagent*	*Observation*
Radicals reactive to dilute acid	1. Acetate H_3COO^-	dilute HCl or dil H_2SO_4	1. Pungent gas
	2. Nitrite NO_2^-		2. Brown vapor
	3. Carbonate CO_3^{2-}		3. Colorless & odorless gas
	4. Sulphite SO_3^{2-}		4. Pungent gas
	5. Sulphide S^{2-}		5. Colorless gas with rotten egg smell
Radicals reactive to conc acid	1. Chloride Cl^-	conc H_2SO_4	1. Pungent colorless gas
	2. Bromide Br^-		2. Colorless pungent gas
	3. Iodide I^-		3. Dark voilet fumes
	4. Nitrate NO_3^-		4. light brown gas
Acid unreactive	1. Sulphate SO_4^{2-}	dil HNO_3 + $BaCl_2$	1. Curdy white precipitate
	2. Phosphate PO_4^{3-}		2. Canary yellow precipitate

8-8-2 Radicals reactive to dilute mineral acid

The radicals of this group react with dilute acid and liberate gases. The group radicals viz., CH_3COO^-, NO_2^-, CO_3^{2-}, SO_3^{2-}, S^{2-}, basing on their solubility can be subdivided into water soluble group (CH_3COO^- & NO_2^-) and volatile group (CO_3^{2-}, SO_3^{2-}, S^{2-})

(A) WATER SOLUBLE GROUP ! It contains acetate and nitrite ions. Accetate and nitrite salts are, in general, highly soluble in water. On reaction with dilute sulphuric acid, acetate liberates acetic acid while nitrite liberates brown colored nitrogen dioxide.

$$CH_3COO^- + H^+ \longrightarrow CH_3COOH$$

$$NO_2^- + H^+ \longrightarrow HNO_2$$

$$2HNO_2 \longrightarrow H_2O + NO + NO_2$$

Acetate (CH_3COO^-) (i) Acetate is best identified by its reactions with ferric chloride giving a deep red coloration in cold condition and red precipitate on heating

$$Fe\,Cl_3 + CH_3COONa \longrightarrow Fe\,(CH_3COO)_3 + 3\,NaCl$$

$$Fe\,(CH_3COO)_3 + 2H_2O \xrightarrow{\text{heat}} \underset{\text{red precipitate}}{Fe(OH)_2(CH_3COO)^- + 2\,(H_3COOH)^+}$$

The red precipitate formed is of a basic ferric acetate.

(ii) Action of ethyl alcohol on acetate in the presence of conc. H_2SO_4 generates sweet smelling ethyl acetate

$$CH_3COONa + C_2H_5OH \xrightarrow{H_2SO_4} CH_3COOC_2H_5 + H_2O$$

Amyl alcohol if used instead of ethyl alcohol gives a characteristic, better distinguishable odour due to formation of amyl acetate.

Nitrite **(NO_2^-)** *(i) Nitrite is best characterized by its azodye test. To an acidified solution of sodium nitrite 1 mg of aniline hydrochloride and 5mg of β-napthol dissolved in NaOH are added. Brilliant orange precipitate indicates nitrite.*

(ii) Nitrite and nitrate exhibit the popular brown ring test. To perform this test conc. H_2SO_4 is added to the salt solution and is slightly warmed. Golden brown gas evolves, immediately the test solution is cooled and freshly prepared acidified ferrous sulphate is added from the sides. At the junction of the two liquids the characteristic brown ring due to $[Fe(NO)]SO_4$ is formed. This compound is unstable and disappears either on shaking or heating the test tube.

$$NaNO_2 + H_2SO_4 \longrightarrow NaHSO_4 + HNO_2$$

$$3HNO_2 \longrightarrow H_2O + HNO_3 + 2NO$$

$$FeSO_4 + NO \longrightarrow [Fe(NO)]SO_4$$

(iii) In alkaline medium zinc reduces nitrite to ammonia. Ammonia can be detected by its characteristic smell and alkaline property (turning exposed red litmus paper blue).

$$3Zn + 5OH^- + NO_2^- \longrightarrow 3\,(ZnO_2)^{2-} + NH_3 + H_2O$$

B. Volatile Group : This group contains carbonate, sulphide and sulphite. All the three volatile radicals can be differentiated by the following reactions carbonates and sulphites react with barium chloride to give white precipitates soluble in dil HCl, where as there is no such reaction with sulphides

$$Na_2CO_3 + BaCl_2 \longrightarrow BaCO_3 + 2NaCl$$

$$Na_2SO_3 + BaCl_2 \longrightarrow BaSO_3 + 2NaCl$$

The carbonate and sulphite may then be differentiated by their reaction with permanganate. Potassium permanganate is decolorized by sulphites and sulphides but carbonates do not show any reaction.

$$2KMnO_4 + 5H_2SO_3 \longrightarrow K_2SO_4 + 2MnSO_4 + 2H_2SO_4 + 3H_2O$$

$$2KMnO_4 + 5H_2S + 6HCl \longrightarrow 2KCl + 2MnCl_2 + 8H_2O + 5S$$

Carbonate ($CO3^{--}$): (i) Carbonates readily react with acids liberating carbondioxide.

$$CO_3^{2-} + 2H^+ \longrightarrow CO_2 + H_2O$$

(ii) CO_2 is readily identified from its property of turning lime water and baryta water milky.

$$Ca(OH)_2 + CO_2 \longrightarrow CaCO_3 + H_2O$$

$$Ba(OH)_2 + CO_2 \longrightarrow BaCO_3 + H_2O$$

(iii) Carbonate solution when treated with barium chloride precipitates barium carbonate

$$BaCl_2 + CO_3^{2-} \longrightarrow BaCO_3 + 2Cl^-$$

Sulphite ($SO3^{--}$): (i) When sulphite salts are treated with dilute acid, they liberate pungent SO_2 gas.

$$SO_3^{2-} + 2H^+ \longrightarrow SO_2 + H_2O$$

(ii) SO_3^2 is oxidized to SO_4^2 in the presence of oxidizing agents like dichromate and permanganate and turn green and colorless respectively.

$$Cr_2O_7^{2-} + 3\,SO_3^{2-} + 8H^+ \longrightarrow 2Cr^{3+} + 4H_2O + 3SO_4^{2-}$$

$$2MnO_4^- + 6H^+ + 5\,SO_3^- \longrightarrow 2Mn^{2+} + 5SO_4^{2-} + 3H_2O$$

Sulphide (S^{--}) : (i) In the presence of dilute acids sulphide salts liberate foul smelling H_2S gas.

$$S^{2-} + 2H^+ \longrightarrow H_2S$$

When a paper dipped in lead acetate is kept in the path of H_2S it turns black due to formation of PbS.

$$Pb^{2+} + H_2S \longrightarrow PbS + 2H^+$$

(ii) H_2S gas is a reducing agent like SO_2 gas. Therefore, similar to sulphite, oxidizing agents like dichromate and permanganate get reduced by acidified sulphide and turn green and colorless respectively.

$$Cr_2O_7^{2-} + 3H_2S + 8H^+ \longrightarrow 2Cr^{3-} + 3S + 7H_2O$$

$$2MnO_4^- + 5H_2S + 6H^+ \longrightarrow 2Mn^{2+} + 5S + 8H_2O$$

8-8-3 Radicals reactive to concentrated mineral acid

All the halides and nitrate constitute this group. On treatment with concentrated mineral acid pungent gases are evolved from the salts. These gases are identified from their characteristic tests.

Chloride (Cl^-): (i) The chloride salt when treated with concentrated H_2SO_4 liberates pungent hydrogen chloride gas.

$$Cl^- + H_2SO_4 \longrightarrow HSO_4^- + HCl$$

This gas produces dense white fumes from a glass rod dipped in ammonia and exhibit acidic properties (turning blue litmus red).

$$NH_3 + HCl \longrightarrow NH_4Cl$$

(ii) When a chloride salt is treated with manganese dioxide (MnO_2) and conc H_2SO_4, pungent smelling chlorine gas is liberated. This gas is reducing and turns the filter paper dipped in dichromate from light yellow to green.

$$MnO_2 + 4HCl \longrightarrow MnCl_2 + Cl_2 + 2H_2O$$

(iii) When silver nitrate is added to the salt solution curdy white precipitate of AgCl is formed. This precipitate disappears with the addition of few drops of ammonia and reappears on acidification with nitric acid.

$$Cl^- + Ag^+ \longrightarrow AgCl$$

$$AgCl + 2NH_3 \longrightarrow [Ag(NH_3)_2]Cl$$

$$[Ag(NH_3)_2]Cl + 2HNO_3 \longrightarrow AgCl + 2NH_4NO_3$$

Treatment of Br^- and I^- with Ag^+ give colored precipitates of AgBr and AgI respectively, the former precipitate is *slightly soluble* in ammonia while the latter is *completely insoluble.*

(iv) Chloride salt, when heated with potassium dichromate crystals and concentrated sulphuric acid produces red fumes of chromyl chloride CrO_2Cl_2. These vapors get cooled on the sides of the test tube and form red oily drops

$$Cr_2O_7^{2-} + Cl^- + 6H_2SO_4 \longrightarrow 2CrO_2Cl_2 + 6HSO_4^- + 3H_2O$$

When the chromylchloride vapors are passed into sodium hydroxide they leave a yellow solution of chromate.

$$CrO_2Cl_2 + 4OH^- \longrightarrow CrO_4^{2-} + 2Cl^- + 2H_2O.$$

Bromide and iodide on treatment with $Cr_2O_7^{2-}$ do not form any chromyl salt but get oxidized to generate corresponding elements.

$$6Br^- + Cr_2O_7^{2-} + 14\ H^+ \longrightarrow 3Br_2 + 2Cr^{3+} + 7H_2O$$

Bromide (Br^-) : (i) Bromide radical reacts with conc H_2SO_4 and liberates colorless, pungent hydrogen bromide and small quantities of brown colored bromine vapor.

$$Br^- + H_2SO_4 \longrightarrow HSO_4^- + HBr$$

$$2HBr + H_2SO_4 \longrightarrow Br_2 + SO_2 + 2H_2O$$

(ii) Light yellow colored AgBr is precipitated, when bromide solution is treated with silver nitrate solution. The precipitate dissolves in large amounts of ammonia.

$$Br^- + Ag^+ \longrightarrow AgBr$$

$$AgBr + 2NH_3 \longrightarrow [Ag(NH_3)_2]Br$$

(ii) On treating the bromide salts with a mixture of manganese dioxide and conc H_2SO_4, brown vapors of bromine are evolved that bleach the litmus color and turn the starch paper orange red.

$$Br^- + H_2SO_4 \longrightarrow HBr + HSO_4^-$$

$$MnO_2 + 4HBr \longrightarrow MnBr_2 + Br_2 + 2H_2O$$

(iii) Bromide radicals are readily oxidized by chlorine gas dissolved in water (chlorine water) to give orange red bromine vapors.

$$2Br^- + Cl_2 \longrightarrow 2Cl^- + Br_2$$

Iodide (I^-) : (i) Iodide salt on treatment with conc H_2SO_4 liberates large proportion of violet colored iodine vapors and small amounts of hydrogen iodide. The violet vapors when passed over starch paper turn it blue.

$$I^- + H_2SO_4 \longrightarrow HSO_4 + HI$$

$$H_2SO_4 + 2HI \longrightarrow I_2 + SO_2 + 2H_2O$$

(ii) The above reaction mixture in the presence of manganese dioxide liberates only iodine.

$$2I^- + MnO_2 + 2H_2SO_4 \longrightarrow I_2 + Mn^{2+} + 2H_2O + 2SO_4^{2-}$$

(iii) $AgNO_3$ when added to the iodide solution precipitates yellow colored AgI. This precipitate is neither soluble in ammonia nor dil HNO_3

$$Ag^+ + I^- \longrightarrow Ag\ I$$

(iv) Iodine is readily formed when chlorine water is added to iodide solution.

$$2I^- + Cl_2 \longrightarrow 2Cl^- + I_2$$

(v) Scarlet red HgI_2 is initially precipitated by the addition of iodide solution to few drops of mercuric chloride. On continuing the addition, the precipitate dissolves due to formation of complex ion $[HgI_4]^{2-}$

$$HgCl_2 + 2I^- \longrightarrow HgI_2 + 2Cl^-$$

$$HgI_2 + 2I^- \longrightarrow [HgI_4]^{2-}$$

Nitrate (NO_3^-) : When conc H_2SO_4 is added to a nitrate salt and slightly heated, dark brown vapors of nitrogen dioxide and pungent smelling HNO_3 vapors are generated.

$$NO_3^- + H_2SO_4 \longrightarrow HSO_4^- + HNO_3$$

$$4HNO_3 \longrightarrow 4NO_2 + O_2 + 2H_2O$$

(ii) Nitrate is also identified by the brown ring test discussed under nitrite. Nitrate salt when heated with conc H_2SO_4 in the presence of copper turnings liberate dark brown NO_2 and the reaction mixture turns blue due to the formation of cupric nitrate.

$$2NO_3^- + Cu + 4H^+ \longrightarrow Cu^{2+} + 2NO_2 + 2H_2O$$

This test is not exhibited by the other group members.

(iii) Nitrate solution when boiled with sodium hydroxide and zinc dust produces pungent smelling ammonia, which can be identified from its conventional chemical tests.

$$NO_3^- + 4Zn + 7OH^- \longrightarrow NH_3 + 4[ZnO_2]^{2-} + 2H_2O$$

8-8-4 Radicals unreactive to mineral acids

The radicals of this group, viz., SO_4^{2-} and PO_4^{3-} do not react with mineral acids even under heating conditions.

Sulphate (SO_4^{2-}): (i) The sulphate solutions on treatment with barium chloride give a curdy white precipitate.

$$BaCl_2 + SO_4^{2-} \longrightarrow BaSO_4 + 2Cl^-$$

This precipitate is insoluble in dil HCl and dil HNO_3, but partially soluble in conc HCl.

(ii) Sulphate radical readily combines with lead acetate forming a white precipitate of $PbSO_4$.

$$Pb(CH_3COO)_2 + SO_4^{2-} \longrightarrow PbSO_4 + 2CH_3COO^-$$

Phosphate (PO_4^{3-}) : (i) Phosphate radical when heated with ammonium molybdate $(NH_4)_2$ $MoO_4]$ forms yellow crystalline precipitate of $(NH_4)_3[PO_4(Mo_{12}O_{36})]$

$$PO_4^{3-} + 12\ MoO_4^{2-} + 24H^+ + 3NH_4^+ \longrightarrow (NH_4)_3PO_4\ 12MoO_3 + 12H_2O$$

This precipitate is soluble in ammonia and alkali hydroxides.

(ii) Posphate salt solutions combine with various cations giving crystalline precipitates

$$2HPO_4^{2-} + 3Ag^+ \longrightarrow Ag_3PO_4 + H_2PO_4^-$$

$$HPO_4^{2-} + Ba^{2+} \longrightarrow Ba(HPO_4)$$

$$HPO_4^{2-} + Mg^{2+} + NH_3 \longrightarrow [Mg(NH_4)PO_4]$$

$$2HPO_4^{2-} + Fe^{3+} \longrightarrow FePO_4 + H_2PO_4^-$$

8-9 CLASSIFICATION OF CATIONS OR BASIC RADICALS

The basic radicals are divided into six different groups based on their reaction with different group reagents to form their respective precipitates such as chlorides (first group), sulphides (second group and fourth group), hydroxides (third group), carbonates (fifth group). Group sixth is a miscellaneous group and consist of magnesium and ammonium radicals, table 8-5.

Before we discuss the group separation and analysis techniques it is essential to know about the theory behind precipitation of basic radicals as their sulphides and hydroxides basing on their solubility product.

8-9-1 Precipitation of Sulphides : The basic radicals of the II (Pb, Cd, Cu, Hg, Bi, Sb, Sn) and IV (Mn, Zn, Ni, Co) groups are precipitated as their sulphides with hydrogen sulphide. If Kipps apparatus for the production of hydrogen sulphide is not available, the commercially available yellow ammonium sulphide solution or a very dilute solution of sodium sulphide say 10^{-3}M can play its role.

$$M^{2+} + S^{2-} \longrightarrow MS$$

The sulphides of the group II are precipitated in an acidic media such as HCl, while the sulphides of group IV precipitate only in the presence of NH_4Cl and NH_4OH. The solubility product of the sulphides of group II is much lower than the sulphides of group IV. When hydrogen sulphide is passed into the solutions of the various salts, the precipitation of the sulphide can occur only when the ionic product $[M^{2+}][S^{2-}]$ exceeds the solubility product of the sulphide at that temperature. The solubility product (K_{sp}) values of some sulphides of group II and group IV are given in the table 8-6.

Table 8-6 Solubility product of sulphides of group II and group IV

K_{sp} *of group* II		K_{sp} *of group* IV	
PbS	3.4×10^{-28}	MnS	1.4×10^{-16}
CdS	3.5×10^{-28}	ZnS	1.2×10^{-23}
CuS	8.5×10^{-46}	NiS	1.4×10^{-24}
HgS	4.1×10^{-53}	CoS	3.0×10^{-26}

In the presence of an acid, hydrogen sulphide ionization gets suppressed due to an increase of protons liberated from the acid, as shown by the following equilibria :

$$HCl \rightleftharpoons H^+ + Cl^-$$

$$H_2S \rightleftharpoons 2H^+ + S^{--}$$

So there are few sulphide ions available in the solution, which are just enough to reach the solubility product of the sulphides of the II group, causing the precipitating of the PbS, CdS, CuS, etc. However, the solubility product of group IV being at least 100 times large, the required ionic product to match the solubility product is not reached and hence they remain in the centrifugate. But in the presence of NH_4OH, the hydroxyl ions combine with the protons produced by the ionization of H_2S to form unionized water molecules, there by increasing the ionization of H_2S with simultaneous generation of more S^{--} ions in the solution, the ionic product then matches the solubility product of the sulphides of group IV leading to precipitation of MnS, NiS, ZnS etc. Hence it is very essential to underline that, an acidic media is essential for the precipitation of the radicals of group II, while alkaline media is required for the group IV sulphide precipitation although in both the cases hydrogen sulphide is the precipitating reagent.

8-9-2 Precipitation of hydroxides : The III group (Fe, Al, Cr) basic radicals have ammonium chloride and ammonium hydroxide as the group reagents. The hydroxides get precipitated on reaction with the solution of the mixtures or the individual salts. In the precipitation of third group hydroxides, the presence of ammonium chloride, more precisely the ammonium ions, is essential to suppress the ionization of NH_4OH thereby limiting the OH^- concentration in the solution. As OH^- ions fall very low and the solubility product of the hydroxides of aluminium, ferrous, ferric and chromium is reached and these metal ions precipitate out, while the hydroxides of IV and VI group elements viz., zinc, magnesium, manganese, nickel, cobalt etc., which have high solubility product are not precipitated in the presence of the ammonium chloride.

Table 8-5 Groupwise Characteristics of Cations

Group number	*Group reagent*	*Radicals in the group*	*Precipitate formed*	*Characteristics of the precipitate*
I Silver group	dil HCl	Ag^{+} Pb^{2+} Hg_2^{2+}	AgCl (white) $PbCl_2$ (white) Hg^2Cl_2 (white)	Soluble in NH_3 Soluble in hot H_2O insoluble
	$HCl + MNO_3 \longrightarrow MCl + HNO_3$			
II A Copper group	dil HCl + H_2S	Hg^{2+} Pb^{2+} Bi^{3+} Cu^{2+} Cd^{2+}	HgS (black) PbS (black) Bi_2S_3 (brown) CuS (black) CdS (yellow)	All precipitates are insoluble in dil HCl & KOH
	$H_2S + MNO_3 \longrightarrow 2HNO_3 + MS$			
II B Arsenic group	dil HCl + H_2S	Sn^{2+} Sn^{4+} Sb^{3+} Sb^{5+} As^{3+} As^{5+}	SnS (brown) SnS_2 (yellow) Sb_2S_3 (orange) Sb_2S_5 (orange) AS_2S_3 (yellow) As_2S_5 (yellow)	All precipitates are soluble in KOH
	$2M(NO_3)_3 + 3H_2S \longrightarrow M_2S_3 + 6\ HNO_3$			
III Iron group	NH_4Cl + NH_4OH	Al^{3+} Cr^{3+} Fe^{2+} Fe^{3+}	$Al(OH)_3$ (white) $Cr(OH)_3$ (green) $Fe(OH)_2$ (dirty green) $Fe(OH)_3$ (brown)	Soluble in excess NaOH Coagulates on boiling Soluble in acids Soluble in acids
	$M\ Cl_3 + 2NH_3 + 3H_2O \longrightarrow M(OH)_3 + 3NH_4Cl$			
IV Zinc group	$NH_4\ Cl^{+}$ NH_4OH + H_2S	Ni^{2+} Co^{2+} Mn^{2+} Zn^{2+}	NiS (black) Co S (black) Mn S (pink) Zn S (white)	Soluble in hot conc. HNO_3 Soluble in hot conc HNO_3 Soluble in dil HCl & acetic acid Soluble in dilute mineral acids
	$MSO_4 + NH_4\ HS + NH_3 \longrightarrow MS + (NH_4)_2\ SO_4$			
V Calcium group	NH_4Cl^{+} NH_4OH+ $(NH_4)_2\ CO_3$	Ba^{2+} Ca^{2+} Sr^{2+}	$BaCO_3$ (white) $CaCO_3$ (white) $SrCO_3$ (white)	Soluble in dil acetic acid & HCl Soluble in H_2O containing H_2CO_3 Sparingly soluble in ammonium salts of strong acids.
	$MCl_2 + (NH_4)_2\ CO_3 \longrightarrow MCO_3 + 2NH_4Cl$			
VI Alkali group	Miscellaneous reagents	Mg^{2+} Na^{2+} K^{+} NH_4^{+}	$Mg(NH_4)PO_4\ 6H_2O$ (white) $NaMg[UO_2)_3C_2H_3O_2]_9$ H_2O $K_2[PtCl_6]$ (yellow) Brown precipitate NH_2Hg_2O with Nessler's reagent (organge brown)	Soluble in acetic acid and mineral acids Slightly soluble in H_2O

8-9-3 Ammonium radical

Although ammonium radical belongs to sixth group, its analysis is carried out first since salts containing this radical are used as reagents in subsequent analyses.

Ammonium (NH_4^+): (i) When a salt containing ammonium radical is warmed with few drops of NaOH pungent smelling ammonia vapors are evolved. These vapors being alkaline turn red litmus paper blue. These vapors also turn a filter paper dipped in mercurous nitrate black.

$$NH_4^+ + NaOH \longrightarrow NH_3 + Na^+ + H_2O$$

$$Hg_2^{2+} + 2NO_3^- + 2NH_3 \longrightarrow (HgNH_2NO + Hg) + NH_4^+ + NO_3^-$$

(ii) Ammonia and ammonium radical reacts with Nessler's reagent (alkaline $K_2[HgI_4]$) to give a brown precipitate.

$$2K_2HgI_4 + 4\ KOH + NH_4Cl \longrightarrow \left[O \begin{smallmatrix} Hg \\ \\ Hg \end{smallmatrix} NH_2 \right] I + 7KI + KCl + 3H_2O$$

8-9-4 Group I Cations

Lead (Pb^{2+}) :

(i) Lead ions are precipitated in the presence of iodide ions yielding yellow PbI_2. Addition of excess iodide leads to dissolution of this precipitate

$$Pb^{2+} + 2I^- \longrightarrow PbI_2$$

$$PbI_2 + 2I^- \rightleftharpoons (PbI_4)^{2-}$$

(ii) Similarly presence of chromate ions give yellow colored lead chromate precipitate which is insoluble in acetic acid and ammonia but gets dissolved in HNO_3 and alkali hydroxides.

$$Pb^{2+} + CrO_4^{2-} \longrightarrow PbCrO_4 \quad \text{(Yellow)}$$

$$PbCrO_4 + 4OH^- \longrightarrow (PbO_2)^{2-} + 2H_2O + CrO_4^{2-}$$

(iii) Lead radicals readily give black precipitates of PbS on passing H_2S gas through them

$$Pb^{2+} + S^{2-} \longrightarrow PbS$$

Mercurous (Hg_2^{2+}) and Silver (Ag^+) :

(i) The $PbCl_2$ precipitate formed during group precipitation is hot water soluble while AgCl and Hg_2Cl_2 are insoluble. Similar to Pb^{2+} the mercurous and Ag^+ ions also give black and yellow precipitates with S^{2-} and I^- ions.

(ii) Mercurous and silver salts too react with potassium chromate but give precipitates distinguishable through their color.

$$Hg_2(NO_3)_2 + K_2CrO_4 \longrightarrow 2KNO_3 + Hg_2CrO_4 \quad \text{(Brown)}$$

$$2AgNO_3 + K_2CrO_4 \longrightarrow 2KNO_3 + Ag_2CrO_4 \quad \text{(Red)}$$

$$Pb(NO_3)_2 + K_2CrO_4 \longrightarrow 2KNO_3 + PbCrO_4 \quad \text{(Yellow)}$$

(iii) Another test to differentiate Hg_2^{2+} and Ag^+ is through their chloride precipitate. AgCl is readily soluble on addition of ammonia, while Hg_2Cl_2 precipitate turns black on addition of ammonia due to formation of white aminomercuric chloride and black mercury.

$$AgCl + 2NH_3 \longrightarrow \underset{\text{Soluble}}{[Ag(NH_3)_2]^+Cl^-}$$

$$Hg_2Cl_2 + 2NH_3 \longrightarrow \underset{\text{White ppt}}{Hg(NH_2)Cl} + \underset{\text{black}}{Hg} + NH_4Cl$$

8-9-5 Group IIA Cations

Pb^{2+} is not totally precipitated in I group and appears in this group also. The identification methods are same as discussed earlier.

The other II A group radicals can be analysed by treating the salt solution with NaOH. The radicals could be identified from the color of the precipitate formed.

$$HgCl_2 + 2NaOH \longrightarrow 2NaCl + H_2O + HgO \quad \text{(reddish brown)}$$

$$Bi(NO_3)_3 + 3NaOH \longrightarrow 3NaNO_3 + Bi(OH)_3 \quad \text{(White-yellow on boiling)}$$

$$CuSO_4 + 2NaOH \longrightarrow Na_2SO_4 + \underset{\text{blue}}{Cu(OH)_2}$$

$$CdSO_4 + 2NaOH \longrightarrow Na_2SO_4 + \underset{\text{White}}{Cd(OH)_2}$$

Further Hg^{2+}, Bi^{3+}, Cu^{2+} react with KI, while Cd^{2+} is unreactive.

$$HgCl_2 + 2KI \longrightarrow 2KCl + HgI_2 \text{ (red)}$$

The precipitate dissolves in excess of reagent to form $K_2[HgI_4]$, if the reaction is conducted in ammonia-free water with all required precautions & concentrations, and if this solution is made alkaline, it can be used as Nessler's reagent.

$$HgI_2 + 2KI \longrightarrow K_2[HgI_4]$$

$$Bi(NO_3)_3 + 3KI \longrightarrow 3KNO_3 + Bi\ I_3 \text{ (dark brown)}$$

The precipitate dissolves in excess of reagent to form yellow colored K $[BiI_4]$.

$$Bi\ I_3 + KI \longrightarrow KBiI_4$$

Bi I_3 complex decomposes on dilution to form orange colored bismuthyliodide.

$$Bi\ I_3 + H_2O \longrightarrow 2HI + BiOI \text{ (orange)}$$

Copper reacts with I^- to form initially CuI_2, which decomposes to CuI and free iodine. CuI dissolves in excess of reagent to free more iodine and the color finally changes to brown.

$$CuSO_4 + 2KI \longrightarrow CuI_2 + K_2SO_4$$

$$2CuI_2 \longrightarrow 2CuI + I_2$$

Mercuric (Hg^{2+}) : The reactions are similar to Hg_2^{2+}.

Bismuth (Bi^{3+}) :

(i) Bismuth is best and easily identified by adding water to the salt, which then immediately forms a white precipitate due to bismuthyl radical.

$$BiCl_3 + H_2O \longrightarrow BiOCl + 2\ HCl$$

The precipitate is soluble in nitric acid.

(ii) Bismuth radicals are precipitated in ammonia and the precipitate is insoluble even in excess of reagent (Cu^{2+} and Cd^{2+} differ here)

$$Bi^{3+} + 2OH^- \longrightarrow BiO^+ + H_2O$$

(iii) Black colored bismuth metal is precipitated when to an alkaline solution of bismuth salt, sodium stannite is added

$$2Bi^{3+} + 3\ [SnO_2]^{2-} + 6OH^- \longrightarrow 2Bi + 3\ SnO_3^{2-} + 3H_2O$$

Copper (Cu^{2+}) : Copper gives a reddish brown (Chocolate colored) precipitate in the presence of ferrocyanide ions.

$$2\ Cu^{2+} + [Fe(CN)_6]^{4+} \longrightarrow Cu_2\ [Fe\ (CN)_6] \text{ Chocolate colored)}$$

This precipitate is unstable in dilute acids but gives a blue colored solution in the presence of ammonia due to formation of cupric hydroxide

$$Cu_2\ [Fe(CN)_6] + 4OH^- \longrightarrow 2Cu\ (OH)_2 + [Fe(CN)_6]^{4-}$$

Cadmium (Cd^{2+}) : (i) Cadmium ions react with ammonia to give a white precipitate of cadmium hydroxide, when excess ammonia is added the precipitate dissolves forming cadmium tetramine complex.

$$Cd^{2+} + 2NH_3 + 2H_2O \longrightarrow Cd(OH)_2 + 2NH_4^+$$

$$Cd(OH)_2 + 4NH_3 \longrightarrow [Cd\ (NH_3)_4]^{2+} + 2OH^-$$

8-9-6 Group II B radicals

The colours of sulphides given in Table 8-5 will give only a rough indication of nature of the salt.

(Arsenic (As^{3+}) & (As^{5+}) : These two oxidation states don't exist in free state. They can be differentiated by following tests

(i) Passing H_2S gas through a salt solution in dil HCl precipitates to give As_2S_3 while arsenites precipitates only on passing the gas for longer times or on warming

$$2Na_3AsO_3 + 3H_2S \longrightarrow As_2S_3 + 6H_2O$$

The slowness of arsenite in precipitation is explained as due to formation of soluble thioarsenic acid.

$$H_3AsO_4 + H_2S \longrightarrow H_2O + \underset{\text{(soluble)}}{H_3AsO_3S}$$

$$H_3AsO_3S \longrightarrow S + H_3AsO_3$$

$$2H_3AsO_3 + 3H_2S \longrightarrow As_2S_3 + 6H_2O$$

In conc HCl the precipitations occurs through a direct step.

$$2H_3AsO_4 + 5H_2S \longrightarrow As_2S_5 + 8H_2O$$

(ii) To the neutral solutions when $AgNO_3$ is added arsenites (As(III)) give a yellow precipitate while arsenate (As(V)) form a chocolate brown ppt.

$$AsO_3^{3-} + 3\ Ag^+ \longrightarrow Ag_3AsO_3 \quad \text{(Yellow)}$$

$$AsO_4^{3-} + 3Ag^+ \longrightarrow Ag_3AsO_4 \quad \text{(Chocolate brown)}$$

(iii) Arsenite does not form any ppt with magnesia mixture while arsenate forms a white crystalline ppt.

$$HAsO_4^{2-} + Mg^{2+} + NH_3 \longrightarrow Mg(NH_4)\ AsO_4$$

Antimony (Sb^{3+} & Sb^{5+}) :

(i) The sulphides of antimony are orange in color. Antimony pentasulphide disproportionates to trisulphide and sulphur. These sulphides are soluble in alkali hydroxides.

$$Sb_2S_5 \longrightarrow Sb_2S_3 + 2S$$

$$2Sb_2S_3 + 4OH^- \longrightarrow SbO_2^- + 3SbS_2^- + 2H_2O$$

(ii) When the salt solution is heated with either zinc dust or iron filings black antimony is formed.

$$2Sb^{3+} + 3Zn \longrightarrow 2Sb + 3Zn^{2+}$$

$$2Sb^{3+} + 3Fe \longrightarrow 2Sb + 3Fe^{2+}$$

Tin (Sn^{2+} & Sn^{4+}) :

(i) The stannous sulphide (SnS) and stannic sulphide (Sn S_2) are brown and yellow colored respectively.

$$Sn^{2+} + H_2S \longrightarrow \underset{\text{(brown)}}{Sn\ S} + 2H^+$$

$$Sn^{4+} + 2H_2S \longrightarrow \underset{\text{(yellow)}}{Sn\ S_2} + 4H^+$$

SnS is alkali insoluble but Sn S_2 dissolves and forms SnO_3^{2-}. The yellow precipitate of SnS_2 reappears on acidification.

$$3\ SnS_2 + 6OH^- \longrightarrow Sn\ O_3^{2-} + 2SnS_3^{2-} + 3H_2O$$

$$SnO_3^{2-} + 2SnS_3^{2-} + 6H^+ \longrightarrow 3\ SnS_2 + 3H_2O$$

(ii) Stannous ion is oxidized by mercuric chloride to form white precipitate of mercurous chloride. This precipitate turns grey due to formation of Hg when stannous chloride is in larger proportion.

$$Sn^{2+} + 2Hg^{2+} + 2Cl^- \longrightarrow Sn^{4+} + Hg_2Cl_2$$

$$Hg_2Cl_2 + Sn^{2+} \longrightarrow Sn^{4+} + \underset{\text{(grey)}}{2Hg} + 2Cl^-$$

8-9-7 Group III Cations

The Third group radicals are iron (ferrous and ferric), aluminium and chromium. Initial identification could be made from the color of the hydroxide precipitate. The $Al(OH)_3$, $Cr(OH)_3$, $Fe(OH)_2$ and $Fe(OH)_3$ are white, blue-green dirty green and brown respectively.

Ferrous (Fe^{2+}) & Ferric (Fe^{3+}) : (i) Ammonium thiocyanate does not impart any color to Fe^{2+} solution while Fe^{3+} solution turns deep red.

$$Fe^{3+} + SCN^- \longrightarrow [Fe(SCN)]^{3+} \text{ dark red.}$$

(ii) Potassium ferrocyanide interacts with Fe^{2+} and Fe^{3+} giving white and dark blue precipitates respectively.

$$2K^+ + Fe^{2+} + [Fe(CN)_6]^{4-} \longrightarrow \underset{\text{white}}{K_2Fe\,[Fe(CN)_6]}$$

$$K^+ + Fe^{3+} + [Fe(CN)_6]^{4-} \longrightarrow \underset{\text{dark blue}}{KFe\,[Fe(CN)_6]}$$

Aluminium (Al^{3+}) : (i) Aluminium forms a gelatinous white precipitate in the presence of weak base like ammonium hydroxide. This precipitate dissolves in excess of alkali hydroxide as well as in a mineral acid, hence it is an amphotesic hydroxide.

$$2Al^{3+} + 6NH_3 + 6H_2O \longrightarrow 2Al(OH)_3 + 6NH_4^+$$

$$Al(OH)_3 + NaOH \longrightarrow NaAlO_2 + 2H_2O$$

$$Al(OH)_3 + 3HCl \longleftrightarrow AlCl_3 + 3H_2O$$

(ii) The precipitate reappears from aluminate solution either by the addition of ammonium chloride or few drops of mineral acid.

$$NaAlO_2 + NH_4Cl + H_2O \longrightarrow Al(OH)_3 + NH_3 + NaCl$$

$$NaAlO_2 + HCl + H_2O \longrightarrow Al(OH)_3 + NaCl$$

Chromium (Cr^{3+}) : (i) Chromic radical precipitates by the addition of ammonia to form green to grey blue colored $Cr(OH)_3$. This complex dissolves in excess of reagent turning the solution violet.

$$Cr^{3+} + 3NH_3 + 3H_2O \longrightarrow Cr(OH)_3 + 3NH_4^+$$

$$Cr(OH)_3 + 6\,NH_3 \longrightarrow [Cr(NH_3)_6](OH)_3 \text{ (violet)}$$

(ii) Cr^{3+} also forms $Cr(OH)_3$ precipitate by the addition of NaOH. However excess addition of reagent leads to dissolution of precipitate forming chromite radical, which is a distinction from aluminium.

$$Cr^{3+} + 3OH^- \longrightarrow Cr(OH)_3$$

$$Cr(OH)_3 + OH^- \longrightarrow CrO_2^- + 2H_2O$$

(iii) To the above solution when hydrogen peroxide is added it turns yellow duc to the formation of the chromate ion

$$2\,CrO_2^- + 3H_2O_2 + 2OH^- \longrightarrow 2CrO_4^{2-} + 4H_2O$$

8-9-8 Group IV Cations

Except Zn^{2+}, rest of the group members viz., Mn^{2+} (pink), Co^{2+} (dark pink), Ni^{2+} (green) are all colored and so physically distinguishable.

Zinc (Zn^{2+}) : (i) In the presence of hydroxide ions Zn^{2+} gives a white precipitate of $Zn(OH)_2$. This precipitate is soluble in both weak and strong bases.

$$Zn^{2+} + 2OH^- \longrightarrow Zn(OH)_2$$

$$Zn(OH)_2 + 4NH_4OH \longrightarrow [Zn(NH_3)_4]^{2+} + 2OH^- + 4H_2O$$

$$Zn(OH)_2 + 2NaOH \longrightarrow Na_2[ZnO_2] + 2H_2O$$

(ii) By the addition of copper sulphate and ammonium mercuric thiocynate reagents to the Zn^{2+} ions a violet colored precipitate is formed.

$$Zn^{2+} + [Hg(SCN)_4]^{2-} \longrightarrow \underset{\text{White}}{Zn[Hg(SCN)_4]}$$

$$Cu^{2+} + Zn^{2+} + 2[Hg(SCN)_4]^{2-} \longrightarrow Zn[Hg(SCN)_4] + \underset{\text{violet}}{Cu[Hg(SCN)_4]}$$

Manganese (Mn^{2+}) : (i) Addition of sodium hydroxide to Mn^{2+} salt solutions gives white precipitate of $Mn(OH)_2$. After absorbing oxygen from air or on addition of H_2O this precipitate turns brown due to its oxidation to manganic hydroxide

$$Mn + 2OH^- \longrightarrow \underset{\text{White}}{Mn(OH)_2}$$

$$4Mn(OH)_2 + 2H_2O + O_2 \longrightarrow \underset{\text{brown}}{4Mn(OH)_3}$$

(ii) This precipitate readily dissolves in dil HNO_3. When sodium bismuthate or lead dioxide solution is added to this solution it turns purple due to formation of MnO_4^- ion.

$$5NaBiO_3 + 2Mn^{2+} + 14H^+ \longrightarrow 2MnO_4^- + 5Bi^{3+} + 7H_2O + 5Na$$

$$5PbO_2 + 2Mn^{2+} + 4H^+ \longrightarrow 2MnO_4^- + 5Pb^{2+} + 2H_2O$$

Nickel (Ni^{2+}) : (i) Green nickelous hydroxide is precipitated by the addition of NaOH to the salt solution. Unlike hydroxides of Mn^{2+}, Zn^{2+}; $Ni(OH)_2$ is insoluble even in the presence of excess hydroxide but dissolves in ammonia forming greenish blue colored, soluble complex ion $[Ni(NH_3)_6]^{2+}$

$$Ni^{2+} + 2OH^- \longrightarrow Ni(OH)_2$$

$$Ni(OH)_2 + 6NH_3 \longrightarrow [Ni(NH_3)_6]^+ + 2OH^-$$

(ii) On treating ammonical salt solutions with dimethyl glyoxime, scarlet red colored nickel dimethyl glyoxime complex is formed.

$$Ni^{2+} + 2C_4H_6N_2O_2H_2 \longrightarrow Ni\ (C_4H_6N_2O_2H_2)\ (red)$$

Cobalt (Co^{2+}) : (i) Cobalt nitrate reacts in cold with sodium hydroxide to form blue colored $Co(OH)NO_3$ precipitate. On heating this precipitate in excess alkali it turns pink due to formation of $Co(OH)_2$. Cobaltous hydroxide gets easily oxidized by arial oxygen or H_2O_2 to give brown precipitate of cobaltic hydroxide.

$$Co^{2+} + OH^- + NO_3^- \longrightarrow Co(OH)NO_3\ (blue)$$

$$Co(OH)NO_3 + OH^- \longrightarrow \underset{(pink)}{Co\ (OH)_2} + NO_3^-$$

$$4\ Co(OH)_2 + O_2 + 2H_2O \longrightarrow 4\ Co(OH)_3\ (brown)$$

(ii) Cobalt reacts with ammonia solution to form blue precipitate soluble in either excess reagent or in the presence of salt solution containing ammonium ions.

$$Co(NO_3)_2 + NH_4OH \longrightarrow Co(OH)NO_3 + NO_3^- + NH_4^+$$

$$Co\ (OH)\ NO_3 + 6\ NH_4OH \longrightarrow \underset{(brownish\ yellow)}{[Co\ (NH_3)_6]^{2+}} + OH^- + NO_3^- + 6\ H_2O$$

(iii) Cobalt ammine complex on oxidation by arial oxygen turns red.

$$2\ [Co(NH_3)_6]^{2+} + O_2 \longrightarrow \underset{(red)}{2\ [Co(NH_3)_6]^{3+}} + 2O^-$$

(iv) Potassium nitrite reacts in acetic acid media with cobalt salt solution to form yellow precipitate of potassium cobaltnitrite. This test does not undergo with nickel salts

$$Co(NO_3)_2 + 7\ KNO_2 + 2CH_3COOH \longrightarrow K_3\ [Co(NO_2)_6] + 2KNO_3 + 2CH_3COOK + NO + H_2O$$

8-9-7 Group V Cations

The fifth group of qualitative analysis contain the alkaline earth metal ions, Ba^{2+}, Ca^{2+} and Sr^{2+}. These radicals are precipitated by the addition of ammonium oxalate.

$$M^{2+} + (NH_4)_2C_2O_4 \longrightarrow MC_2O_4 + 2NH_{4+}$$

The precipitation is spontaneous with Ca^{2+} and rather slow with Ba^{2+} and Sr^{2+}. BaC_2O_4 is soluble in both hot dil acetic acid and dil mineral acids,

SrC_2O_4 is sparingly soluble in both type of acids, CaC_2O_4 is only soluble in dil mineral acids.

Barium (Ba^{2+}) : On treatment with K_2CrO_4, Ba^{2+} forms yellow precipitate of $BaCrO_4$. This precipitate is insoluble in both acetic acid as well as water (Sr^{2+} and Ca^{2+} chromates are soluble).

$$Ba^{2+} + K_2CrO_4 \longrightarrow BaCrO_4 + 2K^+$$

Strontium (Sr^{2+}) : Strontium chromate is partially soluble water and acetic acid.

Calcium, (Ca^{2+}) : Addition of chromate to dilute Ca^{2+} salt solutions does not lead to any precipitation. The precipitate formed with concentrated solutions is highly soluble in acetic acid.

8-9-10 Group VI radicals

This miscellaneous group contains radicals Mg^{2+}, NH_4^+, K^+ and Na^+. Tests for ammonia are already discussed. There is no common group reagent that precipitates all the radicals.

Magnesium (Mg^{2+}) : (i) Magnesium ions react with ammonium and sodium hydroxides to precipitate $Mg(OH)_2$, insoluble in excess reagents.

$$MgSO_4 + 2NH_4OH \longrightarrow Mg(OH)_2 + (NH_4)_2SO_4$$

$$MgSO_4 + 2NaOH \longrightarrow Mg(OH)_2 + Na_2SO_4$$

(ii) Ammonia free solutions of magnesium react with ammonium carbonate to give magnesium carbonate. After being boiled long time with ammonia or ammonium salts it converts into soluble magnesium bicarbonate

$$4Mg^{2+} + 4\ CO_3^{2-} + H_2O \longrightarrow Mg_4\ (CO_3)_3\ (OH)_2 + CO_2$$

(iii) Magnesium reacts with sodium ammonium phosphate to precipitate out $Mg(NH_4)PO_4 \cdot 6H_2O$. This precipitate is soluble in acetic acid.

$$Na_2(NH_4)PO_4 + MgCl_2 \longrightarrow Mg(NH_4)PO_4 + 2NaCl$$

$$Mg(NH_4)PO_4 + H_2O \longrightarrow Mg^{2+} + HPO_4^{2-} + NH_3 + H_2O$$

(iv) Magnesium forms a complex with oxine (8-hydroxy quinoline) in quantitative amounts.

$$\underset{\text{oxine}}{Mg^{2+} + 2C_9H_6NO^- + 4H_2O} \longrightarrow \underset{\text{(yellow)}}{Mg(C_9H_6NO)_2 4H_2O}$$

Potassium (K^+) : (i) Potassium reacts with sodium cobaltinitrite to give a yellow colored precipitate of potassium sodium cobaltinitrite or potassium cobaltinitrite.

$$2K^{+} + Na_3\,[Co(NO_2)_6] \longrightarrow 2Na^{+}K_2Na[Co(NO_2)_6] \text{ or } K_3\,[Co(NO_2)_6]$$

This reaction should be performed in the absence of iodides, ammonia and ammonium salts.

(ii) Chloroplatinic acid precipitates K^{+} to form yellow colored potassium chloro platinate.

$$2K^{+} + H_2[PtCl_6] \longrightarrow K_2[PtCl_6] + 2HCl$$

Sodium (Na^{+}) : (i) Sodium forms yellow crystalline precipitate of sodium magnesium uranyl accetate with magnesium uranyl acetate, from its concentrated solutions.

$$Na^{+} + 2UO_2(C_2H_3O_2)_2 + Mg\,(C_2H_3O_2)_2 + CH_3COOH \longrightarrow HCl + [Na\,Mg\,(UO_2)_3\,(C_2H_3O_2)_4]$$

(ii) Perchloric acid, chloroplatinic acid and sodium cobatic nitrite do not form any precipitate with Na^{+} ions.

8-10 MULTIPLE CHOICE QUESTIONS

8-10-1 A solution whose concentration is known is referred to as a

A) Titrimetric solution B) Primary solution
C) Standard solution D) Secondary solution

8-10-2 The primary standard substances among the following pairs are :

A) KCl & KOH B) $AgNO_3$ & Na_2CO_3
C) $AgNO_3$ & HNO_3 D) Na_2CO_3 & KOH

8-10-3 The secondary standard substances among the following pairs are :

A) HCl & NaOH B) NaOH & Na_2CO_3
C) $KMnO_4$ & HCl D) NaOH & $AgNO_3$

8-10-4 The volumetric analysis is also referred to as

A) Titrimetry B) Titrationmetry C) Gasimetry D) Standardimetry

8-10-5 The solution which has been standardized is known as

A) Titrate B) Titrant C) Primary Solution D) Basic solution

8-10-6 The solution with unknown concentration whose estimation is being carried out is known as

A) Titrant B) Primary Solution C) Titrate D) Acidic solution

8-10-7 The completion of a volumetric reaction can be visualized at

A) Point of normality B) Point of precipitation
C) Point of equivalence D) Point of complexation

8-10-8 The titration involving oxalic acid and potassium hydroxide is classified as

A) Spectrophotometric titration B) Precipitation titration
C) Complexometric titration D) Neutralization titration

8-10-9 The ionic product of water is

A) 1×10^{-14} B) 1×10^{-13} C) 1×10^{-12} D) 1×10^{-7}

8-10-10 The $[OH^-]$ in pure water is

A) 1×10^{-14} B) 1×10^{-13} C) 1×10^{-12} D) 1×10^{-7}

8-10-11 When the $[OH^+]$ is 0.001M, then $[OH^-]$ will be in pure water.

A) 0.01 B) 10^{-9} C) 0.004 D) 10^{-11}

8-10-12 When $[OH^-]$ is 1×10^{-5} the $[H^+]$ is

A) 10^{-5} B) 10^{-9} C) 10^{-8} D) 10^{-7}

8-10-13 When $[H^+]$ is 1×10^{-14} of HCl then the pOH of the solution is

A) 10 B) 4 C) 8 D) 6

8-10-14 The value of pKw is

A) 14 B) 13 C) 2 D) 0

8-10-15 In the term pH, p stands for

A) potent B) potenz C) para D) prudent

8-10-16 The pH of a solution with [H+] = 8.123 is

A) 7.534×10^{-8} B) 7.534×10^{-8} C) 7.534×10^{-7} D) 7.534×10^{-10}

8-10-17 pKIn is calculated by the expression

A) $pH - \log \frac{[In]}{[HIn]}$ B) $pH + \log \frac{[In]}{[HIn]}$

C) $pH - \log \frac{[HIn]}{[In]}$ D) $\log \frac{[In]}{[HIn]} - pH$

8-10-18 The pOH of pure water is

A) 5 B) 6 C) 7 D) 8

8-10-19 pH =

A) $pK_{In} - \log \frac{[In]}{[HIn]}$ B) $pK_{In} + \log \frac{[In]}{[HIn]}$

C) $pK_{In} + \log \frac{[In]}{[HIn]}$ D) $- pK_{In} + \log \frac{[HIn]}{[In]}$

8-10-20 Litmus has a pH range of

A) 6.2 - 7.5 B) 3.2 - 4.4 C) 10 - 12 D) 5.5 - 7.5

8-10-21 Thymol Blue changes its red color in acidic media to color in basic media

A) Blue B) yellow C) colorless D) violet

8-10-22 Phenol red has a pK_{In} value of

A) 9.7 B) 7.9 C) 9.6 D) 7.8

8-10-23 Phenolthalein can be used in the pH range

A) 8.3 - 10 B) 7.5 - 8.3 C) 6.2 - 7.5 D) 10 - 14

8-10-24 The titration curve of a strong acid strong base gives a shape of

A) S B) P C) D D) F

8-10-25 The equivalence point of an equimolar solution of acetic acid and sodium hydroxide is obtained graphically around pH

A) 11.0 B) 4.8 C) 6.1 D) 8.7

8-10-26 The congugate acid of NH_4OH when treated with HCl is

A) Cl^- B) H^+ C) NH_4^+ D) OH^-

8-10-27 The congugate acid H_3O^+ is

A) weaker than NH_4^+ B) stronger than NH_4^+
C) equally strong to NH_4^+ D) None of the statement is correct

8-10-28 The indicator used for a weak acid - weak base titration is usually

A) Phenol red B) Phenol blue C) Phenolpthalein D) Methyl red

8-10-29 Indicators with pK_{In} 7 are called

A) Neutral indicators B) Basic indicators
C) Acidic indicators D) Universal indicators

8-10-30 The acid indicator amongst them is

A) Phenolpthallin B) Methyl red C) Alizarin -S D) Nitrobenzane

8-10-31 The base indicator amongst these is

A) Methyl orange B) Methyl violet
C) Methyl red D) Alizarin yellow

8-10-32 The standard electrode potential the reaction
$Fe^{3+} + e^- \; Fe^{2+}$ is $E^o =$

A) 9.54 V B) 0.77 V C) 09.15 V D) 1.52 V

8-10-33 The value $E^o = 1.36$ is for the reaction

A) $Cl_2 + 2e^- \quad 2Cl^-$ B) $Br_2 + 2e^- \; 2Br^-$

C) $Cu^{++} + le^- \; Cu^+$ D) $I_2 + 2e^- \; 2I^-$

8-10-34 On titrating SCN^- with Fe^{3+} the end point gives a color of

A) Blood red B) Peacock blue
C) Parrot green D) Duck white

8-10-35 Which one is not a redox amongst the following

A) Diphenylamine B) Phenosafranine
C) Methyl red D) Methylene blue

8-10-36 The indicator used for titrating $C_2O_4H_2$ and $KMnO_4$ is

A) Methyl orange B) Methyl violet
C) Methyl red D) None of them

8-10-37 The indicator used in dichrometry

A) Phenolplthalein B) Diphenyl amine
C) Thymol Blue D) Sulphuric acid

8-10-38 Titrations done with standard iodine solutions are called

A) Redox B) Iodimetry C) Oxidation D) Iodometry

8-10-39 Titrations done with liberated iodine against a reductant like hypo are

A) Redox B) Iodimetry C) Reductant D) Iodometry

8-10-40 Iodine reacts with SO_3^{20} ion

$SO_3^{2-} + I_2 + H_2O$ to form

A) $SO_4^{2-} + 2I^-$ B) $SO_4^{2-} + 2H^+ + 2I^-$
C) $S_4O_6^{2-} + 2I^-$ D) $S + 2H^+ + 2I^-$

8-10-41 Iodine hydrolyses in alkaline solutions $I_2 + OH^-$ to give

A) $I_3^- + H_2O$ B) $HI + O_2$ C) $HOI + I^-$ D) $H_2O_2 + I_3^-$

8-10-42 When silver ions are titrated against ammonium thiocyanate using $FeCl_3$ as indicator, the method is known as

A) Mohr's B) Volhard's C) Turbidometry D) Complexometry

8-10-43 In Mohr's method the indicator is

A) Potassium chromate B) Ferric chloride
C) Alizarin - S D) Thorium Nitrate

8-10-44 Adsorption indicators are used in

A) Complexometry B) Acidmetry
C) Precipitation Titration D) Argentometry

8-10-45 Molecules having two or more donor groups is called

A) Chelate B) Ligand C) Complex D) Colloid

8-10-46 The indicator not used in an EDTA titration

A) Xylenol orange B) Eriochrome Black T
C) Pyrocate col violet D) Methyl violet

8-10-47 Heating any precipitate to a certain temperature with the solvent is known as

A) Distillation B) Digestion
C) Coprecipitation D) Post precipitation

8-10-48 The mechanisum of occulsion results in the formulation of

A) Coprecipitation B) Isomorphic Inclusion
C) Non Isomorphic Inclusion D) Adsorption

8-10-49 The post precipitation is common in the precipitation of

A) Sulphides B) Sulphates C) Chlorides D) Chromates

8-10-50 Barium can be estimated gravimetrically as

A) Sulphate B) Acetate C) Nitrate D) Chloride

8-10-51 Chloride is tested in the filternate during gravimetric estimations using

A) HCl B) $BaBr_2$ C) $AgNO_3$ D) $BaSO_4$

8-10-52 A glass rod with a rubber tip is a

A) Stirrer B) Policeman C) Stopper D) Detector

8-10-53 The reagent added to convert sulphide to sulphate during incineration

A) $BaSO_4$ B) $CaSO_4$ C) Na_2SO_4 D) H_2SO_4

8-10-54 Chlorides on treatment with Conc. H_2SO_4 forms

A) Cl_2 B) HCl C) NaCl D) HClO

8-10-55 The brown ring test of nitrate produces

A) $Fe(NO_3)_2$ B) $Fe(NO)_3$ C) $Fe(NO)SO_4$ D) $Fe(NO_2)\ SO_4$

8-10-56 Acidified $KMnO_4$ on reacting with $SO_3^{=}$ produces $MnSO_4$, on reacting with S it produces

A) $MnCl_2$ B) $MnSO_4$ C) $MnSO_3$ D) MnS

8-10-57 Baryta water

A) $BaCO_3$ B) $CaCO_3$ C) $Ba(OH)_2$ D) $Ca(OH)_2$

8-10-58 AgCl on dissolving in ammonia forms

A) $AgNH_4Cl_2$ B) Ag NO_3
C) $[Ag(NH_3)]Cl$ D) $[Ag\ (NH_3)_2]Cl$

8-10-59 Chromylchloride is

A) $CrOCl_3$ B) CrO_2Cl_2 C) CrO_2Cl D) $CrOCl_2$

8-10-60 The solubility product of group II sulphides is less than that of Group IV. The value 3×10^{-26} is that of

A) Lead sulphide B) Cobalt sulphide
C) Copper Sulphide D) Zinc sulphide

8-10-61 Nessler's reagent is

A) K Hg I_3 B) K Hg_2I_3
C) K_2HgI_4 D) $K_2Hg_2I_4$

8-10-62 Bismuthyl iodide formed by the hydrolyses/decomposition of BiI_3 is

A) BiOI B) BiO_2I C) Bi_2OI D) $BiOI_2$

8-10-63 The yellow tinsulphide is

A) SnS_2 B) SnS C) Sn_2S_2 D) Sn_2S

8-10-64 The substitute for H_2S in qualitative analysis is

A) Fe S B) Ca S C) Na_2S D) MgS

8-10-65 $Al(OH)_3$ + NaOH H_2O +

A) Na AlO_3 B) $NaAlO_4$ C) $NaAlO_2$ D) Na_2AlO

8-10-66 The compound reacted with $Mn(OH)_2$ in the presence of boiling HNO_3 to form permangnate ion is

A) MnO_2 B) Bi_2O_3 C) $KMnO_4$ D) $NaBiO_3$

8-10-67 Cobaltic hexammine ion is

A) $[Co(NH_3)_6]^{2+}$ B) $[Co(NH_3)_6]^{3+}$
C) $[Co(NH_3)_6]^{4-}$ D) $[Co(NH_3)_6]^{4+}$

8-10-68 Oxine is
A) Dimethylglyoxime
B) Acizarin - S
C) 8-hydroxyquinoline
D) Phenolpthalene

8-11 ESSAY & SHORT QUESTIONS

8-11-1 Write short notes on :

A) A primary standard substance B) Ionic product of water

C) pH D) Derive an expression for pKw.

8-11-2 Calculate the pH of a) 0.018M KOH b) 0.0024M HCl c) 0.032×10^{-3} NaOH

8-11-3 What is the importance of indicators in the volumetric analysis ?

8-11-4 Discuss about the choice of indicator in the volumetric analysis.

8-11-5 Write notes on :

A) Titration of a strong acid by a strong base
B) Titration of a weak acid by a weak base

8-11-6 Write notes on

A) Titration of a weak acid & strong base
B) Titration of a strong acid & weak base

8-11-7 Discuss the different theories of indicators

8-11-8 In what way the quinonoid theory is more explanatory than the ionic theory of indicators

8-11-9 **What are redox reactions, what are the different type of redox reactions involved in titrimetry.**

8-11-10 **How are standard electrode potentials involved in volumetric analysis.**

8-11-11 **What are specific indicators and true redox indicators.**

8-11-12 **Write notes on permangnometry**

8-11-13 **Write a note on the advantages & differences present between permangnometry and dichrometry**

8-11-14 **Write a note on the titrations involving ceric sulphate, how does these have an edge over permangnometric titrations ?**

8-11-15 **What is the theory involved in the iodometric titrations ?**

8-11-16 **What are the different method involved in argentometry titrations.**

8-11-17 **Write notes on adsorption indicators**

8-11-18 **Discuss the chemistry involved in the complexometric titrations**

8-11-19 **Discuss the theory of gravimetric analysis**

8-11-20 **What are the optimum conditions required for precipitation ?**

8-11-21 **Write notes on**

A) Co-precipitation

B) Post precipitation

8-11-22 **How are anions and cations classified in groups.**

8-11-23 **Justify the inclusion of CO_3^{2-}, S^{2-}, SO_3^{2-}, NO_2^- as radicals of the same group with suitable reactions.**

8-11-24 **Justify the classification of Cl^-, Br^- and I^- with NO_3^- together with suitable reactions.**

8-11-25 **How is chloride differentiated from bromide ? Discuss with proper equation**

8-11-26 **The II group cations precipitate in the presence of an acid as sulphides whereas the IV group cations does not why ?**

8-11-27 **What is meant by solubility product ? and how is it applied in semimicro analysis.**

8-11-28 **Aluminium, chromium and iron belong to different groups in the periodic table, but are classified together in qualitative analysis. Justify your answer with suitable examples.**

8-11-29 **Give proper reasoning in not more than a sentence or two.**

a) Volumetric analysis are known as titrimetric analysis

b) Indicator is required for an acid base titration

c) No indicator is required in the permangnometric titration.

d) 14 - pH = pOH

e) Phenolpthalien is used in a strong acid - strong base titration.

f) Methyl orange is used in the titration of HCl vs Na_2CO_3.

g) Quinonoid theory is better than ionic theory give the main reason.

h) $K_2Cr_2O_7$ cannot oxidize Mn^{2+}
i) $Cr_2O_7^{2-}$ can react with Cu^{++} but not Cl^-.
j) Substance responsible to impart color when starch reacts with I_3^-
k) $KMnO_4$ is not used as a primary standard.
l) Use of phosphoric acid in the titration of $K_2Cr_2O_7$ against Fe.
m) Difference between iodometry and iodimetry.
n) Purpose of using $SnCl_2$ in the titration of Fe^{++}/Fe^{+++} by dichrometry.
o) α-amylose is not used as an indicator but β-amylose can be used.
p) starches derived from potatoe can be used but not that from corn flours
q) In argentometry Mohr's method is better than Volhard's
r) EDTA as a compound is hexadentate but it acts as a quadridentate ligand
s) Precipitates should be digested before filteration.
t) Similarity between occulsion & coprecipitation
u) Nessicity of adding H_2SO_4 to the incinerated $BaSO_4$ precipitate.
v) Nessicity of performing ammonium radical test before testing other cations.
w) Division of group II cations into A & B groups
x) Addition of HCl & NH_4Cl to different sample solutions of $NaAlO_2$
y) Boiling of $Mn(OH)_2$ with $NaBiO_3$ in HNO_3
z) No common group reagent for VI group radicals.

Answers of 8-11-29

a) The method of titration is involved in volumetric analysis
b) As both the titrate & titrant are colorless
c) Permanganate is a self indicator
d) pH + pOH = 14, since water has equal [H+] = [OH-]
e) The pH range of phenolplhalein is 8.3 to 10 with pK_{In} = 9.6 which is in correspondance to the required values for such a titration
f) The pH range of methyl orange is 3.2 to 4.4 with pK_{In} = 3.5 which is in correspondance to the required values for such a titration.
g) Ionic theory does not explain the formation of color during the titration whereas the quinovoid theory explains it.
h) Mn^{2+} has a higher standard electrode potential than $K_2Cr_2O_7$
i) $Cr_2O_7^{2-}$ has a higher standard electrode potential than Cu^{++} but lower than Cl^-
j) Amylose
k) $KMnO_4$ undergoes autocatalytic decomposition due to heat & light effects.
l) Phosphoric acid lowers the Fe^{3+}/Fe^{2+} electrode potential which makes the reaction between $K_2Cr_2O_7$ & iron easy.
m) In iodometry, the iodine released is titrated but in iodimetry, a standard solution of iodine is used.
n) To reduce Fe^{+++} to Fe^{++}
o) α-amylose starch cannot absorb iodine, whereas β-amylose can
p) pototoe starch contains β-amylose whereas cornflour starch contains α-amylose
q) Mohr's method is a direct method of titrating silver with halides whereas the Volhard's method is an indirect method.
r) EDTA is water insoluble hence to make use of it its disodiumsalt is made hence it becomes quadridentate from hexadentate.

s) For coagulation purpose which results in the increase of size of the precipitate particles
t) Occulsion is a type of coprecipitation in which the soluble impurity is entrapped within the crystal structure of the precipitate
u) To convert any sulphide formed to sulphate
v) As ammonium salt solutions are involved as reagents, therefore it should be checked first.
w) The IIB group radicals are KOH soluble whereas IIA group radicals are KOH insoluble
x) To check the ampholesic character of Al $(OH)_3$
y) The reaction gives permanganate ion
z) It constitutes miscellaneous radicals with different properties.

ANSWERS

8-10-1 C	**8-10-2** B	**8-10-3** A	**8-10-4** A	**8-10-5** B	**8-10-6** C
8-10-7 C	**8-10-8** D	**8-10-9** A	**8-10-10** D	**8-10-11** D	**8-10-12** B
8-10-13 A	**8-10-14** A	**8-10-15** B	**8-10-16** A	**8-10-17** A	**8-10-18** C
8-10-19 B	**8-10-20** D	**8-10-21** B	**8-10-22** B	**8-10-23** A	**8-10-24** A
8-10-25 D	**8-10-26** C	**8-10-27** B	**8-10-28** A	**8-10-29** C	**8-10-30** B
8-10-31 D	**8-10-32** B	**8-10-33** A	**8-10-34** A	**8-10-35** C	**8-10-36** D
8-10-37 B	**8-10-38** B	**8-10-39** D	**8-10-40** B	**8-10-41** C	**8-10-42** B
8-10-43 A	**8-10-44** C	**8-10-45** A	**8-10-46** D	**8-10-47** B	**8-10-48** A
8-10-49 A	**8-10-50** A	**8-10-51** C	**8-10-52** B	**8-10-53** D	**8-10-54** B
8-10-55 C	**8-10-56** A	**8-10-57** C	**8-10-58** D	**8-10-59** B	**8-10-60** B
8-10-61 C	**8-10-62** A	**8-10-63** A	**8-10-64** C	**8-10-65** C	**8-10-66** D
8-10-67	**8-10-68** A				

Part II

Organic Chemistry

1 STRUCTURE OF ORGANIC MOLECULES

1-1 ORIGIN OF ORGANIC CHEMISTRY

Chemical compounds can be classified into two classes, based on their origin, namely inorganic compounds and organic compounds. The chemical compounds having a mineral origin, such as common salt, carbon dioxide etc, are called inorganic compounds, while those obtained from plant or animal sources (i.e. living organisms) such as oils, fats, starch, sugars, resins, spirit, indigo, perfumes, textiles etc. are called organic compounds. The part of chemical science dealing with the synthesis, extraction and identification of these organic compounds was defined as organic chemistry.

1-2 THE VITAL FORCE THEORY

Until 1828 chemists believed that the only source of these organic compounds was living organisms. It was thought to be impossible to synthesize them in the laboratory. In 1815 Berzelius suggested that *a mysterious or vital force is responsible for the formation of organic compounds in living organisms.* This theory is known as **vital force theory.**

In 1828, Friedrich Wohler, a German chemist, accidentally obtained urea, a compound found in urine, from inorganic compounds. During one of his experiments, Wohler evaporated a solution of the inorganic salt ammonium cyanate, NH_4CNO. When he heated the solid, he discovered that another substance was formed, which he analyzed and found to be urea.

$$\underset{\text{Ammonium Cyanate}}{NH_4CNO} \xrightarrow{\text{heat}} \underset{\text{Urea}}{H_2N\text{-}\overset{\overset{\displaystyle O}{\|}}{C}\text{-}NH_2}$$

Once the results of Wohler's synthesis were known, many scientists began to attempt to prepare other organic materials with a good deal of success. For example in 1845, Kolbe had synthesized acetic acid and in 1856 Bertholet had synthesized methane. Gradually, the vital force theory was abandoned, and the years that followed saw the many such organic compounds. Today there are well over six million known organic compounds and new ones are being discovered every day.

1-3 DEFINITION OF ORGANIC CHEMISTRY

As, all the organic compounds were isolated from natural sources or synthesized found to contain carbon as one of the elements, the emphasis for defining the organic compounds was shifted from origin to composition. Thus organic chemistry is defined as the chemistry of carbon compounds. On the basis of this definition, the study of compounds like carbon monoxide, carbon dioxide, carbides, cyanates etc., should be included in organic chemistry, but due to their greater resemblance with inorganic compounds these are studied under inorganic chemistry. Further, since all the organic compounds have been found to be mainly compounds of carbon and hydrogen (called hydrocarbons) or their derivatives, the organic chemistry is defined today as the *chemistry of hydrocarbons and their derivatives.*

1-4 REASONS OF STUDY OF ORGANIC CHEMISTRY AS A SEPARATE BRANCH OF CHEMISTRY

The fundamental laws of chemistry like atomic structure, valence, bonding etc., are applicable to both inorganic as well as organic compounds. Even then organic chemistry is studied as a separate branch of chemistry due to enormous and rapid advancements in this branch of chemistry. The reasons to study organic chemistry as a separate branch of chemistry are given as under;

1. Today there are well over six million known organic compounds and new once are being discovered every day. The great abundance of organic compounds is due to the peculiar property of the carbon atom to combine with itself to form long chains, rings and complex combination of both. This property of carbon atom is called catenation. Since catenation is not shown to such an extent by any another atom, the inorganic compounds are less in number.

2. Due to maximum catenation and tetravalency of carbon it is unique in its properties. It has the tendency to form long open and closed chains. Thus, this gives rise to a very wide range of organic compounds from the simplest like methane to the most complex molecules like proteins. These compounds differ from inorganic compounds markedly.

3. Organic compounds are generally covalent compounds and do not ionize when they dissolved in water. Whereas many inorganic compounds are electrovalent and usually ionize in water. For example, addition of silver nitrate solution to chloroform gives no white precipitate, but all metal chlorides give white precipitate with $AgNO_3$ solution.

4. Organic compounds may be classified into different series called as homologous series. Different members belonging to a homologous series have a characteritic functional group and can be represented by a general formula. The common difference between consecutive members of the series is CH_2. They can be prepared by general methods of preparation and possess similar chemical properties with a regular gradation in their physical properties. This has reduced the study of a large number of organic compounds to the study of a few homologous series.

5. Generally, organic reactions are nonionic in nature and proceed very slowly, while the inorganic reactions are ionic in nature and proceed rapidly. Thus organic reaction are much more complicated than inorganic reactions.

6. Organic compounds are usually soluble in organic solvents, such as ether, alcohol, benzene etc., and insoluble in water, where as inorganic compounds are soluble in water, but insoluble in organic solvents.

7. Very often more than one organic compound is represented by the same molecular formula but exhibit different physical and chemical properties. Such compounds are called isomers and the phenomenon is called a isomerism. Whereas in most of the inorganic compounds, isomerism is not found and hence a single formula generally represents a single compound.

8. Organic solutions are usually non-conductors of electricity whereas solutions of inorganic compounds are good conductrors of electricity.

9. Organic compounds are generally valatile, inflammable and possess low melting and boiling points. Whereas inorganic compounds are generally non-volatile, non-combustible and possess high melting and boiling points.

10. Some organic compounds exhibit the phenomenon of polymerization, while inorganic compounds do not undergo polymerization.

Because of the above characteristics of the organic compounds and their differences with inorganic compounds it is, therefore, necessary to treat organic chemistry as a separate branch of chemistry.

1-5 SOURCE OF ORGANIC COMPOUNDS

The chief sources of organic compounds are plants, animals, coal, natural gas, petroleum and laboratory synthesis.

1. **Plants** : By making use of certain plants, some organic compounds such as carbohydrates, citiric acid, proteins, oils, gums, resins, dyes, alkaloids, tetratic acid, perfumes etc. are obtained. For example, Scheele isolated tetratic acid from grapes.

2. **animals** : Animals are the source of many organic compounds like fats, proteins, urea, honey, vitamins, hormones etc. For example, Rouella isolated urea from human urine and Chevreul isolated cholesterol from animal fats.

3. **Fungi and microorganism** : These are responsible for the production of some organic compounds, such as, alcohols, acids, antibiotics, vitamins, etc. The fermentation is a chemical process in which a complex substances like starch break into simpler compounds by the action of living microorganisms.

4. **Coal** : The destructive distillation of coal yields coke and coal-tar. When coal-tar is subjected to fractional distillation, it yields different organic compounds such as benzene, toluene, napthalene, anthracene, cresol etc.

5. **Petroleum** : Petroleum is the major source for a larger number of organic compounds. It provides us with most of the aliphatic hydrocarbons which are used as fuels in various forms such as petrol, kerosine, diesel,

cooking gas etc., paraffin wax, vaseline, fertilizers and many other petroleum products.

6. **Synthesis** : The products obtained by fractionation of petroleum serve as a starting materials for the synthesis of several organic compounds. For example, drugs, dyes, vitamins, hormones, rubber, explosives, plastics etc. have been successfully synthesized in the laboratory.

1-6 IMPORTANCE OF ORGANIC CHEMISTRY

Organic compounds are encountered constantly in daily life. The components of our body are organic substances. To maintain health we need an adequate supply of organic compounds such as food, vitamins, hormones, enzymes etc. Food without which we can not live, mainly consists of organic compounds, which include carbohydrates, proteins, oils and fats etc. Apart from food the basic needs of our life is fulfilled by organic compounds like cotton, silk, polyester, paper, leather, plastics, dyes, drugs, perfumes etc. Understanding of life processes like digestion and assimilation is possible only after a thorough study of organic chemistry of enzymes and other biologically important molecules. Organic chemistry provides fundamental knowledge of biology and medicine. The science of biochemistry and molecular biology deal with various large organic molecules. an ifea about the use of organic compounds in the service of mankind can be had from their application in day to day life as represented below.

1. Food : Carbohydrates, proteins, fats and oils, vitamins, etc.
2. Textile : Cotton, silk, wool, rayon, nylon, terrylene, etc.
3. Fuel : Wood, coal, alcohol, petrol, diesel etc.
4. Drugs : Antibiotics, aspirin, sulpha drugs, etc.
5. Disinfectants : D.D.T., gammaxene, napthalene, etc.
6. Household and commercial articles : Soap, paper, dyes, paints, varnishes, inks, rubber, gum, etc.
7. Perfumes : Vanillin, camphor, etc.
8. Explosives : Trinitrophenol (picric acid), nitroglycerine, dynamite, trinitrotoluene (T.N.T.) etc.

1-7 CLASSIFICATION OF ORGANIC COMPOUNDS

Organic chemistry is a vast subject and more than six million organic compounds are already known and every year thousands of new compounds are being discovered. For a convenience and systematic understanding of the subject, the large number of organic compounds have mainly been classified into different groups based on (i) Carbon skeleton (ii) Functional groups and (iii) Homology.

1-8 CLASSIFICATION OF ORGANIC COMPOUNDS BASED ON THE CARBON FRAME WORK

Organic compounds are classified on the basis of carbon frame work, i.e., the arrangement of carbon atoms in the molecules as shown by the following flow chart;

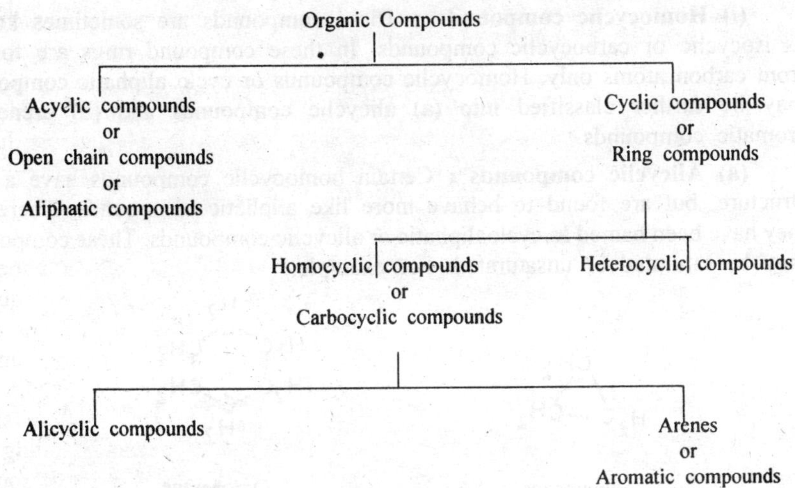

1-8-1 Acyclic, Open Chain or Aliphatic Compounds

The compounds in which carbon atoms are linked together to form an open chain are known as open chain or acyclic compounds. The carbon atoms in open chain compounds may be arranged in a straight chain or in branched chains. In the straight chain compounds, no carbon atom is directly linked to more than two other carbon atoms. In branched chain compounds, a carbon atom may be linked to three or four carbon atoms.

Such compounds are also referred to as aliphatic compounds (Greek aleip-har meaning fat) because the earlier known compounds of this type were either obtained from vegetable and animal fats or they had fat-like character. For example;

$H_3C{-}CH_2{-}CH_2{-}CH_3$ (n-butane) $H_3C{-}CH = CH_2$ (propene) $H_3C{-}CH_2{-}CH_2{-}CHO$ (butyraldehyde)

(straight chain compounds)

$H_3C-CH_2-CH_2-CH_3$ (n-Butane) $H_3C-CH-CH_2$ (propene) $H_3C-CH_2-CH_2$ CHO (butyraldehyde)

(Straight chain compounds)

$(H_3C)_2CH-CH_2-CH_3$ (Isopentane) $H_3C-C(CH_3)_2-CH_3$ (Neopentane)

(Branched chain compounds)

1-8-2 Cyclic Compounds

Cyclic compounds are those in which the atoms are linked to one another, in such a way that closed rings are formed. Cyclic compounds, are further classified into **homocyclic** compounds and **heterocyclic** compounds depending upon the nature of atoms present in the ring.

(i) Homocyclic compounds : These compounds are sometimes known as isocyclic or carbocyclic compounds. In these compound rings are formed from carbon atoms only. Homocyclic compounds or cyclo aliphatic compounds may be further classified into (a) alicyclic compounds and (b) arenes or aromatic compounds.

(a) Alicyclic compounds : Certain homocyclic compounds have a ring structure, but are found to behave more like aliphatic compounds. Therefore, they have been named as cycloaliphatic or alicyclic compounds. These compounds may be saturated or unsaturated. For example;

CH_2, $H_2C—CH_2$ (three-membered ring)

Cyclopropane

H_2C, H_2C, CH_2, H_2C, CH_2, H_2C (six-membered ring)

Cyclohexane

(b) Aromatic compounds : Aromatic compounds contain cyclic structure with alternate arrangement of double and single bonds. Benzene is the parent aromatic compound. these compounds are unsaturated as per the molecular formula, but do not readily undergo addition reactions like typical unsaturated compounds. The name **aromatic compounds** was assigned to them as most of the compounds of this class possess pleasant smell (from the Greek word aroma meaning pleasant smell). The following can be cited as examples of aromatic compounds.

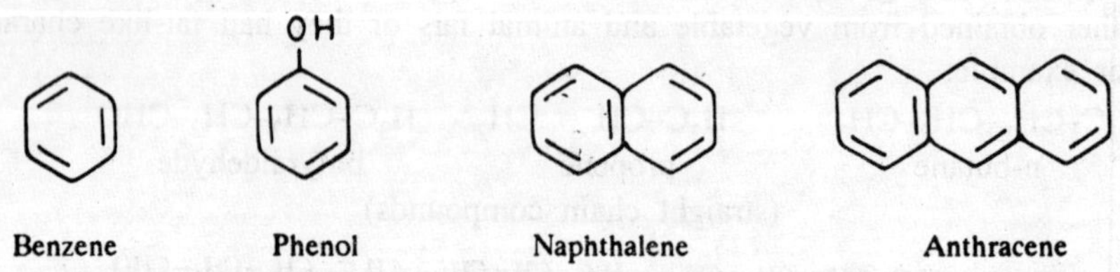

Benzene Phenol Naphthalene Anthracene

(ii) Heterocyclic compounds : Heterocyclic compounds are cyclic compounds in which the ring contains in addition to carbon atoms one or more hetero atoms such as oxygen, nitrogen, sulphur, etc. Heterocyclic compounds may be aromatic or non-aromatic.

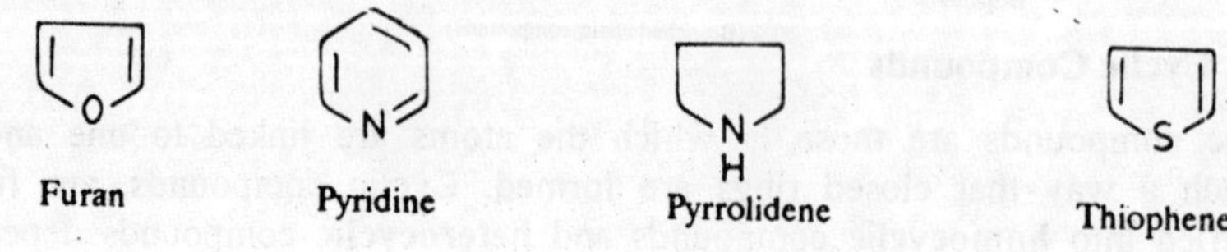

Furan Pyridine Pyrrolidene Thiophene

1-9 CLASSIFICATION OF ORGANIC COMPOUNDS BASED ON FUNCTIONAL GROUPS

According to this type of classification all organic compounds containing a particular functional group are considered as belonging to one class or family. The properties of an organic compound depend upon the functional group(s) present in it.

Classification according to functional groups rationalizes the behaviour of vast number of known organic compounds. It is logical and convenient to divide organic compounds into families with similar properties in the same way that the periodic classification of the elements divided the elements into groups or families having related properties. For example, compounds such as acetic acid (CH_3COOH), Propionic acid (CH_3CH_2COOH), and butyric acid ($CH_3CH_2CH_2COOH$) behave more or less similar. They all contain the same functional group i.e. carboxylic group (–COOH) attached to a saturated carbon atom. So all the compounds containing the carboxylic group form a family of compounds known as carboxylic acids. Thus, *the atom or a group of atoms determining the function or mode of activity (i.e. properties) of a family of organic compounds called the functional group.* Various functional groups are possible in organic compounds. all the organic compounds having the same functional group are said to constitute a particular class of organic compounds.

Some of the common functional groups encountered in organic chemistry and thus the various classes of organic compounds are listed in the table 1-1.

Table 1-1 Classification Based on Functional Group

Functional group	Class of Compounds	Example	Formula
(a) Compounds containing carbon and hydrogen only :			
- C - C -	Alkanes	Ethane	$H_3C - CH_3$
> C = C <	Alkenes	Ethylene	$H_2C = CH_2$
- C = C -	Alkynes	Acetylene	H - C = C - H
(b) Compounds containing carbon			
- C - OH	Alcohols	Methyl alcohol	$H_3C - OH$
- C - O - C -	Ethers	Dimethylether	$H_3C - O - CH_3$
$-\overset{H}{\overset{\vert}{C}} = O$	Aldehydes	Acetaldehyde	$H_3C - \overset{H}{\overset{\vert}{C}} = O$
> C = O	Ketones	Acetone	$H_3C - \overset{O}{C} - CH_3$
$-\overset{O}{\overset{\Vert}{C}} - OH$	Carboxylic acids	Acetic acid	$H_3C - \overset{O}{\overset{\Vert}{C}} - OH$

$-C(=O)-OR$	Estates	Methylacetate	$H_3C-C(=O)-OCH_3$
$-C(=O)-O-C(=O)-$	Anhydrides	Acetic anhydride	$H_3C-C(=O)-O-C(=O)-CH_3$
(c) compounds containing			
$-N(=O)-O$	Nitro compounds	Nitromethane	$H_3C-\overset{+}{N}(=O)-O-$
$-O-N=O$	Nitrites	Methylnitrite	$H_3C-O-N=O$
$-C(=O)-NH_2$	Amides	Acetamide	$H_3C-C(=O)-NH_2$
(d) Compounds containing carbon			
$-NH_2$	Primary amines	Methylamine	H_3C-NH_2
$NH-R$ R^-	Secondary Amines	Dimethylamine	$H_3C-NH-CH_3$
$-N(R)-R$ (R, R, R)	Tertiary amines	Trimethylamine	CH_3 $N -- CH_3$ CH_3 H_3C
$-C\equiv N$	Cyanides	Methycyanide	$H_3C-C\equiv N$
$-\overset{+}{N}\equiv C-$	Isocyanides	Methyl isocyanide	$H_3C-\overset{+}{N}=C^-$
$-N=N-$	Azo compounds	Azobenzene	$C_6H_5-N=N-C_6H_5$
(e) Compounds containing carbon			
-C-X	Halides	Methyl chloride	H_3C-Cl
(f) Compounds containing carbon			
$-C(=O)-X$	Acid halides	Acetyl chloride	$H_3C-C(=O)-Cl$
(g) Compounds containing carbon			
-C-SH	Thiols	mercaptans	Methanethiol
-C-S-C-	Thio ethers	Methylthioether	$H_3C-S-CH_3$

1-10 CLASSIFICATION OF ORGANIC COMPOUNDS BASED ON THE CONCEPT OF HOMOLOGY

The concept of homology is an important simplifying factor in the study of organic chemistry. According to this concept the organic compounds can be divided into different families, known as **homologous series**, on the basis of similarity in structure and chemical properties. A homologous series may thus be defined *as a series of compounds in which the various members have similar structural features and similar chemical properties and the successive members differ in their molecular formulae by* CH_2. The members of a series are known as homologous of each other.

The important characteristics of a homologous series are

(1) Different members of a series can be assigned a general formula. For example homologous series of alkanes represented by C_nH_{2n+2}.

(2) The molecular formula of any two adjacent members of a series differs by CH_2.

(3) The members of a series show similar chemical reactions, since these molecules contain the same functional group.

(4) The physical properties such as melting point, boiling point, density etc., show gradual change with the increase of molecular weight in a series.

(5) Different members of the series can be prepared by the same general methods of preparation.

For instance, let us consider the homologous series of alcohols.

Formula	Name
$CH_3\ OH$	Methyl alcohol
$CH_3\ CH_2\ OH$	Ethyl alcohol
$CH_3\ CH_2\ CH_2\ OH$	Propyl alcohol
$CH_3\ CH_2\ CH_2\ CH_2\ OH$	Butyl alcohol

All these alcohols and also the higher members of the series, can be assigned the general formula C_nH_{2n+1} OH. The molecular formula of each compound differes from that of the preceding and the succeeding members by -CH_2. All these compounds undergo nucleophilic substitution reactions. Their physical properties show a regular gradation. All of them can be prepared by similar methods of preparation.

1-11 STRUCTURAL THEORY OF ORGANIC CHEMISTRY

The structural theory is the basis upon which millions of facts about thousands of individual compounds have been brought together and arranged in a systematic way. It is the basis upon which these facts can be best accounted and understood. The structural theory is the framework of ideas about how atoms are put together to make molecules. The structural theory has to do with the order in which atoms are attached to each other, and with the electrons that hold them together. It has to do with the shapes and sizes of the molecules that these atoms form and with the way that electrons are distributed over them. On the basis of structural formula of a compound we would know the physical properties of the compound, the kind of solvents in which the compound will dissolve, even whether it will be coloured or not, the kind of reagents with which the compound will react and the kind of products that will be formed and whether it will react rapidly or slowly etc.

For the proper understanding of an organic compound, the knowledge of the structure of the atom is very essential.

An atomic orbital is a region of space about the nucleus in which there is a high probability of finding an electron. an electron has a given energy as designated by;

(a) the principal energy level (Quantum number), n, related to the size of the orbital;

(b) The sub level, s, p, d, or f, related to the shape of the orbital;

(c) Except for the s- each sub level has some number of equal energy (degenerate) orbitals differeing in their spatial orientation;

(d) The electron spin, designated by ± 1/2.

The s-orbital is spherical in its shape and the p-orbital consists of two lobes to form dumb-bell shaped structure. The three p-orbitals are labelled p_x, p_y and P_z because they are oriented along the x, y, and z axes, respectively. In a p-orbital there is no chance of finding an electron at the nucleus, (the nucleus is called a nodal point).

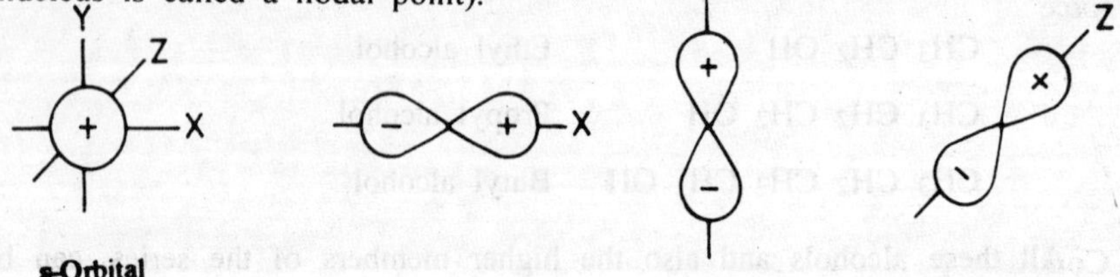

s-Orbital

p-Orbitals

Regions of an orbitals separated by a node are assigned '+' and –' signs. These signs are not associated with electrical or ionic changes. The s-orbital has no node and is usually assigned a '+'.

Any consideration of the structure of molecules must begin with a discussion of chemical bonds, the forces that hold atoms together in a molecule.

1-12 PRINCIPLES UNDERLYING CHEMICAL BONDING

In 1916 two kinds of chemical bond were described. The ionic bond by Walther Kossel and the covalent bond by G.N. Lewis. Both of them based their ideas on the following concept of the atom.

A positively charged nucleus is surrounded by electrons arranged in concentric shells or energy levels. There is maximum number of electrons that can be accommodated in each shell are, two in first shell, eight in the second shell, eighteen in the third shell, and so on. The greatest stability is reached when the outer shell is full, as in the noble gases. Both ionic and covalent bonds arise from the tendency of atoms to attain this stable configuration of electrons. These two ways of formation of a chemical bond may lead to the formation of three types of bonds. (i) electrovalent or Ionic bond (ii) Covalent bond (iii) Co-ordinate covalent bond.

1-12-1 Electrovalent or Ionic Bond

A Ionic bond is formed due to the electrostatic attraction between oppossitely charged stable ions formed by the complete transfer of electron from one atom to another. These oppostiely charged ions usually have stable inert gas configuration. The electropositive elements such as Li, Na, K, Mg, Ca etc. lose their electron(s), while the electronegative elements such as F, O, Cl, Br etc. take up the electrons(s) and get converted into negatively charged ions or anions.

For example, consider the formation of sodium chloride. Sodium atom has ten electrons in its inner shells and one electron in its outer shell or valence shell. The loss of one electrons would leave sodium with full outer shell of eight electrons (**'Ne'** configuration). Chlorine atom has ten electrons in its inner shells and seven electrons in its valence shell. The gain of one electron would give chlorine a full outer shell of eight electrons (**'Ar'** configuration). Sodium chloride is formed by the transfer of one electron from sodium to chlorine. Sodium now bears a positive charge and chlorine bears a negative charge. These Na+ and Cl^- ions are held together by electro static force of attraction to form ionic bond.

$$\text{Na} + \text{energy} \longrightarrow \text{Na}^+ + \bar{e}$$

$$\text{Cl} + e^- \longrightarrow \text{Cl}^- + \text{energy}$$

$$\text{Na}^+ + \text{Cl}^- \longrightarrow \text{Na}^+ + \text{Cl}^-$$

Condition for the formation of ionic bond

(1) Lower the value of ionization energy of an atom greater will be the ease of formation of cation.

(2) Higher the electron affinity of an atom greater the case of formation of the anion from it.

(3) Higher the lattice energy of the resulting ionic compound greater will be the case of its formation.

Characteristics of ionic compounds

(1) Ionic compounds posses high melting and boiling points. This is due to greater amount of thermal energy, required to break- down the inter ionic forces.

(2) They are highly soluble in polar solvents such as water and generally insoluble in non-polar solvents such as ether, benzene, carbon tetrachloride etc.

(3) Ionic compounds are good conductors of electricity in the molten-state and in aqueous solutions, because under these conditions the constituent ions become mobile and able to move under the influence of an electric field.

(4) In solution ionic compounds show ionic reactions which are quite fast and instantaneous.

(5) Ionic bonds are rigid and non-directional in nature. Hence, they do not exhibit space isomerism.

1-12-2 Covalent Bond

A covalent bond is formed by the sharing of a pair of electrons by the two bonded atoms, each contributing one electrons for the process of sharing. These shared electrons contribute to the stability of both the combining atoms as each of them acquires an electronic configuration of nearest inert gas.

For example, consider the formation of hydrogen molecule by the combination of two hydrogen atoms. Each hydrogen atom has only one electron in its 's' orbital, which it contributes in sharing so as to acquire the electron configuration of inert gas helium.

$$H. + H. \longrightarrow H{:}H \text{ or } H\text{-}H$$

When two atoms share a pair of electrons each contributing one electron, the bond thus, formed is a single bond. Similarly, when there is a sharing of two or three electron pairs between the two atoms, the bonds thus formed is known as double or triple bond respectively. For example;

```
      H                 H     H
      •                 •     •
      x                 x     x                 x   x
H• x C x •H       H• x C x   x C x •H      H• x C x   x C x •H
      x                   x   x                 x   x
      •                                         x   x
      H

      H                 H   H
      |                 |   |
  H — C — H         H — C = C — H             H — C≡C — H
      |
      H
```

The covalent bond is typical of the compounds of carbon and it is the bond of chief importance in the study of organic chemistry.

Characteristics of covalent compounds

(1) Covalent compounds are generally gases and liquids, at normal temperature and pressure. They are solids when their molecular weights are high.

(2) Covalent compounds exist as molecules and are held together by weak van der Waals forces. Hence, they possess low melting and boiling points.

(3) They are insoluble in polar solvents like water, but soluble in non polar solvents like ether, benzene etc.

(4) Covalent bonds are directional in nature and hence the phenomenon of isomerism is exhibited in covalent compounds.

(5) As covalent compounds in solutions do not dissociated into ions their solutions are non-conductors of electricity.

(6) Reactions of covalent compounds are usually slow reactions and hence their rates can be conveniently determined.

1-12-3 Co-ordinate Colvalent Bond or Dative Bond

This is a special type of covalent bond in which shared pair of electrons is contributed by one of the two bonded atoms. The atom that contributes the lone pair for bonding electrons is called donor and the one which shares the electrons is called acceptor. The dative bond is shown be means of an arrow (vec pointing away from the donor atom to the acceptor atom). For example consider the formation of NH_{4+} ion.

In ammonia molecule, nitrogen atom has a lone pair of electrons in its valency shell. H^+ can take a pair of electrons to get stability. Thus ammonia donates its unshared pair of electrons to H^+ forming co-ordinate covalent bond in ammonium ion.

Characteristics of co-ordinate covalent compounds

(1) Due to semi-polar nature of the co-ordinate covalent bond, these compounds have melting and boiling points higher than those of the covalent compounds but lower than those of the ionic compounds.

$$H_3N: + H^{\oplus} \longrightarrow \left[H_3N \rightarrow H \right]^{\oplus} \text{ or } NH_4^{\oplus}$$

(2) Generally these compounds are soluble in non-polar solvents, but sparingly soluble in water.

(3) Co-ordinate covalent compounds are stable like covalent compounds.

(4) Co-ordinate covalent bonds are rigid and directional in nature and hence co-ordinate covalent compounds exhibit the phenomenon of isomerism.

(5) These compounds are not ionized in solutions or in fused state. Hence, they do not conduct electricity.

1-13 Types of Overlapping

A covalent bond formed may be explained as due to the overlapping of atomic orbitals. When atomic orbitals of different atoms overlap, potential energy of the system is lowered hence the resulting molecular orbitals are more stable than the corresponding atomic orbitals taking part in overlapping. In general as the outer most orbit possessing s- and p- bonding orbitals, overlapping of orbitals is of three types.

(i) s-s overlapping : Hydrogen atom has a single electron in 1s orbital. During the formation of a stable molecule of hydrogen, the two 's' orbitals of hydrogen atoms overlap with each other forming a covalent bond. The two hydrogen atoms share a pair of electrons, contributed one each by them. The electrons thus shared by the two atoms are located in a region or space between the nuclei of the atoms and hence they are attracted by both the nuclei. As a result of this attraction potential energy of the system decreases. The resulting bond is called as bond and the overlapping is known as s-s overlapping. Since overlapping of the orbitals takes place along the same axis to an appreciable extent, the resulting bond is strong.

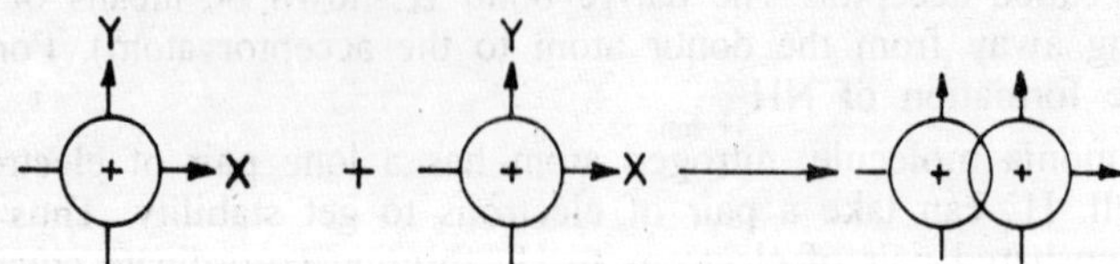

Formation of hydrogen molecules

(ii) s-p overlapping : In hydrogen chloride molecule formation, the 1s orbital of hydrogen and half filled 3p orbital of chlorine are involved in bond formation. These two orbitals overlap and result in s-p molecular orbital. This overlapping is called s-p overlapping and the resulting bond is called s-p bond. Since the overlapping takes place along the same axis, the resulting bond is strong.

P-orbitals can overlap to greater extent than s-orbitals. Hence s-p molecular orbitals are more stable than s-s molecular orbitals.

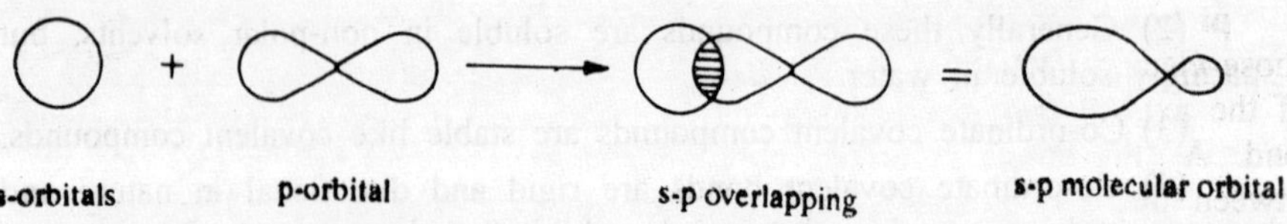

s-orbitals p-orbital s-p overlapping s-p molecular orbital

(iii) p-p Overlapping : It is of two types

(a) p-p head-on or axial overlapping : In this type of overlapping the two p-orbitals of different attoms overlap along the same axis in head on manner resulting in p-p molecular orbital. This type of overlapping is called p-p head-on overlapping.

p-orbital + p-orbital → p-p overlapping ≡ p-p molecular orbital

p-p side on or lateral overlapping : This type of overlapping takes place in unsaturated molecules at double or triple covalent bond. The two "p" orbitals involved remain in the same place and overlap to a very lesser extent and hence the bond so resulted is a very weak bond. This type of overlapping is called p-p lateral overlapping.

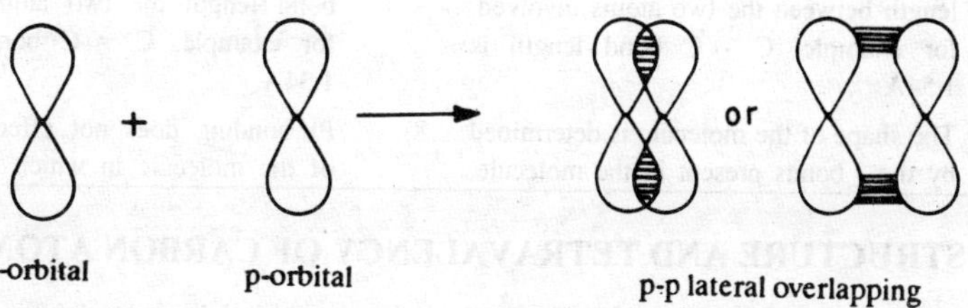

1-13-1 sigma and pi bonds

The above discussion points that a bond is set up due to overlapping at atomic orbitals of the two combining atoms and a strong or weak bond is obtained depending upon the extent of overlapping. Depending upon the nature of overlapping the bonds can be categorised in to sigma and pi bonds.

Sigma bond : A covalent bond formed by the head-on overlapping of orbitals, in which the electron density distribution is concentrated mainly along the inter nuclear axis and cylindrically symmetrical about the axis is called a sigma(σ) bond. This type of bond is, therefore, obtained as a result of maximum overlapping. Sigma bonds are resulted from s-s, s-p and head on p-p overlapping.

pi bond : a covalent bond formed by the lateral overlapping of orbitals, whose electron density is not concentrated along the inter-nuclear axis but lies off the axis in two identical regions on either side of the axis is called Pi(π) bond. A 'Pi' bond is formed only after a sigma bond is already formed between the two atoms. As this overlapping is partial the bond formed is weak.

Table 1-2 Difference between sigma and Pi bonds

	Sigma bond		Pi bond
1)	Sigma bond results from the overlapping of two 's' orbitals or two 'p' orbital along the same axis in end to end manner.	1)	Pi bond results from the lateral or sideways overlapping of two p-orbitals.
2)	In Sigma bond the electron cloud is symmetrical and is concentrated mainly along the inter nuclear axis.	2)	The electron cloud of a p bond is unsymmetrical and is concentrated above and below the plane of the sigma bond.
3)	A sigma bond is formed as a result of maximum overlapping, bond is strong. For example, the energy of a carbon carbon single bond is 80kcal mole-1.	3)	A p bond is formed as a result of partial overlapping and the bond is weak. For example, the energy of a carbon-carbon p bond is about 65 kcal $mole^{-1}$.
4)	The sigma bond electrons are localized electrons.	4)	In p bond electrons are held less firmly and can be delocalized.
5)	Sigma bonds are less reactive	5)	p bonds are more reactive
6)	Sigma bond can exist independently.	6)	p bond can not exist independently, it exists along with a sigma bond.
7)	Sigma bond leads to expected bond length between the two atoms involved, for example, C - C bond length is 1.54A.	7)	p bonding leads to a shortening of the bond length the two atoms involved, for example, C = C bond length is 1.34A.
8)	The shape of the molecule is determined by the - bonds present in the molecule.	8)	Pi bonding does not effect the shape of the molecule in which it occurs.

1-14 STRUCTURE AND TETRAVALENCY OF CARBON ATOM

The atomic number of carbon atom is six and the electronic configuration of carbon in the ground state is $1s^2$, $2s^2$, $2p_x^1$, $2p_y^1$, $2p_z^0$. Because of two unpaired electrons in $2p_x$ and $2p_y$ orbitals a carbon atom might be expected to be only "bivalent" but, carbon atom exhibits tetravelency in almost all of its important compounds. To explain the tetravelency it has been postulated that at the time of bond formation (Which is an energy releasing process), the two electrons in the 2s orbital of carbon get unpaired and one of them is excited to $2p_z$, orbital. Thus the excited state electronic configuration of carbon atom is $1s^2$, $2s^1$, $2p_x^1$, $2p_y^1$, $2p_z^1$. Since, in the excited state, carbon has four half-filled valence orbitals, it can form four covalent bonds (tetravalency of carbon) by overlapping with four different orbitals.

It may be noted that the formation of four bonds by carbon would result in geater lowering of energy, due to formation of highly stable half filled orbitals, than if the carbon were to form only two bonds. This extra lowering of energy more than compensates the little excess energy needed to promote an electron from 2s to 2p orbital. This means that even though the promotion of electron is by itself not favoured energetically, since it takes place with the gain in energy because of formation of half filled orbitals, it ultimately becomes energitically favourable process.

It is evident that the four electrons available for bond formation belong to two different types of orbitals (i.e., *'s'* and *'p'* orbitals), hence the four bonds will be of two different types. In other words the three bonds involving 2px, 2py 2pz orbitals will be of one type and fourth bond involving *'2s'* orbital is spherical in shape. Thus the fourth bond is oriented in no particular direction while the three 2p orbitals are oriented at 90° to one another and pointed in a particular direction. Moreover, since the energy of an electron in *'s'* orbital is less than that from the other three *'2p'* orbital electrons the bonds should be of different strength. But this is contrary to the observation, for instance the four C-H bonds in methane (CH_4) molecule are identical in all respects. The formation of four identical bonds can be explained by a concept called as hybridization.

In other words the *2s* and *2p* orbitals of the carbon atom are not involved as such in bond formation but, firstly, one s orbital and three p orbitals merge, mix or hybridize with each other to form four new identical hybrid orbitals, which then take part in bond formation.

Hybridization : *The phenomenon of mixing of the dissimilar orbitals of an atom having nearly same energies and the redistribution of their energies to form same number of new orbitals of identical shapes and equivalent energies is known as hybridization.*

Conditions to undergo Hybridization

(i) Only orbitals of similar energies and belonging to the same atom or ion can undergo hybridization.

(ii) The number of hybrid orbitals generated is equal to the number of pure orbitals mixed during the hybridization.

(iii) Hybrid orbitals lead to the formation of hybrid bonds which are stronger than the non-hybridized bond or bonds.

(iv) The hybrid orbitals are distributed in space as far apart as possible.

(v) Orbitals undergo the hybridization but not the electrons.

The different types of hybridization, which observed in organic compounds are sp^3, sp^2 and sp hybridization.

(1) sp^3 or tetrahedral hybridization : In this type of hybridization, the 2s orbital and three 2p orbitals are mixed up to give four new identical orbitals known as sp^3 hybrid orbitals. These four new orbitals are neither s nor p in character but are of mixed sp^3 type. These four new orbitals of carbon are oriented in space towards the four corners of a regular tetrahedron. The angle between any two sp^3 orbitals is $109^{\circ}28'$. sp^3 hybridized carbon is also known as tetrahedral carbon.

The sp^3 hybrid orbitals have one fourth s character and three fourths p character. This hybridization takes place during the formation of compounds in which carbon atoms form only single bonds. For example alkanes and their derivatives.

In the formation of methane (CH_4) four sp^3 hybrid orbitals of the carbon atom overlap with 1s orbitals of four hydrogen atoms. This leads to the formation of four C – H bonds of methane. These bonds are sigma bonds.

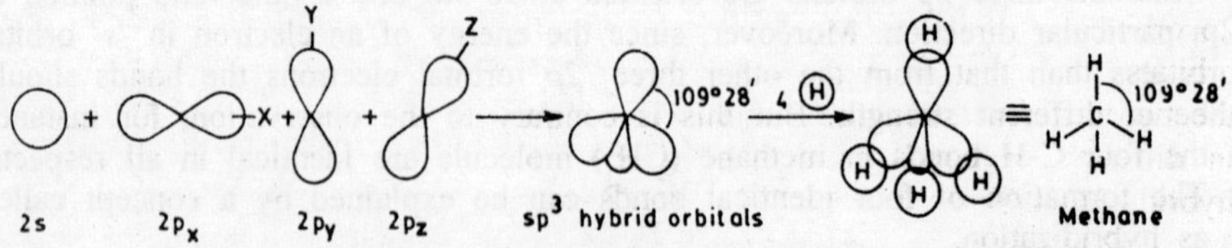

(2) sp^2 or trigonal hybridization : In this type of hybridization, the 2s orbital and two of the three 2p orbitals (spx and 2py) hybridize to form three new identical orbitals known as sp^2 hybrid orbitals. These three sp^2 hybrid orbitals including the carbon atom lie in one plane and are directed towards the corners of an equilateral triangle. The angle between any pair of sp^2 hybrid orbitals is 120°. sp^2 hybridization is therefore known as trigonal hybridization and sp^2 hybrid orbitals as trigonal orbitals.

The sp^2 hybrid orbitals have one third s character and two thirds p character. Their effective size is slightly smaller than that of sp^3 orbitals. This type of hybridization takes place in the formation of compounds containing doubly bound carbon atoms; e.g., compounds containing C = C, C = O, C = N, C = S etc.

In the formation of ethylene, two sp^2 hybridized carbon atoms are involved. Each carbon utilises three sp^2 hybridized carbon atoms are involved. Each carbon utilises three sp^2 orbitals in the formation of a C - C bond and two C - H bonds. C - H bonds are formed by the overlapping of s orbitals of two carbons. The sigma frame work of ethylene molecule is planar *i.e.* the two carbons and four hydrogen atoms linked to them are co-planar *i.e.* present in one plane. In addition, each carbon has an unhybridized p orbital perpendicular to the plane of molecular frame work. The unhybridized pure p orbital on each of the two carbon atoms overlap each other laterally and thus a p bond is formed between the carbon atoms. The H - C - C bond angle in ethylene is 120°.

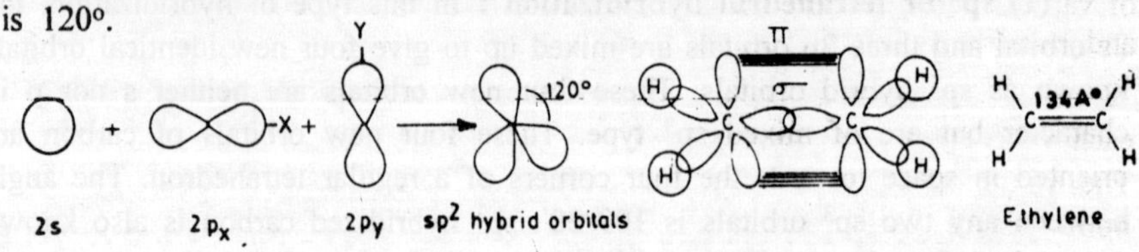

Due to later overlapping of p orbital, the distance between two carbon atoms in ethylene is 1.34Å a and less than the C – C bond distance in ethane, 1.54 Å.

(3) **'sp' or Digonal hybridization** : In this type of hybridization, the 2s orbitals and only one of 2p orbitals ($2p_x$) hybridize to form two new identical orbitals called as diagonal or sp hybrid orbitals. The remaining two 2p orbitals are left in their original state. This gives rise to two identical 'sp' orbitals pointing in opposite directions along a straight line. The angle between the two sp hybrid orbitals is 180°. The unhybridized $2p_y$ and $2p_z$ orbitals lie in different planes at right angles to each other and also to the plane of hybridized orbitals.

The sp hybrid orbitals have one-half 's' character and one-half 'p' character. Their effective size is even smaller than that of sp^2 orbitals. sp hybridization takes place in compounds containing triply bound carbon atoms, *e.g.*, compounds containing C ≡ C, C ≡ N etc.

In the formation of acetylene, one sp hybrid orbital of carbon overlaps with the sp hybrid orbital of the other carbon to form a carbon-carbon sigma bond. Each carbon utilises its second sp hybrid orbital to form a C – H bond by overlapping with s orbital of hydrogen and this gives rise to molecular frame work of acetylene which is linear. The H-C-C bond angle in acetylene is 180°. Overlap of two unhybridized p orbitals of carbon atom with that of the other carbon atom, leads to the formation of two p bonds.

2s + 2 px —X→ sp hybrid orbitals (180°)   H—C—C—H   1.21A° H–C≡C–H Acetylene

Acetylene is thus a linear molecule with p-electron cloud on all four sides of the C-C axis. The triple bond in acetylene consists of a sigma bond and two p-bonds. The C≡C bond distance in acetylene is 1.21 Å.

1-14-1 Parameters of Molecular Structure

For a proper understanding of the nature of chemical bonding, the knowledge of various parameters of molecular structure such as bond lengths, bond angles and bond energies is very much useful.

(1) Bond lengths : The atoms forming a bond cannot come closer to each other than a certain distance where the potential energy is minimum. The minimum distance between the nuclei of two bonded atoms is known as **bond length.** But becuase of the perpetual atomic vibrations, this distance does not remain constant. Thus bond length may be defined as the *average distance between the centres of nuclei of two bonded atoms.*

The bond lengths are usually expressed in Angstrom (Å) units (1Å = 10^{-8} cm). The various bond lengths have been measured by physical methods such as X-ray diffraction, electron diffraction and spectroscopic methods. Some bond lengths are tabulated in table 1-3.

Table 1-3 Bond length of various bonds

Bond	Bond length (Å)
C-C	1.54
C=C	1.34
C≡C	1.21
C-H	1.12
C-O	1.43
C-N	1.47
C-S	1.82
C-F	1.42
C-Cl	1.77
C-Br	1.91
C-I	2.13
O–H	0.97
N-H	1.03

Factors affecting bond lengths

(i) Bond lengths are effected by hybridization. As expected bond length decreases with the increase in s character, because an s- orbital is smaller than a p orbital. For example

sp^3	C-C	1.54 Å
sp^2	C=C	1.34 Å
sp	C≡C	1.21 Å

(ii) The electronegativity of bonding atoms also play an important role in determining the value of bond lengths. The factor electronegativity shortens the carbon double bond length. For example, the C=C bond length in $F_2C = CF_2$ is 1.31 Å instead of 1.34 Å in normal C=C compounds.

(iii) Bond lengths are also effected by delocalization of electrons, which in turn may be due to resonance, hyperconjugation etc. For example, bond length of all the carbon. Carbon bonds in benzene molecule is 1.40 Å, which is in between the carbon-carbon single bond length (1.54 Å) and double bond length (1.34 Å).

Table 1-4

Compound	Angle between	Bond angle in Å
CH_4	H-C-H	109°28′
C_2H_4	H-C-C	120°
C_2H_2	H-C-C	180°
H_2O	H-O-H	104°30′
NH_3	H-N-H	107°
H_2S	H-S-H	92°12′

(2) Bond angles : In the molecules containing more than one bond, the bonds make angles between them. The angle between the lines representing the two bonds is known as bond angle. Since, the atoms in molecules are in constant motion with respect to each other, they are not expected to have a fixed value of a bond angle. Thus, bond angle maye be defined *as the average angle between the direction of two bonds.* Bond angles are measured by X-ray diffraction, spectroscopic methods and dipole moments. The observed bond angles of some of the important compounds are listed in table 1-4.

Factors effecting the bond angles

(1) Bond angles are effected by hybridization of bonding orbitals. As the s character of bonding orbitals increases the bond angle also increase. For example

CH_4	sp^3	109° 28′
C_2H_4	sp^2	120°
C_2H_2	sp	180°

(2) Bond angles are also effected by the repulsion of non-bonding electrons. For example in ammonia (NH_3), three hydrogen atoms overlap with the sp^3 hybrid orbitals and form bonds while the fourth sp^3 hybrid orbital takes a lone pair of electrons, which repel the electrons involved in bond formation, with the result the tetrahedral bond angle of 109° 28′ decreases to 107°. Similarly, water has two lone pairs of electrons, repulsion between lone pair and bond pair electrons is more and hence bond angle is further distorted to 104° 30′.

(3) Bond energies : There are two types of bond energies, (i) Bond dissociation energy (D) and (ii) Bond energy (E).

(i) Bond dissociation energy (D) : *This is the energy required to break a particular bond in a polyatomic molecule, in the gaseous phase, into neutral fragments (free radicals) also in the gaseous phase, i.e.,* bond dissociation energy is the energy required for effecting a homolytic fission of a bond.

$$A - B\ (g) \longrightarrow A\ (g) + B\ (g)$$

(ii) Average bond energy (E) : It is important to note that the bond dissociation energy is characteristic of a particular bond, but not of a particular type of bond. Thus in methane all the four C-H bonds are equivalent but have different values of bond dissociation energies :

CH_4	CH_3 + H	$D(H_3C\text{-}H)$	-104 kcal mole^{-1}
CH_3	CH_2 + H	$D(H_2C\text{-}H)$	-106 kcal mole^{-1}
CH_2	CH + H	D(HC-H)	-106 kcal mole^{-1}
CH	C + H	D(C-H)	-81 kcal mole^{-1}

we must not confuse bond dissociation energy(D) with another measure of bond strnegth called bond energy(E). Actually, the *bond energy is the average value of the various values of bond dissociation energies of the same type of bonds present in the molecule.* Thus the bond energy per C-H bond in methane is 99.2 kcal mole^{-1}. (i.e. 104 + 106 + 106 + 81 = 397 ; 397/4 = 99.2).

Since a diatomic molecule has only one bond, bond dissociation energy(D) is equal to bond energy(E) whereas in polyatomic molecules, bond dissociation energy(D) is different from that of the average bond energy (E). Some bond energies obtained by taking the average from measurements on several polyatomic molecules are listed in Table 1-5.

Table 1-5 Some bond energies

Bond	Energy kcal mole^{-1}	Bond	Energy kcal mole^{-1}
C-H	99	C-F	107
C-C	83	C-Cl	78
C=C	145	C-Br	68
C=C	192	C-I	51
C-O	80	C-S	65
C=O	166	H-H	104
C-N	68	N-H	93
C=N	147	O-H	111
C=N	207	S-H	83

The average bond energy gives an idea about the strength of a particular bond, which intern indicates the stability. Greater the bond energy, more the stability and lesser the reactivity of a bond. As σ-bond is more stronger than p-bond, the bond energy of σ bond is more than p bond. The carbon-carbon bond energy increase with the increase in number of bonds between the two carbon atoms.

Factors effecting the bond energy : Two main factors responsible for the differences in bond energy of a given bond in different compounds are

1. Steric factors : As the steric hindrance increases the bond energy decreases.

2. Angular strain : As the angular strain increases bond energy decreases. For example carbon-carbon bond energy in cyclopropane in lesser than cyclobutane.

A very important fact is that the average bond energies, can be used, in many cases, to compute heats of formation with a fair degree of accuracy. For example, the observed heat of formation of ethyl alcohol from its atoms is 767 kcal mole^{-1}, which is nearly equal to that obtained from the values of its average bond energies (5 C - H, one C-C, one C-O, and one O-H bonds) i.e., 5 x 99 + 83 + 80 + 111 = 769 kcal/mole.

Heat of reaction can also be predicted from average bond energies for example, the reduction of an ethylene to ethane, involving following bond change proceeds with the liberation of 281-249 = 32 kcal mole^{-1} energy which is in very good agreement with the experimental value.

C = C	+	H - H	C - C	+ 2(C - H)
145		104	83	2 × 99
	= 249			= 281

However, the heat of hydrogenation for aromatic compounds calculated on the basis of particular structure for a compound does not always agree with the value measured experimentally for that compound. In such cases, the compound is probably not correctly represented by structure on which the calculation has been based, i.e. the compound is probably a resonance hybrid and thus the extra energy (resonance energy) due to resonance between various possible structures must be considered.

1-15 MULTIPLE CHOICE QUESTIONS

1-15-1 The concept of vital force theory is due to

(A)Bergelius (B) Wholer (C) Avogadro (D) Kolbe

1-15-2 The Scientist who prepared an organic compound for the first time in the laboratory

(A) Belgius (B) Wohler (C) Levoiser (D) Kolbe

1-15-3 Which of the following has shortest bond distance

(A) C-C (B) C=C (C) C=C (D) C-F

1-15-4 Cyclohexane is a typical cxample of

(A) Acyclic compounds (B) Heterocyclic compounds (C) Alicyclic compounds (D) Aromatic compounds

1-15-5 The order of strength of different developings is

(A) *s-s>p-p>s-p* (B) *s-s>s-p>p-p*(C) *s-p>s-s>p-p* (D) *p-p>s-p>s-s*

1-15-6 which of thc bond is wcakest

(A) Ionic bond (B) bond (C) Coordinate bond (D) II-bond

1-15-7 The number of and II bands in benzene are

(A) 12 sigma and 3 pi (B) 6 sigma and 6 pi (C) 6 sigma and 3 pi (D) 12 sigma and 6 pi

1-15-8 C-C bond length in ethane is

(A) 1-54 A° (B) 1.34A° (C)1.20 A° (D) 1.09 A°

1-15-9 Which of the following compound does not involve sp^3 hybridization

(A)H_2O (B) CH_4 (C) NH_3 (D) C_2H_2

1-15-10 Sodium and chlorine atoms in common salt are held together by

(A) Covalent bond (B) Coordinate bond (C) Ionic bond (D) II-bond

1·16 SHORT AND ESSAY-TYPE QUESTIONS

1-16-1 Discuss the geometry of sp^2 and sp hybrid orbitals of carbon. (O.U., Mar. 1990)

1-16-2 Why ethylene is planar molecule and acetylene linear? Comment on their C-C bond length? (O.U., Oct. 1989)

1-16-3 Explain clearly why the bond angle in methane molecule is 109° 28. (O.U., Oct. 1989)

1-16-4 How would you explain the tetravalency of carbon, while its electronic configuration predicts that it would be bivalent ? (O.U., Mar. 1989)

1-16-5 Draw the orbital picture of the acetylene molecule. Justify the statement "A carbon-carbon triple bond is stronger than a carbon-carbon double bond". (O.U., Apr. 1988)

1-16-6 What do you understand by sigma and pi bonds ? Illustrate with examples? (O.U., Oct. 1987)

1-16-7 (a) Define bond energy?

(b) Compare C - C, = C, C = C, with respect to bond length and bond strength. (O.U., Apr. 1987)

1-16-8 What do you understand by the hybridization of orbitals? Explain the tetravelency of carbon using this concept. (O.U., Oct 1985)

1-16-9 Numbering the carbon atoms correctly, point out the type of hybridization of each carbon in the following molecule

$H^2C = CH - CH_2 - CH = CH - CH^3$ (O.U., Apr. 1985)

1-16-10 How are sigma and pi bonds formed in carbon compounds? (O.U., Apr. 1984)

1-16-11 What types of bonds are formed, in organic compounds, by coaxial and lateral (side ways) overlap of carbon orbitals ? Give diagrams. Which one of these is the strongest and why ? (O.U., Oct. 1985)

1-16-12 Explain the type of hybridisation in

(a) Propylene

(b) Propylene (O.U., Oct. 1989)

1-15-13 Explain the tetravelency of carbon and give the hybridization of carbon in the following;

(i) Carbonium ion

(ii) Carbanion

(iii) Acetylene (A.U., Apr. 1988)

1-15-14 (a) Mention the classification and sources of organic compounds.

(b) Write a note on single and multiple bonds. (A.U., Dec. 1989)

1-15-15 Write notes on Bond lengths and Bond angles. (A.U., Dec. 1989)

1-15-16 Give an account on the hybridization and static effects on dissociation constants of carboxylic acids. (A.U., Dec. 1989)

1-15-17 Write notes on (a) Bond angle (b) Resonance. (A.U., Dec. 1990)

1-15-18 (a) What type of hybrid orbitals does the carbon atom make use of in (i) ethane (ii) ethlene and (iii) acetylene.

(b) how does the type of hybridization effect (i) the shape and (ii) C - C bond length in the molecule. (A.U., Oct. 1988)

ANSWERS :

1-15-1 (A) 1-15-2 (B) 1-15-3 (C) 1-15-4 (C) 1-15-5 (B) 1-15-6 (D)
1-15-7 (A) 1-15-8 (A) 1-15-9 (D) 1-15-10 (C)

2

NOMENCLATURE OF ORGANIC COMPOUNDS

2-1 INTRODUCTION

The nomenclature or assignment of names to organic compounds is an important aspect of the study of organic chemistry. This is obviously due to the abundance of organic compounds and the large variety of their molecular structures. There are two main systems for naming of compounds.

(i) Common or trivial system of nomenclature

(ii) IUPAC system of nomenclature.

2-2 COMMON OR TRIVIAL SYSTEM

In the early stages of the development of organic chemistry, when the number of organic compounds known was not very large, the compounds were named on the basis of their history or the source from which they were obtained or on their chemical behaviour. Quite frequently the names chosen had Latin or Greek origin. For example, wood spirit was so called as it was originally obtained by the destructive distillation of wood. It was later on named as methyl alcohol (Greek : Methu = wine and hule = wood), acetic acid derives its name from vinegar (Latin : acetum = vinegar) of which it is the chief constituent, formic acid was so named because it could be obtained by the distillation of red ants (Greek : formicus = red ants), malic acid which was first isolated from apples (Latin : malum = apple). Similarly, citric acid (from citrus plant), urea (from urine) etc. Sometimes, a single compound was given different names e.g., methane was named as "marsh gas" because it is produced in marshes and also as "fire damp" because of its property of catching fire easily (German:fire = fuer and dampf = vapour). These names are known as common or trivial names. One advantage of the common or trivial system is that the names are usually short and easily remembered, but a disadvantage is that a particular compound may have a number of names.

2-3 IUPAC SYSTEM

With the tremendous increase in the number of organic compounds isolated and prepared, it became apparent that it was necessary to systematize the method of nomenclature. The most satisfactory system is one which indicates the structure of the compound. Accordingly, an International Congress of Chemists held in Geneva in 1892 evolved a rational system of nomenclature which ciame to be known as Geneva system. This system is partially modified by a committee, appointed by the International union of chemists, in 1931 and the modified system was referred to as the **IUC System.**

The IUC system, in the light of practical difficulties, has been further revised by the International Union of Pure and Applied Chemistry in 1957. The revised system, which is now known as **IUPAC System**, is the most modern system and is largely used all over the world. The most salient feature of IUPAC system is that a given compound, having a definite structure, can be named in only one manner.

Before discussing the details of IUPAC system of nomenclature, we would briefly survey the common and IUPAC names of various homologous series of aliphatic compounds and their simple members. This would be helpful in understanding the general rules of IUPAC system and their application to more complex compounds. Each systematic name has two or three parts, viz., word root, primary suffix and secondary suffix.

1. Word roots : A owrd root indicates the nature of the basic carbon skeleton. For chains up to four carbon atoms, there are special word roots, while for C_5 onwards the Greek number roots are used as the word roots. The general word root for any carbon chain is alk-

Chain length	*Word root*	*Chain length*	*Word root*
C_1	Meth-	C_6	Hex-
C_2	Eth-	C_7	Hept-
C_3	Prop-	C_8	Oct-
C_4	But-	C_9	Non-
C_5	Pent-	C_{10}	Dec-

2. Primary suffix : A primary suffix is added to the word root to designate saturation or unsaturation in a carbon chain, this is known as primary suffix.

Name of carbon chain	*Primary suffix*	*Genesis name*
Saturated	C - C	-ane
Unsaturated	C = C	-ene
Unsaturated	C ≡ C	-yne

Hydrocarbon containing a hydrogen less than the parent compound incorporates the suffixes -yl or -enyl or -ynyl depending upon, the type of hydrocarbon.

Nature of group	*Primary suffix*	*Genesis name*
Alkane minus one H atom	-yl	alkyl
Alkene minus one H atom	-enyl	alkenyl
Alkyne minus one H atom	-ynyl	alkynyl

3. Secondary suffix : Secondary suffix is added after the primary suffix to indicate the functional group present in the compound, this is known as secondary suffix. When the secondary suffix is added to the primary suffix the terminal 'e' of the latter is usually removed.

Homologous series	*Word root*	*Primary suffix*	*Secondary suffix*	*Genesis name*
Saturated alcohol	alk-	-ane	-ol	alkanol
Saturated aldehyde	alk-	-ane	-al	alkanal
Saturated acid	alk-	-ane	-oic acid	alkanoic acid
Saturated ester	alk-	-ane	-oate	alkanoate

The names of simple members of various homologous series of aliphatic compounds are given below.

A. Alkanes or paraffins (saturated hydrocarbons) : These are saturated hydrocarbons containing only carbon and hydrogen with general formula C_nH_{2n+2}. These are characterized by linking of various carbon atoms to each other through single covalent bonds (C - C). The IUPAC name for this homologous series is alkane and the names of all alkanes will be ending with the suffix -ane. In the case of first few alkanes, the trivial and the IUPAC names are identical.

Common and IUPAC nomenclature of some alkanes

Formula	Common name	IUPAC name
CH_4	Methane	Methane
C_2H_6	Ethane	Ethane
C_3H_8	Propane	Propane
C_4H_{10}	Butane	Butane

For studying the nomenclature of compounds having three or more carbon atoms, it becomes necessary to have a knowledge of the four different types of carbon atoms. Carbon attached to one (or no), two, three or four other carbon atoms are called primary(1°) secondary(2°), tertiary(3°) and quaternary(4°) carbon atoms respectively.

$$\begin{array}{ccccccccc} & & CH_3 & & CH_3 & & & & \\ & & | & & | & & & & \\ H_3C & — & C & — & CH & — & CH_2 & — & CH_3 \\ & & |\ 4^\circ & & 3^\circ & & 2^\circ & & 1^\circ \\ & & CH_3 & & & & & & \end{array}$$

The straight chain hydrocarbons are called normal(n) hydrocarbons.

Branched chain hydrocarbon with the structural

```
          CH3
          |
H3C — CH — (CH2)n — CH3
```

formula are called **iso alkanes** while those having

```
       CH3
       |
H3C — C — (CH2)n — CH3
       |
       CH3
```

are known as **neoalkanes.**

```
                                                       CH3
                                                       |
H3C — CH2 — CH2 — CH2 — CH3              H3C — CH — CH2 — CH3
        n-Pentane                                Isopentane

                 CH3
                 |
          H3C — C — CH3
                 |
                 CH3
             Neopentane
```

B. Alkyl groups or radicals : These are monovalent groups or radicals obtained by the removal of one hydrogen atom from a molecule of an alkane. The general formula of these is C_nH_{2n+1}.

Their names contain the suffix -yl in place of -ane of the names of corresponding alkanes.

Formula	**Corresponding alkane**	**IUPAC name of alkyl group**
CH_3^-	Methane	Methyl
$C_2H_5^-$	Ethane	Ethyl
$C_3H_7^-$	Propane	Propyl
$C_4H_9^-$	Butane	Butyl

An alkyl radical is generally represented by the letter 'R' the alkyl group derived from alkanes having three or more carbon atoms exist in various isomeric structures *e.g.* Butyl group may exist in three isomeric structures.

```
                              CH3
                              |                       |
H3C–CH2–CH2–CH2—      H3C–CH–CH2—      H3C–CH2–CH–CH3
    n-butyl                Isobutyl            sec. butyl
```

Some alkenyl radicals or groups are :

$CH_2 = CH -$ $CH_2 = CH - CH_2 -$

Vinyl Allyl

C. Alkenes or olefins (Unsaturated hydrocarbons) : These are unsaturated hydrocarbons having a double bond between two carbon atoms (C = C). Their names are derived by replacing the suffix *-ane* of the corresponding alkane by *-ene* and therefore the general name alkene. The common names are obtained by adding the suffix *-ylene* to the word root. They are represented by the

General formula : C_nH_{2n} and

Functional group : $> C = C <$

Common and IUPAC nomenclature of some alkenes

Formula	Common name	IUPAC name
C_2H_4	Ethylene	Ethene
C_3H_6	Propylene	Propene
C_4H_8	Butylene	Butene
C_5H_{10}	Pentylene	Pentene

D. Alkynes or acetylenes (Unsaturated hydrocarbons) : These are unsaturated hydrocarbons containing a triple bond between two carbon atoms $- C \equiv C -$

Their names are derived from the corresponding alkane by substituting the suffix - *ane* by *-yne* and the general name as alkynes. They are represented by the

General Formula : C_nH_{2n-2} and

Functional group : $- C \equiv C -$

Common and IUPAC nomenclature of alkynes

Formula	Common name	IUPAC name
C_2H_2	Acetylene	Ethyne
C_3H_4	Allylene or Methyl acetylene	Propyne
C_4H_6	Butylene	Butyne

E. Halogen derivatives : The replacement of one or more hydrogen atoms of an alkane molecule by halogen atom leads to the formation of their halogen derivatives. These derivatives depending upon the number of halogen atoms are classified into mono, di, tri and tetra halo compounds.

(i) Monohalogen derivatives or alkyl halides : These are named as halogen substituted alkanes, the name of halogen being prefixed to the name of alkane involved. In the **trivial or common system,** they are named as halides of the alkyl group present. The monohalogen derivatives may further be classifed into primary, secondary and tertiary alkyl halides depending on the nature of carbon to which the halogen (-X) is attached.

General of formula : $C_nH_{2n+1}X$ or R - X where X is a halogen and 'R' stands for an alkyl group -X characteristic functional group :

R — CH2 — X (**Primary alkylhalide**)

$R-\overset{\displaystyle R}{\overset{|}{C}H}-X$ (**Secondary alkylhalide**)

$R-\underset{\displaystyle R}{\underset{|}{\overset{\displaystyle R}{\overset{|}{C}}}}-X$ (**'Teritiaryalkylhalide**)

Common and IUPAC nomenclature of some monoalkylhalides

Formula	Common name	IUPAC name
CH_3Cl	Methylchloride	Chloromethane
C_2H_5Br	Ethylbromide	Bromoethane
$H_3C-\overset{\displaystyle CH_3}{\overset{\vert}{C}H}-Cl$	Isopropyl chloride	2-Chloropropane
$H_3C-\underset{\displaystyle CH_3}{\underset{\vert}{\overset{\displaystyle CH_3}{\overset{\vert}{C}}}}-Br$	tert-Butylbromide	2-Bromo-2-methyl propane

(ii) Dihalogen derivatives : These are obtained by replacement of two hydrogen atoms by two halogen atoms. If the two halogen atoms are attached to the same carbon atom, the derivative is called an alkylidene halide, whereas it is termed alkylene halide if the two halogen atoms are attached to two different carbon atoms.

General formula : $C_nH_{2n}X_2$

Functional group : (i) Alkylidene halide $-\overset{|}{C}-CX_2$

(ii) Alkylene halide $-\underset{\displaystyle X}{\underset{|}{\overset{|}{C}}}-\underset{\displaystyle X}{\underset{|}{\overset{|}{C}}}-$

Common and IUPAC nomenclature of some dialkylhalides

Formula	Common name	IUPAC name
$H_3C — CHCl_2$	Ethylidene chloride	1, 1- Dichloroethane
$ClH_2C — CH_2Cl$	Ethylene chloride	1,2- Dichloroethane

(iii) Tri and tetrahalogen derivatives : The tri and tetra derivatives are derived by the replacement of three or four hydrogen atoms from the corresponding alkane by three or four halogen atoms respectively.

General formula : $C_n H_{2n-1} X_3$ and $C_n H_{2n-2} X_4$

Characteristic group : X_3 and X_4

Common and IUPAC nomenclature of some tri & tetra -alkylhalides:

Formula	Common name	IUPAC name
$CHCl_3$	Chloroform	Trichloromethane
CHI_3	Iodoform	Triiodomethane
CCl_4	Carbon tetrachloride	Tetrachloromethane

F. Alcohols : These are the compounds containing the hydroxyl group (-OH). They are further classified as mono-, di- and trihydric alcohols based on the presence of one, two, or three hydroxyl groups in the molecule. The suffix for the alcohols is - ol.

(i) Monohydric alcohols : The IUPAC names of monohydric alcohols are derived from the corresponding alkane by replacing the 'e' of alkane by -ol. Hence, they are known as alkanols. These are also further classified into primary, secondary and tertiary alcohols depending on the nature of the carbon to which the alcoholic group is attached.

General formula; $C_n H_{2n+1} OH$ ($C_nH_{2n+2}O$)

Functional group : - OH

Characteristic groups : RCH_2OH, $R—CH(R)OH$ i.e. $R — \underset{|}{\overset{R}{|}}CHOH$, $R — C(R)(R) — OH$

primary alcohol, secondary alcohol, tertiary alcohol

Common and IUPAC nomenclature of some monohydricalcohols

Formula	Common name	IUPAC name
CH_3OH	Methylalcohol	Methanol
C_2H_5OH	Ethylalcohol	Ethanol
$CH_3 — CH_2 — CH_2 — OH$	n-propylalcohol	1-propanol (pri)
$(H_3C)_2CH — OH$	Isopropyl alcohol	2-propanol (sec)
$H_3C — C(CH_3)_2 — OH$	tert. Butyl alcohol	2-methyl-2-propanol (tert)

(ii) Dihydric alcohols or Glycols or alkane diols : The dihydroxy derivatives of alkanes are called as dihydric alcohols or glycols. Since compounds having two hydroxyl groups on the same carbon atom are usually unstable, the two hydroxyl groups of glycols must be attached to two different carbons atoms.

The IUPAC names of the dihydric alcohols are derived by adding the suffix-diol to the name of the corresponding alkane and indicating the positions of the hydroxyl groups. The common names of glycols are derived from the corresponding olefins.

General formula : $C_nH_{2n}\ (OH)_2$ or $(C_n\ H_{2n+2}\ O_2)$

Characteristic group; $> C — OH$

$> C — OH$

Common and IUPAC nomenclature of some dihydricalcohols

Formula	Common name	IUPAC name
$HOCH_2 — CH_2OH$	Ethylene glycol	1,2 Ethanediol
$H_3C — CHOH — CH_2OH$	Propylene glycol	1,2 Propanediol

(iii) Trihydric alcohols : They contain three hydroxyl groups attached to three different carbon atoms in the molecule. In the IUPAC system, they are known as alkane triols, the individual members carrying the suffix *-triol* after the name of the corresponding alkane.

General formula : $C_nH_{2n-1}\ (OH)_3$ or $(C_n\ H_{2n+2}\ O_3)$

Characteristic group : $-\underset{|}{\overset{}{C}}-\underset{|}{C}-\underset{|}{C}$ with OH on each carbon

$$\begin{array}{ccccc} -C & - & C & - & C \\ | & & | & & | \\ OH & & OH & & OH \end{array}$$

Common and IUPAC nomenclature of trihydricalcohols

Formula	Common name	IUPAC name
$HOCH_2 - CHOH - CH_2OH$	Glycerol	1,2,3-Propanetriol

G. Ethers or Alkoxyalkanes : These are derived from alkanes by the replacement of one hydrogen atom by an alkoxy group (-OR), thus the general formula for these becomes R - O - R. When the alkyl groups are same, the ether is called as simple ether and when they are different, it is called mixed ether.

General formula : C_nH_{2n+2} R — O — R (Simple ether) R — O — R (Mixed ether)

Functional group : — O — R
Characteristic group : — O —

According to IUPAC system ethers are called alkoxyalkanes and the particular members are named as alkoxy derivatives of alkanes corresponding to bigger alkyl group. In common system, they are named as dialkyl ethers.

Common and IUPAC nomenclature of some ethers

Formula	Common name	IUPAC name
$H_3C - O - CH_3$	Dimethylether	Methoxy methane
$H_5C_2 - O - C_2H_5$	Diethyl ether	Ethoy cthanc
$H_3C - O - C_2H_5$	Ethyl methyl ether	Methoxy ethane

H. Thioalcohols or thiols or mercaptans or alkanethiols : Thioalcohols are derived by replacing oxygen atom of alcohols by sulphur atom. The secondary suffix used for thioalcohols in IUPAC system is thiol or mercaptan, while the common names are obtained by adding thioalcohol to the name of the alkyl group.

General formula : C_nH_{2n+1} SH or (C_n H_{2n+2} S)
Functional group : - SH

Common and IUPAC nomenclature of some thioalcohols

Formula	Common name	IUPAC name
CH_3SH	Methylthioalcohol or Methylmercaptan	Methane thiol

C_2H_5SH	Ethylthioalcohol or Ethylmercaptan	Ethane thiol

I. Thioethers or Alkyl Sulphides : These compounds contain a sulphur atom (in place of oxygen in ethers) attached to two alkyl groups which may be same or different. According to the IUPAC system they are named by adding the word thioether to the names of the alkyl groups present in the molecule. In trivial system, they are named as sulphides of the alkyl groups.

General Formula : $C_n H_{2n+2}$ R — S — R (Simple thioether) or R — S — R′ (mixed thioether)

Functional group : — SR
Characteristic group : — S —

Common and IUPAC nomenclature of some thioethers

Formula	**Common name**	**IUPAC name**
$H_3C\ S\ CH_3$	Dimethyl sulphide	Dimethyl ether
$C_2H_5\ S\ C_2H_5$	Diethyl sulphide	Diethyl thioether

J. Aldehydes or Alkanals : These are derived from alkanes by the replacement of one hydrogen atom by an aldehydic (-CHO) group. The IUPAC suffix of an aldehyde group is - al and the general name of these compounds is alkanals. The different members are named by replacing the terminal 'e' of the name of corresponding alkane by -al

The common names of aldehydes depend upon the names of acids, to which they may be oxidized the Suffix *ic acid* is replaced by *aldehyde.*

General formula ; $C_n H_{2n+1}$ CHO or ($C_n H_{2n}$ O)

Functional group : — CHO

Common and IUPAC nomenclature of some aldehydes

Formula	**Common name**	**IUPAC name**
HCHO	Formaldehyde	Methanal
H_3C — CHO	Acetaldehyde	Ethanal
H_3C — CH_3 — CHO	Propionaldehyde	Propanal

K. Ketones or Alkanones : These compounds contain the carbonyl group (> C = O) attached to two same or different alkyl groups. The IUPAC suffix for ketones is one and their general name is alkanones. The individual ketones are named by substituting the final '*e*' of the name of corresponding alkane by *-one.* The position of the > C = O group indicated by placing the number of the carbon atom to which oxygen is attached. The common names of ketones are obtained by adding the word ketone to the name of alkyl groups present.

General Formula $C_n H_{2n} O$

Functional group : $> C = O$

Common and IUPAC nomenclature of some ketones

Formula	Common name	IUPAC name
$H_3C-\overset{\overset{O}{\|\|}}{C}-CH_3$	Dimethyl ketone or Acetone	Propanone
$H_3C-\overset{\overset{O}{\|\|}}{C}-C_2H_5$	Ethylmethyl ketone	2-Butanone
$C_2H_5-\overset{\overset{O}{\|\|}}{C}-C_2H_5$	Diethyl ketone	3-Pentanone

L. Carboxylic acids : The compounds containing one or more carboxylic groups (COOH) are known as carboxylic acids. They are classified into mono-, di- and tri carboxylic acids based on the presence of one, two or three carboxylic groups in the molecule.

(i) Mono carboxylic acids or fatty acids or Alkanoic acids : Their IUPAC names are derived from the name of the correspounding alkanes by replacing the final '-e' with -oic acid. The common names are derived from their popular sources.

General Formula : $C_n H_{2n+1}$ COOH or $(C_n H_{2n} O_2)$

Functional group : — COOH

Common and IUPAC nomenclature of some carboxylic acids

Formula	Common name	IUPAC name
HCOOH	Formic acid	Methanoic acid
H_3C COOH	Acetic acid	Ethanoic acid
C_3H_7 COOH	Butyric acid	Butanoic acid
C_4H_9 COOH	Valeric acid	Pentanoic acid

(ii) Dicarboxylic acids or Alkane dionic acids : These compounds contain two carboxylic groups attached to the same or different carbon atoms.

Their IUPAC name is alkane dionic acids. For naming the individual members, the suffix dionic acid is added to the name of the corresponding alkane. In the common system, the names are generally derived from some source of the acid.

General Formula : $C_n H_{2n} (COOH)_2$

Functional group : $>C<\begin{matrix}COOH \\ COOH\end{matrix}$ or $\begin{matrix}>C-COOH \\ | \\ >C-COOH\end{matrix}$

Common and IUPAC nomenclature of some dicarboxylic acids

Formula	**Common name**	**IUPAC name**
HOOC – COOH	Oxalic acid	Ethanedionic acid
HOOC – CH_2 – COOH	Malonic acid	Propane 1, 3-dionic acid
H_2C – COOH | H_2C – COOH	Succinic acid	Butane - 1,4 -dionic acid

M. Acid derivatives : There are four functional derivatives of acids which may be obtained by replacing hydroxyl group of the carboxylic group by other atom or group.

(i) Acid halides or acylhalides : These are obtained by replacing the hydroxyl group (-OH) of the carboxylic group of acids by a halogen atom. Their IUPAC names are obtained by replacing the *'e'* of alkane by *-oyl* halide or they can be named by replacing the suffix -oic acid of the carboxylic acid by -oyl halide. For obtaining the common name, the suffix -ic acid of the common name of the correspounding acid is similarly changed into the suffix -yl halide.

General Formula : C_nH_{2n+1} COX

Functional group : — COX

Common and IUPAC nomenclature of some acid halides

Formula	**Common name**	**IUPAC name**
HCOCl	Formyl chloride	Methanoyl chloride
H_3C — COCl	Acetyl Chloride	Ethanoyl chloride
H_3CCH_2 — COCl	Propionyl chloride	Propanoyl chloride

(ii) Acid amides : They are obtained by replacing the hydroxyl group (OH) of the carboxyl group of acids by $-NH_2$ group. The IUPAC names of individual acid amides are obtained by replacing the final 'e' of the corresponding alkane and adding the suffix - amide or by replacing the suffix -oic acid of the carboxylic acid by amide. For getting the common name, the ending -ic acid of the corresponding acid is replaced by the word amide.

General Formula : C_nH_{2n+1} $CONH_2$

Functional group : — $CONH_2$

Common and IUPAC nomenclature of some acid amides

Formula	Common name	IUPAC name
$HCONH_2$	Formamide	Methanamide
H_3CCONH_2	Acetamide	Ethanamide
$C_3H_7CONH_2$	Propionamide	Propanamide

(iii) Acid anhydrides : They are derived from two (similar or different) carboxylic acids by the elimination of a molecule of water. The IUPAC names of these are obtained by removing the word acid from the IUPAC name of the corresponding acid and adding the word anhydride. The common names are obtained similarly from the common name of the parent acid.

General Formula C_n H_{2n-2} O_3

Functional group : $-\overset{O}{\overset{||}{C}}-O-\overset{O}{\overset{||}{C}}-$ or — CO — O — CO —

Common and IUPAC nomenclature of some acid anhydrides

Formula	Common name	IUPAC name
$H_3C - CO - O - CO - CH_3$	Acetic anhydride	Ethanoic anhydride
$H_5C_2 - CO - O - CO - C_2H_5$	Propionic anhydride	Propanoic anhydride
$H_7C_3 - CO\ O - CO - C_3H_7$	Butyric anhydride	Butanoic anhydride

(iv) Esters : They are derived from carboxylic acids by the replacement of hydroxyl group (-OH) of carboxylic group of - OR'. The IUPAC names of esters are obtained by first writing the name of alkyl part of -OR' group, followed by the IUPAC name of the acid in which the suffix *-ic acid* is replaced by *-ate*. The common names are obtained similarly from the common name of the corresponding acid.

General Formula : C_nH_{2n+1} COOR' or (C_n H_{2n} O_2)

Functional group : — COOR

Common and IUPAC nomenclature of some esters

Formula	Common name	IUPAC name
H $COOCH_3$	Methyl Formate	Methylmethanoate
H_3C — $COOCH_3$	Methyl acetate	Methyl ethanoate
H_3C — $COOC_2H_5$	Ethyl acetate	Ethylethanoate

N. Amines : Alkyl derivatives of ammonia are called amines. These are further classified as primary, secondary and tertiary depending upon the number of alkyl groups attached to nitrogen.

(i) Primary amines : These are monoalkyl derivatives of ammonia. In the IUPAC system primary amines are named by replacing the 'e' of corresponding alkane by amine or by adding prefix amino to the corresponding alkane. The common names are derived according to the alkyl group attached to the nitrogen atom and adding the suffix *amine.*

General Formula : C_nH_{2n+1} NH_2 or (C_n H_{2n+3} N)

Functional group : — NH_2

Common and IUPAC nomenclature of some amines

Formula	Common name	IUPAC name
H_3C — NH_2	Methylamine	Methanamine or Amino methane
$C_2H_5NH_2$.	Ethylamine	Ethanamine or Aminoethane

(ii) Secondary amines : These are dialkyl derivatives of ammonia General Formula : R — NHR

Functional Group : NH —

(iii) Tertiary amines :- These are trialkyl derivatives of ammonia.

General Formula : R_3N

Functional Group : — N —
|

In the IUPAC system secondary and tertiary amine are named as alkyl derivatives of the highest ranking primary amine structure present. Common names are derived according to the alkyl groups attached to the nitrogen atom and adding the suffix amine.

Common and IUPAC nomenclature of some amines

Formula	Common name	IUPAC name
$H_3C - NH - CH_3$	Dimethylamine	N - Methanamine or N-Methyl aminomethane
$H_3C - NH - C_2H_5$	Ethylmethylamine	N-Methylethanamine or N- Methylaminoethane
$(HC_3)_3N$	Trimethylamine	N,N-Dimethylmethanamine or N,N-Dimethylaminomethane
$(HC_3)_2 \quad N - C_2H_5$	Ethyldimethylamine	N,N-dimetylethanamine or N,N- Dimethylaminoethane

O. Alkyl cyanides or nitriles or carbonitriles : These are derived by replacing a hydrogen atom of the alkane or a carboxylic group of the acid by cyanide (-CN) group. Their IUPAC names are obtained by adding suffix nitrile to the name of parent alkane and the common names are derived by adding suffix cyanide to the alkyl group or nitriles of the acids which they give on hydrolysis.

General Formula : C_nH_{2n+1} CN or (C_n H_{2n+1} N)

Functional group : $- C \equiv N$

Common and IUPAC nomenclature of some alkyl cyanides

Formula	Common name	IUPAC name
HCN	Hydrogen cyanide	Methane nitrile
H_3CCN	Methyl cyanide	Ethane nitrile
H_3CCH_2CN	Ethyl cyanide	Propane nitrile
$H_3CCH_2CH_2CN$	Butyl cyanide	Butane nitrile.

P. Alkyl isocyanides or isonitriles or carbylamines : These may be regarded as derivatives of alkane formed by replacement of one hydrogen atom by the -NC group or by replacing carboxylic group of carboxylic acid by the isonitrile (-NC) group. In the IUPAC system, they are named as alkyl carbylamines. In the common system they are either named as alkyl isocyanides or isonitriles of corresponding acids.

General Formula : C_nH_{2n+1} NC C_n $H_{2n-1}N$

Functional group : $- \overset{+}{N} \equiv \overset{-}{C}$ or $-$ NC

Common and IUPAC nomenclature of some alkyl isocyanides

Formula	Common name	IUPAC name
H_3CNC	Methylisocyanide	Methylcarbylamine or Aceto isonitrile
C_2H_5NC	Ethyl isocyanide	Ethylcarbyl amine or Propionoisonitrile

Q. Nitroalkanes or Nitroparaffins : These may be regarded as derived from alkanes by replacing a hydrogen atom by nitro group (-NO_2). The IUPAC and trivial names are generally the same but in the IUPAC names the position of nitro group is also indicated. They are named as the nitro derivatives of the alkanes by adding the prefix nitro before the name of the corresponding alkane.

General Formula : $C_nH_{2n+1}\ NO_2$

Functional group : $-\ NO_2$

Common and IUPAC nomenclature of some nitroalkanes

Formula	Common name	IUPAC name
$H_3C - NO_2$	Nitromethane	Nitromethane
$CH_3CH_2NO_2$	Nitroethane	Nitroethane
$H_3C - CH_2 - CH_2 - NO_2$	Nitropropane	1-Nitropropane

2-4 NOMENCLATURE OF COMPLEX COMPOUNDS

The IUPAC system, is the most rational and widely used system of nomenclature. In the case of complex compounds it has been found rather difficult to search for common names, on the other hand, IUPAC names are found handy and convenient. There are certain rules for naming a complex organic molecule according to IUPAC system.

RULES FOR IUPAC NOMENCLATURE OF ORGANIC COMPOUNDS

2-4-1 Longest Chain Rule

(A) *The longest unbroken chain of carbon atoms in a molecule serves as the parent name for any hydrocarbon or its derivative.*

```
 5      4     3       2      1
H3C - CH - CH2 - CH2 - CH3
       |
      5CH2
       |
      6CH3
```

The parent chain consists of six carbon atoms (not five) and the word root is hex-

```
 5      4      3      2      1
H3C - CH  - CH - CH2 - CH3
       |
      CH2
       |
      CH3
```

Wrong location of parent chain as it is not the longest continuous chain

Caution : The numbering can be done both from right to left and vice versa. But for the sake of uniformity numbering is done here from right to left.

(B) For the compounds containing a functional group, the longest chain must contain the functional group, even if the longest chain rule is violated.

$$\begin{array}{ccccccc} 4 & & 3 & & 2 & & 1 \\ H_3C & - & CH_2 & - & CH & - & CH_2OH \\ & & & & | & & \\ & & & & CH_2 & & \\ & & & & | & & \\ & & & & CH_3 & & \end{array}$$

The parent chain is made up of four carbon atoms (not five) and the word root is but-

$$\begin{array}{ccccccc} 5 & & 4 & & 3 & & \\ H_3C & - & CH_2 & - & CH & - & CH_2OH \\ & & & & \downarrow & & \\ & & & & {}^2CH_2 & & \\ & & & & | & & \\ & & & & CH_3 & & \end{array}$$

Wrong location of parent chain as it does not include the functional group.

(C) While locating the longesty chain, sometimes two equally long chains (but having different number of side chains) appear to be possible. In such cases, the longest chain that carries the larger number of side chains is selected as the parent chain.

$$\begin{array}{ccccccccc} 5 & & 4 & & 3 & & & & \\ H_3C & - & CH_2 & - & CH & - & CH_2 & - & CH_3 \\ & & & & | & & & & \\ & & & & {}^2CH & - & CH_3 & & \\ & & & & | & & & & \\ & & & & {}^1CH_3 & & & & \end{array}$$

The parent chain is made up of five carbons atoms and carried two side chains.

$$\begin{array}{ccccccccc} 5 & & 4 & & 3 & & 2 & & 1 \\ H_3C & - & CH_2 & - & CH & - & CH_2 & - & CH_3 \\ & & & & | & & & & \\ & & & & CH & - & CH_3 & & \\ & & & & | & & & & \\ & & & & CH_3 & & & & \end{array}$$

Wrong location of parent chain as it carried only one side chain.

2-4-2 Lowest Number Rule

(A) The carbon atoms forming the principal chain are numbered from one end to another. The positions of the side-chains are indicated by the numbers of carbon atoms to which these are attached. The numbering is started from the end nearest to the substituents so as to give the lowest numbers possible to the side chains.

$$\begin{array}{ccccccccc} 5 & & 4 & & 3 & & 2 & & 1 \\ H_3C & - & CH_2 & - & CH_2 & - & CH & - & CH_3 \\ & & & & & & | & & \\ & & & & & & CH_3 & & \end{array}$$

Substituent at C_2 (Correct)

$$\begin{array}{ccccccccc} 1 & & 2 & & 3 & & 4 & & 5 \\ H_3C & - & CH_2 & - & CH_2 & - & CH & - & CH_3 \\ & & & & & & | & & \\ & & & & & & CH_3 & & \end{array}$$

Substituent at C_4 (Wrong)

(B) In case, the parent chain has two or more substituents, numbering must be done in such a way that the sum of the locants on the parent chain is the lowest possible.

$$\overset{6}{H_3C} - \overset{5}{CH_2} - \underset{CH_3}{\overset{4}{CH}} - \overset{3}{CH_2} - \underset{CH_3}{\overset{2}{CH}} - \overset{1}{CH_3}$$

Sum of locants : 2 + 4 = 6 (Correct)

$$\overset{1}{H_3C} - \overset{2}{CH_2} - \underset{CH_3}{\overset{3}{CH}} - \overset{4}{CH_2} - \underset{CH_3}{\overset{5}{CH}} - \overset{6}{CH_3}$$

Sum of locants : 3 + 5 = 8 (Wrong)

(C) If different side chains (alkyl groups) are in equivalent positions in relation to the ends of the chain, preference is given to the end where the alkyl group has fewer carbon atoms.

$$\overset{6}{H_3C} - \overset{5}{CH_2} - \underset{H_3C - CH_2}{\overset{4}{CH}} - \underset{CH_3}{\overset{3}{CH}} - \overset{2}{CH_2} - \overset{1}{CH_3}$$

Methyl group is at C_3 (Correct)

$$\overset{1}{H_3C} - \overset{2}{CH_2} - \underset{H_3C - CH_2}{\overset{3}{CH}} - \underset{CH_3}{\overset{4}{CH}} - \overset{5}{CH_2} - \overset{6}{CH_3}$$

Methyl group is at C_4 (Wrong)

(D) If two sets of numbers be equally possible the alphabetical order of the prefixes in the name will decide which set should be used.

$$Cl - \overset{4}{CH_2} - \overset{3}{CH_2} - \overset{2}{CH_2} - \overset{1}{CH_2} - Br$$

Bromo group at C_1 and chloro group at C_4 (Correct)

$$Cl - \overset{1}{CH_2} - \overset{2}{CH_2} - \overset{3}{CH_2} - \overset{4}{CH_2} - Br$$

Bromo group at C_4 and chloro group at C_1 (Wrong)

(E) The name of the substituent is prefix to the name of the parent hydrocarbon and its position on the parent chain is indicated by writing the locant before the prefix : A hyphen (-) is inserted between the locant and the substituent name.

$$\underset{4}{H_3C} - \underset{3}{CH_2} - \underset{2}{\overset{Br}{CH}} - \underset{1}{CH_3}$$

2-Bromobutane

(F) If a substitutent is present more than one times, this is indicated by the prefix di-, tri, tetra, etc. added to the substituents are separated by commas.

$$\underset{4}{H_3C} - \underset{3}{\overset{CH_3}{CH}} - \underset{2}{\overset{CH_3}{CH}}\ \underset{1}{CH_3}$$

2,3-Dimethylbutane

2-4-3 Arrangement of Prefixes

In case more than one group are attached to the parent chain, these should be arranged alphabetically, while doing so, prefixes di and tri etc., are not considered.

$$\overset{7}{H_3C} - \overset{6}{CH_2} - \underset{CH_3}{\overset{5}{CH}} - \underset{CH_3}{\overset{4}{CH}} - \underset{CH_2 - CH_3}{\overset{3}{CH}} - \overset{2}{CH_2} - \overset{1}{CH_3}$$

3-Ethyl-4,5-dimethylheptane

2-4-4 Lowest Number for Functional Group

(A) In case some functional group is present in the chain, the lowest number is to be assigned to it, even if the lowest number rule is violated.

$$\underset{5}{H_3C} - \underset{4}{\overset{CH_3}{CH}} - \underset{3}{CH_2} - \underset{2}{\overset{OH}{CH}} - \underset{1}{CH_3}$$

4-Methyl-2-pentanol (Correct)

$$\underset{1}{H_3C} - \underset{2}{\overset{CH_3}{CH}} - \underset{3}{CH_2} - \underset{4}{\overset{OH}{CH}} - \underset{5}{CH_3}$$

2-Methyl-4-pentanol (Wrong)

(B) The carbon in chain terminating functional groups such as - COOH, - CHO, - $CONH_2$ - CN etc., must always be given number one. The carbon of non-terminal functional group (for example c = 0) may or may not be assigned number one.

$$\overset{5}{H_3C} - \underset{CH_3}{\overset{4}{CH}} - \overset{3}{CH_2} - \overset{2}{CH_2} - \overset{1}{COOH}$$

4-Methylpentanoic acid

$$\overset{6}{H_3C} - \underset{CH_3}{\overset{5}{CH}} - \overset{4}{CH_2} - \underset{O}{\overset{3}{\underset{\|}{C}}} - \overset{2}{CH_2} - \overset{1}{CH_3}$$

5-Methyl-3-hexanone

2-4-5 Naming of Poly Functional Compounds

(A) When a compound containing two are more different functional groups, one of the functional group is choosen as principal group and the remaining functional groups are treated as substituents. The following order of preference is given in the choice of the principal functional group which is to be given the lowest number.

Acids > acid derivatives (except nitriles) aldehydes>nitriles > ketones> alcohols > amines > ethers > alkenes > alkynes.

$$\overset{3}{H_2C} = \overset{2}{CH} - \overset{1}{CH_2} - OH$$

2-propen-1-ol

$$\overset{5}{HC} \equiv \overset{4}{C} - \underset{CH_3}{\overset{CH_3}{\overset{3}{C}}} - \overset{2}{CH} = \overset{1}{CH_2}$$

3,3-dimethyl-1-penten-4-yne

(B) While selecting the parent chain in a poly functional compounds, it should be ensured that it includes the maximum number of functional groups present, including the principal functional group.

$$\overset{5}{H_3C} - \overset{4}{CH} = \overset{3}{\underset{\underset{\overset{2}{CH_2} - \overset{1}{CH_2OH}}{|}}{C}} - CH_2 - CH_2 - CH_3$$

3-Propyl-3-penten-1-ol

(C) In the case of compounds containing more than one functional groups, the ending is suitably modified. Carbon multiple bonds and a second functional group are combined in ending.

$$\underset{4}{H_2C} = \underset{3}{CH} - \underset{2}{\overset{\overset{O}{||}}{C}} - \underset{1}{CH_3}$$

3-Buten-2-one

$$\underset{1}{CH_2} = \underset{2}{CH} - \underset{3}{CH} = \underset{4}{CH_2}$$

1, 3-Butadiene

(D) The halo-, nitro-, nitroso-, azo- and sometimes amino groups are not regarded as functional groups. They are considered as substituents.

(E) In the compounds containing substitutents, side chains and functional groups, first of all mention all substituents in alphabetical manner, then side chains in alphabetical manner and finally functional groups in alphabetical order.

(F) If the chain of carbon atoms selected as the branched chain also contain multiple bonds or functional groups the branched chain in separately numbered.

$$\underset{5}{H_3C} - \underset{4}{\overset{\overset{OH}{|}}{CH}} - \underset{3}{\overset{\overset{CH_2 - OH}{|}}{\overset{CH_2}{\overset{|}{CH}}}} - \underset{2}{CH_2} - \underset{1}{COOH}$$

4-Hydroxy-3-2(2-hydroxyethyl)-pentanoic acid

$$\begin{array}{ccccccccccccccccccc} & & & & & & & & & & CH_3 & & & & & & & & \\ & & & & & & & & & 1 & | & 2 & & 3 & & & & & \\ & & & & & & & & & H_3C & - C - & CH_2 & - & CH_3 & & & & & \\ 10 & & 9 & & 8 & & 7 & & 6 & & | & 4 & & 3 & & 2 & & 1 & \\ H_3C & - & CH_2 & - & CH_2 & - & CH_2 & - & CH_2 & - & C & - & CH_2 & - & CH_2 & - & CH_2 & - & CH_3 \\ & & & & & & & & & & \backslash & & & & & & & & \\ & & & & & & & & & & 1CH_2 & - & 2CH & CH_3 & - & 3CH_3 & & & \end{array}$$

5-(1,1-Dimethylpropyl)-5-(2-methylpropyl)-decane.

On the basis of foregoing discussion, we can sum up the general procedure to be followed in naming a given structural formula.

(1) Locate the longest chain containing the principal functional group and as many of the secondary functional groups and multiples bonds as possible.

(2) Select the word root corresponding to the length of the chain. In the case of alicyclic compounds add the prefix *cyclo* to the word root.

(3) Number the chain from one end so that the various groups get as small number as possible and the principal functional group gets the lowest number.

(4) Add to the word root a suffix (ane,ene,yne) representing the nature of carbon-carbon bonds in the parent chain.

(5) Add suitable prefixes and suffixes, with their respective number of attachment to the parent chain, to denote the number of position of each substituent or functional group.

(6) Write the name of the compound as one word, separating the numbers from the designations of prefixes and suffixes by hyphens and separating the number from each other by commas.

Problem 1 : Name the compound whose structural formula is

$$\begin{array}{ccccccccc} H_2C & — & CH_2 & — & CH_2 & — & C & — & CH_3 \\ | & & & & & & \| & & \\ OH & & & & & & O & & \end{array}$$

The longest chain contains five carbon atoms and no double or triple bond. It is therefore, a derivative of pentane. There are two functional groups - OH and $> C = O$ situated equidistant from the ends. But the latter stands higher in the order of preference. Hence numbering must start from the side, which give the lowest possible number to > C = O. Keto group being the principal functional group will be indicated by the suffix *-one* while hydroxyl group being the secondary functional group will be treated as a substituent and indicated by a prefix hydroxy. Its name therefore, will be

$$\overset{5}{H_3C} - \overset{4}{\underset{\underset{OH}{|}}{CH_2}} - \overset{3}{CH_2} - \overset{2}{\underset{\underset{O}{||}}{C}} - \overset{1}{CH_3}$$

4-Hydroxy-2-pentanone

Problem 2 : Name the compound from its structural formula given below :

$$H_3C - CH_2 - \overset{\overset{Cl}{|}}{CH} - C \equiv C - CH_2OH$$

The longest chain contains six carbon atoms with a triple bond. It is, therefore a derivative of pentyne numbering is to start from the side nearest to the principal functional group i.e., - OH)

$$H_3C - CH_2 - \overset{\overset{Cl}{|}}{CH} - C \equiv C - CH_2OH$$

Hydroxyl group being the functional group will be indicated by the suffix -ol. Triple bond is indicated by the suffix -yne with the root word. Its name, there is 4-chloro-2-hexyn-1-ol.

2-5 WRITING THE STRUCTURAL FORMULA FROM THE GIVEN IUPAC NAME

This can be achieved by the following steps.

(1) Locate the parent alkane from the name and write the number of carbon atoms of this alkane in a straight chain and number them from any end. However, in the case of chain-terminating functional groups, the carbon of the functional group must be included in the parent chain and must be numbered one.

(2) Now locate the suffix which gives information about the nature and number of the functional groups along with their positions and insert on the specified carbon atom (s).

(3) Locate the group(s) or substituent(s) and side chain(s) indicated by suffixes and insert them on the specified carbon atom(s) indicated by prefix.

(4) This gives us the skeletal formula. Lastly, add required number of hydrogen atoms to satisfy four valencies of each carbon atom to get the structural formula.

Let us illustrate these rules by an example.

To write the structural formula of the compound 7-Hydroxy-3-methoxy-4-methyl-5-heptanoic acid.

(i) Parent alkane of the compound is heptane. Write seven carbon atoms in a straight chain and number these.

$$C - C - C - C - C - C - C$$

(ii) The suffix enoic and indicates the presence of functional groups -COOH at the end of the chain (i.e. position one) and a double bond.

(iii) To locate the substituents mentioned in the prefix 7-hydroxy-3-methoxy-4-methyl and the double bond indicated by 5- ene, we attach one -OH group with C7, one $-OCH_3$ group with C3, open CH3 group with C4 and a double bond between C5 and C6. This gives

```
7   6   5   4   3   2     1
C - C - C - C - C - C - COOH

OH          CH3 OCH3
```

(iv) Finally adding hydrogens to satisfy four valencies of each carbon atom, we get the structural formula.

```
  7      6      5      4      3        2        1
H2C — CH = CH — CH — CH  — CH2 — COOH
                       |      |
OH                    CH3   OCH3
```

2-6 COMMON OR TRIVIAL NAMES OF ORGANIC COMPOUNDS

Now-a-days the common names are employed in the case of very familiar and common compounds or very complex compounds (e.g. quinine). The different common names can be classified into six categories.

(1) Functional names : These are obtained from the functional group.

Formula	Functional Group	Functional name
$H_2C = CH_2$	Double bond	Ethylene
$HC \equiv CH$	Triple bond	Acetylene
$H_3C - CO - CH_3$	Ketone	Acetone
$H_3C\ COOH$	Acid	Acetic acid

(2) Substitutive names : In this type, the functional group is considered as the substituent and thus the compounds are considered asthe substituted compound of the parent hydrocarbon.

Formula	Substitutent	Substitutive name
$H_3C\ CH_2\ OH$	Hydroxy group	Hytdroxy ethane
$H_3C\ CH\ NO_2$	Nitro group	Nitroethane

(3) Additive names : These names are adopted where the compound is obtained by the addition of a reagent to an organic compound.

Formula	Reagents	Additive name
$ClH_2C\ CH_2\ Cl$	Ethylene, chlorine	Ethylene chloride
$HOH_2C\ CH_2\ Cl$	Ethylene, hypochlorous acid	Ethylene chlorohydrin
$HOH_2C\ CH_2OH$	Ethylene, water	Ethylene glycol

(4) Replacement names : These names are adopted where one compound is derived by the replacement of an atom by other atom.

Formula	Analogous compound	Replacement name
$H_3C - CH_2SH$	$H_3C\ CH_2OH$ (Ethyl alcohol)	Ethyl thioalcohol
$H_3C - S - CH_3$	$H_3C - O - CH_3$ (dimethlether)	Dimethylthioether

(5) Conjugative names : These names are adopted where two compounds are joined together either directly or through a functional group.

Formula	Conjugative name
$H_3C - \overset{\overset{\large O}{\|\|}}{C} - \overset{\overset{\large O}{\|\|}}{C} - CH_3$	Biacetyl
$H_3C - N = N - CH_3$	Azomethane

(6) Radico functional names : These names are adopted where a functional group is present on the alkyl group (radical)

Formula	Radical, functional group	Radico-functional name
CH_3OH	Methyl, Alcohol	Methyl alcohol
$CH_3CH_2NH_2$	Ethyl, Amine	Ethylamine

Some commonly used observations : There are certain abbreviations in common use for the name of some groups and compounds.

Formula	Name of the group/compound	Abbreviation
$-CH_3$	Methyl	Me
$-C_2H_5$	Ethyl	Et
$-CH(CH_3)_2$	Isopropyl	Pr^i
$-C(CH_3)_3$	tert. Butyl	Bu^t
CH_3CO	Acetyl	Ac
CH_3OH	Methyl alcohol	MeOH
CH_3COCH_3	Acetone	Me_2CO
CH_3COOH	Acetic acid	AcOH
$(CH_3CO)_2O$	Acetic Anhydride	Ac_2O

2-7 MULTIPLE CHOICE QUESTIONS :

2-7-1 IUPAC name of the compound $CH_3\text{-}\underset{CH_3}{C}\text{-}CH_2\text{-}Br$ is

(A) 1-Bromo-2-methyl propane (B) 2-methly-3-bromopropane

(C) both (A) & (B) (D) None of these

2-7-2 The structural formula of 1-hydroxy-2-propanol is

(A) $CH_3\text{-}\underset{\|\atop O}{C}\text{-}\underset{\|\atop O}{C}\text{-}CH_2OH$ (B) CH_3COOH

(C) $CH_3\text{-}\underset{\|\atop O}{C}\text{-}OCH_3$ (D) $CH_3\text{-}\underset{\|\atop O}{C}\text{-}CH_2OH$

2-7-3 IUPAC name of the compound having formula

$$\begin{array}{c} \quad\quad CH_3 \\ \quad\quad || \\ H_3C\text{-}C\text{-}CH{=}CH_2 \\ \quad\quad || \\ \quad\quad CH_3 \end{array}$$

is

(A) 3,3,3-trimethyl-1-propene (B) 1,1,1-dimethyl-2-propene

(C) 3,3-dimethyl-1-butene (D) 2,2-dimethyl-3-butene

2-7-4 IUPAC name of the compound $CH_2{=}CH\text{-}CH(CH_3)_2$ is

(A) 1,1-dimethyl-2-propene (B) 3-methyl-1-butene

(C) 2-Vinylpropane (D) 1-isopropyl ethylene

2-7-5 IUPAC name of

$$\begin{array}{c} \quad\quad O \\ \quad\quad || \\ CH_3\text{-}C\text{-}OC_2H_5 \end{array}$$

is

(A) ehtyl acetate (B) ethyl ethanoute

(C) methylproporate (D) none of these

2-7-6 IUPAC nmae of CH_3 COOH is

(A) Acetic acid (B) methyl carboxylic acid

(C) pyroligneous acid (D) Ethanoic acid

2-7-7 The IUPAC name of the compound

$$\begin{array}{l} HO\text{-}CH_2\text{-}CH_2\text{-}CH\text{-}COOH \\ \quad\quad\quad\quad\quad\quad\quad \backslash \\ \quad\quad\quad\quad\quad\quad\quad CH_3 \end{array}$$

is

(A) 4-hydroxy-2-methyl butane acid (B) 1-hydroxy-3-methyl butane acid

(C) 3-hydroxy-1-methyl butane acid (D) 1-hydroxy-3-carboxy butane

2-7-8 Ester is represented by the general formula

(A) R-COOR (B) ROR

(C) R-O-O-R (D) R-COOH

2-7-9 The IUPAC name of CH2 is

(A) Acetylene (B) Allylene

(C) Ethane (D) Ethyne

2-8 SHORT & ESSAY TYPE QUESTIONS

2-8-1 Define"Homologous series" and give an example of the same ?

(O.U., Mar.1990)

2-8-2 Give the formulae for the following

(A) (i) 2-Methyl-3-ethyl hexane

(ii) 2,3,6 Triethyl heptane

(iii) 2-Methyl-2-pentene -1-ol

(O.U.,Mar.1990)

(b) Name the following compounds as per the IUPAC system?

(i) $H_2C = CH\text{-}CH_2\text{-}CH{=}CH_2$

(ii) $(CH_3)_2\ CH\text{-}CH(CHO)CH_2(CH_3)_2$

(O.U., Mar.1990)

2-8-3 What do you understand bya functional group? What is its significance? Write the functional group of the each of the following classes

(i) a ketone (ii) an acid amide (iii) an ether?

(O.U., Oct.1989)

2-8-4 Write the structural formulae for the following

(i) 3-Chloro - 1iodohexane

(ii) 4-Methyl-3, 3-diethyl-5-isopropyl octane

(iii) 1,1 Dimethy1-2,2- dimethyl butane

(O.U.,Oct 1989)

2-8-5 Write the structural formulae for the following

(i) 5-Bromo -4- methylheptane

(ii) 3-Bethyl-2-propyl-3-butene-1-ol

2-8-6 Name the following compounds on the basis of IUPAC system of nomenclature

```
(I) H3C-CH-CH-CH2-CH-CH3
        |  |        |
        Cl CH3      CH3

(ii) H3C-CH-CH-CH3
         |  |
         Cl OH

        CH3CH3
          \ \
(iii) H3C-C-CH-CH=CH2
          \ \
          CH3CH-CH3
             \
             CH3
```

2-8-7 Write IUPAC name of the following compound

H2C-C=C-CH(CH3)2

(O.U.,Apr.1988)

2-8-8 Write the structural formulae of the following compounds

(a) Hex-1-ene-4-yne

(b) 2-Methyl-buta-1-3-diene

(O.U.,Apr.1988)

2-8-9 What is homology ? Illustrate with two examples ?

(O.U.,Apr.1987)

2-8-10 Write the structural formulae of the following compounds
(i) 4-Isopropyl-6-methyl -decane
(ii) 3,4,4 trimethyl - 1- hexyne?

2-8-11 Write the IUPAC name of the following compound:

$$\begin{array}{ccccc} & & & CH_3 & \\ & & & | & \\ H_3C- & CH- & CH_2- & C- & CH_3 \\ & | & & | & \\ & CH_3 & & CH_2-CH_3 & \end{array}$$

2-8-12 Write the structural formulae of
(i) 4-ethyl-5-methyl-2-hexyne
(ii) 2,3-diiodo -4- octene

(O.U.,Apr.1987)

2-8-13 Write the IUPAC name of the following compound:
CH_3-CH=CH=CH_2-CH_2OH

(O.U.,Apr.1987)

2-8-14 Give the IUPAC names for the following compounds:
(a) CH_3-CH=CH-C=C-H
(b) H_3C-CH_2-CH=CH-CH_3

(O.U.,Nov.1986)

2-8-15 Name the following compounds by the IUPAC method

(a)
$$\begin{array}{l} Cl-CH_2-CH_2-CH-C(CH_3)_2 \\ \qquad\qquad\qquad\quad | \\ \qquad\qquad\qquad CH_3\ C_2H_5 \end{array}$$

(b) L-Butyl alcohol

(O.U.,Mar.1988)

2-8-16 Write the srtuctural formulae for the following :
(a) 3-Chloro-1-Iodohexane
(b) 4-Methyl -3,3-diethyl-5-isoprotyl octane
(c) 1,1-dimethyl butane.

2-8-17 Name the following compounds as per the IUPAC system ?

(a)
$$\begin{array}{ccc} CH_3- & C= & CH-CH_3 \\ & | & \\ & CH_3 & \end{array}$$

(b)
$$\begin{array}{ccc} CH_3- & CH- & CH_2-CH_3 \\ & | & \\ & OH & \end{array}$$

(c)
$$\begin{array}{ccc} & Cl & \\ & / & \\ H_3C- & CH- & CH-CH_3 \\ & \backslash & \\ & & CH_3 \end{array}$$

(O.U., Oct. 1984)

2-8-18 Name the following compounds according to the IUPAC nomenclature ?

(a)
$$H_3C\text{-}CH_2\text{-}\underset{\large CH_2\text{-}CH_3}{\overset{\large CH_3}{C}}\text{-}CH_2\text{-}CH_2\text{-}CH_3$$

(b)
$$H_3C\text{-}\underset{\large CH_3}{\overset{\large CH_3}{C}}\text{-}CH_3$$

2-8-19 Writedown the structural formula of 2,2,4-trymetyl pentane ?

(O.U.,Oct.1984)

2-8-20 Give the IUPAC for the following

(i) C_3H_7COOH
(ii) C_2H_4
(iii) $C_2H_5\,COCH_2\,CH_2\,CH_3$
(iv) $H_3CCHBr\,COOH$
(v) $H_3CCH_2CH{=}\,CHCOOH$

(O.U., Apr.1991)

2-8-21 (a) How is an organic compound is named ?

(b) Write the structure of the following

(i) 2-Chloro-trimethyl-2-pentene
(ii) But-3-ene-1-ol
(iii) 2,4,4-trimethyl-2-pentene
(iv) 3-Methyl-octa-3-ene-5-yne

(N.U.Apr. 1989)

2-8-22 Give IUPAC names for

(I) HCHO
(ii) CH3CHO
(iii) CH3COOH
(iv) CH=CH

(A.U.Apr.1988)

2-8-23 Write a note on IUPAC nomenclature with suitable examples

(A.U.,Dec 1989)

2-8-24 Name the following organic compounds by systematic method of nomenclature.

(a) $H_3C\text{-}CH_2\text{-}CH{=}CH_2$
(b) $CH_2Cl\text{-}CH_2CL$
(c) $C_6H_5N{=}NC_6H_5$
(d) $H_3C\text{-}Ch_2\text{-}\underset{\large O}{\underset{\|}{C}}\text{-}Ch_3$

ANSWERS :

2-7-1 A	**2-7-2** D	**2-7-3** C	**2-7-4** B	**2-7-5** C
2-7-6 D	**2-7-7** A	**2-7-8** A	**2-7-9** D	

3

ALKANES

3-1 INTRODUCTION

The simplest organic compounds, made up of only carbon and hydrogen atoms are known as *hydrocarbons*. In these compounds carbon is tetravalent and hydrogen univalent. Depending upon the frame work of carbon atoms in their molecules, the hydrocarbons may have either open chain or ring structures. The open chain hydrocarbons are called as *aliphatic hydrocarbons*. These are classified into two groups depending on the types of carbon- carbon bonds. Those in which all the four valencies of carbon atoms are satisfied by single bonds to four atoms or groups are known as *saturated hydrocarbons*. The compounds in which carbon atoms are linked together by multiple bonds are known as *unsaturated hydrocarbons*.

Alkanes are open chain saturated hydrocarbons having the general formula C_nH_{2n+2}, *where 'n' represents the number of carbon atoms*. Thus the first member, methane, has the formula CH_4 and the second member ethane has the formula C_2H_6. Since alkanes are saturated hydrocarbons, the carbon-carbon bonds are strong σ bonds comparatively chemically inert and hence are referred as paraffins. (Latin, param = little; affins = affinity).

3-2 OCCURRENCE AND STRUCTURE OF ALKANES

3-2-1 Occurrence

The alkanes are the main constituents of natural gas and petroleum. Petroleum is a complex mixture of alkanes and other hydrocarbons. It usually contains alkanes having upto forty carbon atoms in their molecule. Natural gas from petroleum wells contains only the lower gaseous alkanes like methane, ethane, propane and butane. The neutral wax found near petroleum wells is mixture of higher solid hydrocarbons. Bee wax also contains some solid alkanes. Lower members like methane are also obtained by the natural decay of animals and plants. In fact petroleum is also the product of the decomposition of the complicated organic matter.

3-2-2 Structure of Alkanes

The bonds in alkanes are formed by the overlapping of sp^3 hybridized orbitals of each carbon atom with similar orbitals of other carbon atoms or 's' orbital of hydrogen atoms. As a result all the four bonds linking the carbon atoms are σ bonds. Every C - C and C - H bond has a length of 1.54 A and 1.09 A respectively. The four bonds are directed towards the corners of a regular tetrahedron and are at an angle of $109^\circ 28'$ to each other.

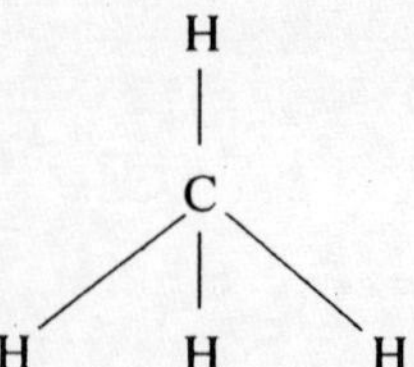

Fig. 3.1 Structure of methane

3-3 NOMENCLATURE

For naming alkanes generally two systems are adopted, viz., 1. common system 2. IUPAC system.

3-3-1 Common System

In this system, the first four members are named by their trivial names. Rest of the members are named by prefixing the greek numerals, viz. indicating the number of carbon atoms in the molecule, to the suffix-*ane* . The common names of a few paraffins are given below.

Molecular formula	*Name*	*Molecular formula*	*Name*
CH_4	Methane	$C_{10}H_{22}$	Decane
C_2H_6	Ethane	$C_{11}H_{24}$	Undecane
C_3H_8	Propane	$C_{12}H_{26}$	Dodecane
C_4H_{10}	Butane	$C_{20}H_{42}$	Eicosane
C_5H_{12}	Pentane	$C_{30}H_{62}$	Triacontane
C_6H_{14}	Hexane	$C_{50}H_{102}$	Pentacontane
C_7H_{16}	Heptane		
C_8H_{18}	Octane		
C_9H_{20}	Nonane		

Usually the saturated hydrocarbons have four different types of carbon atoms in their molecules.

1. Primary (1°) *carbon atom* : A carbon atom attached to one *or no other* carbon atom is called primary (1°) carbon atom.

2. Secondary (2°) *carbon atom* : A carbon atom attached to two other carbon atoms is called secondary (2°) carbon atom.

3. Tertiary (3°) *carbon atom* : A carbon atom attached to three other carbon atoms is called tertiary (3°) carbon atom.

4. Quatenary (4°) *carbon atom* : A carbon atom attached to four other carbon atoms is known as quaternary (4°) carbon atom.

$$\begin{array}{ccccccccc} & & & & CH_3 & & CH_3 & & \\ & & & & | & & | & & \\ H_3C & — & CH_2 & — & CH & — & C & — & CH_3 \\ 1^o & & 2^o & & 3^o & & |\ 4^o & & \\ & & & & & & CH_3 & & \end{array}$$

The hydrogen atom attached to primary, secondary, tertiary carbon atoms are correspondingly named as primary, secondary and tertiary hydrogen atoms.

In case of isomeric alkanes from C_4H_{10} onwards, the following distinctions are made in naming the various isomers. The hydrocarbons having all the carbon atoms in a straight or continuous chain are termed normal or n-hydrocarbons.

e.g. : $H_3C\ CH_2\ CH_2\ CH_3$ is called as n-butane

Branched chain isomers having $\begin{array}{l} H_3C \diagdown \\ \qquad CH - CH_2 \\ H_3C \diagup \end{array}$ structure is called iso,

whereas others containing a quaternary carbon atom i.e.

$$\begin{array}{ccccc} & & CH_3 & & \\ & & | & & \\ H_3C & — & C & — & CH_2 \\ & & | & & \\ & & CH_3 & & \end{array}$$ structure is called as **neo.**

$$H_3C — CH_2 — CH_2 — CH_2 — CH_3$$

n-pentane

$$\begin{array}{l} H_3C \diagdown \\ \qquad CH — CH_2 — CH_3 \\ H_3C \diagup \end{array}$$

Isopentane

$$\begin{array}{ccccc} & & CH_3 & & \\ & & | & & \\ H_3C & — & C & — & CH_3 \\ & & | & & \\ & & CH_3 & & \end{array}$$

Neopentane

The univalent radicals which are fomred by the removal of one hydrogen atom of the paraffins are termed alkyl groups or radicals. The common names of some important alkyl groups are given below.

Structure	Name	Abbreviation
H_3C-	Methyl	Me
H_3C-CH_2-	Ethyl	Et
$H_3C-CH_2-CH_2-$	n-propyl	n-Pr
$(H_3C)_2CH-$	Isopropyl	Pr^i
$H_3C-CH_2-CH_2-CH_2-$	n-butyl	n-Bu
$(H_3C)_2CH-CH_2-$	Isobutyl	iso Bu
$H_3C-CH_2-CH(-)-CH_3$	sec. butyl	sec. Bu
$H_3C-C(CH_3)_2-$	tert. butyl	tert. Bu

3-3-2 IUPAC System

According to this system, the hydrocarbons in which all the carbon atoms are present in continuous chains are named by adding the suffix-*ane.* to the appropriate wordroot.

CH_4	Methane
H_3C-CH_3	Ethane
$H_3C-CH_2-CH_3$	Propane

In case of branched chain alkanes the largest possible continuous chain is selected and the corresponding wordroot followed by the suffix-ane is written down. Next the chain is numbered from that end which gives the lowest sum of numbers for the branching alkyl groups. The names of branching alkyl groups along with their attachment numbers are then prefixed before the wordroot in alphabetical order.

For example.

$$H_3C-\underset{\large |}{\overset{CH_3}{CH}}-CH_2-CH_3$$

1 2 3 4

2-Methylbutane

$$H_3C-\underset{CH_3}{\overset{CH_3}{C}}-CH_3$$

2,2-Dimethylpropane

$$\overset{6}{H_3C}-\overset{5}{CH_2}-\underset{C_2H_5}{\overset{4}{CH}}-\overset{3}{CH_2}-\underset{CH_3}{\overset{CH_3\ 2}{C}}-\overset{1}{CH_3}$$

4 - Ethyl -2, 2-dimethylthexane

3-4 ISOMERISM

The molecular structure of compounds plays an important role in the study of organic chemistry. During this we come across many instances where a given molecular formula represents two are more compounds having different properties. This is evidently due to some difference in the molecular structures of these compounds. *The compounds having same molecular formula but different molecular structures or different properties are called isomers. This phenomenon is known as isomerism.*

Types of Isomerism : Isomerism may be classified into two types

1) Structural isomerism

2) Sterio isomerism

3-4-1 Structural Isomerism

The compounds having the same molecular formula but different structural formulae are called structural isomers, and this phenomenon is called as stuctural isomerism. This isomerism may be further classified into five main types:

(i) Chain isomerism : The compounds having same molecular formula but different chain structures are called chain isomers, and this phenomenon is called as chain isomerism. For example C_5H_{12}.

$$H_3C-CH_2-CH_2-CH_2-CH_3$$

n-pentane

$$H_3C-\overset{CH_3}{\overset{|}{CH}}-CH_2-CH_3$$

isopentane

$$H_3C-\underset{CH_3}{\overset{CH_3}{C}}-CH_3$$

neopentane

n-pentane, isopentane and neopentane are structural isomers of pentane C_5H_{12}. The difference in these isomers is due to the difference in the nature of carbon atoms chain. In n-pentane it is a straight chain whereas in isopentane and neopentane it is a branched chain.

(ii) Position isomerism : *The compounds having the same molecular formula and same chain structure but different positions of substituents are called position isomers and the penomenon is called position isomerism.* For example a molecular formula C_3H_7Cl represents

$H_3C - CH_2 - CH_2 - Cl$ 1-Chloropropane

$H_3C - CH(Cl) - CH_3$ 2-Chloropropane

Both the compounds have straight chain formulae but differ in the position of the chlorine atom, which in the first compound is linked to the end carbon atom and with the middle carbon in the second.

Similarly $H_2C = CH - CH_2 - CH_3$ 1-Butene $H_3C - CH = CH \; CH_3$ 2-Butene

Both have straight formulae but differ in the position of the double bond.

(iii) Functional isomerism : The compounds having the same molecular formula but different functional groups are called functional isomers and this phenomenon is called functional isomerism. For example

Formula	*Functional Isomers*	
C_2H_6O	$H_3C - CH_2 - OH$ Ethyl alcohol	$H_3C - O - CH_3$ Dimethylether
C_3H_6O	$H_3C - CH_2 - CHO$ Propanaldehyde	$H_3C - CO - CH_3$ Acetone
$C_2H_4O_2$	H_3CCOOH Acetic acid	$HCOOCH_3$ Methyl formate
C_3H_9N	$H_3C - CH_2 - CH_2 - NH_2$ n-Propyl amine	$H_3C - CH_2 - NH - CH_3$ Methylethylamine
C_2H_3N	$H_3C - CN$ Methyl cyanide	$H_3C - NC$ Methyl isocyanide

(iv) Metamerism : This type of isomerism exhibited by the members of the same homologous series due to the difference in the nature of alkyl groups attached to the functional group or polyvalent atoms for example $C_4H_{10}O$.

$H_5C_2 - O - C_2H_5$ Diethylether

$H_3C - O - C_3H_7$ Methyl propyl ether

(v) Tautomerism : Tautomerism is a special type of functional isomerism. *This may be defined as a phenomenon by which a single compound exists in two readily interconvertible structures that differe markedly in the relative position of at least one atomic nucleus, generally hydrogen.* The two different structures are known as tautomers of each other. (Greek tauto = same meros = part).

In this type of isomerism both the isomers by omne and the same substance, are always present in equilibrium and it is a **dynamic isomerism.** Tautomerism is due to the oscillation or migration of a mobile hydrogen atom from one polyvalent atom to another with necessary rearrangement of linkages.

Aldehydes, ketones, esters etc., exhibit keto-enol tautomerism.

Example :

$$H_2C(H)-C(=O)-CH_3 \rightleftharpoons H_2C=C(OH)-CH_3$$

Keto-form of acetone Enol-form of acetone

The tautomer containing the carbonyl group is known as the keto-form, while the other containing a hydroxyl group attached to a doubly bonded carbon is termed as enol form. In acetone, keto form changes to the enol form by the migration of a proton to oxygen of the carbonyl group. Then a pair of electrons shifts from the C — H bond to C — C bond.

Similarly alkyl cyanides and nitro compounds also show tautomerism

$$H-C\equiv N \rightleftharpoons C^{-}\equiv N^{+}-H$$

Hydrocyanic acid Isohydrocyanic acid

$$H_3C-N^{+}(=O)-O^{-} \rightleftharpoons H_2C=N^{+}(OH)-O$$

Nitro-form Aci-form

Difference between tautomerism and resonance

Tautomerism should not be confused with resonance the two phenomenon are quite different and the imporatant poi,nts of difference between them can be enumerated as under.

(i) Tautomers involved in tautomerism are **definite** compounds may be separated and isolated whereas canonical forms of a resonance hybrid are **only hypothetical** and can't be isolated.

(ii) Tautomerism involved a change in the position of atom (generally hydrogen), while resonance involved a change in the position of the unshared or π electrons only.

(iii) The two tautomeric forms have different structures or functional groups, whereas the resonating structures have the same functional group.

(iv) Tautomers are in dynamic equilibrium with each other, whereas resonating structures are not in dynamic equilibrium.

(v) Tautomerism involved no change in the bond length, whereas resonance affects the bond length.

(vi) Tautomerism does not lower the energy of the molecule and hence does not play any role in stabilization of the molecule whereas resonance decreases the energy and hence increases the stability of the molecule.

(vii) Tautomerism indicated by the symbol $\rightleftharpoons$ whereas resonance by $\longleftrightarrow$

(viii) Tautomerism may be shown by both planar and non-planar molecules, whereas resonance occurs only in planar molecules.

3-4-2 Stereo Isomerism

The compounds having same molecular formula and same structural formula but different spatial arrangement of atoms or groups are called as stereoisomer and this phenomenon is called as stereoisomerism. The isomerism may be further classified into three types.

(i) Geometrical isomerism

(ii) Optical isomerism

(iii) Conformational isomerism.

(i) Geometrical isomerism (cis-trans isomerism) : The isomers which posses the same structural formula but differ in the spatial arrangement of the groups around double bond or single bond with restricted rotation are known as geometrical isomers and the phenomenon is called as geometrical isomerism. This isomerism is possible in alkenes and cyclo alkanes. This isomerism arises due to restricted rotation of carbon-carbon bond. If similar groups are on the same side of double bond it is called as cis-isomer, or similar groups present diagonally opposite side of the double bond is called a trans-isomer.

For example :

1. Two geometrical isomers with molecular formula C_4H_8

H_3C / CH_3 — C = C — H / H
cis-2-Butene

H_3C / H — C = C — H / CH_3
trans-2-Butene

2. Two geometrical isomers with molecular formula $C_4H_4O_4$

$$\begin{matrix} H & & H \\ & C = C & \\ HOOC & & COOH \end{matrix} \qquad \begin{matrix} H & & COOH \\ & C = C & \\ HOOC & & H \end{matrix}$$

Maleic acid (cis) Fumaric acid (trans)

Geometrical isomerism is possible only when each of the doubly bonded carbon atom has two different groups. The compounds like

$$\begin{matrix} A & & B \\ & C = C & \\ A & & C \end{matrix} \qquad \begin{matrix} A & & B \\ & C = C & \\ C & & B \end{matrix} \qquad \begin{matrix} A & & A \\ & C = C & \\ A & & A \end{matrix}$$

will not exhibit cis as trans isomerism.

(ii) Optical isomerism : This isomerism arises due to lack of symmetry. If four different groups are attached to a single carbon atom. *It exhibits non-super imposable mirror images which have essentially the same physical and chemical properties, except for their effect upon plane polarized light. If one isomer rotates the plane polarized light to the right the other may rotate equally to the left. These are known as* **dextro, 'd' and levo, 'I'** *isomers respectively.* Such a pair of isomers are known as **optical isomers** or **enantiomers** and this phenomenon is known as optical isomerism.

Optical isomers contain an **asymmetric carbon** atom in their molecules i.e. a carbon atom attached to four different atoms or groups. For example

$$\begin{matrix} & CH_3 & \\ & | & \\ H - & C & - OH \\ & | & \\ & COOH & \end{matrix} \qquad \begin{matrix} & COOH & \\ & | & \\ H - & C & - H \\ & | & \\ & COOH & \end{matrix}$$

Lactic acid (Optically active) malonic acid (Optically inactive)

Lactic acid having four different groups (— CH_3, — OH, — COOH, H) on the central carbon atom is optically active, while succinic acid having two similar atoms on the central carbon atom is optically inactive. A compound can exist in enatiomeric forms, if it has an asymmetric carbon atom and is devoid of plane of symmetry, centre of symmetry and alternating axis of symmetry. The compound which does not posses either of these elements of symmetry is asymmetric and hence is optically active and can exist in 'd' and 'l' forms, the number of optical isomers in a molecule containing 'n' number of different asymmetric carbon atoms is given by the relation 2^n.

(iii) Conformational isomerism : *The various configurations of a molecule that arise due to rotation of attached groups along a C — C single bond are known as conformers or rotamers. These are readily inter convertible. This phenomenon is known as conformation of conformational isomerism.*

The term conformation should not be confused with the configuration which relates to those spatial arrangements of the atoms of a molecule that can be changed only by the breaking and making of bonds where as the spatial arrangements in conformation are changed simply by rotation about a single bond.

Representation of conformers : The various methods adopted for projecting a molecule on a planar surface are (i) Fischer projection formula (ii) Sawhorse formula and (iii) Newmann projection formula.

Example : C_2H_6 (CH_3 — CH_3)

The two tetrahedral methyl groups can rotate about the carbon- carbon bond axis yielding several configurations, depending upon the angle of rotation. The angle of rotation is called the dihidral angle. The Fisher and sawhorse formulae for the conformation of ethane can be shown as

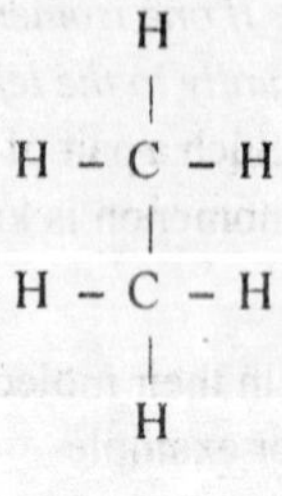

Fischer projection 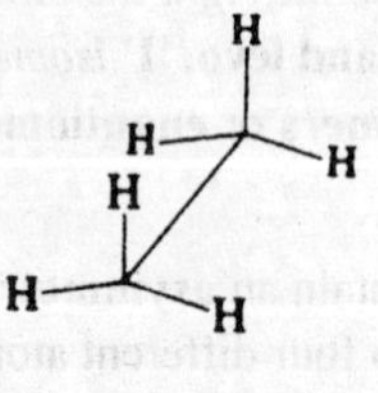 Eclipsed 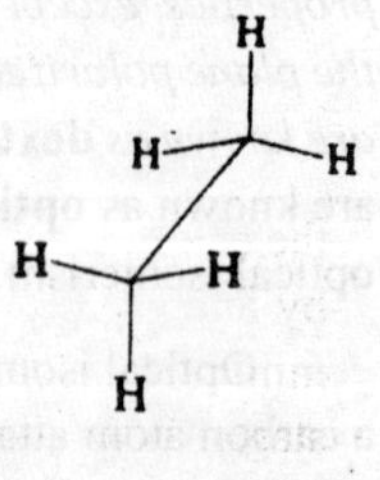 Staggered

Sawhorse formulae

The Newmann projection formulae for the conformation of ethane can be depicted as

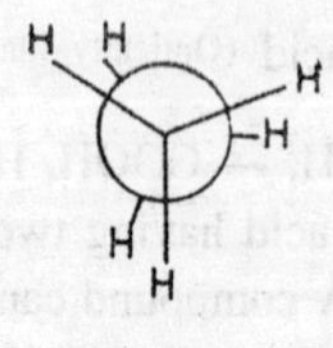

Eclipsed 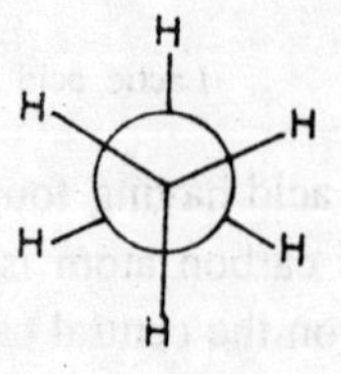 Staggered

Newmann projection formulae

(The bonds of the rear carbon in the eclipsed form are shown slightly displaced simply to make them visible).

The conformer in which the rear methyl group is completely eclipsed and only the one nearer is visible to eye is called as eclipsed form. And the conformer in which the rear methyl group has been rotated upside down and all the six hydrogen atom of the two methyl groups are as far apart as possible is called as staggered form.

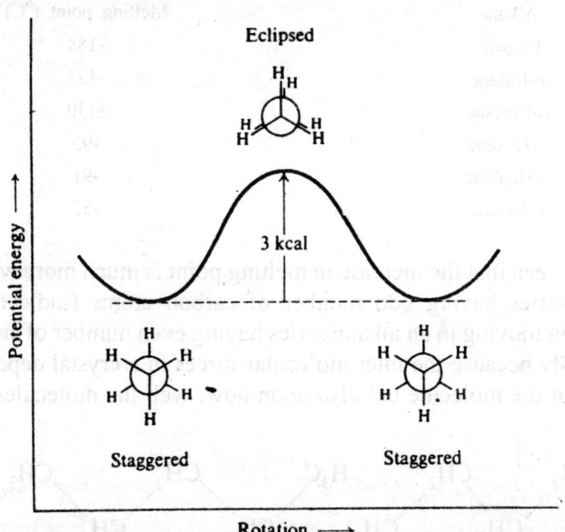

Fig. 3-1 Potential energy diagram of ethane molecule at various dihedral angles.

The rotation of the front methyl group by 60° about the C - C axis converts the eclipsed conformer into the staggered form. If the rotation is continued further by another 60°, a new eclipsed conformer results. Potential energy of ethane is minimum in staggered form, whereas maximum in eclipsed form. Thus the staggered conformer is most stable and has least potential energy. In the eclipsed conformer, the hydrogen atoms attached to each carbon are in direct opposition to each other. This permits minimum separation of the electrons of the six carbon-hydrogen bonds. This eclipsed conformation is, therefore of highest energy and has the lowest stability.

A plot of potential energy of ethane molecule as a function of rotation about carbon-carbon bond (dihedral angle) is shown in figure 3-1.

3-5 PHYSICAL PROPERTIES

Alkanes show a gradation in their physical properties.

3-5-1 Physical State

The lower members of alkanes containing up to four carbon atoms are colorless. The higher normal alkanes having five to seven carbon atoms are colorless liquids and beyond that they are colorless solids at room temperature.

3-5-2 Melting points

The melting points of alkanes do not show a very smooth gradation with the increase in molecular size. For example, let us consider, the melting point of propane to n-octane.

Alkane	Melting point (°C)
Propane	-188
n-Butane	-135
n-Pentane	-130
n-Hexane	-95
n-Heptane	-91
n-Octane	-57

It may be seen that the increase in melting point is much more while moving in an alkane series having odd number of carbon atoms (indicated by **bold** lettering) than in moving in an alkane series having even number of carbon atoms. this is essentially because the inter molecular forces in a crystal depend not only upon the size of the molecule but also upon how, well the molecules fit into the crystal lattice.

H_3C CH_2 CH_2 CH_2 CH_2 CH_3

n-Hexane

(even number of C atoms)

H_3C CH_2 CH_2 CH_2 CH_2 CH_2 CH_3

n-Heptane

(odd number of C atoms)

3-5-3 Boiling points

The boiling points of n-alkanes increase regularly with the increasing number of carbon atoms. This can be explained in terms of intermolecular forces of attraction. Amongst the isomeric alkanes there are small differences in the boiling points. The branched chain isomers have invariably lesser boiling points than the normal compounds and the greater the branching, the lower is the boiling point of branched isomer, for example,

Alkane	Boiling point
n-Pentane	36
isopentane	28

3-5-4 Solubility

Due to nearly same electronegativities of carbon and hydrogen atoms present in alkanes and due to their symmetrical shapes, these molecules are non-polar in character. As such they are insoluble in polar solvents such as water but dissolve freely in non-polar solvents such as ether. The solubilities decrease with the increase in molecular weights.

3-6 CHEMICAL PROPERTIES

Alkanes are quite inert towards common reagents because of their saturated character and thus are fairly unreactive. Carbon to carbon and carbon to hydrogen linkages in alkanes are usually strong because of their pure covalent nature. Thus there are no centers of high or low electron density in the alkanes. This rules out the attack of electrophilic or nucleophilic reagents on these hydrocarbons. As a general rule, therefore most of the chemical reactions of alkanes are free radical substitution reactions.

3-6-1 Halogenation

Alkanes react with chlorine and bromine in the presence of sunlight or UV light or at high temperature resulting in the successive replacement of hydrogen atoms with halogens. Extent of halogenation depends on the amount of halogen used.

Example : Chlorination of methane.

$$CH_4 + Cl_2 \xrightarrow[\text{or high temperature}]{\text{UV light}} \underset{\text{Methyl chloride}}{CH_3Cl} + HCl$$

$$CH_3Cl + Cl_2 \xrightarrow{\text{"}} \underset{\text{Methylene chloride}}{CH_2Cl_2} + HCl$$

$$CH_2Cl_2 + Cl_2 \xrightarrow{\text{"}} \underset{\text{Chloroform}}{CHCl_3} + HCl$$

$$CHCl_3 + Cl_2 \xrightarrow{\text{"}} \underset{\text{Carbon tetrachloride}}{CCl_4} + HCl$$

Bromination proceeds in almost identical manner but is less vigorous than chlorination. Reaction with fluorine is too violent and results in a mixture of carbon, fluorinated alkanes and hydrofluoric acid. Iodine does not substitute directly because reaction is too slow and reversible, but in the presence of oxidizing agents such as iodic acid, nitric acid etc., iodination acid is possible.

$$CH_4 + I_2 \longrightarrow CH_3I + HI$$

$$5HI + HIO_3 \longrightarrow 3I_2 + H_2O$$

Mechanism : Halogenation of alkanes proceed through a free radical mechanism and involves the following steps :

(i) Chain initiation : This is the first step of the reaction and involves the generation of chlorine free radicals by the homolytic cleavage of chlorine molecule. This reaction is an endothermic reaction and requires 58 kcal mole^{-1} of energy, which is supplied either by irradiating it to UV light or by heating the reaction mixture to high temperature.

$$Cl{-}Cl \xrightarrow[\text{or high temperature}]{\text{UV. light}} 2\dot{C}l \quad (\Delta H = +58 \text{ k. cal/mole})$$

(ii) *Chain propagation step :* In this step, the chlorine free radical formed in the first step abstracts a hydrogen atom from alkane molecule and forms an alkyl radical. The alkyl radical on collision with another molecule of chlorine, forms alkyl chloride and a chlorine free radical set free to continue the chain.

(a) $H_3C\ H + Cl. \longrightarrow H_3C. + HCl \quad (\Delta H = -1 \text{ kcal mole})$

(b) $H_3\dot{C} + Cl\ Cl \longrightarrow H_3C\ Cl + \dot{C}l \quad (\Delta H = -23 \text{ kcal mole})$

The steps (a) and (b) are repeated over and over again.

(iii) *Chain termination step :* The above chain propagation reaction stops when any two free radicals combine as follows :

$$\dot{C}l + \dot{C}l \longrightarrow Cl_2$$

$$H_3\dot{C} + \dot{C}l \longrightarrow CH_3Cl$$

$$H_3\dot{C} + \dot{C}H_3 \longrightarrow H_3C\ CH_3$$

Orientation of halogenation : The ease of abstraction of hydrogen atom from the alkane to yield alkyl free radicals has been found to follow the sequence > Tertiary Secondary > Primary

For example :

$$H_3C{-}CH_2{-}CH_3 \xrightarrow{\dot{C}l} H_3C{-}\dot{C}H{-}CH_3 \text{ (More stable)} \xrightarrow{Cl_2} H_3C{-}CH(Cl){-}CH_3$$

2-Chloropropane
Major product

$$H_3C{-}CH_2{-}CH_3 \xrightarrow{\dot{C}l} CH_3{-}CH_2{-}\dot{C}H_2 \text{ (Less stable)} \xrightarrow{Cl_2} H_3C{-}CH_2{-}CH_2Cl$$

1-Chloropropane
Minor product

3-6-2 Nitration

Nitration is replacement of hydrogen by a nitro (NO_2) group. Alkanes especially the long chain members, on heating with nitric acid yield nitroalkanes.

$$R - H + HNO_3 \xrightarrow{\Delta} R\ NO_2 + H_2O$$

Recently developed technique of nitration in vapour phase has made it possible the nitration of lower alkanes in good yields.

$$H_3C - H + HONO_2 \xrightarrow{400^\circ C} H_3C\ NO_2 + H_2O$$

However, at high temperatures partial cleavage of alkanes results in a mixture of nitro alkanes.

$$H_3C{-}CH_2{-}CH_3 \xrightarrow[450^\circ C]{HNO_3} CH_3{-}CH_2{-}CH_2{-}NO_2 + CH_3{-}\underset{\underset{NO_2}{|}}{CH}{-}CH_3 +$$

$$C_2H_5NO_2 + H_3C\ NO_2$$

3-6-3 Sulphonation

When a normal paraffin from hexane onwards is treated with oleum, a hydrogen atom is replaced by a sulphonic acid group. The process of replacement is known as sulphonation. The ease of replacement of hydrogen atom is Tertiary Secondary Primary.

For example :

$$C_6H_{13}\ H + HO\ SO_3H \xrightarrow[400^\circ C]{SO_3} C_6H_{13}SO_3H + H_2O$$

3-6-4 Oxidation

(i) Combustion : Alkanes readily burn with a non-Luminous flame in excess of air of oxygen to give CO_2 and H_2O

$$CH_4 + 2O_2 \longrightarrow CO_2 + 2H_2O + \text{Energy}$$

(ii) Thermal decomposition, Pyrolysis or Cracking : When vapours of higher alkanes are passed through a hot metal tube, it breaks up to yield smaller molecules.

For example :

$$\underset{\text{Propene}}{C_3H_6} + H_2 \longleftarrow \underset{\text{Propane}}{C_3H_8} \longrightarrow \underset{\text{Methane}}{CH_4} + C_2H_4 \longrightarrow \underset{\text{Ethene}}{CH_4 + C}$$

Mechanism of cracking is still obscure. Out of the many mechanisms suggested the one involving formation of free radicals is highly favoured.

$$H_3C{-}CH_2{-}CH_3 \longrightarrow CH_3{-}CH_2{-}\dot{C}H_2 + \dot{H} \longrightarrow H_3C{-}CH = CH_2 + H_2$$

$$H_3C{-}\dot{C}H_2 + \dot{C}H_3 \xrightarrow{\text{Disproportionation}} CH_2 = CH_2 + CH_4$$

3-7 CONFORMATIONS OF ETHANE

The carbon-carbon bond is ethane is a single bond because it is composed of one pair of electrons in the bonding orbital. This bonding orbital allows the free rotation of carbon nuclei without causing any damage to the bond.

To show the three-dimensional character of molecules, many structures for ethane, such as those in Fig. 3.2 can be drawn.

(a) Ball and Stick

(b) Sawhorse structure structure structure I

(c) Newman projection formula

(d) Ball and Stick

(e) Sawhorse structure structure structure II

(f) Newman projection formula

Fig. 3-2 Methods of showing two possible confirmations of ethane; Structure I-A staggered form and Structure II-A eclipsed form

The structures I and II are only two of an unlimited number of possible spatial orientations of the six hydrogen atoms. In arriving at these structures, the hydrogens attached to C_2 have been kept in the same position, while the hydrogens on C_1 have been rotated by 60°

There are several common ways to depict three-dimensional structures on paper. The so-called *Sawhorse structures* represented by I-b and II-e in Fig. 3-2 bear a great resemblance to the ball and stick models, I-a and II-d, of ethane shown in the same figure. I-c and II-f of Fig. 3-2 are called the **Newman projection formulae.** These are especially useful in showing confirmation structures in a simple way. The conventions for drawing Newman projection formulae are as follows.

(i) Sight down a C — C single bond so that the two carbon atoms are superimposed on one other.

(ii) Draw the atoms seen on the first carbon atom of the C C bond and attach them to a common point such that they are 120° apart in the projection.

(iii) The atoms of the second carbon atom, (the atom that is directly behind the first one is not seen) are attached to a circle (they too will be 120° apart in the projection).

The conventions will become more familiar by inspecting molecular models and drawing Newman projection formulae for them. In the Newman projection formula I-c, we notice that the hydrogen atoms on C_1 and C_2 are placed 60° apart, while they are placed at 120° in the projection formula II-c.

Because of the electronic structure of the σ-bond, we can say that there is free rotation about the carbon-carbon single bond. This concept of free rotation, represented for ethane differ only in ithe three-dimensional location of the hydrogen atom in space. These arrangements of atoms as well as an infinite number of others, exist and are converted from one to another through free rotation about the carbon-carbon single bond, we call such structures as conformations.

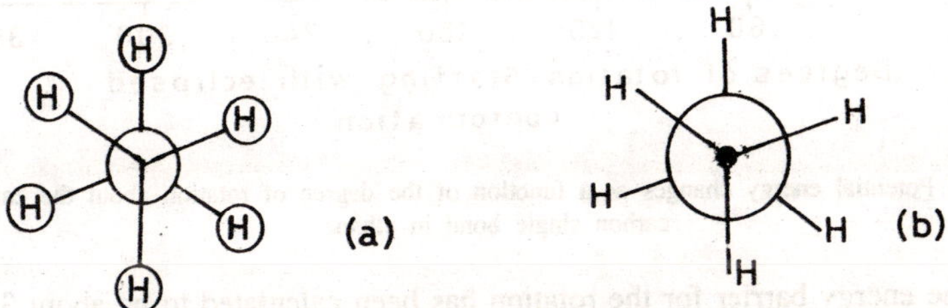

Fig. 3-3 Method for drawing the Newman projection formula for the staggered conformation of ethane.

More generally, any two structures that differ only in their arrangement of atoms in space and are inter-convertible by rotation about single bonds are conformers of each other. The two particular conformations discussed are given the names **staggered and eclipsed** to refer to the three-dimensional arrangement of the atoms. For ethane, the staggered form of conformation, a, in Fig. 3-3 is more stable than any other conformation. In this conformation each hydrogen on C_1 are separated by a maximum possible distance from each hydrogen on C_2. The eclipsed conformation in Fig. 3-3, has the hydrogen atoms on C_1 and C_2, are as close to one another as possible. The end view of ethane in Fig. 5-3 reveals the hydrogens on the first carbon to be completely superimposed or eclipsed with those on the second carbon atom. The myriad of conformations between the staggered and eclipsed forms are called **skew conformations.**

There is continuous rotation about the carbon-carbon single bond with a limitless number of possible conformations. As a result, all possible conformations are in equilibrium with one another. Each slight change in rotation about the single bond produces a new conformation. In most instances, conformation structures cannot be isolated as discrete and stable. We can, however, speak of their relative stabilities. We expect the staggered conformation of ethane I in Fig. 3-1 to be more stable than the eclipsed conformation II and thus preferred because the electron clouds associated with the hydrogen atoms and the bonding electron in the C H σ-bonds are at a maximum distance apart in the staggered form. Electrostatic interactions and perhaps to some extent steric interactions arising out of physical crowding of the atoms are minimised in the staggered conformation. In fact, more ethane molecules, at any instant, are in this staggered conformation than in any other possible conformation.

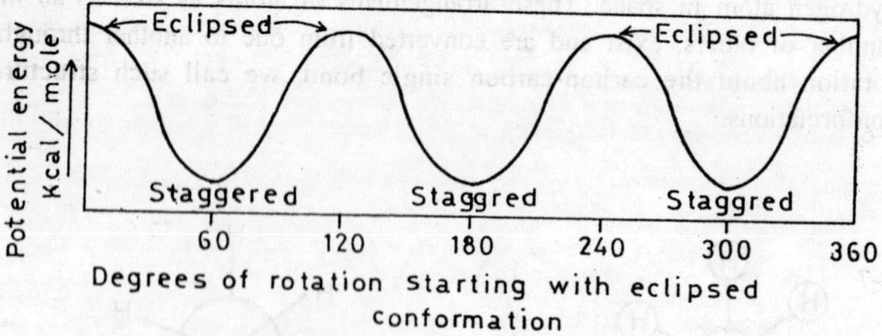

Fig. 3-4 Potential energy changes as a function of the degree of rotation about the carbon-carbon single bond in ethane.

The energy barrier for the rotation has been calculated to be about 3 kcal mole^{-1} in ethane. This amount of energy is required to rotate the hydrogen on the one methyl go from their most stable staggered conformation to their least stable eclipsed conformation.

The energy barrier of 3 kcal mole^{-1} is relatively small at room temperature, where the thermal energy of the surroundings exceeds this value. Hence at RT we may assume that free rotation exists about carbon-carbon single bonds. Ethane interconverts among its various conformers at a rate greater than 10^6 times per second at room temperature. It has been determined that the energy barrier must be greater than 20-25 kcal mole^{-1} to prevent thermal interconversions at room temperature. this renders the respective conformations isolable at sufficiently low temperatures. Even so microwave spectroscopy studies on ethane gas at room temperature confirm this prediction that the preferred conformation is the most stable staggered form.

3-8 MULTIPLE-CHOICE QUESTIONS

3-8-1 Which one of the following has low boiling point.

A) Isohexane B) n-hexane C) Neohexane D) n-heptane

3-8-2 The major product formed by the monobromination of propane is,

A) n-Propyl bromide B) Isopropyl bromide

C) Both D) None

3-8-3 How many types of carbon atoms are present in isohexane

A) 3 B) 2 C) 4 D) 5

3-8-4 The halogen which is most reactive in the halogenation of alkanes under sunlight is

A) Bromine B) Chlorine C) Iodine

3-8-5 Which of the following Isomerism is not possible with the molecular formula C_4H_{10}.

A) Conformational Isomerism B) Geometrical Isomerism

C) Chain Isomerism D) None

3-8-6 Major product formed when propane is subjected to nitration around 400°C

A) 1-Nitropropane B) Nitroethane

C) 2-Nitropropane D) Nitromethane

3-8-7 The most stable conformer of ethane is

A) Eclipsed B) Staggered C) Skew D) None

3-8-8 Alylalcohol is an isomer of

A) Acetone B) Propanal

C) Both D) None

3-8-9 Keto-enol forms of acetylacetone are

A) Geometrical Isomers B) Optical Isomers

C) Tautomers D) Position Isomers

3-8-10 Resorcinol and para Hydroquinone are

A) Functional Isomers B) Chain Isomers

C) Position Isomers D) Metamers

3-8-11 Gauche and anti conformers of n-butane are

A) Tautomers B) Resonance Structures

C) Dynamic Stereoisomers D) Chain Isomers

3-8-12 Salicylaldehyde is isomer of

A) p-hydroxy benzaldehyde B) Benzoic Acid

C) Both D) None

3-8-13 Which of the following exhibits optical activity

A) 1-Bromopropanol B) 2-Bromopropanol

C) Both D) 3-Bromopropanol E) All

3-8-14 Allene is an isomer of

A) Allylene B) Cyclopropene

C) Both D) None

3-8-15 Crotonic acid Exhibits

A) Geometric Isomerism B) Functional Isomerism

C) Chain Isomerism D) Position Isomerism E) All

3-8-16 o, m and p-xylenes are ____________________ Isomers.

3-9 SHORT AND ESSAY-TYPE QUESTIONS

3-9-1 Formulate the mechanism of chlorination of methane. (O.U., Apr. 1990)

3-9-2 Write the structure for all the isomers of the formula C_6H_{14}. Give IUPAC names for all the isomers. Whcih one or those isomers has lowest boiling point and which one has highest boiling point? (O.U., Oct. 1990)

3-9-3 Write the structures of isomeric pentanes. Give their IUPAC name. Which of the isomeric pentanes give only one monochloro-substituted product? (O.U., Oct. 1990)

3-9-4 Give mechanism of the halogenation of alkanes. (O.U., Oct. 1990)

3-9-5 Write Newman projection formulae for different conformers of n-butane. (A.U., Apr. 1990)

3-9-6 What is conformational isomerism? Write in detail with reference to ethane. (A.U., Jun. 1990)

3-9-7 Write a note on petroleum.

3-9-8 Explain the mechanism of halogenation of alkanes.

3-9-9 Write various possible structural formulae of alkanes with M.F. C_6H_{14} and write the IUPAC names.

3-9-10 An alkane with M.Wt. 114 formed only one monochloro substitution product. Suggest a structure for the alkane.

3-9-11 What is conformational isomerism? Write in detail with reference to ethane.

3-9-12 Explain geometrical isomerism, with an apropriate example? (O.U., Mar, 1990)

3-9-13 Define "conformation". Draw the Newman projection formulae of ethane? (O.U., Oct. 1989)

3-9-14 What is meant by conformational isomerism. Illustrate with a suitable example?
(O.U., Mar. 1989)

3-9-15 What do you understand by the term "isomerism"? Distinguish between structural isomerism and stereo isomerism ? Describe, with one example each, any three types of structural isomerism?
(O.U., Oct. 1988)

3-9-16 Explain conformational isomerism with reference to the ethane molecule?
(O.U., Apr. 1988)

3-9-17 Write the structures of the possible isomers in each of the following compounds and indicate the type of isomerism exhibited (a) 2-Butene (b) Pentane?
(O.U., Apr. 1988)

3-9-18 Using appropriate structural formulae illustrate
(a) a pair of position isomers
(b) a pair of conformational isomers
(c) a pair of sturctural isomers
(O.U., Apr. 1987)

3-9-19 Write the different isomers of C_6H_{14}. Give the IUPAC names of these isomers.
(O.U., 1986)

3-9-20 What do you understood by Cis-trans isomerism? Illustrate the same with reference to $C_2H_2Cl_2$?
(O.U., Oct. 1985)

3-9-21 Give an example to show Keto-enol tautomerism?
(O.U., Apr. 1984)

3-9-22 Explain giving the structure, the type of isomerism exhibited by the following.
(i) Propionic acid and methylacetate
(ii) Trimethylamine and n-propylamine
(iii) 1,2-dibromoethane
(O.U., Apr. 1991)

3-9-23 Give one example each for optical isomerism and tautomerism.
(O.U., Apr. 1991)

3-9-24 Define the conformational isomerism and write the possible conformers of n-butane.

3-9-25 Explain the type of isomerism exhibited by the following
(a) Pentane and neopentane
(b) Ethylene chloride and ethylidene chloride
(c) 1,2-dibromoethane
(O.U. Apr., 1989)

ANSWERS

3-8-1 C	**3-8-2** B	**3-8-3** D	**3-8-4** B	**3-8-5** B	**3-8-6** C
3-8-7 B	**3-8-8** C	**3-8-9** C	**3-8-10** C	**3-8-11** C	**3-8-12** C
3-8-13 C	**3-8-14** C	**3-8-15** E			

(c) a pair of structural isomers

(iii) 1,2-dibromoethene

4

CYCLOALKANES

4-1 INTRODUCTION

There are large number of compounds which contain closed rings comprised of carbon atoms only. These compounds are known collectively as *carbocyclic* or *homocyclic compounds.* This group falls in a class of compounds which resemble the aliphatic compounds in many ways, and hence assume the name *alicyclic compounds* (aliphatic cyclic compounds). The saturated hydrocarbons of this family are known as cycloalkanes and are designated by the general formula C_nH_{2n}. Cycloalkanes contain two hydrogen atoms less than the corresponding alkanes. Their general formula is same as the one for alkenes.

4-2 NOMENCLATURE

There are two systems for naming cycloalkanes.

(i) Common system : Since the saturated alicyclic hydrocarbons contain a number of methylene groups joined together to form a ring, they are known as the *polymethylenes,* the number of carbon atoms in the ring being indicated by a Greek or Latin prefix, i.e., *tri, tetra, penta* etc. For example,

Trimethylene

Hexamethylene

(ii) IUPAC system : According to the IUPAC system, the saturated monocyclic hydrocarbons take the names of the corresponding open chain saturated hydrocarbons, preceded by the prefix *cyclo.* For example,

Cyclobutane

Cyclohexane

Substitutents on the ring are named first and then their positions are indicated by numbers such that the carbon bearing the substitutents will have lowest possible number.

Chloro-cyclobutane

1,1-Dimethyl cyclopentane

3-Ethyl-1-methyl cyclopentane

In simple cycloalkanes and cycloalkynes the doubly and triply bonded carbons are considered to occupy position between C_1 and C_2 and in numbering these bonds are given first priority. For example,

3-Ethylcyclo-pentene

1, 3-Cyclohexa-diene

For convenience, aliphatic rings are often represented by simple geometric figures. A triangle for cyclopropane, a square for cyclobutane, a pentagon for cyclopentane, a hexagon for cyclohexane, and so on. It is understood that two hydrogens are located at each corner of the figure unless some other group is indicated. For example,

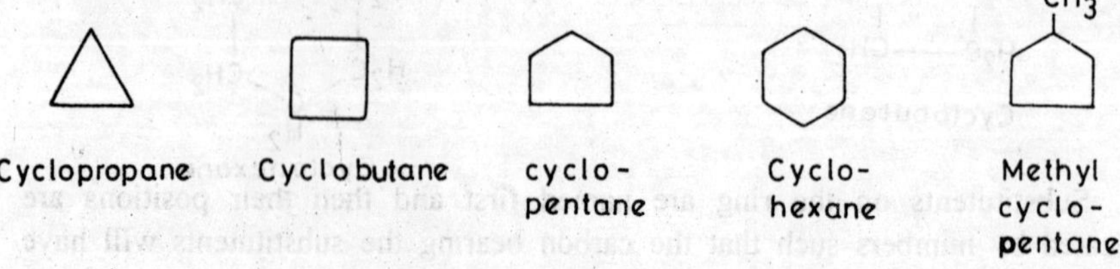

4-3 PREPARATION

Preparation of cyclic compounds from other aliphatic compounds, generally involves two stages.

(a) Conversion of some open-chain compound or compounds into a compound that contains a ring, the process being called cyclization.

(b) Conversion of the cyclic compound thus obtained into the desired compound

There are many methods of forming cycloalkanes, but only a few of the most common are given below.

4-3-1 Freund's Method

Dichloro or dibromo alkanes having terminal halogen atoms, when treated with sodium or zinc produce corresponding cycloalkanes. For example,

$$H_2C(CH_2Br)_2 + Zn \xrightarrow[\Delta]{\text{ethanol}} \text{(cyclo-)}\ H_2C{<}\begin{matrix}CH_2\\|\\CH_2\end{matrix} + ZnBr_2$$

1, 3 - dibromo- propane → Cyclopropane

The method is an extension of Wurtz reaction and may be regarded as an internal Wurtz reaction. It cannot be applied for the formation of a compound containing more than six carbon atoms as the yields are poor.

4-3-2 Perkins method

α, co-dihaloalkanes on condensation with sodio malonic ester give alicyclic/carboxylic esters, which on hydrolysis followed by decarboxylation yield corresponding cycloalkanes.

$$\begin{matrix}H_2C-Br\\|\\H_2C-Br\end{matrix} + NaCH(COOEt)_2 \xrightarrow[\text{(2-steps)}]{NaOEt} \begin{matrix}H_2C\\|\\H_2C\end{matrix}{>}C(COOEt)_2$$

$$\xrightarrow{\text{KOH hydrolysis}} \begin{matrix}H_2C\\|\\H_2C\end{matrix}{>}C(COOH)_2 \xrightarrow[-2\,CO_2]{\Delta} \begin{matrix}H_2C\\|\\H_2C\end{matrix}{>}C{<}\begin{matrix}H\\H\end{matrix}$$

Cyclopropane

4-3-3 Wislicenus method

When calcium or barium salts of dicarboxylic acids are subjected to dry distillation, a cyclic ketone having one carbon atom less, is formed. Such a cyclic ketone on Clemmensen's reduction using Zn-Hg and HCl, forms a cycloalkane.

$H_2C-CH_2-C-O^{\ominus}$ / $H_2C-CH_2-C(=O)-O^{\ominus}$ + Ba^{+2} $\xrightarrow{\Delta}$ H_2C-CH_2 / H_2C-CH_2 $>C=O$ + $BaCO_3$

Cyclopentanone

$\xrightarrow[HCl]{Za-Hg}$ H_2C-CH_2 / CH_2 / H_2C-CH_2

4-3-4 Dieckmann Cyclization

Esters of dicarboxylic acids undergo intramolecular Claisen condensation in the presence of sodium or sodium ethoxide yielding cyclic β-keto ester. Subsequent hydrolysis and heating (decarboxylation) of cyclic β-keto esters gives rise to cycloalkanones, which can be converted into cycloalkanes by the Clemmensen reduction.

For example, cyclopentane can be synthesized from ethyl adipate as given below.

$H_2C-CH_2-COOEt$ / $H_2C-CH_2-COOEt$ $\xrightarrow{NaOEt}$ $H_2C-CH(COOEt)$ / $C=O$ / H_2C-CH_2 $\xrightarrow{H_3O^{\oplus}}$ $H_2C-CH(COOH)$ / $C=O$ / H_2C-CH_2

$\xrightarrow[-CO_2]{\Delta}$ H_2C-CH_2 / $C=O$ / H_2C-CH_2 $\xrightarrow{Zn-Hg/HCl}$ H_2C-CH_2 (C H_2) / CH_2 / H_2C-CH_2 (C H_2)

This reaction proceeds by a mechanism similar to that of Claisen's condensation.

4-3-5 Thorpe-Zeigler's method

When aliphatic α, ω-dinitriles are treated with bases like metal salts of secondary amines, [eg: $LiN(Et)_2$], intramolecular cyclization results in the formation of cyclic imino compounds, which on hydrolysis give the corresponding ketones in good yield.

$$(H_2C)_n\begin{matrix} CH_2-CN \\ CH_2-CN \end{matrix} \xrightarrow{LiN\,Et_2} (H_2C)_n\begin{matrix} CH_2 \\ CH(CN) \end{matrix}C{=}NH \xrightarrow{\text{acid hydrolysis}}$$

$$(H_2C)_n\begin{matrix} CH_2 \\ CH(COOH) \end{matrix}C{=}O \xrightarrow[\Delta]{-CO_2} (H_2C)_n\begin{matrix} CH_2 \\ CH_2 \end{matrix}C{=}O \xrightarrow[HCl]{Zn-Hg} (H_2C)_n\begin{matrix} CH_2 \\ CH_2 \end{matrix}CH_2$$

Cycloalkane

The mechanism of the reaction is similar to that of Dieckmann cyclization, $LiN(Et)_2$ acting as a strong base. Large dilution is required to ensure intramolecular reaction. At present this is the best method of preparing macro cyclic compounds.

4-3-6 Acyloin Condensation

In this method a dicarboxylic ester is treated with metallic sodium in a large volume of benzene or toluene followed by protic solvents. Large dilution is required to ensure intramolecular condensation as against intermolecular reaction. The mechanism of the reaction is as follows :

$$(H_2C)_n\begin{matrix}CH_2-\overset{O}{\overset{\|}{C}}-OEt\\CH_2-\underset{O}{\underset{\|}{C}}-OEt\end{matrix}\xrightarrow{2\,Na}(CH_2)_n\begin{matrix}CH_2-\overset{O^{\ominus}}{\overset{|}{\dot{C}}}-OEt\\CH_2-\underset{O^{\ominus}}{\underset{|}{\dot{C}}}-OEt\end{matrix}$$

$$\downarrow$$

$$(H_2C)_n\begin{matrix}CH_2-\dot{C}=O\\CH_2-\dot{C}=O\end{matrix}+2\,NaOEt\longleftarrow(H_2C)_n\begin{matrix}CH_2-\overset{O^{\ominus}}{\overset{|}{\dot{C}}}-OEt\\CH_2-\underset{O^{\ominus}}{\underset{|}{\dot{C}}}-OEt\end{matrix}$$

$$\downarrow 2\,Na$$

$$(H_2C)_n\begin{matrix}CH_2-C-ONa\\\|\\CH_2-C-ONa\end{matrix}\xrightarrow{H^{\oplus}}(H_2C)_n\begin{matrix}CH_2-C-OH\\\|\\CH_2-C-OH\end{matrix}\xrightarrow[\text{cation}]{\text{Keto-end tantomen-}}(H_2C)_n\begin{matrix}CH_2-C=O\\|\\CH_2-C-OH\end{matrix}$$

acyloin

4-3-7 Reduction of Aromatic Compounds

On catalytic hydrogenation, aromatic compounds yield the corresponding alicyclic compounds. For example,

$$\text{Benzene}+3H_2\xrightarrow[25\text{ atm}]{\text{Ni. }150-250\,°C}\text{Cyclohexane}$$

4-3-8 Diels-Alder Reaction

This reaction involves the addition of an alkene molecule called a *dienophile* to a conjugated *diene* resulting in the formation of an adduct. The adduct is usually a six membered ring. The addition takes place at 1, 4-positions.

$$\begin{matrix}HC=CH_2\\|\\HC=CH_2\end{matrix}+\begin{matrix}CH_2\\\|\\CH_2\end{matrix}\underset{}{\overset{\Delta}{\rightleftharpoons}}\text{Cyclohexene}$$

1, 3 Butadiene (a diene) — Ethylene (a dienophile) — Cyclohexene (adduct)

$$\text{Cyclohexene} \xrightarrow[\Delta]{H_2/Ni} \text{Cyclohexane}$$

Cyclohexane

4-4 PHYSICAL PROPERTIES

(a) The first two members, i.e., cyclopropane and cyclobutane are gases, while higher cycloalkanes are colorless liquids having pleasant odour.

(b) Their B.P., M.P., and specific gravities show a gradual increase with the increase in molecular weight. However the B.P. of cycloalkanes and their specific gravities are higher than those of isomeric alkenes and even corresponding alkanes. This may probably be due to the fact that Zig-Zag structures of alkane molecules are constantly fluttering and making it difficult for the molecules in the solid state, to be closely packed there by lowering their melting points as compared to cycloalkanes where the more rigid ring structure prevents this fluttering to a great extent. Similarly, this fluttering makes the alkane molecules, in the liquid state more stable towards their neighbors. In other words, the intermolecular attractions in alkanes are less than those in corresponding cycloalkanes hence the lower boiling point of the former.

(c) They are insoluble in water but soluble in ether, alcohol, etc.

4-5 CHEMICAL PROPERTIES

Cycloalkanes resemble alkanes in their chemical behaviour. They do not react with acids, bases, oxidizing agents and reducing agents. Like alkanes cycloalkanes undergo substitution reaction. However, cyclopropane and cyclobutane also undergo ring opening reactions and give addition products.

4-5-1 Substitution Reactions

Cycloalkanes react with Cl_2 and Br_2 in the presence of diffused sunlight or ultraviolet light or when heated to give a monochloro or monobromo product respectively.

$$\text{(cyclo-}H_2C\text{–}CH_2\text{–}H_2C\text{)} + Cl_2 \xrightarrow{\text{Light}} \text{(cyclo-}H_2C\text{–}CHCl\text{–}H_2C\text{)} + HCl$$

Cyclopropane Chlorocyclopropane

$$\begin{array}{c} H_2 \\ C \end{array} \quad H_2C\text{—}CH,\ H_2C\text{—}CH_2 \text{ (ring)} + Br_2 \xrightarrow{300^\circ C} \text{(ring with } CH\,Br) + HBr$$

Cyclopentane → Bromocyclopentane

4-5-2 Addition reactions

Besides the free radical substitution reactions that are characteristic of cycloalkanes and of alkanes in general, cyclopropane and cyclobutane ring systems, also undergo addition reactions.

$$\text{Cyclopropane} \xrightarrow{Ni/H_2\ 80^\circ C} H_2\underset{H}{C}-CH_2-\underset{H}{C}H_2$$

$$\text{Cyclopropane} \xrightarrow{Cl_2/FeCl_3} H_2\underset{Cl}{C}-CH_2-\underset{Cl}{C}H_2$$

$$\text{Cyclopropane} \xrightarrow{Conc.\ H_2SO_4} H_2\underset{H}{C}-CH_2-\underset{OH}{C}H_2$$

$$\text{Cyclopropane} \xrightarrow{HX} H_2\underset{H}{C}-CH_2-\underset{X}{C}H_2$$

HX = HCl, HBr or HI

In each of these reactions a carbon-carbon bond is broken, and the two fragments of the reagents appear at the end of the propane chain.

In general, cyclopropane undergoes addition less readily than propylene. For example, chlorination of cyclopropane requires a Lewis acid catalyst to

$$\text{Cyclobutane} \xrightarrow{H_2/Ni\ 120^\circ C} H_2\underset{H}{C}-CH_2-CH_2-\underset{H}{C}H_2$$

$$\text{Cyclobutane} \xrightarrow{HX} H_2\underset{H}{C}-CH_2-CH_2-\underset{H}{C}H_2$$

HX = HI

polarise the chlorine molecule Cyclobutane does not undergo most of the ring opening reactions of cyclopropane, it is hydrogenated, but only under more vigorous conditions than those required for cyclopropane.

Thus cyclobutane undergoes addition less readily than cyclopropane. However, higher cycloalkanes are quite stable and are unaffected by H_2, HI, etc.

In order to explain the ring-opening tendency of cyclopropane and cyclobutane Adolf Von Baeyer proposed a theory, known as *Baeyer's Strain Theory.*

Baeyer's strain theory : Baeyer's strain theory is based on following assumptions :

He based his theory on the classical Le Bel and Van't Hoff's hypothesis that the four valencies of a carbon atom are directed towards the corners of a regular tetrahedron therefore, the normal angle between any pair of bonds of a carbon atom is $109^{\circ}\ 28'$. Baeyer postulated that any deviation of bond angles from the normal tetrahedral angle would impose a condition of internal strain on the ring.

He also assumed that all cyclic rings were planar and thus calculated the angles through which each of the valency bonds was deflected from the normal direction, in the formation of the various rings. This is called the *angle strain* which determines the stability of the ring.

In cyclopropane, the three carbon atoms occupy the corners of an equilateral triangle thus cyclopropane has C - C - C bond angle of 60°. This implied that the normal tetrahedral angle of $109^{\circ}\ 28'$ between any two C - C bonds is compressed to 60° and that each of the two bonds involved is pulled in by $1/2(109^{\circ}28' - 60) = 24^{\circ}44'$. The value $24^{\circ}44$ then represents the angle strain or the deviation through which each bond bends from the normal tetrahedral direction.

In cyclobutane the four carbon atoms occupy the corners of a square. Thus cyclobutane has C - C - C bond angle of 90°. The deviation from the normal tetrahedral angle which each bond shows is, therefore, $1/2\ (109^{\circ}28' - 90) = 9^{\circ}44'$.

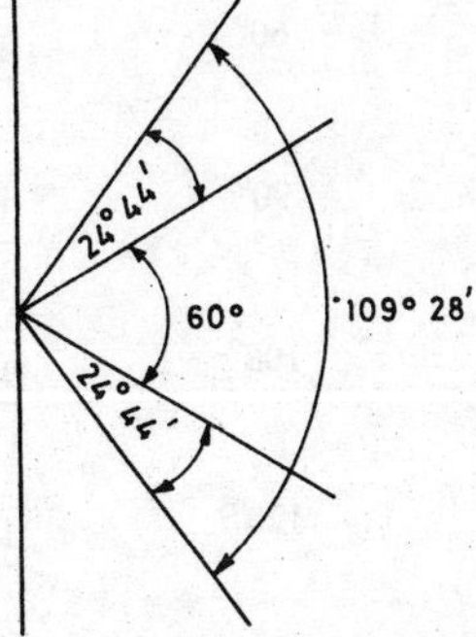

In cyclo pentane the five carbon atoms lie at the corners of a regular pentagon. Thus cyclopentane has C - C - C bond angle of 108°. The deviation from the normal tetrahedral angle which each bond shows is, therefore, $^1/_2$ (109°28′ - 108°) = 0°44′.

In cyclohexane the six carbon atoms occupy the corners of a regular hexagon. Thus cyclohexane has C - C - C bond angle of 120°. The deviation from the normal tetrahedral angle which each bond show is, therefore 1/2 (109°28′ - 120°) = - 5°16′. Similarly, in the case of cycloheptane the deviation from the normal tetrahedral angle is -9°33′. As one proceeds to cyclooctane, cyclononane, etc., the deviation of the bond angles from 109°28′ would become progressively larger, and the molecules would become progressively more strained.

The '+' sign in Table 4.1 indicates that the C-C bonds have to be compressed to satisfy the geometry of the ring. The '-' sign indicates that the C - C bonds have to be widened to satisfy the geometry of the ring. Whether the angle strain is positive or negative, its magnitude determines the extent of strain in the ring. Thus Baeyer considered that rings smaller than cyclopentane or cyclohexane were less stable, it was because of this less stability that the three and four membered rings underwent ring-opening reactions. The success of Baeyer's strain theory can be judged by its agreement with the following facts :

Experimental verification of Baeyer's strain theory (Heats of combustion and relative stabilities of the cyclo alkanes) :

The heat of combustion is the quantity of heat evolved when one mole of a compound is burned to carbon dioxide and water. Let us see if the heats of

Table 4.1 Deviation from the normal tetrahedral angle

Cycloalkane	*Bond angle*	*Deviation from normal tetrahedral angle (angle strain)*
60°	60°	$\frac{109°28' - 60°}{2} = 24°44'$
90°	90°	$\frac{109°28' - 90°}{2} = 9°44'$
108°	108°	$\frac{109°28' - 108°}{2} = 0°44'$
120°	120°	$\frac{109°28' - 120°}{2} = -5°16'$
128°34′	128°34′	$\frac{109°28' - 128°34'}{2} = -9°33'$

combustion of the various cycloalkanes support Baeyer's proposal that rings smaller or larger than cyclopentane and cyclohexane are less stable.

Examination of the data, Table 4.2, for a great many compounds has shown that the heat of combustion of an aliphatic hydrocarbon agrees rather closely with that calculated by assuming a certain characteristic contribution from each structural unit. For stable open chain alkanes each methylene group, $-CH_2-$, contributing 157.4 kcal $mole^{-1}$ to the heat of combustion. A comparison of heats of combustions of cycloalkanes per $-CH_2-$ unit, (Table 4.2), reveals that cyclopropane and cyclobutane rings possess 9 and 6.6 kcal $mole^{-1}$ respectively, more energy per $-CH_2-$ unit. Hence they should be less stable than corresponding alkanes, a fact in agreement with the Baeyer's strain theory. Their tendency to undergo addition reactions involving ring fission is related to this unstability. Its failure to explain the thermochemical data of other ring systems are listed below.

Limitations of Baeyer's strain theory

(i) According to Baeyer's strain theory, cyclopentane should be least strained (more stable) and cyclohexane should be some what strained (less stable). But the heats of combustion data clearly suggests that cyclopentane is quite strained and cyclohexane is perfectly strain free.

(ii) Data on the heats of combustion of various cycloalkanes suggests that rings larger than cyclopentane and cyclohexane are more stable. In fact,

Table 4.2 Heats of Combustion of Cycloalkanes

Cycloalkane $(CH_2)_n$	*Ring size (n)*	*Heat of combustion (kcal/mole)*	*Heat of combustion per* CH_2*(kcal/mole)*	*Total strain energy (kcal/mole)*
Cyclopropane	3	499.8	166.6	27.6
Cyclobutane	4	655.9	164.0	26.4
Cyclopentane	5	793.5	158.7	6.5
Cyclohexane	6	944.5	157.4	0.0
Cycloheptane	7	1108	158.3	6.3
Cyclooctane	8	1269	158.6	9.6
Cyclononane	9	1429	158.8	12.6
Cyclodecane	10	1586	158.6	12.0
Cyclonndecane	11	1742	158.4	11.0
Cyclododecane	12	1891	157.6	2.4
Cyclotridecane	13	2051	157.8	5.2
Cyclotetradecane	14	2204	157.4	0
n-Alkane (reference)	–	–	157.4	0

they are almost equal in stability to the corresponding alkanes, whereas according to Baeyer's theory rings should be considerably strained and therefore less stable.

(iii) According to Baeyer's strain theory, strains of cyclobutane and cycloheptane are very close, but heat of combustion data shows great difference in their stability.

(iv) For rings of larger size, there is no regular increase in strain and rings having more than 14 members are again strainfree.

Basically Baeyer's strain theory is applicable to three and four membered ring systems, but not to rings larger than four membered. Baeyer was wrong for a very simple reason. He assumed that all the rings are Planar. In fact, most cycloalkanes are not planar. For example, cyclohexane thought to be a regular (flat) hexagon having a bond angle of 120° is in fact a nonplanar structure.

Sachse-Mohr concept of strainless rings : In order to account for the stability of cycloalkanes beyond cyclopentane, Sachse and Mohr pointed out that such rings can become absolutely free of strain if all the ring carbons are not forced in to one plane, as was supposed by Baeyer. If a ring assumes a *puckered* condition, the normal tetrahedral angles of 109°28′ are retained

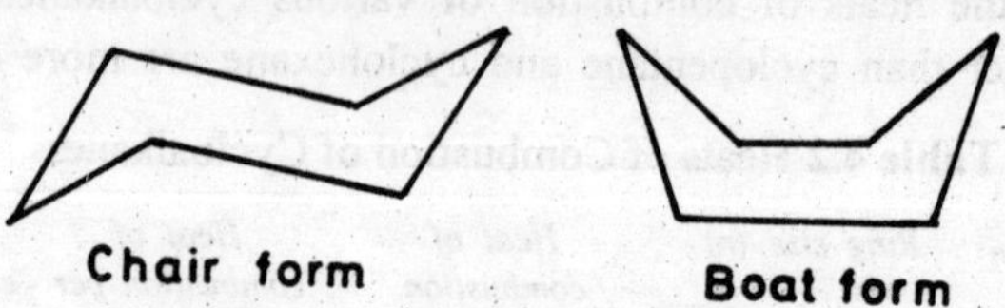

Fig.6.1 Puckered forms of cyclohexane

and as a result, the strain within the ring is relieved. thus cyclohexane can exist in two non-planar strainless forms, namely chair and boat forms, shown in Fig. 6.1

In chair form carbons 1, 3 and 5 lie in one plane and carbons 2, 4 and 6 lie in another plane. In the boat form carbons 1, 2, 4 and 5 lie in the same plane and carbons 3 and 6 are above the plane. Such non-planar strainless rings in which the ring carbon atoms can have normal tetrahedral angles are also possible for larger ring compounds.

Orbital picture of angle strain: What is the meaning of Baeyer's angle strain in terms of the modern picture of the covalent bond? Baeyer's strain theory is based on a mechanical concept of valency and on the assumption that all the rings are planar. This led to conclusions which have now been shown to be wrong. In the first place, physical methods have shown that all rings, except cyclopropane are not planar. Cyclopropane is a three point system and consequently must be planar. In the second place, quantum mechanical calculations do not permit very 'arge, distortions of bond angles. If cyclopropane were an equatorial triangle with bonds directed along the lines joining the carbon atoms, then the ring valency angle of each carbon atom would be 60°. This value is impossible, since the carbon valency angle can never be less than 90° (when they are pure p-orbitals). Further mixture of p- with s-orbitals opens the valency angle. According to Coulson's calculations the smallest carbon valency angle that one can reasonably expect to have is 104°. Coulson has therefore suggested that in cyclopropane, the carbon hybridized orbitals are not pointing one another in the same straight line, and consequently there is a loss of overlap (Fig. 4.2a). It is this loss of overlap that gives rise to instability. The cyclopropane molecule is in a stable of strain due to bent bonds. Applying this argument to cyclobutane (Fig.4.2b), it can be seen that this molecule also has bent bonds, but loss of overlap is less than for cyclopropane and so the former will be more stable than the latter.

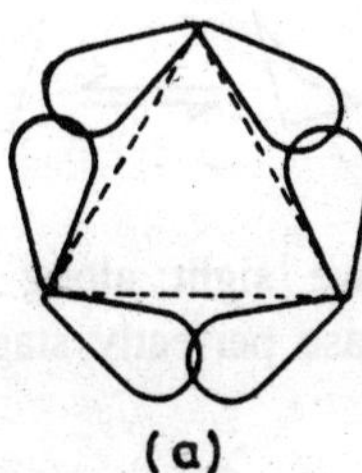
(a)

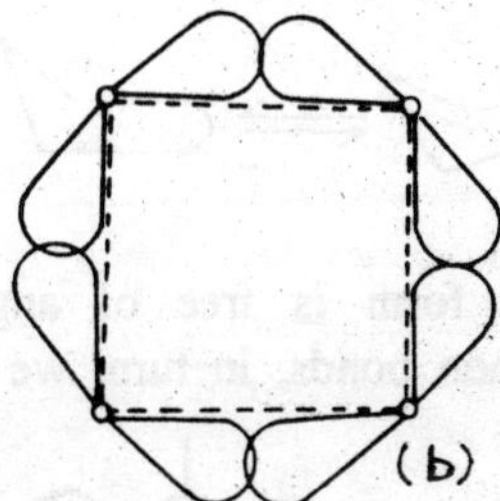
(b)

Fig. 4.2

Factors affecting stability of conformational isomers : To go more deeply into the chemistry of alicyclic compounds, we must use conformational analysis. To begin with let us review the factors that affect the stability of conformational isomers.

Any atom tends to have bond angles that match those of its bonding orbitals; tetrahedral (109°28′) for sp^3 hybridized carbon. Any deviation from the *normal bond angles* is accompanied by angle strain.

Any pair of tetrahedral carbons attached to each other tend to have their bond staggered. That is, any ethane like portion of a molecule tends to be as ethane, to take up a staggered conformation. Any deviation from the staggered arrangement is accompanied by *torsional strain.*

Any two atoms that are not bonded to each other can interact in several ways, depending on their size and polarity, and how closely they are brought together. These non-bonded interactions can be either repulsive or attractive and the result can be either destabilization or stabilization of the conformation.

Non bonded atoms or groups that just touch each other i.e., atoms that are about as far apart as the sum of their van der Waals radii, attract each other. If brought closer together, they repel each other, such crowding together is accompanied by *van der Waals strain* or *steric strain.*

Non-bonded atoms or groups tend to take positions that result in the most favourable *dipole-dipole interactions,* i.e., positions that minimize dipole-dipole repulsions or maximize dipole-dipole attractions. A particularly powerful attraction results from the special kind of dipole-dipole interaction called *the hydrogen bonding.*

All these factors, working together or opposing each other determine the net stability of a conformational isomer.

Conformational analysis of cyclohexane : The cyclohexane ring can assume many shapes. A single cyclohexane molecule is in a continuous state of flexing or flipping into different shapes or conformations. Some of these shapes are shown below :

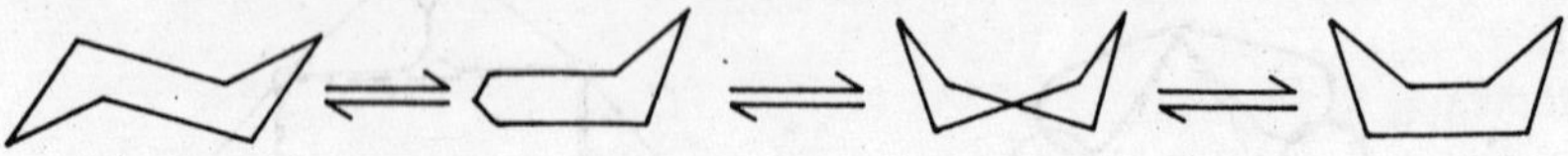

Chair form is free of angle strain. If we sight along each of the carbon-carbon bonds, in turn, we see in every case perfectly staggered bonds.

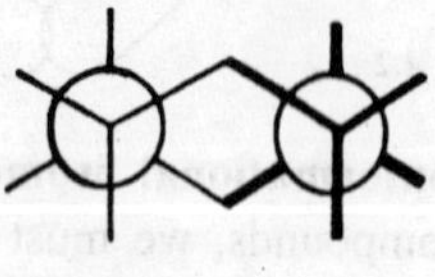

Cyclohexane chairform

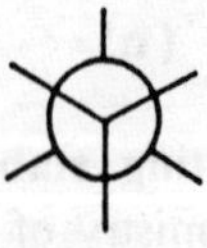

Staggared ethane

The conformation is thus not only free of angle strain but free of torsional strain as well. It lies at an energy minimum. The chair form is the most stable conformation of cyclohexane. Now, let us take chair cyclohexane and flip the end of a bond of molecule to give the boat conformation.

Boat form is also free of angle strain. Sighting along either of two carbon-carbon bonds we set of exactly eclipsed bonds, and hence we except considerable torsional strain.

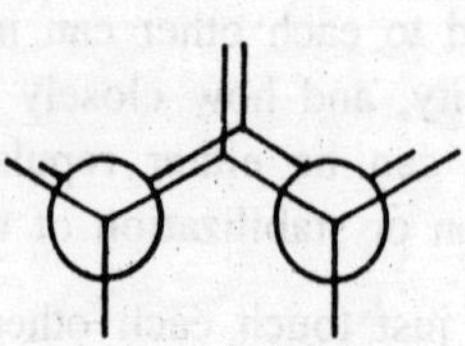

Cyclohexane boat form

Eclipsed ethane

In addition, there is van der Waals strain due to crowding between the *flagpole* hydrogens, which lie only 1.83Å apart, much closer than the sum of their van der Waals radii (2.5Å). The boat conformation less stable (7.1 kcal mole^{-1}) than the chair conformation. It is believed to lie at an energy maximum, and thus not a conformer, but a transition stable between two conformers. Now, what are those two conformers that lie, energetically speaking, on either side of the boat conformation. To see what they are, let us hold a model of the boat conformation. To see what they are, let us hold a model of the boat conformation with the flagpole hydrogens (Ha and Hb) pointing up and look down through the ring. We grasp C_2 and C_3 in the right hand and C_5 and C_6 in the left hand and twist the molecule so that, say, C_3 and C_6 go down and C_2 and C_5 come up. As we do this, H_a and H_b move diagonally apart and we see (below the ring) a pair of hydrogens, H_c and H_d (on C_3 and C_6 respectively), begin to approach each other. When the H_a-H_b distance is equal to the H_c-H_d distance, we stop and examine the molecule. We have minimised the flagpole-flagpole interactions and at the same time have partly relieved the torsional strain at the C_2-C_3 and C_5-C_6 bonds.

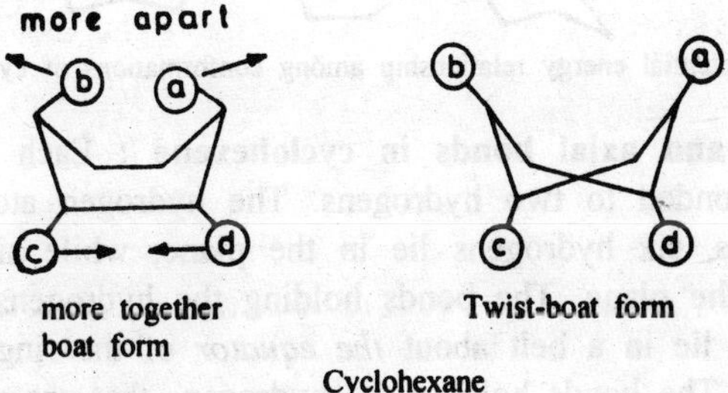

Cyclohexane

This new configuration is the twist boat form. It is a conformer, lying at an energy minimum 5.5 kcal above the chair conformation.

Flagpole bords

Boat cyclohexane

Twist boat cyclohexane

Between the chair form and twist-boat form lies the highest barrier of all, a transition state conformation (the half chair), which, with angle strain and torsional strain, lie about 11 kcal above the *chair* form.

The overall relationships are summarized in Fig. 6.3. Equilibrium exists between the chair and twist-boat forms, with the more stable chair form being favoured.

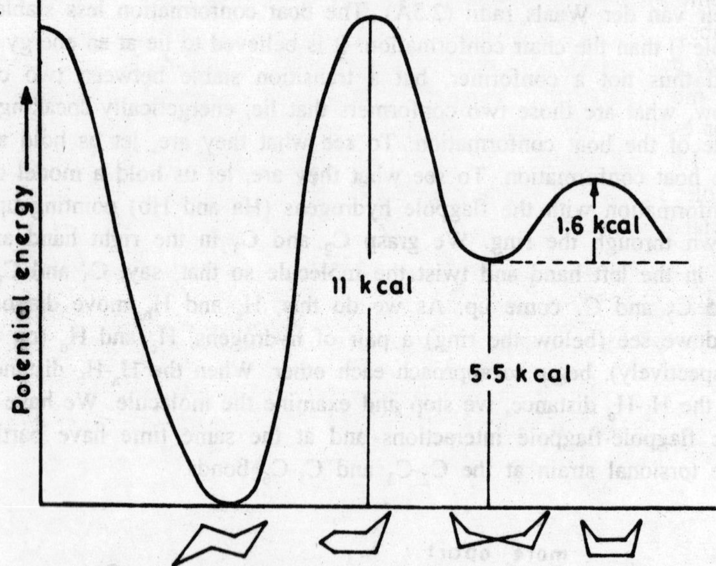

Fig. 6.3 Potential energy relationship among conformations of cyclohexane

Equatorial and axial bonds in cyclohexane : Each carbon atom of cyclohexane is bonded to two hydrogens. The hydrogen atoms occupy two kinds of positions, six hydrogens lie in the plane, while six hydrogens lie above or below the plane. The bonds holding the hydrogens that are in the plane of the ring lie in a belt about *the equator* of the ring, and are called *equatorial bonds.* The bonds holding the hydrogens that are above and below the plane are pointed along *an axis perpendicular to the plane and are called axial bonds.* In the chair conformation each carbon atom has one equatorial bond and one axial bond.

axial hydrogens   Equatorial hydrogens

In the flipping and reflipping between conformations, axial becomes equatorial, while equatorial becomes axial as shown below.

H(a)   H(e)

Conformations of mono substituted cyclohexanes : Let us look at the consequences of axial and equatorial bonds. Cyclohexane rings rapidly flip conformations at room temperature. At any given instant, however, the substitutent of a monosubstituted cyclohexane is either axial or equatorial. The two conformers are not equally stable. The equatorially substituted conformer is more stable than the axially substituted conformer. In methylcyclohexane, for example, the energy difference between axial and equatorial methyl group is 1.8 kcal $mole^{-1}$. At any given instant, 95% of the methyl cyclohexane molecule have the methyl group equatorial, and only 5% have the methyl in axial conformation.

The reason for this energy difference is steric strain due to 1, 3-diaxial interactions. The axial methyl group on carbon 1 suffers steric interference from the axial hydrogens three carbons away (carbon 3 and 5).

Axial methyl (5%)
(less stable)

Equatorial methyl (95%)
(more stable by 1.8 kcal/mol)

What is true for methyl cyclohexane is also true for all other mono substituted cyclohexanes. A substitutent is almost always more stable in an equatorial position than in an axial position. The precise amount of 1, 3-diaxial steric strain depends on the size of the group and table 6.3 lists some values for common substitutents. As expected the amount of steric strain increases through the series $H_3C < H_3C\text{-}CH_2 < (CH_3)_2C - < < (CH_3)_3C$ in parallel with the bulk of the successively larger alkyl groups.

Table 6.3 Steric strain due to 1, 3-diaxial interactions

Substitutent X	*(Strain of one H-X, 1,,3-diaxial interaction kcal $mole^{-1}$)*
-F	0.12
-Cl	0.25
-Br	0.25
-OH	0.5
-C3	0.9
-C2-C3	0.95
-CH(CH3)2	1.1
-C(C3)3	2.7
-C6H5	1.5
-COOH	0.7
-CN	0.1

Strain of one H-X
1, 3 - diaxial interaction

Note that, the values in table 6.3 refer to 1, 3-diaxial interactions with a single hydrogen atom. These values must therefore be doubled to arrive at the amount of strain in a monosubstituted cyclohexane.

4-6-MULTIPLE CHOICE QUESTIONS

4-6-1 Which of the following in most stable

(A) Benzene (B) Cyclohexane (C) Cyclohexene (D) Cyclohexadiene

4-6-2 To which class cyclopentane belongs

(A) Aromatic Compounds (B) Acyclic compounds

(C) Cyclic compounds (D) Heteocyclic compounds

4-6-3 Which of the following rings is most strained

(A) Cyclopriopane (B)Cyclohexane

(C) Cyclooctane (D) Cyclobutane

4-7-4 (i)Write the structural formulae of the two chair conformational isomers of 1-t-batyl - 3 chlorocyclohexane

(ii) Give their appropriate names

(iii) Which of these would be most stable ? Why?

4-7 SHORT AND ESSAY-TYPE QUESTIONS

4-7-1. Write a note on conformation of cyclohexane.

(A.U., Jun. 1990 & O.U., Apr. 1990)

4-7-2. Describe the general methods of preparation of cyclo alkanes.

(A.U., Jun. 1990)

4-7-3. What are conformers? Explain the term conformation analysis.

(A.U., Apr. 1989)

4-7-5 Write an account of potential energy relationship among conformations of cyclo hexane with suitable graph representing potential energy variations.

4-7-6 What are the factors affecting conformational Isomers

4-7-7 Give an account of Baeyer's strain theory and mention its limitations

ANSWERS

4-6-1 (B) 4-6-2 (C) 4-6-3 (A)

5

ALKENES

5-1 INTRODUCTION

Alkenes are the unsaturated aliphatic hydrocarbons that contain one double bond. Alkenes are also known as "olefins". The generic term olefin comes from an old name for ethylene, the first member of the family. Dutch chemists found that the reaction of gaseous ethylene with gaseous chlorine gave on oily liquid, ethylene chloride, so they called ethylene olefant gas (meaning oil making gas). The general formula of alkenes is C_nH_{2n}. Alkenes have two hydrogen atoms lesser than the corresponding alkanes, because two adjacent carbon atoms form a double bond. The characteristic structural feature or functional group is a $>C=C<$ bond with an open chain structure.

5-2 STRUCTURE OF ALKENES

To illustrate the main structural feature of alkenes, let us consider the structure of ethylene, C_2H_4, the simplest member of the alkene family.

In view of the ready conversion of ethylene into ethane, we can reasonably expect certain structural similarities between the two compounds, to understand the difference, let us connect the carbon atoms by a covalent bond, and then attach two hydrogen atoms to each carbon atom. At this stage we find that each carbon atom possesses only six electrons in its valence shell, instead of the required eight to complete the octet and that the entire molecule needs an additional pair of electrons if it is to be neutral. We can solve both these problems by assuming that the carbon atoms can share two pairs of electrons. To describe this sharing of two pairs of electrons, we say that the carbon atoms are joined by a double bond. this carbon-carbon double bond is the distinguishing feature of the alkene structure.

```
    H     H
    :     :
H : C : : C : H
```

```
H  1.10 A°              H
  \                    /
   \     1.34 A°      /
    C ============== C
   /                  \
  /                    \
H      120°             H
```

Ethylene

hybridized orbitals of each carbon with the 1s orbital of a hydrogen. The bonds thus formed are called σ (sigma) bonds as the overlapping of orbitals takes place along their axes. Moreover, since all the three sp^2 orbitals of each carbon lie in the same plane, all the carbon and hydrogen atoms of ethylene are also coplanar. Thus ethylene is a flat molecule in which each carbon is situated at the center of a triangle, at the corners of which are located two hydrogen atoms and the other carbon atoms as is clear from the skeleton of the molecule.

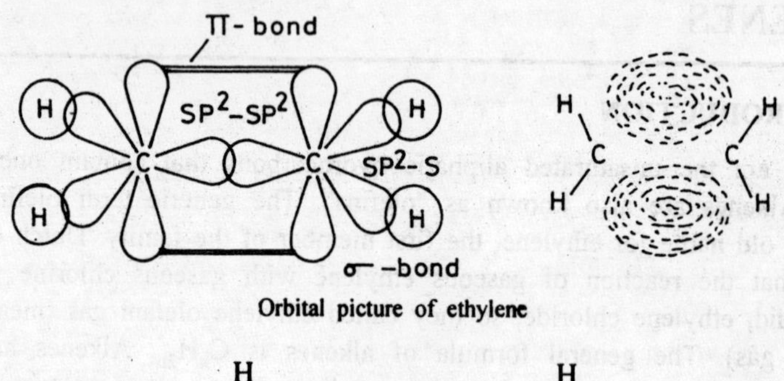

Orbital picture of ethylene

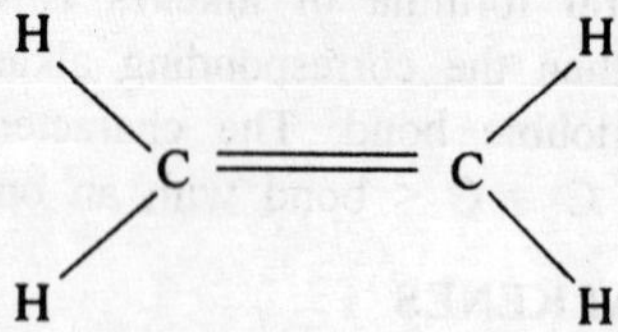

Ethylene molecule shape

5-3 NOMENCLATURE

There are three important systems for naming alkenes.

5-3-1 Common System

The common names of the first few alkenes are derived from the corresponding alkanes by replacing the suffix *'ane'* of the common name of the saturated hydrocarbon by-*ylene*. For example;

Alkene	Common name
$CH_2 = CH_2$	Ethylene
$H_3C — CH = CH_2$	Propylene
$H_3C — C(CH_3) = CH_2$	Isobutylene
$H_3C — CH_2 — CH = CH_2$	α-Butylene
$H_3C — CH = CH — CH_3$	β-Butylene

The isomeric alkenes are distinguished by writing Greek letters α, β, γ etc., as prefixes to indicate the position of double bond at first, second, third etc., carbon atoms respectively.

The radicals derived from alkenes by removal of a hydrogen are shown below with their common names.

$$H_2C=CH-$$ Vinyl

$$H_2C=CH-CH_2-$$ Allyl

$$H_3C-CH=CH-CH_2-$$ Crotonyl

5-3-2 As Derivatives of Ethylene

In this system alkenes are considered to be the derivatives of the ethylene. The names of alkenes are obtained by adding the name of alkyl groups replacing the hydrogen atoms of ethylene to the parent name ethylene.

Formula	Name
$H_3C-CH=CH_2$	Methylethylene
$H_3C-CH_2-CH=CH_2$	Ethylethylene
$H_3C-CH=CH-CH3$	Sym-Dimethylethylene
$H_3C-C(CH_3)=CH_2$	Unsym-Dimethylethylene
$H_3C-C(CH_3)=CH-CH_3$	Trimethyl ethylene
$H_3C-C(CH_3)=C(CH_3)-CH_3$	Tetramethylethylene

The replacement of two hydrogens by two alkyl groups can give rise to two isomers depending on whether the two groups are linked to different or same carbon atom. They are distinguished by writing prefixes sym and unsym which are abbreviations for symmetrical and unsymmetrical.

5-3-3 IUPAC System

According to the IUPAC system of nomenclature, the class suffix of the alkenes (olefins) is ene. The longest carbon chain containing the double bond is chosen as the parent alkene. The name is obtained by changing the suffix-ane of the corresponding alkane into -ene. the position of the double bond and side chains are indicated by numbers, the lowest number possible being given to the double bond, and this is placed before the suffix. the IUPAC system is now the one generally used.

Formula	IUPAC name
$H_2C = CH_2$	Ethene
$H_3C - CH = CH_2$	Propene
$HC_3 - C(CH_3) = CH_2$	2-Methyl propene
$H_2C = CH - CH_2 - CH_3$	1-Butene
$H_3C - CH = CH - CH_3$	2-Butene

The radicals derived from alkenes by removal of one hydrogen are known as alkenyl radicals. The position of double bond being specified by the number of carbon atom preceding the double bond. Their names are obtained by adding -yl to the suffix of alkene. The carbon with the free valency is numbered one. There are three exceptions (see example).

$\overset{3}{H_3C} - \overset{2}{CH} = \overset{1}{CH} -$ Prop-1-enyl

$\overset{5}{H_3C} - \overset{4}{CH}(CH_3)\overset{3}{CH} = \overset{2}{CH} - \overset{1}{CH_2} -$ 4-Methylpent-1-enyl

$CH_2 = CH -$ Ethenyl

$\overset{3}{H_2C} = \overset{2}{CH} - \overset{1}{CH_2} -$ Propene-2-yl

$\overset{2}{H_2C} = \overset{1}{C}(CH_3) -$ 1-methylethenyl

5-4 PREPARATION OF ALKENES

5-4-1 Dehydrohalogenation of Alkylhalides

The name of this reaction describes, what happens to an alkyl halide when it is heated with potassium hydroxide dissolved in alcohol. The alkylhalide loses a hydrogen atom from one carbon atom and a halogen atom from an adjacent carbon atom. The alkyl halides on reluxing with concentrated alcoholic solution of potassium hydroxide yield alkenes. The reaction, then, is called **dehydrohalogenation.**

$$H_2C(H) - CH_2(Cl) \xrightarrow[\Delta]{\text{alcoholic KOH}} H_2C = CH_2 + H_2O + KCl$$

Mechanism : The driving force of the reaction is the polarity introduced in the molecule by the greater electronegativity of the halogen. This is called the inductive effect and can be diagrammed as

$$\begin{array}{ccccc} & H^{\delta+} & & H^{\delta+} & \\ & | & & | & \\ H^{\delta+} — & C & ——— & C & — H^{\delta+} \\ & | & & | & \\ & H^{\delta+} & & X^{\delta-} & \end{array}$$

Depending upon the structure of alkylhalide and some other factors the reaction can take place either by bimolecular elimination (E_2) or unimolecular elimination (E_1) mechanistic pathways.

1. Bimolecular elimination mechanism (E_2) : In dehydrohalogenation the strongly basic hydroxide ($\bar{O}H$) attracts one of the acidic hydrogen atoms on the α-carbon, directly attached to the carbon atom bearing the halogen atom **(I)**. The electron pair that holds the hydrogen to the α-carbon can then contribute to neutralizing the partial positive charge on the halogen bearing carbon atom **(II)**. At the same time, this electron pair, which will ultimately form the α bond, begins to repel the electron pair holding the halogen to the electrophilic carbon **(III)**, further displacing these electrons towards the halogen. The attraction between the halogen with its partial negative charge and the solvent ethanol (called solvation) helps the halogen to pull away from the carbon atom as a solvated halide ion **(IV)**.

$H^{\delta+}$---$\bar{O}R$ — C — $C^{\delta+}$ — $X^{\delta-}$ (α-carbon) I → [$H^{\delta+}$------$\overset{\delta-}{O}R$ — C — C — $X^{\delta-}$ II → $H^{\delta+}$---$\overset{\delta-}{O}R$ — C — $C^{\delta+}$ — $X^{\delta-}$ solvent III] → $>C=C< + H_2O + X^-$ (solvent) IV

The net result of this reaction sequence is

(i) The formation of a water molecule

(ii) The formation of a π bond between the α-carbon and the halogen-bearing Carbon atom, and

(iii) The release of a solvated halide ion.

This mechanism is summarized as follows :

$$H_3C - \underset{\displaystyle H^{\delta+}}{CH} - \underset{\displaystyle Br^{\delta-}}{CH_2} + \bar{O}H \longrightarrow \underset{H}{\overset{H_3C}{>}}C=C\underset{H}{\overset{H}{<}} + H_2O + Br^-$$

1-Bromopropane Propene

It is a single step mechanism. Rate of the reaction depends on concentration of alkylhalide and alcoholic KOH. Hence it is a bimolecular reaction.

Above example gives a typical mechanistic scheme of a bimolecular elimination reaction (E_2). *Primary and secondary alkylhalides are most likely to undergo* E_2 *elimination.*

For example,

$$H_3C - CH_2 - CH_2 - CH_2 - Cl \xrightarrow{\text{Alcoholic KOH}} H_3C\text{-}CH_2\text{-}CH = CH_2$$

1-Chlorobutane (1°) 1-butene

2. **Unimolecular elimination mechanism (E_1)** Elimination can also occur through a carbocation formation. The cation arises through the ionization of the alkylhalide, and the reaction rate is dependent only on the concentration of alkyl halide. This mechanism is called unimolecular elimination (E_1). In the E_1 mechanism, the polar solvent, ethanol, aids in the cleavage of the carbon halogen bond by solvation of both the carbocation and the halide ion (VI). The formation of the carbocation is the rate determining step. In the second step of the reaction the base $\bar{O}H$ attacks on of the electron deficient hydrogen atoms on a carbon atom α- to the positively charged carbon atom (VI). The electron pair holding the proton being attacked by base is simultaneously drawn toward the electron pair that will form the π- bond between the α-carbon atom and the positively charged carbon atom (VII).

$$-\overset{\displaystyle H^{\delta+}}{\underset{|}{\overset{|}{C}}} - \overset{|}{\underset{\displaystyle X^{\delta-}}{C^{\sigma+}}}- \longrightarrow -\overset{\displaystyle H^{\delta+} \leftarrow \bar{O}R}{\underset{+}{C}} \rightarrow \overset{+}{C} \text{------solvent} \quad X^- \text{------solvent} \longrightarrow \;>C=C< + H_2O$$

VI VII

The net result of the E_1 reaction sequence is the same as that of the E_2 reaction except that in the E_1 mechanism the halide ion is dissociated first to give a carbocation, which then looses a proton *Tertiary alkyl halides are most likely to undergo E_l elimination because of the relative ease with which they ionize. The sequence of events in E_l, elimination is given below.*

$$H_3C-\underset{\underset{Br}{|}}{\overset{\overset{CH_3}{|}}{C}}-CH_3 \xrightarrow[\text{(rate determining step)}]{\text{step-1 slow}} H-\underset{\underset{H}{|}}{\overset{\overset{H}{|}}{C}}-\underset{+}{\overset{\overset{CH_3}{|}}{C}}-CH_3$$

$$\xrightarrow[\bar{O}H]{\text{Step-2}} H_2C=\overset{\overset{CH_3}{|}}{C}-CH_3$$

Most of the conditions that favor substitution also favor elimination. For this reason in most instances, when an alkyl halide is to react with a base, there is competition between the two reaction types. We can simplify the problem to some extent by considering the following generatlizations.

(i) Strong bases tend to favor elimination over substitution.

(ii) Since the highly substituted alkenes are more stable, tertiary alkyl halides undergo mostly elimination, substitution occurs only to very small extent.

An interesting feature of dehydrohalogenation is that if an alkyl halide is capable of forming the double bond in any one of the two possible directions, it invariably yields the alkene which has the greater number of substituents across the double bond. This is in accordance with the **Saytzeff's rule** which states that *when alternatives exists, hydrogen in preferentially eliminated from the carbon atom with fever number of hydrogen atoms.* This generalization or the Satyzeff's rule applies to E_1 and most of the E_2 reactions of alkyl halides. Dehydrohalogenation of 2-bromobutane, for instance gives 2-butene as the major product.

$$\underset{\text{2-Bromobutane}}{H_3C-CH_2-\underset{\underset{Br}{|}}{CH}-CH_3} \xrightarrow{\text{alc. KOH}} \underset{\text{2-Butene}}{H_3C-CH=CH-CH_3} + \underset{\text{1-Butene}}{H_3C-CH_2-CH=CH_2}$$

The preferential formation of more substituted alkenes in the above reactions can be correlated with the relative stability of various alkanes. Calculations from heats of combustion and hydrogenation establish that the stability of a double bond is increased by alkyl substitution on the double-bonded carbon atoms. Thus the stabilities of alkenes follow the order :

$$R_2C=CR_2 \qquad R_2C=CHR \qquad RHC=CHR \qquad R_2C=CH_2$$

$$RHC=CHR \qquad RHC=CH_2 \qquad H_2C=CH_2$$

The greater stability of the more substituted alkene is sometimes explained on the basis of hyperconjugation. The order of decreasing ease of elimination of alkyl halides is 3° > 2° > 1°. Further the ease of dehydrohalogenation for different halogens is in the order : Iodide > Bromide > Chloride.

5-4-2 Dehydration of alcohols

Dehydration means the removal of a molecule of water from a reactant molecule. When an alcohol is heated in the presence of concentrated H_2SO_4 or concentrated H_3PO_4, a molecule of water is eliminated and an alkene is formed. The reaction can be summarized as :

$$-\underset{|}{\overset{\overset{H}{|}}{C}}-\underset{\underset{OH}{|}}{\overset{|}{C}}- \xrightarrow[\Delta]{H_2SO_4} >C=C< + H_2O$$

Mechanism : It is also an example of E_1 reaction and its mechanism involves three distinct steps.

In the first step a proton from the acid coordinates with the -OH group forming the protonated alcohol (oxonium ion). This being a highly reactive species, loses readily a molecule of water.

$$-\overset{H}{\underset{|}{\overset{|}{C}}}-\overset{|}{\underset{:\ddot{O}H}{\underset{|}{C}}}- + H_2SO_4 \xrightarrow{\text{Fast}} -\overset{H}{\underset{|}{\overset{|}{C}}}-\overset{|}{\underset{+\,O\,-\,H}{\underset{|}{C}}}- + HSO_4^-$$
(with H on the oxonium O)

Second step involves formation of a carbocation.

$$-\overset{H}{\underset{|}{\overset{|}{C}}}-\overset{|}{\underset{+\,O-H}{\underset{|}{C}}}- \xrightarrow[\text{slow}]{\Delta} -\overset{H}{\underset{|}{\overset{|}{C}}}-\overset{|}{\underset{+}{C}}- + H_2O$$

Carbocation

In the third step the carbocation eliminates a proton from the neighbouring carbon forming the alkene.

$$-\overset{H}{\underset{|}{\overset{|}{C}}}-\overset{|}{\underset{+}{C}}- + \text{Base} \longrightarrow \;>C=C<$$

Alkene

The ease of dehydration of various alcohols has been found to follow the order tertiary > secondary > primary, which conforms to the relative stabilities of the carbocations. The dehydration of tert. butyl alcohol, for instance,. takes place at 90°C in the presence of dilute sulphuric acid, while ethanol is dehydrated at 170°C and that too in presence of concentrated H_2SO_4.

For example,

$$\underset{H}{\underset{|}{H_2C}}-\underset{OH}{\underset{|}{CH_2}} \xrightarrow[170^\circ C]{\text{Conc. } H_2SO_4} H_2C=CH_2 + H_2O$$

$$H_3C-\underset{OH}{\underset{|}{CH}}-CH_3 \xrightarrow[100^\circ C]{60\%\ H_2SO_4} H_3C-CH=CH_2 + H_2O$$

$$H_3C-\overset{CH_3}{\underset{CH_3}{\overset{|}{\underset{|}{C}}}}-OH \xrightarrow[90^\circ C]{20\%\ H_2SO_4} H_3C-\overset{CH_3}{\overset{|}{C}}=CH_2 + H_2O$$

Dehydration of unsymmetrical secondary or tertiary alcohols leads to the formation of two different alkenes.

The dehydration of 2-butanol gives 2-butene and 1-butene

$$\underset{\text{H}}{\underset{|}{H_3C - CH}} - \underset{\text{OH}}{\underset{|}{CH}} - \underset{\text{H}}{\underset{|}{CH}} \xrightarrow{H_2SO_4} \underset{\substack{\text{2-Butene}\\ \text{(Major)}\\ \text{More stable}}}{H_3C - CH = CH - CH_3} + \underset{\substack{\text{1-Butene}\\ \text{(Minor)}\\ \text{Less table}}}{H_3C - CH_2 - CH = CH_2}$$

This is in accordance with the Saytzeff's rule.

Dehydration of an alcohol may also be brought about by passing the alcohol vapors over heated alumina at 350-400°C. Other dehydrating agents which may be employed are P_2O_5 and H_3PO_4.

5-5 PHYSICAL PROPERTIES

(a) Alkenes containing two to four carbon atoms are colorless gases, those containing five to eighteen carbon atoms are liquids, while molecules containing more than eighteen carbon atoms are solids.

(b) They are insoluble in water but are soluble in organic solvents such as ethers, alcohols, and carbon tetrachloride etc.

(c) They are lighter than water, the series has a limiting density less than 0.8.

(d) The melting points and boiling points rise with increasing molecular weight. Further more, branched chain alkenes have lower boiling points than straight chain alkenes.

(e) In general, alkenes have higher melting points than the alkanes with the same carbon skeleton. This is attributed to the stronger attractive forces in alkenes than in alkanes. Due to the mobility of the electrons in the π bond, the double bond is more polarizable than single bond. This permits the ready formation of induced dipoles in alkene molecules which are responsible for stronger intermolecular forces of attraction

(f) Like alkanes, alkenes are generally non-polar, but certain alkenes are weakly polar due to their unsymmetrical geometry. For example, propene, 1-butene have a definite dipole moment.

cis-2-butene, in contrast to trans-2-butene has a small dipole moment because the former is unsymmetrical while the latter is symmetrical. Hence cis-2-butene boils at higher temperature than the trans-2-butene.

5-6 CHEMICAL PROPERTIES

The functional group in alkenes is carbon-carbon double bond (C=C) which is made up of one σ bond and one π bond. The π bond since it is formed by the lateral overalp of p-orbitals is weaker than a σ bond. The energy of a σ bond is 83 kcal mole^{-1}, while that of a π bond is 63 kcal mole^{-1}. the electrons in the weaker π bond are more diffused in space and hence easily accessable to any reactant.

From the above arguments it follows that the weaker π bond would tend to convert itself to two σ bonds, givirng stable addition products. Also, the σ bond in alkene acts as source of electrons. Hence electrophiles which are electron deficient species and can accept electrons, would be attracted by the σ cloud while nucleophiles will be repelled. Therefore, the characteristic reactions of alkenes are the electrophilic addition reactions.

$$\underset{\text{Alkene}}{>C{=}C<} + \underset{\text{Addendum}}{A-B} \longrightarrow \underset{\text{Addition product}}{>\overset{\overset{A}{|}}{C}-\underset{\underset{B}{|}}{C}<}$$

$$\underset{\text{Alkene}}{>C=C<} + \underset{\text{Addendum}}{A^{\delta+}-B^{\delta+}} \longrightarrow \underset{\pi\text{-Complex}}{>C \overset{\overset{B^{\delta+}}{\vdots}\;\overset{A^{\delta+}}{\vdots}}{=} C<} \longrightarrow \underset{\text{carbonium ion}}{>\overset{\overset{A}{|}}{C}-\overset{+}{C}<} \overset{B^-}{\longrightarrow} \underset{\text{Addition product}}{\overset{\overset{A}{|}}{C}-\underset{\underset{B}{|}}{C}<}$$

In the first step of the addition reaction, the peripheral electrons of the π bond attract the electrophilic end of the addendum (A) to form a loose π-complex. The positive end of the addendum buried in the symbol p-election could give positive charge to the symbol p- complex, leaving the nucleophile B. The π- complex then converted into a carbonium ion, which reacts with the nucleophile remainder to form the addition product.

The general reactions of alkenes are given below :

5-6-1 Addition of hydrogen

Alkenes add hydrogen in the presence of a metal catalyst forming the correspoinding alkane. This process known as catalytic hydrogenation proceeds by the addition of a molecule of hydrogen at the carbon-carbon double bond.

$$\underset{\text{Alkene}}{>C{=}C< + H_2} \xrightarrow{\text{Catalyst}} \underset{\text{Alkane}}{>\underset{\underset{H}{|}}{C}-\underset{\underset{H}{|}}{C}<}$$

The hydrogenation of alkene can be effected by passing vapors of alkene and hydrogen over Raney nickel, **Sabatier-Senderens** reduction catalyst, at 200-250°C. However, when the reaction takes place at room temperature, the best catalysts are palladium and platinum. Palladium is normally employed in a very finely divided state supported on an inert material such as charcoal to maximize surface area (Pd/C). Platinum is normally used as PtO_2, a reagent known as **Adams catalyst** after its discoverer, Roger Adams.

Catalytic hydrogenation unlike most other organic reactions is a heterogeneous process rather than a homogeneous one. That is, the hydrogenation reaction occurs on the surface of a solid catalyst particles rather than in solution. For this reason, catalytic hydrogenation has proved very difficult to study for mechanistic purposes. However, empirical observation has revealed that hydrogenation is cis addition, i.e., both hydrogens are added to the double bond from the same direction/face.

Mechanism : The first step in the reaction is absorption of hydrogen onto the catalyst surface. Complexation then occurs between catalyst and alkenes by overlap of vacant metal orbitals with π electrons of alkene. Then hydrogen is inserted into the π-bond, and the saturated product (alkane) diffuses away from the catalyst.

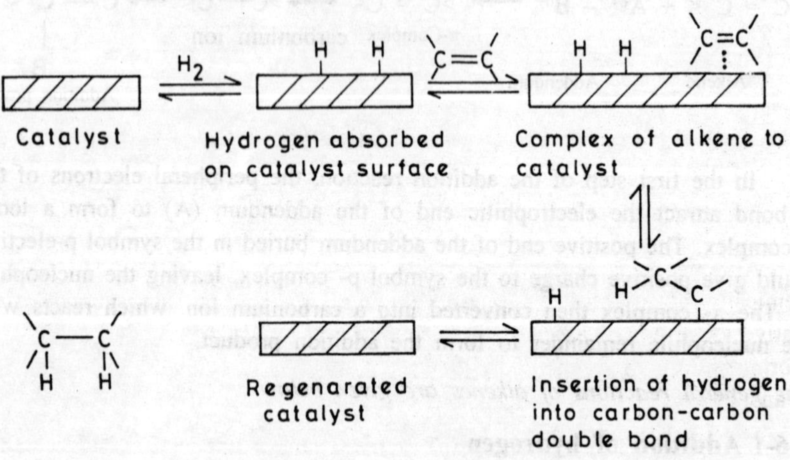

Fig. 5.1 Mechanism of hydrogenation of alkene

Examples

$$\underset{\text{Ethane}}{H_2C = CH_2} + H_2 \xrightarrow[200\text{-}250^\circ]{Ni} \underset{\text{Ethane}}{H_3C - CH_3}$$

$$\underset{\text{Propene}}{H_3C - CH = CH_2} + H_2 \xrightarrow[\Delta]{Pt} \underset{\text{Propane}}{H_3C - CH_2 - CH_3}$$

Heat of hydrogenation and stability of alkenes : Although the cis-trans inter converstion of alkene isomers does not occur spontaneously, it can be made to happen under appropriate experimental conditions. For example, treatment with a strong acid catalyst.

If we were to interconvert cis-2-butene to trans-2-butene and allow them to reach equilibrium, we would find that they are not of equal stability at equilibrium. The ratio of isomers is 76% trans to 24% cis.

H_3C H H_3C CH_3

C=C C=C

H CH_3 H H

trans (76%) cis (24%)

Based on the relationship between equilibrium constants and free energy differences, trans-2-butene is more stable than cis-2- butene by 0.66 kcal mole^{-1}. In general cis alkenes are less stable than their trans isomer because of steric (spatial) strain between the two bulky substituents on the same side of the double bond.

Although it is sometimes possible to inter convert double-bond isomers chemically to establish an equilibrium and thus obtain information about relative stabilities of alkenes, there is an easier way to gain nearly the same information by other methods. One such method is catalytic hydrogenation of an alkene.

Consider the hydrogenation of cis and trans-2-butene. Both the alkenes react with hydrogen to give the same product, n-butane. The heat of hydrogenation for cis isomer, ΔH^o_{cis} = -28.6 kcal mole^{-1}, whereas for trans isomer, ΔH^o_{trans} = -27.6 kcal mole^{-1} (Note that these ΔH^o values are negative since heat is evolved). the difference in the heats of hydrogenation for the 2-butene isomers, 1.0 kcal mole^{-1}, is in good agreement with the energy difference calculated from equilibrium data i.e., 0.66 kcal mole^{-1}. Heats of hydrogenation of some of the alkenes are given in table 5.1

Table 5.1 Heats hydrogenation of some alkenes

Substitution	*Alkenes*	*ΔH^{o} hydrogenation kcal mole^{-}*
Unsubstituted	$CH_2 = CH_2$	-32.8
Monosubstituted	$H_3C\text{-}CH{=}CH_2$	-30.1
Dissubstituted	cis $H_3\ CCH = CH\ \text{-}\ CH_3$	-28.6
	Trans $H_3CCH = CH\ \text{-}\ CH_3$	-27.6
	$(H_3C)_2\ C = CH_2$	-28.4
Trisubstituted	$(H_3C)_2\ C = CH\ \text{-}\ CH_3$	-26.9
Tetrasubstituted	$(H_3C)_2\ C = C\ (CH_3)_2$	-26.6

The data in table 5.1 show that alkenes become more stable with increasing substitution. As a general rule alkenes follow the following stability order.

$R_2C{=}CR_2$ > $R_2C{=}CHR$ > $R_2C{=}CH_2$, $RHC{=}CHR$ (trans) , $RHC{=}CHR$ (cis) > $RHC{=}CH_2$ > $H_2C{=}CH_2$

Tetrasubstituted Trisubstituted Disubstituted

Disubstituted Monosubstituted Unsubstituted

The reason for this observed stability order for substituted alkenes may be due to a stabilizing interaction called hyperconjugation. As the alkyl groups increases the number of hyperconjugation structures increases and stability of alkanes is also increases.

5-6-2 Addition of halogens

When an alkene is treated with chlorine or bromine in carbon tetrachloride, the halogen adds rapidly across the carbon-carbon double bond of the alkene to form 1, 2-dihalo alkane (vicinal dihalide).

$$H_2C = CH_2 + X_2 \xrightarrow{CCl_4} H_2C(X) - CH_2(X)$$

Alkene Halogen 1, 2-Dihaloalkane

For example,

$$H_2C = CH_2 + Br_2 \xrightarrow{CCl_4} \underset{\text{Br}}{\overset{\text{Br}}{H_2C - CH_2}}$$

Mechanism : The addition of bromine to alkene forming 1, 2-dihalo alkenes takes place by the following steps.

Step 1 : When a bromine molecule comes in the proximity of an alkene, the negative π electron cloud of the alkene causes polarization of the bromine molecule, that is, displacement of the pair of electrons joining the bromine atoms towards one of the atoms.

$$Br — Br \longrightarrow \overset{\delta-}{Br} — \overset{\delta+}{Br}$$

Step 2 : The positive end of the polarized bromine molecule is then attracted near to the π cloud to form a low stability π complex.

$$>C = C< \; + \; \overset{\delta-}{Br} — \overset{\delta+}{Br} \longrightarrow >C \overset{Br\delta^-}{\overset{\vdots}{\overset{Br\delta^+}{=}}} C<$$

π - complex

Step 3 : The positively charged bromine atom is then attracted by both carbons of the double bond and the Br-Br bond breaks down, the negatively charged bromine leaves as Br⁻ and cyclic bromium ion is formed.

δ⊖Br

δ⊕Br

>C=C<

π- Complex

⟶

Br⊕

>C — C<

Cyclic bromonium ion

+ Br⊕

Step 4 : Nucleophilic attack of bromide ion, Br^- yields, 1.2 dibromo compound.

$$\text{cyclic bromonium ion } (Br^{\oplus} \text{ bridging } >C-C<) + Br^{\ominus} \longrightarrow >C(Br)-C(Br)<$$

1,2-dibromo compound

It may be noticed that Br^- attacks the cyclic bromonium ion from the opposite (more open) side of the plane of the alkene molecule to form the addition product.

The two-step mechanism of trans-addition of bromine is supported by the fact that if bromination of alkene is done in the presence of chloride ion, a dihalide containing one bromine atom and chlorine atom is produced along with the dibromide. Thus the bromination of ethylene in the presence of NaCl gives both 1, 2-dibromoethane, and 1-bromo 2-chloroethane.

$$H_2C=CH_2 \xrightarrow{Br_2} H_2C\overset{Br^{\oplus}}{—}CH_2 \xrightarrow{Br^{\ominus}} H_2C(Br)-CH_2(Br)$$

1, 2-dibromoethane

$$H_2C\overset{Br^{\oplus}}{—}CH_2 \xrightarrow{Cl^{\ominus}} CH_2(Br)-CH_2(Cl)$$

1-bromo-2-chloroethane

Thus the addition of halogens to alkenes is a trans-addition process in contrast to catalytic hydrogenation which is a cis- addition.

Bromine solution in carbon tetrachloride has orange colour. When it is added drop wise to an alkene, bromine is used up to form colourless vicinal dibromide and the orange color disappears immediately. This is a diagnostic test for the unsaturation in unknown compounds.

Fluorine **can not** be used for the preparation of vicinal difluorides because it causes extensive fragmentation of the alkene molecule.

Iodine reacts slowly with alkenes to form vicinal diiodides which are unstable and eliminate iodine very readily to regenerate the original alkene.

5-6-3 Addition of hydrogen halide

Alkenes react with hydrogen halides by addition across the carbon-carbon double bond to form alkyl halides.

$$>C{=}C< + HX \longrightarrow >\overset{H}{\overset{|}{C}}-\underset{X}{\underset{|}{C}}<$$

Alkene Alkylhalide

where HX = HCl, HBr, HI

Addition of HX to symmetrical alkene : The addition of a hydrogen halide to a symmetrical alkene gives only one product because it does not matter on which carbon of the double bond the halogen atom is placed.

For example,

$$H_2C{=}CH_2 + HBr \longrightarrow \overset{H}{\overset{|}{H_2C}}-\underset{Br}{\underset{|}{CH_2}}$$

Mechanism : Hydrohalogenation of a symmetrical alkene, for example, ethylene is as follows :

Step 1 : The hydrogen halide forms a π-complex which rearranges to give a carbonium ion.

$$H_2C{=}CH_2 + HBr \longrightarrow \underset{\pi\text{-complex}}{[H_2C \overset{\overset{Br^{\sigma-}}{\vdots}}{\overset{H^{\sigma+}}{\vdots}} CH_2]} \longrightarrow \overset{H}{\overset{|}{H_2C}}-\overset{+}{CH_2} + Br^-$$

Step 2 : Alkylhalide is then obtained by the nucleophilic attack of Br^- by the carbonium ion.

$$H_2\overset{+}{C}-\underset{|}{C}H_2 \; (\text{with H on } C) + Br^- \longrightarrow \underset{\underset{H}{|}}{H_2C}-\underset{\underset{Br}{|}}{CH_2}$$

Ethylbromide

As shown in Step 1, only one type of carbonium is possible from a symmetrical alkene. Hence, a single product would be formed in Step 2.

Addition to unsymmetrical alkenes; Orientation of electrophilic addition : When a hydrogen halide reacts with an unsymmetrical alkene, there are two possible addition products depending on the placement on a particular carbon atom. thus propene with HBr can form 1-bromopropane or 2-bromo propane. But it has been established experimentally that 2-bromopropane (isopropyl bromide) is obtained predominantly.

$$H_2C=CH-CH_3 + HBr \longrightarrow H_3C-\overset{\overset{Br}{|}}{CH}-CH_3$$

2-Bromopropane

Markovnikov studied a number of such addition reactions, and suggested a rule to predict the course of reaction for addition of hydrogen halides and other unsymmetrical reagents such as H_2SO_4, HOCl, etc., to unsymmetrical alkenes. The empirical rule named after him as **Markovnikov rule** which states that, *in the addition of unsymmetrical reagents, HX, to unsymmetrical olefins, the negative part of the addendum, HX, is added to the carbon atom of the double bond which is having the least number of hydrogen atoms.*

Validity of Markovnikov's rule can be explained as follows : .

(a) Relative stabilities of carbocations : Markovnikov's rule was purely empirical when it was formulated. However now a theoretical explanation of orientation of addition to olefins can be offered on the basis of relative stability of carbocations. If there is a possibility of formation of more than one carbocations, the addition of the electrophile yields the more stable carbocation.

$$H_3C-CH=CH_2 \xrightarrow{HBr} \begin{cases} H_3C-\overset{+}{C}H-CH_2-H \quad (2^\circ) \\ H_3C(H)-CH-\overset{+}{C}H_2 \quad (1^\circ) \end{cases}$$

The order of stability of carbocations is

Tertiary > Secondary > Primary

Secondary carbocation (isopropyl carbocation) is more stable, predominates and reacts with the Br to form isopropyl bromide.

$$H_3C-\overset{+}{C}H-CH_3 + Br^- \longrightarrow H_3C-\underset{\displaystyle Br}{\underset{|}{CH}}-CH_3$$

(b) Inductive effect of the substitutents : We can also understand the Markovnikov's rule by considering inductive effect of the substituents on the ethylenic bond. In propene, for instance, the polarization of the double bond will be as depicted below due to the electron repelling inductive effect of the methyl group, which will be transmitted through the π bond.

$$H_3C \longrightarrow \overset{\delta+}{C}H = \overset{\delta-}{C}H_2$$

Thus the positive and negative ends of the dipole H-Br will add to the negative and positive ends of the double bond, respectively, giving rise to the isopropyl bromide.

$$H_3C-CH=CH_2 + HBr \longrightarrow H_3C-\underset{\displaystyle Br}{\underset{|}{CH}}-\underset{\displaystyle H}{\underset{|}{CH_2}}$$

Similarly Markovnikov's addition of hydrogen halide can be explained by the electron-donating resonance effect (+ Mesomeric effect) of halogen attached to a double bond, dominating over its electron attracting inductive effect (-I.E.). Thus, 1, 1- dihaloethane is formed from vinyl halide.

$$H_2C{=}CH{-}X: \xrightarrow{H - X} \underset{\displaystyle H}{\underset{|}{H_2C}} - \underset{\displaystyle X}{\underset{|}{CH}} - X$$

Addition of unsymmetrical reagents to unsymmetrical olefin of the type R-CH = CH - R′ : For example in 2-pentene ($H_3C - CH = CH - CH_2 - CH_3$). The two unsaturated carbon atoms are having the same number of hydrogen atoms. According to Marovnikov's rule the addition product should contain nearly equal amounts of the two isomers. Actually 2-bromopentane is formed slightly more than 3-bromopentane.

$$H_3C - CH = CH - CH_2 - CH_3 \xrightarrow{HBr}$$

$$\underset{\text{2-Bromopentane}}{H_3C - \underset{\displaystyle Br}{\underset{|}{CH}} - \underset{\displaystyle H}{\underset{|}{CH}} - CH_2 - CH_3} + \underset{\text{3-Bromopentane}}{H_3C - \underset{\displaystyle H}{\underset{|}{CH}} - \underset{\displaystyle Br}{\underset{|}{CH}} - CH_2 - CH_3}$$

Formation of 2-bromopentane is explained on the basis of relative stabilities of the carbonium ion.

$$H_3CCH = CHCH_3 + HBr \longrightarrow \underset{\text{More stable carbocation (I)}}{H_3C - \overset{+}{CH} - \underset{\displaystyle H}{\underset{|}{CH}} - CH_2 - CH_3}$$

$$H_3CCH = CHCH_3 + HBr \longrightarrow \underset{\text{Less stable carbocation (II)}}{H_3C - \underset{\displaystyle H}{\underset{|}{CH}} - \overset{+}{CH} - CH_2 - CH_3}$$

Since on the basis of hyperconjugation the carbonium ion I is more stable than the II, the addition product will be 2- bromopentane:

$$H_3C - \overset{+}{CH} - \underset{\displaystyle H}{\underset{|}{CH}} - CH_2 - CH_3 \xrightarrow{Br^-} \underset{\text{2-Bromopentane}}{H_3C - \underset{\displaystyle Br}{\underset{|}{CH}} - \underset{\displaystyle H}{\underset{|}{CH}} - CH_2 - CH_3}$$

Formation of the 3-bromopentane is explained on the basis of the inductive effect. Ethyl group is more powerful positive inductive effect group than the methyl group and hence polarization of the double bond takes place in the following way leading to the formation of 3-bromopentane.

$$H_3C - CH = CH - CH_2 - CH_3 \xrightarrow{HBr} H_3C - \underset{\displaystyle H}{\underset{|}{CH}} - \underset{\displaystyle Br}{\underset{|}{CH}} - CH_2 - CH_3$$

3-Bromopentane

Markovnikov's rule has now been replaced by a more general rule, which states that *electrophilic addition to a carbon-carbon double bond involves the intermediate formation of a more stable carbocation.*

Peroxide effect, Anti-Markovnikov addition, Kharash effect : Kharash discovered that the addition of HBr to unsymmetrical alkene in the presence of organic peroxides (R-O-O-R) takes a course opposite to that predicted by Markovnikov rule. *This phenomenon of anti markovnikov addition caused by the presence of peroxide is called Kharash effect or peroxide effect.* Thus when propene reacts with HBr in the presence of a peroxide, the product is mainly 1-bromopropane, whereas in the absence of a peroxide the main product is isopropyl bromide.

$$H_3C - CH = CH_2 + HBr \longrightarrow H_3C - CH_2 - CH_2Br$$

1-Bromopropane
(anti-Markovnikov product)

$$H_3C - CH = CH_2 + HBr \longrightarrow CH_3 - \underset{\displaystyle Br}{\underset{|}{CH}} - CH_3$$

2-Bromopropane
(Markovnikov product)

Anti-Markovnikov addition of HBr to alkene also takes place when the reactants are exposed to U.V. light. This mode of addition is not to be regarded as the reverse process of Markovnikov addition as the two reactions are entirely different in nature. While Markovnikov addition follows ionic mechanism, the anti Markovnikov addition proceeds by free radical mechanism.

Mechanism : Organic peroxides as well as Ultraviolet light are able to initiate the chain reaction which results in the formation of anti-Markovnikov products. Thus propene reacts with HBr to give 1-bromopropane

The mechanism of the reaction can be divided into 3 parts.

(a) Initiation (b) Propogation (c) Termination

(a) Initiation :

(i) Peroxide dissociates to give two alkoxy radicals which attaçhed HBr to form bromine free radical

$$R - O - O - R \longrightarrow 2RO^{\bullet}$$

$$RO^{\bullet} + HBr \longrightarrow ROH + Br^{\bullet}$$

(ii) HBr on photochemical dissociation gives the free radicals $H^{\bullet}$ and $Br^{\bullet}$

$$HBr \longrightarrow H^{\bullet} + Br^{\bullet}$$

(b) Propagation

Bromine free radical formed attacks the alkene molecule giving two possible bromo alkyl free radicals

$$H_3C - CH = CH_2 + Br^{\bullet} \longrightarrow H_3C - \dot{C}H - CH_2Br$$

2° Free radicaal (more stable)

$$H_3C - CH = CH_2 + Br^{\bullet} \longrightarrow H_3C - CH(Br) - \dot{C}H_2$$

1° Free radical (less stable)

The order of stability of the free radicals is 3° 2° 1°. Therefore 2° free radical is formed predominantly.

The more stable 2° free radical reacts with HBr forming 1- bromopropane (anti-Markovnikov's product) and another free radical which propagates the chain reaction the

$$H_3C - \dot{C}H - CH_2Br + HBr \longrightarrow H_3C - CH(H) - CH_2 - Br + Br^{\bullet}$$

1-Bromopropane

(anti markovanikov's product)

(c) Termination

In this step the radicals combine to give stable molecules.

$$Br^{\bullet} + Br^{\bullet} \longrightarrow Br_2$$

It may be noted that HCl and HI do not give anti-Markovnikov's products in the presence of peroxides, and actually form the normal Markovnikov's products. this is because HCl and HI do not undergo free radical addition. With hydrogen chloride the cleavage of HCl bond to give Cl is the initiation step is too unfavourable as HCl bond is stronger than

HBr bond. In case of HI, free iodine $\dot{I}$ is obtained easily since H-I bond is weak. However, the association of iodine radicals to form molecular iodine (I_2) is preferred over formation of a relatively unstable carbon-iodine bond.

5-6-4 Addition of Hypohalous Acids

The addition of chlorine or bromine to alkenes, when carried in the presence of water, has the net result of adding a hypohalous acid (HOX) molecules across the double bond. The compounds produced are known as halohydrin in which a hydroxyl group and a halogen atom are attached to adjacent carbons.

$$>C=C< + HO-X \longrightarrow >\underset{OH}{\underset{|}{C}}-\underset{X}{\underset{|}{C}}<$$

where HOX = HOCl, HOBr

It may be noted that here OH group behaves as the negative part of the addendum ($H^{\delta} - X^{\delta}+$). If HO — X is assumed to add to alkenes the addition follows Markovnikov's rule. For example,

$$H_3C-CH=CH_2 + HOBr \longrightarrow H_3C-\underset{OH}{\underset{|}{CH}}-\underset{Br}{\underset{|}{CH_2}}$$

Mechanism : When an alkene reacts with bromine in the presence of water, water competes with bromide ion as nucleophile and reacts with the bromonium ion intermediate to yield a mixture of dibromide and bromohydrin (bromoalcohol). The net effect is addition of hypobromous acid HO - Br to the alkene.

$$H_3C-CH=CH-CH_3 \xrightarrow[H_2O]{Br_2} H_3C \quad H_3C-CH\overset{\overset{+}{B}}{\frown}CH-CH_3$$

$$\xrightarrow{H_2O} H_3C-\underset{OH}{\underset{|}{CH}}-\overset{Br}{\overset{|}{CH}}-\underset{OH}{\underset{|}{CH_3}}$$

3-bromo —— 2-Butanol

$$\xrightarrow{Br^-} H_3C-\overset{Br}{\overset{|}{CH}}-\underset{Br}{\underset{|}{CH}}-CH_3$$

2, 3-dibromobtane

5-6-5 Addition of sulphuric acid to alkenes (Hydration of alkenes)

Alkenes react with sulphuric acid at room temperature to give an alkyl hydrogen sulphate

$$\underset{\text{Alkene}}{>C=C<} + \underset{\text{Sulphuric acid}}{H_2SO_4} \longrightarrow \underset{\text{Alkylhydrogen sulphate}}{>\overset{\overset{\displaystyle H}{|}}{C}-\overset{\overset{\displaystyle OSO_3H}{|}}{C}<}$$

The mechanism of the addition is similar to that of ionic addition of hydrogen halides and Markovnikov's rule is also applicable for these additions to unsymmetrical alkenes.

For example,

$$H_3C-CH=CH_2 + HOSO_3H \longrightarrow H_3C-\overset{\overset{\displaystyle OSO_3H}{|}}{C}-\underset{\underset{\displaystyle H}{|}}{CH_2}$$

Isopropylhydrogen sulphate

The resulting isopropylhydrogen sulphate upon hydrolysis with warm water gives isopropyl alcohol.

$$H_3C-\overset{\overset{\displaystyle OSO_3H}{|}}{CH}-CH_3 \xrightarrow{HOH} H_3C-CH-\overset{\overset{\displaystyle OH}{|}}{CH_3} + H_2SO_4$$

5-6-6 Addition of water to alkenes (Hydration)

Water adds to the more reactive alkenes in the presence of a catalyst to form alcohols.

$$>C=C< + H_2O \xrightarrow{H^+} \underset{\text{Alcohol}}{>\underset{\underset{\displaystyle H}{|}}{C}-\underset{\underset{\displaystyle OH}{|}}{C}<}$$

This addition also follows Markovnikov's rule.

$$H_3C-CH=CH_2 + H_2O \xrightarrow{H^+} H_3C-\underset{\displaystyle OH}{\underset{|}{CH}}-CH_3$$

Isopropyl alcohol

Mechanism : The mechanism of acid catalysed addition of water to propene, for instance, can be visualized as under :

$$H_2O + H_2SO_4 \rightleftharpoons H_3O^+ + HSO_4$$

$$H_3C-HC=CH_2 + H-\overset{+}{\underset{\displaystyle H}{\underset{|}{O}}}-H \rightleftharpoons H_3C-\overset{+}{CH}--\underset{\displaystyle H}{\underset{|}{CH_2}}$$

$$\Big\Updownarrow H_2O$$

$$H_3C-\underset{\displaystyle OH}{\underset{|}{CH}}-CH_3 \overset{-H^+}{\rightleftharpoons} H_3C-\underset{\displaystyle \overset{+}{O}-H \atop {| \atop H}}{\underset{|}{CH}}-CH_3$$

Isopropyl alcohol

5-6-7 Hydroboration

A useful synthetic procedure in organic chemistry involves addition of borane (BH_3) to carbon-carbon double bond. This process is referred to as **hydroboration.**

$$>C=C< + BH_3 \longrightarrow >\underset{\displaystyle H}{\underset{|}{C}}-\underset{\displaystyle BH_2}{\underset{|}{C}}<$$

Alkyl borane

Borane itself is unknown but its dimer, (B_2H_6), behaves as if it were the hypothetical monomer. Because of the vacant orbital on borane, it acts as an electrophile attacking the π electrons of the alkene and in accordance with Markovnikov's rule. Intramolecular hydride transfer then completes the reaction. Thus the reaction is believed to take place in concerted manner involving a four membered transition state.

$$H_3C - CH = CH_2 + H-BH_2$$

$$\begin{array}{c} H \\ \delta- \quad | \delta+ \\ H \cdots B - H \\ \vdots \qquad \vdots \\ H_3C - \underset{\delta+}{CH} - \underset{\delta-}{CH_2} \end{array} \longrightarrow H_3C - CH_2 - CH_2BH_3$$

Because of the four membered transition state hydroboration is a cis addition. As these reactions do not yield rearranged products, it can be inferred that a free carbocation is not an intermediate in the process.

Alkyl boranes behave like borane and react further with excess of alkene forming di and trialkyl boranes.

$$\underset{\text{n-Propyl borane}}{H_3C - CH - CH_2 - BH_2} \xrightarrow{\text{Propene}} \underset{\text{di-n-Propylborane}}{(H_3C - CH_2 - CH_2)2BH} \xrightarrow{\text{Propene}} \underset{\text{tri-n-Propyl borane}}{(H_3C - CH_2 - (CH_2)_3 B}$$

The rate of addition of borane to alkene decreases with the increase in the alkyl substituents on the double bond.

Trialkyl boranes are used for the synthesis of primary alcohols by reaction with alkaline aqueous solution of H_2O_2

$$(H_3CCH_2CH_2)_3B \xrightarrow{\text{aq. } H_2O_2} (H_3CCH_2CH_2O)_3B \xrightarrow{^-OH} 3CH_2CH_2CH_2OH + B(OH)_3B$$

n-Propyl alcohol

The overall result of this hydroboration is the anti- Markovnikov's hydration on the double bond.

$$H_3C - CH = CH_2 \xrightarrow[\text{(ii) } H_2O_2 / \bar{O}H]{\text{(i) } B_2H_6} \underset{\text{n-Propyl alcohol}}{H_3C - CH_2 - CH_2OH}$$

5-6-8 Oxidation Reactions

Alkenes are more readily oxidized than alkanes, the site of attack being the double bond. The nature of oxidation products depend upon the strength of the oxidizing agent employed.

(i) Epoxidation : Lower alkenes add on to atomic oxygen across the carbon-carbon double bond when reacted with oxygen in the presence of silver catalyst, to form epoxides

$$>C=C< + \frac{1}{2}O_2 \xrightarrow[200\text{-}400^\circ C]{Ag} >C\overset{O}{\frown}C<$$

Epoxide

This oxidation reaction of alkenes giving epoxides is known as epoxidation. Thus

$$H_2C=CH_2 + \frac{1}{2}O_2 \xrightarrow[250^\circ C]{Ag} H_2C\overset{O}{\frown}CH_2$$

Epoxy ethane (ethylene oxide)

A better and more important method of epoxidation is by the treatment of alkenes by peracids which are the organic derivatives of hydrogen peroxide, viz., perform acid, metachloro perbenzoic acid and trifluoroperacetic acid etc.

$$H_2C=CH_2 + Ar-\overset{O}{\overset{\|}{C}}-O-O-H \longrightarrow H_2C\overset{O}{\frown}CH_2 + Ar-\overset{O}{\overset{\|}{C}}-OH$$

metacholorop perbenzoic acid — epoxyethane — metacholoro benzoic acid

$Ar = m\text{-}Cl\,C_6H_4-$

Mechanism : The accepted mechanism of epoxidation of peracids involves simultaneous bond making and bond breaking

$$H_2C=CH_2 + O(H)-O-\overset{\underset{\|}{O}}{C}-Ar \longrightarrow H_2C\overset{O}{\frown}CH_2 + Ar-\overset{O}{\overset{\|}{C}}-OH$$

Ethylene — m-Chloroperbenzoic acid — Epoxyethane — Benzoic acid

$Ar = m\text{-}Cl\,C_6H_4$

(ii) Ozonolysis : The classical reaction for cleaving the carbon- carbon double bond uses ozone. Ozonolysis is carried out in two stages. First, addition of ozone to the double bond to form an ozonide, and second, hydrolysis of the ozonide to yield the cleavage products (i.e. carbonyl compounds). Ozone gas is passed into a solution of the alkene in some inert solvent like CCl_4. Evaporation of solvent leaves the ozonide as a viscous oil. This unstable, explosive compound is not purified, but is treated directly with water, generally in the presence of a reducing agent.

$$R-CH=CH_2 + O_3 \longrightarrow \underset{\text{Ozonide}}{R-HC\overset{O}{\frown}CH_2\ (O-O)} \xrightarrow[A_2O]{Zn} R-\overset{O}{\overset{\|}{C}}-H + \overset{O}{\overset{\|}{C}}H_2$$

Mechanism : The mechanism of ozonide formation is a multi step process. The first step consists of 1, 3-dipolar addition of ozone to double bond forming a molozonide. The molozonide, being unstable, subsequently decomposes into two fragments. Recombination of these fragments in an alternative way yields an ozonide.

$$>C=C< + {}^{\ominus}O-O-O^{\oplus} \longrightarrow \underset{\text{Molozonide}}{>C-C<\ (O-O-O)} \longrightarrow >C=\overset{\oplus}{O}-O^{\ominus} + O=C< \longrightarrow \underset{\text{Ozonide}}{>C\overset{O}{\frown}C<\ (O-O)}$$

Ozonides when treated with a reducing agent like zinc and water or hydrogen in the presence of palladium, are cleaved at the seat of the double bond to give two carbonyl compounds.

$$>C\overset{O}{\frown}C<\ (O-O) \xrightarrow[Zn]{H_2O} >C=O + O=C< + H_2O_2$$

Location of double bond : The overall reaction i.e., fragmentation of the alkene into carbonyl components is termed as ozonolysis and has been found to be of much use for preparative purposes and also for locating the position of a double bond in an unknown compound. This is possible as the cleavage of the double bond produces smaller carbonyl fragments in which a double bonded oxygen is found attached to each of the carbon atoms linked by a double bond. The aldehyde and ketones thus obtained are identified and by locating the number and arrangement of carbon atoms in these products one can work back so as to arrive at the structure of the original alkene.

For example, three isomers having molecular formula C_4H_8 have been characterized as 1-butene, 2-butene and 2-methyl propene respectively on the basis of identification of the aldehydes and ketones obtained by ozonolysis.

$$H_3C - CH_2 - CH = CH_2 \xrightarrow[\text{(ii) } Zn/H_2O]{\text{(i) } O_3} H_3C - CH_2 - \overset{\displaystyle H}{\overset{|}{C}} = O + O = CH_2$$

Propionaldehyde (one mole) formaldehyde (one mole)

$$H_3C - CH = CH - CH_3 \xrightarrow[\text{(ii) } Zn/H_2O]{\text{(i) } O_3} H_3C - \overset{\displaystyle H}{\overset{|}{C}} = O + O = \overset{\displaystyle H}{\overset{|}{C}}\ CH_3$$

acetaldehyde (two moles)

$$H_3C - \overset{\displaystyle CH_3}{\overset{|}{C}} = CH_2 \xrightarrow[\text{(ii) } Zn/H_2O]{\text{(i) } O_3} H_3C - \overset{\displaystyle CH_3}{\overset{|}{C}} = O + O = CH_2$$

acetone (1 more) formaldehyde (1 more)

(iii) Hydroxylation : Hydroxylation of alkenes i.e. the addition of a hydroxyl to each of the two alkene carbons can be carried out with reagents such as potassium permanganate ($KMnO_4$) and Osmium tetroxide (OsO_4) etc.

Potassium permanganate is a common and inexpensive oxidant that reacts with alkenes under carefully controlled conditions in alkaline medium to yield 1, 2-glycols.

$$\underset{\text{alkene}}{>C=C<} \xrightarrow{\text{alk. } KMnO_4} \underset{\text{1, 2-glycol}}{>\underset{OH}{\underset{|}{C}}-\underset{OH}{\underset{|}{C}}<}$$

Mechanism : The reaction occurs through an intermediate cyclic manganate species, which is then hydrolysed to give the 1, 2- glycol.

$$R-CH=CH-R + \underset{\text{Potassium permanganate}}{O_2Mn(=O)OK} \longrightarrow \text{RHC}-\text{CHR (cyclic } -O-Mn(=O)(OK)-O- \text{)} \xrightarrow{\text{hydrolysis}} \underset{\text{1, 2-glycol}}{\underset{HO}{\underset{|}{RHC}}-\underset{OH}{\underset{|}{CHR}}}$$

Hydroxylation with $KMnO_4$ or OsO_4 is a cis **addition** process. This reaction is also the basis of a qualitative test of unsaturation, called **Baeyers test.** When a solution of alkaline permanganate (Baeyer's reagent) is added to an alkene, the double bond is rapidly oxidized and the purple permanganate ion is reduced. Disappearance of purple colour indicates the presence of the carbon carbon double bond.

Hydroxylation of alkenes can also be carried out with osmium tetroxide. In ether solution at room temperature the reagent adds rapidly to an ethylenic

$$RCH=CHR + OsO_4 \longrightarrow \text{RCH}-\text{CHR (cyclic } -O-Os(=O)_2-O- \text{)} \xrightarrow[NaHSO_3,\ H_2O]{\text{hydrolysis}} \underset{\text{1, 2-glycol}}{\underset{OH}{\underset{|}{RCH}}-\underset{OH}{\underset{|}{CHR}}}$$

double bond to form, the intermediate cyclic osmic ester (Osmate) which upon hydrolysis in the presence of $NaHSO_3$ yielding 1, 2-glycol.

5-7 ALKADIENES

Compounds containing two double bonds on the carbon chain, in a molecule are called as dienes or alkadienes. Their general formula is C_nH_{2n-2}. Alkadienes are of three types depending on the position of the double bonds on the carbon chain.

(i) Conjugated dienes : The alkadienes in which the two double bonds are separated by one single bond are called **conjugated dienes**. In these dienes double and single bonds are alternatively present on the carbon chain.

$H_2C = CH — CH = CH_2$ — 1,3-Butadiene

$H_2C = C(CH_3) — CH = CH_2$ — 2-Methyl-1,3 butadiene (isoprene)

Conjugated dienes have a unique chemistry and their properties are sufficiently different from simple alkenes.

(ii) Isolated dienes : *The alkadienes in which the double bonds are separated by more than one single bond are called* **isolated dienes.**

$> C = CH — CH_2 — CH = C <$

For example,

$H_2C = CH - CH_2 - CH = CH_2$ — 1, 4 Pentadiene,

$H_2C = CH - CH_2 - CH_2 - CH = CH_2$ — 1, 5-Hexadiene

In these dienes the two double bonds do not interact, as they are isolated by the intervening CH_2 group(s). the double bonds in isolated dienes behave like independent double bonds or simple olefins. The only difference between the alkenes and isolated dienes is that in dienes the reactions of alkenes are shown twice because of the presence of two double bonds.

(iii) Cumulated dienes or allenes : **Cumulated dienes** are those alkadienes in which the bonds appear on successive carbon atoms and one carbon is a part of the two double bonds.

$> C = C = C <$

For example,

$H_2C = C = CH_2$ — Propadiene (allene)

$\overset{5}{H_3C} — \overset{4}{CH_2} — \overset{3}{CH} = \overset{2}{C} = \overset{1}{CH_2}$ — 1,2-Pentadiene

Allene is the parent compound of the series and hence these are also known as **allenes.** The properties of cumulated dienes are different from both simple alkenes and conjugated dienes. Here the π-systems are not conjugated because they are at right angles to one another. The substituted cumulated dienes have no asymmetric carbon atoms, but they are asymmetric mulecules, i.e., they have no plane of symmetry. The allenes represented by the general formula

$$\begin{matrix} R & & & & H \\ & \diagdown & & \diagup & \\ & C = & C & = C & \\ & \diagup & & \diagdown & \\ H & & & & R \end{matrix}$$

are optically active, because the terminal substituents, R groups, are in different planes

5-7-1 Nomenclature of dienes

The IUPAC nomenclature of alkadienes is same as in case of alkenes except that the numeral **di** is added before the suffix - **ene** to indicate the presence of two double bonds.

For example,

$H_2C = CH - CH = CH_2$ 1, 3-butadiene

$H_2C = CH - CH_2 - CH = CH_2$ 1, 4-pentadiene

$H_2C = C = CH_2$ propadiene

5-7-2 General methods of preparation

Dienes are generally prepared by suitably varying the methods employed for the preparation of simple alkenes. The two most commonly employed methods are given below :

(i) Dehydrogenation of alkenes by passing the vapours over heated catalyst : 1, 3-butadiene may be prepared by passing the vapours of n-butane over heated Cr_2O_3 - Al_2O_3 at 600°C.

$$H_3C - CH_2 - CH_2 - CH_3 \xrightarrow[\text{heat}]{Cr_2O_3 - Al_2O_3} H_2C = CH - CH = CH_2 + 2H_2$$

(ii) Dehydration of diols in the presence of acids : Dehydration of 1, 4-butanediol in the presence of sulphuric acid or phosphoric acid, results in the formation of 1, 3-butadiene.

$$\underset{OH}{\underset{|}{H_2C}} - \underset{H}{\underset{|}{CH}} - \underset{H}{\underset{|}{CH}} - \underset{OH}{\underset{|}{CH_2}} \xrightarrow[\Delta]{H_2SO_4} H2C = CH - CH = CH_2$$

5-7-3 Physical properties

1, 3-butadiene is a colorless gas whereas isoprene is a liquid.

Structure of 1, 3-butadiene : It is a planar molecule and represented as

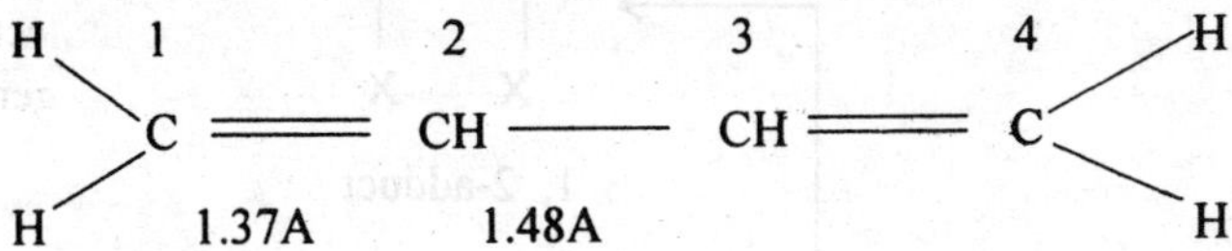

It may be noticed that C_2-C_3 bond is shorter than normal carbon- carbon single bond (1.54A), while C_1-C_2 and C_3-C_4 bonds are somewhat longer than as isolated carbon-carbon double bond (1.34A). In the above formula we are showing its structure as if the two π bonds are independent. Actually, they are not and assume a structure as depicted in figure 5.2. Overlapping is possible between all the four p-orbitals which permits the delocalization of the π electrons about the entire molecule. Each pair of π electrons thus is attracted, not by two, but by all four nuclei. The delocalization of the π electrons adds greatly to the stability of the butadiene molecule.

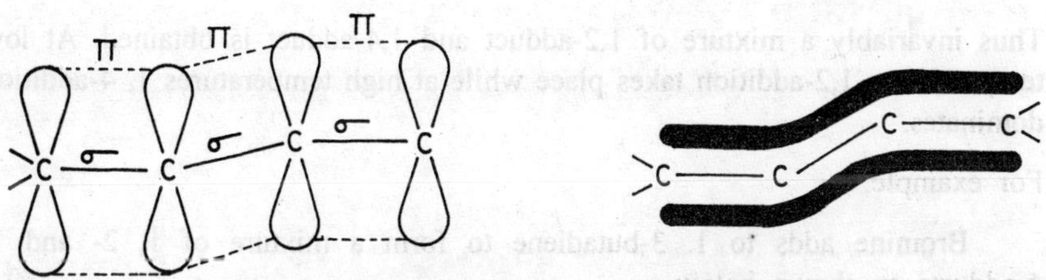

Fig. 5.2 Delocalization of π electrons

5-8 CHEMICAL PROPERTIES

The chemical properties of dienes are essentially due to the presence of two carbon-carbon double bonds. Thus non-conjugated dienes behave in exactly the same manner as simple alkenes except that attaching reagent consumed is twice the amount required for one double bond.

In case of conjugated dienes, however, the properties of the double bonds are considerably modified due to delocalization of π electrons (resonance). Thus conjugated dienes are characterized by their greater stability and the 1, 4- addition reactions. In 1, 4-addition reactions the addition takes place at terminal carbons of the conjugated system. The important reactions of dienes are discussed below.

(a) **Electrophilic addition reactions** : 1, 3-butadiene undergoes addition reactions with reagents such as Cl_2, Br_2, HBr etc. Although it contains two double bonds but only one molecule of the addendum is added either to 1, 2-carbons (1, 2 addition) or to 1, 4-carbons (1, 4-addition).

$$H_2C = CH - CH = CH_2 + X_2 \xrightarrow{\text{1, 2-addition}} \underset{\text{1, 2-adduct}}{H_2C(X) - CH(X) - CH = CH_2}$$

$$H_2C = CH - CH = CH_2 + X_2 \xrightarrow{\text{1, 4-addition}} \underset{\text{1, 4-adduct}}{H_2C(X) - CH = CH - CH_2(X)}$$

Thus invariably a mixture of 1,2-adduct and 1,4-adduct is obtained. At low temperatures, 1,2-addition takes place while at high temperatures 1, 4-addition dominates.

For example,

Bromine adds to 1, 3-butadiene to form a mixture of 1, 2- and 1, 4-adducts as shown below.

$$H_2C = CH - CH = CH_2 \xrightarrow[\text{1,2-addition}]{Br_2,\ \text{Low temp. } (-80^{\circ}C)} H_2C(Br) - CH(Br) - CH = CH_2$$

$$H_2C = CH - CH = CH_2 \xrightarrow[\text{1,4-addition}]{Br_2,\ \text{High temp. } (40^{\circ}C)} \underset{\text{1, 4 - dibromo - 2 - butene}}{H_2C(Br) - CH = CH - CH_2(Br)}$$

The formation of 1, 4-adduct is quite characteristic of 1,3- butadiene and this can be explained by assuming that in the initial electrophoretic attack, the electrophile can add either to C_1 or to C_2. Since addition of the electrophile at C_2 would give rise to an unstable primary localized carbocation, the electrophilic attack takes place at C_1 resulting in the formation of a resonance stabilized allylic cation which also explains the enhanced reactivity of dienes over isolated ethylenic bonds.

Mechanism :

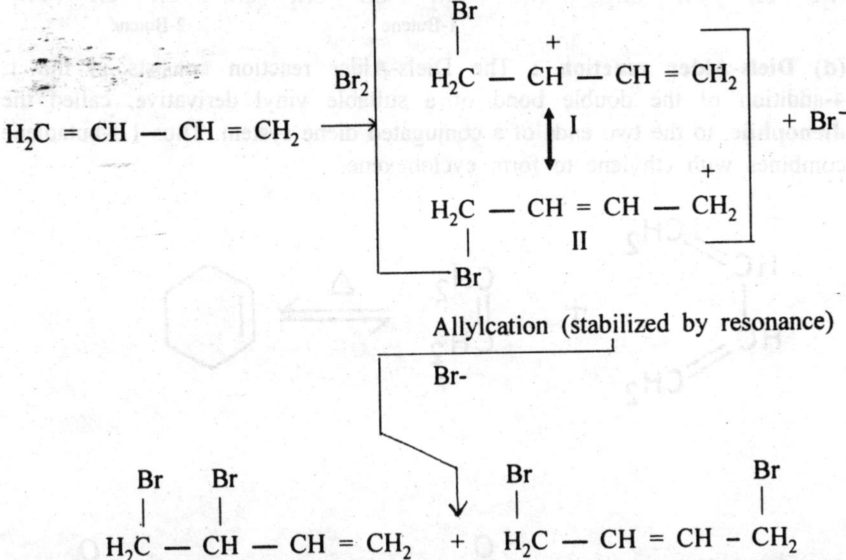

$$H_2C - CH - CH = CH_2 \;(\text{with Br on C1 and C2}) + H_2C - CH = CH - CH_2 \;(\text{with Br on C1 and C4})$$

The positive charge in this intermediate resides primarily on two carbons and either of these positions may be attacked by the nucleophile yielding either the 1,2-or the 1,4-addition product.

It is interesting to note that the products formed are different at different temperatures. At low temperatures, (-80°C), the 1, 2-addition product predominates in the mixture while at higher temperatures (40°), the 1, 4-addition product becomes the major product. The energy of activation of the carbonium ion II is higher for 1, 4-addition than for 1, 2-addition. Therefore 1,4- addition is favored at higher temperatures.

(b) Addition of HX: Addition of hydrogen halides also leads to the formation of two products.

$$H_2C = CH - CH = CH_2 + HX \longrightarrow \left[\overset{H}{H_2C} - \overset{+}{CH} - CH = CH_2 \longleftrightarrow \overset{H}{H_2C} - CH = CH - \overset{+}{CH_2} \right] X^-$$

$$\overset{H}{H_2C} - \overset{X}{CH} - CH = CH_2 + \overset{H}{H_2C} - CH - CH - \overset{X}{CH_2}$$

(c) Addition of hydrogen : 1, 3-Butadiene reacts with hydrogen in the presence of a catalyst to give a mixture of 1-butene and 2- butene.

$$H_2C = CH - CH = CH_2 \xrightarrow{H_2/cat} \underset{\text{1-Butene}}{H_3C - CH_2 - CH = CH_2} + \underset{\text{2-Butene}}{CH_3 - CH = CH - CH_3}$$

(d) Diels-Alder reaction : The Diels-Alder reaction consists in the 1, 4-addition of the double bond of a suitable vinyl derivative, called the **dienophile**, to the two ends of a conjugated diene system. Thus 1,3-butadiene combines with ethylene to form cyclohexene.

$$HC(=CH_2)-HC=CH_2 + CH_2=CH_2 \underset{}{\overset{\Delta}{\rightleftharpoons}} \text{cyclohexene}$$

$$\text{butadiene} + \text{maleic anhydride} \overset{\Delta}{\rightleftharpoons} \text{tetrahydrophthalic anhydride}$$

This reaction is highly stereospecific and occurs exclusively in the cis fashion. The exact mechanism of Diels-alder reaction is still uncertain. It has been suggested that a six membered transition state is formed.

(e) Free radical addition reactions : Like simple alkenes, conjugated dienes also undergo certain free radical addition reactions. Moreover, addition of free radicals can again lead to the formation of both 1, 2- and 1, 4-addition products.

$$H_2C = CH - CH = CH_2 + Cl_2 \xrightarrow[\text{or peroxide}]{h\nu} H_2C - \underset{|}{\overset{Cl}{CH}} - \overset{Cl}{\underset{|}{CH}} - CH_2$$

$$+ \; \overset{Cl}{\overset{|}{H_2C}} - CH = CH - \overset{Cl}{\overset{|}{CH_2}}$$

Mechanism :

(i) Initiation : Formation of free radicals in the presence of light or radical initiator.

$$Cl\ Cl \xrightarrow[\text{peroxide}]{h\nu} 2Cl^{\bullet}$$

(ii) Propagation : Formation of allylic free radical, which is resonance stabilized.

$$H_2C = CH - CH = CH_2 + Cl^{\bullet} \longrightarrow \underset{\text{I}}{CH_2(Cl) - \dot{C}H - CH = CH_2}$$

$$\updownarrow$$

$$\underset{\text{II}}{H_2C(Cl) - CH = CH - \dot{C}H_2}$$

Each of the forms I and II react with Cl_2 to give 1, 2 and 1, 4 products respectively and chlorine free radical.

$$CH_2(Cl) - \dot{C}H - CH = CH_2 + Cl_2 \longrightarrow CH_2(Cl) - CH(Cl) = CH = CH_2 + \dot{C}l$$

$$CH_2(Cl) - CH = CH - \dot{C}H_2 + Cl_2 \longrightarrow CH_2(Cl) - CH = CH - CH_2(Cl) + Cl^{\bullet}$$

(iii) Termination : Radicals combine to give stable molecules, thus terminating the reaction.

5-9 POLYMERIZATION OF ALKENES

Polymerization may be defined as a chemical combination of a number of similar or different molecules to form a single large molecule. Hence a polymer may be defined as a large molecule formed by the repeated combination of small and simple chemical compounds known as the monomers.

$$n > C = C < \longrightarrow \left(\begin{matrix} | & & | \\ C & - & C \\ | & & | \end{matrix} \right)_n$$

Monomer **Polymer**

Ethylene, propylene, and substituted alkenes undergo polymerization readily to form polymers known as plastics. These synthetic plastics can be easily molded into required shapes and possess other properties which enable them to be used for the manufactures of a wide range of articles of common use.

Polymerization reactions are divided into two types :

1. Condensation polymerization
2. Addition polymerization

5-9-1 Condensation polymerization

In condensation polymerization the basic monomer units combine through a condensation reaction by elimination of simple molecules to give a polymer,

Example : Formation of Nylon, terylene etc.

5-9-2 Addition polymerization

In addition polymerization the monomer units combine together to give a new addition product having the same empirical formula but a higher molecular weight. The monomers for addition polymerization are mainly unsaturated compounds.

For example, ethylene polymerizes to give polyethylene or polythene-a plastic. Vinyl chloride polymerizes to give polyvinyl chloride (PVC) used in making irrigation pipes

$$\underset{\text{Ethylene}}{n\ CH_2 = CH_2} \longrightarrow \underset{\text{Polyethylene}}{(CH_2 - CH_2)n}$$

The addition polymerization may be initiated by [a} Free radical or [b] Cationic reagents or [c] Anionic reagents. Such a reagent is known as **initiator** and this determines the type and course of polymerization reaction.

[A] Free radical polymerization : The polymerization of ethylene and substituted ethylenes follow free radical mechanism. The free radical often produced from, R-O-O-R, combines with a monomer in the first step to generate a new free radical. In the second step this free radical combines with another molecule of monomer, and the process continues till a large free radical is obtained. In the third step the previous steps combine to give the polymer. This may be illustrated by taking example of ethylene.

(a) Initiation : Free radicals are produced by the initiator, peroxide.

$$R - O - O - R \longrightarrow 2RO^{\bullet}$$

$$RO^{\bullet} + CH_2 = CH_2 \longrightarrow RO - CH_2 - \dot{C}H_2$$

(b) Propagation : Continued addition of ethylene molecule to the free radical from the previous step produces a large free radical.

$$RO - CH_2 - \dot{C}H_2 + CH_2 = CH_2 \longrightarrow RO - CH_2 - \dot{C}H_2$$
$$- CH_2 - \dot{C}H_2\ RO - CH_2 - CH_2 - CH_2 - \dot{C}H_2 + nCH_2 = CH_2$$

$$RO - CH_2 - CH_2 - CH_2 - CH_2 - (CH_2 - CH_2)_n -$$

(c) Termination : The large free radicals finally combine to give polyethylene.

$$RO(CH_2 - CH_2)_{n+1} - CH_2 - \dot{C}H_2 + \dot{C}H_2 - CH_2 - (CH_2 - CH_2)_{n+1}\ OR \rightarrow RO(CH_2 - CH_2)_{2n+4}$$

Polyethylene

[B] Cationic Polymerization : alkenes polymerize in the presence of H_2SO_4, HF or Lewis acids such as BF_3 or $AlCl_3$ via a carbonium ion mechanism.

Mechanism :

$$H_3C - \underset{}{\overset{CH_3}{\overset{|}{C}}} = CH_2 \xrightarrow{\text{acid}} \left(\underset{\underset{CH_3}{|}}{\overset{\overset{CH_3}{|}}{C}} - CH_2 \right)_n$$

Isobutylene Polyisobutylene

The mechanism of polymerization involves the initial addition of a proton to the alkene forming a carbocation which in the second step adds to another alkene giving rise to a second carbocation.

This carbocation in turn adds to the third molecule of alkene and so on. Thus a chain process sets in resulting in the formation of a polymer which may terminate by the elimination of a proton from the carbocation.

$$(H_3C)_2C = CH_2 + H^+ \longrightarrow (H_3C)_2\overset{+}{C} - CH_3 \xrightarrow{H_2C = C(CH_3)_2} (H_3C)_3C - CH_2 - \overset{+}{C}(CH_3)_2$$

$$(H_3C)_3C\left[CH_2 - C(CH_3)_2\right]_n - CH_2 - \overset{+}{C}(CH_3)_2 \longleftarrow H_3C - C(CH_3)_2 - CH_2 - C(CH_3)_2 - CH_2 - \overset{+}{C}(CH_3)_2 \longleftarrow (H_3C)_2C = CH_2$$

[C] Anionic polymerization : The aniomic polymerization of alkenes can be initiated by a powerful nucleophile. For example; NH_2. The mechanism of anionic polymerization may be formulated as given below :

Nu- + > C = C < ⟶ Nu — C — $\bar{C}$ —

Fast, > C = C <

Nu (—C—C)$_n$ — C — C ⟵ Nu — C — C — C — C —

Carbanion

H^+

Nu (—C—C)$_n$ —C—C—H

The growing chain terminates by any reaction, for example: addition of proton that would destroy the terminal carbanion. Usually the anionic polymerization is not easy to achieve but the presence of a powerful electron attracting group on the alkene facilitates the process.

5-10 RUBBER

Most of the rubber products today are synthesized from alkene derivatives. Rubber is used in homes and industries in a large number of ways. Rubber can be obtained as natural rubber, but now synthetic rubber industries are the major producers of rubber.

5-10-1 Natural rubber

Natural rubber is obtained from nearly 500 different plants, but the main source from which the commercial natural rubber obtained from the tree *Henea braziliencus*. The rubber is obtained from latex collected from the cuts made in the bark of the tree. The latex contains 25%-40% of rubber hydrocarbon. The latex is treated with an acid such as acetic acid which coagulates the crude rubber. The crude rubber is composed of 90%-95% of rubber hydrocarbon. On destructive distillation, rubber gives isoprene as the main product, which led to the consideration that rubber is a polymer of isoprene (C_5H_8) and hence it can be represented as $(C_5H_8)_n$.

$$H_2C = C(CH_3) - CH = CH_2 \xrightarrow{\text{Polymerization}} (CH_2 - C(CH_3) = CH - CH_2)_n$$

Isoprene → Natural rubber

In rubber, various isoprene units are joined in head to tail manner, where the branched end of the isoprene is considered as head. X-ray diffraction studies have shown that in natural rubber the isoprene units are arranged in the cis form, whereas in gutta-percha, isoprene units are arranged in the trans manner.

$$-CH_2-\overset{CH_3}{\overset{|}{C}}=CH-CH_2-CH_2-\overset{CH_3}{\overset{|}{C}}=CH-CH_2-$$

$-H_2C(H_3C)C=C(H)CH_2-CH_2(H_3C)C=C(H)CH_2-CH_2(H_3C)C=C(H)CH_2-$

Rubber (cis-form)

$-H_2C(H_3C)C=C(H)CH_2-CH_2(H_3C)C=C(H)CH_2-CH_2(H_3C)C=C(H)CH_2-$

Gutta-percha (trans-form)

5.10.2 Vulcanization

Raw rubber obtained from latex tapped from the rubber tree does not possess the characteristics of the rubber with which we are familiar. In order to give its strength and elasticity it is vulcanized. In the vulcanization the raw rubber is mixed with small amounts of sulphur or sulphur compounds and heated at a temperature of 150°C.

The sulphur reacts with the polymer molecules forming a cross linked network.

```
  |     |     |     |     |     |
 -X-----X-----X-----X-----X-----X-
  |     |     |     |     |     |
  S-----S-----S-----S-----S-----S
  |     |     |     |     |     |
 -X-----X-----X-----X-----X-----X-
  |     |     |     |     |     |
```

where 'x' represents a monomer unit. The cross linking gives mechanical strength to the rubber.

5.10.3 Synthetic rubbers

(i) Neoprene (Polychloroprene) Rubber : It was the first synthetic rubber manufactured in large scale. The monomer unit is chloroprene which is made from acetylene. Neoprene is particularly resistant to chemical action and is therefore used in making hoses for petrol and oil.

$$2HC\equiv CH \xrightarrow[NH_4Cl]{CH_2Cl_2} \underset{\text{Vinyl acetylene}}{H_2C=CH-C\equiv C-H} \xrightarrow[\text{polymerization}]{HCl} H_2C=\overset{Cl}{\overset{|}{CH}}-C=CH_2$$

$$\underset{\text{Neoprene rubber}}{(CH_2-\overset{Cl}{\overset{|}{CH}}=C-CH_2)_n}$$

(ii) Styrene-Butadiene Rubber : (SBR or Buna S) : It is a copolymer made from styrene and 1, 3-butadiene in emulsion at about 47°C, using potassium persulphate as the initiator.

SBR has its major use in making tyres, shoe soles and in sidewalls of building.

5-11 MULTIPLE-CHOICE QUESTIONS

5-11-1 Which one among the following alkenes is most stable ?
(A) Ethylene (B) 2-butene (C) 2-methyl-2-butene (D) 2, 3- dimethyl-2-butene

5-11-2 Which of the following on treatment with alcoholic KOH gives isobutylene?
(A) tert. Butylchloride (B) isobutyl chloride (C) both (D) none

5-11-3 The major product formed when propylene is reacted with hydrobromic acid is
(A) isopropyl bromide (B) n-propyl bromide (C) none

5-11-4 Which one of the following has 'E' configuration

(A) $(H_2N)(H_3C)C=C(CH_2Cl)(H)$ (B) $(H_3C)(H)C=C(Cl)(CH_3)$ (C) $(H_2N)(NC)C=C(CH_3)(CH_2Cl)$ (A) None

5-11-5 Which among the following gives only acetone on ozonolysis

(A) 2-methyl-2-butene (B) 2, 3-dimethyl-2-butene

(C) 2-m ethyl-1-propene (D) All

5-11-6 Which of the following compounds is formed when isobutylene is subjected to hydroboration followed by treatment with H_2O_2NaOH

(A) Sec. Butylalcohol (B) Isobutyl alcohol

(C) n-Butyl alcohol (D) tert. Butyl alcohol

5-11-7 The major product formed when 1.3 butadiene is treated with one more of Br_2 at high temperature (40°C) is

(A) 1, 4 Dibromo-2-butane (B) 3, 4-Dibromo-2-butene

(C) 1, 2, 3, 4-Tetrabormobutane (D) None

5-11-8 Natural rubber is polymer of

(A) 1, 3-Butadiene (B) Vinyl Chloride

(C) Isopropene (D) Ethylene

5-11-9 Addition of bromine to an alkene is

(A) Cis addition (B) Trans addition (C) Both (D) None

5-11-10 Electrolysis of potassium succinate gives

(A) Ethylene (B) Ethane (C) Propylene (D) Butene

5-12 SHORT AND ESSAY-TYPE QUESTIONS

5-12-1 Formulate the conversion of ethane into ethylene. (O.U., Mar. 1989)

5-12-2 Write the structures of two isomers of an alkene with molecular formulae C_4H_8 and suggest one method for differentiating between them. (O.U., Mar. 1989)

5-12-3 Write a short note on cis trans isomerism. (O.U., Mar. 1989)

5-12-4 Give equations, to show how propene is obtained from :
(i) n-propyl alcohol (ii) isopropyl chloride (iii) Propyne (O.U., Mar. 1990)

5-12-5 Formulate the mechanism of the following addition of HBr to propene in the absence of peroxide and light results in the formation of isopropyl bromide. (O.U., Mar. 1990)

5-12-6 Write the structure or the products obtained on ozonolysis of 3- methyl-2-pentene. (O.U., Mar. 1990)

5-12-7 Identify A, B, C

(i) $H_2C = CH_2 \xrightarrow[HBr]{HI} A \xrightarrow[H_2O]{Mg} B \xrightarrow[Na_2CO_3]{H_2O} C$

(ii) $H_2C = CH_2 \longrightarrow A \xrightarrow[\text{Peroxide}]{} B \longrightarrow C$

(iii) $H_3C - CH = CH_2\text{-}HBr \longrightarrow A$ (K.U., Apr. 1990)

5-12-8 Give the structure of the compound, which gives only acetone on ozonolysis. (O.U., Oct. 1990)

5-12-9 Write notes on diels-Alder reaction (A.U., Jun. 1990)

5-12-10 Describe what is polymerization. (A.U., Apr. 1989)

5-12-11 Complete the following
t-Butyl bromide + $NaOCH_3$ (K.U., Apr. 1989)

5-12-12 Write any two general methods of preparation of alkenes and discuss the mechanism of electrophilic addition in ethylene. (A.U., Dec. 1989)

5-12-13 Explain why dehydration of 1-butanol by conc. H_2SO_4 gives 2- butene instead of 1-butene. (O.U., Apr. 1991)

5-12-14 Give the reasons for 1, 4-addition in butadiene. (S.K.U., Apr. 1990)

5-12-15 Write the industrial significance of rubber. (S.K.U., Apr. 1990)

5-12-16 Give the mechanism of the following reactions :
(a) Dehydrohalogenation (b) Dehydration of alcohols.

5-12-17 How will you prepare the following :
(a) n-Propyl bromide from propylene
(b) 1-Butene from 1, 2-dibromobutane
(c) Allylchloride from propene.

5-12-18 What happens when 1-butene is treated separately with HBr and HOCl.
(i) In presence of and (ii) In absence of peroxides.
State the rule involved in these reactions.

5-12-19 Write down the structural formulae of olefins from which the following products are obtained on ozonolysis

(a) Methyl ethyl ketone and acetaldehyde (b) Acetyaldehyde
(c) Cyclopentanone and acetone (d) Formaldehyde and propionaldehyde

5-12-20 Write short notes on the following
(a) cis-trans isomerism (b) E, Z-configuration

5-12-21 How could one distinguish by qualitative test tube experiments the folllowing:
(a) Propylene and propane (b) n-Butane and 1.3-butadiene

5-12-22 Identify the products in each case

(a) $C_3H_8 \xrightarrow[\text{Monochlorination}]{Cl_2/hv} A \xrightarrow{\text{alc. KOH}} B \xrightarrow{\text{ozonolysis}} C+D$

(b) $C_2H_5OH \xrightarrow{H_2SO_4} A \xrightarrow{H+} B \xrightarrow[\text{Polymerization}]{}$

5-12-23 On Ozonolysis followed by the hydrolysis an organic compound C_5H_{14} (A) gives an aldehyde C_3H_6O (B) and a ketone C_4H_8O (C) Identify the compounds A, B and C. Express the above reactions by means of equations.

5-12-24 Compare the stabilities of ethylene and tetramethyl-ethylene and reason your answer.

5-12-25 Give the products and explain the mechanism involved in the following reactions :

(a) Propene $\xrightarrow[\text{at room temp.}]{Cl_2}$

(b) Propene $\xrightarrow[\text{at} \approx 500^{\circ}C]{Cl_2}$

5-12-26 What happens when,
(a) Propene is treated with HBr in the presence of peroxide.
(b) 2-Butene is subjected to ozoolysis.
(c) Ethylene is treated with Bacye's reagent.
(d) Propylene reacts with diborane and the products is oxidized with alkaline H_2O_2.

5-12-27 What are conjugated dienes ? Discuss the stability of conjugated dienes.

5-12-28 Account for the fact : Addition of bromine to ethylene in the presence of NaCl give not only $BrCH_2CH_2Br$ but also $BrCH_2CH_2CH_2Cl$.

5-12-29 A hydrocarbon of the formula C_9H_{18} absorbs two molecules of hlydrogen when hydrogenated. On oxidation it gives one molecule each of acetic acid propionic acid and acetoacetic acid. Deduce the structure of the hydrocarbon.

5-12-30 Explain the products formed by the addition of bromine to 1, 3- butadiene.

ANSWERS

5-11-1 D	**5-11-2** C	**5-11-3** A	**5-11-4** B	**5-11-5** B	**5-11-6** B
5-11-7 A	**5-11-8** C	**5-11-9** B	**5-11-10** A		

Part III

Physical Chemistry

1

ATOMIC STRUCTURE

1-1 Introduction

The second half of the eighteenth century started with the discovery of oxygen by Joseph Priestly (1774), hydrogen by Henry Cavendish (1766) and introduced chemistry as an experimental science. During 1775-1780, Lavoisier established chemistry as a quantitative science, by proving that in the course of a chemical reaction the total mass of the system is constant. Then nineteenth century saw the rationalization of chemical reactions in the form of laws of definite and multiple proportion, Avogadro's law, etc.

The Twentieth century heralded with the discovery of subatomic particles electrons and protons, by J.J. Thompson in 1897. This soon led to the questions about the internal arrangement of these elementary particles, as an understanding of the internal structure of an atom is expected to answer the question like; why atoms combine?, why some elements are metals and others non- metals?, why some compounds are colored and others colorless?, etc. In response, many models and improvements were suggested from time to time.

1-1-1 Thomson's Atomic Model

J.J. Thomsons proposed a distribution of electrons in an evenly distributed mass and positive charge, like the seeds and flesh in a water melon. This model was rejected with discovery of phenomenon of radio activity and the gold foil x-ray scattering experiment.

1-1-2 Rutherford's Model

E. Rutherford (1911) from his famous scattering experiment of alpha rays by gold foil obtained evidence for the existence of protons as a concentrated positively charged mass known as nucleus. To account for the fact that the oppositely (negatively) charged electrons do not fall in to the nucleus as a result of electro static attraction, he suggested a perpetual motion similar to the planetary rotation around the sun. The centrifugal force, acting outward, due to rotation was expected to be equal and opposite to electrostatic attraction, acting inward. However this perpetual motion of a charged body without loss of energy is against the classical electromagnetic theory. According to this theory a charged body in motion emits it's energy continuously and as a result the radius of the circular orbit steadily decreases finally leading to the collapse of electron in to the nucleus. Further Rutherford's model could not explain the well experimented discontinuous line spectra obtained by Balmer, Lyman and other workers, for the hydrogen atom.

1-1-3 Planck's Quantum Theory

In 1900 Max Planck while analyzing the light emitted by heated objects discovered that the emitted light is non-continuous and is in the form of definite energy packets. This was also found to be true in case of absorption of light. These small packets of energy are known as **quanta** or **photons**. The energy of a quanta depends on the vibrational frequency (in other words temperature) of the atoms of the heated object.

Mathematically the energy of a photon, E, is given by the expression;

$E = h\nu$ where ν is frequency of the oscillator

Further, Planck's quantum theory advocates that the energy is always emitted in integer multiples of hν, for example, hν, 2hν, 3hν..... but never a fraction. This particle aspect of light also received support from the discovery of photoelectric effect, where impact of light on alkali metals eject electrons, the number of ejected electrons showing a direct dependence on the intensity of light used.

1-1-4 Bohr's Model of Hydrogen Atom

Knowledge of hydrogen spectrum and Planck's relationship between energy and frequency of light opened the doors to a new approach of understanding of internal structure of atoms. The primary requisite for having the atomic spectrum is that the atom must be energized, which is normally achieved with an electric discharge. An energized hydrogen atom was found to emit photons corresponding to specific wave length as prescribed by the line spectrum of hydrogen. This indicates that hydrogen atoms can not have random energies but instead are restricted to certain energies as determined by the Ritz combination principle and responsible for the

These energy levels can be represented by the following figure 1-1.

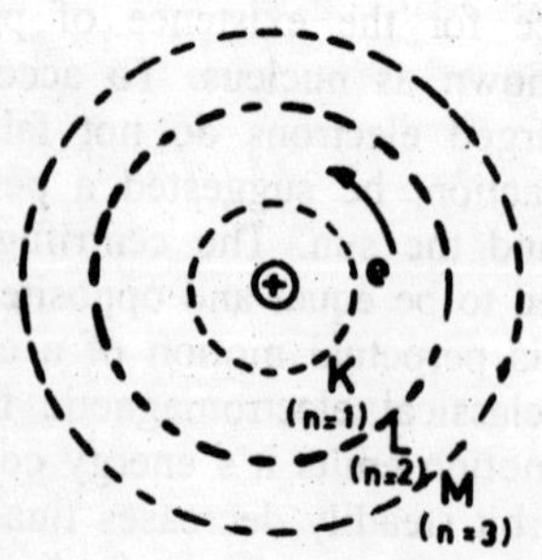

Fig 1-1 Bohr's energy levels

hydrogen spectra. These restricted energies of hydrogen atom could be attained by the revolution of electron in circular orbits around the nuclei. This discrete energy attained is the sum of two fundamental forms of energy viz., kinetic energy obtained by virtue of motion of the electron and potential energy obtained by virtue of the separation (position) of the electron from the proton forming the central nuclei. The consequence of these defined energy levels is that an electron jumps from one energy level to a higher energy level by absorbing a defined amount of energy, which is reemitted in the form of a photons of characteristic frequency. These restricted energy levels are called by Bohr as quantized energy levels.

With these ideas Bohr has proposed the following postulates

Bohr's Postulates:

(i) The model of a hydrogen atom is that of solar system, with the proton forming the central nucleus and electron revolving around the nucleus in stationary orbits of fixed radii.

(ii) Electron does not lose any energy as long as it is in the stationary orbit. This is due to the equal and opposite attraction force operating between the positively charged nuclei and negatively charged electron, which is directed towards the center and the tangential centrifugal force due to the movement of the electron, which is directed away from the center.

(iii) These stationary orbits could be identified by their angular momentum given by the product of mass of the electron, radius of the orbit and the velocity of the electron. Angular momentum is also quantized and these are given by the integral multiples of $h/2\pi$

i.e., $mvr = nh/2\pi$

where h is Planck's constant & n = 1,2,3,...

The integer, n, indicates the orbit or shell in which the electron is located and numbered from the nucleus and is called the principal quantum number. The principal quantum numbers of 1,2,3,4 etc., are also referred by the letters K,L,M,N etc., respectively.

(iv) The larger the principal quantum number or size of the orbit, the greater will be the energy of an electron. The K shell has the lowest possible energy and is called ground state. Other states are referred as excited states.

(v) An electron revolving in a fixed orbit can gain energy and jump to a higher energy level (an outer orbit) and energy is released in the form of light when it falls back to any lower orbit.

(vi) The energy absorbed or released by an atom must be equal to the difference in the energies of the orbit about which the electron transfers. If an electron falls from an energy level E_2 to E_1 the energy of the photon (of frequency ν) emitted is equal to the difference of the energies of these orbits.

$$\Delta E = E_2 - E_1 = h\nu$$

expressing frequency in terms of wave number we have

$$\nu = \frac{E_2 - E_1}{hc}$$

Bohr tested the validity of his postulates by calculating the radii, and energies of the stationary orbits in hydrogen and hydrogen like atoms, He^+, Li^{2+}, etc., having one electron revolving around central nuclei of different masses. His values matched well with the experimental values obtained from different sources.

Success of Bohr's Theory : In spite of the simplicity of the Bohr's theory, it was highly helpful in;

(i) Understanding the internal structure of an atom.

(ii) Predicting the energy and radii of an orbit.

(iii) Providing theoretical validity to the empirical Ritz combination principle.

(iv) Explaining to H spectrum.

(v) Finally in calculating the ionization potential of an atom.

Limitations of Bohr's Theory :

(i) Bohr's theory, though successfully explained the spectra of unielectron atoms and ions like H, He^+, Li^{2+}, Be^{3+} etc, it failed to interpret the spectra of higher atoms having more than one electron and calculate their frequencies.

(ii) Bohr's theory predicts only a single line for a transition between two stationary orbits but using high resolution spectrograph these lines were further split into some more closely spaced lines called **fine structure**. This fine structure suggests that the stationary orbit suggested by Bohr is not a single energy level but has some more closely spaced energy levels.

(iii) No theoretical reasoning was given for the presumption that the angular momentum is equal to whole number multiple of $h/2\pi$

(iv) In the treatment of Bohr's theory it has been assumed that the momentum and position, in a definite orbit, of an electron are known. Heisenberg's uncertainty principle showed that it is impossible to know the exact position and momentum of sub-atomic particles like electrons simultaneously.

(v) The Bohr's theory is essentially pictorial, giving physical significance to the electron as a particle moving in a fixed orbit which is not true in real sense. The modern treatment regards atom purely as a mathematical model without giving any pictorial representation and instead of stationary orbit the probability of finding an electron is discussed.

1-2 THE WAVE NATURE OF ELECTRON

Even ten year after revelation of Bohr's theory and acceptance of its success, scientists were mystified from the concept of quantization. They often found it difficult to understand why stationary orbits of a Bohr atom have some selected energies and why an electron can not opt to stay in orbits with other energy levels. During the same period planck's quantum theory received the acceptance of scientific world and every one is forced to admit a wave, particle (Photon) dual nature of the light with this back ground Louis de Broglie in 1924 argued. With the acceptance of Planck's quantum theory, scientists were forced to admit a *dual nature*, of a photon (an energy particle) and of a wave, to the electromagnetic radiations. In 1924 Louis de Broglie argued that even matter composed of atoms and sub-atomic particles like electrons, protons etc., also exhibit this dual nature. His argument is based on the following considerations;

(i) **Symmetry in Nature :** In nature both the living and non living belonging to a family behave in a similar way irrespective of their origin. As matter and radiation are two fundamental forms of nature they are also expected to exhibit similar characteristics like dual nature.

(ii) **Parallelism between mechanics and optics :** A moving particle always chooses a shortest or least action path. This is also true with a ray of light which always takes a shortest path possible.

(iii) **Bohr's theory and integral rules :** In Bohr's theory the angular momentum was proposed to be equal to integral multiples of h/2!. Similarly he expressed the energy, radii etc, in terms of principal quantum number, which always assumes integral values. But integer rules are quite common in optics suggesting a parallelism between the behavior of electron, a particle, and the radiation.

1-2-1 Wave Length of Matter Waves

De Broglie also deduced the wave length of the matter waves clubbing the Einstein's mass-energy relationship and Planck's theory of energy of a photon of an electromagnetic radiation.

According to Einstein's mass-energy relationship, the energy associated with a particle (here photon) of mass m is given by

$E = mc^2$, where c is the velocity of light,

But from planck's theory the energy of a photon is equal to

$E = h\nu$, where h is Planck's constant, & ν is frequency of light.

Equating both the energy expressions

$$mc^2 = h\nu$$

But ν is equal to c / λ

Therefore $mc^2 = hc/ \lambda$

or $\lambda = h/mc$

If this is applied to a particle like electron moving with a velocity v, c can be replaced with v.

$$\boxed{\lambda = h/mv}$$

This equation is known as **de Broglie's equation** of matter waves. Thus de Broglie's equation combines both the corpuscular nature of the electron reflected from its mass and wave nature shown by the wave length or frequency. The de Broglie's equation successfully bridged the gap existed earlier between matter and radiation. De Broglie's matter-wave relationship may be state as the wave length of a matter wave is inversely proportional to its momentum (mass x velocity).

Thus all the particles in this universe have both particle and wave nature. For larger objects h/mv, there by λ, is insignificant and hence best described by its particle character. For sub-atomic particles like electron which are extremely small h/mv or λ assumes importance and hence the particle character diminishes with a simultaneous increase in the wave character. Davisson and Germer in the USA and G.P. Thomson (son of J.J.Thomson) in England in 1927 actually demonstrated the wave nature of electrons by obtaining the diffraction (an optical character) pattern for electron beams.

1-2-2 Descriptions of An Electron Orbit Using De Broglie's Matter-wave Theory

An electron moving in a circular orbit could form either a **stationary** or **standing wave** or a **non-stationary wave**.

(A) **NON-STATIONARY WAVE** : The non-stationary waves are out of phase and are destroyed due to destructive interference. A maximum in one circuit is cancelled by a minimum of the next. The number of wave lengths are therefore nonintegral.

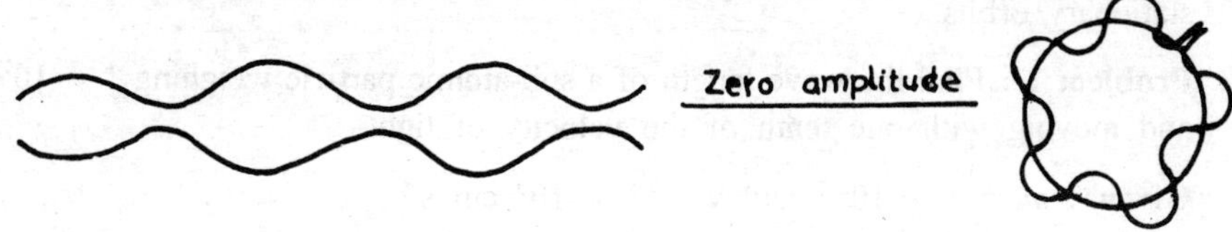

Two waves out of phase leading to destruction

A nonstationary wave

Fig 1-2 Non-stationary waves

(B) STATIONARY WAVES : A stationary wave is continuously in phase. An example of these waves could be found in the vibrations of a string instrument like veena or guitar. When a veena string is plucked and briefly touched in the middle sets a standing wave. They have integral number of wave lengths. The crest and troughs of consecutive circuits match exactly with one another.

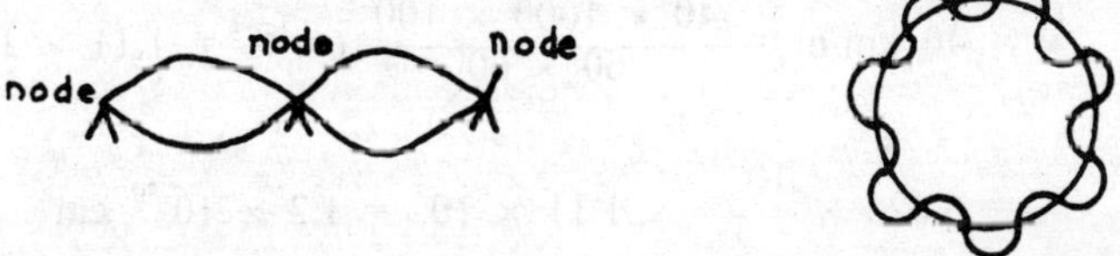

Completely in-phase wave A standing electron wave

Fig 1-3 Stationary waves

Only stationary waves give ever continuing electron waves. The circumference of a stationary wave is equal to integral multiples of wave length. This is an orbit.

Therefore $2\pi r = \text{h } \lambda$ where r is radius of orbit

But according to de Broglie's theory wave length of an electron

$$\lambda = h/mv$$

So $\quad 2\pi r = nh/mv \quad$ or $\quad mvr = nh/2\pi$

This is the same expression for momentum put forward by Bohr in his 3rd postulate. Thus de Broglie has provided a theoretical justification for one of the Bohr's hypothesis and thereby solved the puzzle of quantization of stationary orbits.

Problem 1 : Find the wave length of a sub-atomic particle weighing 1×10^{-27}g and moving with one tenth of the velocity of light.

Given : $m = 1 \times 10^{-27}$ and $v = 3 \times 10^{9}$ cm s^{-1}

$$\lambda = 6.626 \times \frac{10^{-27}}{1} \times 10^{-27} \times 3 \times 10^{9} = 2.2 \times 10^{-9} \text{ cm or } 0.22 \text{ Å}$$

wave length of sub-atomic particle = 0.22 Å

Problem 2 : What must be the velocity of electron, if they display a de Broglie wave length of 1nm. Given Mass of electron = 9.1×10^{-28} g

Solution : Given : $\lambda = 1\text{nm} = 1 \times 10^{-9}$ m $= 1 \times 10^{-7}$ cm

$m = 9.1 \times 10^{-28}$ g

$$v = \frac{h}{m\lambda} = \frac{6.626 \times 10^{-27}}{9.1 \times 10^{-28} \; 1 \times 10^{-7}} = 7.2 \times 10^{-7} \text{ cm s}^{-1}$$

Problem 3 : Calculate wave length associated with a stone weighing 5g and moving with a velocity of 40 km/hr.

Given : $m = 5$g

$$v = 40 \text{ km/hr} = \frac{40 \times 1000 \times 100}{60 \times 60} \text{ cm S}^{-1} = 1.11 \times 10^{3} \text{ cm S}^{-1}$$

$$\lambda = 6.626 \times \frac{10^{-27}}{5} \times /1.11 \times 10^{3} = 1.2 \times 10^{-30} \text{ cm}$$

Problem 4 : A bullet moving with a speed of 1×10^{3} m s^{-1} has a de Broglie wave length of 1×10^{-34} m. calculate the mass of the bullet.

Given : $v = 1 \times 10^{3}$ m S^{-1} = 1×100 cm S^{-1} = 1×10^{5} cm S^{-1}

$\lambda = 1 \times 10^{-34}$ m $= 1 \times 10^{-32}$ cm

$$m = \frac{h}{\lambda v} = \frac{6.626 \times 10^{-27}}{1 \times 10^{-32} \times 1 \times 10^{5}} = 6.626 \text{ g}$$

mass of bullet = 6.626 g

1-3 HEISENBERG'S UNCERTAINTY PRINCIPLE

According to classical mechanics we can predict exactly the position and momentum of any moving particle at any time *t*, when its initial position and momentum are known. Thus it is possible to predict the location of a car running with a constant speed at any time *'t'*, when its initial position and velocity are known. But according to Heisenberg's uncertainty principle, *it is impossible to determine simultaneously both the positions and momentum of a particle on the atomic scale*, in other words, *simultaneous determination of velocity or any related property eg., energy or momentum, and position is not possible for subatomic particles*. This principle is less meaningful for the *macroscopic bodies* which have *predominant particle character* (as discussed by the de Broglie's theory) and mainly useful *for the atomic and subatomic particles* which have *predominantly wave character*.

If Δx is the uncertainty in determining position of an electron and Δp the uncertainty in measuring momentum then uncertainty principle can be mathematically represented as

$$\Delta x \times \Delta p \geq h$$

or more precisely $\quad \Delta x \times \Delta p \geq h/4\pi$

As the momentum *'p'* = mass *'m'* x velocity 'v'

we have $\Delta p = m \times \Delta v$

Hence uncertainty principle in terms of velocity may be represented as

$$\Delta x \times \Delta v \geq h/4\pi m$$

The following example clarifies mathematically why this principle is less meaningful for macroscopic particles.

1. for a particle weighing 1g :

$$\Delta x \times \Delta v \approx \frac{6.62 \times 10^{-27}}{1 \times 4 + 3.14} \approx 0.53 \times 10^{-27}$$

with the modern devices one can measure the position with an accuracy of 1×10^{-7} cm or the uncertainty in *x* can not be greater than this value.

$$\text{therefore, } \Delta v \approx \frac{0.53 \times 10^{-27}}{\Delta x} \approx \frac{0.53 \times 10^{-27}}{1 \times 10^{-27}} \approx 0.53 \times 10^{-20}$$

This uncertainty in the determination of velocity and position in comparison to the size of a macroscopic body is almost negligible. Therefore, macroscopic particles are in the perview of classical mechanics and we are certain in the simultaneous determination of the position and velocity.

2. For A Subatomic Particle :

When we consider a sub-atomic particle like electron having a mass of 9.1 × 10^{-28} g.

$$\Delta x \times \Delta v \approx \frac{6.62 \times 10^{-27}}{9.1 \times 10^{-28} \times 4 \times 3.14} \approx 0.58$$

As the diameter of the electron is 1A° or 1 × 10^{-8} cm the uncertainty in determining its position could not be more than this value.

therefore $\Delta x \approx 1 \times 10^{-8}$ cm

$$\Delta v \approx \frac{0.58}{1 \times 10^{-8}} \approx 0.58 \times 10^{8} \text{ cm s}^{-1}$$

The error in the determination of velocity is quite high. Hence even if we determine the position as accurately as possible the error in the determination of velocity becomes enarmous.

This suggests that in an attempt to determine the velocity or momentum of subatomic particles accurately, we lose the precision in the determination of position and vice-versa.

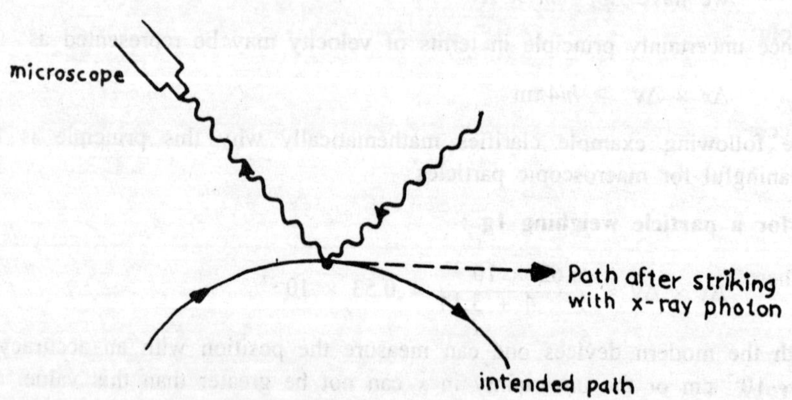

Fig 1-4 Hypothetical equipment to prove uncertainity principle

This fact can also be well illustrated by attempting to design a hypothetical apparatus to achieve our goal of determining both the position and velocity of an electron. Detection of any object requires, some device like light that strikes the object and returns to the detector and passes the information. The K shell, the lowest electronic level, has a diameter of approximately 1A° or 1 × 10^{-10} m. So the wave length of the striking radiation should be having a wave length of 1A° or less than this value like X-rays and gamma-rays. Thus if we hit the electron rotating in a orbit, whose path is under surveillance of a super microscope, with an X-ray photon which is deflected by the electron into the microscope we expect to detect the position and velocity accurately. But in reality as soon as the X-ray photon strikes the electron, it transfers its energy to the electron and instantaneously changes the velocity of electrons, thus bringing an error in the determination of velocity.

This conclusively proves that even using an hypothetical equipment, we will not be able to determine accurately the position and velocity of an electron or any other microscopic particle simultaneously. Applying the Heisenberg's uncertainity principle to the hydrogen atom, we notice that in reality the electron does not revolve in a fixed orbit, i.e., in a well defined path, as Bohr thought. If it did, we could determine the position and momentum of the electrons in violation to uncertainity principle. This limitation of Bohr's concept of stationary orbits was explained by Schrödinger wave mechanics.

Problem 5 : The uncertainity in the moment of a particle is 1×10^{-37} kg m s^{-1} Calculate the uncertainity in the measurement of its position.

(Hint : Direct use of Δp in MKS units could be made by taking Planck's constant $h = 6.626 \times 10^{-34}$ Js.)

Solution: Given: $\Delta P = 1 \times 10^{-37}$ kg m $s^{-1} = 1 \times 10^{-37} \times 1000 \times 100$g cm s^{-1}

$= 1 \times 10^{-32}$ g cm s^{-1}

According to Heisenberg uncertainity principle

$$\Delta x \,.\, \Delta p \geq \frac{h}{4\pi} \text{ or } \Delta x \,.\, \Delta p \approx \frac{h}{4\pi}$$

$$\text{Therefore, } \Delta x \approx \frac{h}{4\pi\ \Delta p} \approx \frac{6.626 \times 10^{-27}}{4 \times 3.14 \times 1 \times 10^{-32}}$$

$$\approx 5.3 \times 10^{4} \text{ cm}$$

Problem 6 : An electron is accelerated to one tenth the velocity of light. If the velocity can be measured with a precession of ± 1%, find the uncertainity in its position. Given mass of electron = 9.1×10^{-28} g

Solution : Given $V = 0.1\ C = 0.1 \times 3 \times 10^{-10}$ cm $s^{-1} = 3 \times 10^{-11}$ cm s^{-1}

$$\frac{\Delta v}{v} = \frac{1}{100}$$

$$\Delta v = \frac{1}{100} \times 3 \times 10^{-11} = 3 \times 10^{-13} \text{ cm s}^{-1}$$

According to Heisenberg principle

$$\Delta x \cdot \Delta p \approx \frac{h}{4\pi} \quad \text{or} \quad \Delta x \cdot \Delta v \approx \frac{h}{4\pi m}$$

$$\Delta x \approx \frac{6.626 \times 10^{-27}}{4 \times 3.14 \times 9.1 \times 10^{-28} \times 3 \times 10^{-13}}$$

$$\Delta x = 1.9 \times 10^{12} \text{ cm}$$

Problem 7 : In the above problem what will be the uncertainity if we consider a proton instead of an electron (mass of proton = 1.67×10^{-27} g).

$$\Delta x \approx \frac{6.626 \times 10^{-27}}{4 \times 3.14 \times 1.67 \times 10^{-27} \times 3 \times 10^{-13}}$$

$$= 1.05 \times 10^{9} \text{ cm}$$

Problem 8 : Calculate the uncertainity in the position of a moving bullet of mass 0.01 kg if the uncertainity in its velocity is $\pm 2 \times 10^{-30}$ m s^{-1}

Solution : Given : = 2×10^{-30} m s^{-1} m = 0.01 kg

$$\Delta x \approx \frac{h}{4\pi m \Delta v} = \frac{6.626 \times 10^{-34}}{4 \times 3.14 \times 0.01 \times 2 \times 10^{-30}}$$

$$\Delta x = 2.64 \times 10^{-3} \text{ m}$$

Problem 9 : A rocket carrying a pay load of 1 ton is moving with a uncertanity in the velocity of $\pm$ 0.02 m S^{-1} Calculate the uncertainity involved in detecting its position. Also find the uncertainity in its momentum.

Solution : Given : m = 1×10^{-27} and v = 3×10^{9} cm s^{-1}

$$\Delta x = \frac{h}{4\pi m \Delta v} = \frac{6.626 \times 10^{-34}}{4 \times 3.14 \times 1000 \times 0.02}$$

$$= 2.64 \times 10^{-36} \text{ m}$$

uncertanity in momentum $\Delta p = m \Delta v = 1000 \times 0.02 = 20$ kg m s^{-1}

Problem 10 : A racing can weighing 750 kg is moving with a velocity of 250 km/hr. If the uncertainity in the measurement of its velocity is 0.1%, find the uncertanity in finding its position at any given instant.

Given : $v = 250 \text{ km/hr} = \frac{250 \times 1000}{60 \times 60} \text{ m s}^{-1} = 69.4 \text{ m s}^{-1}$

$$\frac{\Delta v}{v} = \frac{1}{100} \text{ or } \Delta v = \frac{69.4}{100} = 0.694 \text{ m s}^{-1}$$

m = 750 kg

$$\Delta x \approx \frac{h}{4\pi m\Delta v} = \frac{6.626 \times 10^{-34}}{4 \times 3.14 \times 750 \times 0.694}$$

$$= 1 \times 10^{-30} \text{ m}$$

1-4 THE SCHRöDINGER EQUATION

De Broglie's matter-wave relationship has already ruled out Bohr's firm physical representation of orbits of the revolving electron. Heisenberg's uncertainty principle led to the indeterminacy of position and velocity of an electron. So instead of ascribing a rigid path where we can locate an electron as done in the classical mechanics, various groups of workers attempted for finding the probable location of an electron. In 1926 Erwin Schrödinger and Werner Heisenberg independently introduced entirely new approaches. Schrödinger's approach is called wave mechanics and that of Heisenberg's approach is called matrix mechanics. Schrödinger proved that basically these two approaches are same and they are only two different sides of the same coin known as the quantum mechanics. Schrödinger's method is more popular with the chemists.

From the classical wave mechanics if ψ is the amplitude of a wave moving is a three dimensional space with a velocity of v and frequency ν the wave equation is

$$\frac{\partial^2 \psi}{\partial x^2} + \frac{\partial^2 \psi}{\partial y^2} + \frac{\partial^2 \psi}{\partial z^2} + \frac{4\pi^2 \nu^2 \psi}{v^2} = 0$$

But from non classical de Broglie equation $\lambda = h/mv$ where $v = \lambda\ \nu$

so $\quad 1/\lambda = mv/h = \nu/v$

Substituting in the classical wave equation, the value of ν/v we have;

$$\frac{\partial^2 \psi}{\partial x^2} + \frac{\partial^2 \psi}{\partial y^2} + \frac{\partial^2 \psi}{\partial z^2} + 4\pi^2 \left(\frac{mv}{h}\right)^2 \psi = 0$$

i.e.

$$\frac{\partial^2 \psi}{\partial x^2} + \frac{\partial^2 \psi}{\partial y^2} + \frac{\partial^2 \psi}{\partial z^2} + \frac{4\pi^2 m^2 v^2}{h^2} \psi = 0$$

But kinetic energy 'K' $= {}^1/_2\ mv^2$ or $m^2v^2 = 2mK$

$$\frac{\partial^2 \psi}{\partial x^2} + \frac{\partial^2 \psi}{\partial y^2} + \frac{\partial^2 \psi}{\partial z^2} + \frac{8\pi^2 mK}{h^2} \psi = 0$$

We know that Kinetic energy *'K'*= Total energy *'E'* - Potential energy *'V'*

$$\frac{\partial^2 \psi}{\partial x^2} + \frac{\partial^2 \psi}{\partial y^2} + \frac{\partial^2 \psi}{\partial z^2} + \frac{8\pi^2 m(E-V)}{h} \psi = 0$$

This equation is known as **Schrödinger equation.**

1-4-1 The Interpretation of ψ and ψ^2

In the Schrödinger wave mechanics ψ assumes additional meaning. When a matter-wave is exhibiting exclusively wave nature, in the classical wave mechanics, ψ represents the amplitude and the square of the amplitude ψ^2 gives the intensity of the light. As the intensity of light depends on the number of photons striking an object or the photons density we can say that ψ^2 represents the density of the photons. When the matter-wave has exclusively particle character ψ^2 can be said to represent the density of the particles. But if we consider a case like hydrogen atom where a single electron revolves around the nucleus to talk about the density of particles is meaningless and ψ^2 speaks about *the **probability** of finding an electron at a particular place.* This concept could be better understood by taking an analogy. Suppose we have a piece of chalk and started drawing the cell structure on the black board, after some time we find that the chalk got used up in creating the structure of a cell. The chalk did not disappear but got spread out over the board. We can easily point the places where more chalk is present and where it is thinly spread but to speak of whole chalk and how it is distributed, it is necessary to have the whole drawing. Similarly *to understand the complete space distribution of an electron we must understand the total probability or electron density of an electron. The electron distribution of an orbit does not lie in a sharp well defined circular path as portrayed by Bohr but extends some distance on both sides of the path.*

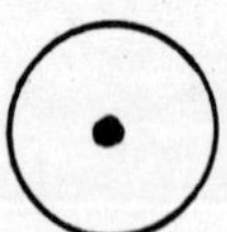

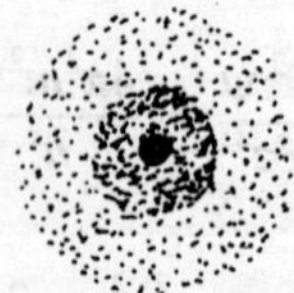

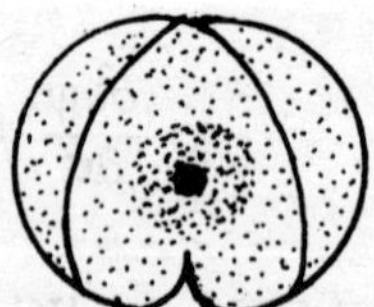

Two dimensional

Bohr's model of H atom

Schrödinger's model of H atom

Fig 1-5 Schrödinger model of H atom

Consider the revolution of an electron present in the K-shell of a hydrogen atom. When we take a number of three dimensional photographs and super impose them on one another we get a dot density photograph as shown in fig 1-5

We find from the diagram that the electron density or probability of finding an electron is very high near the nucleus and maximum at 0.529A^{o} from the nucleus (The Bohr's radius of 1st orbit) but decrease as we go farther away from nucleus.

Schrödinger solved the wave equation for hydrogen and some other simple atoms and actually obtained a set of **wave functions** that describe allowed energies and shapes of the electron waves. Each of these different possible waves is called an **orbital**. Each orbital has a fixed amount of energy and describes a region (**probability**) where the electron could be found. Each wave function is characterized by values known as **quantum numbers**.

1-5 THE QUANTUM NUMBERS

There are number of orbitals or wave functions in which an electron could be located. In order to describe the location and energy distribution it is essential to have the knowledge of quantum numbers.

1-5-1 Principal Quantum Number, '*n*'

This value is same as the '*n*' value of Bohr's atomic orbits or shells but to make a distinction it is given the name **orbital**. This describes the main energy levels present in an atom in which the electron can continue to rotate without losing any energy. Principal quantum number can have only integer values 1, 2, 3, 4...... and the corresponding letter designation is K, L, M, N....

Principal quantum number	1	2	3	4
Letter designation	K	L	M	N

The energy of the orbital increases with increase in the value of *n*. It also describes the radii, the distance of a main energy level,where maximum electron density could be detected, from the nucleus. The maximum number of electrons allowed in a shell are given by the formula $2n^2$. Thus K shell can have only two electrons while L can have $2 \times 2^2 = 8$ electrons. The maximum number of electrons in M and N shells are 18 and 32 respectively.

1-5-2 Azimuthal Or Orbital Quantum Number, '*l*'

The splitting of main lines of H spectra described by Ritz combination principle (or Bohr's theory) gives an idea that a main shell is composed of **sub shells or sub energy levels** each of which is designated by another quantum number *l*. This quantity is called azimuthal quantum number in Bohr's theory or orbital quantum in Schrodinger wave mechanics. For any given shell ***l*** may have values 0,1,2,...(*n*-1).

Thus for $n = 1$ the only value l can assume is $l = 0$. When $n = 2$, l has two values 0 and 1. So to summarize the *total number of values of l or subshells that any given shell has, is simply equal to the value of n of the shell.*

Values of 'l'	0,	1,	2,	3,...
Subshells designation	*s,*	*p,*	*d,*	*f,...*

We are not interested with values of $l > 3$ or f subshell because these are the only levels that are occupied by the electron in atoms in their ground state.

1-5-3 Magnetic Quantum Number '*m*'

The appearance of additional lines in atomic spectra produced when atoms are placed in the magnetic field necessitated the introduction of magnetic quantum number. The subshells were found to have some more orbitals which serve to determine the orientation of an electron in space. The magnetic quantum number **'*m*'** can have all integer values ranging between $-l$ to $+ l$. When $l = 0$, m can have only one value $m = 0$. When $l = 1$, m can have values -1, 0, +1. Total number of values of m are equal to $2l + 1$.

The possible l and m values corresponding to different 'n' values are summarized in Table 1-1

Table 1-1 Possible l and m values corresponding to different 'n' values

Principal quantum number 'n' (main shell)	*Shell desig-nation*	*Azimuthal quantum number 'l'*	*Subshell desig-nation (sub shell)*	*Magnetic quantum number 'm' (orbital)*	*Number of orbitals in a subshell (2l +1)*	*Maximum no. of electrons each orbital*
1	K	0	*1s*	0	1	2
2	L	0	*2s*	0	1	2
		1	*2p*	-1 0 +1	3	6
3	M	0	*3s*	0	1	2
		1	*3p*	-1 0 +1	2	6
		2	*3d*	-2 -1 0 +1 +2	3	10
4	N	0	*4s*	0	1	2
		1	*4p*	-1 0 +1	3	6
		2	*4d*	-2 -1 0 +1 +2	5	10
		3	*4f*	-3 -2 -1 0 +1 +2 +3	7	14

1-5-4 Spin Quantum Number '*s*'

All the above quantum numbers, ***n***, ***l*** and ***m***, came from the solution of the Schrödinger's wave equation. The electron in a shell was found to be fully described by choosing yet another quantum number known as spin quantum number. Uhlenbeck and Goudsmith have proposed that *an electron (like earth) not only rotates around the nucleus (the sun), but also revolves around its* *own axis. This motion is known as spinning of an electron.* A spinning electron has only two options, spinning in a clockwise direction or anti clockwise direction. These are represented by giving a spin quantum number values of $+\frac{1}{2}$ and $-\frac{1}{2}$

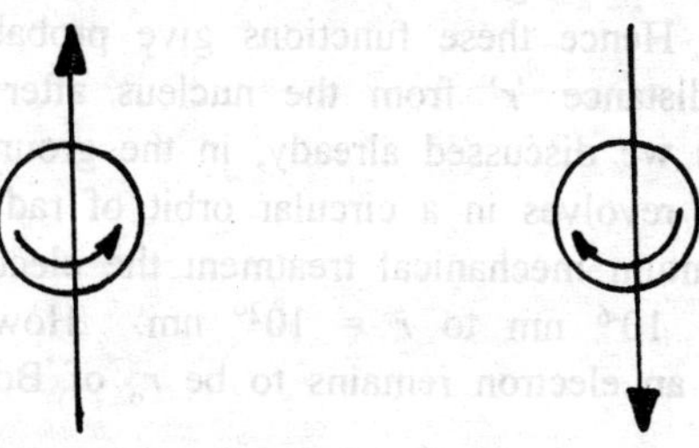

Fig 1-6 Electrons spinning in opposite directions

1-6 SHAPES OF ORBITALS

A wave function of an orbital describes the position of an electron from the nucleus in a three dimensional space. The position of an electron at a given instant may be represented using either cartisian or polar coordinates.

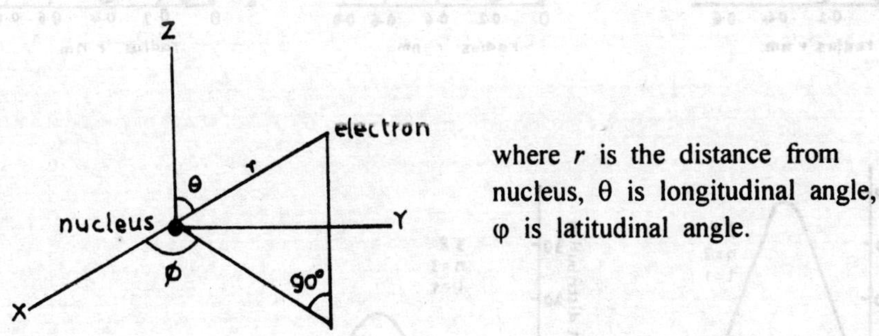

where r is the distance from nucleus, θ is longitudinal angle, φ is latitudinal angle.

Fig 1-7 An electron in a three dimensional space

The wave function (ψ), which describes complete distribution of an electron, is the product of two quantities (i) Radial probability distribution '$R(r)$' and (ii) Angular probably distribution, ϕ (θ, ϕ)

$$\psi (r, \theta, \varphi) = R (r).\phi (\theta, \varphi)$$

(A) RADIAL DISTRIBUTION FUNCTION, *R(r)* : The radial probability distribution defines the most probable distance of an electron from the nucleus. The radial distribution function gives the total probability of finding the electron in a spherical shell. Hence these functions give probability of finding an electron at a given distance 'r' from the nucleus after averaging over its angular variables. As we discussed already, in the ground state the electron in the hydrogen atom revolves in a circular orbit of radius r_o = 0.0529 nm or 0.529 Å . In quantum mechanical treatment the electron could be found any where from r = 10^{-6} nm to r = 10^{29} nm. However the maximum probability of finding an electron remains to be r_0 of Bohr's orbit.

The radial distribution functions of $1s$, $2s$, $2p$, $3s$ and $3p$ orbitals may be given as follows.

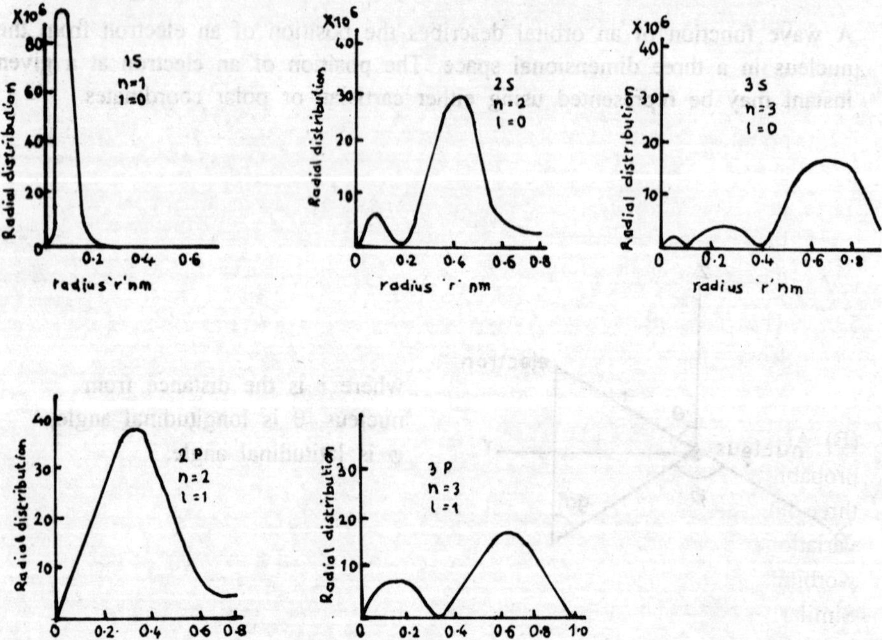

Fig 1-8 Radial distribution functions

From the above figures one can notice

1. *All radial distribution functions have zero probability of finding the electron at the origin, i.e., the nucleus.* This observation over rules the probability of an electron falling in the nucleus on account of loss of energy.

2. *The maximum electron density of a subshell falls both with the increase in the value of n as well as l.* The 1s orbital (n = 1 and l = 0 is highly localized and maximum electron density of 120 × 10^6 is found near the Bohr radius. The less sharp distribution functions with the increase in the distance from nucleus (that is with the increase in the values of n and l), indicate lesser localization of electron cloud at higher r values. This fact also supports the view that the attraction between nucleus and any electron decreases with increase in r.

3. *In the radial distribution curves of higher orbitals have in addition to the main peak, another small peak/s close to the value of r_o, indicating that a smaller but definite probability always lies near r_o.* Thus, 2*s* orbital has two, one small and the other large peak, and 3*s* has three, two small and one main peak. This behavior, known as **penetration**, suggests that the electron in the higher orbital also spends a small portion of it's time near the nucleus.

4. When ever we consider a radial function with more than one peak, either between the smaller and smaller or smaller and larger peaks, a point of zero probability of finding an electron exists. This point is called ***nodal point*** *or simply* ***node***. The node of an electron wave is just like the node of a stationary wave produced on a string instrument like Veena or Guitar. The **nodal plane** is a zero probability plane passing through a node.

5. The radial function does not end abruptly at some particular distance from the nucleus. Instead, it gradually decreases, approaching zero at very large distances.

(B) ANGULAR DISTRIBUTION FUNCTIONS, ϕ (*theta*, φ) : The angular probability distribution function ϕ defines the probability of finding an electron in three dimensional space and is given by the angles θ and φ. These functions indicate variation of l and m values independent of the values of ***n*** and ***r***. Consequently all *s*-orbitals, having same l values but different n and r values, will have same shape. Similarly all p_x-orbitals, p_y-orbitals, etc., will have the same shape irrespective of their n and r values. ϕ^2 (θ, φ) gives the probability of finding an electron in different directions.

While angular wave functions are responsible in deciding the shapes of atomic orbitals, the radial wave functions are responsible in deciding the size. A three dimentional representation of probability of finding an electron in a sub level is done with the help of a **dot density diagram** in which the number of dots per unit area (the dot density) is proportional to the electron density. The dot density diagrams are discussed below.

(i) **THE SHAPES OF *S*-ORBITALS** : *s*-Orbitals, with $l = 0$ and $m=0$, have the same dependence of on ϕ (θ, φ) in a three dimentional space and therefore are spherical. To state in a different way, at any particles distance from the nucleus, the electron density of *s* orbitals is the same regardless of direction. The wave function ψ of *s* orbitals depends exclusively on the radial distribution function $R_{(r)}$. Therefore with the increase in radius '*r*', the diameter of the *s*-orbital increases, retaining its spherical shape, Fig 1-9.

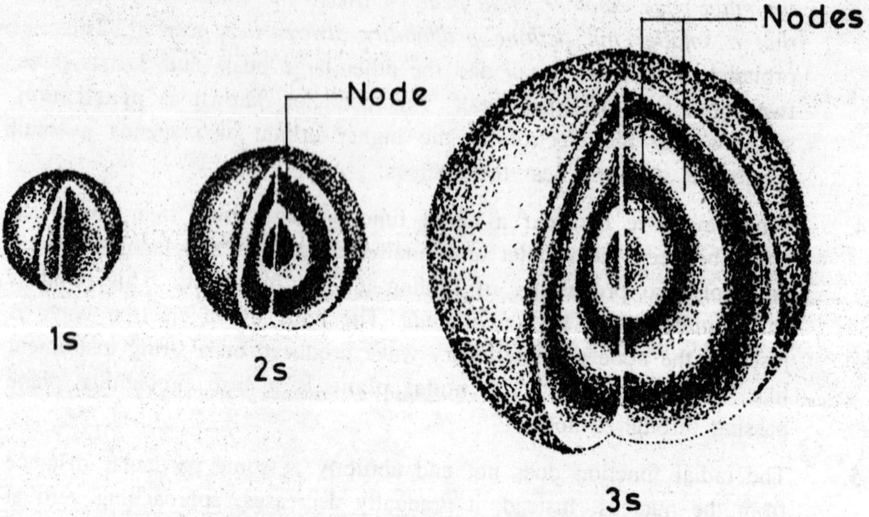

Fig 1-9 A two dimentional representation of electron density distribution in the 1*s*, 2*s* and 3*s* orbitals.

We notice that the dot diagram of the electron wave for 3*s* orbital has two nodes and 2*s* has only one node.

(ii) **THE SHAPES OF *p*-ORBITALS** : The electron cloud of *p*-orbitals, with $l = 1$ and $m = 1, 0, \& -1$, have a pair of lobes (dumb-bell shape) aligned in a straight line Fig 1-10. The electron cloud is not distributed in a spherically symmetrical way as in s-orbital, instead, it is concentrated in two equal-size regions located on opposite sides of the nucleus with their centre coinciding with the centre of the atom. The nucleus lies at the intersection point of the axis of the *p*-orbital and corresponding nodal plane. As expected from the radial distribution curves, Fig 1-8, 2*p* orbital, except for the nodal plane passing through nucleus, has no node, 3*p* and higher orbitals contain nodes. A cross section of 3*p* orbital in given in fig 1-11

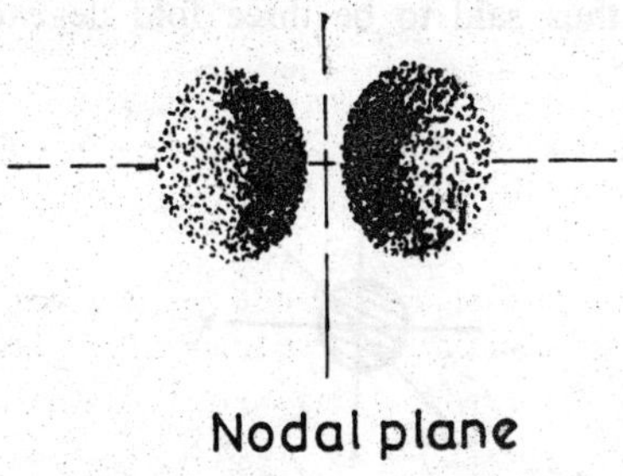

Fig 1-10 Dot density diagram of p orbital

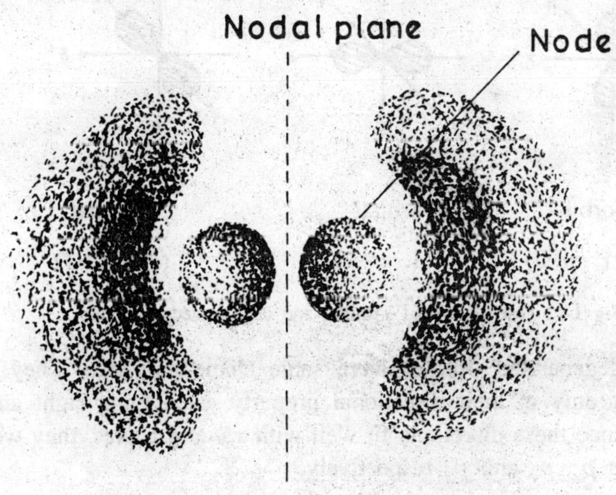

Fig 1-11 A cross section of 3*p* orbital.

The angular and radial functions respectively decide the size and orientation of *p*-orbitals. *p*-Orbitals with larger *n*-values are larger in size.

The orbital quantum number $l = 1$, corresponding to p-orbital, is associated with three magnetic quantum numbers (m = 0, +1 and -1) which designate three orientations or three angular wave functions.

For magnetic quantum number, m = 0, the probability of finding an electron is maximum along z axis. So it is called p_z orbital. For m = 1 the angular distribution is oriented along x axis, hence called p_x orbital. Similarly m = -1 shows angular distribution function along y-axis and termed as p_y orbital. The p-orbital is thus said to be three fold degenerated, Fig 1-12.

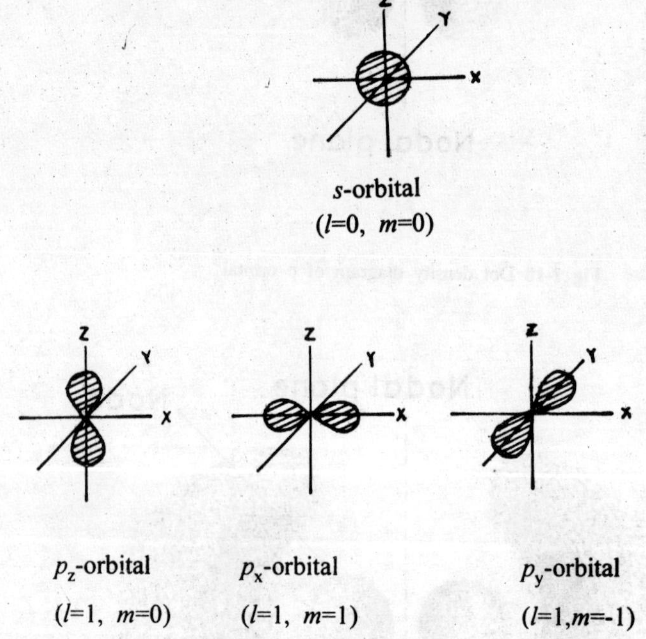

Fig 1-12 Orientations of s-orbital and degenerated p-orbitals

All these degenerated orbitals have same shape and size. They differ from one another only in their directional property, each lying right angles to one another. Since these directions fit well with x,y and z axes, they were assigned the symbols p_x, p_y and p_z respectively.

(iii) THE SHAPES OF d-ORBITALS : The orbital quantum number l = 2 has five magnetic quantum numbers m = -2,-1,0, +1, +2. Corresponding to these five m values, for a given value of n, we have five d-orbitals and the orbital is said to be five fold degenerated. Unlike degenerated p orbitals, d orbitals do not look alike. Four of these d orbitals, d_{xy}, d_{yz}, d_{zx} and $d_{x^2-y^2}$ look alike, each composed of four lobes oriented in different planes. The fifth *d*-orbital, labeled as d_{z^2}, consists of two lobes pointing opposite directions along *z*-axis with a ring shaped structure placed at the centre of these lobes and occupying *xy* plane. The shapes of these orbitals, Fig 1-13, origin purely from mathematical considerations of wave mechanics. Unlike *s* and *p* orbitals, it is rather difficult to get a vivid physical picture of chemical bonding involving *d*-orbitals.

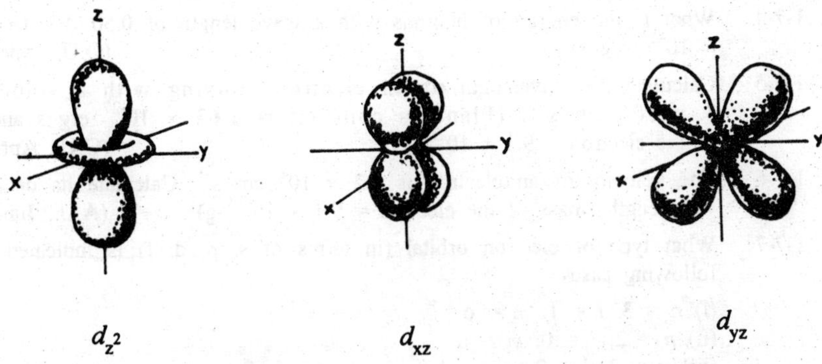

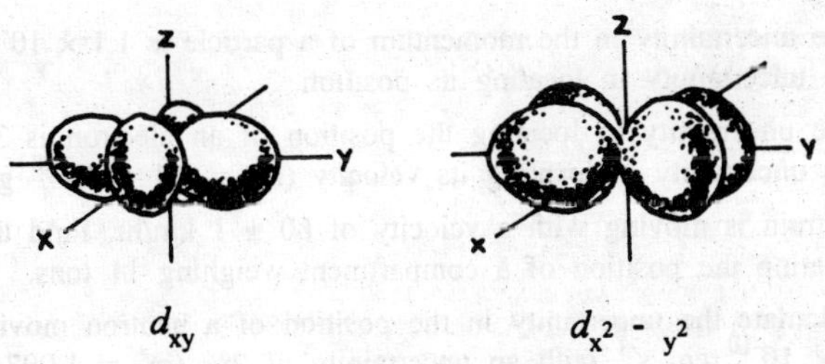

Fig 1-13 Shapes of d orbitals

(iv) **SHAPES OF *f*-ORBITALS** : *f*-orbital is designated by azimuthal quantum number value of $l = 3$. So it has seven magnetic wave functions the angular wave function is said to be seven fold degenerated m = -3, -2,-1,0, +1, +2, +3. Corresponding to these seven m values, for a given value of n, we have seven *f*-orbitals. Due to complexities in reading these three dimensional representations the radial distribution functions and shapes of the f-orbitals are not represented here.

1-7 NUMERICAL PROBLEMS FOR WORKING

1-7-1 Calculate the wave length associated with a particle of one mg, moving with a velocity of 4.5×10^5 cm. $^{-1}$ ($h = 6.63 \times 10^{-27}$ erg sec). (O.U., Oct. 1989)

1-7-2 What is the wavelength associated with a particle of mass 1×10^{-27} g moving with a velocity of 10^5 cm s^{-1} ? If a body of mass 10g moves with the same velocity as the above particle, which would have a greater wave length (Plank's constant $h = 6.626 \times 10^{-27}$ erg.s). (O.U., Mar. 1989)

1-7-3 What is the wave length associated with a helium atom of mass 6.6×10^{-29} g, moving with a velocity of 1.4×10^5 cm s^{-1}? ($h = 6.626 \times 10^{-27}$ erg. sec). (O.U., Oct. 1988)

1-7-4 What is the energy of photons with a wave length of 0.50 A°? ($h = 6.62 \times 10^{-27}$ erg s) (O.U., Apr. 1987)

1-7-5 Calculate the wavelength of an electron, moving with a velocity of 3×10^9 cm s^{-1}. (Planck's constant = 6.63×10^{-27} erg s and mass of the electron = 9.1×10^{-28} g) (O.U., Apr. 1984)

1-7-6 The velocity of an electron is 2.2×10^8 cm s^{-1}. Calculate its de Broglie wavelength (mass of the electron = 9.1×10^{-28} g) (A.U., Jun. 1988)

1-7-7 What type of electron orbital (in terms of s, p, d, f) is indicated in the following cases.

(i) $n = 3, l = 1, m = o$
(ii) $n = 2, l = 0, m = o$
(iii) $n = 3, l = 2, m = +1$
(iv) $n = 4, l = 2, m = -1$
(v) $n = 4, l = 3, m = 0$

1-7-8 What are the n and l values for the orbitals $3p$, $4d$, $2s$, $4f$.

1-7-9 Give the degeneracy of the following subshells $3p$, $2d$, $5f$, $2s$

1-7-10 The uncertainity in the momentum of a particle is 1.1×10^{-5} kg m. Calculate the uncertainity in locating its position.

1-7-11 The uncertanity in locating the position of an electron is 3×10^{-2} m. Find the uncertanity in locating its velocity ($m_e = 9.1 \times 10^{-28}$ g)

1-7-12 A train is moving with a velocity of 60 ± 1 km/hr. Find the uncertainity in locating the position of a compartment weighing 11 tons.

1-7-13 Calculate the uncertanity in the position of a neutron moving at a speed of 2×10^{-10} cm s^{-1} with an uncertainity of 2% ($m_n = 1.007$ amu)

(Hint. 1 amu = $\frac{1}{N}$ g)

1-8 MULTIPLE-CHOICE QUESTIONS

Mark the correct answer

1-8-1 Which electronic level would allow only absorption of photon but not emission

A) $4s$ B) $2p$ C) $3s$ D) $1s$

1-8-2 Maximum number of electrons that can be accommoded in the orbital with $l = 2$ is

A) 2 B) 8 C) 10 D) 14

1-8-3 The four quantum number (n, l, m, s in the same order) of valence electron of rubidium are

A) 4, 0, 0, $\frac{1}{2}$ B) 5, 1, 0 $\frac{1}{2}$ C) 5, 0, 0, $\frac{1}{2}$ D) 5, 0, 1, $\frac{1}{2}$

1-8-4 Trace the impossible arrangement among the following sets of quantum numbers.

A) $n = 1, l = 0, m = 0, s = -\frac{1}{2}$

A) $n = 3, l = 1, m = 2, s = \frac{1}{2}$

A) $n = 4, l = 1, m = 1, s = \frac{1}{2}$

A) $n = 3, l = 1, m = -1, s = \frac{1}{2}$

1-8-5 Which of the following subshells has no node

A) *1s* B) *2s* C) *2p* D) *4f*

1-8-6 Which one of the following subshells is not degenerated

A) *2p* B) *3d* C) *4f* D) *5s*

1-8-7 Which one of the following subshell shows a degeneracy of 5

A) *2p* B) *3d* C) *4f* D) *5s*

1-8-8 Maximum probability of finding an electron along d_{xy} orbital is

A) along x axis
B) at an angle of 45° to x axis
c) at an angle of 60° to x axis
D) at an angle of 90° to x axis.

1-8-9 The d-orbital that has only two lobes is

A) d_{xy} B) d_{zx} C) $d_{x^2-y^2}$ D) d_{z^2}

1-8-10 The main energy level of an electron is designated by

A) Orbital quantum number B) Magnetic quantum number
C) Spin quantum number D) Principal quantum number

1-8-11 After filling the 4d orbital completely, the next incoming electron goes to

A) *4f* B) *5s* C) *5p* D) *5d*

1-8-12 Electron from which of the following orbitals is tightly bound to the nucleus

A) *4s* B) *4p* C) *3d* D) *4f*

1-8-13 Number of nodes present in a 3p orbital are

A) 0 B) 1 C) 2 D) 3

1-8-14 Radial wave function of an orbital decides

A) size of the orbital

B) orientation of the orbital

C) size and orientation of the orbital

D) None of these

1-8-15 The magnetic quantum number 'm' values corresponding to the azimuthal quantum number l = 2 are

A) 1 B) 3 C) 5 D) 7

1-8-16 The f orbital is degenerated by

A) 5 times B) 7 times C) 10 times D) 14 times

1-8-17 Two electrons present in any orbital are distinguished by

A) Principal quantum number
B) Azimuthal quantum number
C) Magnetic quantum number
D) Spin quantum number

1-8-18 The ratio of energy of a red photon (800 n m) to that of violet photon (200 mm) is

A) 4 B) 2 C) $\frac{1}{2}$ D) $\frac{1}{4}$

1-8-19 Failure of Heisenberg's uncertainity principle leads to

A) Description of Bohr's stationary orbits
B) Schrodinger's probability theory
C) De Broglie's wave-particle duality
D) Rutherfords atomic model.

1-8-20 The azimuthal quantum number corresponding to 3d orbital

A) $l = 0$ B) $l = 1$ C) $l = 2$ D) $l = 3$

1-8-21 The maximum number of electrons that a d-orbital can accommodate are

A) 2 B) 6 C) 10 D) 14

1-8-22 The maximum number of electrons that can be found in 1 subshell.

A) 2 B) 8 C) 18 D) 32

1-8-23 Dot density diagram 3p orbital has

A) no node B) 1 node C) 2 nodes D) 3 nodes

1-8-24 Principal quantum number (n) describes

A) radius of the orbit B) radius of the electron
C) no. of nodes of an orbital D) no. of electrons in an orbital

1-8-25 Which one of the quantum numbers is designated by a non integer

A) n B) l C) m D) s

1-8-26 Wave length of the matter wave is given by

A) $mh\ v$ B) h/v C) $\frac{mh}{v}$ D) $\frac{h}{mv}$

1-8-27 According to Hesenberg's uncertainity principle

A) $\Delta x/\Delta p = \frac{h}{4\pi}$ B) $\Delta x \,.\, \Delta p = 0$

C) $\Delta x \,.\, \Delta p \quad \frac{h}{4\pi}$ D) $\Delta x + \Delta p \quad \frac{h}{4\pi}$

1-8-28 The quantum numbers of a valence electron are $n = 3$, $l = 0$, $m = 0$, $s = +\frac{1}{2}$. The atom is

A) N B) Na C) Al D) 0

1-9 SHORT AND ESSAY-TYPE QUESTIONS

1-9-1 What is the de Broglie relationship ? What does it suggest? (O.U., Apr. 1988)

1-9-2 State briefly why Bohr's theory was found inadequate to explain the structure of a hydrogen atom. In what respect is the wave mechanical model of an atom, a refinement over the model suggested by Bohr?

1-9-3 (a) Explain the wave nature of electrons?

(b) Why should the circumference of an electron orbit be an integral of its wave length? (O.U., Nov. 1986)

1-9-4 How is the particle and wave nature duality of the electron explained (O.U., Apr. 1985)

1-9-5 Explain briefly Planck's hypothesis of quantization of energy. (A.U., Apr. 1989 & N.U., Apr. 1989)

1-9-6 (a) It is impossible to determine both the position and velocity of an electron at any instant with high precision.
(b) Define the terms wavelength and wave number. How are they related to each other?
(c) Discuss de Broglie's equation (O.U., Oct. 1985)

1-9-7 State and explain Heisenberg uncertainity principle. Explain why wave nature is observed in small particles only? (A.U., Jun. 1988)

1-9-8 State and explain Heisenberg's uncertainity principle. (A.U., May 1989)

1-9-9 (a) State Hisenberg's uncertainly principle. Why do we neglect it while dealing with the macroscopic particles?
(b) What is an orbital? How many orbitals are there in a d- subshell?
(c) An electron is one of the 4-d orbitals? What could be the possible values of the quantum number n, l, m ? (O.U., Oct. 1987)

1-9-10 Why is it impossible to determine accurately both position and velocity of an electron at any instant (O.U., Apr. 1987)

1-9-11 Explain the fundamental conflict between Bohr's model of hydrogen atom and Heisenberg's uncertainty principle. (O.U., May 1986)

1-9-12 Show that the Heisenberg's uncertainty principle has little significance to macroscopic partilces like moving bullets.

1-9-13 Write the Schrodinger wave equation in cartisan coordinate and explain the terms? (A.U., Sep. 1988)

1-9-14 Write Schrodinger wave equation for general potential energy and explain the terms. (N.U., Oct. 1989)

1-9-15 Give the value of all the quantum number for the electron with the highest energy in the sodium atom. (O.U., Apr. 1985 & K.U., Oct. 1990)

1-9-16 What are quantum numbers? How many quantum numbers does an electron have in an atom? What do the principal quantum number n and the spin quantum number (s) signify? (O.U., Oct. 1984)

1-9-17 (a) What are quantum numbers? (b) How do lines of the Balmer series of hydrogen spectrum? (K.U., Apr. 1984)

1-9-18 An electron is in the 4d-orbital. What possible values can it have for the quantum number n, l, and m? (O.U., Mar. 1989)

1-9-19 Write all quantum numbers of an electron in each of the following orbitals. (a) 2s (b) 2p (c) 3d (O.U., Apr. 1988)

1-9-20 What are quantum numbers? What is the significance of each quantum number?

1-9-21 Write the n, l values corresponding to the 3d and 4s orbital. (O.U., Apr. 1987)

1-9-22 Show the relation between the quantum numbers n, l and m by giving all the values of l and m, when n = 3? (O.U., Oct. 1985)

1-9-23 (a) What are orbitals? Name the different types of orbitals known in elements.

(b) Draw the shapes of orbitals found in hydrogen and carbon atoms.

1-9-24 Give electron configuration of the elements of oxygen famuly. (A.U., Apr. 1989)

1-9-25 Compare and contrast Bohr's stationary orbits and orbitals of the wave mechanical atom as described by Schrodinger.

1-9-26 Write down the Schrodinger equation for the wave mechanical model of an atom. Name three quantum numbers of an electron, which are yielded by this equation. Explain the significance of each quantum number. (O.U., Oct. 1988)

ANSWERS

1-7-1 1.47×10^{-29} cm; **1-7-2** λ of 1×10^{-27} g body is 6.625×10^{-5} cm and of 10g body is 6.626×10^{-33} cm, former has higher wave length **1-7-3** 7.17×10^{-4} cm **1-7-4** 3.97×10^{-8} **1-7-5** 0.243 Å **1-7-6** 3.3 Å **1-7-7** (i) p, (ii) s, (iii) d (iv) d, (v) f **1-7-8** $3p$ - $n = 3$, $l = 1$ $4d$ - $n = 4$, $l = 2$, $2s$ - $n = 2$, $l = 0$, $4f$ - $n = 4$, $l = 3$ **1-7-9** 3, 5, 7, 0 **1-7-10** 4.8×10^{-30} m **1-7-11** 0.19 cm s^{-1} **1-7-12** 1.73×10^{-38} km **1-7-13** 7.94 cm

1-8-1 D	**1-8-2** C	**1-8-3** C	**1-8-4** B	**1-8-5** A	**1-8-6** D
1-8-7 B	**1-8-8** B	**1-8-9** D	**1-8-10** D	**1-8-11** C	**1-8-12** A
1-8-13 C	**1-8-14** A	**1-8-15** C	**1-8-16** B	**1-8-17** D	**1-8-18** D
1-8-19 B	**1-8-20** C	**1-8-21** C	**1-8-22** B	**1-8-23** C	**1-8-24** A
1-8-25 D	**1-8-26** D	**1-8-27** C	**1-8-28** B		

2

IONIC EQUILIBRIA

2-1 THEORIES OF ELECTROLYTES

The ability to conduct electricity is not limited to metals alone. Salts in their fused state, some liquids and liquid solutions are also known to conduct electricity. In some cases solvent possesses poor conducting properties, while a solution made employing the same solvent exhibits excellent conducting properties. For example, pure water is essentially a non-conductor but addition of a pinch of salt to it makes it an excellent conductor. It was also noticed that addition of solutes like acetic acid and aniline to water render it only weakly conducting, while solutes like sugar and alcohol do not bring any change in its conducting properties. These three groups of solutes are termed **strong electrolytes, weak electrolytes** and **non-electrolytes** respectively. These three properties can be demonstrated with the help of a conducting apparatus like the one shown in Fig.2-1

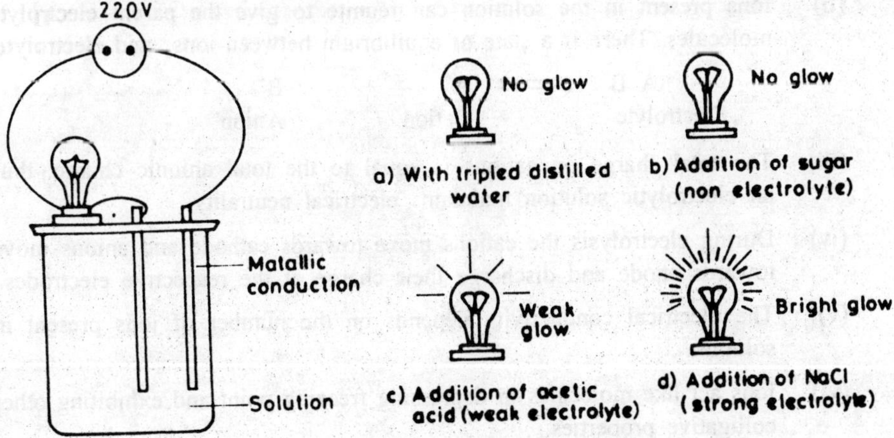

Fig 2-1 Conducting properties of Various electrolytes.

The conducting apparatus is made of two metallic electrodes dipped in pure water/solution and connected to 220V power supply through an incandescent bulb. The glow of the bulb is observed after a sequential addition of sugar, acetic acid and sodium chloride. Bulb does not light with pure water and by the addition of sugar because of their poor conducting properties, glows weakly on addition of acetic acid but glow brightly by the addition of NaCl, a strong electrolyte. How do we explain these variations in electrical conductivity? The nonelectrolytes do not yield any ions. The strong and weak electrolytes break apart to yield ions. In otherwords we say that the electrolytes dissociate into ions (different extent), which are responsible to electrical conductivity.

The electrical conductivity of electrolytes was summarised by Svante Arrhenius in his doctoral thesis submitted in 1884 to the University of Uppsala, Sweden. It is interesting to note that he was nearly denied his doctoral degree but for the support given by van't Hoff and Ostwald, the famous chemists of those days, and was awarded Nobel prize of 1903 for the same findings. His explanation, later known as **Theory of electrolytic dissociation** or **Ionic theory**, not only furnished the basic mechanism of ionic conduction but also paved way for gaining more exact information on the dissociation and ionic equilibria.

2-2 ARRHENIUS THEORY OF ELECTROLYTIC DISSOCIATION

The salient features of this theory are as follows;

(i) When electrolyte molecules are dissolved in water, they split into charged particles called ions. The positively charged ion is called cation and the negatively charged ion is called anion.

(ii) Ions present in the solution can reunite to give the parent electrolyte molecules. There is a state of equilibrium between ions and electrolyte.

$$\underset{\text{Electrolyte}}{AB} \rightleftharpoons \underset{\text{Cation}}{A^+} + \underset{\text{Anion}}{B^-}$$

(iii) The total charge on cations is equal to the total anionic charge, thus an electrolytic solution maintains electrical neutrality.

(iv) During electrolysis the cations move towards cathode and anions move towards anode and discharge their charge at the respective electrodes.

(v) The electrical conductivity depends on the number of ions present in solution.

(vi) Ions act like molecules in depressing freezing point and exhibiting other colligative properties.

(vii) The ratio of the concentration of dissociated molecules to that of total number is termed as **degree of dissociation**. Degree of dissociation (α) increases with increase in dilution.

(viii) At infinite dilution the electrolyte is totally dissociated into ions and therefore the *equivalent conductance*[1] reaches a limiting value(Λ_∞).

(ix) The degree of dissociation is related to equivalent conductance through the expression.

$$\alpha = \frac{\Lambda_v}{\Lambda_\infty} = \frac{\text{(equivalent conductivity of solution at any dilution)}}{\text{(equivalent conductivity of solution at infinite dilution)}} \quad (2.1)$$

2-2-1 Draw-backs of Arrhenius Theory

Although Arrhenius theory was successful in introducing the new concept of ions, it was soon discovered that there are many limitations while interpreting for strong electrolytes and left many observations unanswered. Some of these draw backs are mentioned below.

(i) The degree of dissociation of strong electrolytes obtained from conductometric measurements was found to be different from the value obtained from colligative properties.

(ii) Arrhenius theory assumes ionization of electrolytes only in solution phase, but x-ray diffraction studies of crystalline alkali halides (eg., NaCl) unambiguously proved that they exist in ionized forms even in solid state.

(iii) The molten salts were found to conduct electricity and undergo electrolysis. Arrhenius theory could provide a satisfactory explanation for this conduction but failed to answer what transpires in water.

(iv) Arrhenius theory could not explain the variation of ionic conductivity with concentration of the solution.

(v) This theory also failed to explain the variation of equivalent conductance with dilution in case of NaCl, which exists in fully ionized form even in solid state.

2-2-2 Theory of Weak Electrolytes - Ostwald's dilution Law and Degree of Dissociation of Weak Electrolytes

Since Arrhenius theory didn't face any major opposition in case of weak electrolytes, the concept of chemical equilibrium was readily extended to ionic equilibrium between free ions and undissociated molecules. Ostwald used the Arrhenius theory, in conjunction with the law of mass action to explain the variation of electrical/equivalent conductivity of weak electrolytes with concentration. If AB is a uni-univalent electrolyte, the ionic equilibria may be written as

$$AB \rightleftharpoons A^+ + B^-$$

1 Equivalent conductivity is a measure of conducting capability of one normal solution.

If one mole of electrolyte is dissolved in one litre of water to give 1M solution, and if α is the degree of dissociation then,

Initial conc.	1	0	0
Equilibrium conc.	1-α	α	α

When C moles are dissolved in 1 L solvent

Equilibrium conc.	C (1-α)	C	C

Considering that the ionic equilibrium akin to chemical equilibrium, from the law of mass action we have;

$$K = \frac{[A^+]\,[B^-]}{[AB]}$$

Substituting the equilibrium concentrations of species involved in ionic equilibrium

$$K = \frac{C\alpha \times C\alpha}{C\,(1-\alpha)} = \frac{\alpha^2 C}{(1-\alpha)} \qquad (2.2)$$

This equilibrium constant is often referred to as **dissociation constant or ionization constant**. This constant always has same units as that of concentration but it is customery to drop the same and express as a simple number.

In case of very weak electrolytes, the degree of dissociation is a negligible quantity in comparision with the amount of electrolyte dissolved, i.e.,

$$\alpha << 1 \quad \text{or} \quad 1-\alpha \approx 1$$

Substituting this condition in equation (2.2), we have,

$$\text{or} \quad K = \alpha^2\, C \qquad (2.3)$$

$$\alpha = \sqrt{K/C} \qquad (2.4)$$

If V is the volume of the solution containing 1 mole of the electrolyte, then

$$V = \frac{1}{C} \quad \text{L mol}^{-1}$$

Substituting for 1/C in equation (2.4) we have

For very weak electrolytes $\boxed{\alpha = \sqrt{KV}}$ (2.5)

Using $\alpha = \frac{\Lambda_v}{\Lambda_\infty}$, equation (2.2) changes into

$$K = \frac{C\Lambda_v^2}{\Lambda_\infty\,(\Lambda_\infty - \Lambda_v)} \qquad (2.6)$$

The mathematical expression (2.6) representing the variation of equivalent conductance (Λ_v) with concentration *C*, in other words dilution V and expressions (2.2) to (2.5) representing the variation of degree of dissociation (α) with concentration (*C*) are different forms of **Ostwald's dilution law**. This law is usually interpreted binary (i.e., 1:1) electrolytes which yield only two ions on dissocition.

Ternary electrolytes like oxalic acid, which on dissociation give three ions, are considered to dissociate in two different ionic equilibria and therefore we have two different dissociation constants.

$$\begin{matrix} COOH \\ | \\ COOH \end{matrix} \longrightarrow \begin{matrix} COO^- \\ | \\ COOH \end{matrix} + H^+ \qquad K_1 = 6.46 \times 10^{-2}$$

$$\begin{matrix} COO^- \\ | \\ COOH \end{matrix} \longrightarrow \begin{matrix} COO^- \\ | \\ COO^- \end{matrix} + H^+ \qquad K_2 = 6.17 \times 10^{-5}$$

Similarly phosphoric acid has three dissociation constants.

$$H_3PO_4 \longrightarrow H_2PO_4^- + H^+ \qquad K_1 = 7.08 \times 10^{-3}$$

$$H_2PO_4^- \longrightarrow HPO_4^{2-} + H^+ \qquad K_2 = 6.31 \times 10^{-8}$$

$$HPO_4^{2-} \longrightarrow PO_4^{3-} + H^+ \qquad K_3 = 3.98 \times 10^{-13}$$

2-2-3 Experimental Verification of Ostwald Dilution Law

Though the presence of ions and their chemical reactivity can be felt, it is not possible chemically to establish the extent of dissociation. This is often done employing physical techniques like conductometry and potentiometry. The degree of dissociation, α is given by the ratio of equivalent conductance at any concentration (Λ_v) to that of equivalent conductance at infinite dilution or limiting ionic conductance (Λ_o or Λ_∞).

$$\alpha = \frac{\Lambda_v}{\Lambda_\infty}$$

Λ_v can be readily calculated from a knowledge of electrolytic conductance and concentration of the electrolyte. Λ_∞ can be obtained from the literature ion conductance values. Thus it is possible to calculate degree of dissociation at different concentrations. A constancy of the product of square of degree of dissociation and the concentration {equation (2.2)} verifies the validity of Ostwald's dilution law.

In case of ternary and quaternary electrolytes, the dissociation constants of individual dissociation equilibria are conveniently obtained through potentiometric techniques.

2-2-4 Success of Ostwald's Dilution Law

(i) Ostwald law provides a convenient means of calculating dissociation constants of weak electrolytes.

(ii) A comparative knowledge of dissociation constants of various acids, help us in finding the dissociated forms of the electrolyte present under given experimental conditions.

(iii) A knowledge of dissociation constants of various indicator electrolytes suggest the operative ranges of these compounds.

2-2-5 Limitations of Ostwald's Dilution Law

(i) Ostwald's dilution law is strictly operative with dilute solutions of weak electrolytes. Deviation were noticed at higher concentrations.

(ii) Ostwald's dilution law can not be applied in case of dilute solutions of strong electrolytes.

Problem 1 : The degree of dissociation of aqueous solution of 0.025M phenol is 6.33×10^{-5}. Calculate the dissociation constant of phenol.

Solution : Given $\alpha = 6.33 \times 10^{-5}$

Dissociation constant $K = \dfrac{\alpha^2 C}{1-\alpha}$

Since $\alpha << 1$, $(1-\alpha) \approx 1$

Therefore $K = \alpha^2 C = (6.33 \times 10^{-5})^2 \times 0.025$

$= 1 \times 10^{-10}$

Problem 2 : The equivalent conductivity of 0.1M dichloroacetic acid is 195.36 S cm^2 eq^{-1}. If the ion conductances of the acetate and H^+ ions are 40.9 and 349.82 respectively, find the dissociation constant of the acid.

Solution : Λ (Cl_2CH-COOH) $= \Lambda_+ + \Lambda_-$

$= 349.82 + 40.9$

$= 390.72$ S cm^2

Degree of dissociation $= \dfrac{\Lambda_v}{\Lambda_\infty} = \dfrac{195.36}{390.72}$

$\alpha = 0.5$

Dissociation constant $K = \dfrac{\alpha^2 C}{1-\alpha}$

$$K = \frac{(0.5)^2 \times 0.1}{1-(0.5)} = 0.05 = 5 \times 10^{-2}$$

Problem 3 : The equivalent conductivities of propionic acid at different concentrations are

Λ_v(S cm^2)	14.16	20.0	42.8	58.4
C (mol L^{-1})	1×10^{-2}	0.5×10^{-2}	1×10^{-3}	0.5×10^{-3}

If equivalent conductivity at infinite dilution of propionic acid is 385.62 S cm^2, verify whether the data obtained obeys Ostwald dilution law.

Solution :

C (M)	$\alpha = \frac{\Lambda_v}{\Lambda_\infty}$	$K = \frac{\alpha^2 C}{1-\alpha}$
1.0×10^{-2}	$\frac{14.16}{385.62} = 0.037$	1.35×10^{-5}
0.5×10^{-3}	$\frac{20}{385.62} = 0.052$	1.35×10^{-5}
1.0×10^{-3}	$\frac{42.8}{385.62} = 0.110$	1.38×10^{-5}
0.5×10^{-3}	$\frac{58.4}{385.62} = 0.151$	1.35×10^{-5}

In the above table a constant value of dissociation constant proves the applicability of Ostwald dilution law.

Alternatively, graphical method can be adopted, where, a plot of $\frac{\alpha^2}{1-\alpha}$ versus $1/C$ gives a straight line passing through origin and having a slope equal to dissociation constant.

2-3 ACID-BASE EQUILIBRIA IN AQUEOUS SOLUTION

Acids and bases are among the most common substances found in this world. Actually, the very existence of life is a consequence of various acid-base reactions. In our life system we find them in the digestive and metabolic path ways, in preserving and propagating the hereditary character and in performing various biological processes. We encounter them around us in the form of fruit juices, soda, soft drinks, quick lime, etc. Acids and bases were defined in various ways throughout the history of chemistry. The earliest definitions were purely experimental. For example, acids were defined as the one that have sour taste, turned blue litmus red, liberated H_2 on reacting with metals, had a soapy feel, etc.

2-3-1 Arrhenius Theory of Acids and Bases

In 1890 S. Arrhenius formulated a theory of acids and bases that defines *an acid as a substances that ionizes in water to produce H^+ ions and a base is* *a substance that ionizes in water to produce OH^- ions.* Thus HCl, a typical acid, readily dissociates into H^+ and Cl^- ions.

$$HCl(aq) \longrightarrow H^+(aq) + Cl^-(aq)$$

Arrhenius bases are metal hydroxides such as NaOH.

$$NaOH(aq) \longrightarrow Na^+(aq) + OH^-(aq)$$

When Arrhenius acid and base are brought together they invariably generate water and the process is called neutralization.

$$HCl(aq) + NaOH(aq) \longrightarrow Na^+(aq) + Cl^-(aq) + H_2O$$

The Arrhenius theory also met with success in explaining the catalytic activity of acid/base catalyzed reactions. Arrhenius definition dominated the acid-base chemistry for about three decades and still continues to attract the chemists, knowing well about the limitations of his theory.

2-3-2 Brönsted-Lowry Theory of Acids & Bases

In 1932 Danish chemist J.N.Brönsted and British chemist T. Lowry identified the inadequacy of Arrhenius vision when they found substances that neither produce H^+ nor OH^- ions but still exhibit the acidic or basic characters. For example, liquid ammonia ionizes in nonaqueous media through equilibrium,

$$NH_3 \longrightarrow H^+ + NH_2^- \quad (2.7)$$

where the amide ion, NH_2^-, has strong basic characters akin to OH^- ion produced in aqueous media. Further ammonia vapor and HCl gas free from any solvent were also found to get neutralized, similar to Arrhenius acid-base neutralization.

$$HCl + :NH_3 \longrightarrow NH_4Cl \quad (2.8)$$

According to Brönsted-Lowry theory *an acid is defined as any substance that is capable of donating a proton and base is defined as any substances that in capable of accepting a proton.* In simple terms Brönsted acid is a proton donor and base is a proton acceptor. Unlike the Arrhenius definition Brönsted-Lowry definition do not restrict acid and base behavior to water solutions alone.

In the ammonia-amide equilibrium (2.7), ammonia molecule is a proton donor and hence is an acid. When we start with the reverse reactions of this equilibrium, amide ion readily accepts a proton and hence a base. Thus according to Brönsted theory the relationship between acids and bases is as follows;

$$\boxed{\text{Acid}} \rightleftharpoons \boxed{\text{Base}} + H^+$$

$$\boxed{\text{Base}} + H^+ \rightleftharpoons \boxed{\text{Acid}}$$

It is important to note that a substance which is an acid in an equilibrium could be a base in some other equilibrium. Thus, ammonia, which is an acid in above discussed amide equilibrium, turns into a base in the following equilibrium.

$$\underset{\text{Acid}}{NH_4^+} \rightleftharpoons \underset{\text{Conjugate base}}{NH_3} + H^+ \quad (2.9)$$

So instead of focussing on either an acid or a base, it is more appropriate to discuss what is called a **conjugate acid-base pair** (conjugate means "joined together"). The conjugate base of a Brönsted-Lowry acid is the species left out after removal of the dissociated proton. In the equilibrium (2.9), NH_3 is a conjugate base of ammonium ion. Similarly for the reverse equilibrium, NH_3 is called the base and NH_4^+ as its conjugate acid.

$$\underset{\text{Base}}{NH_3} + H^+ \rightleftharpoons \underset{\text{Conjugate acid}}{NH_4^+} \quad (2.10)$$

Extending this argument for the acid-base neutralization reactions, we have;

$$\underset{\text{Acid}}{HCl} + \underset{\text{Base}}{:NH_3} \rightleftharpoons \underset{\text{Conjugate base}}{Cl^-} + \underset{\text{Conjugate acid}}{NH_4^+}$$

The proton is transferred from HCl to $:NH_3$.

In case of reverse reaction,

$$\underset{\text{Acid}}{NH_4^+} + \underset{\text{Base}}{Cl^-} \rightleftharpoons \underset{\text{Conjugate base}}{:NH_3} + \underset{\text{Conjugate acid}}{HCl}$$

The proton is transferred from NH_4^+ to Cl^-.

Species like NH_3, H_2O and anions like HS^-, $H_2PO_4^-$, HCO_3^-, HPO_4^{2-} can either accept or donate a proton and are said to be **amphoteric**. In particular, the amphoteric nature of water molecules has many important consequences in the acid-base chemistry in aqueous medium.

$$\underset{\text{Acid}}{H_2O} + \underset{\text{Base}}{H_2O} \rightleftharpoons \underset{\text{Conjugate acid}}{H_3O^+} + \underset{\text{Conjugate base.}}{OH^-} \quad (2.11)$$

Boric acid is an oxoacid with the following structure;

$$\begin{array}{c} \text{H} \\ | \\ :\ddot{\text{O}}: \\ | \\ \text{H} - \text{O} - \text{B} - \ddot{\text{O}} - \text{H} \end{array}$$

Although boric acid does not ionize in water to produce H^+ ions the acidic character was proved to be due to its reaction with water .

$$B(OH)_3 \quad + \quad H_2O \rightleftharpoons B(OH)_4^- \quad + \quad H^+$$

The first step of this reaction involves donation of a lone pair of electrons on oxygen in water molecule to the boron atom in boric acid, this then leads to liberation of a proton. This donation of electron pair, in fact, forms the basis of Lewis theory.

2-3-3 Lewis Theory of Acids and Bases

G.N.Lewis, an American chemist, presented his broad based theory of acids and bases in 1932, the same year that Brönsted-Lowry introduced theirs; but the significance was recognized only in 1938. The additional advantage of this theory over Brönsted-Lowry theory is that it brings other reactions involving no proton transfer into the fold of acid-base classification. According to Lewis an acid is an electron pair acceptor and a base is an electron pair donor. Thus in the reaction,

$$\begin{array}{c} \text{Cl} \\ | \\ \text{Cl} - \text{B} \\ | \\ \text{Cl} \end{array} \quad + \quad \begin{array}{c} \text{H} \\ | \\ :\text{N} - \text{H} \\ | \\ \text{H} \end{array} \rightleftharpoons \begin{array}{c} \text{Cl} \quad \text{H} \\ | \quad\quad | \\ \text{Cl} - \text{B} - \text{N} - \text{H} \\ | \quad\quad | \\ \text{Cl} \quad \text{H} \end{array}$$

Lewis acid Lewis base

BCl_3 is called a Lewis acid and NH_3 is called a Lewis base. Other examples of Lewis acid-base reactions are;

$CO_2(g)$	+	$H_2O(l)$	$\rightleftharpoons$	$H_2CO_3(aq)$
$Ag^+(aq)$	+	$2NH_3(aq)$	$\rightleftharpoons$	$Ag(NH_3)_2^+(aq)$
$SnCl_4(s)$	+	$2Cl^-(aq)$	$\rightleftharpoons$	$SnCl_6^{2-}(aq)$
$Ni(s)$	+	$4CO(g)$	$\rightleftharpoons$	$Ni(CO)_4(g)$
$AlF_3(aq)$	+	$NH_3(aq)$	$\rightleftharpoons$	$AlF_3\text{-}NH_3(aq)$
Lewis acid		Lewis base		

2-4 IONIC PRODUCT OF WATER

As already discussed water has a negligible but definite conductance of 3.8 × 10^{-8} S cm^{-1}. This is due to auto ionization or self ionization of water molecules, equation (2.11), which is often simplified and written as

$$H_2O \rightleftharpoons H^+ + OH^- \qquad (2.12)$$

Since the autoionization of water is a reversible process, the equilibrium constant expression takes the form

$$K = \frac{[H^+]\,[OH^-]}{[H_2O]} \qquad (2.13)$$

The pure undissociated water concentration $[H_2O]$ is, of course, constant and is equal to 55.56 mol L^{-1} at 25 °C.

Multiplying both sides of expression (2.13) with this value we get,

$$K_w = K[H_2O] = [H^+]\,[OH^-] \qquad (2.14)$$

where K_w is a constant equal to 1 × 10^{-14} mol^2 L^{-2} at 25 °C and is called **dissociation constant or ionic product of water**. In pure water, the concentrations of hydrogen and hydroxide ions are equal, that is $[H^+] = [OH^-]$. Hence

$$[H^+] = [OH^-] = \sqrt{1 \times 10^{-14}} = 1 \times 10^{-7} \text{ mol L}^{-1}$$

Ionic product expression (2.14) is also obeyed closely by all dilute aqueous solutions. In such solution $[H_2O]$ is essentially unaffected by the chemical reactions that occur. Acidic solutions contain an excess of H^+ ions over OH^- ions and basic solution an excess of OH^- ions over H^+ ions. For example, an acid solution containing 10^{-2} M hydrogen ions has 10^{-12}M $[OH^-]$

$$[OH^-] = \frac{10^{-14}}{10^{-2}} = 10^{-12} \text{ mol L}^{-1}$$

Similarly, when the molar concentration of OH^- is 0.001, the concentration of H^+ ions is

$$[H^+] = \frac{10^{-14}}{10^{-3}} = 10^{-11} \text{ mole L}^{-1}$$

So in aqueous solutions of acids and bases both H^+ and OH^- are present. In acid solution $[H^+]$ is greater than 1 × 10^{-7} M, while in base solution $[H^+]$ is less than 1 × 10^{-7} M.

2-5 THE pH AND pOH SCALE

We often encounter, the acid and base solutions of different concentrations. The acidity and basicity or the strength of a solution is measured and compared in terms of the free hydrogen ion concentrations. For example, a 0.001 M HCl has a hydrogen ion concentration of 0.001 M or 1.0 × 10^{-3} M and a 0.001 NaOH has a hydrogen ion concentration of 1.0 × 10^{-11} M. If we talk within the limits of 1 normal acid to 1 normal base, all the concentrations are always expressed in numbers involving negative exponents. When we have to compare two such numbers, like 1.23×10^{-8} and 9.64×10^{-7}, we first look for the exponents of 10 and then the value before the exponent. This slows down the comparision and puts a non-mathematical man in confusion. To simplify the situation the Danish biochemist S.P.L.Sörensen in 1909 introduced the pH scale, [where p stands for potenz, (German = power) of H^+)

The pH of a solution is defined as negative logarithm of hydrogen ion concentration (mol L^{-1})

$$\boxed{\mathbf{pH = -log[H^+]}} \tag{2.15}$$

Therefore, knowing $[H^+]$ of a solution we can compute its pH or knowing pH we can calculate $[H^+]$ of the solution.

Thus pure water having a $[H^+] = 10^{-7}$ mol L^{-1} has a pH = $-\log 10^{-7} = 7$. pH of an acid solution is always less than 7. For example, the pH of a 10^{-3} M HCl is equal to

$$pH = -\log 10^{-3} = 3$$

Similarly an aqueous solution of a base has always a pH greater than 7.

By common practice, pH scale is restricted to a range 0 to 14, that is to say from $[H^+] = 1$ M to 10^{-14} M or $[OH^-] = 10^{-14}$ M to 1 M or $[H^+] = 1$ M to $[OH^-] = 1$ M. Out side this range, concentrations will not have negative exponents and are simply given in molar concentrations.

To calculate $[H^+]$ from pH, we can use the modified pH expression (2.15) as

$$[H^+] = 1 \times 10^{-pH} \tag{2.16}$$

Extending the same approach to hydroxide ions, it is possible to define pOH as a negative logarithm of hydroxide ion concentration (mol L^{-1}).

$$pOH = -\log [OH^-] \tag{2.17}$$

Similar to interdependence of $[H^+]$ and $[OH^-]$ expressed by equation (2.14), it is possible to arrive a relation between pH and pOH. Taking logarithm on both sides of equation (2.14), we get ,

$$\log K_w = \log [H^+] + \log [OH^-]$$

Multiplying throughout with - 1 gives

$$-\log K_w = -\log [H^+] - \log [OH^-]$$

Extending the concept of pH and pOH, $-\log K_w$ may be replaced by pK_w. Therefore,

$$pK_w = pH + pOH \quad (2.18)$$

Because $K_w = 1 \times 10^{-14}$

$pK_w = -\log 10^{-14} = 14$

Hence

$$pH + pOH = 14 \quad (2.19)$$

Therefore, a knowledge of pH is sufficient to know pOH.

$$pOH = 14 - pH \quad (2.20)$$

We can easily recognize the importance of pH scale by considering the pH values of some common substance, Fig 2-2.

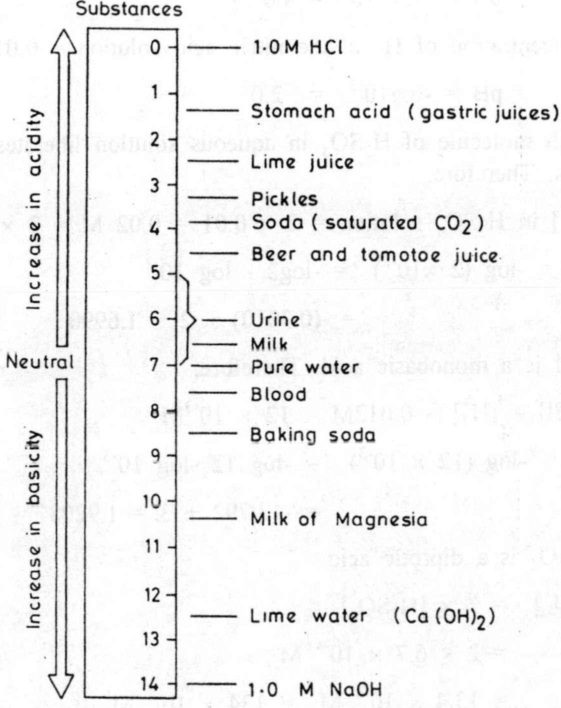

Fig 2-2 The pH scale and pH values of some common substances

While reading the pH scale two factors have to be kept in mind,

(i) *A small change in* pH *means large changes in* [H^+]. Though lime juice and tomato juice seem to differ only by 2 pH units, actually hydrogen ion concentration of former is greater by 100 times over the latter. This means, seemingly small pH change refers to a very large change in the actual hydrogen ion concentration.

(ii) pH *refers to* $[H^+]$ *concentration but not unionized acid concentration.* In case of strong acids, due to 100% dissociation, the $[H^+]$ and acid concentrations are one and the same. But, if we consider a weak acid only a small percentage of molecules exist in dissociated form. Because pH refers to the molar concentration of hydrogen ions, there is no direct, simple relation between pH and unionized acid concentration.

Problem 4 : Calculate the pH of the following acid solutions

(a) 10^{-4}M HCl (b) 0.01M HNO_3 (c) 0.01M H_2SO_4
(d) 0.012M HCl (e) 6.7×10^{-4} M H_2SO_4.

(a) The $[H^+]$ in 10^{-4} M HCl is equal to 10^{-4} M. Therefore,

$$pH = -\log 10^{-4} = 4.0$$

(b) Concentration of H^+ in the nitric acid solution = 0.01 = 10^{-2} M

$$pH = -\log 10^{-2} = 2.0$$

(c) Each molecule of H_2SO_4 in aqueous solution liberates $2H^+$ ions. Therefore,

$[H^+]$ in H_2SO_4 solution = $2 \times 0.01 = 0.02\ M = 2 \times 10^{-2}$ M

$$pH = -\log (2 \times 10^{-2}) = -\log 2 - \log 10^{-2}$$

$$= -(0.3010) + 2 = 1.6990$$

(d) HCl is a monobasic acid. Therefore,

$[HCl] = [H^+] = 0.012M = 12 \times 10^{-3}M$

$$pH = -\log (12 \times 10^{-3}) = -\log 12 - \log 10^{-3}$$

$$= -1.0792 + 3 = 1.9208$$

(e) H_2SO_4 is a diprotic acid.

Therefore $[H^+] = 2 \times [H_2SO_4]$

$= 2 \times 6.7 \times 10^{-4}$ M

$= 13.4 \times 10^{-4}$ M $= 134 \times 10^{-5}$ M

$$pH = \log (134 \times 10^{-5})$$

$$= -\log 134 - \log 10^{-5}$$

$$= -2.1271 + 5$$

$$= 2.8729$$

Problem 5 : Calculate pH and pOH of the following completely ionized bases (a) 0.015 M NaOH, (b) 0.01 M $Ba(OH)_2$

(a) One molecules of NaOH ionizes to give one molecule of OH^-

Therefore $[NaOH] = [OH^-] = 0.015 \text{ M} = 15 \times 10^{-3}\ OH^-$

$pOH = -\log[OH^-] = -\log(15 \times 10^{-3})$

$= -\log 15 - \log 10^{-3} = -1.1761 + 3.0$

$pOH = 1.8239$

$pH = 14 - pOH = 14 - 1.8239 = 12.1761$

(b) Each molecule of $Ba(OH)_2$ on dissociation yields two OH^- ions.

$[OH^-] = 2 \times [Ba(OH)_2] = 2 \times 0.01M = 0.02M = 2 \times 10^{-2}$ M

$pOH = -\log[OH^-] = \log(2 \times 10^{-2})$

$= -\log 2 - \log 10^{-2} = -0.3010 + 2 = 1.6990$

$pH = 14 - 1.6990 = 12.3010$

Problem 6 : Calculate the $[H^+]$ in the solution with pH

(a) 4.032 (b) 8.123

(a) $[H^+] = 1 \times 10^{-pH}$ or $\log[H^+] = -pH$

$\log H^+ = -4.032 = \bar{5}.968$

$[H^+] = \text{anti}\log(\bar{5}.968) = 9.29 \times 10^{-5}$ M

(b) $\log[H^+] = -8.123 = \bar{9}.877$

$[H^+] = \text{antilog}(\bar{9}.877) = 7.534 \times 10^{-9}$ M

Problem 7 : What would be the pH of the mixture formed by mixing 40 mL of 0.015M HCl and 50mL of 0.03M NaOH

Number of moles HCl taken $= \frac{40 \times 0.015}{1000} = 0.6 \times 10^{-3}$ moles

Number of moles of NaOH taken $= \frac{50 \times 0.03}{1000} = 1.5 \times 10^{-3}$ moles .

Since 1 mole of HCl is neutralized by 1 mole of NaOH, 0.6×10^{-3} moles of HCl consumes 0.6×10^{-3} moles of NaOH

Amount of NaOH left unneutralized $= 1.5 \times 10^{-3} - 0.6 \times 10^{-3}$

$= 0.9 \times 10^{-3}$ mole.

Volume of total mixture = 40 + 50 = 90mL

Therefore,

90 mL of solution contains 0.9×10^{-3} moles of NaOH

1000ml of solution contains ?

$= \frac{1000}{90} \times 0.9 \times 10^{-3} = 1 \times 10^{-2}$ M

Concentration of NaOH left out = 1×10^{-2} M

pOH of the mixture = $-\log 10^{-2} = 2.0$

pH of the mixture = $14 - 2 = 12.0$

Problem 8 : To a 50 mL aqueous sodium hydroxide solution of concentration 0.1M, 0.063 g of oxalic acid ($C_2O_4H_2.2H_2O$) is added. Find the pH of the mixture obtained after neutralization.

Molecular weight of oxalic acid = 126

126 g of oxalic acid is equal to 1 g mole

0.063 of oxalic acid is equal to ?

$$\frac{0.063}{126} = 0.5 \times 10^{-3} \text{ moles}$$

Amount of NaOH present in 50 mL of alkali solution = $\frac{50 \times 0.1}{1000}$

$= 5.0 \times 10^{-3}$ mole

0.5 mole of oxalic acid is neutralized by 1.0 mole of NaOH 0.5×10^{-3} mole of oxalic acid is neutralized by ?

$$\frac{0.5 \times 10^{-3}}{0.5} \times 1.0 = 1.0 \times 10^{-3} \text{ moles of NaOH}$$

Amount of NaOH left unneutralized = Total moles - Moles of NaOH of NaOH reacted with oxalic acid

$= 5 \times 10^{-3} - 1 \times 10^{-3}$

$= 4 \times 10^{-3}$ moles of NaOH

Volume of the solution is unchanged since 0.063g of oxalic acid crystals occupy negligible volume. Therefore,

50 mL of mixture contains 4×10^{-3} moles of unneutralized NaOH

1000 mL of mixture contains ?

$= \frac{1000}{50} \times 4 \times 10^{-3} = 8 \times 10^{-2}$ M

Concentration of unneutralized NaOH = 8×10^{-2} M

pOH $= -\log [OH^{-}] = -\log (8 \times 10^{-2})$

$= -0.9031 + 2 = 1.1$

pH $= 14 - 1.1 = 12.9$

Problem 9 : The equivalent conductance of 0.05M solution of propionic acid at 25 °C is 6.32 S cm^2 eq^{-1}. If the Λ_0 of the propionic acid is 385.6 S cm^2 eq^{-1}, calculate the pH of the solution.

Solution : Degree of dissociation of propionic acid $= \dfrac{6.32}{385.6} = 0.0164$

Concentration $[H^+]$ formed as a result of dissociation

$$= 0.0164 \times 0.05 = 0.00082 \text{ M}$$

$$pH = -\log[H^+] \quad = -\log(0.00082)$$

$$pH = 3.086$$

2-6 DISSOCIATION OF ACIDS AND BASES

The strength of an acid or base is defined in terms of fraction of molecules that undergo ionization or dissociation.

$$\text{Percent ionization or dissociation} = \frac{\text{Concentration of ionized acid or base}}{\text{Initial concentration of the acid or base}} \times 100$$

The dissociation capabilities of an acid or a base not only depends on the nature of bonds associated with it but also on the nature of solvent. HCl which ionizes completely when dissolved in water, ionizes partially in nonpolar medium. Similarly acetic acid, a well admitted weak acid in aqueous medium, dissociates 100% when dissolved in aprotic solvent like liquid ammonia. The strength of weak acid and bases are described quantitatively by their **ionization constants.**

2-6-1 Ionization Constants of Weak Acids

Ionization constants of weak acids in aqueous medium are obtained by a small simplification of weak acid dissociation equilibrium law. Our long range goal in the preference of this new quantity over percent ionization is its use in preparing buffers that help to control the pH changes. Suppose a weak acid, HA, dissolved in water is ionizing to give a conjugate base A^-

$$\underset{\text{Weak Brönsted acid}}{HA} + H_2O \rightleftharpoons H_3O^+ + \underset{\text{Conjugate base}}{A^-} \qquad (2.21)$$

The weak acid under consideration could be a neutral molecule like acetic acid or positively charged ion like NH_4^+ or a negatively charged ions like $H_2PO_4^-$. When an acid has more than one ionizable proton like phosphoric acid, H_3PO_4, all the protons are not lost in a single step but are lost under different pH conditions. From the law of mass action the equilibrium constant may be written as;

$$K_{eq} = \frac{[H_3O]\,[A^-]}{[HA]\,[H_2O]}$$

Since the total acid concentration is far lower than $[H_2O]$ and for all practical purposes it may be thought to remain constant. Therefore,

$$K_{eq} \times [H_2O] = \frac{[H_3O^+][A^-]}{[HA]}$$

The product $K_{eq} \times [H_2O]$ is another constant for a given solvent system and a given temperature and is called **acid ionization constant or acid dissociation constant** designated by K_a.

$$K_a = \frac{[H_3O^+][A^-]}{[HA]}$$

Since $[H_3O^+] = [H^+]$, the above expression is also written as

$$\boldsymbol{K_a} = \frac{[H^+][A^-]}{[HA]} \qquad (2.22)$$

Because the ionic product is in the numerator, the larger the K_a stronger is the acid. Following the same approach as for pH, pOH, pK_w, we can define

$$\mathbf{p}\boldsymbol{K_a} = \mathbf{-log}\ \boldsymbol{K_a} \qquad (2.23)$$

pK_a values of some weak and polyprotic acids is given in Table 2-1. The strength of an acid decreases with increase in pK_a.

The ionization constants also allow one to classify the acids into strong, moderate and weak basing on the magnitude of K_a as given below,

$K_a > 1$	**Strong acid**
$K_a = 1$ to 10^{-3}	**Moderate acid**
$K_a < 10^{-3}$	**Weak acid**

Table 2-1 Ionization constants of acid at 25 °C

Weak acids	p*Ka*		*Ionic equilibrium*
Mono basic acids :			
Trichloro acetic acid	0.2	$Cl_3CCOOH \rightleftharpoons$	$Cl_3CCOO^- + H^+$
Dichloro acetic acid	1.30	$Cl_2CHCOOH \rightleftharpoons$	$Cl_2CHCOO^- + H^+$
Bisulphate ion	1.99	$HSO_4^- \rightleftharpoons$	$SO_4^{2-} + H^+$
Chloro acetic acid	2.85	$ClCH_2COOH \rightleftharpoons$	$ClCH_2COO^- + H^+$
Hydrofluoric acid	3.19	$HF \rightleftharpoons$	$F^- + H^+$
Nitrous acid	3.35	$HNO_2 \rightleftharpoons$	$NO_2^- + H^+$
Formic acid	3.74	$HCOOH \rightleftharpoons$	$HCOO^- + H^+$
Benzoic acid	4.14	$C_6H_5COOH \rightleftharpoons$	$C_6H_5COO^- + H^+$
Acetic acid	4.74	$CH_3COOH \rightleftharpoons$	$CH_3COO^- + H^+$

Butyric acid	4.82	C_3H_7COOH	$\rightleftharpoons$	$C_3H_7COO^- + H^+$
Propionic acid	4.85	C_2H_5COOH	$\rightleftharpoons$	$C_2H_5COO^- + H^+$
Hypochlorous acid	7.51	HOCl	$\rightleftharpoons$	$OCl^- + H^+$
Ammonium ion	9.24	NH_4^+	$\rightleftharpoons$	$NH_3 + H^+$
Hydrocyanic acid	9.31	HCN	$\rightleftharpoons$	$CN^- + H^+$
Phenoi	10.0	C_6H_5OH	$\rightleftharpoons$	$C6H_5O^- + H^+$
Polyprotic acids :				
Carbonic acid	6.35	$CO_2 + H_2O$	$\rightleftharpoons$	$HCO_3^- + H^+$
	10.33	HCO_3^-	$\rightleftharpoons$	$CO_3^{2-} + H^+$
Chromic acid	0.75	H_2CrO_4	$\rightleftharpoons$	$HCrO_4^- + H^+$
	6.52	$HCrO_4^-$	$\rightleftharpoons$	$CrO_4^{2-} + H^+$
Oxalic acid	1.27	$H_2C_2O_4$	$\rightleftharpoons$	$HC_2O_4^- + H^+$
	4.27	$HC_2O_4^-$	$\rightleftharpoons$	$C_2O_4^{2-} + H^+$
Phosphoric acid	2.15	H_3PO_4	$\rightleftharpoons$	$H_2PO_4^- + H^+$
	7.20	$H_2PO_4^-$	$\rightleftharpoons$	$HPO_4^{2-} + H^+$
	12.4	HPO_4^{2-}	$\rightleftharpoons$	$PO_4^{3-} + H^+$

2-6-2 Ionization Constants of Weak Bases

Ionization constant of a weak bases in aqueous medium could be obtained in a similar way to that of weak acids.

Suppose a weak base B is dissolved in water, to generate a conjugate acid

$$\underset{\text{Weak Brönsted base}}{B} + H_2O \rightleftharpoons \underset{\text{Conjugate acid}}{BH^+} + OH^- \tag{2.24}$$

A weak base could be a neutral molecule like NH_3 or negatively charged particle like HCO_3^-. From the law of mass action the equilibrium constant may be written as

$$K_{eq} = \frac{[BH^+][OH^-]}{[B][H_2O]} \tag{2.25}$$

Multiplying both sides with $[H_2O]$

$$K_{eq}\,[H_2O] = \frac{[BH^+][OH^-]}{[B]} \tag{2.26}$$

Since total base concentration is far below $[H_2O]$ and for all practical purposes it may be thought to remain constant. Therefore,

$$K_b = K_{eq}\,[H_2O] = \frac{[BH^+][OH^-]}{[B]} \quad (2.27)$$

where K_b is called **base ionization constant or base dissociation constant.**

Following the same practice as for pK_a, we can define pK_b as the negative logarithm of K_b.

$$pK_b = -\log K_b \quad (2.28)$$

pK_b values of some weak bases is given in Table 2-2.

Table 2-2 Ionization constants of some bases at 25 °C

Weak bases	pK_b	*Ion equilibrium*
Ethyl amine	3.25	$C_2H_5NH_2 + H_2O \rightleftharpoons C_2H_5NH_3^+ + OH^-$
Methyl amine	3.35	$CH_3NH_2 + H_2O \rightleftharpoons CH_3NH_3^+ + OH^-$
Ammonia	4.75	$NH_3 + H_2O \rightleftharpoons NH_4^+ + OH^-$
Pyridine	8.77	$C_5H_5N + H_2O \rightleftharpoons C_5H_5NH^+ + OH^-$
Aniline	9.42	$C_6H_5NH_2 + H_2O \rightleftharpoons C_6H_5NH_3^+ + OH^-$
Urea	13.82	$H_2NCONH_2 + H_2O \rightleftharpoons H_2NCONH_3^+ + OH^-$
Oxide ion (O^{2-})	22.0	$O^{2-} + H_2O \rightleftharpoons 2OH^-$
Phosphate ion (PO_4^{3-})	1.66	$PO_4^{3-} + H_2O \rightleftharpoons HPO_4^{2-} + OH^-$
Carbonate ion	3.68	$CO_3^{2-} + H_2O \rightleftharpoons HCO_3^- + OH^-$
Bicarbonate ion	7.59	$HCO_3^- + H_2O \rightleftharpoons H_2CO_3 + OH^-$
Acetate ion	9.24	$CH_3COO^- + H_2O \rightleftharpoons CH_3COOH + OH^-$

Smaller the K_b value or greater the pK_b value weaker is the base.

We can derive another important relation when we multiply the K_a and K_b values of conjugated acid base pair.

For the equilibrium HA $\rightleftharpoons$ $H^+ + A^-$ (2.29)

We know,

$$K_a = \frac{[H^+]\,[A^-]}{[HA]} \quad (2.30)$$

In the equilibrium (2.29) A^- being the conjugate base of Brönsted acid HA, its reaction with water may be represented as

$$A^- + H_2O \rightleftharpoons HA + OH^- \qquad (2.31)$$

$$K_b = \frac{[HA]\,[OH^-]}{[A^-]} \qquad (2.32)$$

K_a and K_b are the ionization constants of conjugate acid-base pair. Now,

$$K_a \times K_b = \frac{[H^+]\,[A^-]}{[HA]} \times \frac{[HA]\,[OH^-]}{[A^-]}$$

Therefore,

$$K_a \times K_b = [H^+]\,[OH^-]$$

The right hand side expression is equal to K_w therefore,

$$\boldsymbol{K_a \times K_b = K_w} \qquad (2.33)$$

(for a conjugate acid-base pair)

Taking logarithms

$$\log K_a + \log K_b = \log K_w$$

Multiplying with -1

$$-\log K_a - \log K_b = -\log K_w$$

or

$$pK_a + pK_b = pK_w$$

or

$$\boldsymbol{pK_a + pK_b = 14} \qquad (2.34)$$

(for a conjugate acid-base pair)

Problem 10 : The pH of a 0.1 M acetic acid is 2.88. Calculate K_a of acetic acid.

Solution : $CH_3COOH \rightleftharpoons CH_3COO^- + H^+$

$$K_a = \frac{[CH_3COO^-]\,[H^+]}{[CH_3COOH]}$$

Because $[H^+] = [CH_3COO^-]$

$$K_a = \frac{[H^+]^2}{[CH_3COOH]}$$

Taking logarithms

$$-2 \log [H^+] = -\log K_a - \log [CH_3COOH]$$

$$2\ pH = pK_a - \log [CH_3COOH]$$

$$2 \times 2.88 = pK_a - \log(0.1)$$

$$pK_a = 5.76 + (-1) = 4.76$$

$$K_a = \text{anti log}(-4.76) = \text{anti log}(5.24) = 1.74 \times 10^{-5}$$

Problem 11 : What will be the relative strengths of two bases (MX-type) of equal concentration with dissociation constants 3.6×10^{-5} and 6.25×10^{-4}

Solution : For solutions of equal strength

$$\frac{\text{Strength of base A}}{\text{Strength of base B}} = \sqrt{\frac{\text{Dissociation constant of base A}}{\text{Dissociation constant of base B}}}$$

$$= \sqrt{\frac{3.6 \times 10^5}{6.25 \times 10^4}} = 0.24$$

Therefore Base B is stronger than base A.

Problem 12 : The pK_a value of hypochlorous acid is 7.51. Find the pK_b of hypochlorite anion.

Solution : Hypochlorous acid dissociates to give hypochlorite anions

$$\underset{\text{Brönsted acid}}{HOCl} \rightleftharpoons H^+ + \underset{\text{Conjugate base}}{OCl^-}$$

For a conjugate acid-base pair

$$pK_a + pK_b = 14$$

Therefore,

$$pK_b \text{ of } OCl^- = 14 - 7.51 = 6.49$$

Problem 13 : 0.01N solution of a weak monobasic acid dissociates to an extent of 10%. Calculate the dissociation constant of acid. (N.U., Apr. 1992)

Solution : Given Percentage of dissociation = 10

It means when we take initially 1 mole, 0.1mole is dissociated on dissolution.

	MA	$\rightleftharpoons$ M^+ +	A^-
Initial	1	0	0
At equilibrium	(1 - 0.1)	0.1	0.1

$$\text{Dissociation constant} = \frac{[M^+]\,[A^-]}{[MA]} = \frac{(0.1)\,(0.1)}{0.9} = \frac{0.01}{0.9}$$

$$K_a = 1.11 \times 10^{-2}$$

Problem 14 : Calculate the pH of 0.0002M NaOH solution assuming complete dissociation. (N.U., Oct. 1991)

Solution : pOH = $-\log[OH^-]$ = $-\log(0.002)$

pOH = 3.7

pH = 14 - pOH = 14 - 3.7

= 10.3

Problem 15 : Calculate the hydrogen ion concentration of 0.01M solution of monobasic acid, whose dissociation constant is 3.5×10^{-12}. (A.U., Apr 1992)

Solution : Given $K_a = 3.5 \times 10^{-12}$

For the equilibrium $HA \rightleftharpoons H^+ + A^-$

$$K_a = \frac{[H^+][A^-]}{[HA]} \text{ and } [H^+] = [A^-]$$

$$3.5 \times 10^{-12} = \frac{[H^+]^2}{[HA]}$$

or

$$[H^+]^2 = 3.5 \times 10^{-12} \times 0.01 = 3.5 \times 10^{-14}$$

$$[H^+] = 1.87 \times 10^{-7}$$

Problem 16 : Assuming complete ionization, calculate pH of 2×10^{-2} M barium hydroxide solution. (O.U., Mar.1992)

Solution : $Ba(OH)_2 \longrightarrow Ba^{2+} + 2OH^-$

$[OH^-]$ produced on complete dissociation of $Ba(OH)_2 = 2 \times 2 \times 10^{-2}$

$= 4 \times 10^{-2}$ M

$$pOH = -\log 4 \times 10^{-2}$$

$$pOH = 1.39$$

$$pH = 14 - 1.39 = 12.61$$

Problem 17 : Calculate the hydrogen ion concentration in 0.1 M acetic acid. Dissociation constant of acetic acid is 1.8×10^{-5} (O.U.,Oct. 1991)

Solution : For the dissociation of acetic acid

$$K_a = \frac{[H+][CH_3COO^-]}{[CH_3COOH]} \text{ and } [H^+] = [CH_3COO^-]$$

$$1.8 \times 10^{-5} = \frac{[H+]^2}{[CH_3COOH]}$$

$$[H^+]^2 = 1.88 \times 10^{-5} \times 0.1 = 1.8 \times 10^{-6}$$

$$[H^+] = 1.34 \times 10^{-3}$$

$[H^+]$ in the acid solution $= 1.34 \times 10^{-3}$ M

2-7 COMMON ION EFFECT

Any particular ionic species in a solution can come from more than one source. Thus in a solution of acetic acid and sodium acetate, the acetate ions are produced due to dissociation of acetic acid and complete ionization of sodium acetate.

$$CH_3COOH \rightleftharpoons CH_3COO^- + H^+ \quad (2.35)$$

$$CH_3COONa \rightleftharpoons CH_3COO^- + Na^+ \quad (2.36)$$

Once the acetate ions are produced, they can not be distinguished whether they are from acetic acid or sodium acetate. Extending the Le Chateliar principle to the above ionic equilibria, we find that presence excess of either H^+ or acetate ions suppresses the dissociation.

The supression of the dissociation of a weak acid or a weak base or suppression of solubility of sparingly soluble salt on the addition of any of its own ions is known as ***common ion effect***. The solutions of strong electrolytes do not exhibit common ion effect as they are fully ionized and have practically no ionic equilibrium expression.

We essentially have four cases that exhibit common ion effect.

(a) Weak electrolyte + Strong electrolyte with common cation.
eg., CH_3COOH + HCl
NH_4OH + NH_4Cl

(b) Weak electrolyte + Strong electrolyte with common anion.
CH_3COOH + CH_3COONa
NH_4OH + NaOH

(c) Weak electrolyte + Weak electrolyte with common cation.
CH_3COOH + HCOOH
NH_4OH + CH_3COONH_4

(d) Weak electrolyte + Weak electrolyte with common anion
CH_3COOH + CH_3COONH_4
NH_4OH + CH_3NH_2OH

In case of (a) and (b), strong electrolyte is completely ionized and therefore, it is sufficient to know the ionization constant of the participating weak electrolyte in order to judge the extent of common ion effect. In case of ionic equilibria (c) and (d), in order to judge the extent of common ion effect we require equilibrium expressions and a knowledge of dissociation constants of both the participating weak electrolytes.

2-8 SOLUBILITY EQUILIBRIA

The solubility equilibria of salts is equally important as conventional chemical equilibria. Salts may be defined as ionic compounds whose cations are any positive ions except H^+ ions and anions are any negative ions except OH^- and O^{2-}

They are soluble in water to different degrees and on dissolution readily yield ions.

The general rules predicting the solubility of ionic compounds in water are;

(i) All alkali metal (Li, Na and K) salts are soluble regardless of counter ions.

(ii) All acetates and nitrates are soluble, regardless of counter ion.

(iii) Most compounds containing Cl^-, Br^-, and I^- are soluble. The exceptions are the lead, silver and mercury (II) salts.

(iv) Most sulphates are soluble. The exceptions are lead, calcium, strontium, mercury (I) and barium salts.

(v) All ammonium salts are soluble.

(vi) All carbonates (CO_3^{2-}), phosphates (PO_4^{3-}), sulfides (S^{2-}) and sulphites (SO_3^{2-}) are insoluble except the group 1A cations and NH_4^+.

These solubility rules, although useful, do not allow us to make any quantitative predictions as to the extent of solubility, dependence of common ion on the ionic equilibria. Solubility product is one such attempt to answer the above questions and has a wide range of applications.

A saturated solution of any substance is in a dynamic solubility equilibrium obeying all principles of chemical equilibrium. In particular the solubility of sparingly soluble salt is of special interest to us. When we add a sparingly soluble salts like AgCl to water, partly it dissolves and most of it is left as undissolved solid mass. Because salts such as AgCl are treated as strong electrolytes, the dissolved portion dissociates 100% to give corresponding ions. Therefore the equilibrium between undissolved man and aquated ions may be expressed as

$$AgCl(s) \rightleftharpoons Ag^+(aq) + Cl^-(aq)$$

The equilibrium constant is written as;

$$K_{eq} = \frac{[Ag^+][Cl^-]}{[AgCl]}$$

or

$$K_{eq}\ [AgCl] = [Ag^+]\ [Cl^-]$$

Applying the convention that the activities or concentration of solids are unity, we can write

$$K_{eq} = [Ag^+]\ [Cl^-]$$

Since the equilibrium constant is a product of molar concentrations of ions of dissolved salt it is called **Solubility product.**

$$K_{sp} = [Ag^+]\ [Cl^-]$$

Similarly for saturated aqueous solution of Ag_2SO_4, we write the solubility product constant as;

$$Ag_2SO_4 \rightleftharpoons 2Ag^+ + SO_4^{2-} \qquad K_{sp} = [Ag^+]^2\ [SO_4^{2-}]$$

When aluminium hydroxide is added to water we have the equation

$$Ag(OH)_3 \rightleftharpoons Al^{3+} + 3OH^- \qquad K_{sp} = [Al^{3+}]\ [OH^-]^3$$

In general solubility product of a compound is the product of the molar concentrations of the constituents ions, each raised to the power of its stoichiometric coefficient in the equilibrium equation.

$$A_xB_y \rightleftharpoons xA + yB \qquad \boldsymbol{K_{sp}} = [A]^x\ [B]^y \qquad (2.37)$$

Solubility products of some sparingly soluble salts is given in Table 2-3.

Table 2-3 Solubility products of some sparingly soluble salts at 25 °C

Salt	K_{sp}	*Salt*	K_{sp}
Fluorides		**Oxalates**	
MgF_2	6.4×10^{-9}	CaC_2O_4	2.3×10^{-9}
CaF_2	4.0×10^{-11}	MgC_2O_4	8.6×10^{-5}
BaF_2	1.7×10^{-6}	BaC_2O_4	1.2×10^{-7}
PbF_2	4.1×10^{-8}	FeC_2O_4	2.1×10^{-7}
		PbC_2O_4	2.7×10^{-11}
Chlorides		**Hydroxides**	
$AgCl$	1.6×10^{-10}		
$PbCl_2$	2.4×10^{-4}	$Mg(OH)_2$	7.1×10^{-12}
$CuCl$	3.2×10^{-7}	$Ca(OH)_2$	6.5×10^{-6}
$HgCl_2$	6.1×10^{-15}	$Fe(OH)_2$	2.0×10^{-15}
Hg_2Cl_2	3.5×10^{-15}	$Fe(OH)_3$	1.1×10^{-36}
		$Mn(OH)_2$	1.2×10^{-11}
Bromides		$Al(OH)_3$	1.8×10^{-33}
$AgBr$	7.7×10^{-13}	$Sn(OH)_2$	5.0×10^{-26}
$HgBr_2$	1.1×10^{-19}	$Ni(OH)_2$	1.6×10^{-14}
$PbBr_2$	2.1×10^{-6}	$Cu(OH)_2$	4.8×10^{-20}
$CuBr$	4.2×10^{-8}	$Zn(OH)_2$	1.8×10^{-14}

		Sulphates	
Iodides		$SrSO_4$	3.2×10^{-7}
AgI	8.5×10^{-17}	$CaSO_4$	2.0×10^{-4}
PbI_2	1.4×10^{-8}	$BaSO_4$	1.1×10^{-10}
CuI	5.1×10^{-12}	$PbSO_4$	6.3×10^{-7}
HgI_2	4.0×10^{-29}		
		Chromates	
Carbonates		$CaCrO_4$	1.0×10^{-4}
$MgCO_3$	3.5×10^{-8}	Ag_2CrO_4	1.9×10^{-12}
$CaCO_3$	9.0×10^{-9}	$BaCrO_4$	2.4×10^{-10}
$SrCO_3$	9.3×10^{-10}	$PbCrO_4$	2.8×10^{-13}
$BaCO_3$	8.1×10^{-9}		
$PbCO_3$	7.4×10^{-14}	**Sulphides**	
		CdS	8.0×10^{-28}
Phosphates		Ag_2S	7.0×10^{-50}
$AlPO_4$	5.8×10^{-19}	Bi_2S_3	1.6×10^{-72}
$Ba_3(PO)_2$	6.0×10^{-39}	NiS	2.0×10^{-28}
$Ca_3(PO_4)_2$	1.2×10^{-26}	HgS	4.0×10^{-54}

2-9 IONIC PRODUCT AND SOLUBILITY PRODUCT

The product of molar concentration of the ions of a dissolved salt, each raised to the stoichiometric coefficient is known as Ionic product. When the Ionic product is less than K_{sp}, the solution is said to be unsaturated. For a saturated solution the ionic product is equal to K_{sp} and a super saturated solution exists when ionic product exceeds K_{sp}

Unsaturated solution	**Ionic product** $> K_{sp}$
Saturated	**Ionic product** $= K_{sp}$
Super saturated	**Ionic product** $> K_{sp}$

When an unsaturated or saturated solution is subjected to any shock treatment or vigorous stirring, it does not lead to precipitation, while a super saturated solution on shaking readily precipitates till ionic product of the solution becomes equal to K_{sp}.

2-9-1 Calculation of Solubility Product

Solubility product calculation mainly involves the estimation of ionic concentrations or solubility of saturated solutions through chemical or instrumental methods.

(a) Volumetric method : The molar concentration of either cation or anion in a saturated solution and at a given temperature can be determined through suitable titrimetric technique. For example, the solubility of $Zn(OH)_2$ can be found from a simple acid-base titration of clear saturated solution.

(b) Gravimetric method : A known volume of saturated solution of a sparingly soluble salt can be evaporated and the weight of the residue corresponds the amount the salt dissolved.

(C) Conductometric method : Since the sparingly soluble salts dissociate 100% and concentrations represent fairly dilute solutions, we have the relation

$$\Lambda_o(\text{salt}) = \Lambda_o(\text{cation}) + \Lambda_o(\text{anion})$$

and $$\Lambda_o(\text{salt}) = \frac{1000\ k}{s}$$

where S is the solubility and k is electrolyte conductivity of the salt. A detailed account of this method is covered under applications of conductance measurements.

2-9-2 The Common Ion Effect and The Solubility

To a clear saturated solution of a sparingly soluble salt, addition of an electrolyte having a common ion was found to result in the precipitation of the sparingly soluble salt. This can be explained taking help of Le Chatelier's principle, which states that addition of a product to a system at equilibrium puts a stress on it and the equilibrium readjusts to remove the stress. Consider a clear saturated solution of silver chloride. Addition of sodium chloride enhances the molar concentration of chloride ions and the stress caused on the ionic equilibrium is relieved by favoring the reverse reaction.

$$AgCl(s) \rightleftharpoons Ag^+(aq) + Cl^-(aq)$$

In the new equilibrium, additional AgCl is precipitated, Fig 2-3. Therefore, the solubility of a sparingly soluble salt is lowered by the presence of a second electrolyte furnishing a common ion.

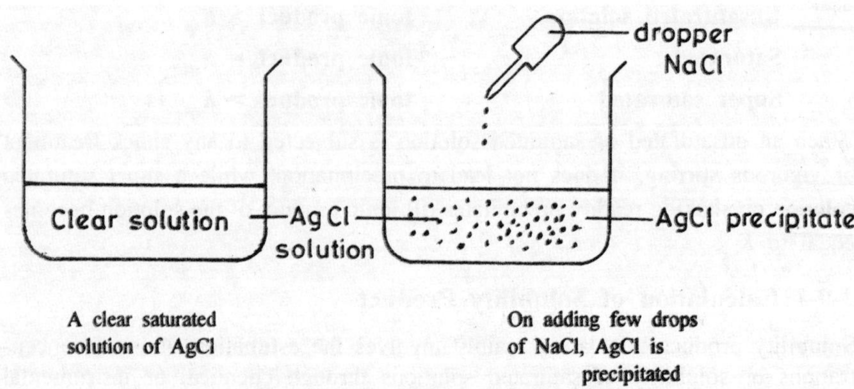

A clear saturated solution of AgCl

On adding few drops of NaCl, AgCl is precipitated

Fig 2-3 Common ion effect

2-9-3 Application of Solubility Product

(i) In the qualitative analysis : The presence and absence of cations and anions in a salt mixture is found exploiting the differences of their solubilities. A basic scheme of analysis is given in Flow chart - I.

FLOW CHART - I

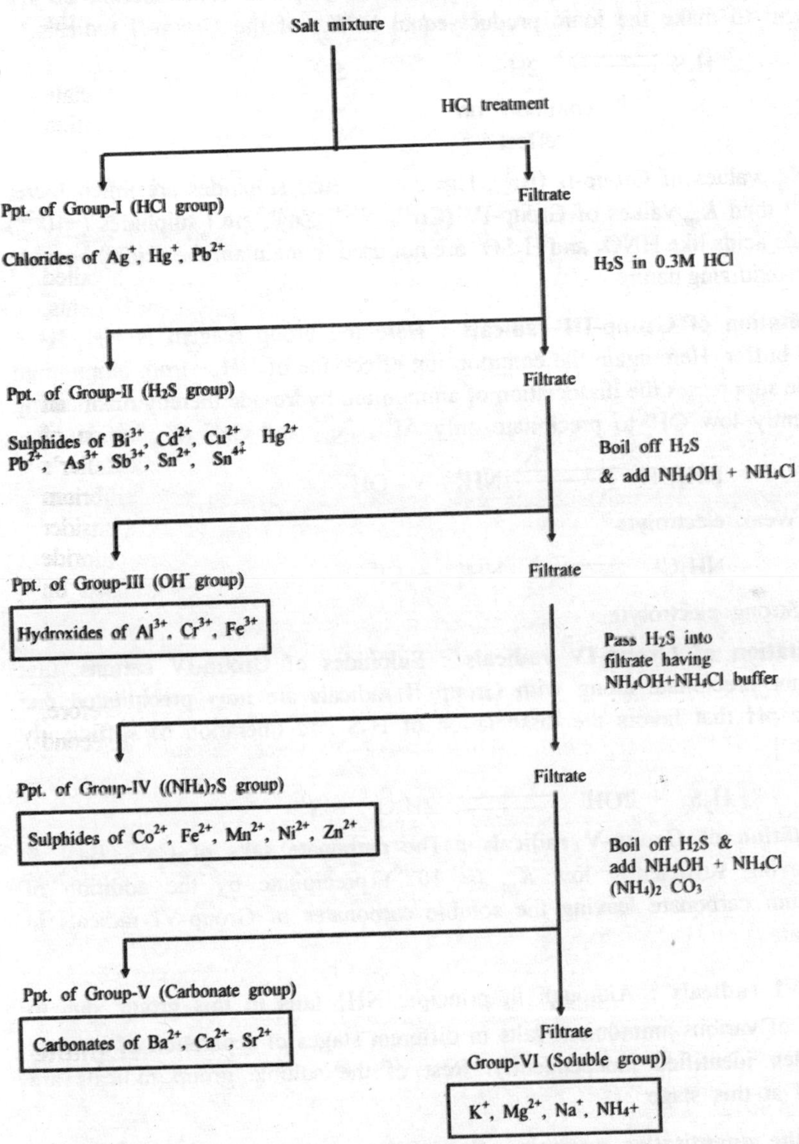

Precipitation of Group-I radicals (Ag^+, Hg^+, Pb^{2+}) : First group radicals are precipitated by the addition of dilute HCl because the ionic product of their chloride salts exceeds the solubility product. Other radicals pass into the filtrate.

Precipitation of Group-II Radicals : For this H_2S in 0.3M HCl acts as group reagent. In the presence of HCl, the dissociation of H_2S is suppressed due to common H^+ ion. Under these conditions the sulphide concentration is just sufficient to make the ionic product equal to K_{sp} of the Group-II radicals.

$$H_2S \rightleftharpoons \underset{\text{common ion effect}}{2H^+} + S^-$$

K_{sp} values of Group-II (Bi^{3+}, Hg^{2+}, Pb^{2+}, etc.) sulphides are much lower ($\approx 10^{-50}$) than K_{sp} values of Group-IV (Co^{3+}, Ni^{2+}, Zn^{2+}, etc.) sulphides ($\approx 10^{-20}$) Minerals acids like HNO_3 and H_2SO_4 are not used to maintain the acidity because of their oxidizing nature.

Precipitation of Group-III radicals : Here the group reagent is NH_4OH + NH_4Cl buffer. Here again the common ion effect due of NH_4+ from ammonium chloride suppresses the dissociation of ammonium hydroxide thereby maintaining sufficiently low OH^- to precipitate only Al^{3+}, Fe^{3+} and Cr^{3+} ($K_{sp} \approx 10^{-35}$).

$$\underset{\text{Weak electrolyte}}{NH_4OH} \rightleftharpoons NH_4^+ + OH^-$$

$$\underset{\text{Strong electrolyte}}{NH_4Cl} \rightleftharpoons NH_4^+ + Cl^-$$

Precipitation of Group-IV radicals : Sulphides of Group-IV cations, that could not precipitate along with Group II radicals are now precipitated due to basic pH that favors the dissociation of H_2S and liberation of sufficiently high $[S^{2-}]$

$$H_2S + 2OH^- \rightleftharpoons 2H_2O + S^{2-}$$

Precipitation of Group-V radicals : The carbonate salts of Ca^{2+}, Ba^{2+} & Sr^{2+}, having sufficiently low K_{sp} ($\approx 10^{-10}$) precipitate by the addition of ammonium carbonate leaving the soluble carbonates of Group-VI radicals in the filtrate.

Group-VI radicals : Although in principle NH_4^+ falls in this group, due to addition of various ammonium salts in different stages of separation of cations, it is often identified independently. Rest of the soluble group radicals are analyzed at this stage.

(ii) In the quantitative analysis : Precipitation reactions are quite common in the quantitative analysis of various metal ions. For example, calcium estimation involves treatment of a soluble calcium salt with ammonium oxalate, where by calcium precipitates as calcium oxalate. The precipitate is purified by washing with ammonium oxalate solution (common oxalate ion reduces the solubility of calcium oxalate). Calcium oxalate is heated strongly and converted into CaO and is weighed.

(iii) In the separation (Fractional precipitation) : The chloride and iodide ions are conveniently separated through fractional separation. AgCl and AgI have solubility product constants (K_{sp}) 1.6×10^{-10} and 8.5×10^{-17} respectively.

$$AgCl\,(s) \rightleftharpoons Ag^+ + Cl^- \qquad K_{sp} = 1.6 \times 10^{-10}$$

$$Ag\,I\,(s) \rightleftharpoons Ag^+ + I^- \qquad K_{sp} = 8.5 \times 10^{-17}$$

Therefore, addition of silver nitrate to iodide + chloride mixture first precipitates AgI(s) and then the precipitation of AgCl(s) follows.

Problem 18 : Milk of magnesia is a suspension of $Mg(OH)_2$ in water commonly used as an antacid. If the solubility product, K_{sp}, for $Mg(OH)_2 = 8.9 \times 10^{-12}$, find the pH of the solution

Solution : $K_{sp} = [Mg^{2+}]\,[OH^-]^2$

Let the solubility of $Mg(OH)_2$ be X

The solubility product is $= X \times (2X)^2 = 4X^3$

$$4X^3 = 8.9 \times 10^{-12}$$

$$X^3 = \frac{8.9 \times 10^{-12}}{4} = 2.22 \times 10^{-12}$$

Solubility $X = \sqrt[3]{2.2\ 10^{-12}} = 1.3 \times 10^{-4}$

Concentration of OH^- in solution $= 2 \times 1.3 \times 10^{-4} = 2.6 \times 10^{-4}$ M

$$pOH = -\log\,[OH^-] = -\log 2.6 \times 10^{-4}$$

$$= 3.585$$

$$pH = 14 - pOH = 14 - 3.585 = 10.415$$

Problem 19 : Calculate the solubility of saturated solutions of $Al(OH)_3$ and $Ba3(PO_4)_2$ having solubility products 2.0×10^{-32} and 6.0×10^{-39} respectively.

Solution : (a) $Al(OH)_3 \rightleftharpoons Al^{3+} + 3OH^-$ $K_{sp} = [Al^{3+}][OH^-]^3 = 2.0 \times 10^{-32}$

If the solubility of $Al(OH)_3$ is X then

$$[Al^{3+}] = X, \quad [OH^-] = 3X$$

Therefore $\quad K_{sp} = (X)\,(3X)^3 = 2.0 \times 10^{-32}$

$$27\,X^4 = 2.0 \times 10^{-32}$$

$$X^4 = \frac{2.0 \times 10^{-32}}{27} = 7.4 \times 10^{-34}$$

$$X = 5.22 \times 10^{-9}$$

Solubility of $Al(OH)_3$ = 5.22×10^{-9} M

(b) If X is the solubility of $Ba_3(PO_4)_2$

$$Ba_3(PO_4)_2 \rightleftharpoons 3Ba^{2+} + 2PO_4 3-$$

$$X \qquad 3X \qquad 2X$$

$$K_{sp} = [Ba^{2+}]^3\ [PO_4^{3-}]^2 = 6.0 \times 10^{-39}$$

$$(3X)^3\ (2X)^2 = 6.0 \times 10^{-39}$$

$$27X^3 \times 4X^2 = 6.0 \times 10^{-39}$$

$$108X^5 = 6.0 \times 10^{-39}$$

$$X^5 = 5.56 \times 10^{-41}$$

$$X = 8.9 \times 10^{-9}$$

Solubility of $Ba_3(PO_4)_2 = 8.9 \times 10^{-9}$ M

Problem 20 : To 100 mL saturated solution of silver chloride (K_{sp} = 1.2 $\times 10^{-10}$), sodium chloride salt is added to give an in solution concentration of 0.0001M . Calculate the amount of precipitate formed.

Solution : Given K_{sp} = $[Ag^+]\ [Cl^-] = 1.2 \times 10^{-10}$

Solubility = $\sqrt{1.2 \times 10^{-10}} = 1.09 \times 10^{-5}$

Solubility of pure saturated solution = 1.09×10^{-5} M

When NaCl is added

Total $[Cl^-] = 1 \times 10^{-4} + 1.09 \times 10^{-5} = 1.109 \times 10^{-4}$

$$[Ag^+] \times 1.109 \times 10^{-4} = 1.2 \times 10^{-10}$$

$$[Ag^+] = \frac{1.2 \times 10^{-10}}{1.109 \times 10^{-4}} = 1.882 \times 10^{-6}$$

Solubility of AgCl after addition of NaCl = 1.882×10^{-6} M

Solubility of AgCl before addition of NaCl= 1.09×10^{-5} M

Amount of AgCl precipitated = $1.09 \times 10^{-5} - 1.882 \times 10^{-6} = 9.02 \times 10^{-6}$M

= $9.02 \times 10^{-6} \times 143.4$ g L^{-1}

= 1.29×10^{-3} g L^{-1}

Amount of AgCl precipitated = 1.3×10^{-4} g per 100 mL

Problem 21 : Calculate the solubility of $Zn(OH)^2$ when dissolved in 1 liter of 1×10^{-5}M NaOH, if its solubility product in water is 8.0×10^{-18}

Solution : Given $[Zn^{2+}]\ [OH]^2 = 8.0\ 10^{-18}$

Let solubility in water be X

$$(X)\ (2X)^2 = 8 \times 10^{-18}$$

$$4X^3 = 8 \times 10^{-18}$$

$$X^3 = 2 \times 10^{-18}$$
$$X = 1.26 \times 10^{-6}.$$

Solubility of $Zn(OH)_2$ in water = 1.26×10^{-6} M

When dissolved in 1×10^{-5} M NaOH

$$[OH^-] = 1 \times 10^{-5} + 2 \times 1.265 \times 10^{-6} = 1.25 \times 10^{-5} \text{ M}$$
$$[Zn^{2+}] [1.252 \times 10^{-5}]^2 = 8 \times 10^{-18}$$
$$[Zn^{2+}] = 5.1 \times 10^{-8}$$

Solubility of $Zn(OH)_2$ in 1×10^{-5} M NaOH = 5.1×10^{-8} M

Problem 22 : The solubility product of Ag_2CrO_4 at room temperature is 9.0 $\times$ 10^{-12}. Calculate the solubility in moles/lit. [N.U., Oct. 1991]

Solution :

$$Ag_2CrO_4(s) \rightleftharpoons 2Ag^+(aq) + CrO_4^{2-}\ (aq)$$
$$K_{sp} = [Ag^+]^2 [CrO_4^{2-}]$$

Let the solubility be X

$$K_{sp} = (2X)^2 X = 9 \times 10^{-12}$$
$$4X^3 = 9 \times 10^{-12}$$
$$X^3 = 2.25 \times 10^{-12}$$
$$X = 1.31 \times 10^{-4}$$

Solubility of Ag_2CrO_4 = 1.31×10^{-4} M

Problem 23 : Solubility of silver chromate is 4.32×10^{-2} g L^{-1}. Calculate the solubility product. At.Wt of Ag = 108, Cr = 52. (O.U., Mar. 1992)

Solution : $Ag_2CrO_4(s) \rightleftharpoons 2Ag^+\ (aq) + CrO_4^{2-}$

Solubility of Ag_2CrO_4 = 4.32×10^{-2} g L^{-1}

Molecular weight of Ag_2CrO_4 = 332

$$\text{Molar solubility} = \frac{4.32 \times 10^{-2}}{332} = 1.3 \times 10^{-4} \text{ M}$$

$$K_{sp} = [Ag^+]^2 [CrO_4^{2-}]$$
$$= (2 \times 1.3 \times 10^{-4})^2 (1.3 \times 10^{-4})$$
$$= 8.79 \times 10^{-12}$$

Solubility product of Ag_2CrO_4 = 8.79×10^{-12}

Problem 24 : The solubility product of magnesium hydroxide at 25 °C is 1.4×10^{-11}. What is the solubility of magnesium hydroxide

Solution : $Mg(OH)_2 \rightleftharpoons Mg^{2+} + 2OH^-$

$$K_{sp} = [Mg^{2+}]\,[OH^-]^2 = 1.4 \times 10^{-11}$$

Let the solubility be X

$$(X)\,(2X)^2 = 1.4 \times 10^{-11}$$

$$4X^3 = 1.4 \times 10^{-11}$$

$$X^3 = \frac{1.4 \times 10^{-11}}{4} = 3.5 \times 10^{-12}$$

$$X = 1.52 \times 10^{-4}$$

Solubility of $Mg(OH)_2$ = 1.52×10^{-4} M

Problem 25 : At 25 °C the solubility product of AgCl is 1.69×10^{-10} M^2. Calculate its solubility (O.U., Oct. 1990)

$$K_{sp} = [Ag^+]\,[Cl^-] = 1.69 \times 10^{-10}$$

If X is the solubility $[Ag^+] = [Cl^-] = X$

Therefore $X^2 = 1.69 \times 10^{-10}$

Solubility $X = \sqrt{1.69 \times 10^{-10}} = 1.3 \times 10^{-5}$ M

Problem 26 : Calculate the concentration of $BaSO_4$ in its saturated solution. The solubility product of $BaSO_4$ is 1×10^{-10}.

Solution : $BaSO_4\,(s) \rightleftharpoons Ba^{2+}\,(aq) + SO_4^{2-}\,(aq)$

$$K_{sp} = [Ba^{2+}]\,[SO_4^{2-}]$$

Suppose the solubility of the salt is S

then $[Ba^{2+}] = S$ and $[SO_4^{2-}] = S$

$$K_{sp} = S^2 = 1 \times 10^{-10}$$

Therefore $S = \sqrt{1 \times 10^{-10}} = 1 \times 10^{-5}$

Solubility of Barium sulphate = 1×10^{-5} M

Problem 27 : Calculate the concentration of aluminium hydroxide present in its saturated solution. The solubility product of $Al(OH)_3$ is 1×10^{-33}.

Solution : $Al(OH)_3\,(s) \rightleftharpoons Al^{3+}\,(aq) + 3OH^-\,(aq)$

Solubility product $K_{sp} = [Al^{3+}]\,[OH^-]^3$

Let the solubility be S

$[Al^{3+}] = S$, $[OH^-] = 3S$

$$(S)\,(3S)^3 = 1 \times 10^{-33}$$

$$S^4 = \frac{1 \times 10^{-33}}{27}$$

$$S = 2.47 \times 10^{-9}$$

Concentration of $Al(OH)_3$ in the saturated solution $S = 2.5 \times 10^{-9}$ M

Problem 28 : The solubility product of calcium sulphate is 9×10^{-6}. Calculate the amount of calcium sulphate precipitated by the addition of 0.01 moles of sodium sulphate to 1 L saturated solution $CaSO_4$ without altering the volume of solution.

Solution : $CaSO_4 \rightleftharpoons Ca^{2+} + SO_4^{2-}$

Let the solubility of $CaSO_4$ after addition of Na_2SO_4 be X.

When 0.01 moles of Na_2SO_4 is added without bringing any change in the total volume

$[SO_4^{2-}] = X + 0.01;$ $\qquad [Ca^{2+}] = X$

$[Ca^{2+}\ [SO_4^{2-}] = K_{sp}$

$(X)\ (X + 0.01) = 9 \times 10^{-6}$

$X^2 + 0.01X = 9 \times 10^{-6}$

X being a very small value, X^2 can be neglected.

$0.01\ X = 9 \times 10^{-6}$

$X = 9 \times 10^{-4}$ M

Now let S be the solubility in the absence of any $NaSO_4$

$[Ca^{2+}] = S;$ $\quad [SO_4^{2-}] = S$

$S^2 = 9 \times 10^{-6}$

$S = 3 \times 10^{-3}$ M

Amount of $CaSO_4$ precipitated $= 3 \times 10^{-3} - 9 \times 10^{-4} = 2.1 \times 10^{-3}$ moles

Problem 29 : How many moles of Ag_2CrO_4 will dissolve in 100 ml. of 0.1M Na_2CrO_4 Neglect the hydrolysis. K_{sp} of Ag_2CrO_4 at 298 K is 2 ℜ10^{-12} $mol^3\ L^{-3}$.

Solution : Na_2CrO_4 is ionized according to equilibrium

$$Na_2CrO_4 \rightleftharpoons 2Na^+ + CrO_4^{2-}$$

1 mole of chromate ions are produced by the complete dissociation of 1 mole of Na_2CrO_4. So 0.1 mole of Na_2CrO_4 in 1 litre solvent completely ionize to give 0.1 mole of chromate ions.

For the equilibrium	$Ag_2CrO_4 \rightleftharpoons$	$2Ag^+$ +	CrO_4^{2-}
Initial	1 mole	0	0
at equilibrium	1-X	$2X$	X

$$[Ag^+]^2\ [CrO_4^{2-}] = (2X)^2\ (X) = 2 \times 10^{-12}$$

$$X^3 = \frac{2 \times 10^{-12}}{4} = 0.5 \times 10^{-12}$$

$$X = 7.9 \times 10^{-5}$$

Total chromate concentration = $7.9 \times 10^{-5} + 0.1 \approx 0.1$

$$[Ag^+]^2\ [CrO_4^{2-}] = 2 \times 10^{-12}$$

$$[Ag^+]^2 = \frac{K_{sp}}{[CrO_4^{2-}]} = \frac{2 \times 10^{-12}}{0.1} = 2 \times 10^{-11}$$

$$[Ag^+] = 4.47 \times 10^{-6}$$

Since 2 moles of Ag^+ is produced by the dissociation of 1 mole of Ag_2CrO_4

Solubility of Ag_2CrO_4 in 1000 mL of Na_2CrO_4 $= \dfrac{4.47 \times 10^{-6}}{2} = 2.24 \times 10^{-6}$

Solubility in100 mL = 2.24×10^{-7} moles

Problem 30 : The solubility of Ag_2CO_3 in water at 298K is 1.26×10^{-4} mol L^{-1}. What is the solubility of this salt in 0.1M $AgNO_3$ and 0.2M Na_2CO_3 solutions, assuming complete dissociation of all the salts?

$$\underset{1\text{-}X}{Ag_2CO_3} \rightleftharpoons \underset{2X}{2Ag^+} + \underset{X}{CO_3}$$

$$K_{sp} = (2X)^2 \cdot X$$

$$K_{sp} = (2 \times 1.26 \times 10^{-4})^2\ (1.26 \times 10^{-4}) = 8.0 \times 10^{-12}$$

Solution : (i) Solubility of Ag_2CO_3 in $AgNO_3$:

Solubility of Ag_2CO_3 in water is negligible when compared to silver ion concentration obtained from $AgNO_3$.

Therefore; $[Ag^+] \approx 0.1M$

$$[Ag^+]^2\ [CO_3^{2-}] = K_{sp}$$

$$[CO_3^{2-}] = \frac{K_{sp}}{[Ag^+]^2} = \frac{8 \times 10^{-12}}{(0.1)^2} = 8 \times 10^{-10}\ M$$

Since 1 mole of CO_3^{2-} is produced by the dissociation of 1 mole of Ag_2CO_3;

solubility of Ag_2CO_3 in $AgNO_3 = 8 \times 10^{-10}$

(ii) Solubility of Ag_2CO_3 in Na_2CO_3

$$[CO_3^{2-}] \approx 0.2M$$

$$[Ag^+]^2 = \frac{K_{sp}}{[CO_3^{2-}]} = \frac{8 \times 10^{-12}}{0.2}$$

$$[Ag^+] = \sqrt{\frac{8 \times 10^{-12}}{0.2}} = 6.3 \times 10^{-6}$$

Solubility of Ag_2CO_3 = $6.3 \times \frac{1}{2} \times 10^{-6} = 3.15 \times 10^{-6}$ M

2-10 NUMERICAL PROBLEMS

2-10-1 Calculate the $[H^+]$ and $[OH^-]$ of a solution of milk of magnetia whose pH is 12.42.

2-10-2 Calculate the pH of the following acid solutions

A) 0.105M HCl B) 0.025 M H_2SO_4

C) 1×10^{-3}M HNO_3 D) 6.8×10^{-4}M H_2SO_4

2-10-3 Calculate the pH of the following base solutions

A) 1×10^{-3}M NaOH B) 0.0096N $Ca(OH)^2$ C) 4.6×10^{-3}M $Ba(OH)_2$

2-10-4 Calculate the pH of 0.01M trichloroacetic acid whose K_a = 0.63.

2-10-5 Calculate the pH of 0.0126M benzoic acid solution. pK_a of benzoic acid = 4.14.

2-10-6 Calculate the pH value of the solution containing 2.5×10^{-7}M of HCl.

[Hint : Since HCl concentration is of the same order as that of H^+ ions generated due of dissociation of pure water, we have to add this value to get total $[H^+]$ in the solution.]

2-10-7 Calculate the pH value of 5.26×10^{-7} M NaOH solution.

2-10-8 Calculate the concentration of NaOH corresponding to a pH of 7.23.

2-10-9 Calculate the pH of a mixture formed by mixing 30 cc of 0.12M NaOH and 40 cc of 0.08M HCl

2-10-10 Calculate the pH of 0.5M solution of methylamine- hydrochloride the pK_b value of methyl amine is equal to 3.36.

2-10-11 Calculate the solubility of chromium hydroxide in its saturated solution. K_{sp} of chromium hydroxide in 1×10^{-17}

2-10-12 How many moles of AgI dissolves in 100 mL of 0.1M KI. K_{sp} of AgI at 298 K is 8×10^{-17} mol^2 L^{-2}.

2-10-13 The solubility of $CuCl_2$ in water at 28K is 2×10^{-7}mol L^{-1}. What is the solubility of this salt in 0.1M $CuSO_4$ and 0.1M HCl solution, assume complete dissociation of all salts.

2-11 MULTIPLE-CHOICE QUESTIONS

2-11-1 pH of a strong acid is 5.0. Its [H+] is equal to

A) 5×10^{-14} B) 5×10^{-5} C) 1×10^{-5} D) 6.9×10^{-5}

2-11-2 The number of moles of hydronium ion in 250 mL of 0.04 M H_2SO_4 is

A) 0.01 B) 0.04 C) 0.1 D) 0.02

2-11-3 The amphoteric character is exhibited by

A) H_3O^+ B) NH_4^+ C) HSO_4^- D) CO_3^{2-}

2-11-4 [H+] in an aqueous solution of aniline is

A) 10^{-7} B) 10^{-7} C) = 10^{-7} D) 10^7

2-11-5 Which one of the following is an Arrhenius acid

A) NH_3 B) BF_3 C) HF D) CH_4

2-11-6 Precipitation occurs when

A) Ionic product = Solubility product

B) Ionic product > Solubility product

C) Ionic product < Solubility product

D) No relation between ionic product and solubility product

2-11-7 Which one of the following is a Lowry-Bronsted base

A) H_3O^+ B) NH_4^+ C) H_2CO_3 D) HCO_3^-

2-11-8 The pH of 0.01 M sodium hydroxide is

A) 2 B) 12 C) 1 D) 0.01

2-11-9 Ostwald's dilution law is applicable to

A) only strong electrolytes B) only weak electrolytes

C) Both strong and weak electrolyte D) None of these

2-11-10 Changes conductance with change in concentration of a strong electrolyte mainly depends on

A) density of the solution B) Extent of ionization salt

C) dielectric constant D) intermolecular attractions

2-11-11 0.1M solution HCN is 0.01% ionized. The ionization constant of the acid is

A) 10^{-3} B) 10^{-5} C) 10^{-7} D) 10^{-9}

2-11-12 0.1M NaOH, 0.01M HCl and pure water are kept in separate vessels. The ionic product of water is

A) Maximum in HCl solution B) Maximum in NaOH solution

C) Maximum in water D) Same in all the three liquids

2-11-13 0.02 moles of NaCl is dissolved in water, the pH of the resultant solution is

A) A greater than 7 B) Less than 7

C) Equal to 7 D) zero

2-11-14 For strong electrolytes Ostwald's dilution law fails because,

A) Specific conductance decreases on dilution

B) Equivalent conductance increases on dilution

C) Conductance ratio is not equal to degree of dissociation

D) is not constant

2-11-15 Which one of the following is correct

A) pH + pOH = 0 B) pH - pOH = 7

C) pH + pOH = 7 D) pH + pOH = 14

2-11-16 The solubility of $CaSO_4$ is 0.005 mole L^{-1} at a certain temperature. The solubility product is

A) 2.5×10^{-5} B) 2.5×10^{-4} C) 2.5×10^{-6} D) 5×10^{-4}

2-11-17 Purification of NaCl by passing HCl is based on

A) Distribution law B) Common ion effect

C) Electrophoretic effect D) Hardy-Schultz rule

2-11-18 The pH of a solution is 8. Its $[OH^-]$ will be

A) 10^{-8} B) 10^{-2} C) 10^{-4} D) 10^{-6}

2-11-19 When H_2S is passed into 0.1M HCl solution

A) Ionization of HCl is suppressed due to common ion effect

B) Ionization of HCl is increased due to common ion effect

C) Ionization of H_2S is suppressed due to common ion effect

D) Ionization of H_2S is increased due to common ion effect.

2-11-20 Ionization of acetic acid decrease in the presence of hydrochloric acid because of ________________

2-11-21 The conjugate base of $HClO_4$ is ________________

2-11-22 The conjugate base of a weak acid is ________________

2-12 SHORT AND ESSAY-TYPE QUESTIONS

2-12-1 Deduce an equation for Ostwald dilution law and discuss its limitations. (S.K.U., Sept. 1990) & (N.U., Apr. 1992)

2-12-2 Explain common ion effect with an example (S.K.U., Sept. 1991) & (N.U., Apr. 1992)

2-12-3 Define solubility product. Write the expression for solubility product of (a) Silver chromate (b) Lead iodide. Write the units. Explain the effect of common ion on solubility of a sparingly soluble salt. (O.U., Oct. 1991)

2-12-4 Derive the relation between degree of dissociation and ionization constant of weak acid. (O.U., Oct. 1991)

2-12-5 Define solubility product. Illustrate two applications of it (S.K.U., Sept. 1990) & (S.K.U., Apr. 1990)

2-12-6 The solubility of a sparingly soluble salt, AB_2 is 5 moles/lit. Derive expression for its solubility product. (N.U., Apr. 1991)

2-12-7 The pH values of equimolar solutions of two acids A and B are 2 and 4 respectively. Which is more acidic ? Explain. (N.U., Apr. 1991)

2-12-8 Describe one example of the application of common ion effect in qualitative analysis. (N.U., Apr. 1991)

2-12-9 (a) What is meant by solubility product? Derive the relation between solubility and solubility product.

(b) How is the principle of solubility product useful in qualitative analysis. (A.U., Jun. 1990 & O.U., Oct. 1990)

ANSWERS

2-10-1 $[OH^-] = 2.63 \times 10^{-2}$ mol L^{-1}, $[H^+] = 3.8 \times 10^{-13}$ mol L^{-1}
2-10-2 (a) 0.98, (b) 1.3, (c) 3.0, (d) 2.87 **2-10-3** (a) 11.0, (b) 10.98, (c) 11.96
2-10-4 pH = 1.1 **2-10-5** pH = 3.02 **2-10-6** 6.46 **2-10-7** pH = 7.79
2-10-8 [NaOH] = 6.98×10^{-8} M **2-10-9** pH = 11.6 **2-10-10** pH = 5.47
2-10-11 1.36×10^{-6} mole L^{-1} **2-10-12** S = 8×10^{-17} mol per 100ml of 0.1M Kl **2-10-13** 1.4×10^{-4}, 1.4×10^{-4}

2-11-1 C **2-11-2** D **2-11-3** C **2-11-4** B **2-11-5** C **2-11-6** B
2-11-7 D **2-11-8** B **2-11-9** B **2-11-10** D **2-11-11** D **2-11-12** D
2-11-13 C **2-11-14** C **2-11-15** D **2-11-16** A **2-11-17** B **2-11-18** D
2-11-19 C **2-11-20** Common ion effect, **2-11-21** ClO, **2-11-22** Strong

3

THE GASEOUS STATE

3-1 STATE OF MATTER

We realize and admire this world with our sensory organs. We are generally wonderstruck at the sight of different forms of nature. All these forms are different expressions of the three fundamental states of matter, *viz., solid, liquid and gas.* These states are basically different aggregation of molecules or atoms and pose no difficulty in their identification.

A solid is identified by its definite volume and shape. A liquid has a definite volume but takes the shape of its container. A gas has no definite volume or shape and simply fills the container. The solid and liquid states, usually referred to as **condensed phases,** are visible but gaseous state is invisible unless the gas is colored.

The state of a substance is determined by two tendencies of matter, the thermal motion and intermolecular attractions, both acting one against the other. With the increase in temperature an ice block melts giving liquid water. Further increase of temperature leads to evaporation of liquid water to give steam. An increase in temperature raises the kinetic energy of molecules. This raise in kinetic energy tends to separate molecules from each other. The intermolecular forces or forces of attraction oppose this tendency towards

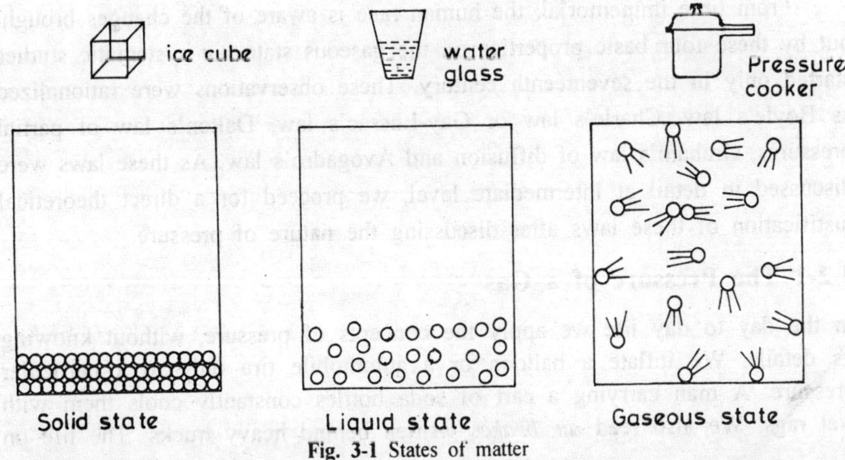

Fig. 3-1 States of matter

separation. In the condensed phases the intermolecular forces of attraction over-weigh the thermal motion, thus retaining their volume.

There is a large difference in the compactness of gases and condensed phases which is revealed by a comparison of densities. For example: 1 g of liquid water at 1 atm pressure and 100°C occupies a volume of ≈ 1 mL while under the same conditions the vapor occupies 1700 mL. (density = 5.9×10^{-4}). But solids have generally slightly higher densities than corresponding liquids. In fact, substances like water that have the slightly higher density in the liquid state than in the solid state were also known. At critical point, solids and liquids have the same density. Therefore the rigidity of solids is not attributed to any more compactness than liquids but to the regular arrangement of the molecules and the nature of bonds between them. The regular arrangement of molecules in a solid maximizes the intermolecular attractions between them and help in possessing a definite shape.

3-2 THE GASEOUS STATE

The most important characteristic of the gaseous state is almost independent behavior of the molecules (atoms in case of noble gases and metal vapors) of one another.

The gas molecules are in a constant random motion. Unlike solids and liquids, gases are relatively simple to understand. Although different gases have different chemical properties they share the same laws covering physical properties. The physical behavior of a gas is fully described by determining four basic properties, viz., the amount of gas, the temperature, the volume and the pressure. In case of solids and liquids it is generally sufficient if we describe the former three properties as pressure plays an insignificant role in deciding the changes affected in them.

From time immemorial, the human race is aware of the changes brought out by these four basic properties on the gaseous state but systematic studies started only in the seventeenth century. These observations were rationalized as Boyle's law, Charle's law or Gay-Lussac's law, Dalton's law of partial pressures, Graham's law of diffusion and Avogadro's law. As these laws were discussed in detail at intermediate level, we proceed for a direct theoretical justification of these laws after discussing the nature of pressure.

3-2-1 The Pressure of a Gas

In the day to day life we apply the concepts of pressure, without knowing its details. We inflate a balloon or a automobile tire only to a particular pressure. A man carrying a cart of soda bottles constantly cools them with wet rags. We also read *air brakes* written behind heavy trucks. The life on this earth itself is at the bottom of a "sea of air" and experiencing a pressure called atmospheric pressure.

Pressure is defined as the force per unit area.

$$\text{Pressure} = \frac{\text{Force}}{\text{Area}} \tag{3.1}$$

We can differentiate pressure and force by considering a pick-axe, Fig 3-2a & 3-2b. It can be conveniently balanced on a palm if held sideways Fig 3-2a, where its weight is distributed over relatively large area of the palm.

$$P = \frac{5 \text{ kg}}{10 \text{ cm}^2} = 0.5 \text{ kg cm}^{-2}$$

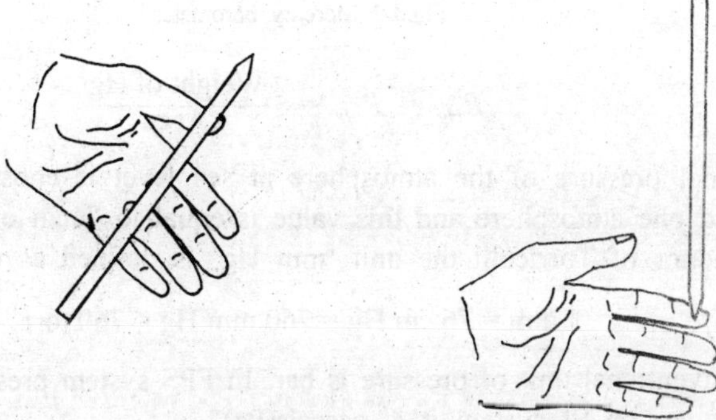

Fig 3-2a pick-axe held horizontally **Fig. 3-2b** Pick-axe held vertically

If the same pick-axe is to be balanced vertically on the tip of a finger it may prove very painful since the total weight is concentrated in a small area, Fig 3-2b.

$$P = \frac{5 \text{ kg}}{0.5 \text{ cm}^2} = 10 \text{ kg cm}^{-2}$$

In both the examples the weight of the pick-axe is equal to the gravitational force and is unchanged.

An Italian scientist and a student of Galileo, Evangelista Torricelli, invented a pressure measuring devise, now known as **barometer.** He filled long glass tubes of different dimensions and inverted them into a bowl of mercury. Torricelli found that the mercury column in these tubes stay suspended to a height of about 76 cm, no matter how long or wide the tube is, Fig 3-3.

The weight of the mercury column represents the force, and the weight applied per unit area gives its pressure. This pressureis equal and opposite to the atmospheric pressure.

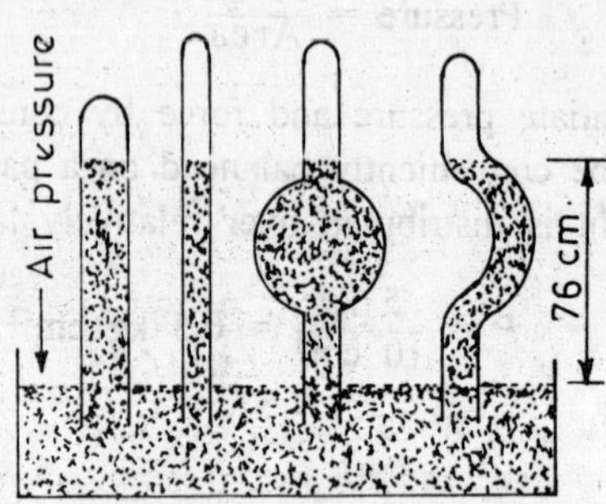

Fig.3-3 Mercury barometer

$$P_{Hg} = P_{air} = \frac{\text{Weight of Hg}}{\text{Area}}$$

The normal pressure of the atmosphere at sea level is chosen as a standard and called one atmosphere and this value is equal to 76cm of Hg or 760 mm Hg. In honor of Torricelli the unit 'mm Hg' is defined as one torr.

$$1 \text{ atm} = 76 \text{ cm Hg} = 760 \text{ mm Hg} = 760 \text{ torr}$$

Other conventional unit of pressure is bar. In FPS system pressure is expressed in lb in^{-2} and in SI system it is pascals(Pa).

$$1 \text{ atm} = 14.7 \text{ lb in}^{-2} = 101325 \text{ Pa} \quad = 1.013 \text{ bar} = 101\,325 \text{ N m}^{-2} \text{ (newton/m}^2\text{)}$$

Problem 1 : What is the height of water column, in meters, that could be supported by normal atmospheric pressure?

Solution : Normal atmospheric pressure supports a column of 76 cm of Hg.

$$\text{Pressure of Hg column} = G \times d_{\text{Hg}} \times h_{\text{Hg}}$$

(where G is acceleration due to gravity, d_{Hg}, the density of mercury and h_{Hg} is the height of the column supported by 1 atm pressure)

Therefore pressure of Hg column $= G \times 13.6 \text{ g cm}^{-3} \times 76$ cm. Similarly, when Hg is replaced by water pressure of H_2O

$$= G \times 1 \text{ g cm}^{-3} \times h_{\text{H}_2\text{O}}$$

Since both these pressures refer to 1 atm pressure are equal.

$$\text{Therefore } G \times 13.6 \times 76 = G \times 1 \times h_{\text{H}_2\text{O}}$$

$$\text{Therefore } h_{H_2O} = \frac{13.6 \times 76}{1} = 1034 \text{ cm}$$

$$= 10.34 \text{ m}$$

(1 atm pressure supports 0.76 m of Hg or 10.34 m of H_2O. The abnormal water column length explains why water manomete are not used for routine pressure measurments)

Problem 2 : Calculate the pressure exerted by a book having dimensions 20 × 10 × 2 cm and weighing 500 g, when it is placed horizontally on its back and upright on its shorter sides.

Solution : Pressure is defined as the force operating per unit area

Force of the body = 500 g

Area covered when placed on its back = 20 × 10 = 200 cm^2

$$\text{Pressure when supported on its back} = \frac{500 \text{ g}}{200 \text{ cm}^2} = 2.5 \text{ g cm}^{-2}$$

Area covered when placed upright on its shorter sides

$$= 2 \times 10 = 20 \text{ cm}^2$$

$$\text{Pressure in the upright position} = \frac{500 \text{ g}}{20 \text{ cm}^2} = 25 \text{ g cm}^{-2}$$

3-3 KINETIC THEORY OF GASES

The general gas laws referred earlier are experimentally well proved and all ideal gases would obey them exactly. They help us to predict the behaviour of gases. Thus we will be in a position to say that the volume doubles at a constant pressure, when the temperature is doubled. Since these laws are empirical they have no answer to questions like "why does the volume of a gas expand upon heating?" An attempt to provide theoretical background for the laws was first made by Bernoulli in 1739. This was further expanded by other scientists like Ludwig Boltzmann, Clark Maxwell, Van der Waals, etc. The kinetic molecular theory of gases (or kinetic theory of gases in short) is a result of the work done by these scientists and is based on the following assumptions.

1. *A gas is composed of point masses called molecules (atoms in case of noble gases and metal vapours) which are separated by distance far greater than their own dimensions and occupy negligible volume.*

This statement suggests that the gas is almost an empty space and when the molecules are pressed hard they occupy a negligible volume in comparison to the total volume of the gas. For example a sample of a gas like oxygen, hydrogen, etc., at STP (760 mm Hg pressure and 0°C temperature are usually referred as STP) occupy a volume of 0.014% leaving 99.986% space merely empty. Similarly 1 mL of water when evaporated occupies 1700 mL at 100°C and 1 atm pressure, suggesting that 99.941% of the total volume of water vapour is vacuum.

2. *The molecules of a gas are in a constant random motion colliding with each other and with the walls of the container.*

Each collision exerts a tiny force on the walls and number of collisions per unit area determine the pressure exerted by the gas molecules on the sides of the container. A molecule travels in straight lines between collisions.

3. *The collisions of gas molecules are perfectly elastic.*

Although some energy exchange between the molecules is not ruled out, the total kinetic energy of a gas always remains constant. No energy is gained or lost by the system because of collisions. Increase in temperature of a gas, increases the thermal motion. A faster moving molecule makes a stronger impact on the walls, there by accounts for a greater pressure.

4. *Gas molecules do not exert any attractive or repulsive forces on one another.*

In condensed phases the molecules are sufficiently close to have a role for internuclear forces. In gases the molecules are widely spaced not to allow any intermolecular attractive or repulsive forces. Even during collisional contact, the high kinetic energy and very small impact time do not allow the attractive or repulsive forces to play any noticeable role.

5. *The motion of gas molecule provides them with kinetic energy. The kinetic energy is found using the expression.* $KE = \frac{1}{2} mu^2$

Each molecule moves with a different velocity and thereby possesses different kinetic energy. The kinetic energy of the system is an algebraic sum of the individual kinetic energies of the molecules. The pressure of a gas reflects only an average kinetic energy of the molecules.

6. *The average kinetic energy of a gas is directly proportional to the absolute temperature of the gas.*

The average kinetic energy at absolute zero or 0 K is equal to zero. So absolute zero refers to a state where the gas molecules would have ceased their motions. At any given instant the gas molecule possesses a wide range of molecular velocities or distribution of kinetic energies, Fig. 4-6. Maxwell and Boltzmann suggested that average kinetic energy $E_K \propto T$. A detailed account of distribution of molecular velocities is given in section 3-4-3.

3-3-1 The Kinetic Gas Equation

The equation of state $(PV = nRT)$ could be derived from a mathematical analysis of the fundamental postulates of the kinetic-molecular theory proposed above. For this consider n gas molecules encased in a cubic container of side l, Fig. 3-4. First we will attempt to get an expression for the pressure on the shaded side of the cube. According to postulate 2, pressure, P, is on account of the collisions of the molecules on the walls of the container with a force, F, per unit area. If a is the area of the side, *Pressure 'P'* $= \frac{F}{a}$

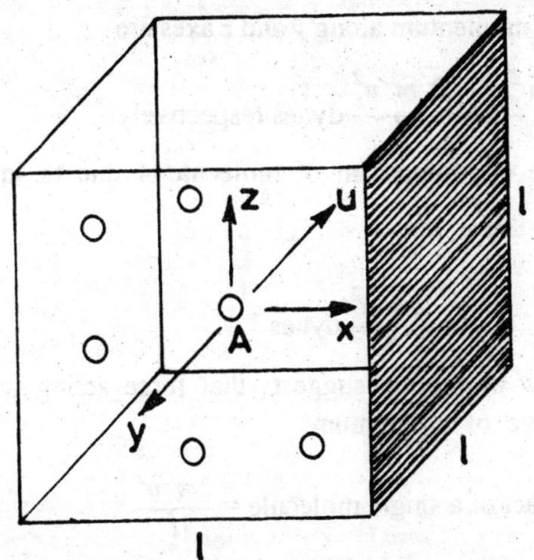

Fig. 3-4 Collisions on the sides of a cube

But, force is a change in momentum with time and momentum is the product of mass and velocity.

The molecules move at random with different velocities in a three dimensional space. If we consider a molecule, A, moving with a velocity, 1 cm per second, its velocity may be resolved along three axes x, y and z parallel to the sides of the cube. The velocity and its components along three axes are related by the expression;

$$u^2 = u_x^2 + u_y^2 + u_z^2 \tag{3.2}$$

When molecule A strikes the shaded wall, the magnitude of the velocity and its components are unchanged but the sign is reversed or direction is reversed. Thus considering the velocity component u_x, after an impact on the shaded wall the molecule rebounces with a velocity $-u_x$. Corresponding momentum change may thus be given as,

Change in momentum for each impact of a molecule $= m\,u_x - (-m\,u_x)$

$$= 2\,mu_x \text{ g cm}^{-1}\text{ s} \tag{3.3}$$

Before the molecule makes a second impact on the opposite wall it has to travel a distance l. So,

$$\text{Number of impacts made by the molecule per sec. along } x \text{ axis} = \frac{u_x}{l} \tag{3.4}$$

Total change in momentum along x-axis in one second due to $\frac{u_x}{l}$ impacts is therefore equal to,

$$2m\, u_x \times \frac{u_x}{l} = \frac{2\ m\ u_x^2}{l}\ g\ cm^{-1}\ \textit{or dynes.}$$

Similarly rate of change of momentum along y and z axes are

$$\frac{2\ m\ u_y^2}{l} \text{ and } \frac{2\ m\ u_z^2}{l} \text{ dynes respectively.}$$

Total rate of change of momentum of molecule A due to impact on all the six sides $= \frac{2m}{l}\,(u_x^2 + u_y^2 + u_z^2)$

$$= \frac{2m\ u^2}{l} \text{ dynes} \tag{3.5}$$

Newton's second law of motion suggests that force acting on a body is equal to the rate of change of momentum.

Therefore force due to impact of a single molecule $= \frac{2m\ u^2}{l}$

But postulate-5 suggests that each molecule in a gaseous system has different velocity. If these velocities are represented by $u_1, u_2, u_3 \ldots\ldots u_n$,

$$\text{Total force due to impact of } n \text{ molecules} = \frac{2m}{l}\,(u_1^2 + u_2^2 + u_3^2 + \ldots.. + u_n^2) \tag{3.6}$$

At this stage we need to introduce new velocity terms called **root mean square velocity** *(C)* and **mean square velocity,** *(C^2)*

$$\text{Mean square velocity } C^2 = \frac{u_1^2 + u_2^2 + u_2^2 + \ldots. + u_n^2}{n} \tag{3.7}$$

$$\text{Root mean square velocity } C = \sqrt{\frac{u_1^2 + u_2^2 + u_3^2 + \ldots. + u_n^2}{n}} \tag{3.8}$$

The root mean square *(rms)* velocity may be better understood by considering an analogy of movement of vehicles on a high way. Consider six vehicles travelling at the following velocities (negative sign indicates movement in an opposite direction).

Velocity (km/hr)	*Speed* (km/hr)	*(Velocity)*^{2}
60	60	3600
50	50	2500
50	50	2500
-40	40	1600

-50	50	2500
20	20	400
Σ Velocity = 90	Σ Speed = 270	Σ Velocity^2 = 13100

$$\text{Average velocity } \bar{u} = \frac{\Sigma \text{ Velocity}}{n} = \frac{90}{6} = 15 \text{ km/hr}$$

This value is far lower than the average speed of the cars, i.e. $\frac{270}{6} = 45$ km/hr

$$\text{Mean square velocity } c^2 = \frac{\Sigma \text{ (velocity}^2)}{n} = \frac{13100}{6}$$

$$= 2183.33 \text{ km/hr}^2$$

$$\text{Root mean square velocity } c = \sqrt{\frac{\Sigma \text{ velocity}^2}{n}} = \sqrt{\frac{13100}{6}}$$

$$= 46.73 \text{ km/hr}$$

Therefore, ***rms*** velocity gives a closer value to the average speed. Because gas molecules move in different directions and different velocities, consideration of *rms* velocity in place of average velocity speaks of true situation.

Now rearranging equation (3.7) we have,

$$u_1^2 + u_2^2 + u_3^2 + \ldots + u_n^2 = n\,c^2$$

Therefore equation (3.6) becomes

$$\text{Total force due to impact of } n \text{ molecules } = \frac{2m\,nc^2}{l} \text{ dynes} \qquad (3.9)$$

Since pressure is defined as the force per unit area and total area of the walls of the cube is $6l^2$, we have;

Total pressure exerted by all molecules on all the six sides

$$= \frac{2mnc^2}{l} \times \frac{1}{6l^2} = \frac{mnc^2}{3l^3} \text{ dynes cm}^{-2}$$

But the quantity l^3 is equal to the volume of the cube V of side l.

Therefore,

$$\text{Pressure } 'P' = \frac{mnc^2}{3V} \text{ dynes cm}^{-2}$$

or (3.10)

$$PV = \frac{1}{3}\, mnc^2$$

This expression is known as **kinetic gas equation.** It relates the pressure of a gas to the volume of the container and the number, mass and mean square velocity of the molecules.

3-3-2 Deductions of Gas Laws from Kinetic Gas Equation

Kinetic gas equation (3.10) obtained from the postulates of kinetic theory of gases is of most general nature and readily used in giving theoretical base to the gas laws, hither to were purely empirical. In the following discussion we first state these laws and then give deduction from kinetic gas equation.

(A) BOYLE'S LAW

Robert Boyle (1662), an English chemist, studied the behaviour of gases systematically and quantitatively. His observations were summarized in a simple statement called Boyle's law, which states that *the volume of a fixed amount of gas maintained at constant temperature is inversely proportional to pressure of the gas.*

$$P \propto \frac{1}{V}$$

To put it in another way, for a given amount of gas at constant temperature, the product of pressure and volume is a constant.

$$PV = \text{constant}$$

From postulate 5, the kinetic energy of n molecules is equal to $\frac{1}{2}\ mnc^2$ *and postulate 6 says that this kinetic energy is directly proportional to the temperature.*

Therefore $$\frac{1}{2}\ mnc^2 \propto T$$

or $$\frac{1}{2} mnc^2 = k_1 T$$

where k_1 is a proportionality constant. Rewriting the kinetic gas equation as,

$$PV = \frac{2}{3}\ (\frac{1}{2}\ mnc^2)$$

Thus we have,

$$PV = \frac{2}{3}\ k_1 T \tag{3.11}$$

At constant temperature equation (3.11) becomes,

$$PV = \text{constant}$$

which is the mathematical representation of Boyle's law.

(B) CHARLES' LAW OR CHARLES & GAY-LUSSAC'S LAW

The French scientists Jacques Charles and Joseph Gay-Lussac studied the Temperature-Volume relationship of the gases. Charles' law states that *the volume of a fixed amount of gas maintained at constant pressure is directly proportional to the absolute temperature of the gas.*

$$V \propto T$$

To put it in another way, for a given amount of gas at constant pressure, the volume-temperature ratio is constant.

$$V/T = \text{constant}$$

At any given pressure the plot of volume *vs* temperature is a straight line. By extending the line to zero volume, we find an intercept on the temperature axis to be -273.15°C. For other pressures also we obtain different straight line plots but with the same zero-volume temperature intercept at -273.15°C, Fig. 3-5.

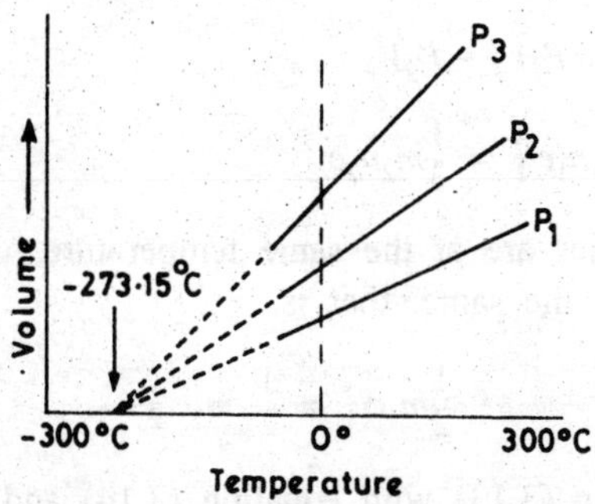

Fig. 3-5 Variation of volume with temperature at constant pressure

Lord Kelvin (1848) identified this point as the lowest attainable temperature, an ideal starting point for the temperature measurements and called it **absolute zero.** In his honour the absolute temperature scale is now called **Kelvin scale** and the temperature as Kelvin (K).

According to this scale the freezing point of water (0°C) is equal to 273.15 K The relationship between K and °C is

$$T\,(K) = t(°C) + 273.15$$

In our regular calculations we use 273 instead of 273.15 in the K and °C conversion.

Rearranging the equation (3.11) we have

$$V = \frac{2}{3} \times \frac{k_1}{P} \times T \qquad (3.12)$$

At constant pressure, $\frac{2k_1}{3P}$ is another constant k_2

Therefore $\quad V = k_2T \quad$ or $\quad V/T = \text{constant}$

which is the mathematical representation of Charles' law

(C) AVOGADRO'S LAW

This volume-amount relationship was enuniciated by the Italian scientist Amedeo Avogadro in 1811. It states that *at constant temperature and pressure, equal volumes of different gases contain the same number molecules.*

$$V \propto n$$

From kinetic theory of gases, for any two gases, indicated by the subscripts 1 and 2, we have the following pressure-volume expressions.

$$P_1V_1 = \frac{1}{3} m_1n_1c_1^2 \text{ and } P_2V_2 = \frac{1}{3}m_2n_2c_2^2$$

At constant temperature $P_1V_1 = P_2V_2$

Hence $$\frac{1}{3} m_1n_1c_1^2 = \frac{1}{3}m_2n_2c_2^2 \quad (3.13)$$

If these two gases are at the same temperature, the kinetic energies of the molecules will be the same, that is,

$$\frac{1}{2}m_1c_1^2 = \frac{1}{2} m_2c_2^2 \quad (3.14)$$

Dividing equation (3.13) with equation (3.14) and rearranging we get,

$$n_1 = n_2$$

Thus Avogadro's law has been derived from kinetic theory for an ideal gas.

(D) GRAHAM'S LAW OF DIFFUSION

The escaping tendencies or diffusion through a fine opening of a container was studied by Thomas Graham (1833). His law governing the diffusion rates may be stated *that at constant pressure and temperature the rates of diffusion of various gases vary inversely as the square roots of their densities* or molecular weights. If u and d are the velocity of diffusion and density of the gas, then

$$u \propto \frac{1}{d} \quad \text{or} \quad \frac{u_1}{u_2} = \frac{d_2}{d_1}$$

Graham's law of diffusion can be deduced readily from the kinetic theory of gases.

At constant pressure and volume from equation (4.13) we have,

$$\frac{1}{3} m_1n_1c_1^2 = m_2n_2c_2^2$$

i.e., $$\frac{c_1^2}{c_2^2} = \frac{n_2m_2}{n_1m_1}.$$

Further, if $n_1 = n_2 = N$, then;

$$\frac{c_1^2}{c_2^2} = \frac{Nm_2}{Nm_1} = \frac{M_2}{M_1} \quad (3.15)$$

where M_1 and M_2 are molecular weights of two gases.

But molar volumes are same at constant temperature and pressure and density (d) is the ratio of molecular weight and molar volume. Therefore,

$$\frac{c_1^2}{c_2^2} = \frac{d_2}{d_1} \quad (3.16)$$

(E) DALTON'S LAW OF PARTIAL PRESSURES

The gas laws apply not only to individual gases but also to misture of gases. When two or more gases that do not react chemically with each ŏther are placed in a container, each gas behaves independently as if it were alone. Thus if we consider the total pressure of gas mixture, each gas contributes a pressure equal to the value it would exert if it were alone in the container, called partial pressure. Dalton generalized this behaviour in his law of partial pressures which states that *the total pressure of a gas mixture is the sum of the partial pressures of all its components.* Partial pressures are represented by lower case p.

$$\text{Total pressure } P = p_1 + p_2 + p_3 + \ldots \quad (3.17)$$

According to kinetic gas theory, if n_1 molecules of a gas – 1, having molecules of mass m_1 and mean square velocity c_1^2 and occupying a volume V, then the pressure exerted by these molecules p_1 is given by the expression;

$$p_1 = \frac{n_1 m_1 c_1^2}{3V} \quad (3.18)$$

Similarly, if n_2 molecules of gas 2 is occupying a volume V, we have;

$$p_2 = \frac{n_2 m_2 c_2^2}{3V} \quad (3.19)$$

Analogous expressions for partial pressures p_3, p_4 ... can be obtained for other gas molecules occupying the same volume. If gases 1, 2, 3... were to be presented as a mixture in a container of volume, V, then Total pressure P of the gas;

$$P = \frac{n_1 m_1 c_1^2}{3V} + \frac{n_2 m_2 c_2^2}{3V} + \frac{n_3 m_3 c_3^2}{3V} \quad (3.20)$$

Upon substituting the inferences drawn from expressions (3.18) and (3.19), etc., in equation (3.20) we have,

$$P = p_1 + p_2 + p_3 + \ldots \quad (3.21)$$

which is the mathematical form of Dalton's law of partial pressures.

3-4 DEDUCTION OF SOME MORE RELATIONS FROM KINETIC THEORY OF GASES

In addition to providing theoretical base to the empirical gas laws, kinetic theory of gases was found useful in predicting velocity of gas molecules and deriving expressions connected to molecular collision, molar heat capacities, distribution of molecular velocities, etc.

3-4-1 Velocity of Gas Molecules

According to kinetic theory of gases, *at a given temperature* even though individual molecules posses different velocities and there by different kinetic energies, the *average kinetic energy* of all molecules having different molecular masses is the same, i.e.,

$$\frac{1}{2} m_1 c_1^2 = \frac{1}{2} m_2 c_2^2 = \frac{1}{2} m_3 c_3^2 = ... \tag{3.22}$$

Therefore, it suggests that heavier molecules move with slower speeds. this can be seen in Table 3-1.

Table 3-1 Molecular velocities of some gases at 25°C

Gas	*Molecular weight*	*Velocity (m s^{-1})*	*Average kinetic energy* $\times 10^{-6}$
H_2	2	1928	7.43
D_2	4	1363	7.43
NH_3	17	665	7.51
N_2	28	515	7.42
CO	28	515	7.42
O_2	32	482	7.43
CO_2	44	411	7.43
Cl_2	71	324	7.45

From the kinetic gas equation, for 1 mole of any gas having Avagadro number of molecules, (i.e., $n = N$)

$$\frac{1}{3} mNc^2 = PV = RT$$

The product *Nm* is the molecular weight, *'M'*, of the gas. Therefore,

$$\frac{1}{3} Mc^2 = PV = RT$$

or $$c^2 = 3RT/M$$

Thermal or *rms* velocity $$c = \sqrt{3RT/M} \tag{3.23}$$

Therefore the root mean square velocity of the gas molecules is directly proportional to the square root of temperature and inversely to the square root of the molecular weight of the gas. Again, replacing *RT* with *PV* in equation (3.23) we have

Thermal velocity $\quad c = \sqrt{3PV/M}$ (3.24)

But *M/V* gives the density of gas, *d*, at temperature *T* and pressure *P* therefore equation (3.24) may be written as;

Thermal velocity $\quad c = \sqrt{3P/d}$ (3.25)

The equation (3.23) and (3.25) provide a means of predicting *rms* velocities of gases, which are difficult to determine experimentally.

3-4-2 Kinetic Energy of an Ideal Gas

The total kinetic energy is the sum of contributions due to constant rapid motion of the particles (translational, vibrational, and rotational). Since vibrational and rotational contributions are very small, entire energy of a molecule may be thought to be possessed in the form of translational kinetic energy. This can be deduced with the help of kinetic gas equation.

For a particle of mass, m_1 and moving with a velocity *u*, the kinetic energy of translation of a molecule = $\frac{1}{2}\, mu^2$. But according to kinetic gas equation *for 1 mole of gas*

$$PV = RT = \frac{1}{3} Nmc^2$$

Therefore $\quad RT = \frac{2}{3}\left(\frac{1}{2} Nmc^2\right)$

$$= (2/3)\, E_k$$

or $\quad \boxed{E_k = (3/2)\ RT}$ (3.26)

Then *average kinetic energy per molecule is,*

$$\varepsilon_k = \frac{E_k}{N} = \frac{3RT}{2N} \tag{3.27}$$

or $\quad \boxed{\varepsilon_k = (3/2)\ kT}$ (3.28)

where $k = R/N$ is called Boltzmann constant and is equal to 1.38×10^{-23} JK^{-1} $molecule^{-1}$ or 1.38×10^{-16} ergs K^{-1} $molecule^{-1}$

In the derivation of kinetic energy relation (3.26), the intermolecular attractions are not considered. This is known as **Maxwell's generalization,** which states that *the kinetic energy of an ideal gas is independent of the nature of the gas and depends only on the absolute temperature.*

$$K.E. \propto T$$

Hence according to Maxwell generalization *at absolute zero the translational motion completely ceases.*

Problem 3 : Calculate kinetic energy of 21 g of nitrogen maintained at 27°C. (Hint : Molecular weight of nitrogen = 28)

Solution :

No. of moles of nitrogen present in 21 g = $\frac{\text{Mass}}{\text{Molecular weight}} = \frac{21}{28} = \frac{3}{4}$

K.E. of one mole of any gas = $\frac{3}{2}\ N\ RT$

Taking R = 8.314 JK^{-1} mol^{-1} & T = 273 + 27 = 300 K we have,

$$\text{K.E.} = \frac{3}{2} \times \frac{3}{4} \times 8.314 \times 300$$

$$= 2.8 \text{ kJ}$$

Kinetic energy of 21 g of N_2 at 27°C is 2.8 kJ.

3-4-3 Distribution of Molecular Velocities

According to postulates 5, kinetic energy of a gas is directly proportional to the mean square velocity of the molecule and Maxwell's generalization says, that it is also directly proportional to the absolute temperature.

$$\text{K.E.} \propto c^2 \quad \text{and} \quad \text{K.E.}\ \alpha\ T$$

Therefore

$$c^2 \propto T \tag{3.29}$$

This mean square velocity relation with temperature of the gas speaks only about an average environment. All the molecules of a gas do not move with the same velocity and the speed of any molecule varies continuously depending on number of collisions per unit time and intermolecular attractions. Maxwell and Boltzman using the theory of probability have evaluated the fraction of molecules having a particular velocity at a given temperature. They found that *the distribution of molecular velocities depends not only on the temperature but also on the molecular weight of the gas* and is given by,

$$\frac{l}{n} \cdot \frac{dn}{du} = 4\pi \left[\frac{M}{2\pi\ RT}\right]^{1/2} e^{-Mu^2/2RT}\ u^2 \tag{3.30}$$

where n is the total number of molecules with molecular weight M and a particular velocity u. The left hand expression $\frac{1}{n}\frac{dn}{du}$ gives probability of finding molecules with a particular velocity u. Equation (3.30) is known as the **Maxwell-Boltzmann distribution of velocity.** The distribution of velocities at two different temperature is given in Fig.3-6.

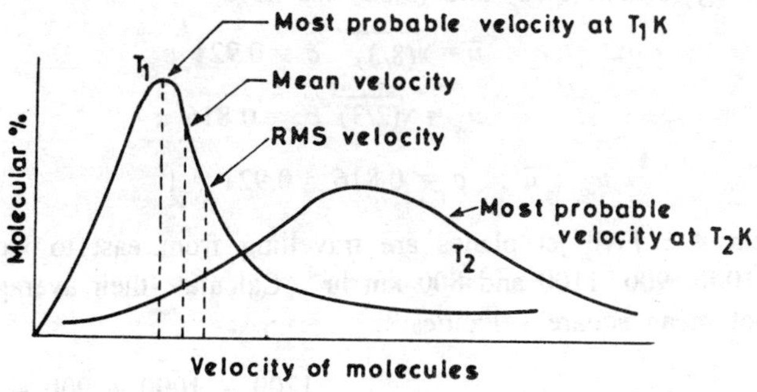

Fig. 3-6 Distribution of velocities

The following conclusion can be drawn from Fig.3-6

(i) At any temperature majority of the molecules lie in a particular velocity zone.

(ii) The percentage of molecules at any given temperature having too small or too high velocities is small.

(iii) The velocity maximum of the curve is called **Most probable velocity** and the fraction of molecules around this have **maximum velocity**.

(iv) With the increase in temperature, say from T_1 *to* T_2 the maximum of the curve shifts to right, corresponding to the increased molecular velocity.

(v) The flattening of the curve at higher temperature indicates a wider distribution of velocities.

In the analogy of vehicular velocities given in section 3-3-1, 50 *kmph* may be said to correspond the most probable velocity and other velocity terms are samller than this value. On differentiation of the equation (3.30) with respect to individual velocities of the molecules and setting the result to zero, as required for a maximum, we get ;

$$u_{\infty} \text{ (most probable velocity)} = \sqrt{2RT/M} \qquad (3.31)$$

From similar consideration the average velocity was found be

$$\text{Average velocity } \bar{u} = \frac{u_1 + u_2 + \ldots + u_n}{n} = \sqrt{\frac{8RT}{\pi M}} \qquad (3.32)$$

From equation (3.23) we already know the *rms* velocity

$$c = \sqrt{\frac{3RT}{M}} \quad (3.33)$$

Comparing (3.31), (3.32) and (3.33) we have,

$$\overline{u} = \sqrt{(8/3)} \quad c = 0.921\ c \quad (3.34)$$

$$u_{\infty} = \sqrt{(2/3)} \ c \ = 0.816\ c \quad (3.35)$$

and $$u_{\infty} : \overline{u} : \ c = 0.816 : 0.921 : 1 \quad (3.36)$$

Problem 4 : Five jet planes are travelling from east to west at velocities 1200, 1000, 900, 1100 and 800 km hr^{-1}. Calculate their average, mean square and root mean square velocities.

Solution : Average velocity is '$\overline{u}$' $= \dfrac{1200 + 1000 + 900 + 1100 + 800}{5}$

$= 1000$ km hr^{-1}

Mean square velocity $c^2 = \dfrac{(1200)^2 + (1000)^2 + (900)^2 + (1100)^2 + (800)^2}{5}$

$= 1.02 \times 10^6$ km^2. hr^{-2}

Root mean square velocity $c = \sqrt{1.02 \times 10^6} = 1010$ km hr^{-1}

Problem 5 : Calculate the rms velocity of CO at 27° C.

Given $T = 27 + 273 = 300$ K, Molecular weight $M = 12 + 16 = 28$
$R = 8.314 \times 10^7$ erg K^{-1} mol^{-1}

$$RMS \text{ velocity} = \sqrt{\frac{3RT}{M}} = \sqrt{\frac{3 \times 8.314 \times 10^7 \times 300}{28}}$$

$= 5.17 \times 10^4$ cm s^{-1}

Problem 6 : Calculate rms velocity of oxygen melecule at NTP. Given that R = 8.313×10^7 ergs K^{-1} mol^{-1} (O.U., Oct. 1987) & (O.U., Oct. 1980)

Solution : Given for oxygen $M = 32$ and at NTP $T = 0°C = 273$ K

rms velocity at NTP $= \sqrt{\dfrac{3 \times 8.313 \times 10^7 \times 273}{32}} = 4.613 \times 10^4$ cm s^{-1}

Problem 7 : Calculate mean velocity of oxygen at 25°C. (O.U., Mar. 1990)

Solution : Mean or average velocity $\overline{u} = \sqrt{8RT/\pi M}$

Given $M = 32$, $T = 25 + 273 = 298$ K and we know $R = 8.314 \times 10^7$ erg K^{-1} mol^{-1}

$$c = \sqrt{\frac{8 \times 8.314 \times 10^7 \times 298}{3.14 \times 32}} = 4.441 \times 10^4 \text{ cm s}^{-1}$$

Problem 8 : At 15°C and 1 atm pressure, the density of a gas is 1.52 g L^{-1}. What are the *RMS* and average velocities?

RMS velocity = $\sqrt{\dfrac{3RT}{M}}$

But for 1 mole of gas $RT = PV$

Therefore *RMS* velocity = $\sqrt{3PV/M} = \sqrt{3P/d}$

Similarly average velocity = $\sqrt{8RT/\pi M} = \sqrt{8P/\pi d}$

Given $T = 15 + 273 = 288$ K, $P = 1$ atm, $d = 1.52$ g L^{-1} and we know $R = 8.314 \times 10^7$ erg K^{-1} mol^{-1}

Pressure = 1 atm = 76 cm Hg = 76 × 13.6 × 981 dynes cm^{-1}

= 1.014×10^6 dynes cm^{-2}

density = 1.52 g L^{-1} = 1.52/1000 g cm^{-3} = 1.52×10^{-3} g cm^{-3}

$$RMS\ \text{velocity} = \sqrt{\frac{3P}{d}} = \sqrt{\frac{3 \times 1.014 \times 10^6}{1.52 \times 10^{03}}} = 4.474 \times 10^4 \text{ cm s}^{-1}$$

$$\text{average velocity} = \sqrt{\frac{8P}{\pi d}} = \sqrt{\frac{8 \times 1.014 \times 10^6}{3.14 \times 1.52 \times 10^6}} = 4.123 \times 10^4 \text{ cm s}^{-1}$$

Problem 9 : Calculate *rms* velocity of oxygen molecules at a temperature of 30°C are pressure of 750 mm Hg. (O.U., Apr. 1988)

Since temperature and *M* are known the pressure data is not required.

$$rms\ \text{velocity} = \sqrt{\frac{3RT}{M}} = \sqrt{\frac{3\ 8.314 \times 10^7 \times 303}{32}} = 4.86 \times 10^4 \text{ cm s}^{-1}$$

Alternatively without taking the standard *R* value this problem could be solved as follows;

$$rms\ \text{velocity} = \sqrt{\frac{3RT}{M}} = \sqrt{\frac{3VP}{M}}$$

$$P = 750 \text{ mm Hg} = 75 \times 13.6 \times 981 \text{ dynes cm}^{-2}$$

A gas at 76 cm Hg pressure (P_1), 273 K (T_1) temperature occupies a volume of 22400 cm^3 (V_1).

Volume occupied at 30°C (T_2) and 75 cm Hg pressure (P_1) is

$$\frac{P_1 V_1}{T_1} = \frac{P_2 V_2}{T_2}$$

$$\frac{76 \times 22400}{273} = \frac{75 \times V_2}{(273 + 30)}$$

$$V_2 = \frac{76 \times 22400 \times 303}{273 \times 75} = 25193 \text{ cm}^3$$

Volume occupied by gas at 30°C and 750 mm Hg = 25193 cm^3

$$rms \text{ velocity} = \frac{3PV}{M} = \frac{3 \times 75 \times 13.6 \times 981 \times 25193}{32}$$

$$= 4.86 \times 10^4 \text{ cm s}^{-1}$$

Problem 10 : At what temperature will the *rms* velocity of a gas be half its value at STP.

Solution : For any given gas

$$rms \text{ velocity at STP} = \sqrt{\frac{3R\ (273)}{M}}$$

$$rms \text{ velocity at STP} = \sqrt{273} \times \sqrt{3R/M}$$

Say, at temperature T the rms velocity is half the value at STP

$$rms \text{ velocity at T°K} = \sqrt{3RT/M}$$

$$= \sqrt{T} \times \sqrt{3R/M}$$

$$\frac{rms\ velocity\ at\ T°K}{rms\ velocityz\ at\ STP} = \frac{1}{2} = \frac{\sqrt{T} \times \sqrt{3R/M}}{\sqrt{273} \times \sqrt{3R/M}}$$

Therefore $\sqrt{T} = \dfrac{\sqrt{273}}{2}$

or T = 68.25 K

So at 68.25 K the *rms* velocity is half the value at STP

3-4-5 Mean Free Path And Molecular Diameter

The molecular velocities presented in Table 3-1 indicated that the molecules move with velocities either comparable to sound (≈ 300 m s^{-1} at room temperature) or much higher. With this we expect to get the smell of a perfume as soon as the perfume bottle is opened in some corner of a hall but the experience is contrary to this expectation. We get the aroma only after a lapse of definite noticeable time. This is because of the small but finite size of the molecules which result in numerous intermolecular collisions. At each collision a molecule deviates from its original path. In other words *each molecule travels in zig-zag paths of different lengths,* Fig. 3-7.

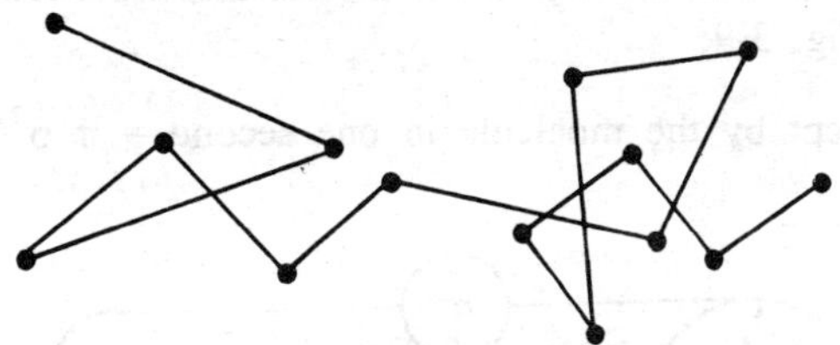

Fig. 3-7 Zig-Zag motion of a molecule

The average distance travelled by a molecule between two successive collisions is called **mean free path.** Mathematically the mean free path λ may be represented as,

$$\lambda = \frac{d_1 + d_2 + d_3 + \ldots + d_n}{n}$$

where d_1, d_2, *etc., give the distance travelled between each collision and n the number of collisions a molecule under goes. Its value depends on the size, velocity and number of molecules present in a given volume of gas. In addition its value also depends on a new molecular parameter called* **collision diameter.** *Collision diameter represents the closest distance between the centers of two molecules taking part in the collision,* Fig.3-8.

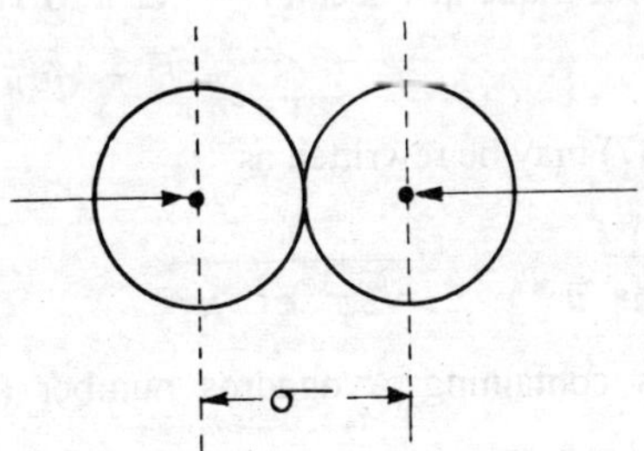

Fig.3-8 The collision diameter

If Z is the number of collisions that a molecule undergoes in one second (called **collision number)** and λ the mean free path, then the total distance travelled in 1 second, i.e., average velocity $\bar{u}$, is given by the product $Z \times \lambda$

Therefore $\overline{u} = Z\,\lambda$ or $\lambda = \dfrac{\overline{u}}{Z}$ (3.37)

The mean free parth of the molecules in a gas can be calculated by considering movement of only one molecule, the others being stationary, through a cylinder of diameter 2σ. If l is the distance travelled by this molecule in one second, Fig. 3-9;

Total volume swept by the moelcule in one second $= \pi\,\sigma^2 l$

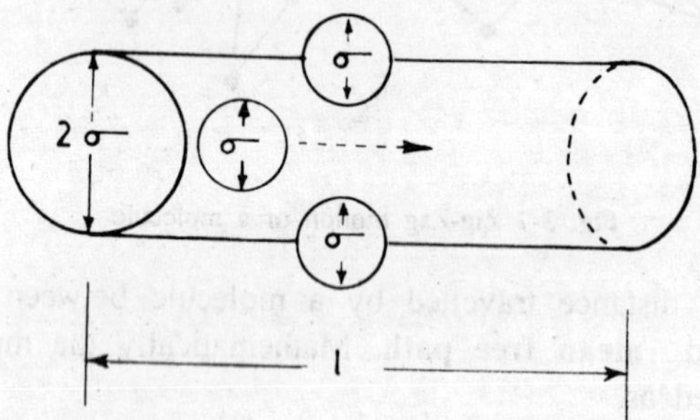

Fig. 3-9 Volume swept by a single molecule in 1 second

If n^* is the number of molecules per unit volume, then;

Number of collisions that a molecule undergoes per second $= \pi\,\sigma^2\, l.\, n^*$ (3.38)

Since l is the distance travelled in one second, it is nothing but the average velocity u.

In a dynamic model where all the molecules are in constant motion the expression (3.38) has to be multiplied with $\sqrt{2}$. Therefore,

Total number of collisions made in 1 second $= \sqrt{2}\,\pi\,\sigma^2 l\, n^*$

$$= \sqrt{2}\,\pi\,\sigma^2\,\overline{u}\,n^* \quad (3.39)$$

Therefore equation (3.37) may be rewritten as

$$\lambda = \frac{\overline{u}}{\sqrt{2}\,\pi\,\sigma^2 n^*\,\overline{u}} = \frac{1}{\sqrt{2}\pi\,\sigma^2\,n^*} \quad (3.40)$$

For an ideal gas containing Avogadros number (N) of molecules in 1 cc, we have;

$$P \times I = n\,RT$$

But the number of moles $n = n^*/N$, where n gives the number of molecules present in 1 cc.

$$P = (n^*/N)\,RT$$

We also know Boltzmann constant $k = R/N$

Therefore

$$P = n^*kT \qquad \text{or} \qquad n^* = P/(kT)$$

Substituting the value of n^* in the expression (3.40)

$$\lambda = \frac{kT}{\sqrt{2}\,\pi\,\sigma^2 P} \tag{3.41}$$

Hence mean free path is directly proportional to the temperature and inversely proportional to the pressure and square of the collisional diameter.

At ordinary temperature and pressure conditions the mean free path is almost equal to the molecule diameter. This results in large number of collisions per unit time, and thereby accounts for the slow movement of the gas molecules in spite of very high molecular velocity.

3-4-6 Principle of Equipartition of Energies

The average kinetic energy of a molecule of mass m and mean square velocity c^2 is

$$\text{K.E.} = \frac{1}{2}mc^2$$

The mean square velocity along any direction will have a velocity component along the three axes and are related through the expression;

$$c^2 = \bar{u}_x^2 + \bar{u}_y^2 + \bar{u}_z^2$$

Multiplying the equation (3.42) with the 1/2 m we get,

$$\frac{1}{2}mc^2 = \frac{1}{2}mc_x^2 + \frac{1}{2}mc_y^2 + \frac{1}{2}mc_z^2$$

or

$$\text{K.E}_a = \text{K.E}_x + \text{K.E}_y + \text{K.E}_z$$

Thus average kinetic energy K.E_a is the sum of the kinetic energy components along three axes.

But the molecules are always in a random motion and do not have any preference to any direction of motion. Therefore all the kinetic energy components may be assumed equal.

$$\boxed{\text{K.E}_x = \text{K.E}_y = \text{K.E}_z} \tag{3.43}$$

This conclusion is called **Principle of Equipartition of Energy** and states that *each energy term in the particular component contributes equally to the total energy.*

But from equation (3.26) total translational kinetic energy per mole is equal to (3/2)RT. So the value in each direction should be $\frac{1}{2}$RT per mole or $\frac{1}{2}$kT per single molecule.

This gives another method of stating equipartition parinciple : *The total energy of a molecule is divided equally among the three degress of freedom (axes), the amount being $\frac{1}{2}$ kT for each.*

$$(K.E.)_x = (K.E.)_y = (K.E.)_z = \tfrac{1}{2}\ RT \text{ per mole}$$

or

$$(K.E.)_x = (K.E.)_y = (K.E.)_z = \tfrac{1}{2}\ kT \text{ per molecul}_e$$

3-4-7 Heat Capacity and Kinetic Theory

The heat *capacity is defined as the quantity of heat required to raise the temperature of a particular mass of a substance by one degree.* When the weight of the material is 1 gram, the heat capacity is called the **specific heat capacity** or simply specific heat and when it is equal to its molecular weight it is referred to as **molar heat capacity.** Product of specific heat and molecular weight of any molecule gives molar heat capacity.

Molar heat capacity = specific heat × molecular weight.

Heat capacities are conventionally measured in calories. *A* ***calorie*** *is defined as the amount of heat required to raise the temperature of 1 gram of water by one degree centigrade.* Because of uncertainty, in isotopic composition of water, the calorie is now equated to 4.184 joules (SI system) or 4184 × 10^7 ergs (CGS system).

Heat can be supplied to any system keeping either volume constant or pressure constant. The former is called as **heat capacity at constant volume** and the latter is called as **heat capacity at constant pressure.** The specific heat capacities at constant volume and constant pressure are designated by c_v and c_p respectively and the corresponding molar hear capacities are designated C_p and C_v respectively.

In case of solids and liquids raising the temperatures by one degree will not bring any recognizable changes in its pressure and volume. Therefore C_p and C_v are nearly the same for these substances. Incase of gaseous systems any change in temperature brings appreciable changes in its pressure and volume. Therefore C_p and C_v of gaseous systems differ significantly.

Molar heat capacity at constant volume, C_v : It is the amount of heat required to raise the temperature of 1 mole of gas through 1°K when the volume is kept constant. This process of heating from any temperature T to $T+1$ is accompanied by an increase in kinetic energy. According to kinetic theory of gases and principle of equipartition of energy, the translational kinetic energy of 1 mole of gas in equal to $\frac{3}{2}$ RT

$$E_k = \frac{3}{2} RT$$

But thermodynamic principles suggest that at constant volume it is impossible to derive any pressure-volume work and all the energy then goes only to raise the kinetic energy. Further, if we assume negligible changes in the other energies like potential energy and vibrational energy.

$$E_k = E_{total} = \frac{3}{2} RT \quad (3.44)$$

Total energy of one mole of gas at $(T + 1) = {}^3/_2 R (T + 1)$

When the temperature is changed by 1° from T to $(T + 1)$, then total energy change at constant volume, is,

$$\Delta E_k = {}^3/_2 R(T + 1) - {}^3/_2 RT$$

$$\Delta E_k = {}^3/_2 R$$

Therefore change in energy (i.e. heat) per degree raise in temperature at constant volume is,

$$\boxed{C_v = {}^3/_2 R} \quad (3.45)$$

Thus molar heat capacity at constant volume of any gas should be equal to $\frac{3}{2} R$.

$$C_v = \frac{3}{2} R = \frac{3}{2} \times 2 = 3 \text{ cal} \qquad (\textit{since } R \approx 2 \text{ cal})$$

$$= 3 \times 4.2 = 12.6 \text{ J}$$

This statement is true only for monoatamic gases like He, Ne, Ar, metal vapors which exhibit ideal behaviour with negligible or no inter atomic attractions and vibrational transformations and the heat supplied in totally utilized in increasing their kinetic energy.

In diatomic and polyatomic molecules, the heat supplied not only enhances the translational energy but also other molecular energies like vibrational and rotational energies of molecules. Therefore more heat is needed to compensate this loss, given by the compensation factor x.

For di and polyatomic molecules $C_v = \frac{3}{2} R + x$

The value of correction factor x depends on the nature of molecule.

Molar heat at constant pressure C_p : *It is the amount of heat needed to raise the temperature of 1 mole of gas through 1°K, when the pressure is kept constant.* When a gas is heated at constant pressure, the heat supplied not only increases the kinetic energy but also increases the volume of the system. This increase in volume of the system gives it a capability to do pressure-volume work, w on the surroundings.

Therefore for an ideal gas

$$C_p = E_k + w$$

$$= \frac{3}{2}R + w \qquad (3.46)$$

The value of external pressure-volume work done by the expanding gas may be obtained as follows

Suppose 1 mole of a gas is present in a cylinder fitted with a friction less piston and occupying a volume V at a pressure P and temperature T. Now heating the cylinder to $T + 1$ degree at constant pressure brings a volume change from V to $V + \Delta V$.

Let the area of cross action of the piston be A cm^2 and displacement of the piston due to expansion be Δl.

Work done w = External force × distance moved

but External force = External Pressure × area of the piston.

$$w = P \times A \times dl$$

$$= P \times \Delta V \; (\therefore \Delta V = A \times dl)$$

For 1 mole of an ideal gas $PV = RT$ (i)

When the temperature changes from T to $(T + 1)$ the volume changes from V to $(V + \Delta V)$

Therefore at the new state $P(V + \Delta V) = R(T + 1)$

$$PV + P\Delta V = RT + R \qquad \text{(ii)}$$

(ii) – (i) gives $P\Delta V = R$

Hence $w = R$

Substituting the value of w in equation (3.46) we have

$$C_p = \frac{3}{2}R + R = \frac{5}{2}R \qquad (3.47)$$

Extending the argument used for C_v in case of di and polyatomic gases

For di and polyatomic gases $C_p = \frac{5}{2}R + x$

Difference of C_p and C_v : Comparison of equation (3.45) and (3.47) indicates that the molar heat capacity at constant pressure is always greater than the molar heat capacity at constant volume.

$$C_p > C_v$$

This γ is an interesting quantity that gives a measure of how far a gas is ideal. Monatomic inert gases like He, Ar, etc., are almost free from interatomic attractions and obey strictly all the postulates of kinetic theory and were found to have a γ value of 1.66. The diatomic gas molecules like oxygen are not perfectly elastic and were found to have a γ value of 1.40. This $\frac{C_p}{C_v}$ ratio further falls away as the gas molecule becomes more and more complex. γ values of few gases are given in the Table 3-3

Table 3-3 γ-Values of some gases

Gas	*Atomicity*	γ
Helium		1.659
Argon	Monatomic	1.664
Krypton		1.665
Hydrogen		1.407
Oxygen	Diatomic	1.402
CO		1.398
CO_2	Triatomic	1.296
NH_3		1.248
C_2H_4 (Ethylene)	Polyatomic	1.249

3-5 DEVIATION FROM IDEAL BEHAVIOUR

To satisfy the ideal behaviour, a gas is supposed to pass mainly two acid tests.

(i) Molecules in gaseous state should not have attractive or repulsive forces.

(ii) Volume of the gas molecules must be negligibly small compared to that of the container.

When a gas satisfies the above conditions it automatically also follows the gas laws under all conditions. But deviations were reported at higher pressures and low temperatures. For example, gases condense to give liquids at low temperature indicating the existence of intermolecular forces. So ideal behaviour is restricted only to a limited pressure and temperature ranges, where the intermolecular forces are negligible and gas molecules occupy negligible volume.

Deviation from Boyle's Law :

The deviation of gases from ideal behaviour, specially Boyle's law, can be well illustrated by plotting *PV/RT* versus *P* plots, Fig. 3- 10, for three real gases at a given temperature. For 1 mole of any ideal gas at any given temperature the ratio *PV/RT*, called **compressibility factor,** should be unity and no variation is expected with the change in pressure. But the curves in Fig. 3-10 reveal that;

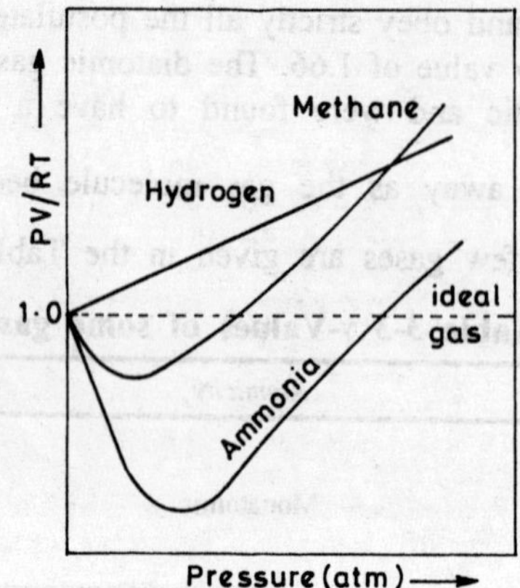

Fig. 3-10 PV/RT versus P curves at 0°C

(i) For real gases the compressibility factor is unity only at moderately low pressures (≈ 10 atm).

(ii) Significant deviations were noticed at higher pressure.

(iii) The compressibility factor for gases like hydrogen increases linearly with increase in pressure. Experimental observations indicate that this type gases can not be condensed under ordinary temperature by the application of any amount of pressure.

(iv) Gases like methane and ammonia which can be easily condensed at moderate temperatures were found to show an initial fall of compressibility factor, followed by an irreversible raise.

To get further understanding of compressibility curves, *PV/RT* values for nitrogen were also plotted against the pressure, Fig. 3-11, at different temperatures.

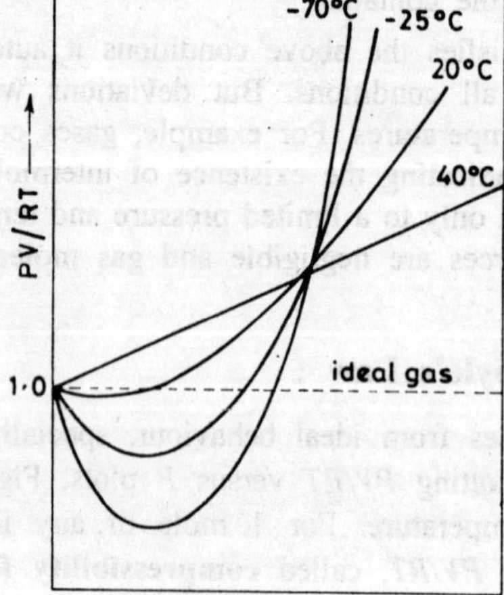

Fig.3-11 *PV/RT vs P* curves of N_2 at different temperatures

The following conclusions can be drawn from the compressibility curves shown in Fig.3-11.

(i) The curves change their shape with change in temperature.

(ii) At relatively high temperature the compressibility factor increases with increases in pressure.

(iii) Around room temperature nitrogen obeys Boyle's law with considerable accuracy.

(iv) As the temperature appraoches the boiling point of liquid nitrogen, (–195°C) the initial dip in the isotherm seems to be more pronounced.

The temperature at which a real gas behaves like an ideal gas over an appreciable range of pressure is called **Boyle's temperature.** The Boyle's temperature for nitrogen is around 50°C, while it is –150°C and – 240°C for hydrogen and helium respectively. Hence at ordinary temperatures the isotherm of nitrogen reaches a minimum before it begins to raise. Hydrogen and helium show a constant raise in compressibility factor with the raise in pressure at ordinary temperature and show a dip only below their Boyle's temperature.

Deviation from Charle's Law :

The changes in shapes of compressibility curves of nitrogen, Fig. 3-11 with changes in temperature also reflect a departure from Charle's law in addition to Boyle's law. Further Charle's law predicts coefficient of volume expansion should be 1/273 at all pressures and temperatures but deviations were noticed. Similar to Boyle's law discrepancies were found to be greater at higher pressures and lower temperatures.

Deviation from Avogadro's Law :

According to Avogadro's law each mole of a gas at STP should occupy 22.4 liters and further at any temperature and pressure volume occupied by a gas should be independent of nature of the gas. But the experimental data in Table 3-4 stand against this contention.

Table 3-4 Molar volumes of gases at NTP

Gas	H_2	N_2	O_2	CO_2	NH_3	Ethyl chloride
Volume (in liters)	22.427	22.405	22.394	22.264	22.084	21.879

3-5-1 Van der Waals Equation

The non-ideality of gases can be dealt with mathematically, by bringing suitable modifications in the equation of state by taking into account of intermolecular forces and finite molar volumes. A Dutch scientist Van der Waals in 1873 made a successful attempt in this regard and won a Noble prize in 1910 for his studies. Van der Waals introduced correction terms in volume (*accounting for the finite volume occupied by the molecule)* and pressure (*accounting for the inter molecular attractions and repulsions*).

Volume Correction (Correction Due to Finite Molecular Volume) :

At low pressures the density of a gas is extremely small, the volume occupied by the molecules themselves is negligible, when compared to the volume of the container. Therefore measurement of volume at low pressure needs no volume correction. On the contrary at higher pressures, the molecules are quite close and compact and the volume occupied by the molecules themselves becomes a considerable fraction of the total volume of the container.

If V is the volume of the container, the *effective volume* or *ideal volume* available for the movement of the molecules is $V_{ideal,}$ *then*

$$V_{ideal} = (V - nb) \tag{3.49}$$

where n is number of moles of the gas and b is a volume correction factor for 1 mole of the gas.

The constant b, referred to as **excluded volume**, is a characteristic of the gas. The value of the excluded volume, b, was found to be nearly four times the actual volume of the molecules, if they were to be at rest. This higher value of b than actual molecular volume speak of the repulsive forces between the molecules which do not allow them to come closer than a particular distance.

Pressure Correction (Correction due to Intermolecular Attraction) :

Real gases in addition to having a finite volume also have intermolecular attractions. Consider a system having moles of gaseous molecules. Each molecule is attracted by its neighbouring molecules and the molecular attraction fades away after few Angstroms. A molecule in the bulk (interior of the vessel) experiences equal pull from all directions. These equal and opposite forces annul each other and ultimately any molecule in the bulk experiences a zero resultant pull. Fig.3-12.

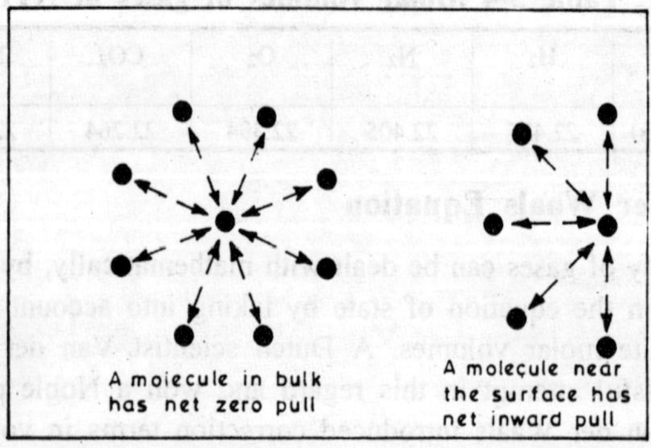

Fig. 3-12 Intermolecular forces of attraction

If the same molecule approaches a wall of the container, it experiences a force imbalance. It assumes a net inward pull. This inward pull reduces the momentum/force of impact and softens the impact made by this molecule against the wall. The overall effect is lowered gas pressure as compared to a gas with no forces of attraction. Further the lowering of gas pressure may be due to interception of a gas molecule by another fast moving molecule and changing the direction. If P is the observed pressure and p is the correction term.

$$P_{\text{ideal}} = P_{\text{observed}} + p \tag{3.50}$$

The two factors responsible for lowering of pressure are proportional to;

(a) The number of molecules per unit volume (molecular density) present in the bulk that are participating in the inward pull.

(b) The number of molecules per unit volume that are responsible for interception of molecular directions, since each of the above factors individually bear a direct proportionality to the molar density, we have;

Pressure correction factor $p \propto d^2$

But $d \propto n/V$

Therefore $p \propto n^2/V^2$

or $$p = \frac{an^2}{V^2} \tag{3.51}$$

where a is a proportionality constant called **coefficient of molecular attraction.** Substituting the value of pressure correction factor p in the expression we have,

$$\mathbf{P_{ideal}} = P_{observed} + \frac{an^2}{V^2} \tag{3.52}$$

Real Gas Equation :

Real gas equation may be obtained by substituting the corrected volume and pressure value from equations (3.49 and 3.52) in the ideal gas equation PV = nRT.

$$\left(\frac{P + an^2}{V^2}\right)(V - nb) = nRT \tag{3.53}$$

This equation is known as **Van der Waals equation.** For 1 mole of a gas per unit volume (n = 1) equation (3-53) reduces to

$$\left(P + \frac{a}{V^2}\right)(V - b) = RT \tag{3.54}$$

3-5-3 Significance of Van der Waals constants :

Table 3-5 lists the Van der Waals constants (*a* & *b*) of some of the gases.

A comparsion of known physical behaviour of a gas and magnitude of Van der Waals constants leads to the following generalizations.

(i) The value of *a* is a measure of how strongly a given type of molecules attract one another. Smallest *a* value of helium indicate weakest intermolecular attractions there by very difficult liquifiability.

(ii) Carbon tetrachloride having *a* value of 20.4 is in accordance with the high liquifiability and high density of the gas.

Table 3-5 Van der Waals constants of some common gases

Gas	*a* (atm L^2 mol^{-2})	*b* (L $mole^{-1}$)
He	0.034	0.0237
Ne	0.211	0.0171
Xe	4.19	0.0266
H_2	0.244	0.0266
Cl_2	6.49	0.0562
N_2	1.33	0.0391
O_2	4.17	0.0318
NH_2	4.17	0.0371
H_2O	5.46	0.0305
CCl_4	20.4	0.138
SO_2	6.71	0.0546
HCl	3.67	4.0408

(iii) Easily liquefiable gases like SO_2, NH_3, H_2O, etc., have relatively higher a value than permanent gases like O_2, H_2, N_2, etc.

(iv) The *b* value was found to maintain a rough correlation with the molecular size. Thus He and H_2 having comparable masses have comparable *b* values, while the bulkier CCl_4 has almost 5 times larger *b-value.*

(v) The constant value of *b* over a wide range of temperatures (not shown in the table) suggest the incompressibility of gas molecules.

3-5-2 Units of Van der Waals Constants *a* & *b*

The Van der Waals constant *a* is related to the pressure correction factor *p* through the expression,

$$p = \frac{an^2}{V^2}$$

On rearranging

$$a = \frac{pV^2}{n^2}$$

when pressure is expressed in atmospheres, volume in liters and n in moles, then units of a are.

$$\text{Units of } a = \frac{(atm)\ (lit)^2}{(mole)^2} = \text{atm L}^2\ \text{mol}^{-2} \tag{3.56}$$

Similarly b is related to the volume correction factor V through the expression,

$$V = nb \qquad \text{or} \qquad b = V/n \tag{3.57}$$

Substituting the units of V and n

$$\text{Units of } b = \frac{\text{Litre}}{\text{mole}} = \text{L mol}^{-1} \tag{3.58}$$

Problem 11 : Using Van der Waals equation, calculate the pressure exerted by 2 moles of a neon gas occupying 300 cc at 15°C. The Van der Waals constants for neon are: a = 0.210 atm^2 mol^{-2}, b = 0.01709 L mol^{-1} (O.U., Oct. 1988)

Solution : Van der Waals equation for n moles is

$$(P + \frac{an^2}{V^2})(V - nb) = nRT$$

Given a = 0.2107 atm L^2 mol^{-1}, b = 0.01709 L mol^{-1}

n = 2 moles, V = 300 cm^3 = 300/1000 = 0.3 L,

T = 15 + 273 = 288 K.

Substituting these values in Van der Waals equation

$$(P + \frac{0.2107 \times 2^2}{(0.3)^2})(0.3 - 2 \times 0.01709) = 2 \times 0.08206 \times 288$$

$$(P + 9.3644)(0.3 - 0.03418) = 47.2666$$

$$(P + 9.3644) = \frac{47.2666}{0.3 - 0.03418} = 177.8142$$

$$P = 177.8142 - 9.3644 = 168.5 \text{ atm}$$

3-5-4 Applicability of Van der Waals Equation :

Van der Waals equation was proved to be in good agreement over a wide range with the pressure, volume, concentration and temperature changes of a gaseous state. Its applicability is discussed below under different pressure and temperature conditions.

(i) At low pressures : At low pressure P is small and V is large. Expanding the Van der Waals equation (3.54), we have,

$$PV - Pb + (a/V) - (ab/V^2) = RT \qquad (3.59)$$

or $PV = RT + Pb - a/V + ab/V^2$ (3.60)

When P is small and V is large, the terms Pb and ab/V^2 can be neglected

Therefore equation (3.60) becomes

$$PV = RT - a/V \qquad (3.61)$$

Dividing equation (3.61) with RT

$$\frac{PV}{RT} = 1 - \frac{a}{VRT} \qquad (3.62)$$

Since a, V, R and T are all positive, under low pressure conditions, the compressibility factor (PV/RT) is always less than unity. With an increase in pressure, the terms Pb and ab/V^2 start assuming importance, thereby accounting for a dip in the plot of PV/RT *vs*, P, Fig.4-11.

(ii) At extremely low pressures : At extremely low pressures, the volume is very large making even the a/V term in the equation (3.61), negligible. Van der Waals equation then reduces to simple ideal gas equation, $PV = RT$. So at extremely low pressure conditions the distinction between real and ideal gases disappears and all gases obey ideal gas equation.

(iii) At high pressures : At high pressures P is large and V is extremely small. Under these conditions the last two volume terms of equation (3.60) being of opposite sign and approximately of equal magnitude, cancel each other. The Van der Waals equation then becomes,

$$PV = RT + Pb$$

or

$$\frac{PV}{RT} = \frac{Pb}{RT} \qquad (3.63)$$

(iv) At extremely high temperatures : PV/RT vs P curves of nitrogen at different temperatures show that at considerably high temperatures the real gases gradually turn to ideality.

At any given pressure and extremely high temperature V is very large. Under these conditions $\frac{a}{V}$, $\frac{a}{V^2}$ and volume correction factor b in equation (3.60) can be neglected, reducing the Van der Waals equation into an ideal gas equation.

(v) At ordinary temperatures : At ordinary temperatures both pressure and volume, make their contribution to the deviations from ideal gas behaviour.

At ordinary temperature and at low pressure $\sqrt{\frac{a}{V^2}}$ (pressure correction term) dominates the van der Waal equation, while b plays an important role at high pressures and ordinary temperatures. However, for permanent gases like helium, hydrogen, neon, etc., the values of a (related to molecular attractions) are so small that essentially b (related to molar volumes) predominates. Hence, neglecting the a/V term, the van der Waals equation (3.60) may be written as,

$$PV = RT + Pb + ab/V^2$$

Since all the terms are positive, a plot of PV/RT versus P has no dip at ordinary temperatures and rises constantly.

(vi) Boyle's temperature : The temperature at which a real gas obeys Boyle's law is known as Boyle temperature. Boyle temperature T_B is related to the van der Waals constants through the expression.

$$T_B = \frac{a}{bR}$$

3-5-5 Limitations of Van der Waals Equation

Although Van der Waals equation is satisfactory over a wide range of temperatures and pressures, deviations were noticed at extremely high pressures and extremely low temperatures. These deviations were attributed to variations in van der Waals constant, which were earlier thought to be temperature independent.

3-6 THE CRITICAL STATE

The liquids are characterized by their subdued thermal motions and increased inter molecular attractions. So gases can be expected to liquify by a decrease in temperature and/or increase in pressure. The beginning of 19th century saw the gases like ammonia, sulphur dioxide, hydrogen sulphide, etc., in the liquid form. Encouraged by these attempts many scientists tried to liquify gases like hydrogen, oxygen and nitrogen, but they were unsuccessful even by the application of 3000 atm pressure. Hence there gases were given a name **permanent gases**. In 1869 T. Andrews made a detailed studies of the pressure - volume - temperature relationships of carbon dioxide, which gave out the conditions for liquefication of permanent gases. So we first discuss Andrews observations on carbon dioxide and then go for generalization of these studies.

3-6-1 Isotherms of Carbon Dioxide

Andrews measured the volumes of carbon dioxide gas at different pressures under constant temperature conditions. The experimental data so obtained was plotted to give pressure *vs* volume **isotherms**, Fig. 3-13.

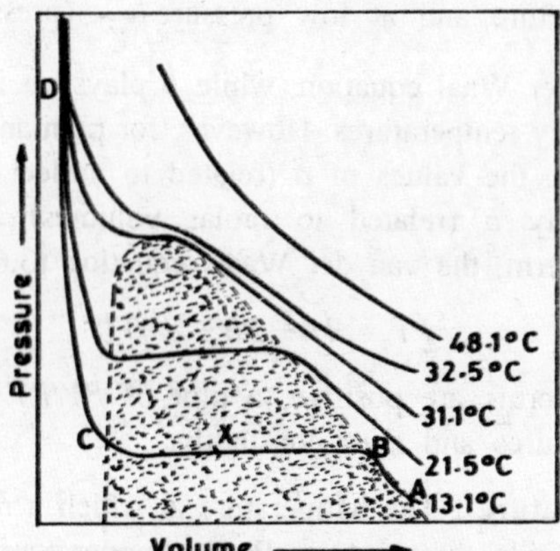

Fig. 3-13 Isotherms of Carbon dioxide

Isotherms at 13.1°C : The lowest pressure represented by *A* indicates a complete gaseous state of CO_2. With the increase in pressure the volume decreases along *AB*. At *B* any attempt to pressurize the liquid only gives more amount of liquid, without bringing any change in the pressure. Point *C* indicates completion of liquefaction of CO_2. Since liquid is not easily compressed any attempt to decrease the volume of the liquid at *C* only rises the pressure to a very high value as indicated by the steep rising compressibility curve *CD*.

The constant pressure represented by *BC* is the vapor pressure of the liquid (49.8 atm for CO_2) where both liquid and gas phases co-exist. At any point *X* on the liquid-vapor equilibrium curve *BC* the composition of liquid to vapor is given by the ratio *BX/CY*.

Isotherm at 21,5°C : The pressure volume curve at 21.5°C is similar to the isotherm at 13.1°C, except that the horizonal portion where liquid and vapor coexist is shorter.

Isotherm at 31.1°C : In this isotherm the horizontal portion has reduced to a point. The dashed boundary, joining the ends of horizonal portions, meets at this point. This was found to correspond a *limiting temperature* above which, no mater how high the applied pressure be, liquefaction will not occur. The gas at this temperature is said to be in a **critical state**. This maximum temperature at which a gas can be liquified is known as **critical temperature.** The pressure just required to cause liquefaction at the *critical temperature* is called the **critical pressure.** If a liquid is heated in a sealed transparent tube, at the critical temperature the meniscus separating the liquid and gaseous phases disappears and there would be no distinction between the liquid and gaseous states, i.e., both the liquid and its vapor will have same density.

Isotherm at 32.5°C : This isotherm is just above the critical temperature. A visible nonlinearity of the curve indicates significant departure from the ideal behaviour. This isotherm, as well as the isotherms lying above critical temperature, show no liquid vapor equilibrium.

Isotherm at 48.1°C : This isotherm is much higher than the critical temperature. This *P-V* curve is almost linear indicating absence of any deviation from ideal behaviour.

3-6-2 Critical Constants

The temperature, pressure and volume corresponding to critical point are called **critical constants.** They are characteristics of the gas. Table 3-6 gives the critical constants of some of the gases.

Table 3-6 Critical constants of some substances

Substance	*B.P. at 1 atm (°C)*	*M.P. at 1 atm (°C)*	*Density at STP* (g L^{-1})	T_c(°C)	P_c (atm)	*Density of liquid at critical pt.* (g m^{-1})
H_2	-252.8	-259.14	0.0899	-239.9	12.8	0.031
O_2	-192.8	-218.4	1.429	-118.8	49.7	0.430
N_2	-195.8	-209.86	1.251	-147.1	33.5	0.311
NH_2	33.35	-77.7	0.721	132.4	111.5	0.235
SO_2	-10	-75.46	2.927	157.2	77.7	0.52
H_2O	100	0.00	999.9	374.0	217.7	0.40

Critical Temperature : *The maximum temperature at which a gas can be liquified is called the critical temperature.* Critical temperature should not be confused with boiling point where the liquid and vapor are at equilibrium at a given pressure. The boiling point lowers with lowering of external pressure. Critical temperature is generally higher than the boiling point at one atmospheric pressure. The critical temperature data given in Table 3-5 also removes the myth of so called permanent gases. Gases like NH_3, SO_2 etc. are easily liquefiable at room temperature just by the application of pressure since their T_c values are far higher than room temperature, while T_c values of permanent gases are far lower than room temperature. It is possible to liquify the permanent gases by subjecting them to initial cooling below their T_c values and then applying required pressure.

Critical pressure : *The pressure required to liquify a gas at critical temperature is defined as its critical pressure,* P_c. These are also characteristic of the gas under consideration. Their values are generally of the order of 10-100 atm for gases and much higher for vapors (The term vapor refers to a gaseous substance, when its temperature is below T_c, i.e., readily liquefiable by simple application of pressure. eg., water, ammonia at room temperature).

Critical Volume : The volume occupied by 1 mole of the gas at its critical temperature and critical pressure is defined as its critical volume V_c. For example 1 mole of CO_2 at 31.1°C (critical temperature) and 72.9 atm (critical pressure) occupies a volume of 95.65 cm^3, making the value its critical volume.

3-6-3 Determination of Critical Constants

Andrews method of plotting *P-V* isotherms for a number of temperatures and joining the ends of horizontal portions to meet at critical point is the most accurate method of determining critical constants. Other simple methods are also available to determine the critical constants separately.

(A) DETERMINATION OF CRITICAL TEMPERATURE AND CRITICAL PRESSURE

Cagniard de la tour's apparatus, Fig 3-14, is widely used to determine T_c and P_c. It uses the principle that the densities of liquid and vapor are same at critical state. The apparatus consists of a *J* shaped tube. The shorter end is blown in the form of a wide bulb. The bulb is initially half filled with the experimental liquid. The rest of the apparatus is filled with mercury leaving a small vacuum portion on the top of the long end of the tube. The long end acts as a mercury manometer.

The bulb containing the experimental substance is immersed in a thermostatic bath initially maintained at low temperature. At this stage the liquid vapor boundary should be clearly visible. Now the temperature of the thermostatic bath is gradually raised such that the experimental substance at every state is in thermal equilibrium with the bath. At a particular temperature the liquid-vapor boundary vanishes and the contents of the bulb look perfectly homogenous. The temperature is noted as the critical temperature (T_c). At this stage pressure corresponding to the vapor in the bulb, read from mercury manometer of the long end of the tube, gives the critical pressure (P_c) of the substance.

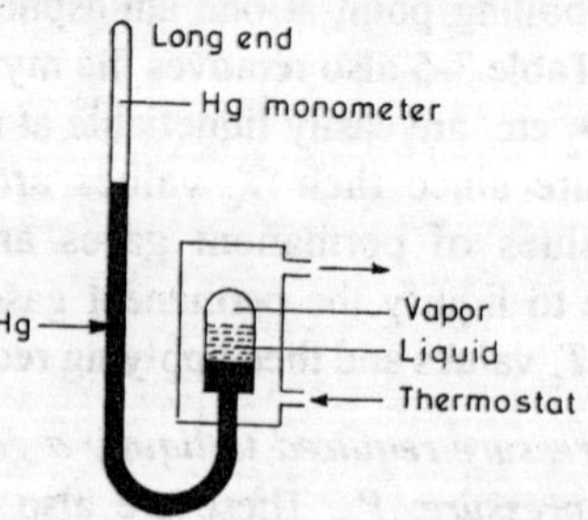

Fig.3-14 Tour's apparatus for the determination of T_c & P_c

(B) DETERMINATION OF CRITICAL VOLUME

Since all the gases deviate significantly at critical state from ideal behaviour, conventional ideal gas equation cannot be used for the calculation of critical volume. Determination of critical volumes (V_c) lie on the rule of L. Cailleted and E. Mathias (1886), known as **law of the rectilinear diameter.** According to this law, *the mean value of densities of any substance in the state of liquid and of saturated vapor at the same temperature is a linear function of the temperature.* Suppose if d_L is the density of the liquid and d_v is the density of the satured vapour then according to law of rectilinear diameter,

$$\boxed{\frac{1}{2}(d_L + d_V) = a + bt} \tag{3.64}$$

where a and b are constants and t is the temperature.

Because these densities correspond to saturated vapor, d_L and d_V they are also called **ortho baric densities.**

The experimental method involves taking a liquid of mass m in an evacuated, graduated sealed tube and heating the liquid to a particular temperature. The volumes V_L and V_v of the liquid and vapor respectively are read at the temperature with the help of calibrations on the ture. Then the mass of the liquid taken is related to the densities with the relation.

$$m = V_L d_L + V_V d_V \tag{3.65}$$

The experiment is repeated with different masses of the liquid and a set of simultaneous equations are obtained. Solving any two simultaneous equations gives the densities d_L and d_V at the experimental temperature. Repeating this method at different temperatures gives a set of d_L and d_V values at the respective temperatures. Mean densities are obtained by taking an average of these two densities.

$$\text{Mean density} = \frac{d_L + d_V}{2} \tag{3.66}$$

Now the orthobaric densities and mean densities are plotted against corresponding temperature to obtain curves as shown in the Fig. 3-15.

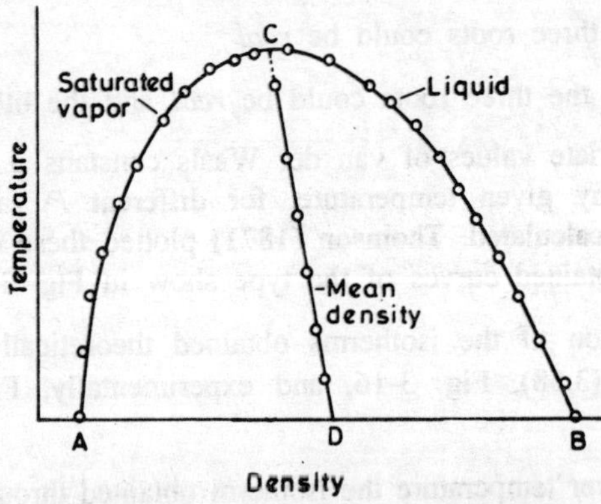

Fig. 3-15 Determination of critical volumes

The points on curve AC correspond to saturated vapor and on BC are those of liquid at various temperatures. C is an intersection point of orthobaric density curves and mean density curve *(CD)*. The density corresponding to C gives the critical density and a reciprocal of this value gives the critical volume per gram. If we are interested with critical molar volumes it is sufficient to divide the molecular weight of the gas with the critical density.

$$\text{Critical molar volume} = \frac{\text{Molecular weight}}{\text{Critical density}} \tag{3.67}$$

3-6-4 The Van der Waals Equation and the Critical State

For one mole of gas per unit volume the van der Waals equation is

$$\left(\frac{P + 1}{V^2}\right)(V - b) = RT$$

on expanding we have,

$$PV - Pb + \frac{a}{V} - \frac{ab}{V^2} = RT$$

Multiplying throughout with V^2,

$$PV^3 - PbV^2 + aV - ab = RTV^2$$

or

$$PV^3 - (RT + Pb)\,V^2 + aV - ab = 0$$

Dividing the above equation with P

$$V^3 - \frac{RT + b}{P}\,V^2 + \frac{a}{P}\,V - \frac{ab}{P} = 0 \tag{3.68}$$

This is a cubic equation in terms of V. In this equation a, b and R are constants. For a given n, P and T values such equation will always have three roots, i.e., three values of V. There are two possibilities for these roots.

(i) All the three roots could be *real*

(ii) *One* of the three roots could be *real* and the other *two imaginary*.

For appropriate values of van der Waals constans a and b and the gas constant R at any given temperature, for different P values corresponding volumes can be calculated. Thomson (1871) plotted these calculated V values against P and obtained curves of the type show in Fig. 3-16.

A comparison of the isotherms obtained theoretically through van der Waals equation (3.68), Fig. 3-16, and experimentally, Fig.3-13, reveal the following facts.

(i) At lower temperature the isotherm obtained through both techniques are similar, except for the replacement of horizontal portion of the experimental isotherm by a sigmoid type (S-shaped) curves in Fig. 3-16.

(ii) The isotherm at lower temperature in a limited pressure range have three volumes (representing three real roots) at each pressure. For example at a pressure represented by M the three volumes are at C, F and B.

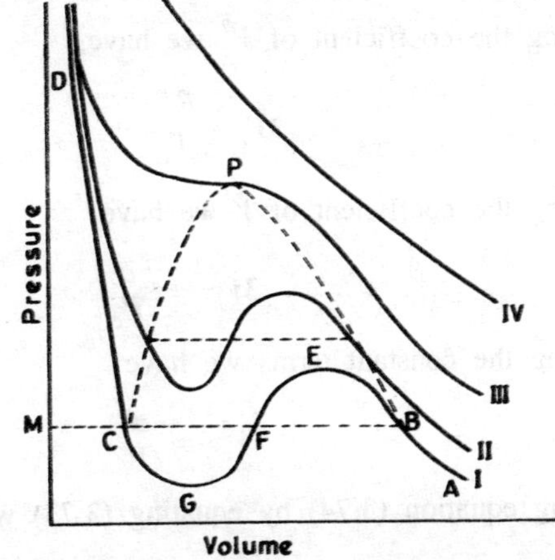

Fig. 3-16 Theoretical isotherms based on van der Waals equation

(iii) In the isotherm represented by III, the three solutions of volume, V, (real values) merge into a single point P.

(iv) The isotherms lying above the limiting curve III will have only one real value of V and other values are imaginary.

(v) The curve segments BE and CG of the isotherm I were realized experimentally under abnormal conditions known as **super saturated vapors** and **super heated liquids** respectively. This gave an unquestionable theoretical support to isotherm in terms of van der Waals equation.

The single point P, where all the three real values of volumes merge is identical with critical point of Fig. 3-13. Under this condition it is possible to write:

$$(V - V_c)^3 = 0 \tag{3.69}$$

having three real roots of $V = V_c$.

On expanding equation (3.69), we get a cubic equation in V as represented below:

$$V^3 - (3V_c)V^2 + (3V_c^2)\,V - (V_c^3) = 0 \tag{3.70}$$

At critical point $T = T_c$ and $P = P_c$

Substituting these values in van der Waals equation (3.68)

$$V^3 - \left(\frac{RT_c}{P_c} + b\right) V^2 + \frac{a}{P_c} V - \frac{ab}{P_c} = 0 \quad (3.71)$$

At critical point the cubic equations (3.70) and (3.71) must be identical. It follows, therefore, that the coefficients of the various volume terms in both the equations must be identical.

Equating the coefficient of V^2 we have,

$$3V_c = \frac{RT_c}{P_c} + b \quad (3.72)$$

Equating the coefficient of V we have,

$$3V_c^2 = \frac{a}{P_c} \quad (3.73)$$

Equating the constant terms we have

$$V_c^3 = \frac{ab}{P_c} \quad (3.74)$$

Dividing equation (3.74) by equating (3.73) we get,

$$\frac{V_c}{3} = \frac{ab}{P_c} \cdot \frac{P_c}{a} = b$$

or

$$\boxed{V_c = 3b} \quad (3.75)$$

Substituting the value $V_c = 3b$ in equation (3.73) we get

$$3(3b)^2 = \frac{a}{P_c}$$

or

$$\boxed{P_c = \frac{a}{27\,b^2}} \quad (3.76)$$

Now substituting the values of V_c and P_c in equation (3.72), we get,

$$3(3b) = \frac{RT_c}{(a/27b^2)} + b$$

$$9b = RT_c \frac{27\,b^2}{a} + b$$

or

$$RT_c \frac{27b^2}{a} = 9b - b = 8b$$

Therefore,

$$\boxed{T_c = \frac{8a}{27\,bR}} \quad (3.77)$$

A knowledge and proper substitution of van der Waals constants in equation (3.75), (3.76) and (3.77) give the values of critical constants and vice-versa.

Another interesting application of van der Waals equation can be obtained by combining the three equation (3.75), (3.76) and (3.77) and is written as, ,

$$\boxed{P_c V_c = \frac{3}{8} RT_c} \qquad (3.78)$$

Problem 12 : Calculate the critical constants of hydrogen for which the van der Waals constants are

$$a = 0.245 \text{ atm L}^2 \text{ mol}^{-1}, \qquad b = 0.266 \text{ L mol}^{-1}$$

Solution : We know

$$V_c = 3b,\ P_c = \frac{a}{27\,b^2} \text{ and } T_c = \frac{8a}{27\,bR}$$

$$V_c = 3b = 3 \times\Re\ 0.0266 = 0.0798 \text{ L}$$

$$P_c = \frac{a}{27\,b^2} = \frac{0.245}{27 \times (0.0266)^2} = 12.82 \text{ atm}$$

$$T_c = \frac{2a}{27bR} = \frac{8 \times 0.245}{27 \times 0.0266 \times 0.08206} = 33.26 \text{ K}$$

Problem 13 : The van der Waals constants of 1 g molecule of a gas are $a = 3.6 \times 10^6$ atm mL^2 mol^{-2} and b = 42.8 mL mole^{-1}. Calculate critical temperature of the gas. (O.U., Apr. 1991)

Solution : Given $a = 3.6 \times 10^6 \text{ atm mL}^2 \text{mol}^{-2} = \frac{3.6 \times 10^6}{10^6} \text{ atm L}^2 \text{mol}^{-2}$

$$b = 42.8 \text{ mL mol} = \frac{42.8}{1000} \text{ L mol}^{-1} = 0.0428 \text{ L mol}^{-1}$$

$$R = 0.08206 \text{ L atm K}^{-1} \text{ mole}^{-1}$$

$$T_c = \frac{8\,a}{27\,bR} = \frac{8 \times 3.6}{27 \times 0.0428 \times 0.08206} = 303.7 \text{ K}$$

Problem 14 : Calculate the van der Waals constants of SO_2. Given the critical density and critical temperature are 0.52 g mL^{-1} and 157.2°C respectively.

Solution : Critical density $= \frac{\text{molecular weight}}{\text{Critical volume}}$

$$\text{Critical volume "}V_c\text{"} = \frac{64}{0.52 \times 1000} = 0.123 \text{ L}$$

But $\qquad V_c = 3b$

$$\text{Therefore } b = \frac{V_c}{3} = \frac{0.123}{3} = 0.041 \text{ L mol}^{-1}$$

$$T_c = \frac{8a}{27bR} = \frac{8a}{27 \times 0.041 \times 0.08206}$$

$$a = \frac{(157.2 + 273) \times 27 \times 0.041 \times 0.08206}{8}$$

$$a = 4.885 \text{ atm L}^2 \text{ mol}^{-2}$$

So van der Waals constants are $a = 4.885$ atm L^2 mol^{-2}

$b = 0.041$ L mol^{-1}

3-6-5 Law of Corresponding States

In the interpretation of van der Waals equation, uncertainties creep in due to selection of proper a and b values and their temperature variation. So van der Waals recommended the usage of ratio for volume, pressure and temperature with corresponding critical values. Thus,

$$\pi = \frac{P}{P_c}; \phi = \frac{V}{V_c} \text{ and } \theta = \frac{T}{T_c} \tag{3.79}$$

or

$$P = \pi P_c, V = \phi V_c \text{ and } T = \theta T_c$$

Substituting the above values of P, V and T in the van der Waals equation (3.54), we have;

$$\left(\pi P_c + \frac{a}{(\phi V_c)^2}\right)(\phi V_c - b) = R\theta T_c$$

Now expressing P_c, V_c and T_c in terms of a, b and R, we get,

$$\left(\pi \frac{a}{27b^2} + \frac{a}{9b^2\phi^2}\right)(3b\phi - b) = \frac{8aR}{27bR}\theta$$

Taking $a/27b^2$ and b common from LHS terms

$$\frac{a}{27b^2}\left(\pi + \frac{3}{\phi^2}\right) b(3\phi - 1) = \frac{8a}{27b}\theta$$

$$\boxed{\left(\pi + \frac{3}{\phi^2}\right)(3\phi - 1) = 8\theta} \tag{3.80}$$

The van der Waals equation (3.80) has many advantages as it does not involve the constants *a, b* and *R*. The quantities π, ϕ and θ are called reduced pressure, reduced volume and reduced temperature respectively. This equation forms the basis of **Law of corresponding states** which states that *any two or more substances that have the same reduced temperature and reduced pressure will have the same reduce volume.* The two gases are then said to be in their corresponding states. Similarly any two gases having same reduced volume and reduced temperature have the same reduced pressure.

3-6-6 Joule-Thomson Effect

It is amply clear that from van der Waals equation the gas molecules experience intermolecular attractions. But most of the conclusions are either purely theoritical or postulates made to explain some experimental observations. J.P. Thomson and W. Thomson (Lord Kelvin) in 1852 made an attempt to provide a direct proof of molecular attractions through a simple experiment. When a real gas at a pressure P_1 is allowed to expand to a pressure P_2 work is done by the gas molecules in breaking the intermolecular forces. If this process is carried out under adiabatic conditions where no heat is allowed to enter or leave the system, energy required for the work is supplied through cooling of the system itself. This phenomena of expansion of a compressed gas through a throttle (regulatory) valve into a low pressure zone (known as throttle expansion), under adiabatic conditions, leading to temperature changes across the valve is called **Joule-Thomson effect.** To demonstrate this effect Joule and Thomson employed a long tube, thermally insulated from outside, Fig. 3-17.

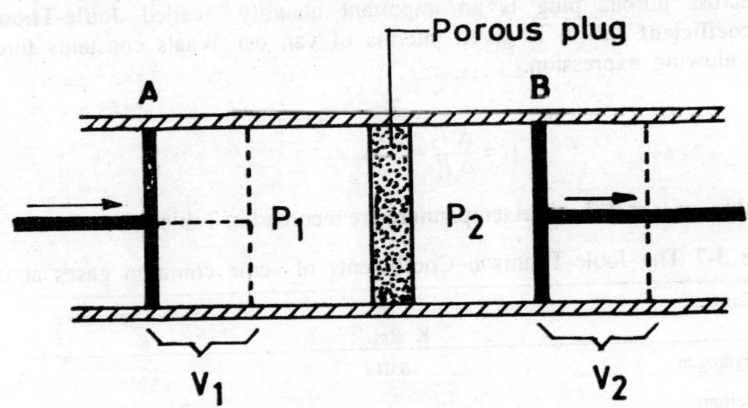

Fig. 3-17 Expansion of gas through porous plug (Joule-Thomson experiment)

In the middle of the tube a porous plug (made of cotton or wool) was placed to permit slow passage of gas from high pressure (P_1) region to low pressure (P_2) region. Two pistons, *A* and *B*, are fitted on either side of porous plug. Both compartments were also fitted with temperature and pressure measuring devices to detect any minor variations of these parameters.

Now piston *A* is pushed inside slowly by a volume *V*, with a simultaneous pulling out of piston *B* such that the pressures P_1 and P_2 of both the compartments respectively remain unaltered. As $P_1 > P_2$, to keep the pressures unaltered, there is a overall expansion of the volume of the gas. The temperature changes across the plug were recorded for different volume changes at different initial temperature and pressure conditions.

Experimental observations :

(i) all the gases show a change in temperature across the plug

(ii) The change in temperature is directly proportional to the difference in pressure on both sides of porous plug or in other words the total volume change of the gas

(iii) At sufficiently low initial temperatures (before throttle expansion) all gases experience a fall in temperature in movirng from high pressure region to low pressure region.

(iv) At moderate initial temperature conditions all easily liquidiable gases (eg., methane, ammonia, CO_2, etc.) suffer a fall in temperature while so called permanent gases (H_2, He, etc.) record a rise in temperature.

(v) Every gas at a characteristic initial temeprature above which the gas shows heating effect and below this temperature it shows cooling effect. This temperature is called the **inversion temperature**. The permanent gases have far lower inversion temperatures, Table 3-7.

(vi) The ratio of change in temperature to the corresponding change in pressure across porous plug is an important quantity called **Joule-Thomson coefficient 'μ'**. It is given interms of van der Waals constants through following expression.

$$\mu = \frac{\Delta T}{\Delta P} = \frac{\frac{2a}{RT} - b}{C_p}$$

The values of μ at 0°C initial temperature are recorded in Table 3-7

Table 3-7 The Joule-Thomson Coefficients of some common gases at 0°C

Gas	μ K atm^{-1}	*Inversion temperature* *K*
Hydrogen	-0.013	193
Helium	-	100
Oxygen	+0.366	-
Nitrogen	0.333	~650
Carbondioxide	0.73	-

The *positive* μ value suggests that the gas shows *cooling effect* at the experimental initial temperature and hence liquifiable. Higher the μ value higher is the liquifiability. the *negative* μ value (eg. Hydrogen) indicates that the gas shows *heating effect* and is not liquifiable at the given initial temperature. to liquify these permanent gases one has to first cool them below their inversion temperature and then subject the gas to **throttle expansion**.

Application of Joule-Thomson Effect : Liquifaction of gases employing Joule-Thomson effect is a widely used industrial process. Pictet (1877) used this technique for the first time in solidifying carbon dioxide and for liquifaction of oxygen. Hampson (1897) and Linde (1895) developed this into a full fledged industrial process for the liquifaction of various gases. In this a gas at an initial pressure of about 250 atm is allowed to steam through a throttle and the gas so cooled is employed to lower the initial temperature of incoming gas. For example when nitrogen at 25°C is expanded from an initial pressure of 200 atm to a final pressure of 1 atm, the gas is cooled by 31°C to a final temperature of -6°C. If this cooled gas is used to lower the initial temperature of incoming compressed gas, the next expansion will yield a still lower final temperature. If this process is repeated over and over in a cyclic operation the temperature falls enough to liquiefy the gas.

Problem 15 : The reduced pressure and volume of ammonia are 1 and 5 respectively at 465°C. Calculate its critical temperature.

Solution : Given reduced pressure $\pi = 1$

reduced volume $\phi = 5$

and $T = 465 + 273 = 738$ K

Van der Waals reduced equation of state is written as

$$\left(\pi + \frac{3}{\phi^2}\right)(3\phi - 1) = 8\theta$$

Now substituting the values of the π and φ, we have

$$\left(1 + \frac{3}{5^2}\right)(3 \times 5 - 1) = 8\theta$$

Therefore $8\theta = \{1 + (3/25)\}(15-1)$

$8\theta = 15.68$

$\theta = 1.96$

But $\theta = T/T_c$ or $T_c = T/\theta$

$T_c = 738/1.96 = 376.5$ K

So critical temperature of ammonia = 376.5 K

Problem 16 : Determine the pressure exerted by 64 g of oxygen at 27°C and occupying a volume 0.5L. Given critical constants of oxygen are T_c = –118.8°C, P_c = 49.7 atm and V_c = 74.42 mL.

Solution : Reduced Temperature $\theta = \frac{T}{T_c} = \frac{27 + 273}{-118.8 + 273} = \frac{300}{154.2}$

$$= 1.95 \text{ K}$$

$$\text{Reduced volume } \phi = \frac{V}{V_c} = \frac{0.5}{(74.42/1000)} = 6.72 \text{ L}$$

$$\text{We know } \left(\pi + \frac{3}{\phi^2}\right)(3\phi - 1) = 8\theta$$

$$\left(\pi + \frac{3}{(6.72)^2}\right)(3 \times 6.72 - 1) = 8 \times 1.95$$

$$\pi = \frac{8 \times 1.95}{(3 \times 672 - 1)} - \frac{3}{(6.72)^2} = 0.75$$

But critical pressure P_c = 49.7 atm

Therefore pressure exerted by 1 mole of oxygen $= \pi P_c$

$= 49.7 \times 0.8$

$= 37.3$ atm

64g of oxygen is 64/32 = 2 moles

Taking 2 moles in the same volume the pressure exerted is 37.3 × 2 = 74.6 atm

3-7 NUMERICAL PROBLEMS

3-7-1 Seven cars participating in a car race having velocities 185, 238, 196, 227, 240, 234 and 230 km hr^{-1}. Calculate the average and *rms* velocities of these cars.

3-7-2 Under normal pressure conditions, a Hg manometer reads 76cm. Find its reading when the pressure is lowered to 0.01 torr.

3-7-3 Calculate the kinetic energy of 42g of CO at -23°C.

3-7-4 A gas mixture taken in a closed container contains 0.2 moles of CO_2, 0.001 mole of NH_3 and 0.05 moles of H_2 moving with velocities 3×10^4, 5×10^4 and 7×10^4 cm s^{-1} respectively. Calculate the *rms* velocity of the gas.

[Hint: When a gas mixture contains n_1, n_2 ... molecules of different gases moving with different velocities u_1, u_2 .... then *rms* velocity of the mixture is

$$rms \text{ velocity} \quad c = \frac{n_1u_1^2 + n_2u_2^2....}{n_1 + n_2 +....}$$

3-7-5 Calculate the *rms* and average velocities of water vapor at 25°C and 1 atm pressure.

3-7-6 Calculate *rms* and average velocities of hydrogen whose molar density at -239.9°C and 12.8 atm pressure is 0.031 g cm^{-3}.
[Hint : 1 atm = 1.014×10^6 dynes cm^{-2}]

3-7-7 Calculate *rms* velocity of nitrogen at 27°C and 1 torr pressures.

3-7-8 The van der Waals constants for xenon gas are a = 4.19 atm L^2 and b = 0.0237 L mol^{-1}.

Calculate the temperature of 1 mole of the gas occupying a volume of 250 cc at a pressure of 5 atm.

3-7-9 Calculate the pressure of oxygen occupying a volume of 1235 cm^3 at 27°C. Given that van der Waals constants for oxygen are a = 1.36 atm L^2 mol^{-2} and b = 0.0318 L mol^{-1}

3-7-10 Calculate the critical constants of oxygen from the van der Waals constants given in the above problem.

3-7-11 Calculate the van der Waals constants of ammonia whose critical density and critical temperature are 0.235 g m^{-1} and 132.4°C respectively.

3-7-12 The van der Waals constants of 48g of oxygen are a = 0.36 atm L^2 mol^{-2}, b = 0.0318 L·mol^{-1}. Calculate its critical temperature.

3-7-13 The reduced pressure and volume of hydrogen are 5 and 0.36 respectively at -268.8°C. Calculate its critical temperature.

3-7-14 The van der Waals constants of chlorine are a = 6.49 atm L^2 mol^{-2}, b = 0.0566 L mol^{-1}. Find its critical constants.

3-7-15 Determine the pressure of 0.05 moles of a gas occupying 235 cc at 127°C. Given T_c = -50°C, P_c = 53 atm and V_c = 100 mL.

3-8 MULTIPLE-CHOICE QUESTIONS

Mark the Correct Answers

3-8-1 At critical point
(A) The density of liquid is greater than its vapor
(B) The density of gas is gerater than its liquids
(C) The densities of gas and vapor are equal
(D) None of the above.

3-8-2 Kinetic theory of gases does not consider
(A) Intermolecular attractions,
(B) Molar volumes of the gases
(C) Critical volumes of the gases
(D) Increase in K. E. of molecules with temperature

3-8-3 Average kinetic energy of a gas increases with increase in
(A) Pressure (B) Volume (C) Mass (D) Temperature

3-8-4 Critical pressure of a gas is equal to

(A) $3b$ (B) $8b/a^2$ (C) $(8a)/(27bR)$ (D) $a/(27b^2)$

3-8-5 γ value of 1.66 is obtained for

(A) Real gases above critical temperature
(B) Real gases below critical temperature
(C) Monatomic ideal gases
(D) Monatomic real gases

3-8-6 Van der Waals constant a is maximum in

(A) H_2 (B) He (C) Cl_2 (D) CCl_4

3-8-7 Van der Waals constant b is minimum for

(A) H_2 (B) He (C) Cl_2 (D) CCl_4

3-8-8 Temperature at which a real gas behaves like an ideal gas over a wide range of pressure is called,

(A) Critical temperature (B) Boyle temperature
(C) Reduced temperature (D) Ideal temperature

3-8-9 The B.P of a liquid whose critical temperature is 327°C

(A) 127°C (B) 218°C (C) 490.5°C (D) 900°C

3-8-10 A gas molecule of mass 1×10^{-23} g and moving with a velocity of 5×10^4 cm s^{-1} has a kinetic energy of

(A) 1.25×10^{-14} g $cm^2 s^{-2}$ (B) 5×10^{-19} g cm s^{-1}
(C) 5×10^{42} g^2 cm s^{-1} (D) 5×10^{-14} g cm^2 s^{-2}

3-8-11 Van der Waals equation has

(A) Three real roots or two imaginary and one real root
(B) Two real roots and two imaginary roots
(C) Three imaginary and two real roots
(D) Two imaginary and three real roots

3-8-12 A gas behaves most ideally at

(A) High temperature and high pressure
(B) High temperature and low pressure
(C) Low temperature and high pressure
(D) Low temperature and low pressure

Fill in the blanks :

3-8-13 A gas can not be liquified above its ____________ temperature through application of any amount of pressure.

3-8-14 The translation motion ceases at ____________

3-8-15 Pressure is defined as ____________ acting on unit area

3-8-16 C_p-C_v for ideal gas in equal to ____________

3-8-17 The real pressure is _______________ the ideal gas pressure

3-8-18 The volume correction factor in van der Waals equation is designated by _______________

3-8-19 Reduced temperature is the ratio between _______________ and _______________

3-8-20 The average distance travelled between two successive collisions is called _______________

3-8-21 The critical temperature of CO_2 is _______________

3-9 ESSAY & SHORT-TYPE QUESTIONS :

3-9-1 What are the fundamental assumptions of kinetic theory of gases? (O.U., Oct. 1991)

3-9-2 Derive the expression for gas equation from kinetic consideration. (O.U., Oct. 1991, Oct. '90 & Oct. '89)

3-9-3 Deduce Graham's law of diffusion from kinetic equation. (O.U., Apr. 1989 & A.U., Apr. 1989)

3-9-4 Starting from van der Waals equation derive to values of critical constants in terms of van der Waals constant *a* and *b*. (A.U., Jun. 1990 & N.U., Apr. 1991)

3-9-5 State the law of corresponding states and deduce the equation

$$\pi + \frac{3}{\phi^2}\ (3\phi - 1) = 8\ \theta$$ (N.U., Apr. 1991 & O.U., Mar. 1989)

3-9-6 Define critical constants. Deduce an expression relating the critical constants with van der Waals constants. (N.U., Nov. 1990 & A.U., Jan 1989 & SKU., Apr. 1990)

3-9-7 Write briefly about the principle of equipartition of energy. (N.U., Apr. 1990 & SKU., Apr. 1990)

3-9-8 Why C_p is more than C_v for an ideal gas? (O.U., Apr. 1991)

3-9-9 What are ideal and real gases? Give reasons for the deviations from ideal behaviour of gases? How are they rectified in van der Waals equation ? (N.U., Apr. 1990, K.U., Apr. 1990 & A.U., Dec. 1989)

3-9-10 Define collision number and collision diameter. (O.U., Apr. 1990, A.U., Dec. 1989)

3-9-11 What is *RMS* velocity? (A.U., Jan. 1990)

3-9-12 What are the units of van der Waals constants *a* and *b*? (K.U., Apr. 1990)

3-9-13 Derive van der Waals equation of state (O.U., Mar. 1992)

3-9-14 Write van der Waals equation in terms of reduced parameters. (N.U., Apr. 1989)

3-9-15 Write an account of law of corresponding states.

3-9-16 Explain why mercury is more suitable as a barometer liquid than water.

3-9-17 Where will it be easy to drink water with a straw, at the sea coast or on the top of a mountain? Answer with proper reasoning.

3-9-18 Explain the Maxwell-Boltzmann's theory of distribution of molecular velocities.

3-9-19 Define mean free path. Why is the mean free path of a gas is inversely proportional to its density?

3-9-20 Write the van der Waals equation for real gases. Explain clearly the meaning of corrective terms of pressure and volume.

ANSWERS

3-7-1 c = 221.43 km hr^{-1}, c^2 = 222.35 km hr^{-1} **3-7-2** 0.76 cm Hg **3-7-3** 3.12 kJ mol^{-1} **3-7-4** 4.64 × 10^4 cm s^{-1} **3-7-5** c = 8.57 × 10^4 cm s^{-1}, $\bar{u}$ = 7.90 × 10^4 cm s^{-1} **3-7-6** *rms* vel = 6.37 × 10^5 cm s^{-1} Aver. vel = 1.016 × 10^4 cm s^{-1} **3-7-8** 19.7K **3-7-9** 21.35 atm **3-7-10** V_c = 95.4 mL, T_c = -118.8°C, P_c = 49.7 atm **3-7-11** a = 2.7 atm L^2 mol^{-2}, b = 0.241 L mol^{-1} **3-7-13** T_c = 15 K **3-7-14** V_c = 16.98 mL, P_c = 75.03 atm T_c = 414.02 K **3-7-15** P = 490 atm

3-8-1 C **3-8-2** A **3-8-3** D **3-8-4** D **3-8-5** C **3-8-6** D

3-8-7 B **3-8-8** B **3-8-9** A **3-8-10** A **3-8-11** A **3-8-12** B

3-8-13 Critical **3- 8-14** 0 K **3-8-15** Force **3-8-16** γ

3-8-17 less than the **3-8-18** b **3-8-19** Temperature, Critical temperature

3-8-20 Mean free path **3-8-21** 31.1°C.

4

SOLUTIONS

4-1 INTRODUCTION

A solution may be defined as a homogenous mixture of two or more substances. They could be gaseous, liquid or solid. The component present in greatest quantity and determines the state of solution is called the **solvent.** All the component/s, other than solvent, present in a solution is (are) called **solute(s)**. In the condensed phases the molecules are held together through inter molecular attractions between solute and solvent molecules. The solute-solute and solvent-solvent interactions are different from solute-solvent interactions. Obeying the phrase *like dissolves like*, the polar substances (e.g., NaCl) dissolve predominantly in polar solvents (e.g., water) and nonpolar substances (e.g., methane) show their preference to non polar solvents (e.g., benzene). Table 4-1 list some of the typical types of solutions.

Table 4-1 Types of solutions

Dispersing phase (solute)	*Dispersion medium (solvent)*	*State of solution*	*Examples*
Solid	Solid	Solid	Alloys (Brass, Bronze, etc.)
Solid	Liquid	Liquid	Sugar water
Liquid	Solid	Solid	Hydrates ($CuSO_4$ $8H_2O$)
Liquid	Liquid	Liquid	Alcohol in water
Gas	Solid	Solid	H_2 gas in palladium
Gas	Liquid	Liquid	Soda water (CO_2 in water)
Gas	Gas	Gas	Air, Natural gas (contrains lower hydro carbons)

In addition to types mentioned in Table 4-1, there could be floor crossing and solid and a liquid or a solid and a gas or solutes of all phases can together dissolve in a solvent and form a solution. For example the soft drinks are generally made by dissolving solid sugar, carbon dioxide gas and some tastemakers in water.

4-2 SOLUTION CONCENTRATION

Amount of solute present in a fixed amount of solvent is called the concentration. In qualitative sense it deals with the crowded nature of solute particles in the dispersion medium. Different concentration units used by chemists have their particular advantages and disadvantages.

$$\text{Mass percent of solute} = \frac{\text{Mass of solute}}{\text{Mass of solute + Mass of solvent}} \times 100\%$$

$$= \frac{\text{Mass of solute}}{\text{Mass of solution}} \times 100\%$$

Problem 1 : Calculate the mass percent of the solution when 14g of sugar is dissolved in 76g of water.

Solution : Mass percent $= \frac{14}{14 + 76} \times 100\% = 15.5\%$

(ii) Volume Percent : Similar of mass percent, volume percentage is given by the formula

$$\text{Volume percent} = \frac{\text{Volume of solute}}{\text{volume of solute + Volume of solvent}} \times 100\%$$

$$= \frac{\text{Volume of solute}}{\text{Volume of solution}} \times 100\%$$

(iii) Parts per Million (*PPM*) : This unit is used to express extremely small concentrations.

$$ppm = \frac{\text{Mass of solute}}{\text{Mass of solution}} \times 10^6$$

Thus *ppm* may be defined as 1mg of substance present in 1000g of solution.

Ppm is popularly used in expressing minute impurities present in a substance and very small spectroscopic and chromatographic concentrations.

Problem 2: The WHO reports indicate that the average DDT accumulation level of the human body is 2 g/kg fat tissue. Express this value in *ppm.*

Amount of DDT present in 1 kg of the fat tissue = 2 g = $2/10^3$ kg

Mass of solution of DDT in the fat tissue = $1 + 2/10^3 = 1$

$$ppm = \frac{(2)/10^3}{1} \times 10^6$$

$$= 2000 \; ppm$$

(iv) Normality : This is one of the widely used concentration unit in the older books on chemistry.

$$\text{Normality } (N) = \frac{\text{Number of gram equivalents of solute}}{\text{Liters of solution}}$$

Molarity has almost replaced this unit in the present literature.

(v) Molarity : Weight of Avogadro number of atoms or molecules expressed in grams gives gram molecules weight. Molarity is defined by the expression;

$$\text{Molarity } (M) = \frac{\text{Number of moles of solute}}{\text{Liters of solution}} \text{ mol L}^{-1}$$

$$= \frac{\text{Weight / Molecular weight}}{\text{Liters of solution}} \text{ mol L}^{-1}$$

Since molarity involves volume term, *it is a temperature dependent quantity.* Conventionally, temperature is mentioned only when molarity is estimated at a temperature other than the room temperature.

Molarity of solutions prepared by weighing pure substances into solvent can be estimated from the above formula. Because several common concentrated acids are actually prepared by dissolving corresponding gases in water, their dilute solutions are prepared by proper dilution with the solvent using the formula,

$$M_i \times V_i = M_f \times V_f$$

where M_i & V_i refer to initial molarity and initial volume and M_f and V_f refer the corresponding final values. Approximate molarities of some concentrated acids are presented in Table 4-2.

Table 4-2 Approximate molarities of concentrated acids

Acid	*Name*	Molarity
HCl	Hydrochloric acid	11.5
H_2SO_4	Sulphuric acid	18
HNO_3	Nitric acid	15
H_3PO_4	Phosphoric acid	15
CH_3COOH	Acetic acid	17.5

Problem 3 : Calculate the molarity of HCl solution when 10mL of concentrated HCl is diluted to 100mL.

Solution : Given : Molarity of concentrated HCl = 11.5 M

$M_i = 11.5$, $V_i = 10$

$M_f = ?$, $V_f = 100$

$11.5 \times 10 = M_f \times 100$

$M_f = 11.5 \times 10/100 = 1.15$ M

Molarity of diluted solution = 1.15 M

(vi) **Molarity :** This *temperature independent* concentration term is defined as the number of moles of solute dissolved in 1 kg of solvent.

$$\text{Molality } (m) = \frac{\text{Moles of solute}}{\text{Mass of solvent in kg}} \text{ mol kg}^{-1}$$

In case of aqueous solutions because the density of water is approximately 1 g mL^{-1}, there is very little difference between its molarity and molality.

(vii) **Mole Fraction :** The ratio of the number of moles of one of the component to the total moles of the components in a solution is called the mole fraction.

For example, if n_1 moles of component "A" and n_2 of component "B" are mixed to give a solution then,

$$\text{Mole fraction of "A"} = \frac{n_1}{n_1 + n_2}$$

$$\text{Mole fraction of "B"} = \frac{n_2}{n_1 + n_2}$$

Sum of mole fractions of all the components is always equal to unity. Similar to molality, mole fraction is also independent of temperature variations.

Problem 4 : Calculate the molarity, molality and mole fraction of a solution of 1g of benzoic acid in 100mL acetone. Given that the density of acetone is 0.8 g/L.

Solution : Molecular weight of benzoic acid (C_6H_5COOH) = 122

1g is dissolved in 100 mL acetone

or

10g dissolved in (1000mL = 1L)

$$\text{Molarity} = \frac{10/122}{1} = 8.2 \times 10^{-2}\ \text{M}$$

Weight of 100mL of acetone = 100 × 0.8 = 80g = 0.08 kg

Therefore 10/122 moles of benzoic acid is dissolved in 0.8 kg

$$\text{Molality} = \frac{10/122}{0.8} = 0.102\ \text{m}$$

$$\text{Number of moles of acetone} = \frac{80}{\text{Mol. wt. of acetone}} = \frac{80}{58} = 1.37\ \text{moles}$$

$$\text{Number of moles benzoic acid} = \frac{10}{122} = 0.082\ \text{moles}$$

$$\text{mole fraction of benzoic acid} = \frac{0.082}{0.082 + 1.37} = 0.057.$$

4-3 SOLUTIONS OF GASES IN LIQUIDS

When we talk about the solutions of gases in liquids, the first thing that strikes our mind is the carbonated water in the form of soda or soft drinks and the fancy advertisements shown in television to promote their sale. Solutions of gases in water play a vital role in the chemistry of life. Entire aquatic life is supported by the small amounts of oxygen dissolved in water. Two factors control the solubility of a gas in a given solvent. They are (i) Temperature and (ii) Pressure.

4-3-1 Effect of Temperature on the Solubility

An increase in temperature invariably reduces the solubility of a gas. The carbonated water when taken out of a refrigerator is completely a clear solution, but even a slight warming results in the evolution of many CO_2 gas bubbles.

Reduction of oxygen solubility in warm waters has a direct consequences on the aquatic life called *thermal pollution.* When hot waters from the cooling towers of nuclear or thermal plants or industrial boilers are allowed to enter water resources like lakes, the temperature of the lake raises by few degrees. This increase in temperature of the surroundings increase the metabolism of acquatic animals, there by consumption of oxygen. At the same time warming up of lakes reduces the solubility of oxygen thus posing a threat to survival of fishes.

4-3-2 Effect of Pressure on the Solubility of gases and Henry's Law

External pressure has negligible effect on the solubility of liquid in a liquid, or solid in a liquid, or solid in a solid, but the solubility of gases in liquids and solids is greatly influenced by changes in external pressure. William Henry (1775-1836), a British chemist, made elaborate stidies on the solubility of gases in liquids and first stated a quantitative relation between gas solubility in liquid called **Henry's law**. Henry's law states that *at any given temperature the solubility of a gas in a liquid is proportional to the pressure of the gas over the solution.*

$$C \propto P$$

or

$$C = kP$$

where C is the molar concentration, P is the external pressure and k is a proportionality constant called ***Henry's law constant.***

An illustration showing how Henry's law is obeyed by solutions of typical gases in water is shown in Fig. 4-1.

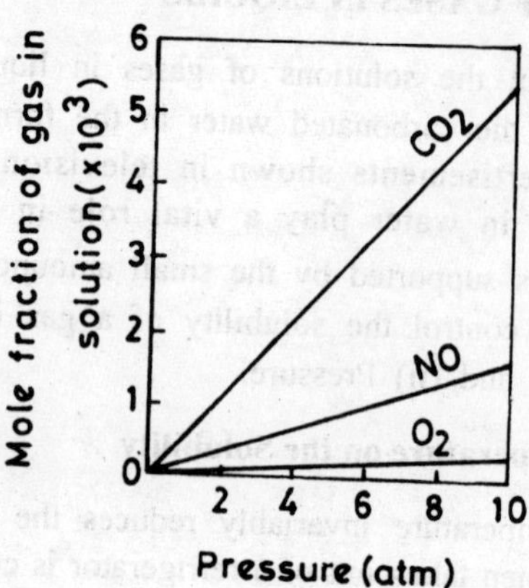

Fig. 4-1 Henry's Law

For a given gas and liquid pair at a given temperature, *k* values remain constant. When P = 1 atm, C is numerically equal to k.

A list of Henry's law constants at 30ºC in water are given in Table 4-3.

Table 4-3 Henry's Law Constants at 30°C

Gas	*k(mm Hg/Mole fraction of solution)*
Argon	3.02×10^{-7}
Carbon dioxide	0.139×10^{-7}
Helium	11.1×10^{-7}
Hydrogen	5.51×10^{-7}
Nitrogen	6.68×10^{-7}
Oxygen	3.52×10^{-7}

***When** we consider a solution of two or more gases, the concentration of any gas was found to bare a direct proportionality with its partial vapor pressure, instead of the total pressure over the solution.*

Mechanism of Henry's Law : Henry's Law could be explained on the basis of kinetic theory of gases. According to this theory pressure is a consequence of collisions of molecules on a surface. In addition we also know that vapor pressure is a result of a *dynamic equilibrium*, were the escaping tendency of liquid molecules is exactly equal to the molecules entering the liquid state. With the increase in the pressure of a gas, the number of collisions on an unit area of a liquid increases. They in turn facilitate the trapping of gas molecules till a new equilibrium is established between molecules in a gas and solution states. This explains why the solubility of a gas increases with pressure. Since the equilibrium constant of the dynamic equilibrium is a function of temperature, Henry's law constant changes with changes in temperature.

Limitations of Henry's Law

(i) Since kinetic theory of gases is strictly obeyed only at Low pressure and High temperature (ideal gas behaviour), Henry's law is obeyed mainly under low pressure and high temperature conditions.

(ii) If dissolved gas reacts with water/solvent higher solubilities can result. Henry's law is not obeyed by gases that are interacting chemically with the solvent.

$$NH_3 + H_2O \rightleftharpoons NH_4OH$$
$$CO_2 + H_2O \rightleftharpoons H_2CO_3$$

Therefore NH_3 which fails to obey Henry's law in aqueous solutions, obeys in benzene where it does not undergo any chemical interaction.

Problem 5 : At room temperature and pressure, oxygen is soluble to an extent of 1.23×10^{-3} mol L^{-1}. Find the solubility when its partial pressure over water is 1000 torr. Vapor pressure of water at 25°C is 24 torr.

Solution : Given : T = 25°C Total pressure = 760 torr

$$P_{total} = p_{H_2O} + p_{H_2}$$
$$760 = 24 + P_{O_2}$$
$$P_{O_2} = 736 \text{ torr}$$

According to Henry's law $C_{O_2} = k \times P_{O_2}$

$$\text{Henry's Law constant} = \frac{C_{O_2}}{P_{O_2}} = \frac{1.23 \times 10{-}3}{736}$$
$$k = 1.67 \times 10^{-7} \text{ mol L}^{-1} \text{ torr}^{-1}$$

When partial pressure is 1000 torr

$$\text{Solubility} = 1.67 \times 10^{-6} \times 1000$$
$$= 1.67 \times 10^{-3} \text{ mol L}^{-1}$$

Problem 6 : Carbondixodie is soluble in water at 20°C and 1 atm external pressure to an extent of 169 mg per 100 g of water. Calculate the solubility of CO_2 when the pressure raised to 2 atm. Vapor pressure of water at 20°C is 20 torr.

Solution : Solubility of CO_2 at 20°C = 169 mg per 100 g of water

$$\textbf{Henry's Law constant} = \frac{C}{p} = \frac{169}{(760-20)} = \frac{169}{740}$$

When external pressure is raised to, 2 atm

Partial pressure of CO_2 at 20°C = (760 × 2) – 20 = 1500 mm Hg

$$\text{Solubility} = \frac{169}{740} \times 1500 = 342.6 \text{ mg per } 100 \text{ g of water}$$

4-4 PHYSICAL PROPERTIES OF DILUTE SOLUTION

When solids or liquids are dissolved in solvents, the solutions formed exhibit a variety of properties which differ markedly from the solvent. These properties can be divided into two classes.

(A) Properties dependent on the nature of solvent and solute : Propêrties like density, electrical conductivity, color, taste, etc., fall in the category. For example pure water is a very poor conductor of electricity but addition of sodium chloride imparts the solution, good conducting properties. Similarly addition of sugar to pure water does not help in increasing the electrical conductivity.

(B) Properties which depend only on the concentration of solute : These properties are called **Colligative properties**. *They depend only on the number of molecular particles present and does not depend on their chemical nature.* There are four important colligative properties,

(i) Lowering of vapor pressure

(ii) Elevation of boiling point

(iii) Depression of freezing point

(iv) Phenomenon of Osmosis

In the following chapter we discuss only about the lowering of vapor pressure, its quantification by **Raoult's law** and related principles.

4-4-1 Raoult's Law

A pure solvent like water, benzene, acetone, acetic acid, ether, etc., has a fixed vapor pressure at any given temperature. It was observed that a solution of non-volatile solute in these solvents also has a vapor pressure, but always less than the pure solvent. This *fall in vapor pressure of a solvent by the addition of a non-volatile solvent is called* ***Lowering of vapor pressure***. This fall in the vapor pressure was found to increase with increase in the concentration of the solute. This is illustrated in Fig. 4-2.

In 1881 the French chemist, F.M. Raoult proposed a quantitative relationship between the lowering of vapor pressure and its concentration. This relationship is called **Raoult's law**. It states that *the partial pressure of a solvent over a solution, (P), is equal to the vapor pressure of the pure solvent, (P^o), times the mole fraction of the solvent in the solution (X_1).*

$$\boxed{P = X_1 P^o} \quad (4.1)$$

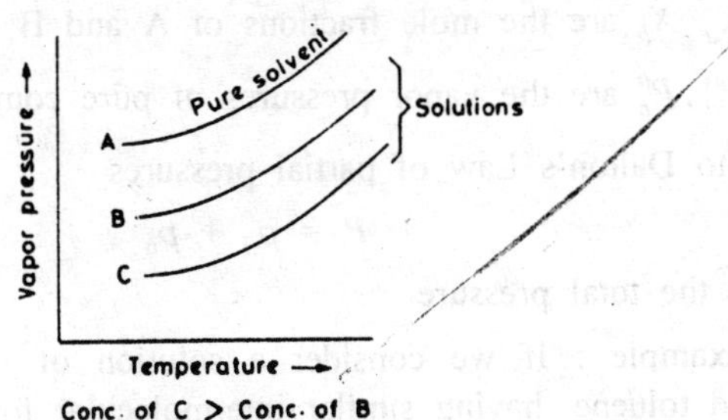

Fig. 4-2 Raoults Law

If the solution contains only one solute with mole fraction X_s,

$$X_1 + X_s = 1$$

or

$$X_1 = 1 - X_s \tag{4.2}$$

Substituting the values of X_1 in equation (4.1)

$$P = (1 - X_s)\, P^o$$

or

$$\boxed{\Delta P = (P - P^0) = X_s\, P^0} \tag{4.3}$$

Therefore lowering of vapor pressure is directly proportional to mole fraction of the solute, X_s.

4-4-2 Mechanism of Raoult's law

Raoult's law can be considered on the basis of molecular hypothesis. In case of pure solvent, the rate of evaporation depends on the number of molecules available on a unit surface of the liquid. When solutions of different non-volatile solutes in the same solvent is considered, the molecular density (number of molecules per unit volume) of the solvent molecules on a unit surface area decreases, there by the rate of evaporation. Since vapor pressure is the result of a dynamic equilibrium between a vapor and corresponding liquid, a fall in the rate of evaporation can be equated to a fall in the vapor pressure. Further increase in the mole fraction of the solute decreases the vapor pressure. This shows that solvent molecules have lesser tendency to leave solution than pure solvent to become vapor.

In case both the solvent and solute are *volatile*, the vapor pressure of the solution is simple algebraic sum of partial pressures of the components. If A and B are any two such components, according to Raoult's law

$$p_a = X_a\, P_a^o$$

$$p_b = X_b\, P_b^o$$

where p_a, p_b are the partial pressures of A and B

X_a, X_b are the mole fractions of A and B

P_a^o, P_b^o are the vapor pressures of pure components.

According to Dalton's Law of partial pressures

$$P = p_a + p_b$$

where P is the total pressure.

For example : If we consider a solution of two nonpolar molecules benzene and toluene, having similar intermolecular forces, we have;

$$P_{\text{total}} = p_{\text{benzene}} + p_{toluene}$$

On plotting P_{total}, p_{benzene} and p_{toluene} against; mole fraction of benzene, X_{benzene}, we get Fig 4-3.

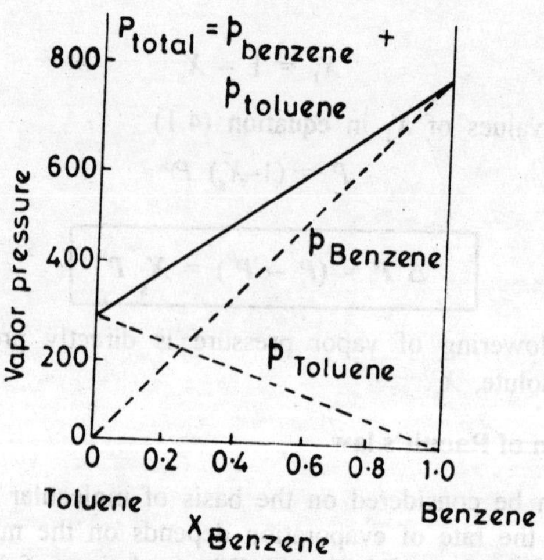

Fig. 4-3 Verification of Raoult's law in Benzene – Toluene system.

4-4-3 Limitation of Raoult's Law

(i) Raoult's law is strictly obeyed by dilute solutions of non- volatile, non-dissociating liquids, e.g., sugar or urea in water.

(ii) Raoult's law is also obeyed by solutions containing volatile components, provided both solvent and solute are of the same nature, e.g., Benzene + Toluene; Cyclohexane + CCl_4.

(iii) Inorganic solutes such as acids, bases and salts which dissociate in water do not obey Raoult's law. They show increased depression of the vapor pressure due to greater number of molecular species produced on account of dissociation.

Problem 7 : Liquid A (mol. wt. 46) and liquid B (mol. wt. 18) fonn an ideal solution. At 20°C the vapor pressure of pure liquid A and pure liquid B are 440 and 17.5 mm of Hg respectively. Calculate the vapor pressure of the solution of A in B containing 0.2 mole fraction of A. (O.U., Mar 1992)

Solution : Given : p^o_A = 440 mm Hg, p^o_B = 17.5 mm Hg

$X_A = 0.2 \; X_B = 1 - 0.2 = 0.8$

$p_A = X_A \; P^o_A = 0.2 \times 440 = 88$ mm Hg

$p_B = X_B \; P^o_B = 0.8 \times 17.5 = 14$ mm Hg

Vapor pressure of solution $P = p_A + p_B$

Vapor pressure of solution of A in B = 88 + 14 = 102 mm Hg

Problem 8 : A solution of two liquids A and B exhibits ideal behaviour. The mole fraction of A is 0.4. The vapor pressure of the pure component A is 0.5 atm and that of B is 0.3 atm. Calculate the partial vapor pressure of a and B is solution. (O.U., Apr. 1987)

Solution : given $X_A = 0.4, \; X_B = 1 - 0.4 = 0.6$

$p^o_A = 0.5, \quad p^o_B = 0.3 \; atm$

Partial vapor pressure of A = $p_A = X_A p^o_A$

$p_A = 0.4 \times 0.5 = 0.2$ atm

Partial vapor pressure of B = $p_B = X_B \; p^o_B$

$p_B = 0.6 \times 0.3 = 0.18$ atm

4-4-4 Ideal solution & Their characteristics

Any solution that obeys Raoult's law is defined as an **ideal solution.** An ideal solution is characterized by following properties.

(i) An ideal solution should obey Raoult's Law

(ii) The enthalpy of solution, ΔH_s, of an ideal solution should be zero, $\Delta H_s = 0$

(iii) The total volume of solution should be algebric sum of volumes of individual components mixed. $\Delta V_{mix} = 0$

(iv) Physical properties like density, refractive index, viscosity, etc., of a solution is an average of the individual components taken in their molar ratios.

4-4-5 Real solutions

Solutions that do not obey Raoult's are considered as *non-ideal* or *real solutions.* The formation of these solutions are accompanied by volume and enthalpy changes.

Designating two volatile components A and B forming nonideal solutions, we can classify the real solutions into two types basing on the nature of deviation from Raoult's law.

Type I – Positive deviation from Raoult's law : In the pure state, components and A and B have intermolecular attractions between like molecules, i.e., between A-A or B-B. When A and B are mixed to give a solution, due to different molecular forces between A and B molecules, the presence of B will affect the escaping tendency of A and *vice versa.* If the attractions between A molecules is much stronger than B molecules, (A-A >> B-B), the A-B intermolecular forces become stronger than B-B.

In such case A molecules increase the escaping tendency of B molecules present in the solution. The net result is an increase of partial pressure of B than expected by Raoult's law, Fig. 4-4. Under these conditions the total vapor pressure of a solution is greater than the sum of the vapor pressures predicted by Raoult's law at the same concentration. Such behaviour is known as **Positive deviation** from Raoult's law. The heat of solution in these solution is *positive endothermic.*

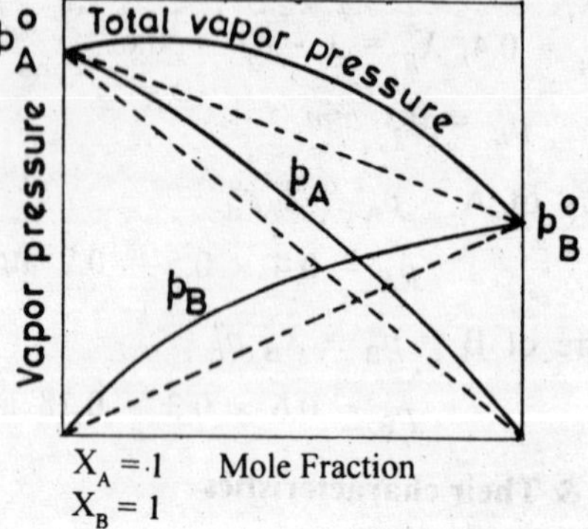

Fig. 4-4 Positive deviation from Raoult's law

If both the components are polar, e.g., mixture of alcohols, ether + acetone, or both non polar, e.g., benzene + toluenee, hydrocarbon mixtures, CCl_4 + heptane, the positive deviation from ideal behaviour is negligible and the solutions may be considered ideal for all practical purposes. *Positive deviation are distinct when one of the component is polar and the other non polar.*

Type – II Negative deviation from Raoult's Law : Negative deviation from Raoult's law is noticed when the intermolecular forces between A and B are stronger than individual components.

$$A\text{-}B >> A\text{-}A \text{ or } B\text{-}B$$

Generally this situation arises when the attractions between A and B molecules is to some extent chemical in nature, i.e., compound formation takes place between A and B. Under these conditions fewer molecules than expected by Raoult's law enter the vapor phase and the total vapor pressure of the solution is smaller than the sum of the vapor pressures predicted by Raoult's law, Fig. 4-5.

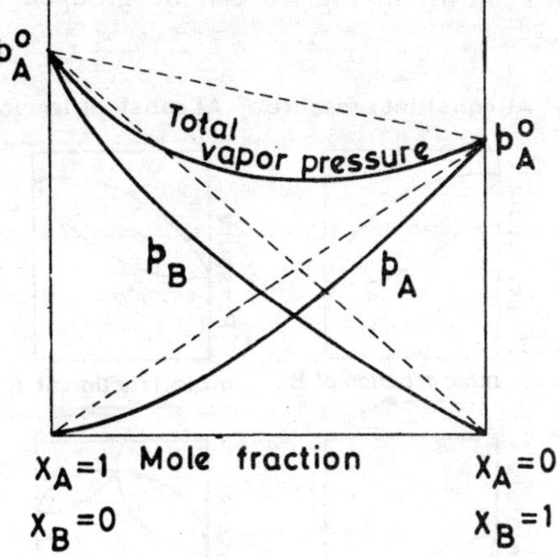

Fig. 4-5 Negative deviation from Raoult's law

In Type II systems mixing of components produces heat, i.e., ΔH_s = –ve. Systems falling under this category are., acetic acid-pyridine, ethyl ether - chloroform, phenol - aniline, sulphuric acid - water, hydrochloric acid- water.

An examination of Fig.4-4 and 4-5 indicate that at very dilute concentrations, approaching either pure A (X_a = 1) or pure B (X_b = 1), the experimental vapor pressures of solutions are not much different from Raoult's law values. This suggests that irrespective of positive or negative deviation, *all very dilute solutions invariably obey Raoult's law.* These solutions are therefore called ideal dilute solutions.

4-5 FRACTIONAL DISTILLATION

For solutions of volatile liquid components, whose individual vapor pressures are different, the composition of the vapor is different from the composition of liquid with which it is in equilibrium. If such a solution is boiled in a open vessel (i.e., at constant pressure), the highly volatile liquid which has low vapor pressure, escapes faster and the liquid-liquid system gets concentrated with the low boiling liquid. the consequence of this process is of importance and serve a useful method of separation of two or more volatile liquids. *The process of heating a liquid mixture and condensing the resultant vapor and enriching the volatile component is called **fractional distillation.*** Similar results could also be obtained by keeping the temperature constant and lowering the pressure and cooling the vapors. From a knowledge of compositions of liquid and vapor at equilibrium, at a given pressure and different tempeatures, it is possible to draw temperature *vs* composition curves, Fig.4-6(a). Similarly at any given temperature it is possible to obtain vapor pressure-composition curves, Fig.4-6b.

All the curves shown in Fig.4-6 can be grouped into three main types.

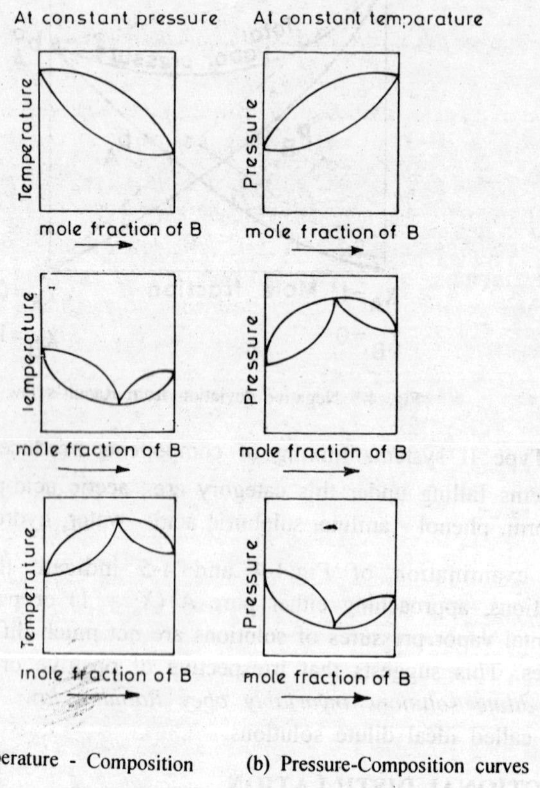

(a) Temperature - Composition (b) Pressure-Composition curves

Fig. 4-6 Fractional distillation curves

Type I : Curves (i) and (ii) represent this category. The vapor pressure varies regularly with composition without showing any maximum or minimum. The boiling point also varies regularly with the composition. Type I systems are close to ideal behaviour. Example : Benzene - toluence, CCl_4 - benzene.

Type II : Curves (iii) and (iv) represent this category. The vapor pressure-composition curves have a maximum. The mixture corresponding to this vapor pressure maximum boils at a minimum temperature than any other mixture. Example : CS_2 - acetone, benzene - ethyl alcohol, chloroform - ethyl alcohol.

Type III : Curves (v) and (vi) represent this category. The vapor pressure - composition curves have a maximum. The mixture corresponding to this vapor pressure maximum boils at a minimum temperature than any other mixture.

Example : Chloroform-acetone, pyridine - acetic acid, water - hydrochloric acid.

4-5-1 Fractional Distillation of Binary Miscible solutions of Type I

Let A and B be the components of a binary solution of type I. The boiling temperature - Composition curves for liquid and vapor phases of the mixtures of A and B are show in Fig. 4-7.

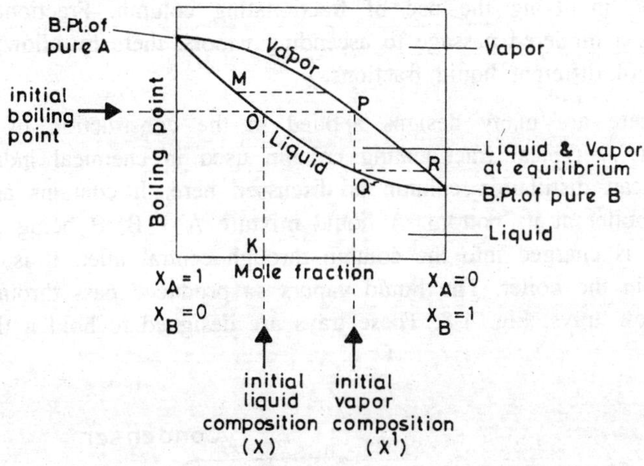

Fig. 4-7 Fractional distillation diagram of Type I

The curve marked "Liquid" gives the temperature at which liquids of different composition attain a vapor pressure of 1 atm, or boil. The curve marked "Vapor" gives the composition of vapor in equilibrium with a liquid at its boiling point. Now considering a liquid with composition X is heated, it traces a path along KO, till its vapour pressure is equal to 1 atm i.e., till it reaches "O" lying on the "liquid" curves. Just at "O" it boils and the composition of vapour coming off at the boiling point will have a composition X_1. This is indicated by the horizontal Tie line OP. If this vapor is cooled and collected, the liquid has a composition X_1.

Since the vapour condensate has a composition X_1, richer in B, than the original liquid with composition X, the residual liquid will become richer in A. If the vapor condensate with composition X_1 is distilled again, the distillate will get further richer in B and acquires a composition corresponding to "R". Repeating this distillation process couple of times, the final distillate will be pure B.

Now let us go back to our new residual liquid richer in A. Let it have a composition represented by point M. Since it is richer in "A, it boils at a higherr temperature. On heating this liquid the vapors coming off will once again be richer in B. As indicated by the tie line MN and the residual liquid get further close to pure A. Therefore boiling the residual liquid ultimately leaves pure A in the residual liquid.

Thus it is possible to separate the two components of Type I mixture by fractional distillation. Though theoretically the process of heating and collecting vapors and using for second distillation etc., sounds well, in practice it is quite cumbersome. Fractional distillation is usually carried out in one operation involving the use of fractionating column. Fractionating column provides a hindered passage to ascending vapors, there by allowing a gradual cooling of different liquid fractions.

There are many designs applied in the construction of fractionating columns. A typical fractionating column used in chemical industries called bubble cap distillation column is discussed here. It contains an electrically heated boiler at its bottom. A liquid mixture A + B, B being more volatile than A, is charged into the column through central inlet. It is heated to its boiling in the boiler. The liquid vapors so produced pass through a number of shallow trays, Fig. 4-8. These trays are designed to hold a thin layer of

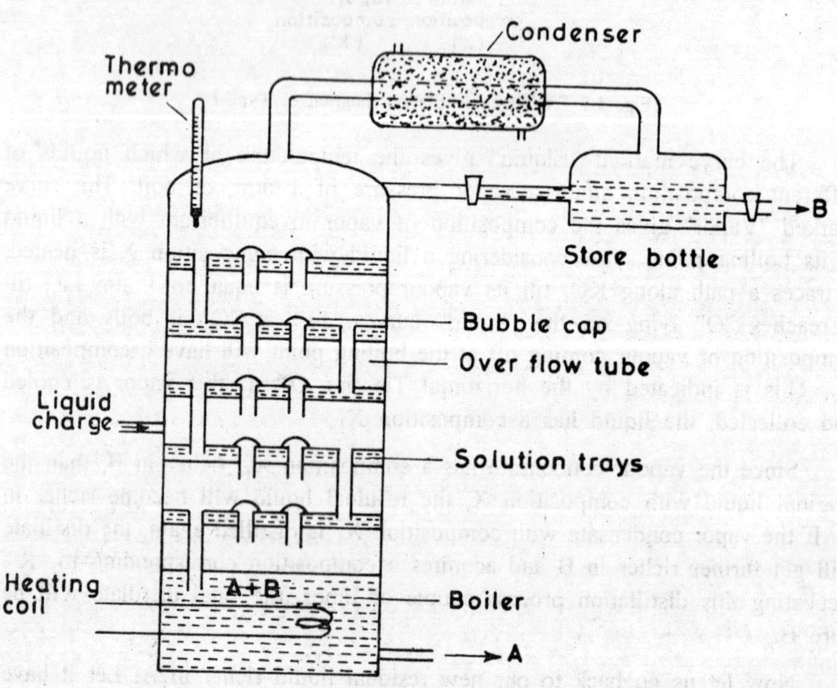

Fig. 4-8 Bubble cap fractionating column

the liquid that condenses in them, allowing excess liquid to over flow in to the plates below. The ascending vapors of the mixture are allowed to rise, to upper plates only after bubbling through liquid films in the lower plates. The ascending vapor are cooled gradually by the liquid layers in the trays. The vapor reaching the top most tray is rich in the low boiling or highly volatile component B. It is condensed and allowed to flow back down the column. This descending liquid rich in B comes in contact with the fresh ascending vapors and rob of their less volatile component A and reach the boiler. Repetition of this cycle ultimately leave pure more volatile component "B" in the tower and less volatile component "A" in the boiler. In this way an almost complete separation of the components of a binary mixture can be achieved.

In the laboratories, special glass columns, Fig.4-9, based on the principle of hindered flow of upward gases and free flow of descending liquid is employed.

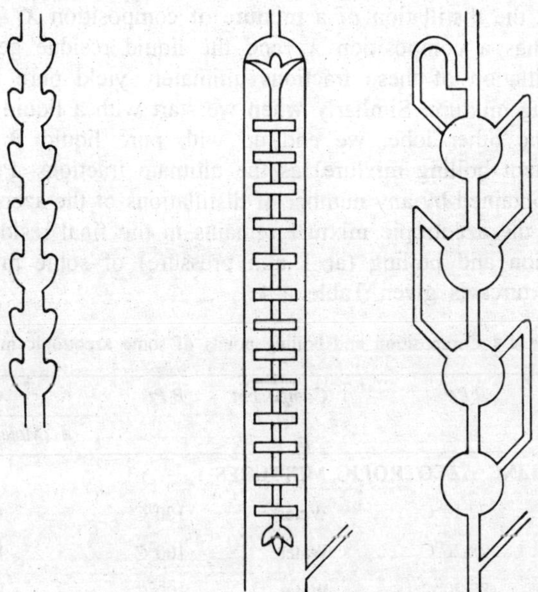

Fig. 4-9 Some glass fractionating columns

4-5-2 Fractional distillation of Binary Miscible solutions of Type II

The vapor pressure - Composition curves of these mixtures show a vapor pressure maximum or these mixtures show a boiling point minimum, Fig. 4-10. At this point the liquid and vapor have the same composition. At this point, called **azeotropic point**, the mixture of components A and B boil at a single temperature and distill completely as if it were a single substance, although no compound formation between A and B has occurred.

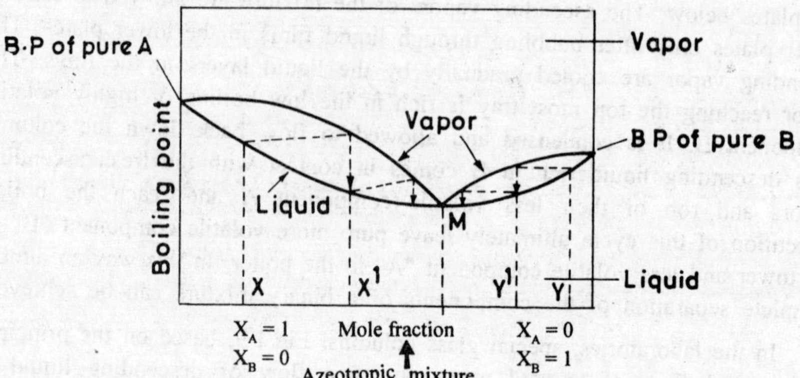

Fig. 4-10 Fractional distillation diagram of Type II

Consider the distillation of a mixture of composition *X*. The distillate of first fraction has a composition X_1 and the liquid residue gets richer in A. Repeated distillation of these fractions ultimately yield pure liquid A and a constant boiling mixture. Similarly when we start with a liquid of composition *Y*, lying in the other lobe, we end up with pure liquid B and azeotropic mixture (constant boiling mixture) as the ultimate fractions. Pure components can never be obtained by any number of distillations of the azeotropic mixtures. In either case the azeotropic mixture remains in the final residue in the flask. The composition and boiling (at 1 atm pressure) of some minimum boiling azeotropic mixtures is given Table 4-4.

Table 4-4 Composition and boiling points of some azeotropic mixtures

Component A	*B.Pt.*	*Component B*	*B.Pt*	*Azeotropic mixture*	
				A (Mole %)	*B.Pt*
MINIMUM BOILING AZEOTROPIC MIXTURES :					
Ethyl acetate	77.10°C	Water	100°C	69.5	70.40°C
Ethanol	78.26°C	Water	100°C	89.4	78.15°C
Benzene	80.20°C	Water	100°C	91.2	69.25°C
Chloroform	61.20°C	Water	100°C	97.2	56.12°C
CCl_4	76.50°C	Ethanol	78.26°C	60.3	64.95°C
MAXIMUM BOILING AZEOTROPIC MIXTURES :					
HCl	–20°C	Water	100°C	20.24	108.50°C
Chloroform	61.20°C	Acetone	56.5°C	80	64.70°C

4-5-3 Fractional distillation of binary miscible solutions of Type III

Solutions showing negative deviation from Raoult's law give a maximum boiling mixture of a fixed composition, Fig. 4-11.

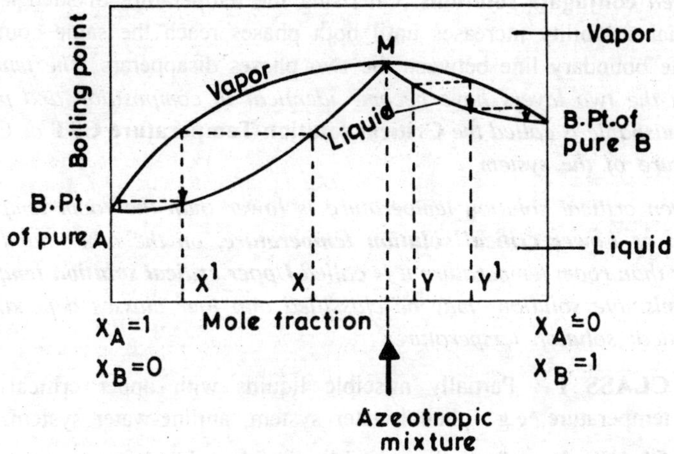

Fig. 4-11 Fractional distillation diagram of type III

Extending the same argument as for type II system we can draw the following conclusions in case of type III systems.

(i) The ultimate distillation fraction of left lobe yields two fractions - a constant boiling mixture with maximum boiling point in the distillate and pure liquid A in the residual liquid.

(ii) The ultimate distillation fraction of right lobe also yields two fractions - a constant boiling mixture with maximum boiling point in the distillate and pure liquid B in the residual liquid.

In either case a maximum boiling azeotropic mixture is obtained in the distillate. Physical data of two of the azeotropes of this class are presented in Table 4-4.

4-6 PARTIALLY MISCIBLE LIQUIDS

All the while we were discussing only about completely miscible liquids. But in reality majority of liquid pairs are only partially miscible. The phenomenon of partial miscibility arises by mixing markedly different liquids that exhibit large positive deviations from ideal behaviour. For example, when ethanol is added to water at room temperature it dissolves in all proportion without the formation of two separate layers. On the other hand when ethanol is replaced by phenol, we notice an initial solubility soon heading for seapration of liquid mixture into two separate layers. a continuous monitoring of concentration of both the liquids from complete miscibility to partial miscibility suggest the arrival of saturation of limit. The layer rich in phenol is a saturated solution of water in phenol, while the other layer is a saturated solution of phenol in water. Such *solutions of different compositions coexisting with one another are termed* **conjugate solutions**. On rising the temperature of such a mixture, the mutual solubility increases until both phases reach the same composition, where the boundary line between the two phases disapperars. *The temperature at which the two layers have become identical in composition and physically indistinguishable is called the* **Critical solution Temperature CST** *or Consolute temperature of the system.*

When critical solution temperature is lower than the room temperture it is known as ***Lower critical solution temperature,*** *on the other hand when it is higher than room temperature it is called* ***Upper critical solution temperature.*** *Partial miscible solutions may be classified into four classes depending upon their critical solution temperatures.*

CLASS I : Partially miscible liquids with upper critical solution temperature, e.g., phenol-water system, aniline-water system.

CLASS II : Partially miscible liquids with lower critical solution temperature, e.g., triethylamine-water system, 1-methyl piperidine-water system.

CLASS III : Partially miscible liquids with upper and lower critical solution temperatures, e.g., m-toluidine-glycerol system, nicotine water system.

CLASS IV : Partiablly miscible liquids with no critical solution temperature, e.g., ethyl ether-water system.

CLASS I : Partially miscible liquids with Upper CST : When roughly equal amounts of aniline and water are mixed at room temperature they give two separate layers-upper layer is saturated with aniline in water while the lower layer with water in aniline. The composition of both these layers are quite different and may be determined experimentally at any given temperature. Similar measurements are possible at other temperatures also. The variation of mutual solubility for this liquid pair with temperature is illustrated in Fig. 4.12.

The area enveloped by the dome shaped curve represent a biphasic liquid-liquid equilibria. All systems whose composition lie out side this dome shaped curve consist of only one liquid layer.

If we mix aniline and water to give a mixture with composition of X and shake it well, it immediately separates into two layers with composition represented by A and B lying on the tie line. The ratio of the distances AX and BX gives the relative amounts of both the immiscible layers. By the addition of sufficient amount of water corresponding to composition A, the two phase equilibria vanishes to give a single layer of unsaturated solution of aniline in water. Similarly addition of excess of aniline to represent the composition of point *b* give an unsaturated solution of water in aniline.

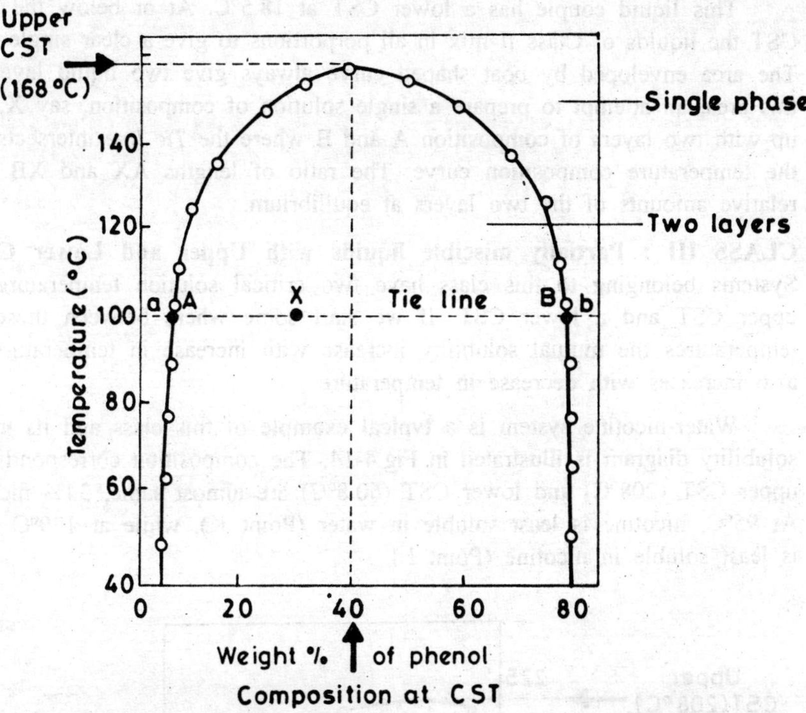

Fig. 4-12 Partially miscible liquids with upper CST

CLASS II : Partially miscible liquids with Lower CST : Although above discussed systems with *upper CST* are of more common occurrence, there are few systems with *lower CST.* In these systems the mutual solubility decreases with temperature. Water- triethylamine is one such system whose temperature-composition curve is illustrated in Fig. 4-13.

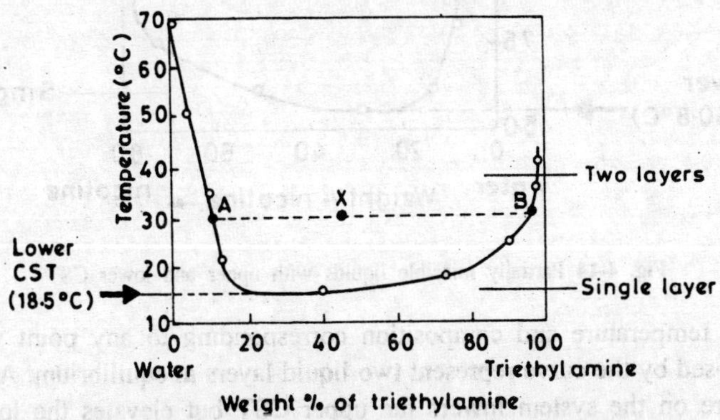

Fig. 4.13 Partially miscible liquids with lowerr CST

This liquid couple has a lower CST at 18.5°C. At or below the lower CST the liquids of Class II mix in all porportions to give a clear single layer. The area enveloped by boat shaped curve always give two liquid layers. In this area, an attempt to prepare a single solution of composition, say X, ends up with two layers of composition A and B where the *Tie line* intersects with the temperature composition curve. The ratio of lengths AX and XB gives relative amounts of the two layers at equilibrium.

CLASS III : Partially miscible liquids with Upper and Lower CST : Systems belonging to this class have two critical solution temperatures, an upper CST and a lower CST. If we start some where between these two temperatures the mutual solubility increase with increase in temperature and also increases with decrease in temperature.

Water-nicotine system is a typical example of this class and its mutual solubility diagram is illustrated in Fig.4-14. The composition corresponding to upper CST (208°C) and lower CST (60.8°C) are almost same, 34% nicotine. At 95°C, nicotine is least soluble in water (Point K), while at 130°C water is least soluble in nicotine (Point L).

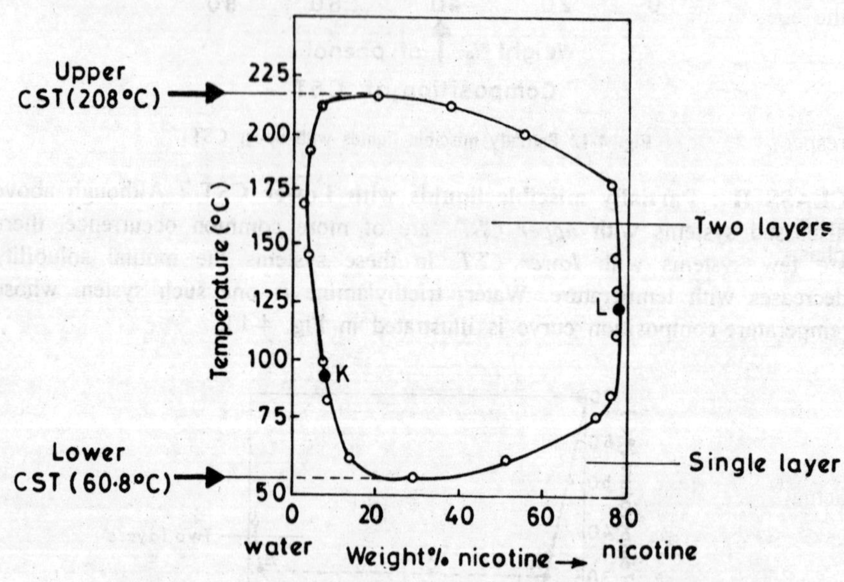

Fig. 4-14 Partially miscible liquids with upper and lower CST

The temperature and composition corresponding to any point within the area enclosed by the curve represent two liquid layers at equilibrium. Application of pressure on the system lowers the upper CST but elevates the lower CST. At a suitable external pressure the upper CST and lower CST merge into a single point and the two layer equilibrium vanishes. Therefore at sufficiently high pressures the liquids are completely miscible at all temperatures in all proportions.

Class IV : Partially miscible solutions without CST : Ethyl ether-water system forms a typical example of this class. This system has neighter an upper nor leads critical solution temperature. Lowering the temperature of the mixture, leads to solidification of the liquid and any attempt to raise the temperature results in the evaporation of the liquid. The two liquids are therefore only partially soluble in each other at all the temperatures over which the solution exists.

4-7 COMPLETELY IMMISCIBLE LIQUIDS

Theoretically there are no two liquids that are absolutely immiscible. The mutual solubility of liquid pairs like mercury and water or chlorobenzene and water is so small that they are regarded as completely immiscible liquid pairs. Their coexistence does not affect the properties of either liquids. Thus the vapour pressure of any liquid may be regarded as independent of other liquid/s.

Therefore the total vapor pressure (P) is the sum of vapour pressure of the constituent liquids.

$$P = p_A + p_B \tag{4.4}$$

where p_A and p_B are the vapor pressures of constituents A and B respectively at the same temperature condition.

Vapour pressure of any substance at a given tiemperature is directly porportional to the number of moles of the substance present in the vapor phase.

$$p_A \propto n_A \text{ and } p_B \propto n_B$$

or

$$\frac{n_A}{n_B} = \frac{p_A}{p_B} \tag{4.5}$$

But number of moles of any substance is equal to the ratio between its actual weight and its molecular weight.

$$n_A = \frac{w_A}{M_A} \text{ and } n_B = \frac{w_B}{M_B} \tag{4.6}$$

where w_A and w_B are the weights of A and B *in the vapor.*

Substituting for n_A and n_B in equation (4.5)

$$\frac{w_A}{M_A} \times \frac{M_B}{w_B} = \frac{p_A}{p_B}$$

or

$$\frac{w_A}{w_B} = \frac{p_A}{p_B} \times \frac{M_A}{M_B} \qquad (4.7)$$

A pure liquid boils when its vapor pressure is equal to the external pressure (1 atm in case of normal boiling point).

Since both the constituent liquids together contribute to the overall vapor pressure than either liquid making their contribution singly, we can infer that a mixture of two or more immiscible liquids of this class must boil at a lower temperature than the constituent liquids. Further, since the vapor pressure of a liquid is independent of amount of a liquid, the liquid mixture must have a single boiling point till at least one of the constituent is distilled off.

These principles gave birth to a new technique called **steam distillation** and used for separation and purification of organic compounds. This process is useful when,

(i) The organic compound is immiscible or almost immiscible in water.

(ii) The compound has high boiling point.

(iii) The compound either dissociates or undergoes some chemical transformation at normal boiling point.

The process of steam distillation involves passing steam through an organic liquid that is immiscible in water. In principle, this is similar to boiling the organic liquid with water forming an immiscible mixture. Bubbling steam in place of boiling with water helps in keeping the liquid mixture in a state of agitation. Under these conditions the total vapor pressure over the mixture is equal to the sum of the pressure of steam (P_s) and organic liquid (P_o)

$$P = P_s + P_o \qquad or \qquad P_o = P - P_s$$

This explains why the organic liquid-water mixture boils at a low temperature than the boiling points of the either liquids. For example, water and iodobenzene at 760mm (1 atm) pressure boil at 100°C and 188°C respectively, while its mixture boils at 98°C. At the boiling point of the mixture the vapor pressure of water and iodobenzene are 712 mm Hg and 48mm Hg respectively.

Total pressure = 712 + 48 = 760 mm Hg

The molecular weights of water and iodobenzene are 18 and 204 respectively. This data is useful in finding the relative amounts of both the constituents in the vapor and hence in the distillate.

$$\frac{W_{IB}}{W_{H_2O}} = \frac{P_{IB}}{P_{H_2O}} \times \frac{M_{IB}}{M_{H_2O}} = \frac{48 \times 204}{712 \times 18} = 0.764$$

Therefore 0.764g of iodobenzene distills off with 1g of steam passed through the system. In a reverse way *determination of composition of the distillate allows one to calculate the molecular weight of unknown liquid.*

A typical steam distillation set up is illustrated in Fig 4-15.

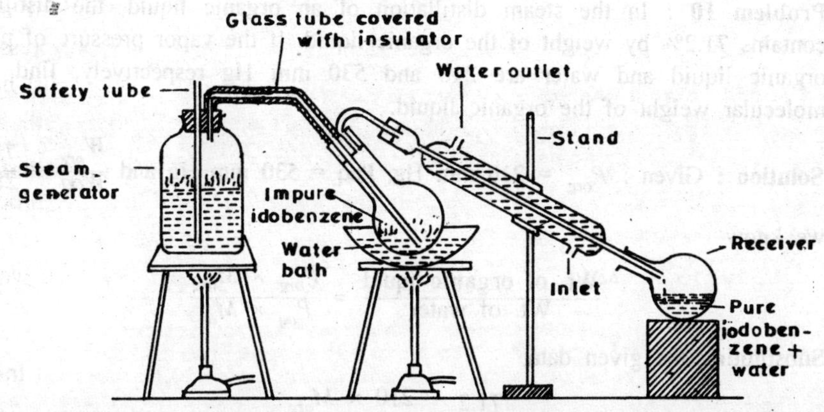

Fig 4-15 Steam distillation of iodobenzene

The flask containing impure liquid to be purified is first heated on a water bath. Now steam from the steam generator is slowly passed into the impure liquid. Steam carrying vapours of pure organic liquid are cooled by the condenser and collected in the receiver. The distillate in the receiver separates out into an organic layer and an aqueous layer. At the end of the distillation the pure liquid is separated from water with the help of a separating funnel. Traces of water dissolved or sticking to the organic liquid are then removed by shaking with anhydrous calcium chloride or any other dehydrating agent.

Problem 9 : Nitrobenzene can be distilled with steam under 1 atm pressure at a temperature of 99.2°C. Calculate the amount of steam necessary of distil 100g of nitrobenzene. The vapor pressure of water at 99.2°C in 739 mm Hg.

(O.U., Mar. 1990)

Solution : Given : P = 760 mm Hg, W_{NB} = 100 g

$p_{H_2O} = 739 \qquad M_{NB}\ (C_6H_5NO_2) = 123 \qquad M_{H_2O} = 18$

$$p_{NB} = P - p_{H_2O} = 760 - 739 = 21 \text{ mm Hg}$$

General expression for weight ratio of two immiscible is

$$\frac{W_A}{W_B} = \frac{p_A M_A}{p_B M_B}$$

$$\frac{100}{W_{H_2O}} = \frac{21 \times 123}{739 \times 18}$$

$$W_{H_2O} = \frac{739 \times 18 \times 100}{21 \times 12} = 515 \text{ g}$$

Weight of steam required to distill 100 g of intro benzene = 515g

Problem 10 : In the steam distillation of an organic liquid, the distillate contains 71.2% by weight of the organic liquid. If the vapor pressure of pure organic liquid and water are 210 and 530 mm Hg respectively, find the molecular weight of the organic liquid.

Solution : Given : P_{org} = 210 mm Hg, Paq = 530 mm Hg and $\frac{W_{org}}{Waq} = \frac{71.2}{28.8}$

we know

$$\frac{\text{Wt. of organic liquid}}{\text{Wt. of water}} = \frac{P_{org} \times M_{org}}{P_{aq} \times M_{aq}}$$

Substituting the given data,

$$\frac{71.2}{28.8} = \frac{210 \times M_{org}}{530 \times 18}$$

$$\text{Molecular weight of organic liquid} = \frac{71.2 \times 530 \times 18}{28.8 \times 210} = 112.3$$

Problem 11 : Nitrobenzene ($C_6H_5NO_2$) can be steam distilled under 1 atm pressure at 372.2 K. If the vapor pressure of water at 372.2 K in 739 mm, calculate the weight percent ratio of nitrobenzene.

Solution : Given : P = 1 atm = 760 mm Hg, p_{H_2O} 739 mm at 372.2 K

M_{H_2O} = 18 $M_{\text{nitrobenzene}}$ = 123

we know P = $p_{org} + p_{H_2O}$ or $p_{org} = p - p_{H_2O}$

Vapour pressure of nitrobenzene = 760 - 739 = 21 mm Hg

Substituting this data in equation (4.7)

$$\frac{W_{NB}}{W_{H_2O}} = \frac{21}{739} \times \frac{123}{18} = 0.194$$

$$\text{Weight percent of nitrobenzene} = \frac{0.194}{0.806} \times 100$$

Weight percent of nitrobenzene = 24%

Problem 12 : Terpinene (M.Wt = 136) forms immiscible mixtures with water and can be steam distilled under 1 atm pressure and 97°C. If pure terpinene has a vapor pressure of 110 mm, calculate the amount of steam needed to collect 0.5g of the pure sample from its crude.

Solution : Given : $M_{\text{terpinene}} = 136, \quad M_{H_2O} = 18$

$P_T = 110 \text{ mm} \qquad P_{H_2O} = 760 - 110 = 650 \text{ mm}$

$Wt = 0.5 \text{ kg}$

Substituting these values in equation (4.7)

$$\frac{0.5}{W_{H_2O}} = \frac{110 \times 136}{650 \times 18}$$

$$W_{H_2O} = \frac{0.5 \times 650 \times 18}{110 \times 136} = 0.391$$

Problem 13 : A mixture of chlorobenzene and wate rboil at 90.3°C at an external pressure of 740 mm. The vapor pressure of water at 90.3°C is 630.1 mm. Calculate the weight composition of the distillate.

(N.U., Apr. 1990)

Given : $\quad P = 740 \text{ mm Hg} \qquad p_{H_2O} = 630.1 \text{ mm Hg},$

$p_{\text{chlorobenzene}} = 18 \qquad M_{\text{chlorobenzene}} = 112.5$

$$\frac{\text{Wt. of chlorobenzene}}{\text{Wt. of water}} = \frac{109.9 \times 112.5}{630.1 \times 18} = 0.92$$

Therefore weight composition of distillate = 0.92

4-8 NERNST'S DISTRIBUTION LAW

Nernst (1891) observed that *when a solute is added to two immiscible or almost immiscible liquids and shaken thoroughly and left aside until the two liquids separate, the molar concentration of the solute in both the layers bear a fixed ratio at any given temperature.* This observation holds goods as long as the solute does not associate or dissociate or undergo any structural transformation in either of the solvents. Further, this relation is satisfactory, for only dilute solutions. The ratio of molar concentrations, called **distribution constant** is independent of amount of solute taken initially. For example, when a solute is placed between an organic solvent-water pair the distribution coefficient may be mathematically expressed as;

$$\text{Distribution or partition coefficient } K_D = \frac{C_{org}}{C_{aq}} \tag{4.6}$$

Distribution law or Partition law may thus be stated that *a solute distributes itself between two immiscible or almost immiscible liquid pair in such a way that the ratio of the concentrations of the solute in the two solvents is always constant at constant temperature, provided the solute does* *not suffer any chemical transformation and remains in the some molecular state in either solvent.*

When iodine is placed between an immiscible solvent pair water and carbon tetrachloride, its distribution pattern is expressed in Fig. 4-16.

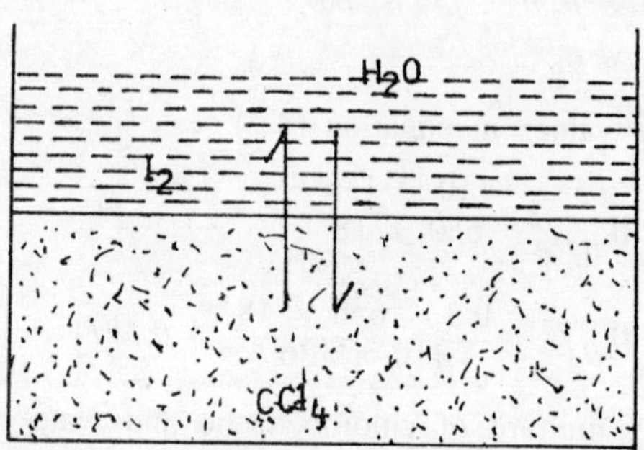

Fig. 4-16 Distribution of for I_2 between CCl_4 & H_2O

4-8-1 Condition for the validity of Distribution Law

The essential conditions for the distribution law to hold good may be summarized as;

(i) **Mutual immiscibility :** The solvent pair should be immiscible or almost immiscible and the extent of mutual solubility should not alter by the addition of solute.

(ii) **Constant temperature :** Solubility of a solute in both the solvents is not changed to the same extent with a change in temperature. Hence it is necessary to keep the temperature constant during the estimation of the partition coefficient.

(iii) **Low concentration of the solute :** Distribution law is strictly obeyed only by dilte solutions. Hence this law holds best throughout the concentration range, if the solute is sparingly soluble in both the solvents.

(iv) **Molecular state :** The solute should exist in the same molecular state in either solvent; that is, the solute should not undergo association or dissociation or any structural deformation or involve in any chemical combination with the solvent. In case of any association or dissociation in any of the solvent, the distribution ratio expression should be properly modified to accommodate these changes.

(A) WHEN THE SOLUTE UNDERGOES ASSOCIATION IN ONE OF THE PHASES : In case of distribution of benzoic acid between benzene and water, in benzene layer benzoic acid exists in the form of a dimer. To accommodate this fact the distribution ratio expression (4.8) has to be modified as follows.

Consider a solute A which associated in solvent I giving polymeric species A_n and exists in normal molecular state in solvent II, Fig. 4-17. If total concentration of solute in solvent I is C_1 and α is the degree of association, then

	nA $\rightleftharpoons$	A_n
Initially	C_1	0
At equilibrium	C_1 (1-α)	$C_1(\alpha n)$

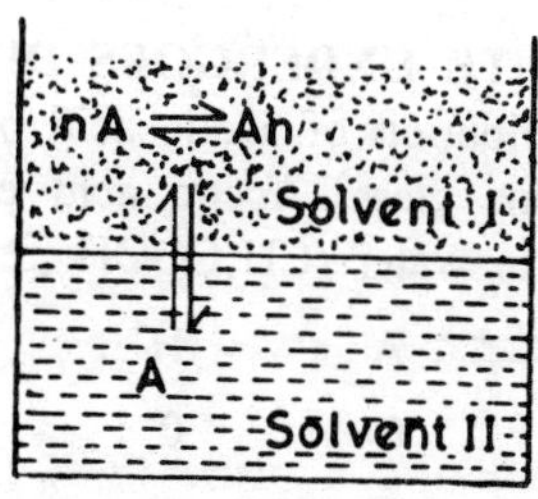

Fig. 4-17 Association in one of the solvent

Equilibrium constant of association equilibrium $= \frac{[A_n]}{[A]^n}$

$$K_{eq} = \left[\frac{C_1(\alpha / n)}{C_1\ (1-\alpha)^n}\right]$$

or
$$C_1\ (1-\alpha) = \left[\frac{C_1\alpha}{K_{eq}n}\right]^{1/n}$$

Now, distribution coefficient

$$K'_D = \frac{C_1\,(1-\alpha)}{C_2} = \frac{(C_1\ \alpha\ / K_{eq}\ n)^{1/n}}{C_2}$$

Since α, K_{eq} and n are constants at a given temperature, their ratio will be another constant.

$$K_D' = \text{constant.}\ \frac{(C_1)^{1/n}}{C_2}$$

or
$$\frac{(C_1)^{1/n}}{C_2} = \frac{K_D'}{\text{constant}} = K_D \qquad (4.9)$$

Therefore, for this typc of system $\frac{(C_1)^{1/n}}{C_2}$ will be constant. Taking logarithms on both sides, we have,

$$\frac{1}{n}\log C_1 = \log K_D + \log C_2$$

$$\log C_1 = n \log C_2 + n \log K_D \quad (4.10)$$

Hence, a plot of $\log C_1$ vs $\log C_2$ gives a straight line having a slope n.

For the benzoic acid example discussed above, association constant of benzoic acid in benzene is $n = 2$ and the distribution coefficient may be given by the expression,

$$K_D = \frac{\sqrt{C_{benzene}}}{C_{water}} \quad (4.11)$$

(B) WHEN THE SOLUTE UNDERGOES DISSOCIATION IN ONE OF THE SOLVENT : Distribution of acetic acid between benzene and water is a typical example of this class. Acetic acid exists in the normal state is benzene and dissociates in water ($CH_3COOH \rightleftharpoons CH_3COO^- + H^+$), Fig. 4.18

	HA $\rightleftharpoons$	A^- +	H^+
Initially	C_2	0	0
At equilibrium	$C_2(1\text{-}\alpha)$	$C_2\alpha$	$C_2\alpha$

$$K_{eq} = \frac{(C_2\,\alpha)^2}{C_2\,(1-\alpha)} \text{ or } C_2\,(1-\alpha) = \frac{(C_2\,\alpha)^2}{K_{eq}}$$

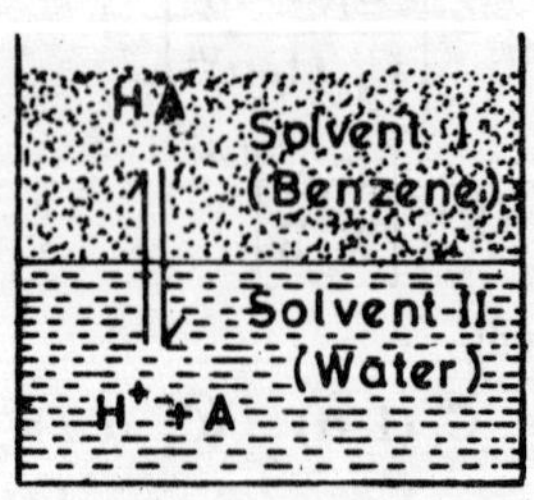

Fig. 4-18 Dissociation in one of the solvent

Now, distribution coefficient, K_D', may be written as

$$K_D' = \frac{C_1}{C_2(1-\alpha)} = \frac{C_1}{C_2^2\,\alpha^2 / K_{eq}} = \frac{C_1}{C_2^2} \cdot \frac{K_{eq}}{\alpha^2}$$

$$K_D' = \frac{C_1}{C_2^2} \text{ . constant}$$

$$\frac{C_1}{C_2^2} = \frac{K_D'}{constant} = K_D \qquad (4.12)$$

Therefore when a molecule dissociates into two fragments the distribution coefficient is given by $\frac{C_1}{C_2^2}$. A more general form of the expression (4.12) is,

$$K_D = \frac{C_1}{C_2^n} \qquad (4.13)$$

Hence, a plot of log C_1 vs log C_2 gives a straight line with a slope equal to n.

4-8-2 Applications of Distribution Law

1. To study the molecular aggregation : As mentioned in the force gone discussion, association or dissociation in one or both the solvents could be easily identified from the constancy of the distribution law expression used. For a comparison some of the standard distribution law expressions are listed in Table 4-5.

Table 4-5 Some standard distribution law expressions

Total concentration		*Distribution law*	*Molecular state*	
Solvent-I	*Solvent-II*	*expression*	*Solvent-I*	*Solvent-II*
C_1	C_2	$\frac{C_1}{C_2}$	Normal	Normal
C_1	C_2	$(C_1)^{1/n}/C_2$	Associated	Normal
C_1	C_2	$C_1/[C_2\ (1-\alpha)]$	Normal	Dissociated
C_1	C_2	$\frac{C_1\ (1-\alpha)}{C_2\ (1-\beta)}$	Dissociated	Dissociated
C_1	C_2	$\frac{C_1}{C_2\ \alpha}$	Normal	Solvated

2. In the extration of organic compounds : A knowledge of distribution coefficients is helpful in deciding the solvent to be chosen for best extraction yields. For example, when a synthetic or natural compound is present in an aqueous extract, we have an option of extracting with wide range of organic solvent like benzene, ether, carbon tetrachloride, carbon disulphide, ethyl acetate, chloroform, etc. A knowledge of distribution coefficient of the organic solute allows us to select the solvent to get optimum extraction. In the extraction process, cost factor of the solvent needs little attention, since after extraction it is fully recoverable by simple distillation for subsequent usage.

The second factor that needs attention is the volumes of the extracting solvent to be used. Best extraction results are obtained by using the given solvent in small amounts and multiple extractions instead of a single lot extraction. This fact is demonstrated in the following example.

Let W g of solute present in 1 litre of water need be extracted with 1 litre of ether. Further let the distribution ratio be 2. Ether can be used in one lot or small multiple lots.

(a) When ether is used in one lot :

Amount of substance extracted - X g

Amount of substance left in water = $(W - X)$ g

Concentration of substance is aqueous layer = $(W - X)/1000$ Conc. of substance in ether layer = $X/1000$

$$\text{Therefore } K_D = 2 = \frac{(X/1000)}{(W-X)/1000} = \frac{X}{W-X}$$

$$2W - 2X = X$$

or

$$X = 2/3\ W$$

This means that around 67% of the original substance is extracted when used in one lot.

(b) When used in two installments of 500mL each:

Let amount extracted in first instalment = X_1

Amount left in aqueous layer = $W - X_1$

Concentration of substance in ether layer = $X_1/500$

Concentration of substance in aqueous layer = $(W-X_1)/1000$

$$K_D = 2 = \frac{X_1/500}{(W-X_1)/1000} = \frac{2X_1}{W-X_1}$$

$$2W - 2X_1 = 2X_1$$

$$X_1 = \frac{1}{2}W$$

That is 50% of the substance is extracted in the first extraction.

Amount left in aqueous layer after 1st extraction = $\frac{1}{2}W$

Let amount extracted in to ether in 2nd extraction = X_2

Amount left in water after 2nd extraction = $\frac{1}{2}W - X_2$

$$K_D = 2 = \frac{X_2/500}{(\frac{1}{2}W - X_2)/1000} = \frac{2X_2}{\frac{1}{2}W - X_2}$$

or

$$W - 2X_2 = 2X_2$$

$$X_2 = \frac{1}{4} W$$

Total amount extracted with 1 Litre ether in two installments of 500 mL

$$\text{each} = \frac{1}{2} W + \frac{1}{4} W = \frac{3}{4} W$$

That is 75% of the solute is extracted by extracting in two installments, which is around 8% larger than the extraction amount in one instalments (67%). Extending this argument we can state that further increase of instalments allow us to extract more amount of solute.

Therefore the conditions for efficient extraction may be stated as:

(i) Multiple extraction using smaller amounts of extracting solvent allow us to achieve maximum extraction.

(ii) Extracting solvent should have maximum distribution ratio.

3. Distribution indicators : Iodine solutions are usually titrated with hypo employing starch indicator. If the iodine concentration in an aqueous solution is too small it becomes very difficult to detect the disappearance of blue color at the end point. Under these circumstances few drops of carbon disulphide is added to the iodine solution. Because of very high K_D value ($K_D = 400$), iodine immediately gets distributed in favor of CS_2 and turn the shining organic liquid drops deep blue. This facilitates the detection of presence of very weak solutions of iodine in the solution and make it possible to titrate. Here CS_2 is called a **distribution indicator**.

4. Study of chemical equilibria : Distribution law is successfully employed to recognize the presence of complex ions and to estimate the equilibrium constants involving them. For example, consider equilibrium,

$$KI + I_2 \rightleftharpoons KI_3$$

A linear variation of partition coefficient with the concentration of KI added to aqueous layer, for the distribution of iodine between CCl_4 and water, suggests a positive interaction between I^- and I_2 to give a complexion I_3^-, Fig 4-19.

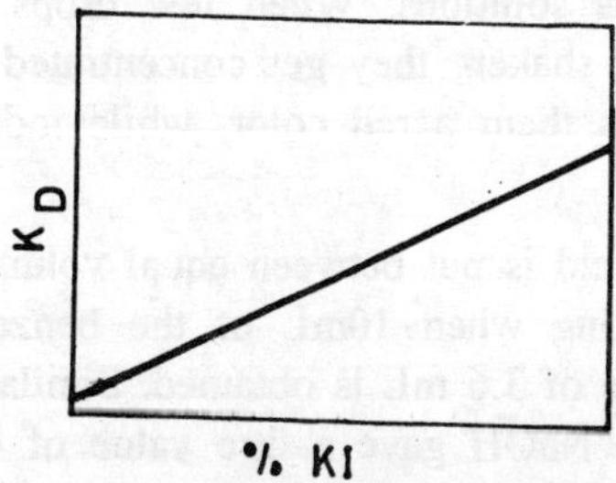

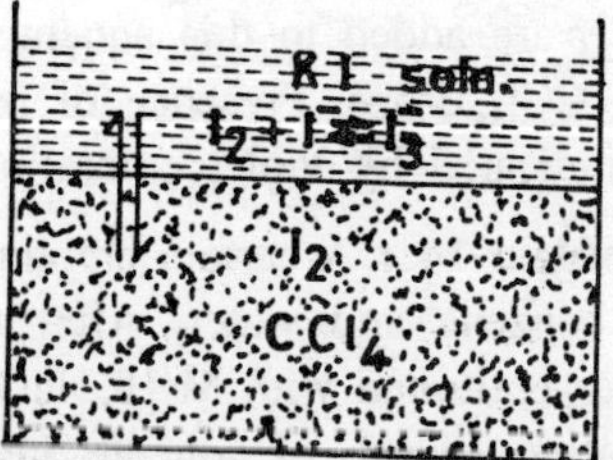

Fig. 4-19 Variation of K_D with % KI

The intercept of such a plot gives the partition coefficient of I_2 between water (0% KI) and CCl_4. Let *a* mol L^{-1} of iodine is distributed between water and CCl_4. Further let the concentration of iodine in CCl_4 layer be *b* mol L^{-1}.

$$K_D = \frac{\text{Concentration of free } I_2 \text{ in aq.layer}}{\text{Concentration of } I_2 \text{ in} CCl_4}$$

$$K_D = \frac{\text{Concentration of free } I_2 \text{ in aq.layer}}{b}$$

Therefore concentration of free I_2 in water = bK_D mol L^{-1}

Now at any concentration *"c"* of KI let the concentration of iodine in aq layer be *"d"*.

The value *"d"* is actually sum of contributions due to complex ion KI_3 and free iodine on account of distribution in pure water.

Total $I_2 = d \quad = [KI_3] + [\text{free } I_2]$

$\quad\quad\quad\quad = \{KI_3\} + b\ K_D$

Equilibrium concentration of KI_3 in aqueous layer = $d - b\ K_D$

Equilibrium concentration of KI in aqueous layer = $c - (d - b\ K_D)$

Equilibrium concentration of I_2 in aqueous layer = $b\ K_D$

$$K_{eq} = \frac{[KI_3]}{[KI]\,[I_2]} = \frac{(d - b\,K_D)}{[c - (d - bK_D)](bK_D)}$$

Different [KI] values and corresponding $[KI_3]$ can be obtained from the Fig (4.19). This allows the calculation of average K_{eq}.

Similar estimation of equilibrium constants are possible for the following equilibria;

$$KBr + Br_2 = KBr_3$$

$$CuSO_4 + 4NH_3 = [Cu(NH_3)_4]SO_4$$

5. Identification of bromide or iodide : Bromide and iodide salts are oxidized to bromine and iodine respectively by the addition of few drops of chlorine water. The oxidized forms, Br_2 or I_2, both give brown colored solutions and are difficult to distinguish in dilute aqueous solutions. When few drops of CCl_4 are added to this aqueous solution and shaken, they get concentrated in the CCl_4 drops. Presence of bromine imparts them a red color, while iodine is indicated by violet coloration.

Problem 14 : A certain amount of benzoic acid is put between equal volumes of benzene and water. After thorough shaking when 10mL of the benzene layer is titrated with M/10 NaOH a titre value of 3.6 mL is obtained. Similarly titrating 20 mL of aqueous layer with M/50 NaOH gave a titre value of 4.2 mL. Calculate the partition coefficient of this system.

Let C_B be the concentration of benzoic acid in benzene layer.

$$10 \times C_B = (M/10) \times 3.6$$

$$C_B = \frac{3.6}{100}\ M = 0.036\ M$$

Let C_W be the concentration of benzoic acid in aqueous phase.

$$20 \times C_w = (M/50) \times 4.2$$

$$C_w = \frac{4.2}{1000}\ M = 0.0042\ M$$

$$\text{Distribution coefficient of benzoic acid} = \frac{\sqrt{C_B}}{C_W} = \frac{\sqrt{0.036}}{0.0042} = 45.2$$

Problem 15 : The partition coefficient of $FeCl_3$ between HCl and ether is 20, the solubility being in favor of ether. If 10mL of HCl layer containing 1g of $FeCl_3$ is extracted twice with 5mL portions of ether, what is the amount of $FeCl_3$ extracted in the first and second extractions. Also find the amount of $FeCl_3$ extracted if all the 10mL is used in one lot.

Since the solubility is in the favor of ether,

$$\text{Partition coefficient, } K_D = \frac{\text{Conc. of } FeCl_3 \text{ in ether}}{\text{Conc. of } FeCl_3 \text{ In water}}$$

For 1st extraction : Let X_1 be the amount of $FeCl_3$ extracted by 5 mL of ether

$$20 = \frac{(X_1/5)\text{ g mL}^{-1}}{(1-X_1)/10\text{ g mL}^{-1}}$$

$$20 \times (1 - X_1)/10 = X_1/5$$

$$2 - 2X_1 = X_1/5$$

$$10 - 10X_1 = X_1$$

$$11X_1 = 10 \text{ or } X_1 = 0.91$$

Amount of $FeCl_3$ extracted in the 1st extraction = 0.91 g

Amount of $FeCl_3$ left in the acid solution = $1 - \frac{10}{11}$ = 0.09 g

For 2nd extraction : Let X_2 be the amount of $FeCl_3$ extracted by 5ml ether

$$20 = \frac{X_2/5}{(0.09-X_2)/10}$$

$$0.18 - 2X_2 = \frac{S_2}{5}$$

$$0.9 = 11X_2 \text{ or } X_2 = \frac{0.9}{11} = 0.08\text{ g}$$

Amount of $FeCl_3$ extracted in the second extraction = 0.08 g

When extracted with one lot of 10 mL : Let X_3 be the amount extracted by 10 mL

$$20 = \frac{(X_3 / 10)}{(1-X_3)/10} = \frac{X_3}{1-X_3}$$

$$20 - 20X_3 = X_3$$

$$X_3 = \frac{20}{21} = 0.95 \text{ g}$$

Amount of $FeCl_3$ extracted when extracted with 10 mL of ether in one lot = 0.95 g

This is less than the amount of $FeCl_3$ extracted (0.98 g) with the same volume of ether used in two lots.

Problem 16 : The molar concentrated of iodine distributed between CCl_4 and water are found to be as under.

$[I_2]$ in CCl_4 layer $\times 10^2$ =	4.8	9.6	12.0	17.6
$[I_2]$ in water layer $\times 10^4$ =	5.7	11.3	14.1	20.7

Show that the results are in accordance with distribution law and also comment on the molecular state iodine in both the layers.

Solution : If I_2 is in the same molecular state, the ratio of concentrations of iodine in both the layers $K_D = \dfrac{\text{Conc. of } I_2 \text{ in } CCl_4}{\text{Conc. of } I_2 \text{ in } H_2O}$, be constant.

$C_{CCl_4} \times 10^2$	4.8	9.6	12.0	17.6
$C_{H_2O} \times 10^4$	5.7	11.3	14.1	20.7
$K_D = \dfrac{C_{CCl_4}}{C_{H_2O}}$ =	84.2,	85.0,	85.1,	85.0

This indicates that the results are in accordance with distribution law and I_2 exists in normal state in both the layers.

Problem 17 : A monobasic acid is distributed between benzene and water. It exists in its normal state in benzene while dissociates in water. If the concentrations in benzene and aqueous layers are 1.44 g L^{-1} and 12 g L^{-1} and distribution coefficient is 0.16. Calculate the degree of dissociation and dissociation constant of the acid.

Solution : Given : C_{aq} = 12 g L^{-1} , C_{org} = 1.44 g L^{-1}

Since the distribution coefficient is a fraction, 0.16, the higher concentration term should appear in the denominator. When it dissociates in the aqueous layer, we have,

$$K_D = \frac{C_{org}}{C_{aq}\,(1-\alpha)} = \frac{1.44}{12(1-\alpha)} = 0.16$$

$$0.12 = 0.16\,(1-\alpha)$$

$$0.12 = 0.16 - 0.16\,\alpha$$

$$0.16\,\alpha = 0.04$$

$$\alpha = \frac{0.04}{0.16} = 0.25$$

Degree of dissociation, $\alpha = 0.25$

$$\text{Dissociation constant } K = \frac{\alpha^2 C}{1-\alpha} = \frac{(0.25)^2}{1-0.25} \times 12 \text{ g L}^{-1}$$

$$K = 1 \text{ g L}^{-1}$$

Problem 18 : In the distribution of a solute between water and chloroform, the following results are obtained.

C_1 in water	:	0.0160	0.0237
C_2 in chloroform	:	0.3380	0.7530

Calculate the molecular state of the solute in chloroform with the help of the data. (N.U., Apr. 1991)

Solution : Given :

C_1 in water	0.0160	0.0237
C_2 in chloroform	0.3380	0.7530
$\frac{C_1}{C_2}$	4.73×10^{-2}	3.15×10^{-2}
$\frac{C_1}{\sqrt{C_2}}$	2.75×10^{-2}	2.73×10^{-2}

Since C_1/C_2 ratios are varying the solute is not in the same molecular state in both the solvents. The constancy of $C_1/\sqrt{C_2}$ suggests that the solute dimerizes in chloroform.

Problem 19 : Experiments in the distribution of phenol between water and chloroform gave the following results :

Conc. in the aqueous Solution (C_1)	0.094	0.163	0.254
Conc. in the Chloroform (C_2)	0.254	0.761	1.850

What conclusion can be drawn from these results concerning the molecular condition of phenol in chloroform solution : **(N.U., Nov. 1990)**

C_1	0.094	0.163	0.254
C_2	0.254	0.761	1.850
$\frac{C_1}{\sqrt{C_2}}$	1.87×10^{-1}	1.87×10^{-1}	1.87×10^{-1}

The constancy of $C_1/\sqrt{C2}$ *suggests that phenol dimerizes in chloroform.*

4-9 NUMERICAL PROBLEMS

4-9-1 5 g of non volatile paraffin with molecular formula $C_{20}H_{28}$ is dissolved in 50 g of benzene. If the vapor pressure of pure benzene at the boiling point of the solution in 300 torr, find the vapour pressure of the solution.

[Hint : First calculate moles of paraffin and benzene, followed by determination of mole fraction. Since paraffin is a non- volatile solute, the vapor pressure of solution is equal to partial vapour pressure of solvent]

4-9-2 Ammonia is soluble in water at 20°C and 760 torr pressure to an extent of 518 g per 100 g of water. Calculate the solubility of ammonia when the barometer reads 1200 torr. Vapour pressure of pure water at 20°C is 20 torr.

4-9-3 At 50°C the vapour pressure of pure CCl_4 and pure chloroform are 317 torr and 526 torr respectively. Calculate the vapour pressure of a mixture prepared by mixing 20 g of CCl_4 and 30 g of $CHCl_3$ at 50° C.

4-9-4 20 g of an non-volatile organometallic compound of molecular mass 300 is dissolved in 50 g of chloroform at 50°C. If the vapour pressure of the pure solvent at 50°C is 526 torr, calculate the vapour pressure of the solution.

4-9-5 Heptane (C_7H_{16}) and octane (C_8H_{18}) have vapour pressures 791 and 352 torr respectively at 1100°C. Predict the pressure at which a mixture containing 50 g of each will boil at 100°C.

4-9-6 The following results are obtained for a study of distribution of benzoic acid between benzene and water.

Benzene layer (mol L^{-1})	$C_1 \times 10^2$	1.56	4.95	8.35	19.5
Water layer (mol L^{-1}),	$C_2 \times 10^3$	3.29	5.79	7.49	11.4

If benzoic acid exists in its normal state in water, establish the molecular state of benzoic acid in benzene.

4-9-7 The distribution coefficient in favour of water for the distribution of mercuric chloride in water and benzene is 13. If 10 mL of water containing 0.1 g of $HgCl_2$ is extracted with two lots of 25 mL of benzene each, find the total amount of $HgCl_2$ extracted. Compare this result with the extraction carried out by a single instalment of 50 mL of benzene.

4-9-8 An organic liquid, completely immiscible with water, is steam distilled under normal pressure conditions. At the distillation temperature of the mixture, the partial pressure of water is 725 torr. If the weight percent of the liquid in the distillate is 40, find the molecular weight of the organic liquid.

4-9-9 Aniline has a vapour pressure of 45 torr at the steam distilation temperature of 98°C. If the total vapour pressure at the temperature is 755 torr, predict the vapour composition of aniline.

4-10 MULTIPLE-CHOICE QUESTIONS

Mark the correct answer

4-10-1 Out of the following substance, maximum change in a colligative property at a given concentration, in aqueous medium is brought by,

A) Starch B) NaCl C) $CuSO_4$ D) $AlCl_3$

4-10-2 An ideal solution of a liquid is prepared by mixing equal mole fractions of two volatile liquids A and B. Then the composition of vapour,

A) is equal to 0.5
B) contains more of A
C) mole fraction of volatile vapors could be different
D) contains more of B

4-10-3 The partial vapour pressure of a liquid in a true solution is given by the expression

A) $X_A p_A^o$ B) X_A/p_A^o C) p_A^o/X_A D) $(X_A/X_B)\ p_A^o$

4-10-4 Purification by steam distillation is possible for

A) Ethyl alcohol B) Acetone C) Acetic acid D) Chlorobenzene

4-10-5 Nicotine-water system is a typical example for
A) Mixtures with upper CST
B) Mixtures witth lower CST
C) Mixtures with upper and lower CST
D) Mixtures that have no CST

4-10-6 Distribution expression when the solute dissociates in one of the solvent,

A) $\frac{C_1}{C_2}(1-\alpha)$, B) $\frac{C_1}{C_2}$, C) $\frac{\sqrt{C_1}}{C_2}$, D) $\frac{C_1\alpha}{C_2(1-\alpha)}$

4-10-7 Henry's law is obeyed by

A) Liquid-liquid equilibria B) Liquid-solid equilibria
C) Gas- liquid equilibria D) None of the above

4-10-8 A postive deviation is observed for the mixtures of

A) Water-Acetaldehyde (B) Water-Mercury
C) Water-Benzene D) Water-Sulphuric acid

4-10-9 Henry's law is strictly applicable at
A) High pressure of low temperature
B) Low pressure and low temperature
C) High pressure and high temperature
D) Low pressure and high temperature

4-11 SHORT AND ESSAY-TYPE QUESTIONS

4-11-1 State Raoult's law and derive the Raoult's equation. (O.U., Mar. 1992)

4-11-2 The binary mixture of ethanol and water can not be separated into pure components by fractional distillation, explain, giving temperature-composition diagram. (O.U., Mar. 1992)

4-11-3 State and explain Henry's law. What are its limitations?
(K.U., Apr. 1989, SKU., Apr. 1990, Sep. 1991, O.U., Oct 1991),

4-11-4 How the pressure of a gas affects its solubility in a liquid? (SKU., Sept. 1991)

4-11-5 Write note on

(a) Critical solution temperature

(b) Critical phenomena (A.U., Mar 1990)

4-11-6 What are azeotropes? Give two examples. Why an azerotrope is called a mixture and not a compound? (SKU., Apr. 1990, N.U., Apr. 1991 & A.U., Mar 1990)

4-11-7 What is vapour pressure? Write the equation for the determination of heat of vaporization of a liquid. (O.U., Oct. 1990)

4-11-8 Give the conditions for a solution to be considered as an ideal one. (O.U., Oct. 1990)

4-11-9 Write a note on steam distillation (SKU., Apr. 90, O.U., Mar. 1990)

4-11-10 Describe fully the distillation of a mixtures of two liquids which show a maximum in the B.P. Curve of their mixtures. (O.U., Oct. 1989)

4-11-11 State the distribution law. Mention any two of its applications. (A.U., Dec. 1989, N.U., Apr. 1990, O.U., Oct. 1989 & N.U., Apr. 1991)

4-11-12 State Henry's law. What is the effect of temperature? (K.U., Apr. 1990)

4-11-13 State Raoult's law. Explain the principles involved in the fractional distillation of completely miscible liquid mixtures (A.U., Jan. 1989)

4-11-14 Describe two partially miscible liquids whose mutual solubilities vary with temperature. (N.U., Nov. 1990)

ANSWERS

4-9-1 291 torr **4-9-2** 826g per 100g of water **4-9-3** p = 455 torr
4-9-4 453 torr **4-9-5** 584.6 torr **4-9-6** Exists in benzene in dimeric form
4-9-7 In two extractions = 0.03g In single extraction = 0.028g **4-9-8** 248.5
4-9-9 32.75% by weight of water.

4-10-1 D **4-10-2** C **4-10-3** A **4-10-4** D **4-10-5** C

4-10-6 A **4-10-7** C **4-10-8** A **4-10-9** D

Part IV

Practical Chemistry

VOLUMETRIC ANALYSIS

1.
ACIDIMETRY TITRATIONS

ACID-BASE REACTIONS

1-1 PREPARATION OF STANDARD SODIUM CARBONATE SOLUTION AND STANDARDIZATION OF HYDROCHLORIC ACID SOLUTION

Aim : *To prepare a standard solution of sodium carbonate and titrate it for the standardization of hydrochloride acid.*

Chemical Reaction : $Na_2CO_3 + 2HCl \longrightarrow 2NaCl + H_2O + CO_2$

1-1-1 Preparation of Standard Na_2CO_3 Solution (0.05 M)

Weigh accurately about 0.53 g of pure anhydrous Na_2CO_3 into a clean weighing bottle and transfer it to a 100 mL standard flask fitted with a funnel. Dissolve the solid in minimum quantity of distilled water and then make up the solution upto the mark. Shake thoroughly for about a minute or two.

$$X = \text{Wt. of } Na_2CO_3$$

$$\text{Molarity of } Na_2CO_3 = \frac{X}{M} \times 10. \quad M = \text{Mol. Wt. of } Na_2CO_3 \text{ (106)}$$

1-1-2 Standardization of Hydrochloric Acid Solution

Clean the burette with water and rinse it with distilled water and then rinse it with supplied hydrochloric acid solution. Now fill the burette with the hydrochloric acid solution. If air bubbles are formed they are removed by opening the stop cock. The burette is clamped to the burette stand in a vertical position. Note down the initial burette reading.

Clean the pipette with water and rinse it with distilled water and then rinse it with sodium carbonate solution. Pipette out 20 mL of standard Na_2CO_3 solution into the clean conical flask which is cleaned with water and rinsed with distilled water only. Add one or two drops of methyl orange indicator to this. Now the solution in the burette is slowly run down into the conical flask and the conical flask is rotated in a swirling motion. While approaching the end point, indicated by longer persistence of pink colour, addition of HCl should be done cautiously and carefully. Light permanent pink colour is obtained which indicates the end point. On the completion of the titration note down the final burette reading without parallax error.

Repeat the titration for concurrent readings. Tabulate the readings and calculate the molarity of HCl.

Sl. No.	*Volume of Na_2CO_3 taken*	*Burette reading*		
		Initial reading	*Final reading*	*Volume of HCl run down (Difference)*
1.				
2.				
3.				

Calculations :

$$\frac{M_1 V_1}{n_1} = \frac{M_2 V_2}{n_2}$$

M_1 = Molarity of Na_2CO_3 (from 1-1-1); M_2 = Molarity of HCl

V_1 = Volume of Na_2CO_3 (20 mL); V_2 = Volume of HCl (B.R.)

n_1 = Mole equivalence of Na_2CO_3 (= 1) n_2 = Mole equivelance of HCl (= 2)

So $M_2 = \frac{V_1 M_1 n_2}{n_1 V_2}$

Amount of the HCl = Molarity x M.Wt

= M_2 x 36.5 g L^{-1}.

1-2 DETERMINATION OF CARBONATE AND BICARBONATE IN A GIVEN MIXTURE

Chemical Reactions :

$Na_2CO_3 + 2HCl \longrightarrow 2NaCl + H_2O + CO_2$ (Total reaction)

$Na_2CO_3 + HCl \longrightarrow NaHCO_3 + NaCl$ (Phenolphthalein)

$NaHCO_3 + HCl \longrightarrow NaCl + H_2O + CO_2$ (Methyl Orange)

Principle : When phenolphthalein is used as an indicator it decolourizes as soon as sodium carbonate is half neutralized by HCl. The neutralization of the total bicarbonate (bicarbonate that was present initially in the mixture and the bicarbonate formed after half neutralization of carbonate present in the mixture) is monitored using methyl orange indicator.

1-2-1 Preparation of Standard Sodium Carbonate

Same as in experiment 1-1-1

1-2-2 Standardization of Hydrochloric Acid

Same as in experiment 1-1-2

1-2-3 Estimation of Na_2CO_3 and $NaHCO_3$ in a given mixture

Pipette out 20 ml of Na_2CO_3 and $NaHCO_3$ mixture into clean conical flask add one or two drops of phenolphthalein to it (better results are obtained with thymol blue-cresol red mixed indicator). It imparts pink colour. Note down the initial reading of the burette. Titrate the solution as described in 1-1-2, against hydrochloric acid until pink colour disappears. Note down the final reading of the burette. Let the volume be V_1 mL.

Now add one or two drops of methyl orange indicator to the same flask and add HCl till the contents turn yellow and note down the final reading. Let this volume be V_2 mL. Repeat the titrations until constant volumes of HCl are obtained. Record the values in a tabular form.

Sl. No.	*Volume Na_2CO_3 $NaHCO_3$ Mixture*	*Burette reading using Phenolpthalein*		V_1mL	*Burette reading using methyl orange*	V_1mL	*(V_2 - V_1)*
		Burette reading Initial	*Final*				
		R_1	$R2$	$R_2 - R_1$	*Final* R_3	$R_3 - R_2$	

Calculations :

I. The volume V_1 corresponds to half neutralization of Na_2CO_3.

$$Na_2CO_3 + HCl \longrightarrow NaHCO_3 + NaCl$$

Hence $2V_1$ is the volume of HCl required for the total neutralization of Na_2CO_3 in the mixture.

$$\frac{M_1V_1}{n_1} = \frac{M_2V_2}{n_2}$$

So $$M_2 = \frac{V_1M_1n_2}{n_1V_2}$$

M_1 = Molarity of HCl (from 1-1-2) M_2 = Molarity of Na_2CO_3

n_1 = Mole equivalence of HCl = 1 n_2 = Mole equivalence of Na_2CO_3 = 1

V_1 = $2V_1$ (Volume of HCl × 2) V_2 = Volume of Na_2CO_3 (20 mL)

II. Volume of HCl corresponding to $NaHCO_3$ alone = $(V_2 - 2V_1)$

$$\frac{M_3V_3}{n_3} = \frac{M_4V_4}{n_4}$$

Molarity of bicarbonater $M_4 = \frac{M_3V_3n_4}{n_3V_4}$

M_3 = Molarity of HCl (from 1-1-2 as M_1) M_4 = Molarity of $NaHCO_3$

$V_3 = V_2 - 2V_1$ (Volume of HCl) (B.R.) V_4 = Volume of mixture (20 mL)

n_3 = mole equivalence of HCl = 1 n_4 = mole equivalence of $NaHCO_3$ = 1

Amount of the Na_2CO_3 = Molarity × M. Wt.

= $M_2 \times 106$ g L^{-1}.

Amount of $NaHCO_3$ = $M_4 \times 84$ g L^{-1}.

1-3 ESTIMATION OF AMMONIUM SALT

Principle : Ammonium salts hydrolyse in the presence of NaOH to give ammonia. The ammonia so produced is removed from the reaction mixture by boiling the sample. From a knowledge of initial concentration of the alkali used and concentration of the unused alkali, NaOH concentration that is used up could be calculated which gives the concentration of ammonium salt.

Chemical Reactions :

$$NH_4Cl + NaOH \longrightarrow NaCl + H_2O + NH_3$$

$$HCl + NaOH \longrightarrow NaCl + H_2O$$

1-3-1 Preparation of Standard HCl Solution

Described in 1-1-2

1-3-2 Preparation of Standard NaOH Solution

Approximately 1.0 g (5 pellets) is dissolved in 250 mL of water and 20 mL of this solution is pipetted into a conicai flask. 2 Drops of phenolphthalein indicator is added to this solution.

A clean burette is rinsed with 0.1 M standard HCl solution and the alkali solution is titrated till it decolourizes. Titre value is noted and repeated for concurrent readings.

Calculations :

$$\frac{M_1V_1}{n_1} = \frac{M_2V_2}{n_2}$$

So $M_2 = \frac{V_1 M_1 n_2}{n_1 V_2}$.

M_1 = Molarity of HCl (from 1-3-1) M_2 = Molarity of NaOH

V_1 = Volume of HCl (B.R.) V_2 = 20 mL

n_1 = Mole equiv. of HCl (= 1) n_2 = Mole equiv. of NaOH (= 1)

1-3-3 Estimation of Ammonium Ion Concentration

Pipette out 20 mL of the unknown ammonium salt solution into a conical flask and add 100 mL of 0.1 M NaOH solution to it. Place a funnel in the neck of the flask in order to prevent any spilling, boil the solution until the vapour does not turn the paper moistened with mercurous nitrate solution black. Now bring the conical flask to room temperature and add two drops of methyl red indicator and titrate with standard 0.1 M HCl solution. Repeat the process for concurrent readings.

Calculations :

I. Titre value for 100 mL NaOH without participating in reaction with ammonium salt = V_1

II. Titre value of NaOH after participating in reaction with 20 mL of unknown ammonium salt solution = V_2

III. Titre value corresponding to 20 mL of ammonium salt solution = $V_3 = V_1 - V_2$

$$\frac{M_3 V_3}{n_3} = \frac{M_4 V_4}{n_4}$$

Molarity of ammonium salt $M_4 = \frac{M_3 V_3 n_4}{n_3 V_4}$

M_3 = Molarity of HCl (from 1-1-2 as M_1) M_4 = Molarity of NH_4^+

$V_3 = V_1 - V_2$ = Volume of HCl V_4 = Volume of ammonium salt (20 mL)

n_3 = Mole equivalence of HCl (= 1) n_4 = mole equiv. of NH_4^+ (= 1)

Concentration of ammonia = Molarity × Mol. Wt.

$= M_4 \times 17 \text{ g L}^{-1}$.

2.

REDOX TITRATIONS – I

PERMANGANOMETRY

Permanganate is a powerful and commonly available oxidizing agent with a reduction potential of 1.51V in acid medium. Since it is often contaminated with manganese dioxide and also undergoes auto-decomposition it can not be taken as a primary standard and needs to be standardized using sodium oxalate or arsenious oxide as primary standards.

2-1 PREPARATION OF STANDARD SODIUM OXALATE SOLUTION AND STANDARDIZATION OF POTASSIUM PERMANGANATE SOLUTION

Chemical Reactions :

$$2KMnO_4 + 3H_2SO_4 \longrightarrow K_2SO_4 + 2MnSO_4 + 3H_2O + 5O$$

$$5C_2H_2O_4^* + 5[O] \longrightarrow 5H_2O + 10CO_2$$

2-1-1 Preparation of Standard Sodium Oxalate Solution

Weigh about 0.67 g of pure sodium oxalate and then transfer it into a clean 100 mL standard flask fitted with a funnel. Dissolve the solid in minimum quantity of distilled water and then about 20 mL of dilute H_2SO_4 make up the solution up to the mark. Shake thoroughly for about one or two minutes

Molarity of sodium oxalate $= (X/M) \times 10$

X = Weight of sodium oxalate

M = M.Wt. of sodium oxalate (134)

2-1-2 Standardization of Potassium Permanganate Solution

Clean the burette with distilled water and rinse it with the given $KMnO_4$ solution and then fill it with the prepared permanganate solution. Note the initial reading in the burette. Rinse the pipette with sodium oxalate solution and pipette out 20 ml of sodium oxalate solution into a clean conical flask

---

* Sodium oxalate reacts with sulphuric acid to form oxalic acid which reacts with the liberated oxygen from the permanganate, therefore in almost all reactions, oxalic acid is given instead of sodium oxalate.

washed with distilled water. Add about 15 mL of dil. H_2SO_4 to it and heat up to 70-80° (i.e. till a bubble appears in the solution). Then add $KMnO_4$ from the burette into the warm solution of the mixture of sodium oxalate and sulphuric acid with continuous shaking in a swirling motion till the end point is reached, indicated by the appearnace of a permanent light pink color, is reached. Note the final reading of the burette. Repeat the titration till two concurrent readings are obtained. Tabulate the readings.

Note : *This titration cannot be carried out in presence of acids like HNO_3 and HCl, because the anions of these acids react chemically with $KMnO_4$ liberating NO_2 and Cl_2 gases respectively.*

Sl. No.	*Volume of sodium oxalate solution*	*Burette Readings*		
		Initial Reading	*Final Reading*	*Volume of $KMnO_4$ consumed (Difference)*

Calculations :

$$\frac{M_1V_1}{n_1} = \frac{M_2V_2}{n_2}$$

So $$M_2 = \frac{M_1V_1n_2}{n_1V_2}$$

M_1 = Molarity of sodium oxalate; (from 2-1-1) M_2 = Molarity of $KMnO_4$

V_1 = Volume of sodium oxalate (20 mL) V_2 = Volume of $KMnO_4$ (B.R.)

n_1 = mole equivalence of $Na_2C_2O_4$ (5) n_2 = Mole equivalence of $KMnO_4$ (2)

2-2 DETERMINATION OF FERROUS

Chemical reaction : $5Fe^{+2} + MnO_4^- + 8H^+ \longrightarrow Mn^{+2} + 5Fe^{+3} + 4H_2O.$

2-2-1 Preparation of Standard Sodium Oxalate Solution :

Same as described in 2-1-1.

2-2-2 Standardization of Potassium Permanganate Solution

Same as described in 2-1-2

2-2-3 Estimation of Ferrous

Rinse a clean burette well with standard $KMnO_4$ solution and fill it with the same solution. Remove the air bubbles, if trapped, by opening the stop cock. Note down the initial burette reading.

Pipette out 20 mL of the Ferrous solution into a clean conical flask and add 10 mL of dil. H_2SO_4. Titrate it against the permanganate solution until the light pink colour persists after through shaking. Note down the final burette reading. Repeat the titration for the concurrent readings.

Sl. No.	Volume of sodium oxalate solution	Burette Readings		
		Initial Reading	Final Reading	Volume of $KMnO_4$ consumed (Difference)

Calculations :

$$\frac{M_3V_3}{n_3} = \frac{M_4V_4}{n_4}$$

So $$M_4 = \frac{V_3M_3n_4}{n_3V_4}$$

M_3 = Molarity of $KMnO_4$ (from 2-1-2) ; M_4 = Molarity of Ferrous iron

V_3 = Volume of $KMnO_4$ (B.R.) V_4 = Volume of Ferrous iron (20mL)

n_3 = Mole equivalence of $KMnO_4$ (1); n_4 = Mole equivalence of Ferrous iron (5)

Amount of Fe^{2+} = Molarity × At. Wt.

$= M_4 \times 56 \text{ g L}^{-1}$.

2-3 ESTIMATION OF CALCIUM

Principle : Calcium forms sparingly soluble oxalate which is separated by filtration and brought back in to solution by dissolving the precipitate in dilute sulphuric acid. The oxalate due to calcium is titrated with standard potassium permanganate.

Chemical Reaction :

$$Ca^{2+} + \begin{matrix} COONH_4 \\ | \\ COONH_4 \end{matrix} \longrightarrow \begin{matrix} COO \\ | \\ COO \end{matrix}\!\!>Ca + 2NH_4^+$$

$$\begin{matrix} COO \\ | \\ COO \end{matrix}\!\!>Ca + H_2SO_4 \longrightarrow CaSO_4 + H_2C_2O_4$$

$$2MnO_4 + 5H_2C_2O_4 + 6H^+ \longrightarrow 2Mn^{2+} + 10CO_2 + 8H_2O$$

2-3-1 Preparation of Standard $KMnO_4$

Already described in 2-1-2.

2-3-2 Estimation of Unknown Calcium

Pipette out 20 ml of the unknown calcium solution into a beaker. Add excess of ammonium oxalate (say 100 ml of 0.1 M solution) to precipitate the calcium as calcium oxalate. Now add dilute ammonia (1:1) until the solution is neutral or faintly alkaline. Decant the filtrate through a Whatman 40 filter paper and then wash the precipitate repeatedly with small quantities of ice cold water. Wash the bulk of the precipitate into a conical flask and dissolve it in dilute H_2SO_4. Heat the conical flask to 70°C (i.e. till a bubble appears in the solution) and titrate with standard potassium permanganate taken in a burette till a light pink color appears which is the end point. Repeat the experiment to check the accuracy with another sample.

Sl. No.	*Volume of calcium solution*	*Burette Readings*		
		Initial Reading	*Final Reading*	*Volume of $KMnO_4$ consumed (Difference)*

Calculations :

$$\frac{M_1V_1}{n_1} = \frac{M_2V_2}{n_2}$$

So $$M_2 = \frac{M_1V_1n_2}{n_1V_2}$$

M_1 = Molarity of $KMnO_4$ (from 2-1-2); M_2 = Molarity of Calcium solution

V_1 = Volume of $KMnO_4$ (B.R.) V_2 = Volume of Calcium (20mL);

n_1 = Mole equivalence of $KMnO_4$ (1); n_2 = Mole equivalence of oxalate (5)

Amount of Ca^{2+} = Molarity × At.. Wt.

$= M_2 \times 40.08$ g L^{-1}

3.

REDOX TITRATIONS – II

DICHROMETRY

Potassium dichromate is not so powerful an oxidizing agent as permanganate (standard reduction potential $\simeq 1.3$ V in acid medium) but has some advantages over latter. $K_2Cr_2O_7$ is a primary standard and is not reduced by cold, dilute hydrochloric acid. Potassium dichromate is therefore of particular value in the estimation of iron in iron ores.

3-1 DETERMINATION OF FERRIC IRON

3-1-1 Preparation of Standard Potassium Dichromate Solution

Weigh accurately about 0.49 gm of pure anhydrous $K_2Cr_2O_7$ and dissolve in a clean 100 ml standard flask. Make up the solution upto the mark. Shake the solution throughly.

$$\text{Molarity of } K_2Cr_2O_7 = \frac{X}{M} \times 10$$

X = Weight of $K_2Cr_2O_7$ $\qquad$ M = M. Wt. of $K_2Cr_2O_7$ (294)

3-1-2 Determination of Ferric Iron

Chemical reaction

$$Cr_2O_7^{-2} + 6Fe^{+2} + 14H^+ \longrightarrow 6Fe^{+3} + 2Cr^{+3} + 7H_2O$$

Principle : Ferric ion is first reduced to ferrous ion with the help of stannous chloride.

$$2FeCl_3 + SnCl_2 \longrightarrow 2FeCl_2 + Sn\ Cl_4$$

The excess unreacted stannous chloride is oxidized by the addition of a few drops of mercuric chloride solution.

$$SnCl_4 + 2HgCl_2 \longrightarrow SnCl_4 + Hg_2Cl_2$$

The Precipitated mercurous chloride settles down. Now the solution is titrated with standard $K_2Cr_2O_7$ solution and concentration of the ferric ion is estimated.

Procedure : Pipette out 20 mL of Ferric iron solution into a clean conical flask. Add 5 mL of concentrated HCl. Now add stannous chloride solution drop wise until the solution becomes colourless. Cool the solution under tap water. Now add, drop by drop, mercuric chloride solution until a white silky precipitate is formed. Avoid adding excess of $SnCl_2$. If excess has been added a grey precipitate is formed, reject the solution and start the reduction process once again. Dilute the solution to about 100 mL with distilled water then add 5mL concentrated sulphuric acid and 2 mL of phosphoric acid* (or 5mL of sulphuric and phosphoric acid mixture) and 4 to 5 drops of sodiumdiphenylamine sulphonate indicator. Note down the initial burette reading. Titrate the solution against standard potassium dichromate solution until violet colour persists. Note down the final burette reading. Repeat the titration for concurrent readings.

Sl. No.	*Volume of Fe^{2+} solution*	*Burette Readings*		
		Initial Reading	*Final Reading*	*Volume of $K_2Cr_2O_7$ consumed (Difference)*

Calculations :

$$\frac{M_1V_1}{n_1} = \frac{M_2V_2}{n_2}$$

So
$$M_2 = \frac{M_1V_1n_2}{n_1V_2}$$

M_1 = Molarity of $K_2Cr_2O_7$ (from 3-1-1) M_2 = Molarity of Ferric iron

V_1 = Volume of $K_2Cr_2O_7$ (B.R.) V_2 = Volume of Ferric iron (20 mL)

n_1 = Mole equivalence of $K_2Cr_2O_7$ (6) n_2 = Mole equivalence of Fe^{3+} (1)

Amount of Fe^{3+} = Molarity × M.Wt. of Fe

$= M_2 \times 56$ g L^{-1}.

3-2 DETERMINATION OF FERROUS AND FERRIC IRON IN THE GIVEN MIXTURE

Principle : The ferrous iron is directly estimated by titrating with standard $K_2Cr_2O_7$ solution. Then in another sample ferric iron is reduced to ferrous iron by the addition of $SnCl_2$ and the total ferrous iron is estimated. Subtracting from this, the value of ferrous iron before reduction, value corresponding to ferric iron is obtained.

3-2-1 Preparation of Standard Potassium Dichromate Solution

Same as given in 3-1-1.

3-2-2 Determination of free Fe^{2+} in a Mixture

Chemical reaction :

$$Cr_2O_7^{-2} + 6Fe^{+2} + 14H^+ \longrightarrow 6Fe^{+3} + 2Cr^{+3} + 7H_2O$$

* Note : Addition of Phosphoric acid lowers the oxidation potential of Ferric - Ferrous couple by forming a complex $[Fe(HPO_4)]^+$ and facilitates the oxidation of Fe^{2+}

Procedure : Pipette out 20 mL of the given mixture solution into a clean conical flask and add 10 mL of dil. H_2SO_4 and dilute the solution to about 75 mL by adding distilled water. Now add 2 mL of phosphoric acid and 4 to 5 drops of sodiumdiphenylamine sulphonate indicator. Titrate the solution against the standard potassium dichromate solution until violet purple colour appears. Repeat the titrations for concurrent readings. Tabulate the readings.

Sl. No.	*Volume of Fe^{2+} + Fe^{3+} solution*	*Burette Readings*		
		Initial Reading	*Final Reading*	*Volume of $K_2Cr_2O_7$ consumed (Difference)*

$$\frac{M_1V_1}{n_1} = \frac{M_2V_2}{n_2}$$

So Molarity (M_2) of Fe^{2+} in mixture $= \dfrac{M_1V_1n_2}{V_2n_1}$

M_1 = Molarity of $K_2Cr_2O_7$ (from 3-1-1) M_2 = Molarity of Fe^{+2} in the mixture

V_1 = Volume of $K_2Cr_2O_7$ (B.R.) V_2 = Volume of the mixture (20 mL)

n_1 = Mole equivalence of $K_2Cr_2O_7$ (1); n_2 = Mole equivalence of Fe^{+2} (6)

Amount of free Ferrous iron = molarity of ferrous iron × atomic weight of Fe
= $M_2 \times 56$ g L^{-1}.

3-2-3 Determination of Ferric Iron in the Mixture

Same as given in 3-1-2.

Calculations :

$$\frac{M_3V_3}{n_3} = \frac{M_4V_4}{n_4}$$

So $M_4 = \dfrac{M_3V_3n_4}{n_3V_4}$

M_3 = Molarity of $K_2Cr_2O_7$ (from 3-1-1); M_4 = Molarity of total iron

V_3 = Volume of $K_2Cr_2O_7$ (B.R.) V_4 = Volume of the mixture (20 mL);

n_3 = Mole equivalence of $K_2Cr_2O_7$ (1); n_4 = Mole equivalence of Fe^{2+} (6)

Amount of total iron (Fe^{3+} + Fe^{2+}) = $M_4 \times 56$ g L^{-1}.

Amount of Ferric iron = Amount of total iron - Amount of free Ferrous iron in the mixture.

4. REDOX TITRATIONS – III

IODOMETRY

4-1 STANDARDIZATION OF SODIUM THIOSULPHATE SOLUTION

4-1-1 Preparation of Standard Potassium Dichromate Solution

Same as given in 3-1-1.

4-1-2 Standardization of Sodium Thiosulphate Solution

Chemical Reactions :

$$Cr_2O_7^{-2} + 14H^+ + 6S_2O_3^{-2} \longrightarrow 2Cr^{+3} + 3S_4O_6^{-} + 7H_2O$$

$$3\ (S_2O_3^{2-} + I_2 \longrightarrow S_4O_6^{2-} + 2I^-)$$

Procedure : Pipette out 20 mL of standard $K_2Cr_2O_7$ solution into an iodination flask and add 3 mL of conc. HCl followed by a pinch of sodium bicarbonate. Close the flask and allow the effervescence to subside. To this add one test tube full of 10% potassium iodide solution and keep it in dark for five minutes, so that the iodine is liberated by the oxidation of I^- by $K_2Cr_2O_7$. Titrate the solution against sodium thiosulphate solution (Hypo) till the solution becomes yellowish-green. Then add 2-3 mL of starch indicator and continue the titrations till the blue colour changes to light green. Repeat the titration for concurrent readings.

$$\frac{M_1V_1}{n_1} = \frac{M_2V_2}{n_2}$$

$$\text{So Molarity of hypo } M_2 = \frac{M_1V_1n_2}{n_1V_2}$$

M_1 = Molarity of $K_2Cr_2O_7$ (from 3-1-1); M_2 = Molarity of hypo (?)

V_1 = Volume of $K_2Cr_2O_7$ (20 mL) V_2 = Volume of hypo (B.R.)

n_1 = Mole equiv. of $K_2Cr_2O_7$ (1) n_2 = Mole equiv. of hypo (6)

4-2 DETERMINATION OF COPPER

4-2-1 Preparation of Standard $K_2Cr_2O_7$ Solution

Same as in experiment 2-1-1.

4-2-2 Standardization of Hypo

Same as in experiment 4-2-2.

4-2-3 Determination of Copper

Chemical Reactions :

$$2\,Cu^{2+} + 4\,I^- \longrightarrow 2\,CuI + I_2$$

$$I_2 + 2\,Na_2S_2O_3 \longrightarrow 2\,NaI + Na_2S_4O_6$$

Procedure : Pipette out 10 mL of given copper sulphate solution and add dil. NaOH dropwise until a slight turbidity appears. Add dil. acetic acid until it dissolves. Dilute the solution to about 50 mL and add 10 mL of potassium iodide solution and shake it. Fill a clean burette with $Na_2S_2O_3$ solution and titrate the copper sulphate solution until brown colour fades. Add at this stage 1 mL of starch solution. Continue the addition of hypo, until blue colour commences to disappears. Repeat the titrations for concurrent readings.

S No.	*Volume of* $K_2Cr_2O_7$	*Burette Readings*		
		Initial Reading	*Final Reading*	*Volume of* $Na_2S_2O_3$ *(Difference)*

Calculations :

$$\frac{M_3V_3}{n_3} = \frac{M_4V_4}{n_4}$$

$$\text{Molarity of } Cu^{2+}\ M_4 = \frac{M_3V_3n_4}{n_3V_4}$$

M_3 = Molarity of hypo (from 4-1-2); M_4 = Molarity of Cu^{+2} (?)

V_3 = Volume of hypo (B.R.) V_4 = Volume of Cu^{+2} solution (10 mL)

n_3 = Mole equivalence of hypo (2) n_4 = Mole equivalence of Cu^{+2} (2)

Amount of copper = Molarity of Cu^{+2} × At. Wt. of copper

$= M_4 \times 63.5\ g\ L^{-1}$.

5.

REDOX TITRATION - IV

CERIMETRY

Ceric ion is a powerful oxidizing agent and *titration procedure employing this oxidant is called* ***cerimetry.*** In the present experiment ferrous ions are oxidised by ceric ions.

$$Ce^{4+} + Fe^{2+} \longrightarrow Ce^{3+} + Fe^{3+}$$

5-1 ESTIMATION OF FERROUS IRON

5-1-1 Preparation of standard (0.01N) ferrous ammonium sulphate solution : Weight accurately about 0.39 g of A.R. grade ferrous ammonium sulphate (equivalent weight = 392.13) and tranfer it into 100 mL standard flask. Dissolve the solid in 1N H_2SO_4 and make upto the mark with the same acid

Molarity of $Fe^{2+} = \frac{X}{392.13} \times 10$ x = wt of Ferrous ammonium sulphate

5-1-2 Preparation of standard (0.01N) ferrous ammonium sulphate solutions : Weigh about 0.33 g of ceric sulphate, $Ce(SO_4)_2$ or 0.63 g of ceric ammonium sulphate $(NH_4)_4$ $[Ce(SO_4)_4]$. $2H_2O$ and transfer it into a clean 100 mL standard flask. Dissolve the solid in IN H_2SO_4 and make up to the mark with the acid solution.

5-1-3 Standardization of ceric solution : Fill the bottle with the ceric solutions. Take 20 mL of standard ferrous solution into a conical flask, add 10 mL of 1N sulphuric acid and one or two drops of ferroin indicator. Titrate with the ceric sulphate solution to the first sharp color change. Repeat the titration for the concurrent readings.

Molarity of Ceric solutions $M_2 = \frac{M_1 V_1}{V_2}$

M_1 = Molarity of ferr ous solution

V_1 = Volume of ferrous solution taken (20 mL)

V_2 = Titre value (Volume of Ce^{4+} solutions)

5-1-4 Estimations of Fe^{2+} in a given solutions : Repeat the above titration. This time Ce^{4+} concentration is known and Fe^{2+} solution is unknown.

Molarity of Fe^{2+} $M_1 = \frac{M_2 V_2}{V_1}$

6.

PRECIPITATION TITRATION

6-1 ESTIMATION OF ZINC

Principle : Zinc ions in neutral or acid medium react with potassium ferrocyanide solution to form very sparingly soluble potassium zinc ferrocyanide.

$$3Zn^{2+} + 2K_4\,[Fe(CN)_6] \rightleftharpoons K_2Zn_3[Fe(CN)_6]_2 \downarrow + 6K^+$$

3,3′ - dimethylnaphthidine* or 3,3′ dimethylnaphthidine disulphonic acid is used as an internal indicator.

6-1-1 Preparation of 0.05 M potassium ferrocyanide : Weigh out ≈ 2.1 g of A.R. potassium ferrocyanide and ≈ 0.03 g potassium ferricyanide in 100 mL distilled water.

6-1-2 Preparation of standard zinc solution : Weigh approximately about 0.65 g A.R. grade zinc metal (atomic wt = 65.38) or 0.81 g of zinc oxide (Mol wt = 81.38), transfer into a 100 mL standard flask and dissolve the solid in dilute sulphuric acid and make up to the mark.

$$\text{Molarity of } Zn^{2+} = \frac{X}{65.38 \text{ or } 81.38} \times 10$$

6-1-3 Standardisation of potassium ferrocynide : Pipette out 20 mL of zinc solution into a conical flask, add 2 drops of 3,3′-dimethylnaphthadine indicator. Titrate this solution with ferrocyanide solution slowly and with vigorous shaking till the first permanent color changes from purple-red to greyish-green. Repeat the titration for concurrent readings.

$$\text{Concentration of Ferrocyanide } M_2 = \frac{M_1 V_1 n_2}{V_2 n_1}$$

M_1 = Molarity of Zinc solution

V_1 = Volume of Zinc solution = 20 mL

n_1 = Mole equiv. of Zn^{2+} = 3

V_2 = Titre value

n_2 = Mole equiv. of ferrocyanide = 2

6-1-4 Estimation of Zn^{2+} in a given solution : The above titration is repeated.

---

* This indicator is prepared by dissolving 0.2g of the solid, while warming, in 100 mL of glacial acetic acid.

7.

COMPLEXOMETRY

EDTA TITRATIONS

Principle : Ethylenediaminetetracetic acid is largely used as the disodium salt. It is one of the potential ligands and hence EDTA is employed in the estimation of metal ions by complexometric technique. EDTA serves as a hexadentate ligand and is represented by the following structure

```
        :O:                                   :O:
        ||                                    ||
 ⁻:Ö—C CH2                              CH2C —Ö:⁻
            \                          /
             :NCH2CH2N:
            /                          \
 ⁻:Ö—C CH2                              CH2C—Ö:⁻
        ||                                   ||
        :O:                                  :O:
```

Structure of EDTA

In tnese titrations the moşt widely used indicator is Eriochrome Black T also called Erio T. This dye has two phenolic groups and has pK_a values at 6.3 and 11.5 for the respective hydroxylic groups. Erio T should not be used in strong acid solutions because of its dimerization. In the pH range 7-11 metal salts show a colour change from blue to red.

$$M^{2+} + HD^{2-} \longrightarrow MD^- + H^+$$

HD^{2-}: blue Dye		MD^-: red Metal-Dye complex	

Addition of EDTA to a solution of metal ions, complexed weakly with the dye (MD^-), results in their replacement to form a more stable metal EDTA complex, consequently the colour changes sharply from red to blue at the end point.

$$\underset{\text{red}}{MD^-} + EDTA \longrightarrow M\text{-}EDTA + \underset{\text{blue}}{HD^{2-}}$$

It is necessary to maintain the pH of the solution at pH = 10, so as to detect a sharp end point.

7-1 ESTIMATION OF MAGNESIUM

Chemical Reactions : $Mg^{2+} + HD^{2-} \longrightarrow MgD^- + H^+$

$MgD^- + EDTA \longrightarrow Mg - EDTA + HD^{2-}$

7-1-1 Preparation of Standard EDTA Solution

Disodiumdihydrogenethylenediaminetetra-acetate can be used as a primary standard. Weigh accurately around 3.7g of disodium salt of EDTA into a clean 100mL standard flask fitted with a funnel. Dissolve the solid in minimum quantity of distilled water and then make up the solution to the mark. Shake thoroughly for about a minute or two.

$$\text{Molarity of } Na_2EDTA = \frac{X}{M} \times 10$$

X = Wt. of Na_2EDTA, M = Mol. Wt. of Na_2EDTA (372.25)

7-1-2 Estimation of Magnesium present in an Unknown Solution

Rinse a clean burette well with standard Na_2EDTA solution and fill it with the same solution. Remove the air bubbles, if trapped, by opening the stop cock. Note down the initial burette reading.

Pipette out 20mL of the unknown Mg^{2+} solution into a conical flask, dilute it by adding about 50mL of distilled water, add 5 mL of buffer pH = 10* followed by 3-4 drops of Eriochrome Black T indicator solution**. Titrate this mixture with standard Na_2EDTA solution present in the burette until the colour changes from red to blue. Since the complexation is not instantaneous better end point is obtained by warming the solution to about 40°C. Note down the final burette reading. Repeat the titration for the concurrent readings.

Calculation :

$$\frac{M_1 V_1}{n_1} = \frac{M_2 V_2}{n_2}$$

So $$M_2 = \frac{M_1 V_1 n_2}{n_1 V_2}$$

M_1 = Molarity of Na_2 EDTA (5-1-1); M_2 = Molarity of Mg^{2+}

V_1 = Volume of Na_2 EDTA (B.R.) V_2 = Volume of Mg^{2+}; (20 mL)

n_1 = Mole equivalence of Na_2 EDTA (1) n_2 = Mole equivalence of Mg^{2+} (1)

Amount of Mg^{2+} = M_2 × At. Wt. of Mg

= M_2 × 24.32 g L^{-1}.

---

* Prepared by dissolving 17.5 g amonium chloride in 142 mL concentrated ammonia (sp. gr = 1.88) and diluting to 250 mL by the addition of water.

** Prepared by dissolving 0.2g of dye in 15 mL of triethanolamine and 5 mL of absolute alcohol.

7-2 DETERMINATION OF COPPER

Chemical Reactions : $Cu^{2+} + HD^{2-} \longrightarrow CuD^- + H^+$

$CuD^- + EDTA \longrightarrow Cu^- EDTA + HD^{2-}$

7-2-1 Preparation of Standard EDTA Solution

Same as given in 7-1-1.

7-2-2 Estimation of Copper

Repeat the procedure given in 7-1-2, taking copper solution in place of magnesium solution.

Calculation :

$$\frac{M_1V_1}{n_1} = \frac{M_2V_2}{n_2}$$

So $$M_2 = \frac{M_1V_1n_2}{n_1V_2}$$

M_1 = Molarity of Na_2 EDTA (from 7-1-1); M_2 = Molarity of Cu^{2+}

V_1 = Volume of Na_2 EDTA (B.R.); V_2 = Volume of Cu^{2+} (20 mL);

n_1 = Mole equivalence of Na_2 EDTA (1); n_2 = Mole equivalence of Cu^{2+} (1)

Amount of Cu^{2+} = M_2 × At. Wt. of Cu

= M_2 × 63.54 g L^{-1}.

7-3 Estimation of Nickel

Nickel forms a very stable complex with EDTA. Any excess EDTA added may be back titrated with zinc ion solution using Erio T indicator.

7-3-1 Preparation of standard EDTA solution : same as given in 7-1-1

7-3-2 Estimation of Nickel using Erio-T as indicator (back titration method)

Take 20 mL of the almost neutral solutions of nickel into a conical flask and add a slight excess of 0.1M EDTA solution (25-30 mL). Dilute the solution to about 200 mL, add 4 mL of pH 10 buffer solution and a few drops of Erio-T indicator. Titrate the excess of EDTA with standard 0.1 M Zinc sulphate solution until the color changes from blue to wine red.

Every mL difference between the volume of 0.1 M EDTA added and the 0.1 M Zinc ion solution used in the back-titration corresponds to 5.871 mg of Ni

7-4 ESTIMATION OF ZINC USING ERIO-T INDICATOR

A similar procedure as in the case of estimation of magnesium (7-1-2) is adopted here.

7-5 DETERMINATION OF HARDNESS OF WATER

Principle : Hardness of water is due to the presence of dissolved salts of Ca^{2+} and Mg^{2+} ions. The bicarbonate salts on heating, decompose to give water insoluble carbonates and hence are said to impart only temporary hardness.

$$Ca\ (HCO_3)_2 \longrightarrow CaCO_3 + H_2O + CO_2$$

Other soluble salts of calcium and magnesium which cannot be separated by simple heating techniques impart permanent hardness to the water.

7-5-1 Preparation of Standard EDTA Solution

Standard EDTA solution is prepared by adopting the procedure given in 7-1-1.

A. ESTIMATION OF TOTAL HARDNESS OF WATER

7-5-2 Estimation of Total Hardness of Water

Take 50 mL of a water sample into a conical flask and add 3 mL of the pH = 10 buffer solution and 3-4 drops of Eriochrome Black T indicator. Titrate this solution with standard 0.01 M EDTA solution taken in a burette. Titrate until the colour changes from wine red to purple blue.

If water examined has no magnesium salts, add 0.1 mL of 0.1M magnesium-EDTA complex solution prior to the addition of indicator. (The 0.1M magnesium EDTA complex could be prepared by mixing equal volumes of 0.2M magnesium and 0.2M EDTA solutions). The addition of magnesium EDTA complex ensures a colour change at the end point.

Calculation :

Hardness is generally expressed in terms of calcium carbonate, as $CaCO_3$ has a molecular weight of 100, 1mL titre value of 0.01 M EDTA is equivalent to 1.000 mg of $CaCO_3$. If V is the titre volume, V mg of $CaCO_3$ is present in 50mL of water sample taken.

$$\text{So Concentration of } CaCO_3 = \frac{V \times 1000}{50} = 20 \text{ mg L}^{-1}.$$

B. ESTIMATION OF PERMANENT HARDNESS OF WATER

7-5-3 Estimation of Permanent Hardness

Place 250mL water in a beaker and boil it gently for 30 minutes. Cool the beaker and we notice insoluble salt particles floating in the water. Filter the water into a volumetric flask and make up the volume to the mark. *Never wash the filter paper.* Now take 50mL of the made up solution and repeat the process given in 7-3-2. Express the permanent hardness as mg L^{-1} of $CaCO_3$.

C. ESTIMATION OF TEMPORARY HARDNESS OF WATER

7-5-4 Estimation of Temporary Hardness

We get the temporary hardness from the difference of values obtained for total hardness and permanent hardness.

Temporary hardness = (Total hardness - Permanent hardness) mg L^{-1}.

GRAVIMETRIC ANALYSIS

1 ESTIMATION OF NICKEL AS THE DIMETHYLGLYOXIMATE

Faintly acidic solution of nickel is precipitated by the addition of an alcoholic solution of dimethylglyoxime (H_2DMG)

$$Ni^{2+} + 2H_2DMG = \underset{\text{red ppt.}}{Ni(HDMG)_2} \downarrow + 2H^+$$

This precipitate is soluble in mineral acids (including acid liberated in the reaction), in hot water, in more than 50% alcohol and concentrated ammonia, but insoluble in dilute ammonia solutions.

Therefore complete precipitation is effected by the addition of dilute ammonia. The precipitate is digested, filtered, through sintered crucible, washed till free from chloride ions, dried at 80-100° C and weighed as such.

1-1 Procedure : Dilute the given nickel solution to 200 mL in a 400 mL beaker by the addition of distilled water. Add 1 mL of dil HCl, heat to 70-80° C and add 30-35 mL of 1% alcoholic dimethylglyoxime solution*. Immediately add dilute ammonia solution dropwise directly to the solution (not down the beaker walls) till a distinct smell of ammonia persists. Allow to stand on the steam bath for 20-30 minutes. Allow the thick red precipitate to cool and settle and test the clean supernatant liquid for completion of precipitation by adding few drops of H_2DMG.

Filter the precipitate through a previously weighed Gooch or sintered glass crucible, wash the precipitate three to four times with cold water and dry it at 110 - 120° C for 45-50 minutes. Allow to cool in a descicator and weigh.

1-2 Calculations :

Weight of empty crucible = w_1

Weight of crucible + precipipate = w_2

---

* Prepared by dissolving 1g dimethylglyoximate in 100 mL alcohol and then filtered to remove turbidity, if any.

Weight of precipitate $x = w_2 - w_1$

288.69 g of $Ni(HDMG)_2$ has 58.69g of Ni.

Hence, weight of Nickel = 0.2033 × weight of precipitate (x)

2 ESTIMATION OF BARIUM AS CHROMATE

2-1 Procedure

Dilute the given solution to 200 mL. Add 1 mL of 6N acetic acid and 10 mL neutral ammonium acetate to the solution, heat to boiling and treat with slight excess of dilute ammonium chromate* (added dropwise from a burette). After allowing the solution to cool and settle, test supernatant liquid for completion of precipitation by adding few drops of the reagent.

Filter the precipitate through a previously weighed sintered glass crucible, wash with hot water until 1 mL of washings give scarcely any reddish-brown coloration with neutral silver nitrate solution. Dry to constant weight at 120°C

2-2 Calculations :

Weight of empty crucible = w_1

Weight of crucible + precipitate = w_2

Weight of precipitate $x = w_2 - w_1$

253.326 $BaCrO_4$ contains 137.33 g of Ba

Hence weight of Barium = 0.542 × weight of precipitate (x)

---

* The ammonium chromate in prepared by dissolving 10g of pure ammonium dichromate in 100 mL water and adding dilute ammonia solution until the color of the solution is clear yellow.

APPENDIX

APPENDIX A

IMPORTANT INORGANIC ACIDS AND THEIR ANIONS

Formula	*Name*	*Anion*	*Name*
H_2CO_3	Carbonic acid	HCO_3^- CO_3^{2-}	hydrogen carbonate ion (bicarbonate ion) carbonate ion
HNO_3	Nitric acid	NO_3^-	nitrate ion
HNO_2	Nitrous acid	NO_2^-	nitrite ion
H_2SO_4	Sulphuric acid	HSO_4^- SO_4^{2-}	hydrogen sulphate ion (bisulphate ion) sulphate ion
H_2SO_3	Sulphurous acid	HSO_3^- SO_3^{2-}	hydrogen sulphite ion (bisulphite ion) sulphite ion
H_3PO_4	Phosphoric acid (Ortho phosphoric acid)	$H_2PO_4^-$ HPO_4^{2-} PO_4^{3-}	dihydrogen phosphate ion monohydrogen phosphate ion phosphate ion
$HClO_4$	Perchloric acid	ClO_4^-	perchlorate ion
$HClO_3$	Chloric acid	ClO_3^-	chlorate ion
$HClO_2$	Chlorous acid	ClO_2^-	chlorite ion
HClO	Hypochlorous acid	ClO^-	hypochlorite ion
HCl	Hydrochloric acid	Cl^-	chloride ion

APPENDIX B

PREFIXES FOR FRACTIONS AND MULTIPLES OF SI UNITS

Prefix	*Symbol*	*Factor*	*Prefix*	*Symbol*	*Factor*
tera	T	10^{12}	centi	C	10^{-2}
giga	G	10^{9}	milli	m	10^{-3}
mega	M	10^{6}	micro	μ	10^{-6}
kilo	k	10^{-3}	nano	n	10^{-9}
hecta	h	10^{2}	pico	p	10^{-12}
deca	da	10	femto	f	10^{-15}
deci	d	10^{-1}	atto	a	10^{-18}

APPENDIX C
CONVERSION TABLE

Physical quantity	*Unit*	*SI equivalent*
Length	angstrom (Å)	10^{-1} nm
	inch (in)	0.025m
	mile	1.60934km
	international nautical mile	1.852 km
Area	square inch (in^2)	645.16 mm^2
Volume	cubic inch	1.63871 x 10^{-5} m^3
Mass	pound (p)	0.453 592 37 kg
Force	Newton	(N = m kg 5^{-2} = Pa m^{-2})
	dyne (dyn)	10^{-5} N
	poundal	0.138 255 N
	pound force	4.448 22 N
Pressure	Pascal	(Pa = Nm^{-2})
	atomosphere (atm)	101.325 Pa
	bar (bar)	10^5 Pa
	torr (torr)	133.322 Pa
	p.s.i.	6894.76 Pa
Energy	Joule	(J = Nm)
	calorie (cal)	4.184 J
	B.t.u.	1055.06 J
	electron Volt (eV)	1.6021 × 10-19 J
	erg	10^{-7} J
	L atm	101.325 J
	Wave number (cm^{-1})	1.986 × 10^{-23} J
Power	horse power	745.700 W

APPENDIX D
GENERAL PHYSICAL CONSTANTS AND THEIR VALUES

Constant	*Symbol*	*Value*	*multiply* (SI)	*multiply* (cgs)
Speed of light	c	2.997 923	$10^8 ms^{-1}$	10^{10} cm s^{-1}
Elementary charge	e	1.602 1917	10^{-19} C	10^{-20} $cm^{1/2}$ $g^{1/2}$
Avogadro consitant	N_A	6.022 169	10^{23} mol^{-1}	10^{23} mol^{-1}
Atomic mass unit	amu	1.660 531	10^{-27} kg	10^{-24} g
Electron rest mass	m^e	9.109 558	10^{-31} kg	10^{-28} g
Proton rest mass	m^p	1.672 614	10^{-27} kg	10^{-24} g
Neutron rest mass	mn	1.674 954	10^{-27} kg	10^{-24} g
Faraday constant	F	9.648 670	10^4 C mol^{-1}	10^3 $cm^{1/2}$ $g^{1/2}$ mol^{-1}
Planck constant	h	6.626 196	10^{-34} Js	10^{-27} erg s
Boltzmann constant	k	1.380 620	10^{-23} J K^{-1}	10^{-16} erg K^{-1}
Gas constant	R	8.314 33	J K^{-1} mol^{-1}	8.314 33 × 10^7 erg K^{-1} mol^{-1} 1.989717 cal K^{-1} mol^{-1} 0.082056 L atm K^{-1} mol^{-1}

APENDIX E

ELECTRONIC CONFIGURATION OF ELEMENTS

Atomic number	*Element*	*Electron configuration*
1	H	$1s^1$
2	**He**	$\mathbf{1s^2}$
3	Li	[He]$2s^1$
4	Be	$-2s^2$
5	B	$-2s^2\ 2p^1$
6	C	$-2s^2\ 2p^2$
7	N	$-2s^2\ 2p^3$
8	O	$-2s^2\ 2p^4$
9	F	$-2s^2\ 2p^5$
10	**Ne**	$\mathbf{1s^2\ 2s^2\ 2p^6}$
11	Na	[Ne] $3s^1$
12	Mg	$-3s^2$
13	Al	$-3s^2\ 3p^1$
14	Si	$-3s^2\ 3p^2$
15	P	$-3s^2\ 3p^3$
16	S	$-3s^2\ 3p^4$
17	Cl	$-3s^2\ 3p^5$
18	**Ar**	$\mathbf{1s^2\ 2s^2\ 2p^6\ 3s^2\ 3p^6}$
19	K	[*Ar*] $4s^1$
20	Ca	$-4s^2$
21	Sc	$-3d^3\ 4s^2$
22	Ti	$-3d^2\ 4s^2$
23	V	$-3d^3\ 4s^2$
24	Cr	$-3d^5\ 4s^1$
25	Mn	$-3d^5\ 4s^2$
26	Fe	$-3d^6\ 4s^2$
27	Co	$-3d^7\ 4s^2$
28	Ni	$-3d^8\ 4s^2$
29	Cu	$-3d^{10}\ 4s^1$
30	Zn	$-3d^{10}\ 4s^2$
31	Ga	$-3d^{10}\ 4s^2\ 4p^1$
32	Ge	$-3d^{10}\ 4s^2\ 4p^2$
33	As	$-3d^{10}\ 4s^2\ 4p^3$
34	Se	$-3d^{10}\ 4s^2\ 4p^4$
35	Br	$-3d^{10}\ 4s^2\ 4p^5$
36	**Kr**	$\mathbf{-1s^2\ 2s^2\ 2p^6\ 3s^2\ 3p^6\ 3d^{10}\ 4s^2\ 4p^6}$

37	Rb	[*Kr*] $5s^1$
38	Sr	$-5s^2$
39	Y	$-4d^1\ 5s^2$
40	Zr	$-4d^2\ 5s^2$
41	Nb	$-4d^4\ 5s^1$
42	Mo	$-4d^5\ 5s^1$
43	Tc	$-4d^5\ 5s^2$
44	Ru	$-4d^7\ 5s^1$
45	Rh	$-4d^8\ 5s^1$
46	Pd	$-4d^{10}$
47	Ag	$-4d^{10}\ 5s^1$
48	Cd	$-4d^{10}\ 5s^2$
49	In	$-4d^{10}\ 5s^2\ 5p^1$
50	Sn	$-4d^{10}\ 5s^2\ 5p^2$
51	Sb	$-4d^{10}\ 5s^2\ 5p^3$
52	Te	$-4d^{10}\ 5s^2\ 5p^4$
53	I	$-4d^{10}\ 5s^2\ 5p^5$
54	**Xe**	$\mathbf{1s^2\ 2s^2\ 2p^6\ 3s^2\ 3p^6\ 3d^{10}\ 4s^2\ 4p^6\ 4d^{10}\ 5s^2\ 5p^6}$
55	Cs	**[Xe]** $6s^1$
56	Ba	$6s^2$
57	La	$5d^1\ 6s^2$
58	Ce	$4f^1\ 5d^1\ 6s^1$
59	Pr	$4f^3\ 6s^2$
60	Nd	$4f^3\ 6s^2$
61	Pm	$4f^5\ 6s^2$
62	Sm	$4f^6\ 6s^2$
63	Eu	$4f^7\ 6s^2$
64	Gd	$4f^7\ 5d^1\ 6s^2$
65	Tb	$4f^9\ 6s^2$
66	Dy	$4f^{10}\ 6s^2$
67	Ho	$4f^{11}\ 6s^2$
68	Er	$4f^{12}\ 6s^2$
69	Tm	$4f^{13}\ 6s^2$
70	Yb	$4f^{14}\ 6s^2$
71	Lu	$4f^{14}\ 5d^1\ 6s^2$
72	Hf	$4f^{14}\ 5d^2\ 6s^2$
73	Ta	$4f^{14}\ 5d^3\ 6s^2$

74	W	$4f^{14}\,5d^{4}\,6s^{2}$
75	Re	$4f^{14}\,5d^{5}\,6s^{2}$
76	Os	$4f^{14}\,5d^{6}\,6s^{2}$
77	Ir	$4f^{14}\,5d^{7}\,6s^{2}$
78	Pt	$4f^{14}\,5d^{9}\,6s^{1}$
79	Au	$4f^{14}\,5d^{10}\,6s^{1}$
80	Hg	$4f^{14}\,5d^{10}\,6s^{2}$
81	Tl	$4f^{14}\,5d^{10}\,6s^{2}\,6p^{1}$
82	Pb	$4f^{14}\,5d^{10}\,6s^{2}\,6p^{2}$
83	Bi	$4f^{14}\,5d^{10}\,6s^{2}\,6p^{3}$
84	Po	$4f^{14}\,5d^{10}\,6s^{2}\,6p^{4}$
85	At	$4f^{14}\,5d^{10}\,6s^{2}\,6p^{5}$
86	**Rn**	$\mathbf{1s^{2}\,2s^{2}\,2p^{6}\,3s^{2}\,3p^{6}\,3d^{10}\,4s^{2}\,4p^{6}\,4d^{10}}$ $\mathbf{4f^{14}\,5s^{2}\,5p^{6}5d^{10}\,6s^{2}\,6p^{6}}$
87	Fr	**[Rn]** $7s^{1}$
88	Ra	$-7s^{2}$
89	Ac	$-6d^{1}\,7s^{2}$
90	Th	$-6d^{2}\,7s^{2}$
91	Pa	$-5f^{2}\,6d^{1}\,7s^{2}$
92	U	$-5f^{3}\,6d^{1}\,7s^{2}$
93	Np	$-5f^{4}\,6d^{1}\,7s^{2}$
94	Pu	$-5f^{6}\,7s^{2}$
95	Am	$-5f^{7}\,7s^{2}$
96	Cm	$-5f^{7}\,6d^{1}\,7s^{2}$
97	Bk	$-5f^{8}\,6d^{1}\,7s^{2}$
98	Cf	$-5f^{10}\,7s^{2}$
99	Es	$-5f^{11}\,7s^{2}$
100	Fm	$-5f^{12}\,7s^{2}$
101	Md	$-5f^{13}\,7s^{2}$
102	No	$-5f^{14}\,7s^{2}$
103	Lr	$-5f^{14}\,6d^{1}\,7s^{2}$
104	Unq	$-5f^{14}\,6d^{2}\,7s^{2}$
105	Unp	$-5f^{14}\,6d^{3}\,7s^{2}$
106	Unh	$-5f^{14}\,6d^{4}\,7s^{2}$
107	Uns	$-5f^{14}\,6d^{5}\,7s^{2}$
108	Uno	$-5f^{14}\,6d^{6}\,7s^{2}$
109	Une	$-5f^{14}\,6d^{7}\,7s^{2}$

QUESTION PAPERS

OSMANIA UNIVERSITY
MARCH/APRIL 1994
(UNDER COMMON CORE SYLLABUS SCHEME)
PAPER I - CHEMISTRY

Answer all question from Section A and B.

Answer all the bits of a question together.

SECTION - A

I. Explain the general trends in the following properties of alkali metals.

a) Electrode potentials b) Basic characters of Hydroxides

c) Hydration energies. d) Flame colouration e) Ionization energies

OR

Discuss the periodicity of the following properties among the elements O, S, Se, Te

a) Oxidation States b) Nature of oxides
c) Electro negativity d) Ionisation energies e) Electron affinity.

II. a) What are pseudo halogens ? How do they resemble halogens in their properties ?

b) Write a not on hydrides of Nitrogen family elements (V B)

OR

a) Name the different fuel gases ? Give their composition and uses. How is water gas prepared ?

b) Write a note on hydrides of Carbon family-elements (IV B)

III. Explain the type of isomerism exhibited by the following compounds whose formulae are given in brackets.

a) The alcohol, ethanol (C_2H_6O)
b) The unsaturated dibasic acid, maleic acid ($C_4H_4O_4$)
c) The manobasic hydroxy acid, Lactic acid ($C_3H_6O_3$)
d) The alkene, Butene (C_4H_8)

OR

Explain with examples
a) The mechanism of free radical substitution
b) Primary, secondary and tertiary carbon atom
c) Hyperconjugation
d) Carbonium ion

IV a) Derive the expression for the equilibrium Constant Kp and Kc for the following reaction.

$$PCl_5 \rightleftharpoons PCl_3 + Cl_2$$

b) 2 moles of HI were heated in a sealed tube at 440°C till the equilibrium was established. It was found to be 22% dissociated. Calculate the equilibrium constant for the dissociation reaction.

OR

Explain why

a) A colloidal solution is not precipitated on the addition of an electrolyte in presence of gelatine.

b) Precipitation of colloidal solutions is favoured by electrolytes.

c) Alum is used in town water supply.

d) A colloidal solution exhibits Brownian movement.

e) The colour of the sky is blue.

SECTION - B

V. What are interhalogen compounds? How are they classified? Discuss the geometry and hybridisation in ClF_3.

OR

What are important Oxyacids of Phosphorous? Mention the oxidation state of Phosphorus in them? Give the structural formula of Phosphorous acid.

VI. What are Carbides? How are they classified? Give one example in each case.

OR

Discuss the structure of diborane.

VII. Describe the preparation and important properties of Borax? How it is used in qualitative chemical analysis.

OR

How is caustic soda manufactured by Castner-Kellner process? Give the important industrial uses.

VIII Write a short note on super phosphate of lime.

OR

Explain why

a) Water is a liquid whereas H_2S is a gas.

b) Chlorine has the highest electron affinity but Fluorine has the highest oxidising power.

IX. a) Write IUPAC name of the following compounds.

(i)

$$CH_3-\underset{\underset{CH_3}{|}}{\overset{\overset{H}{|}}{C}}-CH=CH-CH_2-\underset{\underset{O}{||}}{C}-OH$$

(ii)

$$CH_3-\underset{\underset{CH_3}{|}}{CH}-\underset{\underset{NO_2}{|}}{CH}-CH=CH_2$$

b) What is mesomeric effect? Explain how the mesomeric effect of chlorine affect the C-Cl bond length in $CH_2=CHCl$

OR

a) Write the structural formulae of the following.

(i) 4-Iodo-3-nitropent-1-ene.

(ii) 1-Bromo-2-Penten-3-ol.

b) What is a functional group? Illustrate you answer with the formulae and the name of any three examples.

X. Explain with examples, intermolecular and intramolecular hydrogen bonding and their effect on boiling points.

OR

Explain with examples nucleophilic and electrophilic reaction.

XI. Taking oxygen as an example, explain Hund's rule and Pauli's exclusion principle

OR

What do you understand by the dual nature of matter? Formulate de Broglie equation.

XII. Explain Lechatelier's principle and its application.

OR

Write a brief not on Freundlich adsorption isotherm.

OSMANIA UNIVERSITY

FACULTY OF SCIENCE

B.Sc. I Year Examination
March/April 1995

Subject : **CHEMISTRY** Paper : **I**

Time : **3 HOURS** Marks : **100**

Answer all Questions from Sections A&B
Answer all the bits of a question together

SECTION - A (4x15=60)

1. a) What are Clathartes ? Why Helium and Neon do nor form Clathrates? Give important uses.
   b) Why does Iodine show Basic Properties ? Name two slats in which Cationic to these acids ?

OR

   a) How are the two Persulphuric acids prepared ? Wht formulae have been assigned to these acids ?

II a) Describe the bonding in F_2 molecule using a molecular orbital energy level diagrm.
   b) What are Indicators ? Discuss the theory of Indicators. Explain the use of acid-base indicators in volumetric analysis.

OR

   a) What are Pseudo halogens ? Give examples. Compare their properties with those of halogens.
   b) Give the reactions of diborane with ammonia and chlorine.

III. a) What do you understand by inductive effect ? Explain the higher acidity of chloroacetic acid compared to acctic acid .
   b) What are conformational isomers ? Draw Newman and Sawhorse formulae for the conformations of ethane.

OR

   a) Give the mechanism of addition of HBr to propene.
   b) Write the structures of isomeric pentanes. Give their IUPAC names.

IV a) What is ostwald's dilution law? How it is experimentally verified ? Discuss its limitations.
   b) Formic acid is 4.5% dissociated in O.IN solution at 20°C, Calculate the dissociation constant of the acid.

OR

a) Discuss the principles of fractional distillation of miscible liquid pairs.

b) What is meant by the term solubility product ? Give one important application in analytical chemistry.

SECTION - B

(8x5=40)

V. What are Silicones ? How are they formed ?

OR

Write short note on the use of H_2S in qualitative analysis.

VI. What is meant by the pH of a solution ? Calculate pH value of solution having Hydrogen ion concetration $4x10^{-5}$ gram ions per litre.

OR

What are carbides ? How are they classiffied ?

VII. Give the hybridization and structure of CIF_3

OR

Describe the structure of tetrasulfer tetranitride (S_4N_4)

VIII. How does the Lewis acid character of different Boran halides vary? Explain the trend.

OR

How is the paramagnetic behaviour of oxygen molecule explained by the moleculer orbital concept ?

IX. Write the Mechanism for the trans addition of bromine to carbon - carbon double bond.

OR

Write a note on the mechanism off chlorination of methane.

X. Write a note on Dieckmann's method.

OR

What are dienes ? Give the different types of dienes with examples.

XI State Nernst distribution law ? What are its limitations .

OR

Explain the classification of partially miscible liquid pairs on the basis of critical solution temperature.

XII What is uncertainity Principle ? Explain how it leads to the concept of probability.

OR

State and explain Raoult's law the vapour pressures.

TELUGU VERSION

అన్ని ప్రశ్నలకు సమా ధానములు వ్రాయుము

ఒక ప్రశ్నలోని అన్ని భాగములు ఒకే చోట వ్రాయవలెను.

విభాగము - ఎ **(4x15=60)**

1 ఎ) పంజర సమ్మేళనములనగానేవి ? హీలియమ్, నియన్ లు ఎందుచేత ఈ సమ్మేళనలు ఏర్పరచవు, వీని ఉపయోగములను వ్రాయుము.

బి) అయోడిన్ ఎందుచేత క్షారధర్మాలను ప్రదర్శిస్తుంది. ఈ ధర్మాలను ప్రదర్శించే ఏవైనా రెండు లవణాల పేర్లను వ్రాయండి.

లేదా

ఎ) పెర్ సల్ఫ్యూరిక ఆమ్లాలని తయారుచేసే రెండు పద్ధతుల్ని వ్రాయండి. సాంకేతికము **(formulae)** ద్వారా ఈ ఆమ్లాలని గుర్తిస్తారు.

బి) అణు ఆర్బిటాల్ సిద్ధాంతశక్తి స్థాయి పటము ఆధారంగా F_2 అణువులో గల బంధాన్ని వివరించండి.

2. సూచికలు అనగానేవి ? సూచికా సిద్ధాంతాల్ని విశదీకరించండి. ఘన పరిమాణాత్మక విశ్లేషణములో ఆమ్ల-క్షార సూచికలు ఏ విధంగా ఉపయోగపడతాయి

లేదా

ఎ) సూడోహేలోజినల్లు అనగా నేవి ? ఉదాహరణల నివ్వండి వాని ధర్మాలను హేలోజినల ధర్మాలతో పోల్చండి.

బి) అమ్మోనియా, క్లోరిన్‌లతో డైబోరేన్ పొందే చర్యలను వ్రాయండి.

3. ఎ) ప్రేరేపక ప్రభావము (**inductive effect**) అనగానేవి ? కోరో ఎసిటికామ్లము 1 ఎసిటికామ్లము కంటే ఎందుచేత అధిక ఆమ్లతను కలిగి ఉంటుంది.

బి) అనురూపాత్మక సాదృశకాలు అనగానేవి ? ఈథేన్ యొక్క న్యూమన్ సాహర్ (**Sawhorse**) సాదృశక ఫార్ములా లను వ్రాయండి.

లేదా

ఎ) **HBr** ప్రోపేన్ ల సంకలనము యొక్క చర్యా విధానాన్ని వ్రాయండి.

బి) సదృశ పెంటేన్‌ల నిర్మితి (**structures**) వ్రాయుము.

4. ఎ) ఆస్వాల్డ్ విలీనతా సిద్ధాంతమును విశదీకరించుము. దీన్ని ప్రయోగాత్మకంగా ఏ విధంగా నిర్ధారిస్తారు. దాని అవధులను తెల్పుము.

బి) **0.1 N** ఫార్మిక్ ద్రావణము 4.5% వియోజనము పొందుతుంది. ఈ ఆమ్లము యొక్క వియోజన స్థిరాంకాన్ని గణించండి.

లేదా

ఎ) మిళితమయ్యే స్వభావంకల ద్రవాల అంశికస్వేదన సూత్రాన్ని విశదీకరించండి.

బి) ద్రావణీయతా లబ్దమనగానేమి ? విశ్లేషణాత్మక రసాయన శాస్త్రములో ఈ లబ్దము యొక్క ఒక ముఖ్య వుపయోగాన్ని వ్రాయండి.

విభాగము - బి (4x5=40)

5. సిలికోన్‌లు అనగానేమి ? ఇవి ఏ విధంగా ఏర్పడతాయి ?

లేదా

H_2Sను గుణాత్మక విశ్లేషణములో ఏ విధంగా ఉపయోగించుతారో క్లుప్తంగా తెలపండి.

6. ద్రవము యొక్క pHదేన్ని తెలుపుతుంది ? H^+అయానుల గాఢత $4x10^{-5}$ గ్రాము అయాన్/లీటరు గల ఒక ద్రవము pH విలువను గణించండి.

లేదా

కార్బయిడ్‌లు అనగానేవి ? వాటిని ఏ విధంగా వర్గీకరిస్తారు.

7. ClF_3 యొక్క సంకరీకరణము, నిర్మాణములను గూర్చి వ్రాయుము.

లేదా

టెట్రాసల్ఫర్ టెట్రా నైట్రేట్ (S_4N_4) యొక్క నిర్మాణమును విశదీకరించుము.

8. బోరాన్‌హెలైడ్‌ల లూయిస్ ఆమ్లలక్షణాలు ఏ విధంగా మారుతాయి. ఆమ్ల లక్షణ క్రమాలను విశదీకరించండి.

లేదా

ఆమ్లజని యొక్క పారా అయస్కాంత ధర్మాలు అణు ఆర్బిటాల్ సిద్ధాంతం ద్వారా ఏ విధంగా వివరిస్తారో విశదీకరించండి.

9. కార్బన్-కార్బన్ ద్విబంధానికి బ్రోమిన్‌ను ట్రాన్స్ సంకలనము చేసే చర్యా విధానాన్ని వ్రాయండి.

లేదా

మిథేన్ కోరనీకరణం చేసే విధానాన్ని వ్రాయండి.

10. డిక్‌మాన్ పద్ధతిని గూర్చి వ్రాయండి.

లేదా

డయిన్‌లు అనగానేవి ? వేర్వేరు రకాల డయిన్‌లను ఉదాహ రణ పూర్వకంగా తెలపండి.

11. నెర్నస్ట్ వితరణ సిద్ధాంతాన్ని వివరించండి దాని యొక్క అవధులను తెల్పండి

లేదా

సందిగ్ధ ద్రవ ఉష్ణోగ్రత ఆధారంగా పాక్షిక మిశ్రణీయ ద్రవయుగ్మాలను ఏ విధంగా వర్గీకరిస్తారో విశదీకరించండి.

12. అనిశ్చితత్వ సూత్రము అనగానేవి ? ఇది సంభావ్యతా నియమానికి ఏ విధంగా దారితీస్తుంది.

లేదా

బాష్ప పీడనములకై రేల్ట్ నియమాన్ని నిర్వచించి విశదీకరించండి.

ANDHRA UNIVERSITY MARCH 1994

PART II- CHEMISTRY

(Common Core Scheme)

Paper I-Inorganic, Organic and Physical

(Time : 3 Hours)

Maximum : 100 Marks

Part I

విభాగము I

(Marks : 6x10 =60)

Answer six questions choosing any two from Section A, Section B and Section C

సెక్షన్ A నుండి రెండు Bనుండి రెండు మరియు C నుండి రెండు ప్రశ్నలకు తీసుకుని మొత్తము ఆరు ప్రశ్నలకు జవాబులు వ్రాయుము

Section A

సెక్షన్ A

1. (a) Draw the 'MO' diagrams for B_2 and He_2+

   (b) Discuss the preperation and structures of Borazole

   (a) B_2 మరియు He_2+ లకు అణు ఆర్బిటాల్ శక్తి స్థాయి చిత్ర పటాలను గీయండి.

   (b) బోర్‌జోల్‌ని తయారు చేయు పద్ధతి మరియు నిర్మాణమును చర్చించండి.

2. (a) Explain why alkalimetals are strongest reducing agents and have strong ionic nature.

   (b) Give a brief note on silicones.

   (a) క్షారలోహాలు మిక్కిలి బలమైన క్షయకరణకాలుగాను, బలమైన అయానిక స్వభావాన్ని కలిగి ఉండుటకు గల కారణాలను వివరించండి.

   (b) సిలికోన్‌లను గురించి సంక్షిప్తంగా వ్రాయండి.

3. (a) Explain Phosphonitrilic halides.

   (b) Write preperation and structures of peroxy acids of sulphur.

   (a) ఫాస్ఫొనైట్రిక్ హైలైడ్‌లను గూర్చి వివరింపుము.

   (b) సల్ఫర్ పెరాక్సి ఆమ్లాలను తయారు చేయు పద్ధతి మరియు వాటి నిర్మాణములను వ్రాయండి.

4. (a) Describe the theory of indicators.

   (b) Explain Complexometric titrations

   (a) సూచికల సిద్ధాంతాన్ని వర్ణించండి.

   (b) కాంప్లెక్సోమెట్రిక్ అంశమానములను వివరింపుము.

Section B

సెక్షన్ B

1. (a) Define p^{ka} and how it is related to the strength of acids.

   (b) Write the reaction mechanisms of electrophilic substitution and nucleophilic substitution.

(a) p^{K_a} ని నిర్వచించి అది ఆమ్లాల బల తీవ్రతతో ఎటువంటి సంబంధాన్ని కలిగి ఉండునో తెల్పుము.

(b) ఎలక్ట్రోఫిలిక్ మరియు న్యూక్లియోఫిలిక్ ప్రతిక్షేపణ చర్యల యొక్క సంవిధానంను గూర్చి వ్రాయండి.

2. (a) Define isomerism and explain stereo isomerism with suitable example

(b) Explain the IUPAC rules in naming the organic compounds.

(a) అణుసాదృశ్యం అననేమి ? ప్రాదేశిక సాదృశ్యమును సోదాహరణంగా వివరించండి.

(b) కర్బన సమ్మేళనాల పేర్లు సూచించుటకు IUPAC నియమాలను వివరింపుము.

3. (a) Explain the basicity of aliphatic and aromatic amines.

(b) What is inductive effect and give its applications ?

(a) అలిఫాటిక్ మరియు అరోమాటిక్ అమైన్ల క్షారత్వాన్ని వివరింపుము.

(b) ప్రేరేపక ప్రభావము అననేమో తెల్పి దాని అనువర్తనాలను వ్రాయండి.

4. (a) Distinguish the reactivity of alkanes, alkenes and alkynes.

(b) What is hydrogen bonding and how it influence the boiling point and solubility of organic compounds ?

(a) ఆల్కేనులు, ఆల్కీనులు, ఆల్కైన్‌ల చర్యా శీలతలలోని వ్యత్యాసాలను తెల్పండి.

(b) హైడ్రోజన్ బంధం అనగా నేమి ? సేంద్రియ పదార్థాల భాష్పీభవన ఉష్ణోగ్రతలు, ద్రావణీయలపై హైడ్రోజన్ బంధము యొక్క ప్రభావాన్ని తెల్పండి.

Section C

సెక్షన్ C

1. (a) State and explain Heisenberg uncertainity principle.

(b) Defene solubility product and write its application.

(a) హైసన్ బర్గ్ అనిశ్చితత్వ నియమాన్ని నిర్వచించి వివరించండి.

(b) ద్రావణీయతా లబ్ధాన్ని నిర్వచించి దాని అనువర్తనాన్ని తెల్పండి.

2. (a) Define the terms inversion temperature and critical solution temperature.

(b) Give an account of distribution law and its application.

(a) విలోమ ఉష్ణోగ్రత మరియు సందిగ్ధ ద్రావణ ఉష్ణోగ్రతలను నిర్వచించి వివరించండి.

(b) వితరణ నియమములను గురించి తెల్పి దాని అనువర్తనాన్ని వ్రాయండి.

3. (a) What is meant by law of corresponding states ? Deduce reduced equation of state.

(b) Explain the nature of partially miscible liquid mixtures with an example.

(a) అనురూప స్థితుల నియమము అనగానేమి ? క్షయీకృత స్థితి సమీకరణమును ఉత్పాదించండి.

(b) పాక్షిక మిశ్రమనీయ ద్రవ యుగ్మ మిశ్రమముల స్వభావాన్ని సోదాహ రణంగా తెల్పండి.

4. (a) Describe steam distillation method for the seperation of immiscible liquids.

(b) Derive critical constants from van der Waals' equation

(a) నీటి ఆవిరి స్వేదనం ద్వారా అమిశ్రమనీయ ద్రవ మిశ్రమాలను వేరువరుచు పద్ధతిని వర్ణించండి.

(b) వాండర్ వాల్ సమీకరణము నుండి సందిగ్ధ స్థిరాంకాలను రాబట్టండి.

Part II

విభాగము II

(Marks : 10 x 4=40)

Answer All questions

Each question carries 4 marks

అన్ని ప్రశ్నలకు జవాబులు వ్రాయుము

ప్రతి ప్రశ్నకు 4 మార్కులు

1. Discuss the reactivity of corbonates and bicarbonates of alkalimetals
క్షారలోహ కార్బోనేట్‌లు , బై కార్బోనేట్‌ల చర్యాశీలతను వివరించండి.

2. What is meant by inert electron pair effect ? Explain with two examples.
జడ ఎలక్ట్రాను జంట ప్రభావం అనినేమి ? రెండు ఉదాహరణలతో వివరింపుము.

3. Describe the borax-bead test with two examples.
బోరాక్స్-బీడ్ పరీక్షను రెండు ఉదాహరణలతో వర్ణించండి.

4. Write short notes on silicates.
సిలికేట్‌లను గురించి సంక్షిప్తంగా వ్రాయండి.

5. Write about the projections of organic structural formulae.
సేంద్రియ సమ్మేళనాల నిర్మాణాత్మక ఫార్ములాల ప్రక్షేపణలను గూర్చి వ్రాయండి.

6. Write the structures for the following :-
(a) β-hydroxy butyric acid.
(b) 2-propen-1-ol.
క్రిందివానికి నిర్మాణములు వ్రాయండి :-
(a) β హైడ్రాక్సి బ్యూటిరిక్ ఆమ్లము
(b) 2-ప్రొపీన్-1- ఓల్.

7. What is meant Marcovnikoff's rule ? Explain with suitable example.
మార్కొనికఫ్ నియమము అనినేమో సోదాహరణంగా వివరింపుము.

8. Describe the liquifaction of air by claude method.
'క్లాడ్' పద్ధతిలో గాలిని ఎట్లు ద్రవీకరణంగావిస్తార్ వర్ణించండి.

9. Expalin the reasons for the deviation of gases from ideal behaviour.
వాయువులు ఆదర్శ ప్రవర్తననుండి విచలనాలను ప్రదర్శించుటకు గల కారణాలను వివరించండి

10. Write Schrödinger wave equation and explain the significance of ψ and ψ^2.
ష్రోడింగర్ తరంగ సమీకరణమును వ్రాసి, ψ, ψ^2ల ప్రాముఖ్యతలను వివరించండి.

(94-M)

ANDHRA UNIVERSITY
PART II- CHEMISTRY
(Common Core Scheme)
Paper I – Inorganic, Organic and Physical

(Time : 3 Hours) **Maximum : 100 Marks**

Part I (Marks : 6x10 =60)

Answer six questions choosing any TWO from Section A, Section B and Section C

SECTION A

1. (a) **Explain coprecipitation and post precipitation by giving one example for each.**

   అనుబంధ అవక్షేపము మరియు తరువాత అవక్షేపములు ఒక్కొక్క ఉదాహరణముతో వివరింపుము.

   (b) **Draw MO diagram for C_2 and NO**

   C_2 మరియు NOలకు అణు ఆర్బిటాల్ శక్తి స్థాయి చిత్రపటాలను గీయండి.

2. (a) **Write briefly on pseudohalogens**

   మిథ్యాహాలోజనుల గూర్చి క్లుప్తముగా వ్రాయుము.

   (b) **What are metallic carbonyls? Give their preperation and structure.**

   లోహ కార్బోనైల్ ఏమిటో తెల్పండి ? వాటిని తయారుచేయు విధానాన్ని మరియు నిర్మాణాన్ని వ్రాయండి.

3. (a) **Explain hydration in alkali metals.**

   క్షారలోహాలలో హైడ్రేషన్ వివరించుము.

   (b) **Write short notes on the anamalous behaviour of lithium.**

   లిథియం యొక్క ప్రత్యేక ప్రవర్తనను సంక్షిప్తంగా వివరించండి.

4. (a) **Write short notes on noble gases.**

   ఉత్కృష్ట వాయువులను గురించి సంక్షిప్తంగా వ్రాయండి.

   (b) **Explain the principles involved in redox titrations.**

   ఆక్సీకరణము–క్షయకరణ అంశ మాపనములలో ఇమిడియున్న సూత్రములను గురించి వివరించండి.

SECTION B

5. (a) **What is meant by carnium ions? Explain the relative stability of primary, secondary and tertiary carbonium ions.**

   కార్బొనియం అయాన్‌లు అంటే ఏమిటి ? ప్రైమరి, సెకండరి, టెర్షరి కార్బొనియం అయాన్‌ల యొక్క సాపేక్షిక స్థిరత్వాన్ని విశదీకరించండి.

   (b) **Explain inductive effect examples.**

   ప్రేరేపక ప్రభావాన్ని ఉదాహరణలతో వివరించుము.

6. (a) **Explain geometrical isomerism of C=C compounds with an example.**

   C=C అణువులోని క్షేత్ర సాదృశ్యాన్ని ఒక ఉదాహరణముతో వివరించుము.

(b) What is hyper conjuction ? Explain with two examples.
అతి సంయుగ్మము అంటే ఏమిటో రెండు ఉదాహరణముల ద్వారా వివరించండి.

7. (a) Explain with suitable examples bond angle, bond dissociation and bond length in organic molecules.
కర్బన పదార్థాల అణువులలో, బంధకోణం, బంధ విఘనశక్తి, బంధదైర్ఘ్యం గూర్చి సోదాహరణంగా విశదీకరించండి.

(b) Explain with Examples substitution, adition and elimination organic reactions.
ప్రతిక్షేపణ, సంకలన, విలోపన కర్బన రసాయన చర్యలను ఉదాహరణములతో వివరించండి.

8. (a) Discuss the steric effects on the dissociation constants of carboxylic acids.
కార్బాక్సిలిక్ ఆమ్లాల విఘటన స్థిరాంకాల పై ప్రాదేశిక ప్రభావాన్ని చర్చించండి.

(b) What is meant by hydrogen bonding ? Explain how it influences the boiling point and solubility of organic compounds.
హైడ్రోజన్ బంధం అంటే ఏమిటి ? కర్బన పదార్థాల బాష్పీభవన ఉష్ణోగ్రతలు, ద్రావణీయలపై హైడ్రోజన్ బంధము యొక్క ప్రభావాన్ని తెల్పండి.

9. (a) State law of corresponding states. Deduce reduced equation of state.
అనురూప స్థితుల నియమాన్ని నిర్వచింపుము. క్షయీకృత స్థితి సమీకరణమును ఉత్పాదించండి.

(b) Define solubilityproduct. Explain it with an example.
ద్రావణీయతాలబ్ధమును నిర్వచించుము. ఒక ఉదాహరణముతో దీనిని వివరింపుము.

10. (a) Define critical constants
సందిగ్ధ స్థిరాంకాలను నిర్వచించుము.

(b) State and explain Heisenberg's uncertainity principle.
హైసన్‌బర్గ్ అనిశ్చితా సూత్రాన్ని నిర్వచించి, విశదీకరించండి.

11. (a) Discuss the liquifaction of air by Claude method.
'క్లాడ్' పద్ధతిలో గాలిని ఎట్లు ద్రవీకరణం గావిస్తారో వర్ణించండి.

(b) Write Schrodinger's wave equation and explain the terms in it.
ష్రోడింగర్ తరంగ సమీకరణం వ్రాసి దానిలోని పదాలను వివరించండి.

12. (a) Discuss Joule-Thomson effect.
జౌల్- థామ్‌సన్ ప్రభావమును గురించి వివరించుము.

(b) Write notes on shapes of atomic orbitals.
పరమాణు ఆర్బిటాల్ యొక్క ఆకృతులను గురించి సంక్షిప్తంగా వ్రాయండి.

PART II – (10x4 = 40 marks)

Answer ALL questions.

Each question carries 4 marks

13. **Boron chloride exists as monomer, while aluminium chloride exists as a dimer. Explain.**
బోరాన్ ట్రైక్లోరైడ్ మోనామర్ రూపంలో వుంటుంది. కాని అల్యూమినియం క్లోరైడ్ డైమర్ గా ఉంటుంది. కారణాన్ని వివరించండి.

14. **Name and write the formulae of four oxyacids of chlorine.**
నాలుగు క్లోరిన్ ఆక్సీ ఆమ్లాల పేర్లను, వాటి రసాయనిక ఫార్ములాలను వ్రాయండి.

15. **Explain Nessler's test for detecting ammonium.**
అమ్మోనియంను గుర్తించు నెస్లర్స్ పరీక్ష వివరింపుము.

16. **Write the preperation and structure of hydrazoic acid.**
హైడ్రోజోయిక్ ఆమ్లము తయారు చేయుటను, మరియు నిర్మాణమును వ్రాయండి.

17. **Explain the acid behaviour of phenol.**
ఫినాల్ యొక్క ఆమ్ల స్వభావాన్ని గురించి వివరింపుము.

18. **Give an account on the projections of organic structural formulae.**
కర్బన సమ్మేళనాల నిర్మాణాత్మక ఫార్ములాల ప్రక్షేపణలను గూర్చి వ్రాయండి.

19. **Write the structures of the following :**
    - (a) **3-methyl-octa-3-ene-5-yne**
    - (b) **4-chloro-3-methyl-butyl-2-ene-1-ol.**

ఈ క్రింది సమ్మేళనాల ఫార్ములాను వ్రాయండి.
    - (a) **3-మిథైల్- ఆక్టా-3- ఈన్-5-ఐన్.**
    - (b) **4-క్లోరో-3-మిథైల్-బ్యుట్-2- ఈన్-1- ఓల్**

20. **Describe steam distillation method for the separation of immiscible liquids.**
నీటి ఆవిరి స్వేదనం ద్వారా అ మిశ్రమనీయ ద్రవ మిశ్రమాలను వేరువరుచు పద్ధతిని వర్ణించండి.

21. **What is common ion effect ? Mention its application.**
ఉమ్మడి అయాను ప్రభావ మనగానేమి ? దీని ఉపయోగమును తెలుపుము.

22. **Discuss distribution law and its applications.**
వితరణ నియమము మరియు దాని ఉపయోగములను గురించి చర్చించండి.